COMPUTATIONAL STAR FORMATION

IAU SYMPOSIUM No. 270

COVER ILLUSTRATION: Star fomation simulation

This image shows the distribution of gas and stars in two galaxies ~50 Myr after their encounter. Cold and hot gas are shown in brown and blue colors, respectively. Young, intermediate-age, and old stars are shown as white, yellow, and red colored points, respectively. A number of star clusters are formed between two galaxies. The formation process of these star clusters is "bottom-up", quite different from the conventional "monolithic" picture of star cluster formation.

Credit: Takayuki Saitoh (Division of Theoretical Astrophysics/National Astronomical Observatory of Japan) & Takaaki Takeda (4D2U project/National Astronomical Observatory of Japan)

INTERNATIONAL ASTRONOMICAL UNION
UNION ASTRONOMIQUE INTERNATIONALE

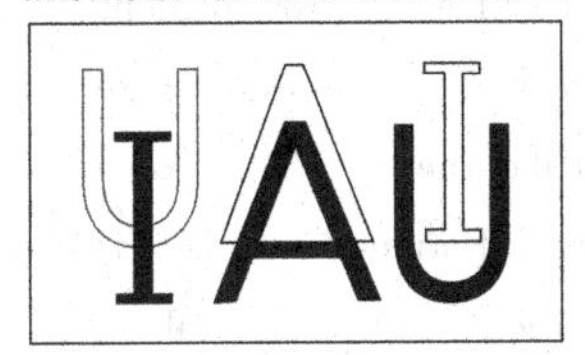

COMPUTATIONAL STAR FORMATION

PROCEEDINGS OF THE 270th SYMPOSIUM OF THE INTERNATIONAL ASTRONOMICAL UNION HELD IN BARCELONA, CATALONIA, SPAIN MAY 31 - June 4, 2010

Edited by

JOÃO ALVES
Institute of Astronomy, University of Vienna, Austria

and

BRUCE G. ELMEGREEN
IBM Research Division, Yorktown Heights, New York, USA

and

JOSEP M. GIRART
Institut de Cincies de l'Espai, (CSIC-IEEC), Barcelona, Catalonia, Spain

and

VIRGINIA TRIMBLE
Department of Physics and Astronomy, University of California, Irvine, USA; Las Cumbres Observatory, Goleta, California, USA

Shaftesbury Road, Cambridge CB2 8EA, United Kingdom

One Liberty Plaza, 20th Floor, New York, NY 10006, USA

477 Williamstown Road, Port Melbourne, VIC 3207, Australia

314–321, 3rd Floor, Plot 3, Splendor Forum, Jasola District Centre, New Delhi – 110025, India

103 Penang Road, #05–06/07, Visioncrest Commercial, Singapore 238467

Cambridge University Press is part of Cambridge University Press & Assessment, a department of the University of Cambridge.

We share the University's mission to contribute to society through the pursuit of education, learning and research at the highest international levels of excellence.

www.cambridge.org
Information on this title: www.cambridge.org/9780521766432

First published 2011

A catalogue record for this publication is available from the British Library

ISBN 978-0-521-76643-2 Hardback

Table of Contents

Presentations

Session I. Historical Introduction

Session II. Individual Star Formation: Observations

Session III. Individual Star Formation: Theory

Session IV. Formation of Clusters: Observations

Session V. Formation of Clusters: Theory I

Session VI. Formation of Clusters: Theory II

Session VII. Numerical Methods: MHD

Session VIII. Numerical Methods: Radiative Dynamics

Session IX. Tuesday Evening: Local Star Formation Processes

Session X. Star Formation Feedback I

Session XI. Star Formation Feedback II

Session XII. Star Formation Feedback III

Session XIII. Star Formation on Galactic Scales I

Session XIV. Star Formation on Galactic Scales II

Session XV. Special Purpose Hardware

Session XVI. Thursday Evening: Computational Methods

Session XVII. Radiation Diagnostics of Star Formation

Session XVIII. Large Scale Star Formation

Session XIX. Cosmological Star Formation

Session XX. Computational Star Formation

Preface

IAU Symposium 270 had its origins at the Fall 2008 JENAM meeting in Vienna, where three of the four present editors participated in a session on star formation that proved too short to clarify either all of the problems or what progress was being made on them. We decided that a larger meeting would be useful, particularly one with an emphasis on numerical simulations and comparisons with observations. Fortunately, the IAU executive committee agreed.

Star formation is complex, involving unknown initial conditions and poorly understood physical processes, such as supersonic turbulence, magnetic diffusion and reconnection, radiation transfer of background and young stellar light, and cooling by collisional excitation and decay of transient molecules and dust particles, all operating in a medium with rapidly changing substructures spanning 20 orders of magnitude in density. It is a violent storm of collapse into filaments, clumps, disks, and protostars, with equally violent energy release in the form of jets, winds, and heat, plus ionization when the most massive stars appear. Yet viewed at various embedded stages through infrared, mm, and radio telescopes, the result of this activity is a fairly regular assortment of young stars and protostars, with a power law distribution of separations and a power law distribution of masses, both extending from the largest scales and masses down to minimum values where the motions become subsonic. By the time these stars are visible to the eye in the night sky, the process is mostly over, the dense gas has dispersed, the jets have calmed, and the dense young clusters have started to disperse.

What lies between the dispersed gas before star formation and the dispersed gas after star formation, minus the few percent that has turned into stars, is the concern of theoreticians and observers at this conference. After 50 years of exponential growth in the speed, storage, and capacity of computers, we are at a stage where many of the formerly unimaginable processes involved with star formation can be studied with some realism. These processes include cloud formation in galaxies, cloud turbulence and collapse, disk and binary star formation, pre-stellar jets and winds, the effects of ionization, and star cluster evolution. Remarkably, simulators get about the same results as observers: power law structures and mass functions are reproduced in computers, filaments, clumps and disks are all present, the timescale for star formation comes out about right, and the overall efficiency of turning gas into stars is also right.

Still there are many details that need to be evaluated. In fact, the first two decades of simulations look almost too good in retrospect. When realistic heating and windy feedback are included, the stellar mass function sometimes changes in seemingly unacceptable ways. The full complexity of magnetic processes is not yet modeled either. Different magnetic field configurations could affect the binary fraction and disk sizes. Processes such as ion-molecule-radiation chemistry that determine the ionization fraction and rate of diffusion are not in computer codes, nor is magnetic reconnection. Radiative transfer through complex gas structures has barely begun. There is still a lot to do.

IAU Symposium No. 270 was convened to bring us up to date on the state of our field. We selected Barcelona (the capital of Catalonia) because of its beautiful climate, famous architecture, and friendly citizens, and we were not disappointed. The Symposium was attended by 220 scientists from 31 countries between May 31st and June 4th, 2010. Almost all of our time was packed with talks and discussions, even in two evening sessions when tapas and drinks were provided to keep us going. We visited the Mare Nostrum Supercomputer Center and had a wonderful excursion to the Codorniu and Freixenet

Caves winery with a sumptuous banquet at the Masia Torreblanca. We were all saddened to hear that one of the evening entertainers, Lluis Barrera Torné, leader of the acrobatic troupe "Colla Castellera Xicots de Vilafranca" that thrilled us with towering human pyramids, met an untimely death in a motorcycle accident only 2 months later.

It is a great pleasure to acknowledge the support of the IAU and all of the hard work by the SOC and LOC. We are deeply grateful to our invited and contributing speakers who wrote this book. The posters from the conference may be viewed at the Cambridge University Press website.

November 2010

Bruce G. Elmegreen, Virginia Trimble, João Alves, and Josep Miquel Girart
SOC co-chairs (pictured below at the reception)

THE SCIENTIFIC ORGANIZING COMMITTEE

THE LOCAL ORGANIZING COMMITTEE

Acknowledgements

The symposium was sponsored by the IAU and coordinated by the IAU Division IV Stars, Division VI Interstellar Matter, Division VII Galaxies & the Universe, proposed by the IAU Commission No. ?? (??).

Funding by the
International Astronomical Union,
Centro Astronómico Hipano-Alemán Calar Alto (CAHA),
Ministerio de Ciencia e Innovación,
Consejo Superior de Investigaciones Científicas (CSIC),
AGAUR - Generalitat de Catalunya,
and the Universitat de Barcelona
are gratefully acknowledged.

FACULTAT
DE
BIOLOGIA

Participants

Tom **Abel**, KIPAC/Stanford, USA tabel@stanford.edu
David **Acreman**, University of Exeter, U.K. acreman@astro.ex.ac.uk
Emilio **Alfaro**, Instituto de Astrofísica de Andalucía -CSIC, Spain emilio@iaa.es
Richard **Allison**, University of Sheffield, U.K. r.allison@sheffield.ac.uk
Felipe **Alves**, Institut de Ciències de l'Espai (IEEC-CSIC), Spain oliveira@ieec.uab.es
João **Alves**, University of Vienna, Austria joao.alves@univie.ac.at
Catarina **Alves de Oliveira**, Laboratoire d'Astrophysique de Grenoble, France Catarina.Oliveira@obs.ujf-grenoble.fr
Loren **Anderson**, Le Laboratoire de Astrophysique de Marseille, France loren.dean.anderson@gmail.com
Philippe **André**, CEA/SAp Saclay, France pandre@cea.fr
Hector **Arce**, Yale University, USA hector.arce@yale.edu
Bastian **Arnold**, Department of Astronomy, University of Vienna, Austria bastian.arnold@univie.ac.at
Jane **Arthur**, Centro de Radioastronomía y Astrofísica, UNAM, Mexico j.arthur@crya.unam.mx
Dori **Arzoumanian**, CEA Saclay, France doris.arzoumanian@cea.fr
Joana **Ascenso**, Centro de Astrofísica da Universidade do Porto, Portugal jascenso@astro.up.pt
Dominique **Aubert**, Observatoire Astronomique de Strasbourg, France dominique.aubert@astro.unistra.fr
Aycin **Aykutalp**, Kapteyn Astronomical Institute, Netherlands aykutalp@astro.rug.nl
John **Bally**, University of Colorado at Boulder, USA john.bally@colorado.edu
Shantanu **Basu**, University of Western Ontario, Canada basu@astro.uwo.ca
Matthew **Bate**, University of Exeter, U.K. mbate@astro.ex.ac.uk
Christophe **Becker**, Laboratoire d'Astrophysique de Grenoble, France beckerc@ujf-grenoble.fr
Maite **Beltrán**, Osservatorio Astrofisico di Arcetri, Italy mbeltran@arcetri.astro.it
Frank **Bigiel**, UC Berkeley, USA bigiel@astro.berkeley.edu
Thomas **Bisbas**, Academy of Sciences of the Czech Republic, Czech Republic spxtb@astro.cf.ac.uk
John **Bochanski**, MIT, USA jjb@mit.edu
Ian **Bonnell**, University of St Andews, U.K. iab1@st-and.ac.uk
Frederic **Bournaud**, CEA Saclay, France frederic.bournaud@cea.fr
Jerome **Bouvier**, Laboratoire d'Astrophysique de Grenoble, France jbouvier@obs.ujf-grenoble.fr
Herve **Bouy**, European Space Agency, Spain hbouy@sciops.esa.int
Chris **Brook**, University of Central Lancashire, U.K. cbbrook@uclan.ac.uk
Stephanie **Bush**, Harvard Smithsonian Center for Astrophysics, USA sbush@cfa.harvard.edu
Gemma **Busquet**, Departament d'Astronomia i Meteorologia, Universitat de Barcelona, Spain gbusquet@am.ub.es
Jonathan **Carroll**, University of Rochester, USA johannjc137@gmail.com
Damien **Chapon**, CEA-Saclay, France damien.chapon@cea.fr
Hui-Chen **Chen**, Institute of Astronomy, National Central University, Taiwan huichen@astro.ncu.edu.tw
Rumpa **Choudhury**, Indian Institute of Astrophysics, India rumpa@iiap.res.in
Pedro **Colin**, Centro de Radioastronomía y Astrofísica, Mexico p.colin@crya.unam.mx
Benoît **Commerçon**, MPIA, Germany benoit@mpia-hd.mpg.de
Paul **Cornwall**, University of Kent , U.K. pc247@kent.ac.uk
Timea **Csengeri**, CEA-Saclay, Service d, France timea.csengeri@cea.fr
Nicola **Da Rio**, Max Planck Institute for Astronomy, Germany dario@mpia-hd.mpg.de
James **Dale**, Czech Academy of Sciences, Czech Republic jim@ig.cas.cz
Wolfgang **Dapp**, The University of Western Ontario, Canada wdapp@astro.uwo.ca
Massimo **De Luca**, Observatoire de Paris and Ecole Normale Supérieure., France massimo.de.luca@lra.ens.fr
Lise **Deharveng**, Laboratoire d'Astrophysique de Marseille, France lise.deharveng@oamp.fr
Lauriane **Delaye**, CEA/IRFU/SAp, France lauriane.delaye@cea.fr
Vasily **Demichev**, Space Research Institute, Russian Federation vademichev@gmail.com
Clare **Dobbs**, MPE, Germany cdobbs@mpe.mpg.de
Gustavo **Dopcke**, Institut für Theoretische Astrophysik - Universitat Heidelberg, Germany gustavotche@gmail.com
Gaspard **Duchêne**, University of California Berkeley, USA gduchene@berkeley.edu
Dennis **Duffin**, McMaster University, Canada duffindf@mcmaster.ca
Bruce **Elmegreen**, IBM T.J. Watson Research Center, USA bge@us.ibm.com
Barbara **Ercolano**, Institute of Astronomy, University of Cambridge, U.K. be@ast.cam.ac.uk
Robert **Estalella**, Universitat de Barcelona, Spain robert.estalella@am.ub.es
Neal **Evans**, The University of Texas at Austin, USA nje@astro.as.utexas.edu
Samuel **Falle**, University of Leeds, U.K. sam@amsta.leeds.ac.uk
Christoph **Federrath**, ZAH, MPIA Heidelberg, Germany chfeder@ita.uni-heidelberg.de
Laura **Fissel**, University of Toronto, Canada fissel@astro.utoronto.ca
Pau **Frau**, Institut de Ciències de l'Espai (CSIC-IEEC), Spain frau@ice.cat
Josep M. **Girart**, Institut de Ciències de l'Espai (CSIC-IEEC), Spain girart@ice.cat
Philipp **Girichidis**, Institut für theoretische Astrophysik, Germany girichidis@ita.uni-heidelberg.de
Simon **Glover**, Zentrum für Astronomie Heidelberg, Germany sglover@ita.uni-heidelberg.de
Oleg **Gnedin**, University of Michigan, USA ognedin@umich.edu
Gilberto **Gómez**, CRyA - UNAM, Mexico g.gomez@crya.unam.mx
Alyssa **Goodman**, Harvard-Smithsonian Center for Astrophysics, USA agoodman@cfa.harvard.edu
Simon **Goodwin**, University of Sheffield, U.K. s.goodwin@sheffield.ac.uk
Eva **Grebel**, Astronomisches Rechen-Institut, Germany grebel@ari.uni-heidelberg.de
Matthias **Gritschneder**, The Kavli Inst. for Astronomy & Astrophysics, Peking Univ., China gritschneder@kiaa.pku.edu.cn
Vasilii **Gvaramadze**, Sternberg Astronomical Institute, Moscow State University, Russian Federation vgvaram@mx.iki.rssi.ru
Michal **Hanasz**, Centre for Astronomy, Nicolaus Copernicus University, Poland mhanasz@astri.uni.torun.pl
Tomoyuki **Hanawa**, Chiba University, Japan hanawa@cfs.chiba-u.ac.jp
Elizabeth **Harper-Clark**, Canadian Institute for Theoretical Astrophysics, Canada h-clark@cita.utoronto.ca
Tim **Harries**, University of Exeter, U.K. th@astro.ex.ac.uk
Patrick **Hennebelle**, Ecole normale supérieure and Observatoire de Paris, France patrick.hennebelle@ens.fr
G. **Hensler**, Institute of Astronomy, Austria gerhard.hensler@univie.ac.at
Seyit **Hocuk**, Kapteyn Astronomical Institute, Netherlands seyit@astro.rug.nl
David **Hubber**, Department of Physics and Astronomy, University of Sheffield, U.K. D.Hubber@sheffield.ac.uk
Georgios **Ioannidis**, University of Kent, U.K. gi8@kent.ac.uk
Izaskun **Jiménez-Serra**, Harvard-Smithsonian Astrophysical Observatory, USA ijimenez-serra@cfa.harvard.edu
Viki **Joergens**, MPIA, Germany viki@mpia.de
Peter **Johansson**, University Observatory Munich, Germany pjohan@usm.lmu.de
Marc **Joos**, Ecole Normale Supérieure and Observatoire de Paris, France marc.joos@lra.ens.fr
Mika **Juvela**, University of Helsinki, Finland mika.juvela@helsinki.fi
Simon **Karl**, University Observatory Munich, Germany skarl@usm.uni-muenchen.de
Chang-Goo **Kim**, Seoul National University, South Korea kimcg@astro.snu.ac.kr
Jongsoo **Kim**, Korea Astronomy and Space Science Institute, U.K. jskim@mrao.cam.ac.uk
Sungsoo **Kim**, Kyung Hee University, South Korea sungsoo.kim@khu.ac.kr
Timothy **Kinnear**, University of Kent, U.K. tk218@kent.ac.uk
Maria **Kirsanova**, Institute of Astronomy, Russian Academy of Sciences, Russian Federation kirsanova@inasan.ru
Ralf **Klessen**, Zentrum für Astronomie Heidelberg, Germany rklessen@ita.uni-heidelberg.de
Michael **Knight**, University of Kent, U.K. mk296@kent.ac.uk
Vera **Könyves**, Service d'Astrophysique, CEA/Saclay, France vera.konyves@cea.fr
Alexei **Kritsuk**, University of California, San Diego, USA akritsuk@ucsd.edu
Pavel **Kroupa**, Argelander Institute for Astronomy, Bonn, Germany pavel@astro.uni-bonn.de
Diederik **Kruijssen**, Astronomical Institute Utrecht, Leiden Observatory, Netherlands kruijssen@astro.uu.nl
Mark **Krumholz**, UC Santa Cruz, USA krumholz@ucolick.org
Voldymyr **Kryvdyk**, Taras Shevchenko National University of Kyiv , Ukraine kryvdyk@univ.kiev.ua
Rolf **Kuiper**, Max-Planck-Institut für Astronomie, Germany kuiper@mpia.de

Maria **Kun**, Konkoly Observatory, Hungary kun@konkoly.hu
Richard **Larson**, Yale University, USA richard.larson@yale.edu
Pak Shing **Li**, University of California at Berkeley, USA psli@astro.berkeley.edu
Oliver **Lomax**, School of Physics & Astronomy, Cardiff University, U.K. Oliver.Lomax@astro.cf.ac.uk
Tuomas **Lunttila**, University of Helsinki, Finland tuomas.lunttila@helsinki.fi
Mordecai-Mark **Mac Low**, Department of Astrophysics, American Museum of Natural History, USA mordecai@amnh.org
Masahiro **Machida**, National Astronomical Observatory of Japan, Japan masahiro.machida@nao.ac.jp
Jun **Makino**, National Astronomical Observatory of Japan, Japan makino@cfca.jp
Johanna **Malinen**, University of Helsinki, Finland johanna.malinen@helsinki.fi
Marie **Martig**, CEA Saclay, France marie.martig@cea.fr
Josep-Maria **Masqué**, Departament d'Astronomia i Meteorologia, Universitat de Barcelona, Spain jmasque@am.ub.es
Silvano **Massaglia**, Department of General Physics - University of Turin, Italy massaglia@ph.unito.it
Jacques **Masson**, ENS Cachan, France jacques.masson@ens-cachan.fr
Tomoaki **Matsumoto**, Hosei University, Japan matsu@hosei.ac.jp
Leonid **Matveyenko**, Space Research Institute, Russian Federation lmatveenko@gmail.com
Christopher **McKee**, UC Berkeley, USA cmckee@astro.berkeley.edu
Andrew **McLeod**, Cardiff University, U.K. Andrew.McLeod@astro.cf.ac.uk
Jingqi **Miao**, School of Physical Sciences, Univ. of Kent, U.K. j.miao@kent.ac.uk
Milica **Milosavljevic**, Institute for Theoretical Astrophysics , Germany milica@ita.uni-heidelberg.de
Vincent **Minier**, CEA Saclay, France vincent.minier@ce.fr
Nickolas **Moeckel**, Institute of Astronomy, U.K. moeckel@ast.cam.ac.uk
Joe **Monaghan**, Monash University, Australia joe.monaghan@sci.monash.edu.au
Anthony **Moraghan**, Korea Astronomy and Space Science Institute, South Korea ajm@kasi.re.kr
Estelle **Moraux**, Laboratoire d'Astrophysique de Grenoble, France emoraux@obs.ujf-grenoble.fr
Kazuhito **Motogi**, Department of Cosmosciences, Graduate School of Science, Hokkaido University, Japan motogi@astro1.sci.hokudai.ac.jp
Kazutaka **Motoyama**, National Astronomical Observatory of Japan, Japan motoyama@cfca.jp
Taishi **Nakamoto**, Tokyo Institute of Technology, Japan nakamoto@geo.titech.ac.jp
Fumitaka **Nakamura**, National Astronomical Observatory of Japan / Niigata University, Japan fnakamur@ed.niigata-u.ac.jp
Dylan **Nelson**, Harvard Smithsonian Center for Astrophysics, USA dnelson@cfa.harvard.edu
Tetiana **Nikolaiuk**, The National Academy of Fine Arts and Architecture, Ukraine kryvdyk@univ.kiev.ua
Åake **Nordlund**, Niels Bohr Institute, Denmark aake@nbi.dk
Michael **Norman**, San Diego Supercomputer Center, UCSD, USA mlnorman@sdsc.edu
Evangelia **Ntormousi**, University Observatory Munich, Germany eva@usm.uni-muenchen.de
Stella **Offner**, Harvard-Smithsonian Center for Astrophysics, USA soffner@cfa.harvard.edu
Mayra **Osorio**, Instituto de Astrofísica de Andalucía, Spain osorio@iaa.es
Eve **Ostriker**, University of Maryland, USA ostriker@astro.umd.edu
James **Owen**, Institute of Astronomy, U.K. jo276@ast.cam.ac.uk
Paolo **Padoan**, ICC - University of Barcelona, Spain ppadoan@ucsd.edu
Marco **Padovani**, Institut de Ciències de l'Espai (CSIC-IEEC), Spain padovani@ice.cat
Aina **Palau**, Institut de Ciències de l'Espai (CSIC-IEEC), Spain palau@ice.cat
Jan **Palouš**, Astronomical Institute, Academy of Sciences of the Czech Republic, Czech Republic palous@ig.cas.cz
Ross **Parkin**, Université de Liège, Belgium parkin@astro.ulg.ac.be
Yaroslav **Pavlyuchenkov**, Institute of astronomy of the RAS, Russian Federation pavyar@inasan.ru
Stephanie **Pekruhl**, University Observatory Munich, Germany pekruhl@usm.uni-muenchen.de
Veli-Matti **Pelkonen**, IPAC/Caltech, USA vmpelkon@ipac.caltech.edu
Stoyanka **Peneva**, Institute of Astronomy, Bulgarian Academy of Science, Bulgaria speneva@astro.bas.bg
Antonio **Pereyra**, Observatorio Nacional (ON/MCT), Peru pereyra@on.br
Paolo **Persi**, IASF-Roma/INAF, Italy paolo.persi@iasf-roma.inaf.it
Jan **Pflamm-Altenburg**, Argelander-Institute for Astronomy, Germany jpflamm@astro.uni-bonn.de
Julian **Pittard**, The University of Leeds, U.K. jmp@ast.leeds.ac.uk
Leila **Powell**, CEA-Saclay, France leila.powell@cea.fr
Daniel **Price**, Monash University, Australia daniel.price@sci.monash.edu.au
Joaquin Patricio **Prieto Brito**, Pontificia Universidad Catolica, Chile jpprieto@astro.puc.cl
Ralph **Pudritz**, Origins Institute, McMaster University, Canada pudritz@physics.mcmaster.ca
Roberto **Raddi**, CAR/STRI - University of Hertfordshire, U.K. r.raddi1@herts.ac.uk
Alireza **Rahmati**, Leiden Observatory, Netherlands rahmati@strw.leidenuniv.nl
Ramprasad **Rao**, Academia Sinica Institute of Astronomy and Astrophysics (ASIAA), USA rrao@asiaa.sinica.edu.tw
Darren **Reed**, University of Zurich, ITP, Switzerland reed@physik.uzh.ch
Florent **Renaud**, Observatoire Astronomique de Strasbourg, France florent.renaud@astro.unistra.fr
Boyke **Rochau**, Max-Planck-Institut für Astronomie, Germany rochau@mpia.de
Javier A. **Rodón**, Laboratoire d'Astrophysique de Marseille, France jarodon@oamp.fr
Hazel **Rogers**, University of Leeds, U.K. phy5hr@leeds.ac.uk
Carlos **Román-Zñiga**, Centro Astronómico Hispano Alemán, Spain croman@caha.es
Takayuki **Saitoh**, National Astronomical Observatory of Japan, Japan saitoh.takayuki@nao.ac.jp
Álvaro **Sánchez-Monge**, Universitat de Barcelona (IEEC-UB), Spain asanchez@am.ub.es
Hsi-Yu **Schive**, Department of Physics, National Taiwan University, Taiwan b88202011@ntu.edu.tw
Daniel **Seifried**, Institute of Theoretical Astrophysics, University of Heidelberg, Germany dseifried@ita.uni-heidelberg.de
Evgeni **Semkov**, Institute of Astronomy, Sofia, Bulgaria, Bulgaria esemkov@astro.bas.bg
Rahul **Shetty**, Institut für Theoretische Astrophysik, Germany rshetty@ita.uni-heidelberg.de
Maryam **Shirazi**, Leiden Observatory, Leiden University, Netherlands shirazi@strw.leidenuniv.nl
Michal **Simon**, Stony Brook University, USA michal.simon@sunysb.edu
Jan **Skalicky**, Dep. of Theoretical Physics and Astrophysics, Faculty of Science, Masaryk Univ., Czech Republic skalicky@physics.muni.cz
Michael **Smith**, University of Kent, U.K. m.d.smith@kent.ac.uk
Rory **Smith**, Universidad de Concepción, Chile rsmith@astro-udec.cl
Rowan **Smith**, Institut fur Theoretishce Astrophysik, Universitat Heidelberg, Germany rowan@ita.uni-heidelberg.de
Valeriy **Snytnikov**, Novosibirsk State University, Boreskov Institute of Catalysis SB RAS, Russian Federation snyt@catalysis.ru
Marco **Spaans**, Kapteyn Astronomical Institute, Netherlands spaans@astro.rug.nl
Volker **Springer**, Heidelberg Institute for Theoretical Studies, Germany volker@mpa-garching.mpg.de
Dimitris **Stamatellos**, Cardiff University, U.K. D.Stamatellos@astro.cf.ac.uk
Jörgen **Steinacker**, Max-Planck-Institut für Astronomie, Germany stein@mpia.de
Greg **Stinson**, Central Lancashire, U.K. gsstinson@uclan.ac.uk
Curtis **Struck**, Iowa State Univ., USA curt@iastate.edu
Hajime **Susa**, Department of Physics Konan University, Japan susa@konan-u.ac.jp
Kei **Tanaka**, Tokyo Tech, Japan kt503i@geo.titech.ac.jp
Mauricio **Tapia**, Instituto de Astronomia, UNAM-Ensenada, Mexico mt@astrosen.unam.mx
Elizabeth **Tasker**, McMaster University, Canada taskere@mcmaster.ca
Ovidiu **Tesileanu**, University of Bucharest, Romania ovidiu.tesileanu@gmail.com
Romain **Teyssier**, CEA Saclay and University of Zurich, Switzerland romain.teyssier@cea.fr
John **Tobin**, University of Michigan, USA jjtobin@umich.edu
Kengo **Tomida**, The Graduate University for Advanced Studies/National Astronomical Observatory of Japan, Japan tomida@th.nao.ac.jp
L. Viktor **Toth**, Department of Astronomy Eötvös University, Hungary l.v.toth@astro.elte.hu
Pascal **Tremblin**, CEA Saclay, France pascal.tremblin@cea.fr
Sandra **Treviño Morales**, Centro de Radioastronomía y Astrofísica, Mexico s.trevino@crya.unam.mx
Virginia **Trimble**, Univ. of California & Las Cumbres Observatory, USA vtrimble@uci.edu
Andrea **Urban**, Jet Propulsion Laboratory, California Institute of Technology, USA andrea.urban@jpl.nasa.gov
Sergey **Ustyugov**, Keldysh Institute of Applied Mathematics, Russian Federation Sergey.Ustyugov@gmail.com

Sven **Van Loo**, University of Florida, USA svenvl@astro.ufl.edu
Aristodimos **Vasiliadis**, Centre for Star and Planet Formation, Denmark armivas@gmail.com
Enrique **Vázquez-Semadeni**, Centro de Radioastronomía y Astrofísica, UNAM, Morelia, Mexico e.vazquez@crya.unam.mx
Erika **Verebelyi**, Eotvos Lorand University, Hungary everebelyi@elte.hu
Wolfgang **von Glasow**, University Observatory Munich, Germany glasow@usm.uni-muenchen.de
Eduard **Vorobyov**, The Institute for Computational Astrophysics, Saint Mary's University, Canada vorobyov@ap.smu.ca
Keiichi **Wada**, Kagoshima University, Japan wada@cfca.jp
James **Wadsley**, McMaster University, Canada wadsley@mcmaster.ca
Stefanie **Walch**, Cardiff University, U.K. stefanie.walch@astro.cf.ac.uk
Carsten **Weidner**, School of Physics & Astronomy, University of St Andrews, U.K. Carsten.Weidner@st-andrews.ac.uk
Daniel **Whalen**, Carnegie Mellon University, USA dwhalen@lanl.gov
Emma **Whelan**, Laboratoire d'Astrophysique de Grenoble, France whelane@obs.ujf-grenoble.fr
Anthony **Whitworth**, School of Physics & Astronomy, Cardiff University, U.K. A.Whitworth@astro.cf.ac.uk
Dmitry **Wiebe**, Institute of Astronomy of the RAS, Russian Federation dwiebe@inasan.ru
Nicole **Wityk**, University of Western Ontario, Canada nwityk@uwo.ca
Richard **Wunsch**, Astronomical Institute, Academy of Sciences of the Czech Republic, Czech Republic richard@wunsch.cz
Masako **Yamada**, Institute of Astronomy and Astrophysics, Academia Sinica, Taiwan masako@asiaa.sinica.edu.tw
Jason **Ybarra**, University of Florida, USA jybarra@astro.ufl.edu
Jincheng **Yu**, Shanghai Astronomical Observatory, Chinese Academy of Sciences, China yujc@shao.ac.cn
Olga **Zakhozhay**, Main Astronomical Observatory NAS of Ukraine, Ukraine zkholga@mail.ru
Manuel **Zamora**, CRyA-UNAM, Mexico m.zamora@crya.unam.mx
Sergey **Zamozdra**, Chelyabinsk State University, Russian Federation sezam@csu.ru
Nicola **Schneider-Bontemps**, CEA Saclay, France nicola.schneider-bontemps@cea.fr
Imma **Sepúlveda**, Universitat de Barcelona, Spain inma@rocknrock.com
Hans **Zinnecker**, Astrophysical Institute Potsdam and SOFIA Science Center, Germany hzinnecker@aip.de

Computational Star Formation
Proceedings IAU Symposium No. 270, 2011
J. Alves, B. G. Elmegreen, J. M. Girart, V. Trimble, eds.

doi:10.1017/S1743921311000093

Historical Perspective on Computational Star Formation

Richard B. Larson
Department of Astronomy, Yale University,
New Haven, CT 06520-8101, USA
email: richard.larson@yale.edu

The idea that stars are formed by gravity goes back more than 300 years to Newton, and the idea that gravitational instability plays a role goes back more than 100 years to Jeans, but the idea that stars are forming at the present time in the interstellar medium is more recent and did not emerge until the energy source of stars had been identified and it was realized that the most luminous stars have short lifetimes and therefore must have formed recently. The first suggestion that stars may be forming now in the interstellar medium was credited by contemporary authors to a paper by Spitzer in 1941 in which he talks about the formation of interstellar condensations by radiation pressure, but then oddly says nothing about star formation. That may be because, as Spitzer later told me, when he first suggested very tentatively in a paper submitted to The Astrophysical Journal that stars might be forming now from interstellar matter, this was considered a radical idea and the referee said it was much too speculative and should be taken out of the paper. So Spitzer removed the speculation about star formation from the published version of his paper.

But the idea apparently got around anyway, and it was soon developed further by Whipple in a paper that credited Spitzer for the original suggestion. Whipple says in a footnote that although his work was first presented in 1942, its publication was delayed by "various circumstances" until 1946. By that time, the idea that stars are forming now in the interstellar medium had evidently become respectable enough to be published in The Astrophysical Journal, and Whipple's paper may be the first published presentation of it. In 1947, Bok & Reilly called attention to the compact dark clouds in the Milky Way that later became known as Bok globules, and they suggested that these dark globules might be prestellar objects and might form stars, referencing the papers by Spitzer and Whipple. This suggestion was controversial at the time, and it remained so for many years. But in 1948 Spitzer, in an article in Physics Today, laid out what are essentially modern ideas about star formation in dark clouds, and he pointed specifically to the dark globule Barnard 68 as a possible prestellar object, or 'protostar' as he called it.

By the 1950s, the theory of star formation had become a popular subject and many papers were written on it. The most influential one was probably a 1953 paper by Hoyle that introduced the concept of hierarchical fragmentation, whereby a cloud is assumed to collapse nearly uniformly until at some point separating or fragmenting into smaller clouds, which then individually collapse nearly uniformly and repeat the process. The idea of hierarchical fragmentation remained influential for a long time in theoretical work, even though the assumption of uniform collapse was later disproven by numerical calculations.

Numerical work on star formation began in a serious way in the 1960s, and I came into the picture in 1965 when the problem of protostellar collapse was suggested to me by my thesis advisor Guido Münch at Caltech. Originally I had grandiose ideas about calculating galaxy formation, but Guido was skeptical and said "before you try to understand how a galaxy forms, why don't you try to understand how one star forms?" He also suggested that I talk to Robert Christy, who had recently used numerical techniques to study stellar pulsation, and see if I could use similar techniques to calculate the collapse of an interstellar cloud to form a star. I thought that this sounded like an interesting and challenging project, and I went to talk to Christy, a nuclear physicist who had worked in the nuclear weapons program at Los Alamos. He thought that my calculation might be feasible, and he handed me some reprints and preprints, among which was a recently declassified report from the Livermore National Laboratory presenting a numerical method for doing gas dynamics with radiation and shocks that had originally been developed to calculate powerful explosions in the Earth's atmosphere. I realized that I could use some of the same techniques for the star formation problem, and I also recognized in this report the origin of what became the most widely used method for calculating stellar evolution, the 'Henyey method', which had been derived from the same Livermore bomb code by taking out the hydrodynamics. Many of the numerical techniques later used in astrophysics thus had their origins in nuclear weapons research, perhaps not surprisingly given that a nuclear explosion may be the closest terrestrial counterpart to astrophysics, involving similar physical processes.

When I began work on the protostellar collapse problem in late 1965, I had no idea what I would find or how far I would get, but I thought that even a start on the problem would be worthwhile. Along the way I wrote and tested two completely independent codes, Lagrangian and Eulerian, each with its advantages and disadvantages, and I tried as far as possible to replicate my results with both codes to increase my confidence in them. About a year later in late 1966, I completed my first calculation that had started with something like a Bok globule and ended with a pre-main sequence star. The basic result was that the collapsing cloud became so centrally condensed that only a tiny fraction of its mass at the center first attained stellar density, becoming a 'stellar core' that continued to grow in mass by accretion until eventually acquiring most of the initial cloud mass. The essential implication of this was that star formation is largely an accretion process. This was clearly an important result, and I realized that I still had a lot of work to do to demonstrate its correctness and robustness, so I spent another year running more cases and varying the assumptions and approximations involved. Eventually, after much testing, I acquired considerable confidence in my results, and I presented them in my thesis in 1968. In my thesis defense I was careful to note that my calculation was still an idealized case assuming spherical symmetry and neglecting rotation and magnetic fields, which seemed unlikely to be realistic. But one of my examiners, I think it was Peter Goldreich, said "don't be so apologetic, this is a good calculation and you should publish it."

Thus encouraged, I published my results in 1969 and presented them at meetings. They attracted considerable interest, but also received a lot of flak and criticism. There followed about a decade of debate and controversy over whether my results were correct, with some studies yielding conflicting results and with observers producing apparently conflicting observations showing outflows rather than inflows around newly formed stars. But Bok was delighted that I had shown how one of his globules could form a star, and he decided to spend his retirement years as a kind of evangelist for Bok globules. He was vindicated in 1978, when he proudly sent me a photograph he had taken of a dark globule with a Herbig-Haro jet emerging from it, showing that a star had recently formed in this globule. My vindication came in 1980 when two groups, Winkler & Newman and Stahler, Shu, & Taam, published results very similar to mine. More recently, Masunaga & Inutsuka in 2000 considerably refined the spherical collapse calculation and again obtained similar results.

What was learned from all this work that could be credited specifically to the use of numerical methods? Looking back, I think that the most important result of my work might have been the very first one that I found when I got my first collapse code running at the end of 1965. I had written a simple Lagrangian code to calculate isothermal collapse, and the first successful run with this code showed the runaway growth of a sharp central peak in density. I plotted the density distribution logarithmically and noticed that it was approaching a power-law form with $\rho \propto r^{-2}$, a form similar to that of a singular isothermal sphere, even though the cloud was collapsing almost in free fall. This power-law behavior extended to smaller and smaller radii as the collapse continued. Although this result was unexpected, I realized that it could be understood qualitatively in terms of the inward propagation of a pressure gradient from the boundary, and I later found an asymptotic similarity solution showing this behavior and was able to show that the numerical solution was evolving toward it, giving me increased confidence in the result. I also later learned that at about the same time Michael Penston had been doing similar work and finding similar results, and he independently derived the same similarity solution. This 'Larson-Penston solution', as it has been called, has been perhaps the most enduring result of that early work, and similar asymptotic similarity solutions have been found for a variety of other more realistic collapse problems, including non-isothermal and non-spherical collapse and even collapse with rotation and magnetic fields.

Concerning collapse with rotation, I tried in 1972 to calculate the collapse of a rotating cloud with axial symmetry, but this time I got it wrong. My numerical resolution in 2 dimensions, limited by the computers then available, turned out to be inadequate to follow the development of a sharp central density peak, and my calculation showed instead the formation of a ring. Later when we got a bigger computer, I repeated the calculation with a finer grid and got a smaller ring, causing me to wonder whether the ring might go away completely with infinite resolution. The first person to get it right was Michael Norman, and in 1980 Norman, Wilson, & Barton showed that when sufficient care is taken to ensure adequate resolution at the center, the result is not a ring but a centrally condensed disk that evolves in a quasi-oscillatory fashion toward a central singularity. This result was later confirmed in more detail in 1995 by Nakamura, Hanawa, & Nakano, who also derived an asymptotic similarity solution similar in form to the Larson-Penston solution describing the evolution of the disk toward a central singularity. Finally in 1997, Basu showed that a similar asymptotic similarity solution describing evolution toward a central singularity can be derived even when a magnetic field is included in addition to rotation and when ambipolar diffusion is properly included in the calculation.

What these results show is that in all of these cases, star formation begins with the runaway development of a central singularity in the density distribution. This conclusion now seems to be universal, and even in more realistic 3-dimensional simulations of the formation of systems of stars, the formation of each simulated star or 'sink particle' always begins with the sudden appearance of a near-singularity in the density distribution in a place where local collapse is occurring. This might now seem an unsurprising result because stars are essentially mass points or singularities on the scale of interstellar clouds, so that the formation of a star must involve the development of a near-singularity in the density distribution. But this result was not anticipated before the numerical calculations were done by Penston and me, and also by Bodenheimer & Sweigart at about the same time. Even though earlier studies, notably the work of Hayashi & Nakano in 1965, had shown a tendency for collapsing clouds to become increasingly centrally condensed, no one had anticipated the runaway development of a density singularity, and it took computers to discover this result (computers which at the time had far less computing power than your cell phone.) So this seemingly universal feature of star formation can be regarded as a true discovery of numerical work, and as an example of how computation can discover qualitatively new phenomena.

A second apparently universal feature of star formation that has become clear from much computational work over the years is that, when no artificial symmetries are imposed and fully 3-dimensional behavior is allowed to occur, we are immediately in the realm of chaotic dynamics, because only the very simplest physical systems show regular and predictable behavior. Newton famously solved the 2-body problem but failed to solve the 3-body problem because it exhibits chaotic behavior, a phenomenon that is now understood largely on the basis of computational work. Even the restricted 3-body problem, where the third body is massless, is chaotic and can show exceedingly complex and unpredictable behavior. Three-body interactions are almost certainly very common in star formation, and in my 1972 paper on collapse with rotation I had speculated that in reality the result might often be the formation of a triple system that decays into a binary and a single star, yielding binaries and single stars in roughly the right proportions. Such unstable and chaotic behavior is in fact often seen in 3D simulations of the formation of systems of stars, even the first crude ones that I made in 1978, and it is not surprising because as more mass accumulates into the near-singularities or 'sink particles', the system becomes increasingly like a gravitational n-body system whose dynamics is well known to be chaotic. In addition to chaotic gravitational dynamics, another source of chaotic behavior that can be important in star formation is the development of fluid-dynamical turbulence in star-forming clouds.

Because of these effects, even the simplest extension of star formation modeling from one star forming in isolation to two stars forming in a binary system involves chaotic dynamics. Not only is the gravitational dynamics of the gas circulating around the forming stars intrinsically chaotic, but the gas flow can become turbulent, in which case there are two sources of chaotic behavior in the system. Gravitational and MHD instabilities in the gas orbiting around the forming stars might introduce yet additional sources of chaotic behavior. As a result, the formation of a binary system is not a deterministic or predictable process in its details – every calculation will produce a different result. Therefore we can only hope to predict the statistical properties of binary systems. Large 3D simulations are beginning to be able to do this, and they have already yielded some realistic-looking results for the distributions of binary properties, including a very wide spread in separations resulting from the chaotic dynamics. Similar considerations also apply to predicting stellar masses – we can't predict the mass of an individual star, whose accretion history may be very chaotic and irregular, but we might be able to predict the IMF of a large ensemble of stars if we can include enough of the relevant physics. Again, large numerical simulations are beginning to be able to address this problem. Of course, extensive computations are needed to do these things, and powerful computers are required; computers with the power of cell-phone processors are no longer adequate.

These examples illustrate that, in my view, the most valuable contributions that computing can make to science are not numbers but new discoveries and insights. So I hope that the participants in this meeting who are doing computational work on star formation keep this in mind, and I look forward to learning about many new discoveries made by computational work.

References

Basu, S. 1997, *ApJ*, 485, 240
Bodenheimer, P. & Sweigart, A. 1968, *ApJ*, 152, 515
Bok, B. J. & Reilly, E. F. 1947, *ApJ*, 105, 255

Bok, B. J. 1948, in *Centennial Symposia, Harvard Observatory Monographs No. 7* (Cambridge, Mass.: Harvard College Observatory), p. 53
Bok, B. J. 1978, *PASP*, 90, 489
Christy, R. F. 1966, *ApJ*, 144, 108
Hayashi, C. & Nakano, T. 1965, *Prog. Theor. Phys.*, 34, 754
Hoyle, F. 1953, *ApJ*, 118, 513
Jeans, J. H. 1902, *Phil. Trans. Roy. Soc.*, 199, 49
Jeans, J. H. 1929, *Astronomy and Cosmogony* (Cambridge: Cambridge Univ. Press)
Larson, R. B. 1969a, *MNRAS*, 145, 271
Larson, R. B. 1969b, *MNRAS*, 145, 297
Larson, R. B. 1972, *MNRAS*, 156, 437
Larson, R. B. 1978, *MNRAS*, 184, 69
Masunaga, H. & Inutsuka, S. 2000, *ApJ*, 531, 350
Nakamura, F., Hanawa, T., & Nakano, T. 1995, *ApJ*, 444, 770
Newton, I. 1692, letter to Bentley quoted by Jeans (1929), p. 352
Norman, M. L., Wilson, J. R., & Barton, R. T. 1980, *ApJ*, 239, 968
Penston, M. V. 1966, *Roy. Obs. Bull.*, No. 117, 299
Penston, M. V. 1969, *MNRAS*, 144, 425
Spitzer, L., Jr. 1941, *ApJ*, 94, 232
Spitzer, L., Jr. 1948a, in *Centennial Symposia, Harvard Observatory Monographs No. 7* (Cambridge, Mass.: Harvard College Observatory), p. 87
Spitzer, L., Jr. 1948b, *Physics Today*, Vol. 1, No. 5, p. 7
Spitzer, L., Jr. 1949, *ASP Leaflet*, No. 241
Stahler, S. W., Shu, F. H., & Taam, R. E. 1980a, *ApJ*, 241, 637
Stahler, S. W., Shu, F. H., & Taam, R. E. 1980b, *ApJ*, 242, 226
Whipple, F. L. 1946, *ApJ*, 104, 1
Whipple, F. L. 1948, in *Centennial Symposia, Harvard Observatory Monographs No. 7* (Cambridge, Mass.: Harvard College Observatory), p. 109
Winkler, K.-H. A. & Newman, M. J. 1980a, *ApJ*, 236, 201
Winkler, K.-H. A. & Newman, M. J. 1980b, *ApJ*, 238, 311

Computational Star Formation
Proceedings IAU Symposium No. 270, 2011
J. Alves, B. G. Elmegreen, J. M. Girart, V. Trimble, eds.

doi:10.1017/S174392131100010X

Historical perspective on astrophysical MHD simulations

Michael L. Norman[1,2]

[1]Center for Astrophysics and Space Sciences
[2]San Diego Supercomputer Center
La Jolla, CA 92093, USA
email: mlnorman@ucsd.edu

Abstract. This contribution contains the introductory remarks that I presented at IAU Symposium 270 on "Computational Star Formation" held in Barcelona, Spain, May 31 – June 4, 2010. I discuss the historical development of numerical MHD methods in astrophysics from a personal perspective. The recent advent of robust, higher-order accurate MHD algorithms and adaptive mesh refinement numerical simulations promises to greatly improve our understanding of the role of magnetic fields in star formation.

Keywords. ISM: star formation, ISM: magnetic fields, methods: numerical

1. Introduction

It is a distinct pleasure to be invited to speak to you today about numerical MHD simulations of star formation. Moreover it is a great honor to speak second following Richard Larson, whom I consider the founder of computational star formation. As I will relate, his research influenced me in ways he is probably unaware of, and it is nice to have the opportunity to tell that story. I must admit this is the first historical perspectives talk I have been asked to give which means I must be getting old. On the other hand I cannot deny that I have been meddling in computational star formation on and off for 35 years now and have a few reminiscences and battle scars to relate. In this short contribution I do not attempt to be comprehensive about the given topic, but rather describe my personal experiences developing and applying numerical MHD methods to problems of interest, including star formation.

2. Caltech coincidences

Before I do that I must relate a couple of strange coincidences that occurred to me when I was an undergraduate at Caltech which in hindsight foreshadowed my graduate research at Livermore. First, as a new freshman I wandered into Millikan Library–a Caltech landmark–to browse the astronomy and physics library. I saw a shelf filled with beautifully bound red volumes, and picked one off the shelf at random to see what they were. I picked Richard Larson's PhD thesis which I would later, as a graduate student, study in great detail. At the time though I didn't understand anything and could barely comprehend how a PhD thesis came into existence. I flipped through it, impressed with the graphs and equations, and put it back on the shelf. The second foreshadowing occurred when I was a sophomore or junior. I did a term paper on supernova explosions for Peter Goldreich's class on the interstellar medium. In the process I ran across a paper in the Astrophysical Journal written by Jim Wilson on numerical simulations of neutrino-driven iron core collapse supernova explosions. Jim would later become my PhD thesis advisor and suggest a topic in star formation that would eventually bring me into contact with Richard Larson's early research.

3. Livermore Years

I did my PhD thesis on numerical star formation under the supervision of Jim Wilson at the Lawrence Livermore National Laboratory from 1975 to 1980. Jim was one of the true pioneers of numerical astrophysics (Centrella *et al.* 1985), and I was fortunate to have him as my supervisor. He was absolutely fearless when it came to tackling a new problem numerically. This was due to the fact that in the 1960s he had developed 2D multiphysics codes to simulate the internal operations of nuclear weapons, which gave him an encyclopedic knowledge of hydrodynamics and MHD, neutronics and radiative transfer, plasma physics, nuclear reactions, etc. In the late 60s Jim became interested in astrophysics and started to work on core collapse supernovae, relativistic stars, magneto-rotationally driven jets, and, somewhat later, numerical general relativity. In the 1970s Jim had assisted David Black and Peter Bodenheimer at UC Santa Cruz to develop a 2D hydro code which they applied to axisymmetric, rotating, protostellar cloud collapse simulations (Black & Bodenheimer 1975, Black & Bodenheimer 1976). They found the collapse produced a gravitationally bound ring, confirming a result published by Richard Larson in 1972. Jim suggested I look at the stability of this ring to nonaxisymmetric perturbations using a 3D self-gravitating hydro code he had written. I said OK. He gave me two boxes of IBM punch cards and said get to work. I did, and two years later I had my first publication (Norman & Wilson 1978).

For my PhD thesis I developed a new 2D, axisymmetric, Eulerian hydro code to study rotating protostellar cloud collapse. Years later this code would become the basis for the first ZEUS code. I showed that the self-gravitating ring seen by Larson (1972) and Black & Bodenheimer (1976) was a numerical artifact produced by spurious transport of angular momentum (Norman, Wilson & Barton 1980). I presented this result, and the truncation error analysis it was based on, at the 1979 Santa Cruz star formation summer school. Larson, Black, and Bodenheimer were in the audience. Here I was, an unknown graduate student, telling the big names in the field that their results were incorrect in front of the star formation community. Afterwards Richard was very gracious about it.

That work taught me an important lesson about numerical simulations which I have never forgotten and young researchers should not forget: that numerical errors masquerade as physics, and that one needs not to take numerical results at face value. A high level of skepticism needs to be applied to any new and interesting result, because it may simply be wrong. The code may simply be doing the best it can under difficult circumstances. Numerical star formation, with its vast range of scales, is a very difficult problem. This I learned reading Richard Larson's thesis.

4. Protostars and Planets, Tucson, 1978

I became aware of the importance of magnetic fields to star formation when I attended the first Protostars and Planets meeting in Tucson, Arizona in January 1978. That is where I met Richard Larson for the first time. All the big names were there, including George Field, Hannes Alfvén, and Joe Silk. Chaisson and Vrba talked about magnetic field structures in dark clouds. Field talked about conditions in collapsing clouds, and John Scalo talked about the stellar mass spectrum. A combative young astrophysicist by the name of Telemachos Mouschovias presented theoretical models of magnetically supported clouds, and of how ambipolar diffusion would lead to gravitational instability

GMCs → clouds → cores → disks → jets and outflows

Object	Process	B important?	Simulation
Jets and outflows	Hydro-magnetic wind	yes	2 ½ D 3D
Proto-stellar disk	magnetic viscosity MRI	yes	2 ½ D 3D
Cores	magnetic support magnetic braking	maybe	2D, 2 ½ D, 3D
Clouds	MHD turbulence	at some level	3D, 3D AMR
GMCs	Parker instability MRI	at some level	3D, 3D AMR

importance

numerical difficulty

Figure 1. Magnetic fields and star formation. We understand star formation as a sequence of related objects and phenomena involving self-gravity, magnetic fields, and turbulence. The importance of magnetic fields seems to increase with decreasing length scale, while the difficulty of numerical modeling the relevant systems increases with increasing length scale because of the lack of simplifying symmetries.

once a critical mass to magnetic flux ratio was exceeded. This work is exceedingly well known now, but in 1978 it was still rather new. One of my strongest recollections of the conference was the Q & A after Alfvén's talk. Mouschovias and Alfvén were in violent agreement about the fundamental importance of magnetic fields to star formation, but seemed to agree on nothing else. That evening I presented a 16mm movie of my 3D hydrodynamic ring fragmentation instability simulations to a receptive audience. But by then I was convinced I was solving the wrong equations, and that what was really required was 3D MHD simulations with ambipolar diffusion and self-gravity, a tall order. In fact, this was what Jim Wilson suggested I work on for my thesis, but I got side-tracked on the 2D axisymmetric work and then decided it was time to graduate. Nonetheless, the takeaway that astrophysical fluid dynamics is fundamentally MHD, not HD, was strongly impressed on me.

Fig. 1 summarizes the current view of the star formation process, and the role magnetic fields are thought to play. We tend to organize the subject around objects at different length scales, proceeding from the largest (giant molecular clouds or complexes) to the smallest (protostars). In between are clouds, cloud cores, protostellar accretion disks and jets. Magnetic fields appear to be important at all these scales, and at some scales fundamental. In the last column I list the minimum useful computational model to study these objects. The importance of magnetic fields seems to increase with decreasing length scale, while the difficulty of numerical modeling increases with increasing length scale because of the lack of simplifying symmetries.

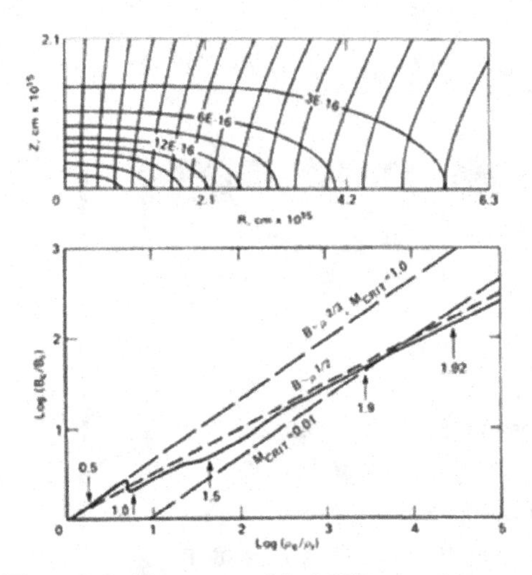

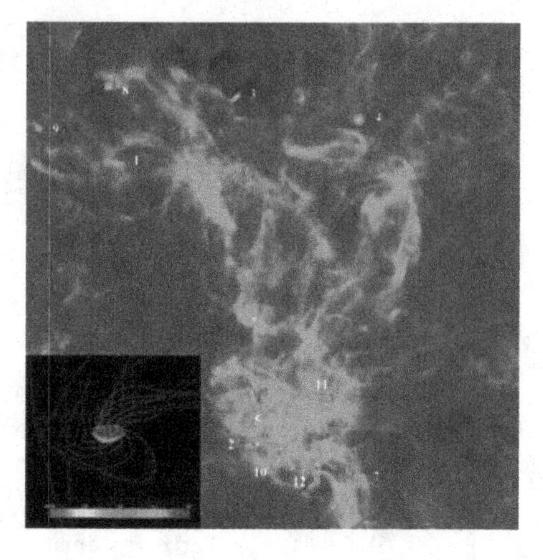

Figure 2. Progress with MHD simulations of star formation. Left: flattened cloud core and central B-rho relation in a 2D non-rotating magnetized collapse simulation (from Scott & Black 1980). Right: self-gravitating cores in a 3D simulation of super-Alfvénic turbulence. Inset: magnetic field topology in a core (from Li *et al.* 2004).

5. Astrophysical Jets

After graduation, my career took a decade-long detour into simulations of astrophysical jets. It was this application, not protostars that got me seriously and permanently involved in developing numerical MHD methods. The VLA had just come online and was producing spectacular radio maps of extragalactic radio jets like those of Cygnus A which were undeniably magnetized. Hydromagnetic launching mechanisms were being proposed by Blandford & Payne (1982) for radio jets, and by Pudritz & Norman (Colin) (1986) and Shibata & Uchida (1985) for protostellar jets. My first simulations of radio jets were purely hydrodynamic, carried out with an improved version of my thesis code. But by 1986 I had incorporated magnetic fields. Working with University of New Mexico radio astronomer Jack Burns and his graduate student David Clarke, I applied this code to magnetically-confined supersonic jet models of extragalactic radio sources (Clarke, Norman & Burns 1986).

6. Evolution of Numerical MHD

6.1. *Early Days*

The development and application of numerical MHD to problems in star formation lagged HD simulations by more than a decade because the simplest nontrivial problem is 2D axisymmetric, where as the early hydrodynamic work could be done in 1D spherical symmetry (e.g., Larson 1969, Westbrook & Tarter 1975). Although Mouschovias had already published by the mid 1970s 2D static models of magnetically supported clouds, it was not until 1980 that the first dynamic MHD simulation was published. Scott & Black (1980) simulated the gravitational collapse of a non-rotating cloud threaded by a uniform magnetic field. They used a first order upwind scheme (donor cell) to evolve the poloidal flux function, ensuring divergence-free poloidal fields. They showed that collapse produces flattened cores as expected, and that the central density and magnetic field scale as $B_c \propto \rho_c^{1/2}$ (Fig. 2a).

Motivated by the recently discovered jets from young stellar objects, Shibata & Uchida (1985) carried out 2-1/2D axisymmetric MHD simulations of hydromagnetically-driven disk wind models. The difference between a 2D and a 2-1/2D simulation is that in rotating axisymmetric systems, toroidal velocity and magnetic components are also evolved. Their so-named sweeping magnetic twist mechanism rediscovered much earlier work by LeBlanc & Wilson (1970) in which rotation efficiently coverts poloidal B-fields into toroidal B-fields, producing what is in effect a coiled magnetic spring that uncoils along the rotation axis due to magnetic pressure, launching a jet. They evolved all three components of B using the second order Lax-Wendroff method, stabilized with artificial viscosity. Such an approach is not guaranteed to maintain divergence-free B-fields.

Clarke, Burns & Norman (1989) performed 2-1/2D MHD simulations of extragalactic radio jets using the original code called ZEUS. The code evolved the poloidal flux function and the toroidal component of the magnetic field using 2nd-order upwind finite differences. This ensures divergence-free magnetic fields, as can easily be demonstrated. The poloidal flux function is defined $a_\phi = rA_\phi$, where r is the cylindrical radius and A_ϕ is the magnetic vector potential. We then have $B_r = -\frac{1}{r}\frac{\partial a_\phi}{\partial z}, B_z = \frac{1}{r}\frac{\partial a_\phi}{\partial r}$. By virtue of the axisymmetry of the toroidal field B_ϕ it is evident that

$$\nabla \cdot \vec{B} = \frac{\partial B_z}{\partial z} + \frac{1}{r}\frac{\partial r B_r}{\partial r} + \frac{1}{r}\frac{\partial B_\phi}{\partial \phi} = +\frac{1}{r}\frac{\partial^2 a_\phi}{\partial z \partial r} - \frac{1}{r}\frac{\partial^2 a_\phi}{\partial r \partial z} + 0 = 0.$$

Faraday's law for evolving the magnetic field becomes

$$\frac{\partial B_\phi}{\partial t} + \frac{\partial}{\partial r}(B_\phi v_r) + \frac{\partial}{\partial z}(B_\phi v_z) = r\vec{B} \cdot \nabla \Omega$$

$$\frac{\partial a_\phi}{\partial t} + v_r \frac{\partial a_\phi}{\partial r} + v_z \frac{\partial a_\phi}{\partial z} = 0,$$

where $\Omega = v_\phi / r$. These equations were evolved in ZEUS using a second-order monotonic upwind scheme alongside the hydrodynamic equations, with the Lorentz force term constructed from first and second difference of B_ϕ and A_ϕ. This was a very neat, stable, and reasonably accurate scheme for 2-1/2D MHD simulations. However it could not be generalized to 3D, and therefore a divergence-free method working directly with the components of B had to be found.

6.2. *Constrained Transport*

Fortunately, in 1988 Evans & Hawley solved half the problem when they introduced the Constrained Transport (CT) method. CT solves the magnetic induction equation in integral form and uses a particular centering of the magnetic and velocity field components in the unit cell so as to transport vector B through a 3D mesh in a divergence-free way. For a recent exposition of this see Hayes *et al.* (2006). I say they solved only half the problem because what they addressed was how to treat the kinematics of magnetic fields, not their dynamics. As we discuss below, an accurate and stable treatment of the dynamics of magnetic fields requires judicious choices for how the EMFs and Lorentz force terms are evaluated.

6.3. *ZEUS and Sons*

In 1987 University of Illinois grad student Jim Stone and I set out to build a version of the Clarke-Norman ZEUS code that evolved (B_z, B_r, B_ϕ) in a divergence-free way using CT (Evans visited NCSA in 1987 and told us about it). We figured if we could make this work in 2-1/2D, it could easily be generalized to 3D. The end result of this effort was a code called ZEUS-2D (Stone & Norman 1992a,b), developed by Jim, and a

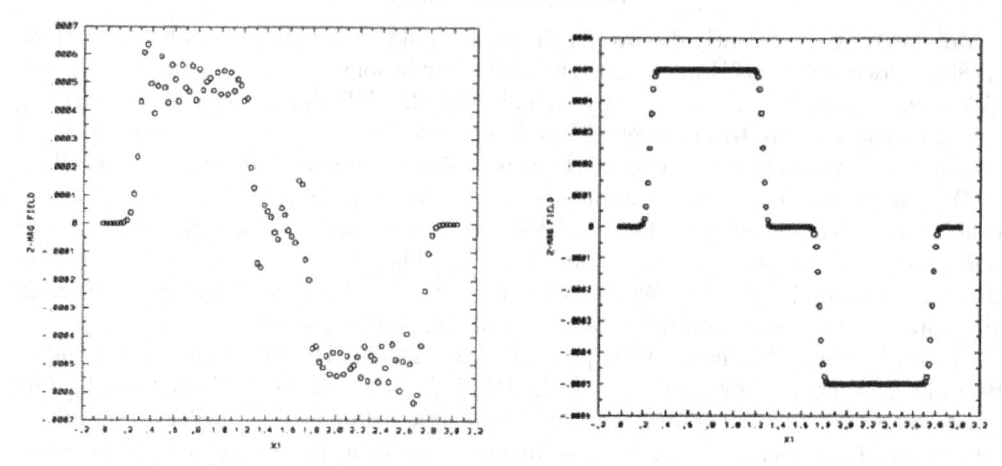

Figure 3. Importance of upwinding. Shear Alfvén wave test problem. *Left:* computed using original CT algorithm of Evans & Hawley (1988); *Right:* computed using MOC-CT algorithm of Stone & Norman (1992b). From Stone & Norman (1992b).

code called ZEUS-3D, developed by Clarke who became my postdoc in 1988. Jim and I were motivated to improve on the hydromagnetic disk wind simulations of Shibata & Uchida (1985). When we tried CT as described by Evans & Hawley (1988), it failed miserably to stably evolve the torsional Alfvén waves generated when the rotating disk starts twisting the initial poloidal field (see Fig. 3). The reason for this is that the EMFs used in the vanilla CT scheme were not upwind in the Alfvén wave characteristics, but rather were computed using simple centered differences and averages. Jim and I came up with a different way to calculate the EMFs using a Method of Characteristics approach specifically for Alfvén waves. The resulting hybrid scheme we called MOC-CT. It worked beautifully on the torsional Alfvén wave problem (Fig. 3b) and for 2-1/2D simulations magnetized accretion disks (Stone & Norman 1994).

Using ZEUS-2D, we accidently discovered the MRI in 1989 (Norman & Stone 1990) but didn't realize the significance of what we were seeing. Several years later, with the pioneering work of Balbus and Hawley, we realized what we had computed was the axisymmetric channel solution of the MRI (Stone & Norman 1994). This anecdote is the counterexample to what I said above about being skeptical of numerical results. In this case, the simulations contained a discovery which we failed to recognize. It was present in the simulation because we had improved the algorithm to the point where we didn't need excessive amounts of artificial viscosity to damp numerical instabilities.

In 1989, David Clarke began a 3D implementation of MOC-CT algorithm which resulted in the ZEUS-3D code. However he encountered explosive numerical instabilities when he applied it to magnetized extragalactic jet models. The lobes of these jets exist in a state of super-Alfvénic turbulence, which, as we later learned from simulations of molecular cloud turbulence, is very tough on numerical schemes. Hawley and Stone eliminated the numerical instability using the tried and true method of adding numerical dissipation (Hawley & Stone 1995). This fix was incorporated into ZEUS-3D, and at that point it was publicly released via the Laboratory for Computational Astrophysics (LCA) website.

Subsequently, ZEUS-3D became widely used for many different kinds of applications including some of the earliest work on decay rates in turbulent molecular clouds (MacLow *et al.* 1998). The stabilized MOC-CT of Hawley and Stone made its way into other code implementations by Hawley, Stone, Gammie, Eve Ostriker, and others who have done important work on the MRI, molecular cloud turbulence, protostellar accretion disks,

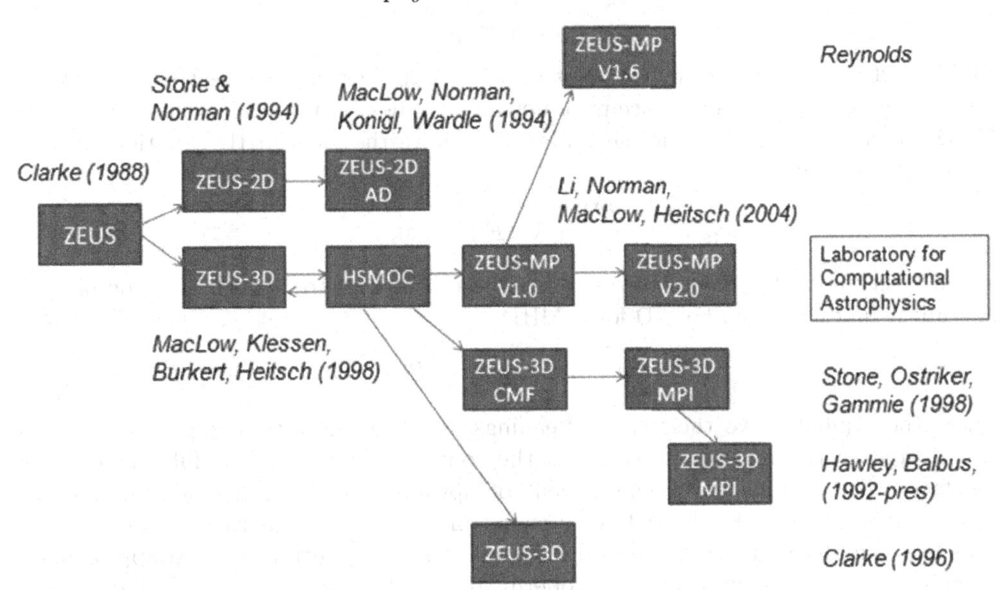

Figure 4. The "ZEUS diaspora". ZEUS' MHD algorithms have made their way into a number of code implementations. A few citations to significant contributions to computational star formation are included.

and Galactic ISM dynamics. The LCA developed its own MPI-parallel version of ZEUS-3D called ZEUS-MP (Norman 2000) which is now in its version 2.0 release (Hayes *et al.* 2006). Fig. 4 shows the "ZEUS diaspora" to the best of my knowledge.

To cite just a few significant applications of ZEUS to computational star formation, I would mention MacLow *et al.* (1998) and Stone, Ostriker & Gammie (1998) on turbulence decay rates in molecular clouds, Heitsch, MacLow & Klessen (2001) on self-gravitating molecular cloud turbulence, Gammie *et al.* (2003) and Li *et al.* (2004) on gravitationally bound core formation in turbulent molecular clouds, and MacLow *et al.* (1995) incorporating ambipolar diffusion into ZEUS.

6.4. *A solar physicist gets involved*

The field of computational star formation was enlivened when Paolo Padoan went to Copenhagen for his PhD research in the mid-90s. There he joined forces with Åke Nordlund, a prominent solar physicist who not surprisingly was in possession of a 3D compressible MHD code called the STAGGER code. The codes solved the ideal MHD equations in non-conservative form on a 3D staggered mesh of size 128^3, using higher-order finite differences, and stabilized using artificial viscosity. The divergence-free condition on the magnetic field was not enforced. Using this code they carried out isothermal compressible MHD turbulence-in-a-box simulations ignoring gravity and other effects. Varying the Alfvén Mach number, they showed that a super-Alfvénic model more closely match a variety of molecular cloud observations than a trans-Alfvénic model (Padoan & Nordlund 1999). This set the stage for their turbulent fragmentation theory of star formation (Padoan & Nordlund 2002) which has been very influential in the field. They argued that there is a direct link between the mass function of gravitationally bound cores and the statistical properties of super-Alfvénic turbulence. Pakshing Li, Mordecai Mac-Low, Fabian Heitsch and myself verified this claim with the first 512^3 simulation of self-gravitating, super-Alfvénic MHD turbulence using ZEUS-MP (Li *et al.* 2004).

6.5. *The Rise of Upwind Schemes*

The last 20 years have witnessed a lot of algorithmic development activity in what is generically called higher-order accurate upwind schemes (or Godunov schemes) for ideal MHD. While details differ, the basic idea is to write the ideal MHD equations in fully conservative form:

$$\frac{\partial \vec{U}}{\partial t} + \nabla \cdot \vec{F}(\vec{U}) = 0,$$

where U is the vector of unknowns, and F is the flux vector, which is a complicated non-linear function of U. For 3D ideal MHD

$$U^T = (\rho, \rho v_x, \rho v_y, \rho v_z, \mathcal{E}, B_x, B_y, B_z)$$

where the symbols have their usual meanings and $\mathcal{E}$ is the total energy. Schemes use the divergence theorem to update U in the control volume cells by differencing F on the faces. The entire burden and benefit of upwind schemes is to find accurate and stable representations for F that are upwind in all the wave characteristics of MHD. This is accomplished through the use of Riemann solvers, both exact and approximate, of which there are many available. Modern MHD codes implementing upwind schemes are built in a modular fashion, mixing and matching half a dozen basic ingredients in different ways. Within the conservation law solver, these are: order accuracy of the spatial interpolation, order of accuracy of the temporal integration, choice of Riemann solver, choice of monotonicity-preserving flux limiters, and directional splitting versus unsplit. Three basic approaches to maintaining the divergence free condition are: 1) ignore it; 2) clean it (elliptic, hyperbolic); 3) prevent it (constrained transport). Varying these choices leads to hundreds of potential combinations, not all of which have been explored. In the following I give a very brief survey of existing methods.

Zachary & Colella (1992) developed an exact solver for the MHD Riemann problem. Ryu *et al.* (1995, 1998) developed 2D and 3D ideal MHD codes based on the Total Variation Diminshing (TVD) method. Dai & Woodward (1994, 1998) generalized the Piecewise Parabolic Method (PPM) to ideal MHD. Balsara (1998a,b) developed a linearized Riemann solver and improved TVD schemes for adiabatic and isothermal MHD. Balsara & Spicer (1999) incorporated Constrained Transport (CT) into a TVD MHD scheme. Powell *et al.* (1999) introduced elliptic divergence cleaning. Londrillo & del Zanna (2000) introduced further improvements to a TVD MHD scheme. Dedner *et al.* (2002) introduced hyperbolic divergence cleaning. Gardner & Stone (2005) developed an improved PPM+CT scheme and stressed the importance of directionally unsplit schemes. Popov & Ustyugov (2008) introduce a PPM on a local stencil (PPML) and married it to CT. Ustyugov *et al.* (2009) added stability improvements to PPML+CT and demonstrated its application to super-Alfvénic turbulence.

6.6. *AMR MHD*

The severe dynamic range requirements to resolve gravitational collapse in multi-dimensions requires nested grids or adaptive mesh refinement (AMR) to attack. In pioneering work, Dorfi (1982) first introduced static hierarchically refined grids into calculations of isolated cloud collapse, including magnetic fields, self-gravity and rotation. Using a simple finite difference scheme he showed that magnetic breaking is 10 times as efficient in the perpendicular rotator case as in the aligned rotator case. He found that bar-like structures are produced by the collapse and breaking. Hierarchical nested grid codes were also developed by Ziegler & Yorke (1997), Tomisaka (1998), and Machada *et al.* (2005, 2005)

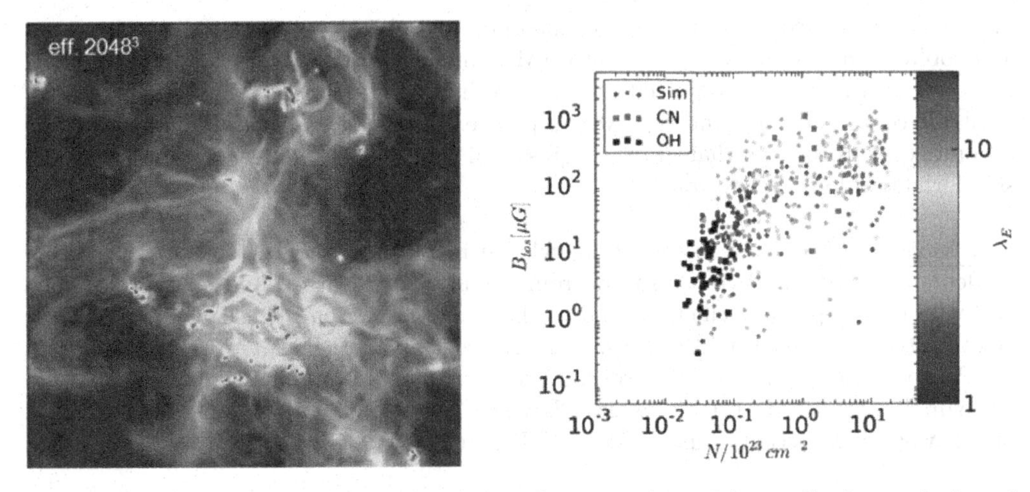

Figure 5. AMR MHD simulation of turbulent fragmentation, with an effective resolution of 2048^3. *Left:* projected gas density in the 10 pc box. *Right:* comparison of simulated and observed Zeeman measurements in dense cores. From Collins *et al.* (2010b).

and applied to isolated core collapse of ever increasing dynamic range. The last authors showed that disk fragmentation is sensitive to magnetic field strength and inclination.

True AMR MHD has only come onto the scene recently due to the numerical challenges involved. It builds on progress made with AMR hydro codes developed by Berger & Colella (1989), Bryan & Norman (1997), Truelove *et al.* (1998), Fryxell *et al.* (2000), and Teyssier (2002). The first AMR MHD code was the RIEMANN code by Balsara (2001). This was followed by the FLASH code (Linde 2002), the NIRVANA code (Ziegler 2005), the RAMSES code (Fromang *et al.* 2006), and the ENZO code (Collins *et al.* 2010a). The last of these, developed by my graduate student David Collins, is the result of a 5 year effort to marry a higher order upwind scheme for ideal MHD with CT on a block structured adaptive mesh. We tried a quite a number of different conservation law solvers, Riemann solvers, and CT strategies before we found one stable enough to deal with the rigors of super-Alfvénic turbulence.

7. Results, Finally!

Fig. 5 shows results ENZO-MHD applied to turbulent fragmentation with self-gravity with an effective resolution of 2048^3 (Collins *et al.* 2010b). On the left is projected gas density through the box at 0.75 free-fall times. On the right is a scatter plot of the simulated LOS magnetic field strength versus the gas column density in bound cores overlaid on Zeeman observations. Color coding indicates the ratio of gravitational to magnetic energy in the cores. The simulation is in good agreement with the observations, lending further support to the turbulent fragmentation picture.

8. Brand New Day

The AMR MHD simulation shown above was beyond my wildest dreams when I was a graduate student listening to Mouschovias and Alfvén square off at Protostars and Planets I. After all these years, I think the field of numerical MHD is finally where we want it to be. With AMR MHD we are finally solving the problem Nature hands us, not some reduced problem (although we have learned a tremendous amount solving reduced problems.) Admittedly, we still have to incorporate ambipolar diffusion, dust, chemistry and cooling, and radiative transfer into AMR simulations. However, the progress being made on all these fronts reported at this meeting is very encouraging. The arsenal of available codes and the number of young people engaged in their development and use encourage me to believe that great progress will continue to be made in computational star formation for years to come.

Acknowledgements. I gratefully acknowledge my former and current star formation students, postdocs, and collaborators from whom I have learned and enjoyed so much and who have kept me challenged and in the game: Tom Abel, Dinshaw Balsara, Greg Bryan, David Clarke, David Collins, John Hayes, Fabian Heitsch, Alexei Kritsuk, Hui Li, Pakshing Li, Shengtai Li, Mordecai MacLow, Åke Nordlund, Brian O'Shea, Paolo Padoan, Jim Stone, Matt Turk, Sergey Ustyugov, Rick Wagner, and Dan Whalen. Some of this work was partially supported by NSF grant AST-0808184.

References

Balsara, D. S. 1998a, *ApJS*, 116, 119
Balsara, D. S. 1998b, *ApJS*, 116, 133
Balsara, D. S. 2001, *J. Comp. Phys.*, 174, 614
Balsara, D. S. & Spicer, D. S. 1999, *J. Comp. Phys.*, 149, 270
Berger, M. J. & Colella, P. 1989, *J. Comp. Phys.*, 82, 64
Black, D. C. & Bodenheimer, P. 1975, *ApJ*, 199, 619
Black, D. C. & Bodenheimer, P. 1976, *ApJ*, 206, 138
Blandford, R. D. & Payne, D. 1982, *MNRAS*, 199, 883
Bryan, G. L. & Norman, M. L. 1997, *Computational Astrophysics, 12th Kingston Meeting on Theoretical Astrophysics*, ASP Conference Series No. 123, eds. D. A. Clarke & M. J. West., 363
Centrella, J., LeBlanc, J., Bowers, R., & Wheeler, J. (eds.) 1985, *Numerical Astrophysics. Proceedings of a symposium in honor of James R. Wilson* (Portola Valley: Jones & Bartlett)
Clarke, D. A., Norman, M. L., & Burns, J. O. 1986, *ApJL*, 311, L63
Clarke, D. A., Burns, J. O., & Norman, M. L. 1989, *ApJ*, 342, 700
Collins, D. C., Xu, H., Norman, M. L., Li, H., & Li, S. 2010a, *ApJS*, 186, 308
Collins, D. C., Padoan, P., Norman, M. L., & Xu, H. 2010b, arXiv:1008.2402
Dai, W. & Woodward, P. R. 1994, *J. Comp. Phys.*, 115, 485
Dai, W. & Woodward, P. R. 1998, *ApJ*, 494, 317
Dedner, A., Kemm, F., Kroener, D., Munz, C.-D., Schnitzer, T., & Wesenberg, M., 2002, *J. Comp. Phys.*, 175, 645
Dorfi, E., 1982, *A&A*, 114, 151
Evans, C. R. & Hawley, J. F. 1988, *ApJ*, 592, 203
Fromang, S., Hennebelle, P., & Teyssier, R. 2006, *A&A*, 457, 371
Fryxell, B., Olson, K., Ricker, P., Timmes, F. X., Zingale, M., Lamb, D. Q., MacNeice, P., Rosner, R., Truran, J. W., Tufo, H. 2000, *ApJS*, 131, 273
Gammie, C. F., Lin, Y.-T., Stone, J. M., & Ostriker, E. C. 2003, *ApJ*, 547, 280
Gardner, T. A. & Stone, J. M. 2005, *J. Comp. Phys.*, 205, 509
Hawley, J. F. & Stone, J. M. 1995, *Comp. Phys. Comm.*, 89, 127
Hayes, J. C., Norman, M. L., Fiedler, R. A., Bordner, J. O., Li, P.-S., Clark, S. E., ud-Doula, A., & Mac Low, M.-M. 2006, *ApJS*, 165, 188
Heitsch, F., MacLow, M., & Klessen, R. 2001, *ApJ*, 547, 280
Larson, R. B. 1969, *MNRAS*, 145, 271
Larson, R. B. 1972, *MNRAS*, 156, 437
LeBlanc, J. M. & Wilson, J. R. 1970, *ApJ*, 161, 541
Li, P.-S., Norman, M. L., MacLow, M., & Heitsch, F. 2004, *ApJ*, 605, 800

Linde, T. 2002, *Annual APS March Meeting, March 18 - 22, 2002*, abstract #F3.005
Londrillo, P. & del Zanna 2000, *ApJ*, 530, 508
MacLow, M., Klessen, R., Burkert, A., & Smith, M. D. 1998, *Phys. Rev. Lett.*, 80, 2754
Machada, M. N., Matsimoto, T., Hanawi, T., & Tomisaka, K. 2005, *MNRAS*, 362, 382
Machada, M. N., Matsimoto, T., Hanawi, T., & Tomisaka, K. 2006, *MNRAS*, 645, 1227
MacLow, M.-M., Norman, M. L., Konigl, A., & Wardle, M. 1995, *ApJ*, 442, 726
Norman, M. L. 2000, Arthur, J., Brickhouse, N., & Franco, J. (eds.), *Astrophysical Plasmas: Codes, Models, and Observations RMxCA*, 9, 66
Norman, M. L. & Stone, J. M. 1990, Beck, R. (ed.) 1990, *Galactic and Intergalactic Magnetic Fields. IAU Symposium 140* (Dordrecht, Netherlands, Kluwer)
Norman, M. L. & Wilson, J. R. 1978, *ApJ*, 224, 497
Norman, M. L., Wilson, J. R., & Barton, 1980, *ApJ*, 239, 968
Padoan, P. & Nordlund, A. 1999, *ApJ*, 526, 279
Padoan, P. & Nordlund, A. 2002, *ApJ*, 576, 870
Popov, M. V. & Ustyugov, S. D. 2008 2008, *Comp. Math. Math. Phys.*, 48 (3), 477
Powell, K. G., Roe, P. L., Linde, T. J., Gombosi, T. I., & de Zeeuw, D. L. 1999, *J. Comp. Phys.*, 154, 284
Pudritz, R. E. & Norman, C. A. 1986 1986, *ApJ*, 301, 571
Ryu, D., Jones, T. W., & Frank, A. 1995, *ApJ*, 452, 785
Ryu, D., Miniati, F., Jones, T. W., & Frank, A. 1998, *ApJ*, 509, 244
Scott, E. H. & Black, D. 1980 1980, *ApJ*, 239, 166
Shibata, K. & Uchida, Y. 1985, *PASJ*, 37, 31
Stone, J. M. & Norman, M. L. 1992a, *ApJS*, 80, 753
Stone, J. M. & Norman, M. L. 1992b, *ApJS*, 80, 791
Stone, J. M. & Norman, M. L. 1994, *ApJ*, 433, 746
Stone, J. M., Ostriker, E., & Gammie, C. F. 1998, *ApJL*, 508, L99
Teyssier, R. 2002, *A&A*, 385, 337
Tomisaka, K., 1998, *ApJ*, 502, 163
Truelove, J. K., Klein, R. I., McKee, C. F., Holliman, J. H., Howell, L. H., Greenough, J. A., & Woods, D. T. 1998 *ApJ*, 495, 821
Ustyugov, S. D., Popov, M. V., Kritsuk, A. G., & Norman, M. L. 2009, *J. Comp. Phys.*, 228, 7614
Westbrook, C. K. & Tarter, C. B. 1975, *ApJ*, 200, 48
Zachary, A. L. & Colella, P. 1992, *J. Comp. Phys.*, 99, 341
Ziegler, U. 2005, *A&A*, 435, 385
Ziegler, U. & Yorke, H. A. 1997, *Comp. Phys. Comm.*, 101, 54

Robert Estallela (arm raised) and Michael Norman (right) among others at the banquet

Computational Star Formation
Proceedings IAU Symposium No. 270, 2011
J. Alves, B.G. Elmegreen, J. M. Girart & V. Trimble, eds.

doi:10.1017/S1743921311000111

Modeling the prestellar cores in Ophiuchus

Ant Whitworth[1], Dimitri Stamatellos[1] and Steffi Walch[1]

[1]School of Physics & Astronomy, Cardiff University, 5 The Parade, Cardiff CF24 3AA, UK
emails: A.Whitworth@astro.cf.ac.uk, Dimitris.Stamatellos@astro.cf.ac.uk,
Stefanie.Walch@astro.cf.ac.uk

Abstract. We present the first results from a project to model the prestellar cores in Ophiuchus, using initial conditions constrained as closely as possible by observation. The prestellar cores in Ophiuchus appear to be evolving in isolation — in the sense that the timescale on which an individual prestellar core collapses and fragments is estimated to be much shorter than the timescale on which it is likely to interact dynamically with another core. Therefore it is realistic to simulate individual cores separately, and this in turn makes it feasible (a) to perform multiple realisations of the evolution of each core (to allow for uncertainties in the initial conditions which persist, even for the most comprehensively observed cores), and (b) to do so at high resolution (so that even the smallest protostars are well resolved). The aims of this project are (i) to address how best to convert the observations into initial conditions; (ii) to explore, by means of numerical simulations, how the observed cores are likely to evolve in the future; (iii) to predict the properties of the protostars that they will form (mass function, multiplicity statistics, etc.); and (iv) to compare these properties with the properties of the observed pre-Main Sequence stars in Ophiuchus. We find that if the observed non-thermal velocities in the Ophiuchus prestellar cores are attributed to purely solenoidal turbulence, they do not fragment; they all collapse to form single protostars. If the non-thermal velocities are attributed to a mixture of solenoidal and compressive turbulence, multiple systems form readily. The turbulence first generates a network of filaments, and material then tends to flow along the filaments, at first into a primary protostar, and then onto a compact accretion disc around this protostar; secondary protostars condense out of the material flowing into the disc along the filaments. If the turbulence is purely solenoidal, but part of the non-thermal velocity dispersion is attributed to solid-body rotation, then again multiple systems form readily, but the pattern of fragmentation is quite different. A primary protostar forms near the centre of the core, and then an extended accretion disc forms around the primary protostar, and eventually becomes so massive that it fragments to produce low-mass secondaries; these frequently end up in hierarchical multiple systems.

Keywords. star formation, binary formation, prestellar cores, turbulence

---

1. Introduction

At an estimated distance between 120 and 140 pc, Ophiuchus is one of the nearest sites of ongoing, mildly clustered star formation, and it has been extensively observed at many different wavelengths. The aim of this project is to explore whether these observations can be used to inform the initial conditions for simulations of the evolution of the observed prestellar cores in Ophiuchus, and to compare the resulting protostars with the observed population of protostars and pre-Main Sequence stars in Ophiuchus. Such a project is timely because recent estimates of the mass function of the prestellar cores in Ophiuchus (Motte *et al.* 1998, Johnstone *et al.* 2000, Simpson *et al.* 2010) indicate that it is very similar in shape to the stellar initial mass function, but shifted to higher masses by a factor of order three. This in turn implies that the efficiency of star formation in a prestellar core is $\sim 33\,\%$. It is therefore appropriate to ask how the core mass function maps into the stellar mass function. Since this is a highly non-linear mapping, the only way to explore it is through numerical simulations.

The work of André *et al.* (2007) suggests that the prestellar cores in Ophiuchus evolve more-or-less in isolation. In other words, an individual core collapses and fragments to form a protostar, or a small-N system of protostars, on a timescale that is sufficiently short that the process is complete before the core has time to interact with one of the other cores in the cloud. Therefore we can model an individual core with up to 10,000 SPH particles per $5\,\mathrm{M}_{\mathrm{JUPITER}}$. Since protostars significantly less massive than this are unlikely to form, this means that all protostars are extremely well resolved.†

2. Initial conditions

By combining observations of dust continuum emission (Motte André & Neri, 1998; Johnstone *et al.*, 2000; Stanke *et al.*, 2006; Simpson *et al.*, 2010) with observations of molecular-line emission, in particular N_2H^+ $(1-0)$ emission (André *et al.*, 2007) and $HCO^+(1-0)$ emission (Simpson *et al.*, 2010), and with detailed radiation transport modeling (Stamatellos *et al.*, 2007; Simpson *et al.*, 2010), we have estimates – for 43 cores – of the mass, the extent on the sky, the velocity dispersion along the line of sight, the ambient radiation field, and the temperature. Strictly speaking, this is the dust temperature, but it is probably also close to the gas-kinetic temperature. We have most of this information for a further 30+ cores (also in Ophiuchus).

This does not specify the initial conditions exactly. First, even if we constrain the 3D shape of a core to be a triaxial ellipsoid, we do not know the extent along the line of sight. However, Goodwin, Whitworth & Ward-Thompson (2004) suggest that this may not be critical, and therefore we start by taking all the cores to be spherically symmetric, with radius equal to the geometric mean of the projected minor and major axes.

Second we do not know the detailed density profile of a core. For simplicity, we assume that all cores have the density profile of a critical Bonnor-Ebert sphere, since many low-mass prestellar cores appear to approximate to this form. However, that we are not assuming that a core is in hydrostatic balance. In general, the cores are thermally supercritical, i.e. they do not have sufficient thermal pressure to support themselves. We note that the self-gravitational potential energy, moment of inertia, and full width at half maximum for optically thin emission from a critical Bonnor-Ebert sphere are given by

$$-\,\Omega_{\mathrm{CORE}} = \frac{0.732\,G\,M_{\mathrm{CORE}}^2}{R_{\mathrm{CORE}}}, \tag{2.1}$$

$$I_{\mathrm{CORE}} = 0.283\,M_{\mathrm{CORE}}\,R_{\mathrm{CORE}}^2, \tag{2.2}$$

$$D_{\mathrm{FWHM}} = 0.802\,R_{\mathrm{CORE}}. \tag{2.3}$$

Third, we do not know the split of non-thermal energy between (i) statistically isotropic random turbulence, (ii) ordered global rotation, and (iii) ordered in- or out-flow. Further we do not know how the turbulence is split between solenoidal and compressive modes,

† We are here using the term protostar to embrace any object that forms by gravitational instability, on a dynamical timescale, and with an approximately uniform interstellar elemental composition – as distinct from planets, which form by core accretion, on a much longer timescale and with an initially fractionated composition.

whether the rotation is solid-body or differential, and whether the in- or out-flow is homologous or isotropic. These appear to be the most critical issues, and these are the issues that we focus on here.

3. Numerical method

We use the recently developed SPH code SEREN (Hubber *et al.* 2010). Radiative cooling and the associated radiation transport are treated using the method of Stamatellos *et al.* (2007). Sink particles are introduced where the density rises above $\rho_{\rm SINK}$, provided that the condensation at this location is gravitationally bound, and the divergences of the acceleration is negative. We use $\rho_{\rm SINK} = 10^{-9}\,{\rm g\,cm^{-3}}$, so that all protostars are well into their Kelvin-Helmholtz contraction phase before they are replaced with sink particles. We do not include MHD effects (either ideal or non-ideal).

4. Ophiuchus Core B2-MM8

As an illustration of the results, we discuss just two cores. The first core is Ophiuchus B2-MM8. This core is embedded in clump B2. Its estimated mass is $M_{\rm CORE} = 0.77\,{\rm M_\odot}$, and its estimated radius is $R_{\rm CORE} = 0.032\,{\rm pc}$ (corresponding to a full-width at half maximum in optically thin dust continuum emission of 0.026 pc). Its estimated temperature is $T_{\rm CORE} = 10\,{\rm K}$. The estimated ratios of thermal and non-thermal energy to gravitational potential energy are $\alpha_{\rm THERM} = 0.07$ and $\alpha_{\rm NONTH} = 0.11$. The core is therefore presumed to be both thermally and non-thermally supercritical. We have simulated the evolution of this core with the non-thermal energy divided between purely solenoidal turbulence and solid-body rotation in the proportions $1:2$ and $2:1$. The solenoidal turbulence is assumed to have a power spectrum $P_k \propto k^{-4}$. In both cases the central low-angular momentum parts of the core collapse to form a primary protostar, and then an extended accretion disc forms around this. In the case where only one third of the non-thermal energy is attributed to rotation, the disc fragments after about 32 kyr, to form five protostars, and these end up in an hierarchical quintuple system with separations ranging from $\sim 2\,{\rm AU}$ to $\sim 200\,{\rm AU}$. In the case where two thirds of the non-thermal energy is attributed to solid-body rotation, the disc takes longer to form and is more extended. After about 37 kyr, the disc quickly fragments, producing six further protostars; three are ejected, one becomes a close companion to the primary at $\sim 20\,{\rm AU}$, and the remaining two form a close binary system orbiting the primary at $\sim 170\,{\rm pc}$. We emphasise that in all cores that we have simulated with purely solenoidal turbulence, there is no fragmentation. When the turbulence is purely solenoidal, the addition of ordered rotation is essential for fragmentation to occur, and it often leads to the formation of hierarchical multiples.

5. Ophiuchus Core A-MM8

The second core to be discussed is Ophiuchus A-MM8. This core is embedded in clump A. Its estimated mass is $M_{\rm CORE} = 1.28\,{\rm M_\odot}$, and its estimated radius is $R_{\rm CORE} = 0.025\,{\rm pc}$ (corresponding to a full-width at half maximum in optically thin dust continuum emission of 0.020 pc). Its estimated temperature is $T_{\rm CORE} = 11\,{\rm K}$. The estimated ratios of thermal and non-thermal energy to gravitational potential energy are $\alpha_{\rm THERM} = 0.03$

and $\alpha_{\rm NONTH} = 0.02$. The core is therefore presumed to be both thermally and non-thermally supercritical. We have simulated the evolution of this core with the non-thermal energy attributed entirely to turbulence, with a random mix of solenoidal and compressive modes and a power spectrum $P_k \propto k^{-4}$. In this case the collapse of the core leads to the formation of a network of filaments, along which material flows into the central regions. There a primary protostar forms at ~ 10 kyr and grows by accreting the material that flows in along the filaments. There are three main filaments, and because these filaments are not aligned, the inflow brings in material with high angular momentum relative to the primary. This material quickly forms a compact disc (diameter of order 30 AU) around the primary, and the disc then fragments to produce a secondary protostar at ~ 11.4 kyr. The two protostars then compete for the same inflowing material and quickly grow to have similar mass. This is because, as long as the secondary has lower mass, it is better positioned to accrete the inflowing high-angular momentum material (Whitworth *et al.* 1995). At about 13.7 kyr, a third protostar condenses out of the circumbinary disc, but it is quickly driven into a wide and extended orbit, so that it accretes more slowly and remains a somewhat lower-mass object. At ~ 25 kyr, a three-body exchange interaction occurs, leading to the disruption of the original binary; one component is ejected as a single, and the other recoils as a binary system with the lower-mass object.

6. Conclusions

We conclude that the division of non-thermal energy amongst different modes is critical to the outcome of the collapse of a prestellar core, even though the amounts of non-thermal energy are usually quite small compared with the gravitational potential energy. With purely solenoidal turbulence, at the low levels typical of the cores in Ophiuchus (i.e. $\alpha_{\rm NONTH} \lesssim 0.1$), no multiple systems are formed, and therefore we conclude that the non-thermal energy is not usually in purely solenoidal turbulence. We have considered two alternative divisions of the non-thermal energy, which both yield multiple systems, but in different ways. If some of the energy is attributed to ordered solid-body rotation, a massive and extended disc forms around the primary protostar; this then fragments to produce additional protostars, some of which are ejected, and some of which end up in an hierarchical multiple system. If some of the energy is allocated to compressive rather than solenoidal modes, close intermediate-mass binaries form by fragmentation of compact unrelaxed discs that grow from the filaments in which they are embedded; the components often evolve towards equal mass, $q \sim 1$.

ACKNOWLEDGEMENTS. AW and DS gratefully acknowledge the support of a rolling grant from the UK Science and Technology Funding Council (Ref. ST/H001530/1). AW and SW gratefully acknowledge the support of a Marie Curie Research Training Network (Ref. MRTN-CT-2006-035890). The simulations reported here were performed on the MERLIN supercomputer of the Cardiff Advanced Research Computing Centre.

References

André, Ph., Belloche, A., Motte, F., & Peretto, N. 2007, *A&A*, 472, 519
Goodwin, S. P., Whitworth, A. P., & Ward-Thompson, D. 2004, *A&A*, 414, 633
Hubber, D. A., Batty, C. P., McLeod, A., & Whitworth, A. P. 2010, *A&A*, submitted
Johnstone, D., Wilson, C. D., Moriarty-Schieven, G., Joncas, G., Smith, G., Gregerson, E., & Fich, M. 2000, *ApJ*, 545, 327
Motte, F., André, Ph., & Neri. R. 1998, *A&A*, 336, 150

Simpson R, *et al.* 2010, *MNRAS*, submitted
Stamatellos, D., Whitworth, A. P., & Ward-Thompson, D. 2007, *MNRAS*, 379, 1390
Stamatellos, D., Whitworth, A. P., Bisbas, T., & Goodwin, S. P. 2007, *A&A*, 475, 37
Stanke, T., Smith, M. D., Gredel, R., & Khanzadyan, T. 2006, *A&A*, 447, 609
Whitworth, A. P., Chapman, S. J., Bhattal, A. S., Disney, M. J., Pongracic, H., & Turner, J. A. 1995, *MNRAS*, 277, 727

Neil Evans, Ralph Pudritz, and Richard Larson

Computational Star Formation
Proceedings IAU Symposium No. 270, 2011
J. Alves, B.G. Elmegreen, J. M. Girart & V. Trimble, eds.

doi:10.1017/S1743921311000123

Low-mass Star Formation: Observations

Neal J. Evans II[1]

[1]Department of Astronomy, The University of Texas at Austin, 1 University Station, C1400
Austin, TX 78712-0259, USA
email: nje@astro.as.utexas.edu

Abstract. I briefly review recent observations of regions forming low mass stars. The discussion is cast in the form of seven questions that have been partially answered, or at least illuminated, by new data. These are the following: where do stars form in molecular clouds; what determines the IMF; how long do the steps of the process take; how efficient is star formation; do any theories explain the data; how are the star and disk built over time; and what chemical changes accompany star and planet formation. I close with a summary and list of open questions.

Keywords. stars: formation, infrared: ISM, submillimeter

1. Introduction

Recent large-scale surveys of nearby molecular clouds have provided a solid statistical basis for addressing some long-standing questions about the formation of low-mass stars. Ideally, one would like unbiassed surveys at wavelengths ranging from radio to X-ray of all molecular clouds within some radius of the Sun. Radio and X-ray surveys probe non-thermal and transient events; millimeter continuum surveys probe the mass and structure of the dust, thereby tracing the gas, the far-infrared contains the luminosity information for embedded stages, the mid-infrared probes the disk, the near-infrared probes the inner disk and star, and the visible and ultraviolet probe the star and ongoing accretion. Together, these define the most basic properties of the molecular cloud material and the forming star. While we will concentrate on the continuum in this review, spectroscopic information provides vital complementary information (e.g., see § 8). Again, a wide wavelength range for spectroscopy is ideal.

The most significant recent additions to our arsenal include large scale surveys of the nearby clouds in the Gould Belt in the mid-infrared with Spitzer (Evans *et al.* 2003, Evans *et al.* 2009, Allen *et al.* 2010 and references therein) and in the millimeter continuum (Enoch *et al.* 2006, Young *et al.* 2006, Enoch *et al.* 2007, Enoch *et al.* 2008, Enoch *et al.* 2009). The Spitzer-based surveys will soon by joined by a complete survey with Herschel spanning the wavelengths between mid-infrared and millimeter (André *et al.* 2010 and André in this volume). Sometime later, the same clouds will be surveyed completely in millimeter continuum with SCUBA-2 on the JCMT (Ward-Thompson *et al.* 2007).

2. Where do Stars Form in Molecular Clouds?

Spitzer surveys of 20 clouds in the Gould Belt, now including nearly all clouds within 500 pc, have provided a much more complete data base of Young Stellar Objects (YSOs) (Evans *et al.* 2009, Güdel *et al.* 2007, Rebull *et al.* 2010, Allen *et al.* 2010). These surveys have clearly established that star formation is not evenly distributed over molecular clouds. Instead, star formation proceeds in a clustered manner, concentrated in regions of high extinction (Jørgensen *et al.* 2008, Evans *et al.* 2009). Using the criterion of Lada & Lada (2003) of 1 $M_\odot$ pc^{-3}, 91% of the stars in the c2d survey (Evans *et al.* 2009) and 75% of those in a larger sample, including Taurus and all Gould Belt clouds, but excluding Orion, form in a clustered environment (Bressert *et al.* 2010). Clustering is a matter of degree, with a broad range of surface densities and no clear sign of separate "clustered" and "distributed" modes. Furthermore, the degree of clustering is stronger in younger SED classes, such as Class I and pre-stellar cores, suggesting that the Class II sources have dispersed slightly since their formation.

Using only the Class I and Flat SED classes to avoid any dispersal issues, Heiderman *et al.* (2010) have found evidence for a threshold gas surface density above which star formation is much faster. This threshold lies roughly at $A_V = 8$ mag, or 120 $M_\odot$ pc^{-2}. A very similar threshold was identified by Lada (2010) from an independent analysis. Since most of the mass of molecular clouds lies below this threshold, at least in local clouds, a threshold may yield insight into why star formation efficiencies are so low in molecular clouds (see § 5).

3. What Determines the IMF?

This age-old question has been revivified by catalogs of prestellar cores that show a mass function somewhat similar to that of stars but shifted to higher masses (e.g., Motte *et al.* 1998, Johnstone *et al.* 2000, Enoch *et al.* 2008, Sadavoy *et al.* 2010). Similar distributions were found for starless, but unbound and hence not prestellar (see di Francesco *et al.* 2007 for definitions) cores in the Pipe Nebula (Lada *et al.* 2008). These results are consistent with a picture in which the IMF is set by the process of core formation and favor a picture in which about 1/3 of the core mass is incorporated into the star. However, arguments have been advanced that some cores will evolve faster [either the less massive (Clark *et al.* 2007) or the more massive (Hatchell & Fuller 2008)], and hence the current core mass function is not consistent with a one-one mapping onto the IMF. Swift & Williams (2008) have explored other issues, and Reid *et al.* (2010) have argued that existing measurements of the clump mass spectrum are highly compromised by limitations of the observational techniques, such as sensitivity, spatial resolution, and spatial filtering of large scale emission. While these theoretical issues will be hard to resolve, the statistics should at least improve as the Herschel Gould Belt surveys become available (for a preview, see article by André in this volume).

4. How Long Do the Stages of the Process Take?

The major evolutionary organizing principle for star formation has been the Class system, first introduced by Lada (1987) and extended by Greene *et al.* (1994) and André *et al.* (1993), using the shape of the SED, measured in various ways to establish a putative evolutionary scheme from prestellar cores to Class 0 to Class I to Flat SED to Class II to Class III. Recently, Robitaille *et al.* (2006) has usefully distinguished the SED Class from the physical Stages, which proceed from Prestellar to Stage 0 (most of mass still in envelope) to Stage I (most of mass in central star and disk but still substantial envelope) to Stage II (envelope gone, but substantial disk) to Stage III (disk mostly gone, mass in star). The history of this system has been reviewed by Evans *et al.* (2009). The connection of Class to Stage is generally accepted, but orientation effects can easily confuse a neat identification of Class with Stage: for example, a face-on Stage 0 source may be classified as a Class I source, while an edge-on Stage I source could have a Class 0 SED.

The method of assigning durations to these Classes is to count the numbers in each Class, relating the numbers to the durations by pinning all durations to one that is assumed to be known. One has to further assume that star formation has been continuous over a time longer than the duration of the class used to calibrate ages and that other variables, such as mass, are not important. With these caveats in mind, the results of the c2d studies of nearby clouds with Spitzer led to longer durations for Class I and Class 0 than had previously been derived. The durations are all pinned to a timescale of 2 Myr for the Class II phase, based on studies of infrared excess frequency in clusters of known ages. That age is uncertain by ± 1 Myr and should be thought of as half-life, rather than a fixed duration for each individual star. Using the latest, but still preliminary, statistics from the combined c2d plus Gould Belt projects (3124 YSOs in 20 clouds), the Class I duration would be 0.55 ± 0.28 Myr and the Flat SED lasts 0.36 ± 0.18 Myr, where the uncertainties are due to the uncertainty in the Class II duration. In this calculation, Class 0 is included with Class I. If they are separated using the T_{bol} criterion, the Class 0 phase lasts 0.16 Myr (Enoch *et al.* 2009, Evans *et al.* 2009), substantially longer than earlier estimates. Using the millimeter surveys together with the infrared to separate out the prestellar cores, Enoch *et al.* (2008) found a duration of 0.43 Myr for prestellar cores once the mean density of the core exceeded the detection threshold of about $n = 2 \times 10^4$ cm^{-3}. Most cores had higher mean densities, and more careful comparison suggested a duration roughly 3 times the free-fall time (Enoch *et al.* 2008). This is another area that should be revolutionized by Herschel observations of the Gould Belt clouds.

5. How Efficient is Star Formation?

The c2d and Gould Belt surveys provide a uniform analysis of a much more complete sample of YSOs. Eventually the Taurus cloud and some other clouds should be brought into this analysis. Currently, we use the same 3124 YSOs (N_{YSOs}) in 20 clouds discussed in § 4. These surveys are about 90% complete to about 0.05 $L_\odot$. The total mass of YSOs is calculated from $M_* = N_{YSOs} \times 0.5$ $M_\odot$, where 0.5 $M_\odot$ is the mean stellar mass. The star formation rate is then $SFR = M_*/t_{II}$ $M_\odot$ Myr^{-1}, where t_{II} is the duration of the Class II phase in Myr. The YSO counts are incomplete for Class III objects, so we are effectively computing a star formation rate averaged over the last 2 Myr. For efficiencies, we compare the mass in stars to the mass in the molecular cloud, measured from extinction mapping over the same region in which YSOs are counted (generally $A_V \geqslant 2$ mag). For clouds with millimeter continuum maps, the mass in dense gas can be computed from the sum over masses of all the dense cores.

Based on the sum of all stellar and cloud masses for all 20 c2d and Gould Belt clouds, about 3% of the mass is in YSOs younger than 2 Myr and the mass in dense cores ranges from 2% to 5% for the three clouds with complete maps. In contrast, the total mass in YSOs is very similar to that in dense cores, suggesting that star formation is rapid and efficient once the dense cores have formed. Clearly, star formation is not very efficient for the clouds as a whole, but the efficiency of star formation is quite high once dense cores have formed.

While not efficient in an absolute sense, star formation in the local clouds is much more efficient than would be predicted from the Kennicutt-Schmidt relations used in extragalactic studies. The relations predict the star formation rate surface density from the gas surface density (e.g., Kennicutt 1998). While a few of the least active clouds in the Gould Belt sample lie on the extragalactic relation, almost all lie well above, most by an order of magnitude (Evans *et al.* 2009). This discrepancy has been explored in detail by Heiderman *et al.* (2010), who find evidence for a threshold surface density, as discussed above. They suspect that the measures of gas surface density in other galaxies, based almost entirely on CO emission, include large amounts of gas below the threshold (§ 2).

6. Do Any Theories Explain the Data?

The existing data provide some reality checks for star formation theories. For example, the most commonly used picture of low-mass star formation is inside-out collapse (Shu 1977). In this picture, collapse begins at the center of a singular isothermal sphere, and a wave of infall propagates outward at the sound speed (c_s). The end of infall is less defined in this model, but it is usually assumed that, when the infall wave reaches the outer boundary of the core, the remainder of the core falls in. The time for this is equal to the time taken for the infall wave to reach the outer boundary. (For a more correct treatment of the evolution of a core with a distinct boundary, see Vorobyov & Basu 2005a). For a kinetic temperature, $T_K = 10$ K, $c_s = 0.19$ km s^{-1}. If we associate the time indicated above for the Class I phase of 0.55 Myr with the final infall of the last bit of envelope, the wave of infall would have reached the outer boundary in 0.55/2 Myr (as noted above, the infall wave reaches the outer boundary in half the total time of the embedded phase). The corresponding radius would be $r_{out} = c_s t_I/2 = 0.055$ pc, roughly consistent with many of the core size distributions (e.g., Enoch *et al.* 2008) and the mean separation of YSOs in clusters of 0.072 ± 0.006 pc (Gutermuth et. al 2009).

The density distribution of the initial state is set by the temperature in the simplest version of this model. Thus, the mass available in the mean duration of a Class I source is also set. If a fraction f ends up in the star, the resulting stellar mass is $M_\star = 0.86f$ M$_\odot$. If $f = 0.3$, as suggested above, the resulting mass is 0.26 M$_\odot$, near the mode of the IMF. Thus a simple picture of cores undergoing inside-out collapse meets some general consistency checks.

However, serious problems appear when we compare the luminosity function of the YSOs in the c2d sample to predictions for the Shu model. The Shu model predicts a mass infall rate of $\dot{M}_{inf} = 1.6 \times 10^{-6}$ M$_\odot$ yr^{-1}. The radiation released by this infall onto an object of 0.08 M$_\odot$, at the boundary between brown dwarfs and stars, and radius of 3 R$_\odot$ produces a luminosity $L_{acc} = 1.6$ L$_\odot$. Most (59%) of the YSOs in the c2d clouds lie below this luminosity (Dunham *et al.* 2010). More generally, the distribution of sources in the T_{bol}-L_{bol} plane (Evans *et al.* 2009 is very poorly represented by tracks that follow the Shu solution and use radiative transfer to calculate the SED as a function of time (Young & Evans 2005).

These issues are not peculiar to the Shu model. In fact, almost all other models feature faster infall and will produce a still larger discrepancy with the data. One is faced with the problem of decreasing the typical (observed) accretion rate onto the star, while still removing the envelope in about 0.5 Myr and building the star.

7. How Are the Star and Disk Built over Time?

The problem of low luminosities and a very large spread (at least three orders of magnitude) in luminosity must be telling us about the process of growth of the disk and star. Because the vast majority of the gravitational energy is released when matter accretes onto the star, rather than onto the disk, storing matter in the disk is an obvious option. Since there is no reason that the accretion rate from disk to star should be synchronized with the infall rate onto the disk, it is very plausible that material accumulates in the disk until an instability causes a fast matter transfer to the star, followed by a slower rebuilding of the disk from the envelope. This picture of episodic accretion explains the high incidence of low luminosity values (little or no accretion onto the star) and the large spread in luminosity (small numbers of sources caught in a high-accretion state).

Indeed, Kenyon *et al.* (1990) suggested this solution to the problem of low luminosity, which was apparent even with IRAS data. The Spitzer data has only exacerbated the problem, and the episodic accretion solution is even more attractive. Furthermore, simulations of the flow from envelope to disk to star show instabilities and rapid variations in accretion onto the star (Vorobyov & Basu 2005b, Vorobyov & Basu 2006). Observationally, there is direct evidence in the form of FU Orionis events (Hartmann & Kenyon 1996), outflow morphologies showing multiple ejection events (e.g., Lee *et al.* 2007), and studies of low luminosity sources with extended outflows. These last studies show that minimum mean luminosities needed to drive the observed outflows exceed the current luminosity of the YSO (Dunham *et al.* 2006, Dunham *et al.* 2010a).

To put these ideas on firmer footing, Dunham *et al.* 2010b explored a simple model of episodic accretion, including full 2D radiative transfer with outflows removing matter from a Terebey *et al.* (1984) rotating collapse model. He could reproduce reasonably well the observed luminosity distribution and the distribution in the T_{bol}-L_{bol} plane.

If accretion onto the star is episodic, there are substantial consequences. First, the connection between Stages and Classes becomes even more tenuous, as the luminosity flares can move an object in one Stage back and forth across Class boundaries. Second, the luminosity is not an indicator of stellar mass until nuclear burning dominates accretion luminosity. Third, there may be long-term consequences for the star, causing incorrect estimates of stellar ages (Baraffe *et al.* 2009) and possibly the low lithium abundances in young stars (Baraffe *et al.* 2010). Finally, the initial conditions for planet formation may be determined by the timing of the last accretion event: if it happens just before the last of the envelope falls onto the disk, the star will start with a low-mass disk; conversely, if the disk was close to its maximum stable mass when the last of the envelope accretes, the star would have a massive disk.

8. What Chemical Changes Accompany Star and Disk Formation?

While we have focused on the macroscopic changes as material flows from envelope to disk to star, microscopic changes to the chemical state occur during the process. Infrared spectroscopy from the ground and from Spitzer have revealed a rich spectrum of ices on dust grains before (Knez *et al.* 2005) and during (Boogert *et al.* 2008, Pontoppidan *et al.* 2008, Öberg *et al.* 2008, Bottinelli *et al.* 2010) the infall phase. The size distribution of dust grains shifts to larger grains in molecular clouds, as compared to the diffuse interstellar medium (Flaherty *et al.* 2007, Chapman *et al.* 2009). A new model of dust opacities in dense cores that incorporates these aspects, especially the ice features, is needed to fit SED data that include mid-infrared spectroscopy (e.g., Kim *et al.* 2010).

The flip side of the formation of ices is severe depletion of the gas phase species via freeze-out onto grains (for reviews, see Ceccarelli *et al.* 2007, van Dishoeck 2009). This can reach extreme levels in prestellar cores. As a forming star increases in luminosity, central heating can evaporate some ices, producing a complex pattern of abundance (Jørgensen *et al.* 2004). To interpret molecular line observations correctly, one should ideally use fully self-consistent chemo-dynamical models (Lee *et al.* 2004, Evans *et al.* 2005, Chen *et al.* 2009) that track the chemistry during collapse. Episodic accretion models further complicate the picture, as the changes in luminosity can subject the grains to the equivalent of freeze-thaw cycles that may drive greater differentiation and complexity. For a concrete example, CO frozen on grains may be converted to CO_2 (Pontoppidan *et al.* 2008, which does not evaporate as easily as the CO; repeated cycles can systematically deplete the CO in the envelope (Kim *et al.* 2010).

When the envelope material falls onto the disk, some ices may evaporate (Watson *et al.* 2007) and outflows will drive further thermal and chemical changes along the outflow walls, which are nicely traced by very high-J CO lines observable with Herschel PACS spectroscopic observations (van Kempen st al. 2010a, van Kempen st al. 2010b).

9. Summary and Open Questions

The main points are summarized here.

(*a*) Stars form mostly in clustered environments, but the degree of clustering varies smoothly over a wide range.

(*b*) The core mass function is similar to the initial mass function of stars, but shifted to higher masses. If the core mass function maps to the IMF, the shift suggest an efficiency of about 1/3. However, there are many caveats in this discussion.

(*c*) The timescales for the Class 0 and Class I SED classes are longer than previously thought. Based on number counts, they are roughly 0.16 Myr for Class 0 and 0.55 Myr for Class I.

(*d*) Efficiencies for star formation over the last 2 Myr are low ($\sim 3\%$) in nearby clouds, but quite high ($> 25\%$) in dense cores.

(*e*) Star formation rate surface densities are much higher for a given gas surface density that would be predicted by Kennicutt-Schmidt relations.

(*f*) Both the discrepancy with the Kennicutt-Schmidt relations and the low absolute efficiency are likely related to a surface density threshold for efficient star formation that is not met by the great majority of the cloud mass in nearby clouds, nor in normal galaxies.

(*g*) A simple inside-out collapse picture for a core is consistent with much of the data on sizes and masses, but no model with uniform accretion reproduces the distribution of YSOs in the T_{bol}-L_{bol} plane. Episodic accretion is strongly indicated.

(*h*) Complex chemical changes accompany the large-scale flows of matter, as material moves from gas to ice and back. Simple models of abundances that are constant, either in time or space, can be misleading.

The most significant open questions are summarized here.

- What sets the mass of stars? Is it the core mass function, or is it feedback?
- How do brown dwarfs form? A core forming a brown dwarf would have to be very small and very dense to be bound.
- What controls the threshold for star formation? While a surface density threshold is indicated by the data, this may be a secondary indicator for a more complex set of variables, including volume density, turbulent velocity, magnetic field, ionization, etc.
- And finally a question that has been with us since the early days of molecular clouds: why is star formation so inefficient?

I would like to acknowledge the collaboration of A. Heiderman, M. Dunham, L. Allen, D. Padgett, and E. Bressert in preparing this review. Support for this work, part of the *Spitzer* Legacy Science Program, was provided by NASA through contracts 1224608 and 1288664 issued by the Jet Propulsion Laboratory, California Institute of Technology, under NASA contract 1407. Support was also provided by NASA Origins grant NNX07AJ72G and NSF grant AST-0607793 to the University of Texas at Austin.

References

Allen, L., Koenig, X., Gutermuth, R. & Megeath, T. 2010, Bulletin of the American Astronomical Society, 41, 559

André, P., Ward-Thompson, D., & Barsony, M. 1993, Astrophys. J., 406, 122

André, P., *et al.* 2010, Astr. Ap., 518, L102

Baraffe, I., Chabrier, G., & Gallardo, J. 2009, Astrophys. J. (Letters), 702, L27
Baraffe, I. & Chabrier, G. 2010, Astr. Ap., in press
Boogert, A. C. A., *et al.* 2008, Astrophys. J., 678, 985
Bottinelli, S., *et al.* 2010, arXiv:1005.2225
Bressert, E., *et al.* 2010, submitted to MNRAS.
Ceccarelli, C., Caselli, P., Herbst, E., Tielens, A. G. G. M., & Caux, E. 2007, Protostars and Planets V, 47
Chapman, N. L., Mundy, L. G., Lai, S.-P., & Evans, N. J. 2009, Astrophys. J., 690, 496
Chen, J.-H., Evans, N. J., Lee, J.-E., & Bourke, T. L. 2009, Astrophys. J., 705, 1160
Clark, P. C., Klessen, R. S., & Bonnell, I. A. 2007, Mon. Not. Roy. Astr. Soc., 379, 57
di Francesco, J., Evans, N. J., II, Caselli, P., Myers, P. C., Shirley, Y., Aikawa, Y., & Tafalla, M. 2007, in Protostars and Planets V, ed. B. Reipurth, D. Jewitt, & K. Keil (Tucson: Univ. Arizona Press), 17
Dunham, M. M., *et al.* 2006, Astrophys. J., 651, 945
Dunham, M. M., Evans, N. J., Terebey, S., Dullemond, C. P., & Young, C. H. 2010b, Astrophys. J., 710, 470
Dunham, M. M., Evans, N. J., II, Bourke, T. L., Myers, P. C., Huard, T. L., & Stutz, A. M. 2010a, submitted to Astrophys. J.
Enoch, M. L., Evans, N. J., Sargent, A. I., & Glenn, J. 2009, Astrophys. J., 692, 973
Enoch, M. L., Evans, N. J., II, Sargent, A. I., Glenn, J., Rosolowsky, E., & Myers, P. 2008, Astrophys. J., 684, 1240
Enoch, M. L., Glenn, J., Evans, N. J., II, Sargent, A. I., Young, K. E., & Huard, T. L. 2007, Astrophys. J., 666, 982
Enoch, M. L., *et al.* 2006, Astrophys. J., 638, 293
Evans, N. J., II, *et al.* 2003, Pub. Astr. Soc. Pacific, 115, 965
Evans, N. J., *et al.* 2009, Astrophys. J., 181, 321
Evans, N. J., II, Lee, J.-E., Rawlings, J. M. C., & Choi, M. 2005, Astrophys. J., 626, 919
Flaherty, K. M., Pipher, J. L., Megeath, S. T., Winston, E. M., Gutermuth, R. A., Muzerolle, J., Allen, L. E., & Fazio, G. G. 2007, Astrophys. J., 663, 1069
Greene, T. P., Wilking, B. A., André, P., Young, E. T., & Lada, C. J. 1994, Astrophys. J., 434, 614
Güdel, M., Padgett, D. L., & Dougados, C. 2007, Protostars and Planets V, 329
Gutermuth, R. A., Megeath, S. T., Myers, P. C., Allen, L. E., Pipher, J. L., & Fazio, G. G. 2009, Astrophys. J. Suppl., 184, 18
Hartmann, L. & Kenyon, S. J. 1996, Ann. Rev. Astr. Astroph., 34, 207
Hatchell, J. & Fuller, G. A. 2008, Astr. Ap., 482, 855
Heiderman, A., Evans, N. J., II, Allen, L. E., Huard, T., & Heyer, M. 2010, submitted to Astrophys. J.
Johnstone, D., Wilson, C. D., Moriarty-Schieven, G., Joncas, G., Smith, G., Gregersen, E., & Fich, M. 2000, Astrophys. J., 545, 327
Jørgensen, J. K., Johnstone, D., Kirk, H., Myers, P. C., Allen, L. E., & Shirley, Y. L. 2008, Astrophys. J., 683, 822
Jørgensen, J. K., Schöier, F. L., & van Dishoeck, E. F. 2004, Astr. Ap., 416, 603
Kennicutt, R. C., Jr. 1998, Astrophys. J., 498, 541
Kim, H. J., Evans, N. J., II, Dunham, M. M., Chen, J.-h., Lee, J.-E., Bourke, T. L., Huard, T. L., Shirley, Y. L., & De Vries, C. 2010, submitted to Astrophys. J..
Knez, C., *et al.* 2005, Astrophys. J. (Letters), 635, L145
Lee, J.-E., Bergin, E. A., & Evans, N. J., II 2004, Astrophys. J., 617, 360
Lee, C.-F., Ho, P. T. P., Hirano, N., Beuther, H., Bourke, T. L., Shang, H., & Zhang, Q. 2007, Astrophys. J., 659, 499
Kenyon, S. J., Hartmann, L. W., Strom, K. M., & Strom, S. E. 1990, Astron. J., 99, 869
Motte, F., Andre, P., & Neri, R. 1998, Astr. Ap., 336, 150
Lada, C. J. 1987, IAU Symp. 115: Star Forming Regions, 115, 1
Lada, C. J. & Lada, E. A. 2003, Ann. Rev. Astr. Astroph., 41, 57
Lada, C. J., Lombardi, M., & Alves, J. F. 2010, submitted.

Lada, C. J., Muench, A. A., Rathborne, J., Alves, J. F., & Lombardi, M. 2008, Astrophys. J., 672, 410
Öberg, K. I., Boogert, A. C. A., Pontoppidan, K. M., Blake, G. A., Evans, N. J., Lahuis, F., & van Dishoeck, E. F. 2008, Astrophys. J., 678, 1032
Pontoppidan, K. M., *et al.* 2008, Astrophys. J., 678, 1005
Rebull, L. M., *et al.* 2010, Astrophys. J. Suppl., 186, 259
Reid, M. A., Wadsley, J., Petitclerc, N., & Sills, A. 2010, arXiv:1006.4320
Robitaille, T. P., Whitney, B. A., Indebetouw, R., Wood, K., & Denzmore, P. 2006, Astrophys. J. Suppl., 167, 256
Sadavoy, S. I., *et al.* 2010, Astrophys. J., 710, 1247
Shu, F. H. 1977, Astrophys. J., 214, 488
Swift, J. J. & Williams, J. P. 2008, Astrophys. J., 679, 552
Terebey, S., Shu, F. H., & Cassen, P. 1984, Astrophys. J., 286, 529
van Dishoeck, E. F. 2009, Astrophysics in the Next Decade, 187
van Kempen, T. A., *et al.* 2010a, Astr. Ap., 518, L128
van Kempen, T. A., *et al.* 2010b, Astr. Ap., 518, L121
Vorobyov, E. I. & Basu, S. 2005a, Mon. Not. Roy. Astr. Soc., 360, 675
Vorobyov, E. I. & Basu, S. 2005b, Astrophys. J. (Letters), 633, L137
Vorobyov, E. I. & Basu, S. 2006, Astrophys. J., 650, 956
Ward-Thompson, D., *et al.* 2007, Pub. Astr. Soc. Pacific, 119, 855
Watson, D. M., *et al.* 2007, Nature, 448, 1026
Young, C. H. & Evans, N. J. 2005, Astrophys. J., 627, 293
Young, K. E., *et al.* 2006, Astrophys. J., 644, 326

Computational Star Formation
Proceedings IAU Symposium No. 270, 2011
J. Alves, B. G. Elmegreen, J. M. Girart, V. Trimble, eds.

doi:10.1017/S1743921311000135

Formation of massive stars

Maria T. Beltrán[1]

[1]INAF, Osservatorio Astrofisico di Arcetri,
Largo E. Fermi 5, 50125 Firenze, Italy
email: mbeltran@arcetri.astro.it

Abstract. The formation of high-mass stars represents a challenge from both a theoretical and an observational point of view. Here, we present an overview of the current status of the observational research on this field, outlining the progress achieved in recent years on our knowledge of the initial phases of massive star formation. The fragmentation of cold, infrared-dark clouds, and the evidence for star formation activity on some of them will be discussed, together with the kinematics of the gas in hot molecular cores, which can give us insights on the mechanism leading to the birth of an OB star.

Keywords. ISM: molecules, stars: formation, radio lines: ISM

---

1. Introduction

How do massive ($>8\ M_\odot$) stars form? Do they form like lower-mass stars or in a fundamentally different way? These crucial questions have profound implications for many areas of astrophysics including high-redshift Population III star formation, galaxy formation and evolution, galactic center environments and super-massive black hole formation, star and star cluster formation, and planet formation around stars in clusters.

High-mass stars are rare objects that form very fast. What is more, the closest high-mass star-forming regions are located further away than the nearest low-mass ones. They form deeply embedded in clusters, where the extinction is very high and there is strong interaction with the environment and feedback from other recently formed stars. All this makes the study of the formation of massive stars challenging, due to the difficulty of tracing the primordial configuration of the molecular cloud, and the impossibility of studying the earliest stages of massive star-formation at optical or even infrared wavelengths.

From a theoretical point of view, massive stars represent a challenge because of their shorter evolutionary timescales. For massive protostars, the Kelvin-Helmholtz timescale is shorter than the free-fall timescale. Therefore, massive stars reach the Zero Age Main Sequence still deeply embedded in the parental cloud and with ongoing accretion. According to theoretical predictions, at this stage, their powerful radiation pressure appears to be strong enough to halt the infalling material (Kahn 1974; Wolfire & Cassinelli 1987), which should inhibit further growth of the star beyond the limit of 8 $M_\odot$ (although recent 3-D radiation hydrodynamic simulations by Krumholz *et al.* (2009) indicate that radiation pressure may not be a problem). In order to solve this problem, competitive accretion has been proposed (Bonnell & Bate 2002), that predicts that initially a molecular cloud fragments in low-mass cores of a Jeans mass $\sim$0.5–1 $M_\odot$, which form stars that compete to accrete mass from the common gas reservoir. Protostars located near the center of the gravitational potential well accrete at a higher accretion rate because of a stronger gravitational pull. Core accretion (McKee & Tan 2002) may also be a viable theoretical solution to overcome the radiation pressure with the ram pressure of the infalling material. In this turbulent accretion model, stars form via monolithic collapse of a massive core fragmented from the natal molecular cloud. Massive star formation

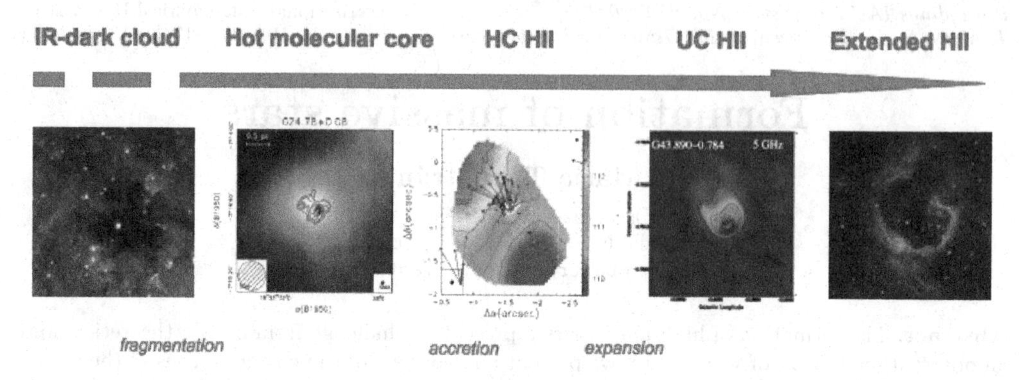

Figure 1. Evolutionary sequence for high-mass stars: from cold IRDCs to chemically rich HMCs to HII regions.

is therefore, in this picture, a sort of scale-up version of low-mass star formation, with collapse, accretion, rotation, and outflows.

Due to the short timescales needed to form a massive star, it is very difficult to trace the earliest stages of its formation and define an evolutionary sequence for high-mass stars, similar to the one defined for low-mass star. However, a reasonable hypothesis may be that the formation process starts with the fragmentation of a cold cloud, sometimes dark at IR wavelengths (IRDC). Subsequent infall and heating inside each fragment (core) eventually leads to the formation of chemically rich hot molecular cores (HMCs), which are the cradle of high-mass stars. Accretion continues onto high-mass protostars that gain mass. The UV-radiation from the young embedded star will develop a bubble of ionized gas, known as hypercompact HII (HC HII) region that eventually will start expanding, becoming first an ultracompact HII (UC HII) region and later on, an extended or giant HII region. Eventually, the gas is dissipated by the ionized winds, exposing the newly formed OB star (see Fig. 1).

In this contribution I will focus on the observational aspects of the earliest stages (IRDCs and HMCs) of massive star formation, which are crucial to understand how massive stars form, and I will review the observational evidence for rotation, outflows, and collapse in massive star-forming regions.

2. Infrared dark clouds

Massive stars form in parsec scale high-density clumps inside cold molecular clouds that when seen in absorption at IR wavelengths against the bright background are called IRDCs. These IRDCs were first detected in absorption at 8 μm up to 21 μm with ISO and MSX (Egan *et al.* 1998; Carey *et al.* 1998), and later on, they have been seen in absorption at 24 μm, sometimes up to 70 μm, with *Spitzer* (Rathborne *et al.* 2006), and more recently with *Herschel* (Peretto *et al.* 2010). Millimeter and submillimeter continuum observations have revealed extended cold dust emission associated with IRDCs (e.g. Schuller *et al.* 2009; Rosolowsky *et al.* 2010). These clouds have typically masses of 10^3–10^4 $M_\odot$, sizes of 1 to 5 pc, densities of $\sim 10^5$ cm^{-3}, and temperatures lower than 20 K.

Temperature structure. The knowledge of the initial stages of massive star formation, and in particular of IRDCs has experienced a significant step forward thanks to the advent of IR and submillimeter telescopes such as *Spitzer* and *Herschel*. In particular, the open time key project Hi-GAL (Molinari & Hi-GAL Consortium 2010), which will observe the inner part of the Galactic Plane with *Herschel* in five bands, from 70 to 500 μm, will allow to construct well constrained spectral energy distributions (SEDs) for these objects, and determine their internal temperature and column density distributions. Using the data obtained during the *Herschel* Science Demonstration Phase, Peretto *et al.* (2010) have developed a simple pixel-by-pixel SED fitting method, and have recovered the spatial variations of both the dust column density and temperature within the IRDCs. Their analysis shows that IRDCs are not isothermal, but the dust temperature decreases significantly from background temperatures of 20–30 K to minimum temperatures of 8–15 K within the clouds. Knowing the temperature and its fluctuations is very important to understand fragmentation in these infrared-dark clouds. In fact, recent hydrodynamical simulations including radiative transfer (Bate 2009; Krumholz *et al.* 2010) show that protostellar heating could reduce significantly fragmentation on small scales.

Core fragmentation. Recent Submillimeter Array (SMA) high-angular resolution observations of IRDCs in the G28.34+0.06 region, have shown that IRDCs fragment into smaller cores with masses of 22 to 64 $M_\odot$ and projected separations of 0.19 pc (Zhang *et al.* 2009). Despite the fact that the SMA observations spatially resolve the Jeans length of 0.05 pc, the core masses are 10 times larger than the Jeans mass of 1 $M_\odot$, for a temperature of 16 K. This suggests that fragmentation may not be controlled just by thermal pressure and gravity, but that other stabilizing factors such as turbulence and maybe magnetic fields or radiative feedback are required to account for the large mass observed. In fact, the observed NH_3 line width in one of the cores is a factor 8 larger than the thermal line width (Wang *et al.* 2008), suggesting that supersonic turbulence might play a crucial role in the core fragmentation.

Star formation activity. One of the hot topics in massive star formation is how pre-stellar are IRDCs. High-sensitivity *Spitzer* and *Herschel* observations have shown that many IRDCs show evidence for active star formation. Some of them show enhanced, slightly extended 4.5 μm emission, called "green-fuzzies" by Chambers *et al.* (2009), and broad SiO line emission (Jiménez-Serra *et al.* 2010) which indicates shocked gas. These shocks could be a remnant of the IRDC formation process, or be produced by decelerated or recently processed gas in large-scale outflows driven either by 8 and 24 μm embedded sources, or by an undetected and wide spread population of lower mass protostars. In some cores bright 8 μm continuum emission, probably from HII regions, has been detected or 3.6 μm emission from unextincted stars. Finally, some cores show bright, compact 24 μm emission that indicates deeply embedded protostars (Chambers *et al.* 2009).

Based on the evidence of active star formation, Rathborne *et al.* (2010) classify the IRDCs in five specific groups: quiescent, intermediate, active, red, and blue cores. Quiescent cores show total absence of star formation, with no significant 3–8 μm nor 24 μm emission. Intermediate cores contain either a "green-fuzzy" or a 24 μm point source, but not both. Active cores contain a "green-fuzzy" and a 24 μm point source, and red and blue cores are associated with bright 8 and 3.6 μm emission, respectively. The dust temperatures and luminosities are higher for cores with active, high-mass star formation than for quiescent cores, as expected if the evolutionary sequence described in Fig. 1 is valid and the quiescent cores have no internal heating source to significantly heat the dust. On the other hand, the mass histograms of active and quiescent cores are similar, which suggests that the associated stellar populations are nearly the same, and that the differences in their temperatures and luminosities are due to different evolutionary stages. In this scenario, the cooler quiescent cores may be the pre-stellar precursors of the warmer and more active protostellar cores. The evolutionary differences between the groups are also seen in the bolometric luminosity versus core dust mass diagram (see Fig. 10 of Rathborne *et al.* 2010), where for a given mass, the luminosity increases from the quiescent to the active cores as a result of an increase of the protostellar activity.

Table 1. Core and infall parameters towards HMCs.

HMC	M_{core} ($M_\odot$)	R (pc)	V_{infall} (km s^{-1})	$\dot{M}_{infall}$ ($M_\odot$ yr^{-1})	Refs.[a]
G10.62−0.38	82	0.02	4.5	3×10^{-3}	1, 2
G24.78+0.08 A1	130	0.02	2	4×10^{-4}–10^{-2}	2, 3
W51 North	90	0.07	4	4×10^{-2}–7×10^{-2}	4
W51e2	140	0.01	3.5	6×10^{-3}	5, 6
G31.41+0.31	490	0.04	3.1	3×10^{-3}–3×10^{-2}	3, 7
G19.61−0.23	415	0.03	4	$>3\times10^{-3}$	2

Notes:[a] References for the core parameters: 1: Keto *et al.* (1988); 2: Beltrán *et al.* (2010); 3: Beltrán *et al.* (2004); 4: Zapata *et al.* (2008); 5: Zhang & Ho (1997); 6: Shi *et al.* (2010) 7: Girart *et al.* (2009)

3. Hot molecular cores

HMCs, the cradles of OB stars, are dense, $n\sim10^7$ cm^{-3}, compact, dusty cores with temperatures in excess of 100 K, and luminosities $> 10^4$ $L_\odot$. These cores are often found in association with typical signposts of massive star formation such as UC HII regions and maser emission of different species. As a result of the evaporation of dust grain mantles by the strong radiation of the deeply embedded early-type star, hot cores exhibit a rich chemistry observable in molecular line emission at (sub)millimeter wavelengths (e.g. Beuther 2007). In these cores, hydrogenated molecules, such as NH_3, HS, oxygen-bearing molecules, such as SiO, SO, CO and isotopomers, deuterated species, complex organic molecules, such as CH_3OH, CH_3CN, $HCOOCH_3$, and pre-biotic molecules, such as CH_2OHCHO (Beltrán *et al.* 2009), are abundant. The fact that so many different molecules are observed towards these regions allows us to estimate the physical parameters, such as temperature, density, and mass, and moreover, to study the kinematics of these young massive stellar objects, making it possible, in some cases, to distinguish infall, outflow, and rotation signatures as we will discuss next.

Infall. Despite the importance of infall to test models of massive star formation, as a result of the difficulty to detect and recognize infall, up to now there are still very few detections. A very effective method of identifying the presence of infalling gas is to observe blue-skewed profiles, also known as red-shifted self-absorption, of optically thick lines (e.g. Wu & Evans 2003), or in case of very bright embedded continuum sources, to observe red-shifted absorption. In the latter case, the red-shifted absorption has been observed against the bright continuum emission of an embedded hypercompact HII region, as for G10.62−0.02 (Keto *et al.* 1988; Sollins *et al.* 2005) and G24.78+0.08 A1 (Beltrán *et al.* 2006), and in others, against the bright optically thick continuum emission from the core center, as for W51 North (Zapata *et al.* 2008), G19.61−0.23 (Wu *et al.* 2009), and G31.41+0.31 (Girart *et al.* 2009). As the continuum source is very bright ($\gtrsim$2000 K), it is easy to observe the colder molecular gas ($\sim$100 K) in absorption against it. If the material surrounding the protostar(s) is not only rotating, but also accreting onto the central star(s), one expects to see absorption at positive velocities relative to the stellar velocity. This is what has been detected towards the hot cores in Table 1.

The infall rates are very high, of the order of 10^{-3}–10^{-2} $M_\odot$ yr^{-1}, and in some cases, could be sufficient to quench the formation of an HII region (Walmsley 1995), or to slow down the expansion of it. In fact, detailed models by Keto (2002) predict that trapped HC HII regions could be long-lived, and that accretion could proceed through them in the form of ionized accretion flows.

Molecular outflows. Molecular outflows trace the earliest phases of protostellar evolution when extinction is high. They are ubiquitous phenomena in star formation, and have broadly been detected towards massive young stellar objects, from B- to O-type stars

Table 2. List of rotating disks and toroids in high-mass (proto)stars

Number[a]	Core	$M_{\rm gas}^{\rm OH94\ b}$ ($M_\odot$)	$M_{\rm gas}^{c}$ ($M_\odot$)	R (pc)	$V_{\rm rot}$ (km s^{-1})	Refs.[d]	$t_{\rm ff}/t_{\rm rot}^{e}$
1	IRAS 20126+4104	0.93	4	0.008	1.3	1,2	0.36
2	Cepheus A HW2	2.1	8	0.0016	3.0	3	0.25
3	IRAS 23151+5912	11	26	0.010	3.0	4	0.27
4	GH2O 92.67+3.07	12[f]	12	0.035	1.2	5	0.18
5	G29.96−0.02	28	28	0.011	1.6	6	0.10
6	G20.08−0.14N	40	95	0.024	3.5	7	0.26
7	NGC 6334I	43	17	0.0014	5.1	8, 9	0.09
8	IRAS 18089−1732	59	45	0.010	4.0	10	0.16
9	G10.62−0.38	82	82	0.016	2.1	6	0.09
10	G28.87+0.07	98	100	0.029	0.5	11	0.03
11	G24.78+0.08 C	98	250	0.040	0.5	12	0.03
12	W51 North	160	90	0.068	1.5	13	0.09
13	G24.78+0.08 A2	163	80	0.020	0.75	12	0.03
14	W51e2	241	140	0.010	2.0	14, 15	0.04
15	G24.78+0.08 A1	264	130	0.020	1.5	12	0.04
16	G23.01−0.41	274	380	0.060	0.6	11	0.03
17	IRAS 18566+0408	304	70	0.034	3.0	16	0.09
18	G19.61−0.23	415	415	0.031	1.0	6	0.03
19	G31.41+0.31	508	490	0.040	2.1	12	0.06
20	NGC 7538S	607	100	0.070	1.35	17	0.04

Notes: [a] Numbers in Fig. 2; [b] Masses estimated assuming the dust opacities of Ossenkopf & Henning (1994); [c] Masses from the literature; [d] References for the core parameters: 1: Cesaroni *et al.* (2007); 2: Cesaroni *et al.* (2005); 3: Patel *et al.* (2005); 4: Beuther *et al.* (2007c); 5: Bernard *et al.* (1999); 6: Beltrán *et al.* (2010); 7: Galván-Madrid *et al.* (2009); 8: Hunter *et al.* (2006); 9: Beuther *et al.* (2008); 10: Beuther *et al.* (2005); 11: Furuya *et al.* (2008); 12: Beltrán *et al.* (2004); 13: Zapata *et al.* (2008); 14: Zhang & Ho (1997); 15: Shi *et al.* (2010); 16: Zhang *et al.* (2007); 17: Sandell *et al.* (2003); [e] The free-fall times $t_{\rm ff}$ have been estimated from the masses obtained assuming the dust opacities of Ossenkopf & Henning (1994) . [f] The mass of this core has been estimated from CS (Bernard *et al.* 1999) and is the value used to derive $t_{\rm ff}$.

with luminosities of up to $10^6\ L_\odot$ (Shepherd & Churchwell 1996). Surveys of massive molecular outflows (Beuther *et al.* 2002; Wu *et al.* 2004; López-Sepulcre *et al.* 2009) have shown that there is a continuity in the correlations between outflow-related quantities, such as, mechanical luminosity, mechanical force, and mass loss rate, and the bolometric luminosity from low-mass to high-mass protostars. The higher the luminosity, the higher the outflow-related quantity. The fact that these correlations hold for a broad range (0.1–$10^6\ L_\odot$) of luminosities suggests that the luminosity of the powering source determines the outflow energetics, and that the driving mechanism is similar for all luminosities. Note that although the possible multiplicity of massive outflows would shift the mechanical force and luminosity of the individual outflows towards lower values, the overall appearance of the correlations would remain the same (López-Sepulcre *et al.* 2009).

Molecular outflows may be crucial in the formation of massive stars. First of all, they can reduce the radiation pressure experienced by the gas in the infalling envelope (Krumholz *et al.* 2005). In fact, the radiation can escape through the cavity opened by them, which allows the infalling material to reach the central protostar. What is more, as predicted by theoretical works, outflows can be an important feedback mechanism in the process of high-mass/cluster formation (e.g. Wang *et al.* 2010). In fact, López-Sepulcre *et al.* (2010) have shown that up to 20–25% of the clump mass can be affected by molecular outflows. Molecular outflows are also important because, as shown by López-Sepulcre *et al.* (2010), they can help to discern whether star formation has already started in a massive core. These authors have found that for a sample of massive clumps, when the surface density Σ is higher than 0.3 g cm^{-2}, close to the critical theoretical value predicted by Krumholz & McKee (2008) to form a high-mass star, the outflow detection rate is 100%. And last but not least, since accretion and outflow are tightly linked, well collimated outflows can provide evidence for the existence of accretion disks around massive stars. In particular, the mass loss rate can provide a rough estimate of the accretion rate onto the star (e.g. Beuther *et al.* 2002), which is a much more difficult parameter to measure.

Rotation. In recent years an ever growing number of massive, rotating structure have been found around high-mass young stellar objects (see Table 2). Most of them have been discovered thanks to velocity gradients perpendicular to the corresponding outflow. These structures have sizes of a few thousands of AUs and masses that range from 4 to about 500 $M_\odot$. Following Cesaroni *et al.* (2006), these rotating structures should be classified into two classes on the basis of the ratio between the mass of the rotating structure and that of the star. When the mass is lower than that of the central star, the object is a centrifugally supported disk. Instead, structures having masses in excess of several 10 $M_\odot$, much greater than the mass of the central star(s), are called toroids and appear to be non-equilibrium structures. Their properties are different from those of disks around low-mass young stellar objects. These toroids may host not just a single star, but a whole cluster. Due to the fact that the gravitational potential is dominated by the massive toroid, Keplerian rotation is not possible on scales of 10^4 AUs. For this reason, it has been proposed that toroids never reach equilibrium and could be transient entities with timescales of the order of the free-fall time, $\sim 10^4$ yr.

This distinction between disks and toroids has been illustrated by Beltrán *et al.* (2010) who have plotted $t_{ff}/t_{\rm rot}$ versus $M_{\rm gas}$ (re-calculated using the dust opacities of Ossenkopf & Henning 1994) for a number of disks and toroids around high-mass (proto)stars found in the literature (see Fig. 2 and Table 2). The free-fall time t_{ff} is proportional to the dynamical timescale needed to refresh the material of the toroid and $t_{\rm rot}$ is the rotational period at the outer radius, $2\,\pi\,R/V_{\rm rot}$, needed to adjust the system to a new equilibrium configuration. If the structure rotates fast, the infalling material has enough time to settle into a centrifugally supported disk. Vice versa, if the structure rotates slowly, the infalling material does not have enough time to reach centrifugal equilibrium and the rotating structure is a transient toroid. Therefore, the higher the $t_{\rm ff}/t_{\rm rot}$ ratio, the more similar should be the rotating structure to a circumstellar disk. In Fig. 2, one sees that the least massive structures, which have masses comparable to or lower than that of the central star have the highest $t_{\rm ff}/t_{\rm rot}$ ratio, while above a core mass of $\sim 10^2$ $M_\odot$ the ratio decreases significantly.

The main conclusion, is that up to now no true disks in Keplerian rotation have been found around the most massive O-type stars, like the one imaged towards the B-type star IRAS 20126+4104 (Cesaroni *et al.* 2005). This could be due to an observational bias: disks could be heavily embedded inside the toroids and their emission difficult to disentangle from that of the large-scale rotating structure with current instrumentation. Another possibility could be that single disks are truncated at small radii (Bonnell & Bate 2005) by interactions with stellar companions in the cluster (but see Ian Bonnell's contribution in this volume), or by photo-ionization (Hollenbach *et al.* 2000), or not detected because too distant.

Magnetic fields. A hot topic in massive star formation is the role of magnetic fields. Recent dust polarization observations of the HMCs G31.41+0.31 (Girart *et al.* 2009), and W51 e2 and e8 (Tang *et al.* 2009) have revealed magnetic field lines perpendicular to the major axis of the core, in the direction of the rotation axis, with a clear hourglass morphology. This suggests that magnetic fields play an important role in the formation of massive stars and could control the dynamical evolution of the cores (see Ramprasad Rao's contribution in this volume). What is more, Girart *et al.* (2009) have also found evidence of magnetic braking in the G31.41+0.31 contracting core.

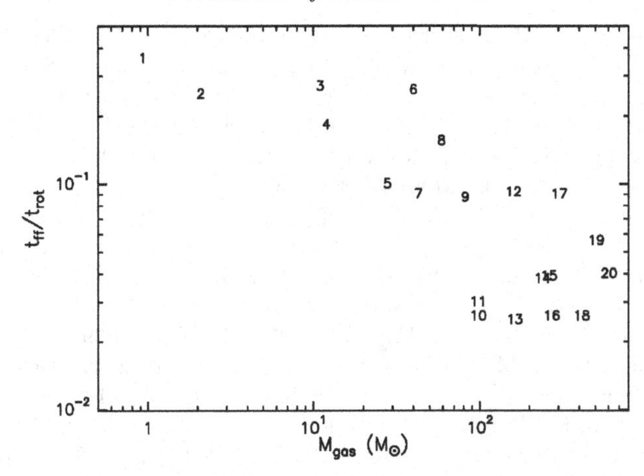

Figure 2. Free-fall timescale to rotational period ratio versus gas mass of known rotating disks and toroids. The numbers correspond to the entries of Table 2 (figure adapted from Beltrán *et al.* 2010). Masses estimated assuming the dust opacities of Ossenkopf & Henning (1994).

More recently, 6.7 GHz methanol maser polarization observations (Vlemmings *et al.* 2010) have determined the full 3-D magnetic field structure around Cepheus A HW2. The field is predominantly aligned along the protostellar outflow and has a strength of 23 mG. According to Vlemmings *et al.* (2010), the magnetic field regulates accretion onto the disk, dominates the turbulent energies by a factor of ~3, and is strong enough to stabilize the massive accretion disk and sustain the high accretion rates needed during massive star formation.

4. Conclusions

High-angular resolution observations of high-mass star-forming regions allow us to start studying in detail fragmentation, infall, accretion, rotation, and outflow towards these regions. However, the available data do not allow us yet to distinguish which accretion-based mechanism, monolithic accretion or competitive accretion, is the responsible for the formation of massive stars. Therefore, beside the important progress made in this field in recent years, there is still a long way to go before our understanding of the formation of all kinds of massive stars, from 8 to ~100 $M_\odot$, will be complete. In particular, although rotating structures are routinely found towards HMCs, no true accretion disk has yet been found around an O-type protostar, which could indicate that the most massive stars for via a different mechanism. In this sense, the advent of ALMA, with its superior sensitivity and resolution, will be crucial. Taking into account that the first call for ALMA early science will be by the end of this year, the search for circumstellar disks around O-type stars could end soon.

References

Bate, M. R. 2009, *MNRAS*, 392, 1363

Beltrán, M. T., Cesaroni, R., Codella, C., Testi, L., *et al.* 2006, *Nature*, 443, 427

Beltrán, M. T., Cesaroni, R., Neri, R., Codella, C., Furuya, R. S., *et al.* 2004, *ApJ*, 601, L187

Beltrán, M. T., Cesaroni, R., Neri, R., & Codella, C. 2010, *A&A*, submitted

Beltrán, M. T., Codella, C., Viti, S., Neri, R., & Cesaroni, R. 2009, *ApJ*, 690, L93

Bernard, J. P., Dobashi, K., & Momose, M. 1999, *A&A*, 350, 197

Beuther, H. 2007, *Proceedings of IAU S237* (Cambridge: Cambridge University Press), p. 148

Beuther, H., Churchwell, E. B. *et al.* 2007, *Protostars & Planets V*, p. 165
Beuther, H., Schilke, P., Sridharan, T. K., Menten, K. M., *et al.* 2002, *A&A*, 383, 892
Beuther, H., Walsh, A. J., Thorwirth, S., Zhang, Q., *et al.* 2008, *A&A*, 481, 169
Beuther, H., Zhang, Q., Hunter, T. R. *et al.* 2007b, *A&A*, 473, 493
Beuther, H., Zhang, Q., Sridharan, T. K., & Chen, Y. 2005, *ApJ*, 628, 800
Bonnell, I. A. & Bate, M. R. 2002, *MNRAS*, 336, 659
Bonnell, I. A. & Bate, M. R. 2005, *MNRAS*, 362, 915
Carey, S. J., Clark, F. O., Egan, M. P., Price, S. D., *et al.* 1998, *ApJ*, 508, 721
Cesaroni, R., Galli, D., Lodato, G., Walmsley, C. M., & Zhang, Q. 2006, *Nature*, 444, 703
Cesaroni, R., Galli, D. *et al.* 2007, *Protostars & Planets V*, p. 197
Cesaroni, R., Neri, R., Olmi, L., Testi, L., *et al.* 2005, *A&A*, 434, 1039
Chambers, E. T., Jackson, J. M., Rathborne, J. M., & Simon, R. 2009, *ApJS*, 181, 360
Egan, M. P., Shipman, R. F., Price, S. D., Carey, S. J. *et al.* 1998, *ApJ*, 494, L199
Furuya, R. S., Cesaroni, R., Takahashi, S., Codella, C., *et al.* 2008, *ApJ*, 673, 363
Galván-Madrid, R., Keto, E., Zhang, Q., Kurtz, S, *et al.* 2009, *ApJ*, 706, 1036
Girart, J. M., Beltrán, M. T., Zhang, Q., Rao, R., & Estalella, R. 2009, *Science*, 324, 1408
Hollenbach, D. J., Yorke, H. W., & Johnstone, D. 2000, *Protostars & Planets IV*, p. 401
Hunter, T. R., Brogan, C. L., Megeath, S. T., Menten, K. M., *et al.* 2006, *ApJ*, 649, 888
Jiménez-Serra, I., Caselli, P., Tan, J. C., Hernández, A. K., *et al.* 2010, *MNRAS*, in press
Keto, E. 2002, *ApJ*, 580, 980
Keto, E. R., Ho, P. T.P., & Haschick, A. D. 1988, *ApJ*, 324, 920
Khan, F. D. 1974, *A&A*, 37, 149
Krumholz, M. R., Cunningham, A. J., Klein, R. I., & McKee, C. F. 2010, *ApJ*, 713, 1120
Krumholz, M. R. & McKee, C. F. 2008, *Nature*, 451, 1082
Krumholz, M. R., McKee, C. F., & Klein, R. I. 2005, *ApJ*, 618, L33
Krumholz, M. R., Klein, R. I. *et al.*, 2009, *Science*, 323, 754
López-Sepulcre, A., Cesaroni, R., & Walmsley, C. M. 2010, *A&A*, in press
López-Sepulcre, A., Codella, C., Cesaroni, R. *et al.* 2009, *A&A*, 499, 811
McKee, C. F. & Tan, J. C. 2002, *Nature*, 416, 59
Molinari, S. & Hi-GAL Consortium 2010, *A&A*, 518, L100
Ossenkopf, V. & Henning, Th. 1994, *A&A*, 291, 943
Patel, N., Curiel, S., Sridharan, T. K., Zhang, Q., *et al.* 2005, *Nature*, 437, 109
Peretto, N., Fuller, G. A., Plume, R., Anderson, L. D., *et al.* 2010, *A&A*, 518, L98
Rathborne, J. M., Jackson, J. M., Chambers, E. T., Stoijmirovic, I., *et al.* 2010, *ApJ*, 715, 322
Rathborne, J. M., Jackson, J. M., & Simon, R. 2006, *ApJ*, 641, 389
Rosolowsky. E., *et al.* 2009 *ApJS*, 188, 123
Sandell, G., Wright, M., & Forster, J. R. 2003, *ApJ*, 590, L45
Schuller, F. *et al.* 2009, *A&A*, 504, 415
Shepherd, D. S. & Churchwell, E. 1996, *ApJ*, 472, 225
Shi, H., Zhao, J.-H., & Han, J. L. 2010, *ApJ*, 710, 843
Sollins, P. K., Zhang, Q., Keto, E., & Ho, P. T.P. 2005, *ApJ*, 624, L49
Tang, Y.-W., Ho, P. T.P., Koch, P. M., Girart, J. M., *et al.* 2009, *ApJ*, 700, 251
Vlemmings, W. H.T., Surcis, G. *et al.* 2010, *MNRAS*, 404, 134
Walmsley, M. 1995, *Rev. Mex. Astron. Astrophys. Conf. Ser. 1*, 137
Wang, P., Li, Z. Y., Abel, T., & Nakamura, F. 2010, *ApJ*, 709, 27
Wang, Y., Zhang, Q., Pillai, T., Wyrowski, F., & Wu, Y. 2008, *ApJ*, 672, L33
Wolfire, M. G. & Cassinelli, J. P. 1987, *ApJ*, 319, 850
Wu, J. & Evans, N. J. 2003, *ApJ*, 592, L79
Wu, Y., Qin, S.-L., Guan, X., Xue, R., *et al.* 2009, *ApJ*, 697, L116
Wu, Y., Wei, Y., Zhao, M., Shi, Y., *et al.* 2004, *A&A*, 426, 503
Zapata, L. A., Palau, A., Ho, P.T P., Schilke, P., *et al.* 2008, *A&A*, 479, L25
Zhang, Q. & Ho, P. T. P. 1997, *ApJ*, 488, 241
Zhang, Q., Sridharan, T. K., Hunter, T. R., Chen, Y., *et al.* 2007, *A&A*, 470, 269
Zhang, Q., Wang, Y., Pillai, T., & Rathborne, J. 2009, *ApJ*, 696, 268

Computational Star Formation
Proceedings IAU Symposium No. 270, 2011
J. Alves, B. G. Elmegreen, J. M. Girart, V. Trimble, eds.

doi:10.1017/S1743921311000147

Pre-main sequence multiple systems

Hervé Bouy[1]

[1]Centro de Astrobiologia, INTA-CSIC, P.O. Box - Apdo. de correos 78, Villanueva de la Cañada Madrid 28691, Spain
email: hbouy@cab.inta-csic.es

Abstract. It is now well established that the majority of young stars are found in multiple systems, so that any theory of stellar formation must account for their existence and properties. Studying the properties of multiple star systems therefore represents a very powerful approach to place observational constraints on star formation theories. Additionally, multiple systems offer other advantages. They provide the most accurate and unambiguous way to measure masses, using orbital fitting and Kepler's laws, and even the stellar radius in the special case of eclipsing binaries. They also allow to compare the properties of 2 coeval objects with different masses, providing important tests for the evolutionary models.

Keywords. Stars: binaries: general, Stars: formation

1. Introduction: Parameters of interest

Multiple systems are very diverse: binaries can be wide or tight, the 2 components can have very different masses, and we even know triple, quadruple and higher order systems. It would be impossible to review all the different properties in just 15 min, and this presentation will focus on three major parameters: the multiplicity frequency, the distribution of mass ratio and of separation.

These various properties depend directly on the mechanisms at work during the early stages of stellar formation and in the subsequent few million years. Early studies of multiplicity focused essentially on the frequency of multiple systems, but the advent of powerful and sensitive instruments allowed astronomers to investigate in more details the statistical properties of multiple systems and in particular their distributions of mass ratio and separation. These properties, and how they depend on the age, the environment and the mass, can in turn be readily compared to theoretical predictions and numerical simulation outputs.

2. Multiplicity vs Environment

Early studies of multiplicity have found a significantly higher rate of multiple systems in young loose associations than in young clusters. Young clusters on the other hand seem to have multiplicity properties similar to those observed for the nearby field population. Early numerical simulations showed that a clustered population can indeed evolve into the field binary distribution through dynamical decay, providing additional evidence that the field population most likely formed in a clustered environment rather than in loose associations.

3. Multiplicity vs Primary's mass

Figure 1 shows the observed binary fraction as a function of spectral type. The results plotted in this figure come from studies with very different levels of sensitivity and

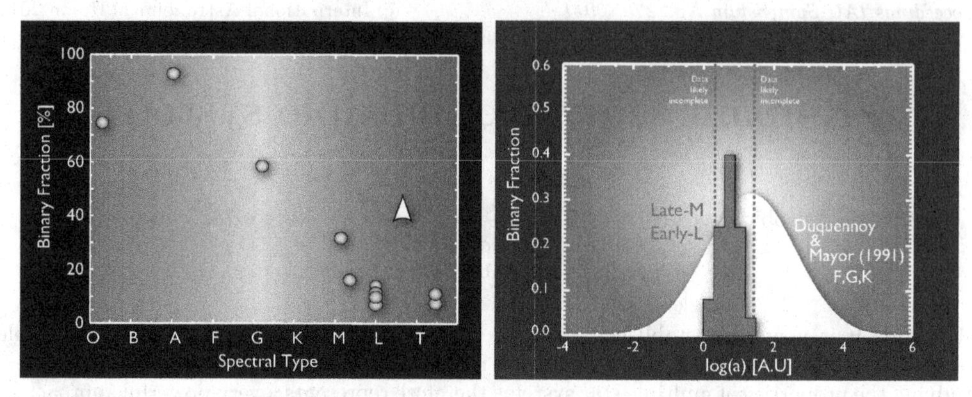

Figure 1. Left Panel: Multiplicity frequency as a function of Spectral Type. Values from Maíz Apellániz (2010), Mason al. (2009), Tokovinin & Smekhov (2002), Duquennoy & Mayor (1991), Siegler *et al.* (2005), Leinert & al. (1997), Fisher & Marcy (1992), Reid & Gizis (1997), Goldman *et al.* (2008), Burgasser *et al.* (2003), Joergens (2006), Bouy *et al.* (2003, 2006), Maxted & Jeffries (2005),Basri & Reiners (2006). **Right Panel:** Distribution of separation for solar mass stars from Duquennoy & Mayor (1991) and very low mass stars Bouy *et al.* (2003).

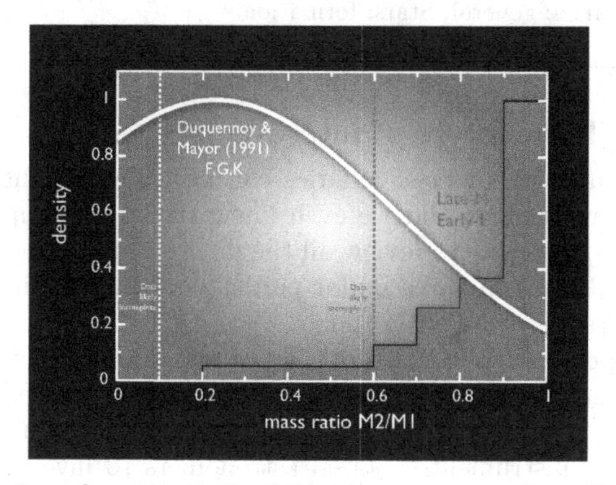

Figure 2. Distribution of mass ratio for field solar mass stars from Duquennoy & Mayor (1991) and very low mass stars Bouy *et al.* (2003).

completeness. For example the measurements obtained for the ultracool objects are missing the closest binaries, and the real binary fraction could add up to 40 or 50%. The value reported at the high mass end is also an upper limit, and the binary fraction is expected to be nearly 100%. A direct comparison is therefore strictly not possible, but this figure nevertheless displays a clear trend of decreasing multiplicity fraction with the mass of the primary.

Figure 1 shows the distribution of separation for solar type stars and very low mass stars and brown dwarfs. The two distributions are gaussian, the very low mass star distribution being a scaled down version of the solar mass one. The much narrower separation range covered to date for very low mass objects prevents a more detailed quantitative comparison.

Finally, although the current surveys for very low mass stars were not sensitive to small mass ratio, the mass ratio distributions for solar type stars and very low mass stars are significantly different, the later one displaying a clear preference for equal mass systems as illustrated int figure 2.

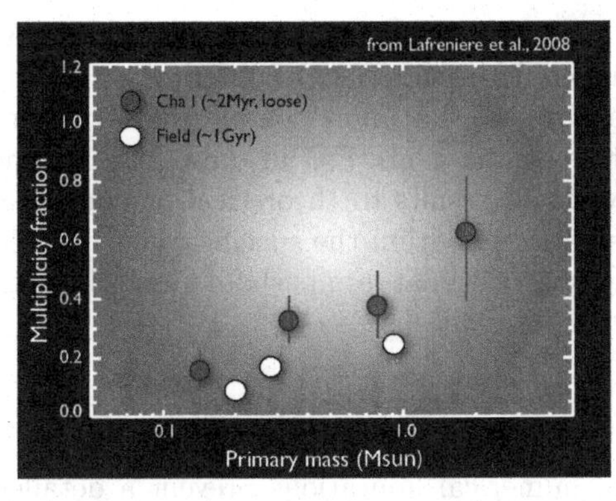

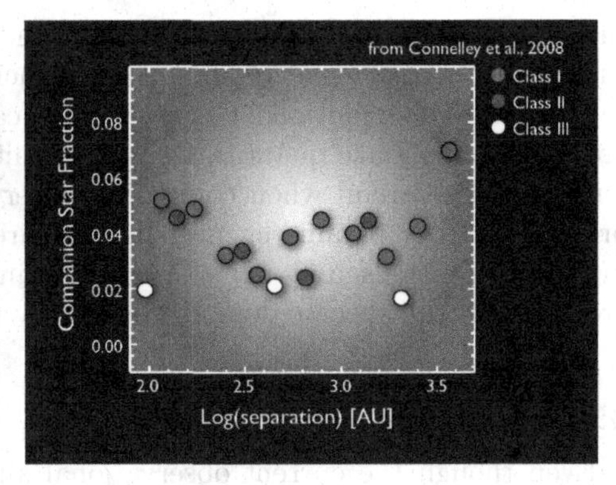

Figure 3. Left Panel: Multiplicity frequency as a function of mass for Cha I members and field solar mass stars. Figure from Lafreniere *et al.* (2008). **Right Panel:** Distribution of separation for class I, II and III. Figure from Connelley *et al.* (2008).

4. Multiplicity vs Age

Most of the multiplicity properties are set in the very early stages of star formation. Dynamical decay plays a very active role in the first few million years. Figure 3 shows the multiplicity frequency of the young loose association Chameleon I compared to that of the solar neighborhood as measured by Lafreniere *et al.* (2008). As explained earlier, the young association has a multiplicity frequency about twice higher than that of field stars, and the multiplicity frequency decreases with decreasing mass. In such a lose association of a few million years, dynamical interactions are unlikely to modify further the multiplicity rate. In a denser cluster, gravitational interactions would on the contrary be more frequent and modify the multiplicity frequency until it becomes similar to that of field stars. We therefore see that the dependence on the age is in fact tightly related to the initial conditions and environment.

Figure 3 shows the distribution of separation of class I, class II and class III sources as reported in Connelley *et al.* (2008). All three are relatively flat over the small separation range covered here and within the uncertainties, and once again this figure illustrates the higher multiplicity frequency among young objects members of loose associations.

The dependence on the age of the distribution of mass ratio has not been studied in great details, mostly because observations of deeply embedded protostars or young stellar objects are challenging with the current instrumentation. The best results have been obtained for high mass stars, which are relatively brighter and easier to observe. At young ages, the distribution of mass ratio for high mass star companions seem to follow random sampling, while at older evolutionary stages, the least massive companions have been ejected and the mass ratio distribution peaks towards unity.

5. Comparing observations to simulations

Since the aim of this conference is to discuss about computational star formation, and therefore to put together observations, theories and numerical simulations, a few words of cautions for both observers and theoreticians should be given.

Interpreting the observations can be a perilous exercise. All the results mentioned in this presentation were obtained within given instrumental limitations. For example, direct imaging using adaptive optics provides similar spatial resolution than speckle interferometry but different limits of sensitivity. Radial velocity surveys are so far limited to relatively bright sources, and often miss the long period companions. None of these surveys cover the entire separation range and mass ratio range of binary populations. As such, differences in measured binary frequencies could be the result of differing separation or mass ratio distributions. General trends can be seen, but more complete observations are required to make quantitative and definitive comparisons. For all these reasons, one must be very careful when comparing observational results to theoretical or numerical predictions. Any parameter must be compared strictly within the specific range covered by the observations, and any article presenting observational results should describe with great details the specific domain covered by the observations.

6. Conclusions and perspectives

Even though the current observational and numerical limitations prevent a detailed and quantitative comparison of observations to numerical predictions, tremendous progresses have been made over the past decade regarding the multiplicity properties of pre-main sequence stars. These have brought crucial informations and shed a new light on our understanding of stellar formation. The development of new instrumentation for both high resolution imaging, high spectral resolution, high astrometric accuracy, as well as the advent of new multi-epoch surveys, give very exiting perspective for this field of research. In this short presentation, I gave only a very shallow overview of the multiplicity properties. Multiple systems are very rich, and many more of their properties are extremely important. The distribution of eccentricity, the properties of their discs, the occurrence of high order multiple systems, and the presence and properties of planets in multiple systems (as we know multiple systems hosting planets) provide very exiting perspectives for the study of stellar formation.

References

Basri, G. & Reiners, A. 2006, *AJ*, 132, 663

Bouy, H., Brandner, W., Martín, E. L., Delfosse, X., Allard, F., & Basri, G. 2003, *AJ*, 126, 1526

Bouy, H., Martín, E. L., Brandner, W., Zapatero-Osorio, M. R., Béjar, V. J. S., Schirmer, M., Huélamo, N., & Ghez, A. M. 2006, *A&A*, 451, 177

Burgasser, Adam J., Kirkpatrick, J. Davy, Reid, I. Neill, Brown, Michael E., Miskey, Cherie L., Gizis, & John E. 2003, *ApJ*, 586, 512

Connelley, Michael S., Reipurth, Bo, Tokunaga, & Alan T. 2008, *AJ*, 135, 2526

Duquennoy, A. & Mayor, M. 1991, *A&A*, 248, 485

Fisher D. & Marcy G. 1992, *ApJ*, 396, 178

Goldman, B., Bouy, H., Zapatero Osorio, M. R., Stumpf, M. B., Brandner, W., Henning, T. 2008, *A&A*, 490, 763

Joergens, V. 2006, *A&A*, 448, 655

Lafrenière, David, Jayawardhana, Ray, Brandeker, Alexis, Ahmic, Mirza, van Kerkwijk, & Marten H. 2008, *ApJ*, 643, 844

Leinert, C., Henry, T., Glindemann, A., & McCarthy, D. W., Jr. 1997, *AJ*, 325, 159

Maíz Apellániz, J. 2010, *A&A, arXiv:1004.5045*

Mason, Brian D., Hartkopf, William I., Gies, Douglas R., Henry, Todd J.; Helsel, & John W. 2009, *AJ*, 137, 3358

Maxted, P. F. L. & Jeffries, R. D. 2005, *MNRAS*, 362, 45

Reid, I. N. & Gizis, J. 1997, *AJ*, 114, 1992

Siegler, Nick, Close, Laird M., Cruz, Kelle L., Martín, Eduardo L., Reid, & I. Neill 2005, *ApJ*, 621, 1023

Tokovinin, A. A. & Smekhov, M. G 2002, *A&A*, 382, 118

Computational Star Formation
Proceedings IAU Symposium No. 270, 2011
J. Alves, B. Elmegreen, J. Girart & V. Trimble, eds.

doi:10.1017/S1743921311000159

Pre-main sequence disks

Gaspard Duchêne[1,2]

[1]Astronomy Department, University of California Berkeley, Berkeley CA 94720-3411, USA
email: gduchene@berkeley.edu

[2]Université Joseph Fourier - Grenoble 1/CNRS, Laboratoire d'Astrophysique de Grenoble (LAOG) UMR 5571, BP 53, 38041 Grenoble Cedex 09, France

Abstract. In this contribution, I briefly review our empirical knowledge of disks around $\lesssim 2\,M_\odot$ pre-main sequence (T Tauri) stars, focusing first on the dichotomic question of their frequency before moving on to some more detailed disk properties (overall orientation, total mass, outer radius). Finally, I conclude with a brief discussion of disks around embedded protostars, which will play in the next few years a major role in testing star formation theory and simulations.

Keywords. stars: formation, circumstellar matter, stars: pre-main sequence

---

1. Introduction

The study of disks around pre-main sequence stars is primarily motivated by our will to understand the initial conditions, processes and timescales associated with planet formation. However, it must be emphasized that the presence and detailed properties of these protoplanetary disks are determined during the star formation process itself, of which they are a natural by-product due to angular momentum conservation. The build-up of mass in a circumstellar disk is indeed set by the accretion history of the object, as well as by dynamical interactions with its immediate neighbours in the early phases of its evolution. Therefore, one can use circumstellar disks as a "final boundary condition" that must be met by star formation theories and simulations, akin to the roles of the IMF and stellar multiplicity (see contributions by J. Ascenso and H. Bouy, this volume). Below, I outline a few key findings that theory must account for.

2. Frequency of disks around pre-main sequence stars

For about three decades, the presence of circumstellar disks around T Tauri stars has been probed through their copious amounts of infrared excess, most recently through a slew of *Spitzer* surveys of nearby star-forming regions. Of particular interest for this discussion are studies of the youngest ($\lesssim 1$ Myr) stellar populations, such as NGC 1333 and NGC 2024, where the proportion of members hosting a circumstellar disks is as high as 80%. Even slightly older (~ 2 Myr) star-forming regions like Taurus-Auriga, the Trapezium Cluster and NGC 2068/71 boast disk frequency rates of 60% and higher (see Hernández et al. 2008 and references therein). Considering that some disks may be disrupted in early dynamical interactions between nearby objects, it is clear that *the large majority ($\gtrsim$90%) of stars form accompanied by a circumstellar disks.*

Breaking down this global disk frequency, it is interesting to test whether the mass of the central star has any impact on the presence of a circumstellar disk, which could be expected via the depth of its potential well. Sufficiently large samples have been probed in several nearby star-forming regions (the Trapezium Cluster, Taurus-Auriga, IC 348, Chamaeleon I) so as to split them in a few stellar mass bins, with very similar results in each region (Hillenbrand et al. 1998; Luhman *et al.* 2010). In short, the disk frequency is typically only 20-30% lower for very low-mass objects than for solar-type stars. Therefore, *even among extremely low-mass objects, at least half are surrounded by a circumstellar disk at an age of ∼1 Myr.* We also note that there is no significant dependence of the disk frequency with stellar mass in the 0.25–2 $M_{\odot}$. Unfortunately, there are too few Herbig AeBe stars (with masses in the 2–8 $M_{\odot}$ range) known in star-forming regions to assess whether this high disk frequency extends to higher masses.

Another factor that is thought to play an important role in the detailed physics of star formation is the type of environment in which stars form. While young dense clusters are characterized by lower multiplicity rates thought to be caused by their violent early dynamical history, the four populations with disk frequencies higher than 80% are all young dense clusters (Hernández et al. 2008). It thus appears that *disk formation and long-term survival are not nearly as strongly (if at all) affected in a densely clustered environment as multiplicity.* The other way in which the environment can affect the formation and survival of circumstellar disks is through the intense UV radiation field imposed by high-mass O-type stars. However, in the Orion Trapezium Cluster, where a handful of high-mass stars are responsible for the well-known "proplyd" phenomenon, the proportion of young stars hosting a disk appears to be roughly constant, possibly even slightly decreasing, as a function of the distance to θ^1 C Ori on a scale of 1–3 pc (Hillenbrand *et al.* 1998). This may be because the UV radiation field, while sufficient to photoevaporate the outer regions of disks, is nonetheless too weak to disrupt whole disks altogether. In any case, this seems to indicate that *the ultraviolet field of high-mass stars does not preclude the formation of protoplanetary disks*, except in the immediate vicinity of individual O stars (e.g., Guarcello *et al.* 2009, Mercer *et al.* 2009).

3. Quantative properties of disks

While assessing the presence of a disk around a pre-main sequence star is a relatively easy task, it is a more complex one to determine the disk bulk properties, such as total mass and global geometry. For one, near and mid-infrared observations only probe the presence of dust within a few AU of the central object. Generally speaking, going beyond this dichotomic assessment requires to make use of a combination of 1) simplistic assumptions, 2) complex radiative transfer modeling, and/or 3) spatially-resolved observations. As a result, using large populations of disks to test theoretical and numerical predictions is plagued by a complex set of biases and is generally a model-dependent endeavor.

Arguably the least ambiguous disk "property" to determine is the disk symmetry axis (i.e., the orientation of its angular momentum vector). Scattered light images and interferometric mapping at millimeter wavelengths have now successfully resolved tens of circumstellar disks in nearby star-forming regions†, enabling global statistical studies. In particular, it is possible to compare the orientation of disks with respect to the local cloud magnetic field, as traced by linear polarization measurements of background stars. Contrary to earlier studies (that were plagued by small sample sizes and selection biases), it appears that the symmetry axis of young star+disk systems are randomly oriented with respect to the local magnetic field in the Taurus-Auriga cloud (Ménard & Duchêne 2004). This suggests that *the orientation of the initial magnetic field has little to no influence on that of the final angular momentum vector of the system, or, alternatively, that there is an almost complete "loss of memory" of any originally preferred orientation during the star formation process.*

† See an up-to-date list at `http://www.circumstellardisks.org`

The orientation of disks within multiple stellar systems is also an interesting tracer of the dynamical evolution inherent to the formation of the systems themselves. In binary systems, the circumstellar disks associated with each component appear to be preferentially aligned with each other, although there is a non-trivial number of exceptions (see Monin et al. 2007 for a review). Intriguingly, this is reminiscent of the relative orientation of orbital planes within triple systems (e.g., Tokovinin 1997). As for the relative orientation of disks and orbital planes, too few systems have been studied to date to reach definitive conclusions. It is worth noting, however, that there are examples of misalignment in the case of circumstellar disks (e.g., Stapelfeldt et al. 1998), i.e., disks around a given component, whereas circumbinary disks seem to be coplanar with their inner binary (e.g., Prato *et al.* 2002). Because the former type of systems is generally characterized by larger separations (a result of selection biases), this may be interpreted as evidence for *a (partial) randomization of the individual angular momentum vectors during the fragmentation of a collapsing core.*

Because disk studies have long focused on the question of planet formation, significant efforts have been placed on assessing their total mass and overall size, which are critical initial conditions. While there is a large object-to-object scatter and considerable uncertainty in the conversion of (sub)millimeter fluxes to total disk masses (due to unknown dust opacities and gas-to-dust ratios), the maximum disk masses over the 0.25–3 $M_\odot$ stellar mass range is on the order of 10% of the stellar mass (see Natta *et al.* 2000 for a review). This conclusion seems to extend all the way into the brown dwarf regime (Bouy et al. 2008; Momose *et al.* 2010). This suggests that *most, if not all, disks around pre-main sequence stars are intrinsically robust against gravitational instabilities.* However, since disk instabilities occurs on much shorter timescales than the typical age of pre-main sequence stars, this should only be considered an inescapable conclusion. Disk fragmentation, if it occurs at all, must take place in a much earlier phase, with the presently observed disk a mere remnant of the initial disk or the result of subsequent continuous accretion from the collapsing core.

Finally, the outer radius of disks, which may be considered a fossil tracer of past dynamical evolution, remains difficult to assess with accuracy because of the limited spatial resolution and sensitivity to faint extended structures of current observations. Typical disk sizes are in the 50–500 AU range (Kitamura *et al.* 2002; Watson *et al.* 2007), with no significant correlation between disk size and stellar mass. Among disks resolved around single very low-mass stars, two are remarkably small (20–40 AU, Luhman *et al.* 2007; Momose et al. 2010) while one is extremely large (1100 AU, Glauser *et al.* 2008), suggesting that *other factors are more important than the stellar mass in determining the disk size.* Because we observe disks long after they form, the nature of these factors remains a speculative topic of discussion.

4. The future: disks around embedded protostars

Ultimately, the study of disk properties, even global ones like disk mass and size, in pre-main sequence stars is of little help to constrain the stellar formation process. While this is in part because of selection biases and instrumental limitations, there is a much more profound issue. By the pre-main sequence phase, a large fraction of the initial disk mass has been accreted or expelled from the system and viscous evolution has spread the disk's outer radius well beyond its original size. In order to use disk properties to test predictions of star formation theory, it is mandatory to move back in time and get as close as possible to star formation itself. Specifically, it is necessary to study disks around embedded protostars.

Class I protostars represent the phase that precedes the T Tauri phase, and are a natural first step in that direction. Yet, only Class 0 sources, in which the central protostar still amounts for a minority of the total mass of the system, are readily comparable to the outcome of numerical simulations. Unfortunately, Class 0 sources are dominated by a massive envelope, and it is particularly challenging to discern their circumstellar disk. In dedicated studies, two remarkably large disks have been resolved and distinguished from the envelope in which they are embedded (150–300 AU, Jörgensen *et al.* 2005; Enoch *et al.* 2009). On the other hand, a first systematic survey of five several Class 0 has found no resolved disk on scales of 70–120 AU (Maury *et al.* 2010), suggesting either much smaller disk sizes or a delayed formation of disks. The picture is still murky at this point, but further studies of disks among Class 0 sources in the upcoming *ALMA* era will yield a clearer picture and, consequently, a more stringent test of star formation theories and simulations.

References

Bouy, H., Huélamo, N., Pinte, C., Olofsson, J., Barrado y Navascués, D., Martín, E., Pantin, E., Monin, J. L., Basri, G., Augereau, J. C., Ménard, F., Duvert, G., Duchêne, G., Marchis, F., Bayo, A, Bottinelli, S., lefort, B., & Guieu, S. 2008, *A&A*, 486, 877

Enoch, M., Corder, S., Dunham, M., & Duchêne, G. 2009, *ApJ*, 707, 103

Glauser, A., Ménard, F., Pinte, C., Duchêne, G., Güdel, M., Monin, J. L., & Padgett, D. 2008, *A&A*, 485, 531

Guarcello, M., Micela, G., Damiani, F., Peres, G., Prisinzano, L., & Sciortino, S. 2009, *A&A* , 496, 453

Hernández, J., Hartmann, L., Calvet, N., Jeffries, R., Gutermuth, R., Muzerolle, J., & Stauffer, J. 2008, *ApJ*, 686, 1195

Hillenbrand, L., Strom, S., Calvet, N., Merril, M., Gatley, I., Makidon, R., Meyer, M., & Skutskie, M. 1998, *AJ*, 116, 1816

Jörgensen, J., Bourke, T., Myers, P., Schöier, F., van Dishoeck, E., & Wilner, D. 2005, *ApJ*, 632, 973

Kitamura, Y., Momose, M., Yokogawa, S., Kawabe, R., Tamura, M., & Ida, S. 2002, *ApJ*, 581, 357

Luhman, K., Allen, P., Espaillat, C., Hartmann, L., & Calvet, N. 2010, *ApJS*, 186, 111

Luhman, K., Adame, M., D'Alessio, P., Calvet, N., McLeod, K., Bohac, C., Forrest, W., Hartmann, L., Sargent, B., & Watson, D. 2007, *ApJ*, 666, 1219

Maury, A., André, P., Hennebelle, P., Motte, F., Stamatellos, D., Bate, M., Belloche, A., Duchêne, G., & Whitworth, A. 2010, *A&A*, 512, 40

Ménard, F. & Duchêne, G. 2004, *A&A*, 425, 973

Mercer, E., Miller, J., Calvet, N., Hartmann, L., Hernández, J., Sicilia-Aguilar, A., & Gutermuth, R. 2009, *ApJ*, 138, 7

Momose, M., Ohashi, N., Kudo, T., Tamura, M., & Kitamura, Y. 2010, *ApJ*, 712, 397

Monin, J. L., Clarke, C., Prato, L., & McCabe, C. 2007, in: B. Reipurth, D. Jewitt, & K. Keil (eds.), *Protostars & Planets V*, p. 395

Natta, A., Grinin, V., & Mannings, V. 2000, in: V. Mannings, A. Boss, & S. Russell (eds.), *Protostars & Planets IV*, p. 559

Prato, L., Simon, M., Mazeh, T., Zucker, S., & McLean, I. 2002, *ApJ* (Letters), 579, L99

Stapelfeldt, K., Krist, J., Ménard, F., Bouvier, J., Padgett, D., & Burrows, C. 1998, *ApJ* (Letters), 502, L65

Tokovinin, A. 1997, *A&AS*, 124, 75

Watson, A., Stapelfeldt, K., Wood, K., & Ménard, F. 2007, in: B. Reipurth, D. Jewitt, & K. Keil (eds.), *Protostars & Planets V*, p. 523

Computational Star Formation
Proceedings IAU Symposium No. 270, 2011
J. Alves, B.G. Elmegreen, J. M. Girart & V. Trimble, eds.

doi:10.1017/S1743921311000160

Morphological Complexity of Protostellar Envelopes

John J. Tobin[1], Lee Hartmann[1], Edwin Bergin[1], Leslie W. Looney[2], Hsin-Fang Chiang[2], Fabian Heitsch[3]

[1]Department of Astronomy, University of Michigan, Ann Arbor, MI 48109; jjtobin@umich.edu

[2]Department of Astronomy, University of Illinois at Champaign/Urbana, Urbana, IL 61801

[3]Department of Physics and Astronomy, University of North Carolina-Chapel Hill, Chapel Hill, NC

Abstract. Extinction maps at 8μm from the Spitzer Space Telescope show that many Class 0 protostars exhibit complex, irregular, and non-axisymmetric structure within the densest regions of their dusty envelopes. Many of the systems have highly irregular and non-axisymmetric morphologies on scales ~1000 AU, with a quarter of the sample exhibiting filamentary or flattened dense structures. Complex envelope structure is observed in regions spatially distinct from outflow cavities, and the densest structures often show no systematic alignment perpendicular to the cavities. We suggest that the observed envelope complexity is the result of collapse from protostellar cores with initially non-equilibrium structures. The striking non-axisymmetry in many envelopes could provide favorable conditions for the formation of binary systems. We then show that the kinematics around L1165 as probed with N_2H^+ are indicative of asymmetric infall; the velocity gradient is not perpendicular to the outflow.

Keywords. stars: formation, dust, extinction, ISM: molecules, ISM: globules

---

1. Introduction

Sphericity and axisymmetry have been standard assumptions on which our theoretical understanding of star formation has rested for some time (e.g. Shu 1977; Terebey *et al.* 1984; Galli & Shu 1993; Hartmann *et al.* 1996). However, it is not clear whether or not envelopes around protostars are accurately described by symmetric models. The shapes of dense cores have been studied on large scales (>0.1 pc) using molecular line tracers of dense gas (Benson & Myers 1989; Myers *et al.* 1991). While the ammonia cores appeared compact and round due to low resolution, associated IRAS sources are often located off-center from the line emission peaks indicating non-axisymmetry. Furthermore, emission from the millimeter-line tracers appears quite complex on scales outside the ammonia emission(Myers *et al.* 1991).

Recently, observations with the *Spitzer Space Telescope* have given a high-resolution view of envelope structure in extinction at 8μm against Galactic background emission (Looney *et al.* 2007; Tobin *et al.* 2010). This method enables us to observe the structure of collapsing protostellar envelopes on scales from ~1000AU to 0.1 pc for the first time with a mass-weighted tracer. In contrast, single-dish studies of envelopes using dust emission in the sub/millimeter regime generally have lower spatial resolution. The continuum emission depends upon temperature and density, while molecular tracers are additionally affected by complex chemistry.

In this contribution, we present IRAC 8μm extinction maps of envelopes around Class 0 protostars. Most of the envelopes in our sample are found to be irregular and non-axisymmetric. We suggest that these envelopes may be important for binary formation

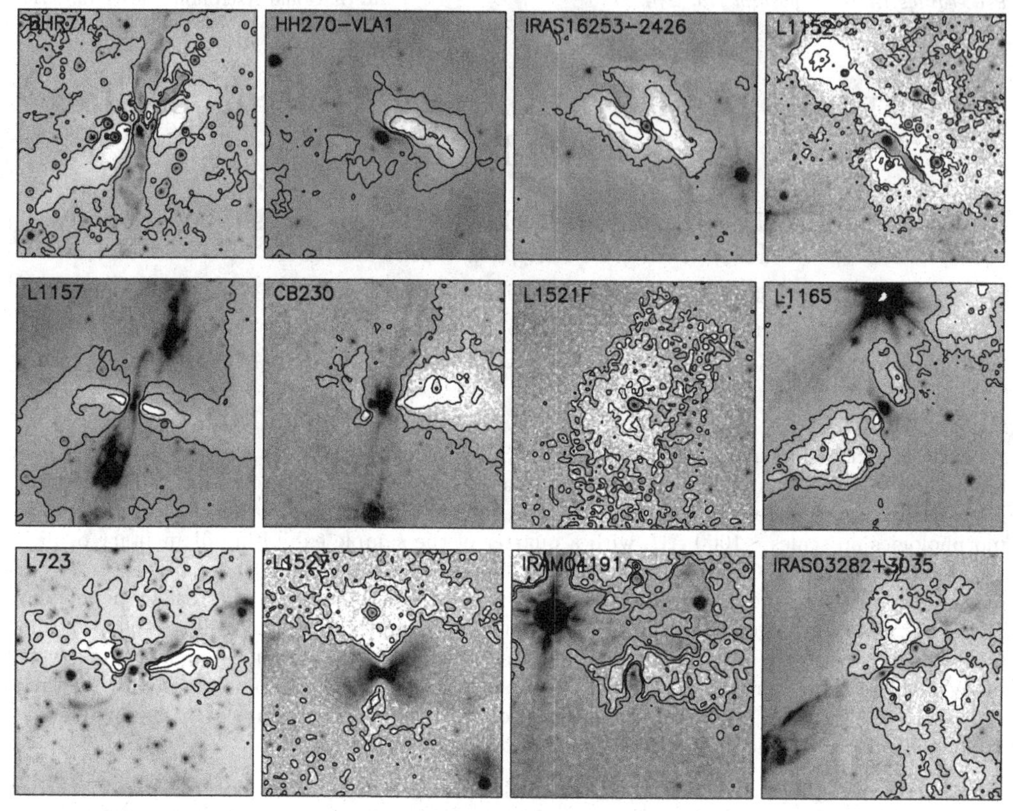

Figure 1. IRAC 8μm images with 8μm optical depth contours overlaid for a selection of the protostars from Tobin *et al.* (2010). From left to right the columns are filamentary envelopes, one-sided envelopes, quasi-symmetric envelopes, and irregular envelopes respectively.

as collapse models have shown that asymmetric structure can induce small-scale fragmentation of the infalling envelope. We also show initial results in analyzing the kinematic structure of these envelopes using the dense gas tracers N_2H^+.

2. Asymmetric Envelopes in Extinction

In Figure 1, we display a selection of 12 systems (out of 22 from Tobin *et al.* (2010)) for which we have detected an envelope in extinction. Each panel shows the 8μm image of the protostar with optical depth contours overlaid. In most images, the outermost extinction contour corresponds to A_V ~5-10 and going up to $A_V \sim 30$.

The most striking feature of the 8μm extinction maps is the irregularity of envelopes in the sample. Most envelopes show high degrees of non-axisymmetry; in most cases, spheroids would be an inaccurate representation of the structure. Some of the most extreme examples have most extincting material mostly on one side of the protostar (e.g. CB230, HH270 VLA1) or the densest structures are curved near the protostar (e.g. BHR71, L723). The structures seen in extinction at 8μm do not seem to be greatly influenced by the outflow. Tobin *et al.* (2010) demonstrated that the outflow cavities of these sources are generally quite narrow and the dense material detected in extinction is often far from the outflow cavities and thus unlikely to be produced by outflow effects.

For convenience, we categorize the systems into 4 groups according to their morphology, though some systems have characteristics of multiple groups. The first column of Figure 1 shows the envelopes that have a highly filamentary/flattened morphologies; column 2

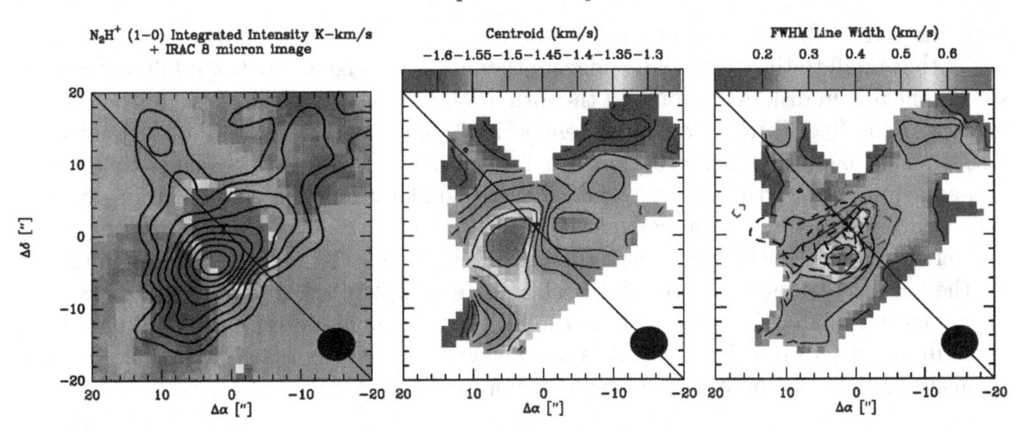

Figure 2. N_2H^+ ($J = 1 \rightarrow 0$) observations for the protostar L1165; in all panels the protostar is marked with an X and the outflow axis is marked with a straight line. Left: IRAC 8μm image (*color scale*) with N_2H^+ ($J = 1 \rightarrow 0$) integrated intensity (*contours*). Middle: Centroid velocity of N_2H^+ emission derived from fitting the hyperfine lines; notice the sharp gradient across the protostar which is not perpendicular to the outflow. Right: Full-width half maximum of N_2H^+ lines; note that the linewidth is greatest at the peak of N_2H^+ emission. The dashed contours in the rightmost panel are the HCO^+ blue (*blue*) and redshifted (*black*) emission.

shows envelopes that have most material on one side of the protostar; column 3 shows the protostars whose envelopes are more or less spheroidal in projection; column 4 shows envelopes that do not strictly fall within the above categories.

3. Kinematic Structure

In addition to morphological structure, the kinematic structure of the envelope resulting from the non-axisymmetric envelope is of great importance for the subsequent evolution of the system. Figure 2 shows the N_2H^+ ($J = 1 \rightarrow 0$) observations for the protostar L1165 from the CARMA millimeter array. The N_2H^+ emission shows that the envelope is elongated along the larger filament consistent with the IRAC 8μm image and the peak emission is offset from the protostar; the envelope also remains filamentary down to small scales. The centroid velocity map shows a strong velocity gradient across the envelope, 0.35 km/s over 2400 AU; though, the linewidth indicates that the gradient could be larger. At present, it is difficult to differentiate between *projected* infall or rotation giving rise to the velocity structure; in either case the infall motions will be complex. Given that the N_2H^+ velocity gradient is $\sim$30° from perpendicular to the outflow and the HCO^+ gradient is perpendicular it seems possible that the angular momentum vector may change with time.

4. Implications of Non-Axisymmetric Collapse

The envelope asymmetries may result from the initial cloud structure. Stutz *et al.* (2009) recently surveyed pre-stellar/star-less cores using 8 and 24μm extinction; their results, and those of Bacmann *et al.* (2000), showed that even pre-collapse cloud cores already exhibit asymmetry. Given the initial asymmetries, the densest, small-scale regions are likely to become more anisotropic during gravitational collapse (Lin *et al.* 1965).

4.1. *Binary Formation*

The smallest scales we observe, $\sim$1000 AU, is where angular momentum will begin to be important as the material falls further in onto the disk; this is the scale we probe with the N_2H^+ observations of L1165. The envelope asymmetries down to small scales imply that infall to the disk will be uneven; therefore, non-axisymmetric infall may play a significant role in disk evolution and the formation of binary systems. Several theoretical investigations (e.g. Burkert & Bodenheimer 1993) showed that collapse of a cloud with just a small azimuthal perturbation can form binary or multiple systems; thus, *large* non-axisymmetric perturbations should make fragmentation even easier. Fragmentation can even begin before global collapse in a filamentary structure (Bonnell *et al.* 1993). Numerical simulations of disks with infalling envelopes (e.g. Kratter *et al.* 2009) informed by the results of this study could reveal a more complete understanding of how non-axisymmetric infall affects the disk and infall process. Asymmetric infall may be taking place in at least L1165. However, it is not known if this system is a binary because the requisite high-resolution observations have not been taken.

4.2. *Initial Conditions*

The envelopes we have shown are not obviously consistent with quasi-static, slow evolution, which might be expected to produce simpler structures as irregularities have time to become damped; one needs initial non-axisymmetric structure to get strong non-axisymmetric structure later on. This raises the question of the role of magnetic fields in controlling cloud dynamics. In some models (e.g. Tassis *et al.* 2007), pre-stellar cores would probably live long enough to adjust to more regular configurations; in addition, collapse would be preferentially along the magnetic field, which would also provide the preferential direction of the rotation axis and therefore for the (presumably magneto-centrifugally accelerated) jets (Basu & Mouschovias 1994). The complex structure and frequent misalignment between collapsed structures and outflows pose challenges for such a picture. Alternatively, this could indicate that the topology of magnetic field is complex and not initially well ordered.

In contrast, more recent numerical simulations (e.g. see review by Ballesteros-Paredes *et al.* 2007; Offner & Krumholz 2009) suggest that cores are the result of turbulent fluctuations which naturally produce more complex structure with less control by magnetic fields amplified by subsequent gravitational contraction and collapse. Thus, the structure of protostellar envelopes indicates that the dynamic, turbulent model of rapid star formation seems more correct. However, recent work has indicated that turbulence can reduce the time needed for ambipolar diffusion to take place which may allow complex structures to form (Basu & Ciolek 2009).

References

Ballesteros-Paredes, J., *et al.* 2007, *Protostars and Planets V*, 63
Bacmann, A., *et al.* 2000, *A&A*, 361, 555
Basu, S., Ciolek, G. E., Dapp, W. B. & Wurster, J. 2009, *New Astronomy*, 14, 483
Bonnell, I. & Bastien, P. 1993, *ApJ*, 406, 614
Burkert, A. & Bodenheimer, P. 1993, *MNRAS*, 264, 798
Galli, D. & Shu, F. H. 1993, *ApJ*, 417, 220
Kratter, K. M., Matzner, C. D., Krumholz, M. R., & Klein, R. I. 2010, *ApJ*, 708, 1585
Lin, C. C., Mestel, L., & Shu, F. H. 1965, *ApJ*, 142, 1431
Looney, L. W., Tobin, J. J., & Kwon, W. 2007, *ApJL*, 670, L131
Offner, S. S. R., & Krumholz, M. R. 2009, *ApJ*, 693, 914
Shu, F. H. 1977, *ApJ*, 214, 488
Stutz, A. M., *et al.* 2009, *ApJ*, 707, 137
Terebey, S., Shu, F. H., & Cassen, P. 1984, *ApJ*, 286, 529
Tobin, J. J., Hartmann, L., Looney, L. W., & Chiang, H.-F. 2010, *ApJ*, 712, 1010

Computational Star Formation
Proceedings IAU Symposium No. 270, 2011
J. Alves, B.G. Elmegreen, J. M. Girart & V. Trimble, eds.

doi:10.1017/S1743921311000172

Fragmentation and dynamics in Massive Dense Cores in Cygnus-X

T. Csengeri[1] and S. Bontemps[2] and N. Schneider[1] and F. Motte[1]

[1]Laboratoire AIM, CEA - INSU/CNRS - Université Paris Diderot, IRFU/SAp CEA-Saclay, 91191 Gif-sur-Yvette, France
email: timea.csengeri@cea.fr

[2]OASU/LAB-UMR5804, CNRS, Université Bordeaux 1, 33270 Floirac, France

Abstract. A systematic, high angular-resolution study of IR-quiet Massive Dense Cores (MDCs) of Cygnus-X in continuum and high-density molecular tracers is presented. The results are compared with the quasi-static and the dynamical evolutionary scenario. We find that the fragmentation properties are not compatible with the quasi-static, monolithic collapse scenario, nor are they entirely compatible with the formation of a cluster of mostly low-mass stars. The kinematics of MDCs shows individual velocity components appearing as coherent flows, which indicate important dynamical processes at the scale of the mass reservoir around high-mass protostars.

Keywords. stars: formation, ISM: kinematics and dynamics

1. Introduction

Two competing scenario is challenged by observations to describe the formation of high-mass stars, a quasi-static model (known also as core accretion model) versus a highly dynamical model, also known as competitive accretion. The former one is a turbulence regulated scenario, where a high level of micro-turbulence balances gravity complementing the thermal pressure (e.g. McKee & Tan, 2002). This scenario describes the formation of high-mass stars as a scaled-up version of the low-mass, quasi-static star-formation process. Accretion rates of up to $\sim 10^{-3}$ $M_{\odot}yr^{-1}$ are reached for these turbulent, massive cores, which is high enough to overcome radiation pressure and enables the protostar to collect a mass larger than 8-10 $M_{\odot}$. Alternatively, dense cores may form and evolve via highly dynamical processes (e.g Vázquez-Semadeni *et al.* 2002; Heitsch *et al.* 2008; Klessen & Hennebelle, 2009), where large-scale turbulent flows create structures by shock-dissipation and, as shown by numerical simulations, supersonic turbulence fragments the gas efficiently in very short time-scales (Padoan *et al.* 2001; Vazquez-Semadeni *et al.* 2007). Gravitationally bound density fluctuations created in this picture of *gravo-turbulent fragmentation* are the seeds for star-formation. Bonnell *et al.* (2001) proposed that in a clustered environment some of these seeds may continue to accrete material from regions which were originally not bound to their protostellar envelope and therefore compete for mass in the central parts of clusters forming high-mass stars. This scenario is introduced as *competitive accretion* (see also Bonnell & Bate, 2006).

We present here a systematic, high angular-resolution study towards five IR-quiet Massive Dense Cores (MDCs) in Cygnus-X. These MDCs lack mid-IR emission (>8μm), but host powerful outflows, therefore must be in an early evolutionary stage. They have masses between 60-200 $M_{\odot}$ with 10× higher volume densities than nearby cores and therefore serve as the best prototypes to observationally constrain the turbulence regulated, quasi-static or a highly dynamical formation scenario. Cygnus X is located at 1.7 kpc and using high angular-resolution IRAM PdBI continuum and molecular line

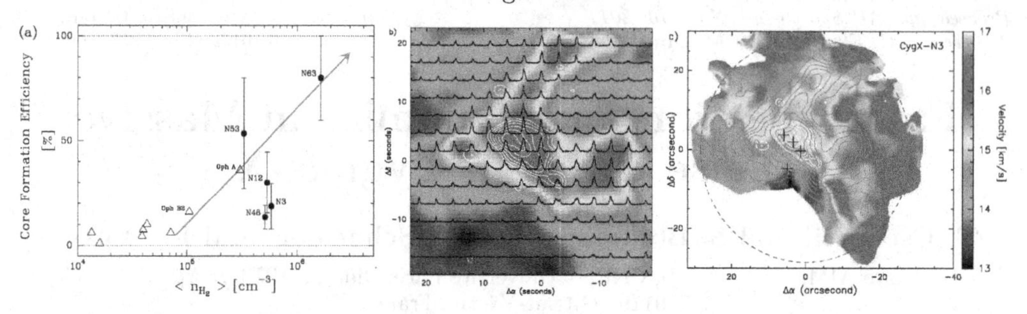

Figure 1. *a)* Core/condensation formation efficiency as a function of the average densities for dense cores of similar sizes (FWHM~0.1 pc) but of different masses in Cygnus MDCs (filled symbols) and in ρ Ophiuchus (open triangles; Motte *et al.* (1998)). A transition from low efficiency cores to possible, single collapsing cores is observed from a few 10^4 to a few 10^6 cm^{-3}. *b)* CygX-N3 is shown as an example of the sample of 5 MDCs, the $H^{13}CO^+$ (J=1–0) integrated intensity maps and spectra map with the 3mm continuum contours in white overlaid, *c)* a map of the position of the spectra peak, which represents the velocity field of the bulk material, crosses mark the position of the four 1mm continuum sources, white contours shows the 3mm continuum emission and grey contours represent the integrated emission of $H^{13}CO^+$.

observations at 1mm and 3mm offers the opportunity of reaching small scales (less than 2000 AU at 1mm and 6000 AU at 3mm) to identify individual collapsing objects. The analysis of the continuum maps allowed us to separate individual protostars (Sect.2.), and the line emission is the perfect probe of the kinematics of the gas in the MDCs (Sect.3.).

2. Fragmentation

These MDCs in Cygnus X are expected to collapse either to form a single (or binary) high-mass star or they might be sub-fragmented into a cluster of mostly low-mass protostars. Bontemps *et al.* (2010) studied the fragmentation properties of these MDCs and found that they are actually sub-fragmented but not in a too large number of fragments. Only the most compact core, CygX-N63, is not sub-fragmented and seems to correspond to a single massive protostar with an envelope mass of 60$M_\odot$. The fragments inside the other cores have sizes and separations similar to low-mass proto-stellar objects in nearby proto-clusters. A total of 23 fragments are resolved in the sample with typical sizes of ~4000 AU and masses between 2 and 55$M_\odot$. Nine of them are found to be massive enough to be precursors of OB stars. We conclude that the level of fragmentation in 4 out of 5 MDCs is higher than in the turbulence regulated collapse scenario, but is not as high as expected in a pure gravo-turbulent scenario where the distribution of mass is dominated by low-mass protostars. In addition we find that the densest MDCs have a large fraction of their total mass in only a few massive fragments (Figure 1 a) showing that they have an exceptionally high core formation efficiency with a clear excess of massive fragments in their central regions which are then proposed to correspond to the expected primordial mass segregation of stellar clusters.

3. Kinematics

To go one step further in understanding the origin of these massive fragments, Csengeri *et al.* (2010) studied the high-density tracers $H^{13}CO^+$ and $H^{13}CN$ at high angular-resolution obtained with the PdBI. We consider that such high-density tracers represent well the common mass reservoir of the high-mass protostars (see the structures in Figure 1 *b*). Thus, there is either a high level of micro-turbulence, which provides sufficient support against gravity and keeps the MDCs in a quasi-static evolution

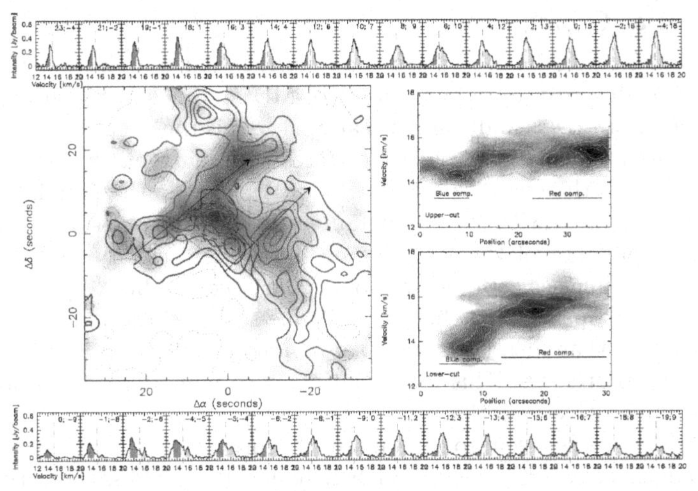

Figure 2. Map of CygX-N3 in $H^{13}CO^+$ obtained with the PdBI and zero-spacings added. The grey scale shows integrated intensity between 14.5-15.5 km s^{-1}. Blue contours correspond to integrated intensity between 13.5-14.5 km s^{-1} and red contours show integrated intensity between 15.5-16.5 km s^{-1}. Contours go from 5×rms noise and are increased by steps of 3×rms noise. Black contour shows the 30% of the peak intensity of the 3mm continuum maps. Fully sampled spectra are extracted in two cuts following two filamentary structures, where the position corresponding to each spectra are shown in the upper right corner. The integration ranges from the blue shifted component are in dark grey and white grey shows the red component. The local rest velocity of the MDC is shown by dashed grey line. The right panels show cuts of spectra represented in position-velocity diagrams for both cuts, respectively.

towards a monolithic collapse. Alternatively, the MDCs may not be in equilibrium and their formation and evolution is mostly driven by dynamical processes. The spectra overlays in Figure 1 *b)* show that all MDCs exhibit complex kinematics in $H^{13}CO^+$ with several individual velocity components, which are separated and dispersed over ~1 to 3 km s^{-1}.

The velocity field of all MDCs is determined by extracting the velocity of the peak intensity of the spectra. We find that three MDCs show dominant organized velocity fields, while in the others no global gradient is seen. Figure 1 *c)* shows the example with a velocity gradient perpendicular to the axis of the continuum fragments. The velocity gradients cover a range of 1.2 - 4.1 km s^{-1} pc^{-1} for the total sample. We calculated the ratio of rotational versus gravitational energy, and we find that even if systematic motions were due to rotation only, the rotational energy would negligible compared to the gravitational (E_{rot}/E_{grav}<0.025). Recent numerical simulations by Dib *et al.* (2010) report a variety of dynamical patterns for cores formed in a turbulent, magnetized and self-gravitating molecular clouds ranging from easily recognizable rotational features to more complex ones.

At high angular resolution the individual spectral components may correspond to coherent structures within the mass reservoir around the high-mass protostars. We studied the distribution of gas by disentangling the individual spectral components as shown for example in Figure 2. Coherent velocity features were extracted by integrating around the local rest velocity and in the blue and red-shifted velocity range, which components are shown as contours. We find sub-filaments, which are perpendicular to the main filament containing the 3 most massive fragments seen in continuum emission. The velocity pattern of these sub-filaments shows velocity drifts and shears within them implying a high level of dynamics with organized motions at these small scales.

4. Discussion & Conclusions

The high angular-resolution observations reveal rich and complex kinematics of dense gas around massive protostars in our sample of MDCs. Since the high mass protostars reside in the center of MDCs and the gravitational well is the deepest there, we interpret the above demonstrated coherent velocity features with filamentary structure (Figure 2) as flows converging to the central part of MDCs. We suggest that the interaction point of these flows is coincident with the continuum peaks, e.g. the high-mass protostars. The relative velocity difference of these small-scale flows and velocity shears may indicate $\sim 10^4$ yr dynamical time-scales for new protostellar seeds to built up. These estimated crossing times are comparable to the free-fall time-scales of the individual protostellar fragments suggesting that such flows may play an important role in building up the final mass of high-mass protostars. This is compatible with numerical simulations with high dynamics and competitive accretion.

The dynamical processes seem to govern also the formation of MDCs as it was recently shown by Schneider *et al.* (2010) for the DR21 filament, a 10 pc-scale very high density massive structure in Cygnus-X. To summarize, Bontemps *et al.* (2010) and Csengeri *et al.* (2010) present a systematic, high angular-resolution study of 5 MDCs in Cygnus-X focusing on their fragmentation properties and the kinematics of the mass reservoir, from which high-mass protostars form. The main findings are:

(*a*) The MDCs of Cygnus-X are found to be sub-fragmented with a total of 23 fragments within 5 MDCs. They have masses up to 55 $M_\odot$ and 9 of them may potentially form high-mass protostars.

(*b*) High angular-resolution maps of high density tracers as $H^{13}CO^+$ reveal a significant substructure of the mass reservoir around the high-mass protostars. In all MDCs several velocity components are found, which are disentangled into small-scale coherent flows with intrinsic velocity gradients and/or velocity shears.

(*c*) The relative difference in velocity position of the flows give dynamical time-scales of the order of the free-fall time-scale for 4 out of 5 MDCs.

(*d*) Therefore we suggest that the evolution of MDCs is driven more by dynamical processes than quasi-static evolution in a turbulent medium.

References

Bonnell, I. A., Bate, M. R., Clarke, C. J., & Pringle, J. E., 2001 *MNRAS*, 323, 785

Bonnell, I. A. & Bate, M. R., 2006 *MNRAS*, 370, 488

Bontemps, S., Motte, F., Csengeri, T., & Schneider, N., 2010 *A&A, in press, ArXiv:astro-ph/0909.2315*

Csengeri, T., Bontemps, S., Schneider, N., Motte, F., & Dib, S. , 2010 *A&A, submitted, ArXiv:astro-ph/1009.0598*

Dib, S., Hennebelle, P., Pineda, J. E., Csengeri, T., Bontemps, S., Audit, E., & Goodman, A. A.. 2010 *ApJ, in press, ArXiv:astro-ph/1003.5115*

Heitsch, F., Hartmann, L. W., Slyz, A. D., Devriendt, J. E. G., & Burkert, A., 2008 *ApJ*, 674, 316

Klessen, R. S. & Hennebelle, P., *ArXiv:astro-ph/0912.0288*

McKee, C. F. & Tan, J. C., 2002 *Nature*, 416, 59

Motte, F., André, P., & Neri R., 1998 *A&A*, 336, 150

Padoan, P., Juvela, M., Goodman, A. A., & Nordlund, Å , 2001 *ApJ*, 553, 227

Schneider, N., Csengeri, T., Bontemps, S., Motte, F., Simon, R., Hennebelle, P., Federrath, C., & Klessen, R. , 2010 *A&A, in press, ArXiv:astro-ph/1003.4198*

Vazquez-Semadeni, E., Shadmehri, M., & Ballesteros-Paredes, J. , 2002 *ArXiv:astro-ph/0208245*

Vázquez-Semadeni, E., Gómez, G. C., Jappsen, A. K., Ballesteros-Paredes, J., González, R. F., & Klessen, R. S., 2007 *ApJ*, 657, 870

Computational Star Formation
Proceedings IAU Symposium No. 270, 2011
J. Alves, B.G. Elmegreen, J. M. Girart & V. Trimble, eds.

doi:10.1017/S1743921311000184

The Formation of Massive Stars

Ian A. Bonnell[1] and Rowan J Smith[2]

[1]SUPA, School of Physics & Astronomy, Universiy of St Andrews, North Haugh, St Andrews, Fife KY16 9SS, UK
email: iab1@st-andrews.ac.uk
[2]Institut fuer Theoretische Astrophysik, Albert-Ueberle-Str. 2, 69120, Heidelberg, Germany

Abstract. There has been considerable progress in our understanding of how massive stars form but still much confusion as to why they form. Recent work from several sources has shown that the formation of massive stars through disc accretion, possibly aided by gravitational and Rayleigh-Taylor instabilities is a viable mechanism. Stellar mergers, on the other hand, are unlikely to occur in any but the most massive clusters and hence should not be a primary avenue for massive star formation. In contrast to this success, we are still uncertain as to how the mass that forms a massive star is accumulated. there are two possible mechanisms including the collapse of massive prestellar cores and competitive accretion in clusters. At present, there are theoretical and observational question marks as to the existence of high-mass prestellar cores. theoretically, such objects should fragment before they can attain a relaxed, centrally condensed and high-mass state necessary to form massive stars. Numerical simulations including cluster formation, feedback and magnetic fields have not found such objects but instead point to the continued accretion in a cluster potential as the primary mechanism to form high-mass stars. Feedback and magnetic fields act to slow the star formation process and will reduce the efficiencies from a purely dynamical collapse but otherwise appear to not significantly alter the process.

1. Introduction

Understanding how massive stars form is important as high-mass stars dominate the luminous, kinematic and chemical output of stars, and thus the evolution of galaxies. Massive star formation is problematic as high-mass stars are rare, and generally form in dense stellar clusters. They rarely if ever form in isolation (De Wit *et al.* 2005) and need to be understood in the context of forming the full mass distribution of stars. In addition, massive stars are commonly found in close binary systems with other high-mass stars.

There are several outstanding issues that need to be addressed in order to build a complete theory of massive star formation (Zinnecker & Yorke 2007). Firstly, there is the issue of their actual formation and whether they form from disc accretion as low mass stars do, or from something more exotic such as stellar mergers. Secondly, there is the question as to what drives high-mass stars to form and at what stage is the mass gathered. This argument involves the possibility that the mass is gathered into a single high-mass but prestellar core that collapses into one stellar system, versus the idea that the mass is gathered during the star formation process due to the combined gravitational potential of the stellar cluster in which the proto-massive star resides. Lastly, we would like to understand the properties of massive stars and how they affect their environment, including any subsequent star formation.

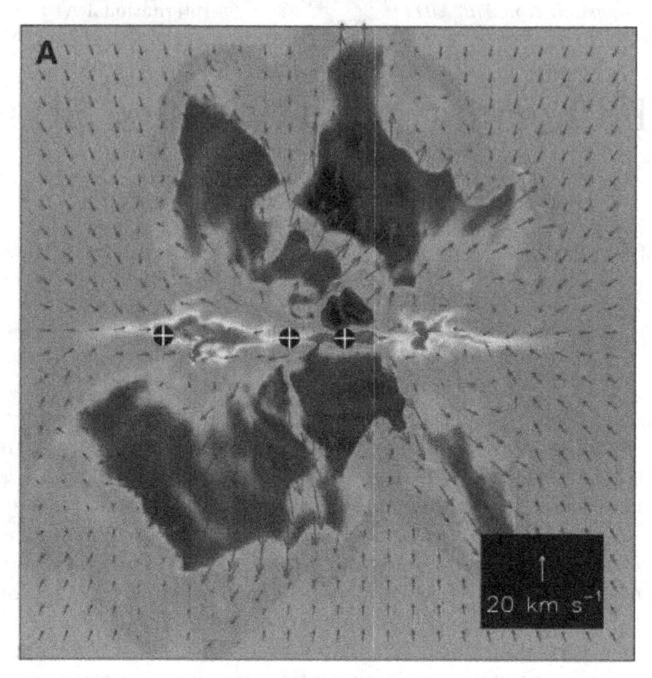

Figure 1. The onset of Rayleigh-Taylor instabilities is shown in a numerical simulation of accretion onto a forming high-mass star. The system is shown edge-on. Note the presence of the low-density bubbles (dark blue) and high density gas (light yellow/red) as well as the arrowed gas velocities (Krumholz *et al.* 2009).

2. Disc Accretion Versus Stellar Mergers

An long-standing problem of massive star formation has been how to circumvent the high radiation pressure from stars with masses greater than 10 solar masses (Khan 1974; Wolfire & Casinelli 1986). Such stars have radiation pressures that can act sufficiently strongly on the dust to reverse the infall, and thus potentially halt mass accretion, limiting the growth of the massive star. Suggestions put forward to circumvent this problem include disc accretion, dust destruction in the inflow (Keto 2003), beaming of the radiation, Raleigh-Taylor instabilities and stellar mergers (Bonnell *et al.* 1998). Several studies (Yorke & Sonnhalter 2002; Krumholz *et al.* 2009; Kuiper *et al.* 2010) have shown that when increasingly realistic physics is included in models, such as frequency dependent treatment of the radiation, gravitational instabilities in the disc and Rayleigh-Taylor instabilities, the formation of a high-mass star can still occur through disc accretion.

Krumholz *et al.* (2009) performed the first 3-D simulation including radiation pressure in the flux-limited diffusion approximation, and showed that the discs that formed were unstable and fragmented to form multiple objects. This moved the source of the radiation pressure away from the centre of mass which aided further accretion. The accretion disc itself was fed by a combination of flow around the radiation-driven bubble and Rayleigh-Taylor instabilities inside from the heavier gas on top of the bubbles (Fig.1).

In addition to the successes of disc accretion in forming high-mass stars, it is also increasingly clear that stellar mergers are unlikely to play a role in the formation of all high-mass stars. This is due to the difficulty in obtaining the necessary physical conditions for stellar mergers to occur on timescales of less than a million years (Bonnell *et al.* 1998; Bonnell & Bate 2002; Davies *et al.* 2006). While accretion onto a clusters stellar core can drive the core into collapse and dramatically increase the stellar density as $n \propto m_*^9$

(Bonnell & Bate 2002), the core itself may dissolve through its gravitational interactions if it becomes uncoupled from the rest of the cluster (Clarke & Bonnell 2009). This occurs as the core is then a small-N system which has a relatively short evaporation timescale due to two-body interactions in which a central binary ejects the other members. To counteract this, the core must be well coupled to the rest of the cluster such that the excess kinetic energy given to individual objects is smoothly transferred to the outer regions of the cluster where the energy can be shared by many stars. This requires a large cluster with a very high central density that smoothly decreases with radius until scales where the relaxation timescale of the cluster is long.

Recent work (Moeckel & Clarke 2010) has indeed shown that a large-N cluster of order 30,000 or more is required for accretion to be able to drive the core to such high stellar densities for collisions to occur. However, mergers amongst binary systems are still a possibility even in smaller-N systems, as close binaries are common and their cross section to encounter other stars is much larger (Bonnell & Bate 2005).

3. Monolithic Massive star Formation

One of the leading models for massive star formation invokes the collapse of a single, high-mass prestellar core to directly form a single high-mass star (McKee & Tan 2003; Banerjee & Pudritz 2007). The appeal in this model arises from it being a scaled-up version of low-mass star formation where stars are observed forming within low mass gas cores. A further attraction is that observations show that the core mass function (CMF) resembles the stellar IMF (Motte *et al.* 1998; Testi & Sargent; Johstone *et al.* 2000), which has lead some to propose a one-to-one correlation between the two, albeit with some efficiency factor (Alves *et al.* 2007; Simpson *et al.* 2008). In order for this model to work, the core must be a completely distinct element of the molecular cloud and must not gain significant mass or fragment during the star formation process.

Smith *et al.* (2009a) have investigated the relation between the CMF and the stellar IMF through numerical simulations. They find that although the CMF does resemble the IMF, the core masses do not map directly into stellar system masses. This arises due to a combination of geometrical effects during core collapse, and varying amounts of subsequent accretion from the surrounding environment which adds significant dispersion to the relation between the core mass and the final stellar mass. This is not very surprising given that the cores in a clustered region are somewhat artificial in that they do not have distinct boundaries but are instead the high-density peaks of a larger mass distribution (Smith *et al.* 2008, Kainulainen *et al.* 2009).

The question of a massive core's fragmentation is one that has been addressed by several studies (Dobbs *et al.* 2005; Krumholz *et al.* 2006, 2010). The symetrical, centrally condensed nature of the core, believed to arise if its pre-collapse evolution is quasi-static, helps suppress any fragmentation (eg. Boss 1993?). The difficulty arises in that the core needs to be turbulently supported in order that the equivalent turbulent Jeans mass is of order the core mass. McKee & Tan modelled this turbulence as an equivalent isotropic sound speed but turbulence is much more complicated and does not support objects isotropically. Instead such objects are rapidly deformed by their internal motions providing the necessary seeds for fragmentation (Dobbs *et al.* 2005). Even centrally condensed cores that are turbulently supported fragment provided that they are able to remain nearly isothermal (Dobbs '*et al.* 2005; Krumholz *et al.* 2006; 2010).

One solution to this problem that has been suggested is the radiative heating from forming stars internal to the cloud (Krumholz refs). This will work on relatively small scales as the temperature due to radiative heating scales as the distance from the source as $t \propto r^{-0.4-0.5}$ (Chakrabarti & McKee 2005). It also requires that the central sources are already present in the core without having previously affected its density distribution in any way. This is likely to be problematic as the core itself must form quasi-statically, over many dynamical timescales if it is sufficiently centrally condensed. The problem aries in that the core should be highly susceptible to fragmentation on a dynamical timescale during this process, hence before any internal sources could have formed.

It is worth noting here that numerical simulations of larger-scale cluster formation, including feedback and magnetic fields, have not been able to form a massive prestellar core. Instead the accumulating gas fragments before it can be assembled into one object inducing the formation of a stellar cluster and subsequent accretion-driven formation of high-mass stars (see below).

3.1. *Critical Surface Density for Massive Star Formation*

Based on the above arguments to suppress fragmentation, Krumholz & McKee 2008 have suggested that there is a critical surface density in order to form high-mass stars. This arises due to the short-distances over which radiative heating is effective which then necessitate high gas densities and hence surface densities. Regardless that this may not be sufficient to suppress fragmentation which would occur during the formation phase of the core, there is a simpler explanation for a critical surface density to form high-mass stars. High-mass stars are not formed in isolation, but in stellar clusters that follow a standard IMF. This necessitates a total mass of order 1000 $M_\odot$. Given a mean stellar mass of ≈ 0.5 $M_\odot$, and assigning a gas density to produce a similar Jeans mass, the system has a size around 1 pc and hence a surface density of order 1 g cm^{-2}. Not surprisingly, such a system has roughly the same properties as the ONC would have during the prestellar stage.

4. Competitive Accretion

The alternative to monolithic collapse is accretion from a larger-scale reservoir onto originally lower-mass cores to form massive stars. In an embedded stellar cluster containing a significant fraction of the total mass in gas, individual stars can accrete from this reservoir but their individual accretion rates depend on their masses, velocities, gas densities and the tidal fields from the other cluster members (Zinnecker 1982; Bonnell *et al.* 1997, 2001; Bonnell & Bate 2006).

Numerical simulations show that molecular clouds fragment down to their thermal or magnetic Jeans masses (Bonnell *et al.* 1991; Klessen *et al.* 1998; Klessen 2001; Bate *et al.* 2003; Jappsen *et al.* 2005; Bonnell *et al.* 2006; Federrath *et al.* 2010). As massive stars form in stellar clusters where the mean stellar mass is of order $0.5 M_\odot$, the fragmentation mass must be of that order. Fragmentation is also highly inefficient in that only a small fraction of the mass is initially in the high density fragments (Klessen *et al.* 1998; Bonnell *et al.* 2003; Smith *et al.* 2009b). Accretion from this large-scale reservoir is then a plausible way to form high-mass stars.

The formation of a stellar cluster from a massive clump of molecular gas commences as as soon as the clump is formed and is gravitationally bound (Smith *et al.* 2009b; Bonnell *et al.* 2010). The internal turbulence fragments the clump into Jeans mass objects which form into small-N subclusters. The overall collapse drives gas and stars together to form a large-N cluster. The gravitational potential of the cluster ensures that gas is funneled down to the centre to be accreted by what will become the most massive star. The combination of the higher gas density, and subsequently the star's higher mass ensures

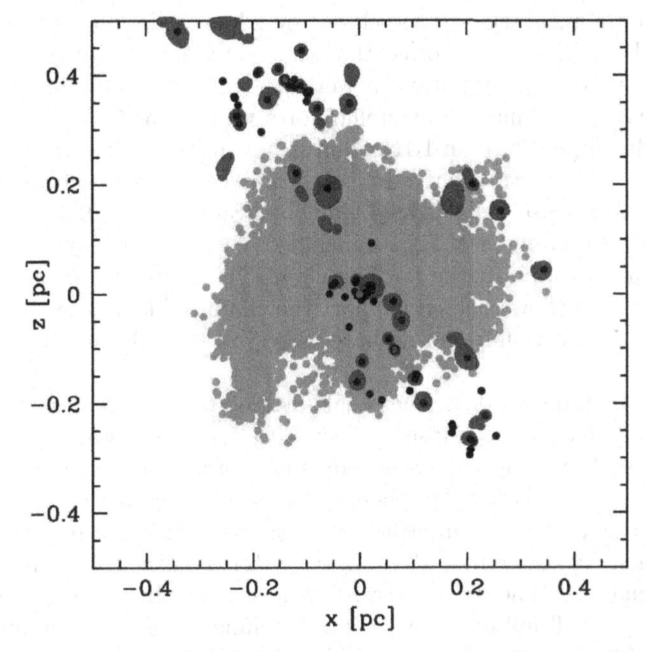

Figure 2. The formation of a stellar cluster from a fragmenting filament. The green dots show the positions of gas which will eventually be accreted by the massive sink (red dot). Black dots show the position of protostars and blue dots show the location of material in bound cores. The gas which will be accreted by the massive protostar is well distributed throughout the clump, but is funneled to the central star by large scale collapse motions (Smith *et al.* 2009b).

that the central object has a high accretion rate and thus becomes a massive star (Fig. 2). This also helps establish an initial mass segregation.

Smith *et al.* (2009b, see also Bonnell *et al.* 2004) show how this results in the simultaneous formation of a massive star and the stellar cluster. The causal relation between the two is that the overall gravitational potential is necessary to gather the mass required for the high accretion rate and thus the formation of the high-mass stars. Forming protoclusters therefore start off as highly fragmented, elongated dispersed objects and evolve towards more well defined centrally condensed spherical objects dominated by the central source. In fact, synthetic maps predict that they will appear less fragmented in continuum maps once fragmentation has occurred and the central temperatures increase due to radiative heating (Smith *et al.* 2009b).

5. Feedback from Massive Stars

Smith *et al.* (2009b) included an overestimate of radiative heating from protostars and did not find any massive prestellar core. The radiative heating does result in a lower fragmentation rate suppressing some of the low-mass stars from forming but cannot halt the ongoing accretion from large scales as the warm gas remains bound to the cluster. Similarly, Urban *et al.* 2010 considered the effect of dust heating on a cluster that was prone to fragmentation, heating reduced fragmentation and increased the characteristic stellar mass.

Dale *et al.* (2005) included ionisation from a central O star and found the overall cluster dynamics largely unaffected as the ionisation found the weakest point in the surrounding cloud through which to escape. This produced a one-sided HII region but did not affect the fragmentation or accretion. Including stellar winds in a forming stellar cluster, Dale & Bonnell (2008) showed that the cloud is partially supported and has a slightly reduced rate of star formation. Nevertheless, the dynamics of the accretion were largely unaffected and no massive prestellar cores were formed. Peters *et al.* (2010) also found that the development of an HII region is not in itself sufficient to stop accretion. It is generally difficult to stop accretion onto massive stars via feedback as it occurs primarily along a few dense well shielded filaments (Smith *et al.* 2010; Wang *et al.* 2010).

A recent result from Wang *et al.* (2010) included both outflows from stars and magnetic fields in an attempt to support the forming cloud. Similar to the Dale & Bonnell result, they found that the outflows helped support the cloud reducing the star formation rate but did not halt the accretion from the shared reservoir that ultimately forms the massive star.

In all numerical studies to date, there has been no occurrence of a massive prestellar core being formed that will collapse directly into a massive star. The reason for this lies in the large scale potential in which massive stars form. Once a stellar cluster or similarly bound gas core is formed, there exists a well defined gravitational potential. Gas falling into this potential can either enter as low density dispersed gas or as high-density, gravitationally bound fragments. In the latter case, the fragments collapse rapidly on their own dynamical timescale to form low-mass stars. In the former case, the gas is tidally unbound and will fall into the potential seeking out a potential minimum defined by an existing object. The accretion of this infalling gas onto the young stars is the competitive accretion process. We cannot envision a way that gas infalling into a well defined potential can form into a gravitationally bound core containing multiple Jeans masses without either being accreted by the stars present in the cluster or fragmenting into low-mass stars (Smith *et al.* 2009b). To date, no feedback processes, even in the presence of magnetic fields (Wang *et al.* 2010), have been found which alter this.

6. Binary systems

It has long been realised that most high-mass stars are members of binary systems and that these systems often comprise two massive components in a tight orbit of < 1 AU. How such systems form is unclear. We expect that massive stars should commonly be in binary systems either through three body capture in the cores of dense clusters or through a disc fragmentation process (Kratner M̂etzner 2006; Krumholz *et al.* 2009). In both cases, the expectant binary system will still be fairly wide of many tens to 100 AU. Hardening such systems is possible through accretion (Bonnell & Bate 2005) or two-body relaxation in clusters (Heggie 1975) although at present it is not certain that any posited method can actually form the closest systems.

7. Conclusions

Massive star formation is now believed to proceed via disc accretion. Stellar mergers may play a minor role in close binary systems or very massive stellar clusters but cannot be the primary mechanism to form massive stars. Radiation pressure does not appear to halt accretion due to to combination of accretion through a disc, and possibly gravitational instabilities in the disc and Rayleigh-Taylor instabilities in the radiation pressure driven bubbles.

Massive prestellar cores will form massive stars, but the question is whether such objects ever exist. Theoretical expectations and numerical simulations both indicate that such cores are unlikely to form without first fragmenting or accreting onto pre-eisting stars. In contrast, clustered accretion appears to be a viable mechanism to form massive stars, unimpeded by either magnetic fields or feedback. It does require that massive stars form in stellar clusters and cannot explain the formation of isolated massive stars without the necessity of a high Jeans mass. To date, simulations of feedback have found that it has a limited effect on the massive star formation, but it does slow the large-scale star formation and reduces the star formation efficiency.

References

Alves J., Lombardi M., Lada C. J., 2007, A&A, 462, L17
André P., Belloche A., Motte F., & Peretto N., 2007, A&A, 472, 519
Banerjee R. & Pudritz, R., 2007, MNRAS, 660, 479
Bate M. R., Bonnell I. A., & Bromm V., 2003, MNRAS, 339, 577
Bonnell, I. A. & Bate, M. R., 2005, MNRAS, 362, 915
Bonnell, I. A. & Bate, M. R., 2006, MNRAS, 370, 488
Bonnell, I. A., Bate, M. R., Clarke, C. J., & Pringle, J. E., 2001a, MNRAS, 323, 785
Bonnell I. A., Bate M. R., & Vine S. G., 2003, MNRAS, 343, 413
Bonnell I. A., Bate M. R., & Zinnecker H., 1998, MNRAS, 298, 93
Bonnell I. A., Clarke C. J., Bate M. R., & Pringle J. E., 2001, MNRAS, 324, 573.
Bonnell I. A., Clarke C. J., & Bate M. R., 2006, MNRAS, 368, 1296
Bonnell I., Martel H., Bastien P., Arcoragi J.-P., & Benz W., 1991, ApJ, 377, 553
Bonnell I. A., Smith R. J., Clark P. C., & Bate M. R., 2010, MNRAS, in press
Chakrabarti S. & McKee C. F., 2005, ApJ, 631, 792
Dale, J. E. & Bonnell, I. A. 2008, MNRAS, 391, 2
Dale, J. E., Bonnell, I. A., Clarke, C. J., & Bate, M. R. 2005, MNRAS, 358, 291
de Wit W. J., Testi L., Palla F., & Zinnecker H., 2005, A&A, 437, 247
Dobbs C. L., Bonnell I. A., & Clark P. C., 2005, MNRAS, 360, 2
Federrath C., Banerjee R., Clark P. C., & Klessen R. S., 2010, ApJ, 713, 269
Heggie, D., 1987, MNRAS, 173, 729
Jappsen A.-K., Klessen R. S., Larson R. B., Li Y., & Mac Low M.-M., 2005, A&A, 435, 611
Kahn, F. D. 1974. A&A 37, 149.
Kainulainen J., Lada C., Rathborne J. M., & Alves J F., 2009, A&A, 497, 399
Keto, E., 2003, ApJ 580, 980986.
Klessen R. S., 2001, ApJ, 556, 837
Klessen R. S., Burkert A., & Bate M. R., 1998, ApJ, 501, L205
Kratter, K. & Matzner, C., 2006, MNRAS, 373, 1563
Krumholz M., Klein R. I., & McKee C. F., 2005, ApJ, 618, L33
Krumholz M. R., Klein R. I., McKee C. F., Offner S. S. R., & Cunningham A. J., 2009, Sci, 323, 754
Krumholz M. R. & McKee C. F., 2008, Natur, 451, 1082
Larson R. B., 2005, MNRAS, 359, 211
McKee, C.F. & Tan, J.C., 2003, ApJ, 585, 850
Motte F., Andre P. & Neri R., 1998, A&A, 336, 150
Nutter D. & Ward-Thompson D., 2007, MNRAS, 374, 1413
Peters T., Banerjee R., Klessen R. S., Mac Low M.-M., Galván-Madrid R., & Keto E. R., 2010, ApJ, 711, 1017
Simpson, R.J., Nutter D., & Ward-Thompson D., 2008, MNRAS, 391, 205
Smith R. J., Clark P. C., & Bonnell I. A., 2008, MNRAS, 391, 1091
Smith R. J., Clark P. C., & Bonnell I. A., 2009, MNRAS, 396, 830
Smith R. J., Longmore, S., & Bonnell I. A., 2009, MNRAS, 400, 1775
Testi L., Sargent A., 1998; ApJL, 508, L91
Urban A., Martel, H., & Evans N. J., 2010, ApJ, 710, 1343
Wang P., Li Z.-Y., Abel T., & Nakamura F., 2010, ApJ, 709, 27

Weidner C. & Kroupa P., 2006, MNRAS, 365, 1333
Weidner C., Kroupa P., & Bonnell I. A. D., 2010, MNRAS, 401, 275
Wolfire, M. G. & Cassinelli, J. P. 1987. ApJ 319, 850867.
Yorke, H. W. & Sonnhalter, C. 2002, ApJ 569, 846.
Zinnecker H., 1982, New York Acad. Sci. Ann., 395, 226
Zinnecker H., Yorke, H., 2007, ARA&A, 45, 481

Computational Star Formation
Proceedings IAU Symposium No. 270, 2011
J. Alves, B.G. Elmegreen, J. M. Girart & V. Trimble, eds.

doi:10.1017/S1743921311000196

Recent Developments in Simulations of Low-mass Star Formation

Masahiro N. Machida[1]

[1]National Astronomical Observatory of Japan, Mitaka, Tokyo 181-8588, Japan
email: masahiro.machida@nao.ac.jp

Abstract. In star forming regions, we can observe different evolutionary stages of various objects and phenomena such as molecular clouds, protostellar jets and outflows, circumstellar disks, and protostars. However, it is difficult to directly observe the star formation process itself, because it is veiled by the dense infalling envelope. Numerical simulations can unveil the star formation process in the collapsing gas cloud. Recently, some studies showed protostar formation from the prestellar core stage, in which both molecular clouds and protostars are resolved with sufficient spatial resolution. These simulations showed fragmentation and binary formation, outflow and jet driving, and circumstellar disk formation in the collapsing gas clouds. In addition, the angular momentum transfer and dissipation process of the magnetic field in the star formation process were investigated. In this paper, I review recent developments in numerical simulations of low-mass star formation.

Keywords. stars: formation, stars: low-mass, brown dwarfs, ISM: clouds, ISM: magnetic fields

1. Introduction

A star is born in a molecular cloud core through gravitational contraction. Molecular clouds that are the initial state of the star formation are frequently observed in various star-forming regions, and we have much information about them. In star-forming regions, we have also observed various objects and phenomena, such as protostars, protostellar jets, bipolar outflows, and circumstellar disks, resulting from the gravitational contraction of molecular cloud cores. Thus, numerous observational studies have allowed us to understand both the initial state and its outcome for star formation. However, observations do not allow us to understand the star-formation process itself (or gravitational contraction phase), because (proto)star formation occurs in dense cloud cores, which are difficult to observe directly, thus necessitating a theoretical approach. In order to understand the star-formation process in a collapsing cloud core, in addition to the self-gravity of the contracting gas, we have to consider the effects of thermal pressure, the Lorentz force, and rotation, which are all intricately interrelated. Therefore, we need detailed numerical simulations to unveil the star-formation process in a dense collapsing cloud core.

However, such detailed numerical simulations are difficult to formulate, because it is difficult to calculate the star-formation process from the molecular cloud core (the prestellar core stage) to protostar formation (the protostellar phase), and ultimately through the termination of the runaway gravitational collapse. One reason these numerical simulations are so difficult is because they have to resolve the spatial scale over ~7 orders of magnitude and the density scale over ~18 orders of magnitude. Molecular clouds have sizes of $\sim 10^5$ AU and densities of $\sim 10^4\ \mathrm{cm}^{-3}$, while protostars have sizes of ~ 0.01 AU and densities of $\sim 10^{22}\ \mathrm{cm}^{-3}$. Thus, we require special numerical techniques, such as AMR (Adaptive Mesh Refinement) and SPH (Smoothed Particle Hydrodynamics), to spatially resolve both the molecular cloud core and the protostar. Now, using these methods, we can unveil the star-formation process by directly calculating the star formation starting at the prestellar core stage. This review summarizes recent developments in numerical simulations of low-mass star formation in the collapsing cloud cores.

2. Outline of Protostar Formation

At first, based on the results of spherically symmetric calculations (Larson 1969, Masunaga & Inutsuka 2000), I briefly outline the low-mass star formation process and its thermal evolution (for details, see Fig. 2 of Masunaga & Inutsuka 2000 along with its explanation). Observations indicate that stars are born in molecular cloud cores that have number densities of $n \sim 10^4\,\mathrm{cm}^{-3}$. After the gravitational collapse begins, the density of the cloud increases with time. In the cloud core, the gas collapses isothermally and remains at $\sim 10\,\mathrm{K}$ until the number density reaches $n \sim 10^{10}\,\mathrm{cm}^{-3}$; at this point, the central region becomes optically thick and the equation of state becomes hard. Then, the first adiabatic core (the so-called first core) with a size of $\sim 1\,\mathrm{AU}$ appears. Subsequent to the formation of the first core, further rapid collapse is induced in a small central part of the first core because of the dissociation of molecular hydrogen when the number density exceeds $n \gtrsim 10^{16}\,\mathrm{cm}^{-3}$. Finally, the gas becomes adiabatic again for $n \gtrsim 10^{21}\,\mathrm{cm}^{-3}$ because of the completion of the dissociation of molecular hydrogen, and a protostar with a size of $\sim 0.01\,\mathrm{AU}$ appears in the collapsing cloud core.

At its formation, the first core has a mass of $\sim 0.1 - 0.01\,M_{\odot}$, while the protostar has a mass of $\sim 10^{-3}\,M_{\odot}$, which corresponds to the Jovian mass. Thus, the massive first core ($\sim 1 - 10\,\mathrm{AU}$) encloses the protostar ($\sim 0.01\,\mathrm{AU}$). A spherically symmetric calculation, which could not include the effect of the rotation, showed that the first core gradually shrinks and disappears in $\sim 10\,\mathrm{yr}$ after the protostar formation. On the other hand, multidimensional calculations, which did include the effect of the rotation, showed that the first core remains long after the protostar formation (Saigo & Tomisaka 2006), evolving into the circumstellar disk in the main accretion phase (Machida *et al.* 2010). In the main accretion phase, the protostar acquires almost all its mass by gas accretion, reaching $\sim 1\,M_{\odot}$. In the subsequent sections, I review recent developments in low-mass star-formation simulations, especially in the gas-collapsing phase.

3. Fragmentation and Binary Formation

Observations have shown that the multiplicity of pre-main sequence stars is larger than that of main-sequence stars in star-forming regions (e.g., Mathieu 1994). Recently, extremely young protostars (i.e., Class 0 protostars) have been observed with radio interferometers (e.g., Looney *et al.* 2000) and wide-field near-infrared cameras (Duchêne *et al.* 2004). These observations showed that stars already have a high multiplicity at the moment of their birth. Thus, we expected that a large fraction of stars are born as binary or multiple systems.

It is considered that rotation causes fragmentation in a collapsing cloud, which then lead to the formation of binary or multiple star systems. Several three-dimensional simulations of the evolution of rotating collapsing clouds have investigated the possibility of fragmentation and binary formation (see, review of Bodenheimer *et al.* 2000 and Goodwin *et al.* 2007). Miyama *et al.* (1984) and Tsuribe & Inutsuka (1999) calculated the evolution of spherical clouds in the isothermal regime with initially uniform density and rigid-body rotation. They found that fragmentation (and thus binary formation) occurs in the isothermal contracting phase ($n < 10^{10}\,\mathrm{cm}^{-3}$) only when the initial cloud is (highly) thermally unstable against gravity (see also Boss 1993). However,

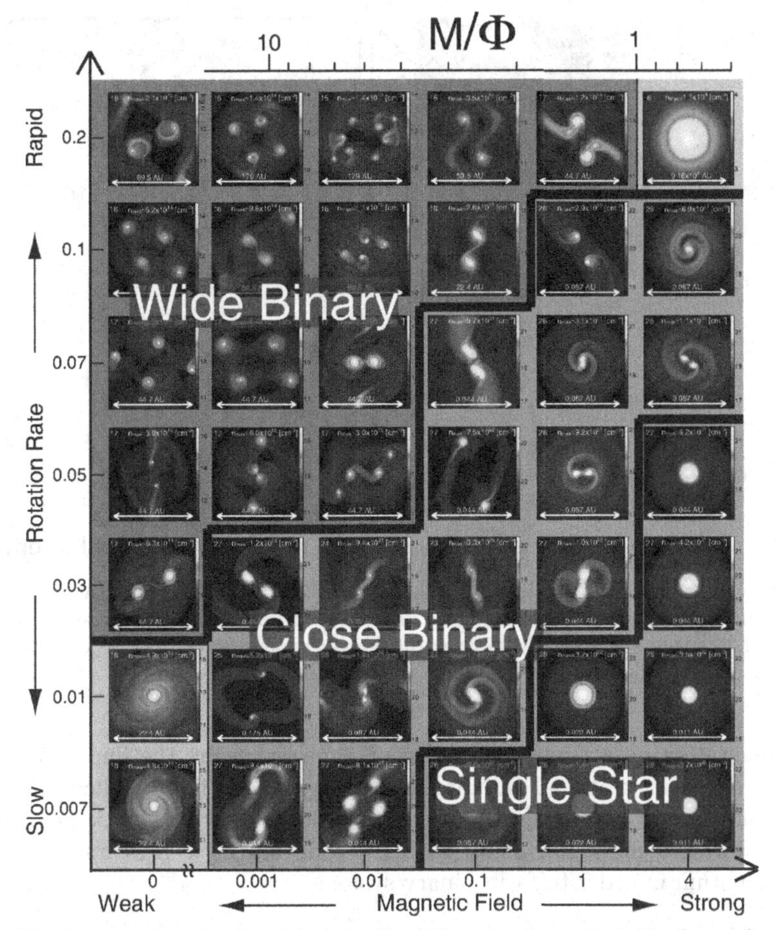

Figure 1. Final states for clouds with initially different magnetic fields (x-axis) and angular velocities (y-axis). The densities (color-scale) on the cross section of the $z = 0$ plane are plotted in each panel. Background colors indicate the following: fragmentation occurs with separation of $> 1\,\mathrm{AU}$, resulting in wide binaries (blue); fragmentation occurs with separation of $< 1\,\mathrm{AU}$, resulting in close binaries (pink); no fragmentation occurs through all phases of the cloud evolution, resulting in single-stars (gray); and the cloud no longer collapses, resulting in no star formation (green).

fragmentation easily occurs after the gas becomes adiabatic ($n > 10^{10}\,\mathrm{cm}^{-3}$, e.g., Matsumoto & Hanawa 2003). Several studies have shown that fragmentation frequently occurs in the adiabatic phase even when the molecular cloud has a small angular momentum and no magnetic field. This is because after the gas becomes adiabatic, it collapses very slowly and the perturbation that induces fragmentation can grow. In addition, in the adiabatic phase, the cloud rotation can form a disk sufficiently thin for fragmentation to occur. On the other hand, recent magnetohydrodynamics simulations have shown that the magnetic field suppresses fragmentation and binary formation (Hosking & Whitworth 2004, Machida *et al.* 2004, Machida *et al.* 2005b, Machida *et al.* 2008a, Hennebelle & Teyssier 2008b and Price & Bate 2007). This is because the angular momentum that could lead to the formation of a disk thin enough for fragmentation to occur is transferred by magnetic braking and the protostellar outflow (see, §4). Thus, in a strongly magnetized cloud, no thin disk appears (Mellon & Li 2009) and, therefore, no fragmentation occurs. Figure 1 shows the rotation and magnetic field conditions under

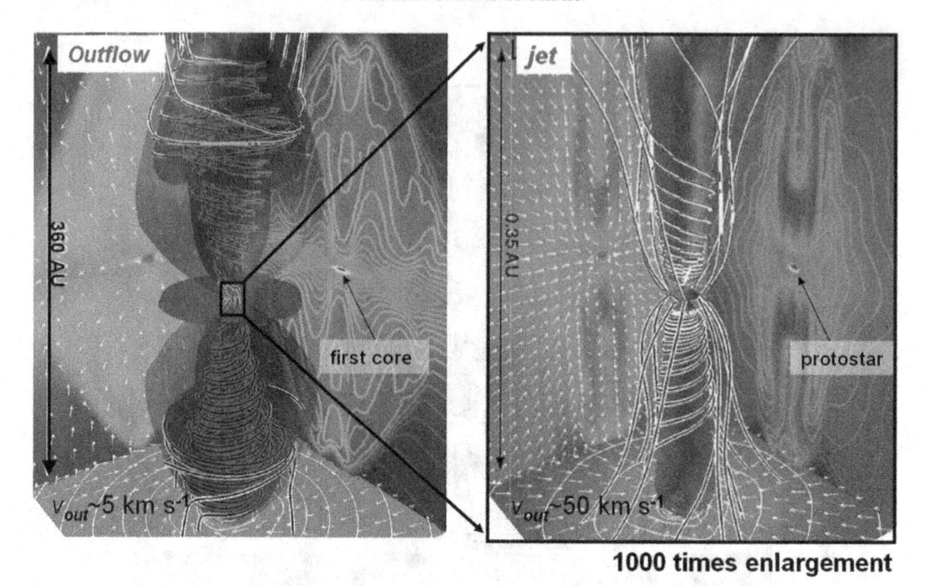

Figure 2. Low-velocity outflow driven by the first core (left panel) and high-velocity jet driven by the protostar (right panel). The magnetic field lines are plotted by black-and-white streamlines. Inside the purple surfaces, the flow is outflowing from the central object (the first core or protostar), and outside the purple surface (in the blue regions), the flow is inflowing to the central object.

which fragmentation occurs with the different panels showing the final state of clouds with initially different rotational and magnetic energies. This figure indicates that a large cloud rotation rate promotes fragmentation, but a large cloud magnetic field suppresses it. In other words, a molecular cloud with a strong magnetic field must have a large angular momentum in order to form binary systems.

4. Protostellar Outflows and Jets

The observations indicate that protostellar outflows are ubiquitous in star-forming regions. Flows originating from protostars are typically classified into two types: molecular outflows observed mainly through line emission from their CO molecules (Arce *et al.* 2006), and optical jets observed through their optical emission (Pudritz *et al.* 2007). Molecular outflows exhibit wide opening angles and slow velocities ($10 - 50\,\mathrm{km\,s^{-1}}$, e.g., Belloche *et al.* 2002), while optical jets exhibit good collimation and high velocities ($100 - 500\,\mathrm{km\,s^{-1}}$, e.g., Bally *et al.* 2007). The observations also indicate that around each protostar, a wide-opening-angle low-velocity outflow encloses a narrow-opening-angle high-speed jet (Mundt & Fried 1983).

Such two-component flows are naturally reproduced in recent star forming simulations, in which the star formation process is calculated from the prestellar stage until protostar formation (Tomisaka 2002, Machida *et al.* 2005a, Machida *et al.* 2006, Machida *et al.* 2008b, Machida *et al.* 2009a, Banerjee & Pudritz 2006, Duffin & Pudritz 2009, Commerçon *et al.* 2010 , and Tomida *et al.* 2010). As described in Figure 2, two nested cores (the first core and protostar) appear in the star formation process, and each core can drive different types of flows. The first core is formed in the low-density region ($n < 10^{12}\,\mathrm{cm^{-3}}$). Thus, a relatively strong magnetic field surrounds the first core, because the first core does not experience Ohmic dissipation. Note that Ohmic dissipation becomes effective within a range of $10^{12} \lesssim n \lesssim 10^{15}\,\mathrm{cm^{-3}}$ (§5). This strong magnetic field can drive a low-velocity outflow by the magneto-centrifugal mechanism (Blandford & Payne 1982). On the other hand, the protostar appears in the high-density region ($n > 10^{21}\ \mathrm{cm}^{-3}$). Thus, an extremely weak magnetic field surrounds the protostar, because such a high-density region experiences Ohmic dissipation. This weak field cannot drive outflow by the magneto-centrifugal mechanism. Instead, the rotation of the protostar generates a strong toroidal field, and the magnetic pressure gradient force can drive a high-velocity flow. As a result, the first core and the protostar drive two flows resulting in a low-velocity outflow surrounding a high-velocity jet as seen in Figure 2.

The different depths of the gravitational potential (or different Kepler velocities) cause the different outflow speeds. The first core has a relatively shallow gravitational potential and drives a relatively slow outflow, while the protostar has a deeper gravitational potential and drives a high-velocity jet. The different driving mechanisms for the outflow and jet cause the difference in the degrees of collimation. The magneto-centrifugal mechanism drives the low-velocity outflow with wide opening angle, while the magnetic pressure gradient force along the rotation axis drives the high-velocity jet with good collimation (Machida *et al.* 2008b). In summary, the different properties of the drivers (the first core and protostar) cause the difference in the properties of these two flows.

5. Angular Momentum and Magnetic Flux Problems

Molecular clouds have rotational energy equal to $\sim 2\%$ of their gravitational energy (e.g., Caselli *et al.* 2002), while they have magnetic energy comparable to their gravitational energy (e.g., Crutcher 1999). Conservation of the angular momentum and the magnetic flux in a collapsing cloud suggests that the rotation and the magnetic field in the cloud gradually increase as the cloud collapses, thus preventing further collapse and protostar formation. However, the rotation and magnetic field strength of the observed protostars indicate that neither the angular momentum nor the magnetic flux is conserved in collapsing clouds. In general, these anomalies are referred to as the "angular momentum problem" and "magnetic flux problem." The former problem is that the specific angular momentum of a molecular cloud is much larger than that of a protostar. The latter problem refers to the fact that the magnetic flux of a molecular cloud is much larger than that of a protostar with equivalent mass. These problems imply that there must be mechanisms for removing angular momentum and magnetic flux from a cloud core. In a collapsing cloud, these two problems are related. Namely, the angular momentum is removed by magnetic effects (i.e., magnetic braking, outflows, and jets), while the magnetic field is amplified by the shearing motion caused by cloud rotation. Hence, the rotation and the magnetic field cannot be treated independently while considering the angular momentum and magnetic flux problems.

Recently, the evolution of the angular momentum and magnetic flux in a collapsing cloud through protostar formation has been investigated (Machida *et al.* 2007 and Duffin & Pudritz 2009). In the collapsing cloud, magnetic braking and protostellar outflow in the magnetically active regions ($n < 10^{12}\ \mathrm{cm}^{-3}$ and $n > 10^{16}\ \mathrm{cm}^{-3}$) remove the angular momentum. By the time a protostar is formed, about 3–4 orders of magnitude of the initial angular momentum have been transferred by such magnetic effects. Simulations suggest that the protostar at its formation has a rotation period of several days, which is comparable to the observations (Herbst *et al.* 2007). In addition, when the number density exceeds $n \gtrsim 10^{12}\ \mathrm{cm}^{-3}$ in the collapsing cloud, the degree of ionization becomes considerably low and Ohmic dissipation (and ambipolar diffusion) removes the magnetic flux. Then, after the density exceeds $n \gtrsim 10^{16}\ \mathrm{cm}^{-3}$, thermal ionization of alkali metals reduces the resistivity and Ohmic dissipation becomes ineffective. By the time a protostar

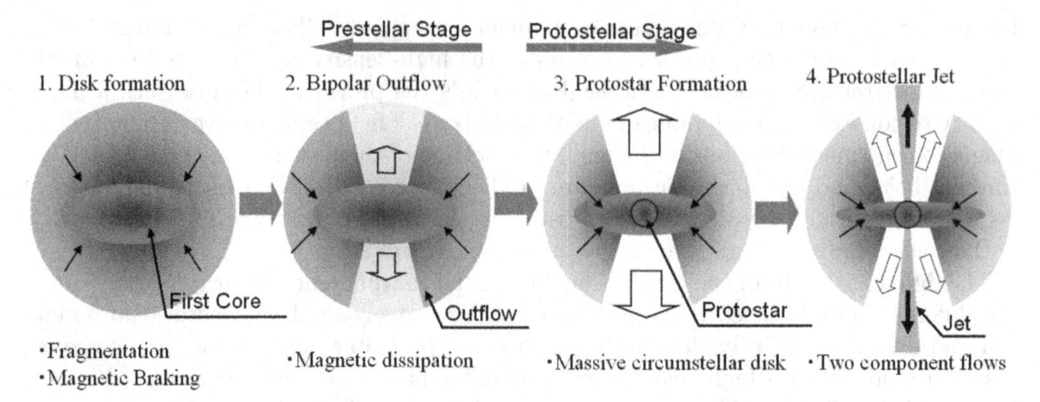

Figure 3. Star formation scenario from the prestellar cloud core through protostar formation.

forms, about 3–5 orders of magnitude of the initial magnetic flux has been removed within the range of $10^{12}\,\mathrm{cm}^{-3} \lesssim n \lesssim 10^{15}\,\mathrm{cm}^{-3}$. Simulations suggest that the protostar at its formation has a sub-kilogauss magnetic field strength, which is also comparable to the observations (Bouvier *et al.* 2007). Thus, recent numerical simulations could resolve the angular momentum and magnetic flux problems in the early phase of the star formation (i.e., until the Class 0 phase). However, to determine the rotation period and magnetic field strength of older protostars (Class I, II, and III phases) and main-sequence stars, we must further investigate their evolution.

6. Circumstellar Disk Formation

Stars form in molecular cloud cores that have nonzero angular momenta, thus, the appearance of a circumstellar disk is a natural consequence of star formation when angular momentum is conserved in the collapsing cloud. In addition, observations have shown the existence of circumstellar disks around protostars. Numerous observations indicate that the circumstellar disks around Class I and II protostars have sizes of $\sim 10 - 1000\,\mathrm{AU}$ and masses of $\sim 10^{-3} - 0.1\,M_\odot$ (e.g., Natta *et al.* 2000). Because the formation sites of the circumstellar disk and protostar are embedded in a dense infalling envelope, it is difficult to directly observe newborn or very young circumstellar disks. Thus, in general, we observe only the circumstellar disks long after their formation, i.e., around Class I or II protostars. Observations also indicate that younger protostars have more massive circumstellar disks (e.g., Natta *et al.* 2000 and Meyer *et al.* 2007). Recently, Enoch *et al.* (2009) observed massive disks with $M_{\rm disk} \sim 1\,M_\odot$ around Class 0 sources, indicating that a massive disk can be present early in the main accretion phase. However, observations cannot determine the real sizes of circumstellar disks, or how and when they are formed. Therefore, both theoretical approach and numerical simulations are necessary to investigate the formation and evolution of circumstellar disks.

In the collapsing cloud, before the protostar forms, the first core appears with a size of $\sim 1 - 10\,\mathrm{AU}$ and mass of $\sim 0.01 - 0.1\,M_\odot$. Recent studies showed that the first core directly evolves into a circumstellar disk after the protostar forms (Bate 1998, Bate 2010, Walch *et al.* 2009, Machida *et al.* 2010, Inutsuka *et al.* 2010). The first core has a disk-like structure at its formation because the first core is supported not only by thermal pressure but also by rotation. Thus, even after the protostar forms (or after the dissociation of molecular hydrogen begins), the first core does not disappear; instead it becomes a Keplerian rotating disk in the main accretion phase. In summary, the protostar is formed inside the disk-like first core. In other words, a massive circumstellar disk with size of > 1 AU already exists at the moment of the birth of the protostar. In the main accretion phase, such a massive disk tends to show fragmentation, subsequently forming a binary companion or gas-giant planet. Thus, recent numerical results support the concept that gravitational instability creates gas-giant planets.

7. Summary

Recent numerical simulations have changed the classical star formation scenario. In Figure 3, I briefly summarize the new star formation scenario, suggested by these recent studies. Gas collapse occurs around a small central part of the molecular cloud. In the collapsing cloud core, the gas becomes adiabatic and the first adiabatic core appears prior to the protostar formation (stage 1). In this stage, fragmentation frequently occurs to form binary or multiple star systems, because the gas collapse slows down and the perturbations that induce fragmentation can grow. In addition, the first core can drive a low-velocity outflow with a wide opening angle, because the rotation timescale becomes shorter than the collapsing timescale and the magnetic field is amplified by the rotation. Then, the amplified magnetic field drives outflow by the magneto-centrifugal mechanism. This flow corresponds to the observed molecular outflow. Also, in this stage, over 99 (or 99.9%) of the angular momentum of the central part of the cloud core is transferred by magnetic effects such as magnetic braking and outflows. The first core increases its mass and density through gas accretion. Then, the magnetic field begins to dissipate through Ohmic dissipation when the central density exceeds $n > 10^{12}\,\mathrm{cm}^{-3}$ (stage 2). Within the range of $10^{12}\,\mathrm{cm}^{-3} \lesssim n \lesssim 10^{16}\,\mathrm{cm}^{-3}$, the magnetic flux is largely removed from the collapsing cloud. In this period, about 3–5 orders of magnitude of the initial magnetic flux of the collapsing cloud is removed. The removal of angular momentum and magnetic flux makes further collapse possible. When the central density exceeds $n \sim 10^{21}\,\mathrm{cm}^{-3}$, the protostar appears (stage 3). At the protostar formation epoch, the protostar is enclosed by the disk-like first core. After the protostar formation, the first core directly evolves into a circumstellar disk with Keplerian rotation. Just after the protostar forms, a high-velocity jet with good collimation appears near the protostar (stage 4). The magnetic field around the protostar is very weak because of Ohmic dissipation. Thus, the high-velocity jet is driven by the magnetic pressure gradient force (or strong toroidal field) that is generated by the rotation of the protostar. In addition, the jet is well collimated, because it propagates along the rotation axis. Moreover, the low-velocity outflow with a wide-opening angle continues to be driven by the circumstellar disk that originated from the first core. Thus, a high-velocity jet is enclosed by a low-velocity outflow after protostar formation. At the protostar formation epoch, the protostar and first core (or the circumstellar disk) have masses of $10^{-3}\,M_\odot$ and $0.01 - 0.1\,M_\odot$, respectively. Thus, in the main accretion phase, the circumstellar disk is more massive than the protostar. Such a massive disk tends to fragment due to gravitational instability, thus creating a binary companion or gas-giant planet in the circumstellar disk.

Recent numerical simulations have unveiled the protostar formation process starting from the prestellar core stage, while protostellar evolution long after the protostar formation (Class I, II and III phases) remains veiled. Further developments or long-term calculations starting from the prestellar core stage are necessary in order to understand the later phases of star formation.

References

Arce, H. G., Shepherd, D., Gueth, F., Lee, C.-F., Bachiller, R., Rosen, A., & Beuther, H. 2007, Protostars and Planets V, 245

Bate, M. R. 1998, ApJL, 508, L95

Bate, M. R. 2010, MNRAS, 404, L79
Belloche A., Andé P., Despois D., & Blinder S. 2002, A&A, 393, 927
Bally, J., Reipurth, B., & Davis, C. J. 2007, in Protostars & Planets V, ed. B. Reipurth, D. Jewitt, & K. Keil (Tucson: Univ. Arizona Press), 215
Banerjee, R., & Pudritz, R. E. 2006, ApJ, 641, 949
Bodenheimer P., Burkert A., Klein R. I., & Boss A. P., 2000, in Mannings V., Boss A. P., Russell S. S., eds, Protostars and Planets IV. Univ. Arizona Press, p. 675
Boss, A. P. 1993, ApJ, 410, 157
Bouvier, J., Alencar, S. H. P., Harries, T. J., Johns-Krull, C. M., & Romanova, M. M. 2007, Protostars and Planets V, 479
Blandford, R. D., & Payne, D. G. 1982, MNRAS, 199, 883
Caselli, P., Benson, P. J., Myers, P. C., & Tafalla, M. 2002, ApJ, 572, 238
Commerçon, B., Hennebelle, P., Audit, E., Chabrier, G., & Teyssier, R. 2010, A&A, 510, L3
Crutcher R. M. 1999, ApJ, 520, 706
Duffin, D. F. & Pudritz, R. E. 2009, ApJL, 706, L46
Duchêne, G., Bouvier, J., Bontemps, S., Andr; P., & Motte, F. 2004, A&A, 427, 651
Enoch, M. L., Corder, S., Dunham, M. M., & Duchêne, G. 2009, arXiv:0910.2715
Goodwin S. P., Kroupa P., Goodman A., & Burkert A., 2007, in Reipurth B., Jewitt D., Keil K., eds, Protostars and Planets V. Univ. Arizona Press, p. 133
Herbst, W., Eislöffel, J., Mundt, R., & Scholz, A. 2007, Protostars and Planets V, 297
Hennebelle, P. & Teyssier, R. 2008b, A&A, 477, 25
Hosking, J. G. & Whitworth, A. P. 2004, MNRAS, 347, 1001
Inutsuka, S., Machida, M. N., & Matsumoto, T. 2010, ApJL, 718, L58
Kalas, P., *et al.* 2008, Science, 322, 1345
Larson, R. B., 1969, MNRAS, 145, 271.
Looney, L. W., Mundy, L. G., & Welch, W. J. 2000, ApJ, 529, 477
Machida, M. N., Tomisaka, K., & Matsumoto, T. 2004, MNRAS, 348, L1
Machida, M. N., Matsumoto, T., Tomisaka, K., & Hanawa, T. 2005a, MNRAS, 362, 369
Machida, M. N., Matsumoto, T., Hanawa, T., & Tomisaka, K. 2005b, MNRAS, 362, 382
Machida, M. N., Inutsuka, S., & Matsumoto, T. 2006, ApJL, 647, L151
Machida, M. N., Inutsuka, S., & Matsumoto, T. 2007, ApJ, 670, 1198
Machida, M. N., Tomisaka, K., Matsumoto, T., & Inutsuka, S., 2008, ApJ, 677, 327
Machida, M. N., Inutsuka, S., & Matsumoto, T. 2008b, ApJ, 676, 1088
Machida, M. N., Inutsuka, S., & Matsumoto, T. 2009a, ApJL, 704, L10
Machida, M. N., Inutsuka, S., & Matsumoto, T. 2010, arXiv:1001.1404
Masunaga, H. & Inutsuka, S., 2000, ApJ, 531, 350
Mathieu, R. D. 1994, ARA&A, 32, 465
Matsumoto T. & Hanawa T., 2003, ApJ, 595, 913
Mellon, R. R. & Li, Z.-Y. 2009, ApJ, 698, 922
Meyer, M. R., Backman, D. E., Weinberger, A. J., & Wyatt, M. C. 2007, Protostars and Planets V, 573
Miyama S. M., Hayashi C., & Narita S., 1984, ApJ, 279, 621
Mundt, R. & Fried, J. W. 1983, ApJ, 274, L83
Natta, A., Grinin, V., & Mannings, V. 2000, Protostars and Planets IV, 559
Price, D. J. & Bate, M. R. 2007, MNRAS, 377, 77
Pudritz, R. E., Ouyed, R., Fendt, C., & Brandenburg, A., 2007, in Protostars and Planets V, eds. B. Reipurth, D. Jewitt, and K. Keil, University of Arizona Press, Tucson, p.277
Reipurth, B. 2000, AJ, 120, 3177
Saigo, K. & Tomisaka, K. 2006, ApJ, 645, 381
Tomida, K., Tomisaka, K., Matsumoto, T., Ohsuga, K., Machida, M. N., & Saigo, K. 2010, ApJL, 714, L58
Tomisaka, K. 2002, ApJ, 575, 306
Tsuribe T. & Inutsuka S. 1999, ApJ, 523, L155
Walch, S., Burkert, A., Whitworth, A., Naab, T., & Gritschneder, M. 2009, MNRAS, 400, 13

Computational Star Formation
Proceedings IAU Symposium No. 270, 2011
J. Alves, B.G. Elmegreen, J. M. Girart & V. Trimble, eds.

doi:10.1017/S1743921311000202

The Luminosity Problem: Testing Theories of Star Formation

Christopher F. McKee[1] **& Stella R. R. Offner**[2]

[1]Departments of Physics and Astronomy, University of California, Berkeley, CA94720, USA
and
Laboratoire d'Etudes du Rayonnement et de la Matière en Astrophysique, LERMA-LRA, Ecole Normale Superieure, 24 rue Lhomond, 75005 Paris, France
email: cmckee@astro.berkeley.edu
[2]Harvard-Smithsonian Center for Astrophysics, 60 Garden St, Cambridge MA 02138, USA

Abstract. Low-mass protostars are less luminous than expected. This luminosity problem is important because the observations appear to be inconsistent with some of the basic premises of star formation theory. Two possible solutions are that stars form slowly, which is supported by recent data, and/or that protostellar accretion is episodic; current data suggest that the latter accounts for less than half the missing luminosity. The solution to the luminosity problem bears directly on the fundamental problem of the time required to form a low-mass star. The protostellar mass and luminosity functions provide powerful tools both for addressing the luminosity problem and for testing theories of star formation. Results are presented for the collapse of singular isothermal spheres, for the collapse of turbulent cores, and for competitive accretion.

1. The Luminosity Problem

Why don't protostars shine more brightly? In a seminal paper, Kenyon *et al.* (1990) identified this luminosity problem, developed an approach to treat it (the protostellar luminosity function), and proposed almost all the solutions to the problem that have subsequently been studied. The luminosity problem is simple to state: The accretion luminosity of a protostar is

$$L_{\rm acc} = f_{\rm acc}\,\frac{Gm\dot{m}}{r_*} = 3.9 f_{\rm acc}\left(\frac{m}{0.25 M_\odot}\right)\left(\frac{2R_\odot}{r_*}\right)\left(\frac{\dot{m}}{10^{-6}\,M_\odot\,{\rm yr}^{-1}}\right)\quad L_\odot, \tag{1.1}$$

where $f_{\rm acc}$ is the fraction of the accretion energy that goes into radiation (Kenyon *et al.* took $f_{\rm acc} = 1$), m is the protostellar mass, and $\dot{m}$ is the accretion rate. Stahler (1988) has calculated the protostellar radius for a variety of masses and accretion rates, including the thermostatic effects of deuterium burning; we find that $r_* \simeq 2R_\odot$ is a typical value for the harmonic mean radius. By comparing the number of embedded sources with the number of T Tauri stars, Kenyon *et al.* (1990) inferred a star formation time $t_f \simeq (0.1 - 0.2)$ Myr in Taurus-Auriga; to build up a star of average mass $(0.5\,M_\odot)$ in this time requires an accretion rate $\dot{m} \simeq (2.5-5)\times 10^{-6}\,M_\odot\,{\rm yr}^{-1}$. On average, the mass of a protostar will be about half its final stellar mass, which implies that the typical luminosity of a protostar that will become a star of average mass is $L \sim (10 - 20)L_\odot$. (The median luminosity of the average mass star is comparable to the median luminosity—see §3 below.) The median luminosity in the Kenyon *et al.* sample is $1.6L_\odot$. One statement of the luminosity problem is that the median protostellar luminosity is observed to be about an order of magnitude less than the expected value.

Kenyon *et al.* (1990) also provided an alternative description of the luminosity problem: Identifying the peak in the observed luminosity function of embedded sources at $0.3\,L_\odot$

as the luminosity of the lowest mass stars, which they took to be $0.1\,M_\odot$ (keep in mind that this was prior to the discovery of brown dwarfs), and estimating the radius of these protostars as $r_* \sim 1\,R_\odot$, they inferred a mass accretion rate of only $10^{-7}\,M_\odot\,\mathrm{yr}^{-1}$. However, very general arguments (Stahler, Shu, & Taam 1980) indicate that the accretion rate due to gravitational collapse should be of order

$$\dot{m} \sim \frac{(c_s^2 + v_A^2 + \sigma_{\rm turb}^2)^{3/2}}{G}, \tag{1.2}$$

$$\geqslant \frac{c_s^3}{G} = 1.4 \times 10^{-6} \left(\frac{T}{10\,\mathrm{K}}\right)^{3/2} \quad M_\odot\,\mathrm{yr}^{-1}. \tag{1.3}$$

There are solutions that give higher accretion rates than this, such as the Larson (1969)-Penston (1969) solution for the collapse of an isothermal sphere of constant density, but there are no solutions that give lower accretion rates, at least prior to the time that the accretion is affected by the outer boundary (Henriksen *et al.* 1997). Since the observed temperature in molecular clouds is $T \sim 10$ K, this again leads to an order of magnitude discrepancy between observation and theory.

Results from the recent *Spitzer* c2d survey of five nearby star-forming molecular clouds (Evans *et al.* 2009; the survey does not include Taurus-Auriga) confirm that protostars have low luminosities. Evans *et al.* (2009) classified the young stellar objects (YSOs) in the traditional class system, in which Class 0 represents protostars with envelope masses greater than the protostellar mass; Class I represents embedded objects with masses greater than the envelope mass; Flat Spectrum objects represent a transitional class; and Class II corresponds to T Tauri stars. Dunham *et al.*(2010) analyzed this sample and found an extinction-corrected median luminosity of 1.5 $L_\odot$. Including 350 μm data for many of the brighter sources, they obtained an extinction-corrected mean of 5.3 $L_\odot$.

The luminosity problem is thus well established observationally. It is fundamental because observations appear difficult to reconcile with some of the the basic premises of star formation theory, that stars form via gravitational collapse (eq. 1.3) and that they radiate the binding energy in the process (eq. 1.1). As the discussion above illustrates, there are two related aspects to the problem: the observed luminosity appears to be less than that theoretically expected, and there is an excess of very low-luminosity sources.

2. Proposed Solutions

Kenyon *et al.* (1990) proposed a number of solutions to the luminosity problem, of which the two major ones are slow accretion and episodic accretion (discussed below). They also suggested that brown dwarfs could alleviate the luminosity problem by providing sources with luminosities less than they considered. Since their work predated the discovery of brown dwarfs, they did not include them. It is now known that brown dwarfs constitute about 20% of stars (Andersen *et al.* 2008), and they do permit some sources to have lower luminosities. We include brown dwarfs in the models described below.

There is an additional effect that reduces the luminosity problem: The hydromagnetic outflows observed from protostars extract kinetic energy from the accreting gas (McKee & Ostriker 2007), reducing the energy radiated. If half the energy lost by the disk is mechanical energy and this in turn is half the total potential energy, then $f_{\rm acc} \simeq \frac{3}{4}$ (Offner & McKee 2010).

2.1. *Episodic Accretion*

Kenyon *et al.* (1990) suggested that the accretion onto the protostar (as opposed to infall onto a circumstellar disk) could be episodic, so that much of the protostellar mass would be accreted in short periods of time, with high luminosities. Such brief, high-luminosity accretion events could be associated with FU Ori outbursts, which have inferred accretion rates of $\dot{m} \sim 10^{-4}\,M_\odot\ \mathrm{yr}^{-1}$. Kenyon *et al.* (1990) pointed out that the 150 YSOs in Taurus Auriga that formed in the last 10^6 yr correspond to a total accretion rate of $0.75 \times 10^{-4}\,M_\odot\ \mathrm{yr}^{-1}$, comparable to that of a single FU Ori object; thus, a signficant fraction of the mass of a protostar might be acquired during FU Ori events.

Hartmann & Kenyon (1996) refined this argument: They cited a rate of star formation within 1 kpc of the Sun of $(5-10)\times 10^{-3}\,M_\odot\ \mathrm{yr}^{-1}$, which is similar to the recent estimate of $8 \times 10^{-3}\,M_\odot\ \mathrm{yr}^{-1}$ by Fuchs *et al.*(2009). At that time there were 5-9 known FU Ori objects within 1 kpc; if each were to accrete at a rate of $10^{-4}\,M_\odot\ \mathrm{yr}^{-1}$, then protostars could gain about 10% of their mass this way. They pointed out that this was a lower limit, since more FU Ori objects would most likely be discovered. That has indeed occurred: Greene *et al.* (2008) count a total of 22 known FU Ori objects, of which 18 are within 1 kpc. Rounding off, we infer that 20 objects accreting at $10^{-4}\,M_\odot\ \mathrm{yr}^{-1}$ would account for 25% of the mass accreted by protostars within 1 kpc.

To elaborate on this result, consider a simple model for episodic accretion, in which protostars are in a high-accretion state for a total time Δt_{high} and in a low-accretion state for the rest of the time, which is effectively the star formation time, $\Delta t_{\mathrm{low}} \simeq t_f$. The fraction of the mass accreted during a high state is

$$F_{\mathrm{high}} = \frac{\dot{m}_{\mathrm{high}}\Delta t_{\mathrm{high}}}{\langle m_f\rangle}. \tag{2.1}$$

We shall adopt parameters appropriate for FU Ori outbursts, but in fact all episodes of high accretion could be included in F_{high}. The total time spent in the high state is

$$\Delta t_{\mathrm{high}} = \frac{\mathcal{N}_{p,\,\mathrm{high}}}{\dot{\mathcal{N}}_*}, \tag{2.2}$$

where $\mathcal{N}_{p,\,\mathrm{high}}$ is the number of protostars in a high state in some volume of the Galaxy and $\dot{\mathcal{N}}_*$ is the star formation rate there. For a mean stellar mass of $0.5\,M_\odot$, the local star formation rate cited above corresponds to a birthrate within 1 kpc of the Sun of 0.016 stars yr^{-1}. If the number of FU Ori objects within 1 kpc is about 20, then $\Delta t_{\mathrm{high}} = 1250$ yr. Inserting Hartmann & Kenyon's estimate that $\dot{m}_{\mathrm{high}} \simeq 10^{-4}\,M_\odot\ \mathrm{yr}^{-1}$, we recover the result given above, $F_{\mathrm{high}} \simeq 0.25$. We also note that the duty cycle of FU Ori outbursts is small: $\Delta t_{\mathrm{high}}/t_f \sim 10^3/(10^{5-6}) < 0.01$. This small value is consistent with the absence of any known FU Ori sources in the Evans *et al.* (2009) sample.

There are two uncertain numbers that entered into this estimate of F_{high}, the number of FU Ori objects, $\mathcal{N}_{p,\,\mathrm{high}}$, and the accretion rate, $\dot{m}_{\mathrm{high}}$. Undoubtedly, more such objects will be discovered within 1 kpc of the Sun in the future; however, it should be noted that several of the bursting objects, such as L1551 IRS5, have luminosities, and therefore accretion rates, an order of magnitude less than the average (Hartmann & Kenyon 1996). Further study will also refine the observed value of the mean accretion rate. Hartmann & Kenyon (1996) infer an accretion rate $\dot{m}_{\mathrm{high}} = 1.9\times 10^{-4}\,M_\odot\ \mathrm{yr}^{-1}$ and a stellar radius $r_* = 5.9\,R_\odot$ for FU Ori itself. However, after the rapid accretion ceases, the star will shrink back to its original size, releasing a comparable amount of energy. A mean value $\dot{m}_{\mathrm{high}} \simeq 1 \times 10^{-4}\,M_\odot\ \mathrm{yr}^{-1}$ thus seems reasonable in this case.

Hartmann & Kenyon (1996) also noted that given 5 known outbursts in 60 years and a star formation rate of 0.01 stars yr^{-1} within 1 kpc of the Sun implies that there are about 10 bursts per object. The more recent data cited above lead to a similar conclusion. Since $\Delta t_{\mathrm{high}} \simeq 1250$ yr, this means that a typical outburst lasts about 100 yr, consistent with the observationally inferred lifetime (Hartmann & Kenyon 1996).

Finally, we note that a potential problem with the FU-outburst solution to the luminosity problem is that these outbursts appear to be located preferentially in regions of low star-formation rates (Greene *et al.* 2008).

2.2. *Slow Accretion ($\dot{m} \lesssim 10^{-6} M_\odot$ yr^{-1})*

As an alternative solution to the luminosity problem, Kenyon *et al.* (1990) suggested two ways to increase the ages of the protostars and therefore reduce the inferred accretion rate: (1) If the star formation rate decreased substantially over the lifetime of the T Tauri stars, then the observed number of Class I sources would require a greater lifetime; or (2) if the lifetime of the T Tauri stars were larger, then the protostellar lifetime would increase proportionately. They rejected the first possibility since it is difficult to understand how star formation can decelerate in 1 Myr when it is in a region with a crossing time of 10 Myr. The second possibility is more plausible, since the estimated lifetime of T Tauri stars has increased in the intervening 20 years. In their recent analysis of protostellar lifetimes, Evans *et al.* (2009) concluded that the lifetime of Class II sources is about 2 Myr, which would imply a Class I lifetime of $t_f \simeq (0.2 - 0.4)$ Myr in Taurus Auriga. The larger of these two values is favored by data in the c2d survey, for which Evans *et al.* (2009) found a Class I lifetime of 0.44 Myr. The total protostellar lifetime, t_f, must also include the time spent in the Class 0 phase, which they found to be $\simeq 0.1$ Myr, so that the total inferred mean lifetime of protostars in these clouds is $\langle t_f \rangle = 0.54$ Myr. This long protostellar lifetime, together with the correction for non-radiative energy losses ($f\_acc = \frac{3}{4}$) and the correction for unseen outbursts (also a factor of $\frac{3}{4}$) effectively resolves the luminosity problem: The accretion rate for a star of average mass is then $\dot{m} \simeq 10^{-6}\, M_\odot$ yr^{-1}, which corresponds to a typical accretion luminosity from equation (1.1) of about $2.2\, L_\odot$, comparable to the observed median luminosity of $1.5\, L_\odot$.

What about the other aspect of the luminosity problem, which is that the inferred accretion rates are much less than expected theoretically? There are two physical effects that reduce the accretion rate below that in standard models, namely, protostellar outflows and the finite size of the protostellar envelope. Bontemps *et al.* (1996) found that protostellar outflow rates vary as the envelope mass and showed that this could be understood if accretion rates declined exponentially with time: $\dot{m} = \dot{m}_0 \exp(-t/t_*)$ (see also Myers *et al.* 1998), where the decay time is $t_* = m_f/\dot{m}_0$ and m_f is the final protostellar mass (ie., the initial stellar mass). This exponential decline in the accretion rate does not capture the reduction due to the initial stage of protostellar outflows (Shibata & Uchida 1985). It is thus plausible when these effects are included, current theories can be consistent with the inferred low accretion rates. However, this will be possible only if the high accretion rates associated with an initial Larson-Penston accretion phase (Henriksen *et al.* 1997, Schmeja & Klessen 2004) do not release a signficant amount of radiative energy.

3. The Protostellar Luminosity Function

3.1. *Previous Work*

Any solution of the luminosity problem must explain the distribution of protostellar luminosities–the protostellar luminosity function (PLF). Kenyon *et al.* (1990) introduced the PLF and considered two cases: (1) the standard isothermal sphere (IS) case (Shu 1977), in which the accretion rate is independent of mass so that the formation time, t_f, is proportional to the final mass, m_f; and (2) a case in which the accretion rate is proportional to m_f, so that t_f is independent of m_f. Fletcher and Stahler (1994a,b) extended this work in the IS case by determining both the protostellar mass function (PMF) and the PLF for pre-main-sequence stars.

Dunham *et al.* (2010) carried out a detailed analysis of the joint distribution of protostellar luminosities and bolometric temperatures. Interestingly, they adopted an accretion rate of $4.6\times10^{-6}\ M_\odot\ \mathrm{yr}^{-1}$, much higher than implied by the protostellar lifetimes inferred by Evans *et al.* (2009). In one series of models, they included outflows, which entrained envelope material and resulted in a core star-formation efficiency of 0.3-0.5. The outflows were not intrinsically collimated, as in the study by Matzner & McKee (2000), so that outflow cavities eventually expanded to an opening angle of 90 deg, terminating accretion. The outflows led to substantial variations in the observed bolometric luminosity and temperature as functions of the inclination angle. Despite the reduction in the accretion rate, the mean luminosities were still too high. The best agreement with observation was obtained by assuming that most of the mass was accreted in FU Ori-type events with accretion rates of $10^{-4}\ M_\odot\ \mathrm{yr}^{-1}$, although the resulting models spent much more time with bolometric temperatures above 10^3 K than is observed. These models required $F_{\mathrm{high}} \sim 0.8$ (Dunham, private communication). This is consistent with observation only if the number of FU Ori sources within 1 kpc of the Sun were about 3 times higher than currently known (see eq. 2.2).

3.2. *The PMF and the PLF*

We follow McKee & Offner (2010) and Offner & McKee (2010) in describing the protostellar mass function (PMF) and the protostellar luminosity function (PLF). Let $\psi(m_f)d\ln m_f$ be the fraction of stars born in the mass range dm_f; ψ is thus the IMF. The PMF is the present-day mass function of the protostars, and it must be consistent with the IMF when the protostellar mass reaches its final value, m_f. We denote the fraction of protostars in the mass range dm that will have final masses in the mass range dm_f by $\psi_{p2}(m,m_f)d\ln m d\ln m_f$; it is the IMF weighted by the time the protostar spends at a given mass,

$$\psi_{p2}(m,m_f) = \left[\frac{(m/\dot{m}(m,m_f))}{\langle t_f\rangle}\right]\psi(m_f), \tag{3.1}$$

where $\langle t_f\rangle$ is the average protostellar lifetime. We have assumed that the accretion rate is a function of m and m_f, which can be readily generalized to allow for high and low accretion states. The PMF is the integral of this function over all possible values of the final mass, ranging from the lower bound on the IMF, m_ℓ, to the upper bound, m_u, subject to the constraint $m \leqslant m_f$:

$$\psi_p(m) = \int_{\max(m_\ell,m)}^{m_u} \psi_{p2}\, d\ln m_f. \tag{3.2}$$

It follows that the fraction of protostars in the mass range dm and the luminosity range dL is

$$\Psi_{p2}(L,m)\, d\ln m\, d\ln L = \psi_{p2}(m,m_f)\, d\ln m\, d\ln m_f. \tag{3.3}$$

The PLF is then

$$\Psi_p(L) = \int_{m_{\min}}^{m_{\max}} d\ln m \Psi_{p2}(L,m), \tag{3.4}$$

$$= \int_{m_{\min}}^{m_{\max}} d\ln m\, \frac{\psi_{p2}[m,m_f(L,m)]}{|\partial\ln L/\partial\ln m_f|}, \tag{3.5}$$

where the limits of integration are such that $m_\ell \leqslant m_f \leqslant m_u$ and $m \leqslant m_f$. The PMF and PLF depend on the history of the mass accretion rate, $\dot{m}(m, m_f)$ (eq. 3.1) *and therefore on the theory of star formation.*

4. Testing Theories of Star Formation

4.1. *Accretion Histories*

We consider three different theories of star formation, plus one variant:

* **Isothermal Sphere** (IS, Shu 1977)—Inside-out collapse of singular isothermal sphere:

$$\dot{m} = \dot{m}_{\rm IS} = 1.54 \times 10^{-6} (T/10\,{\rm K})^{3/2} \quad M_\odot\ {\rm yr}^{-1}. \tag{4.1}$$

The formation time is proportional to the mass, $t_f \propto m$, so this accretion model is valid only for low-mass stars (Shu, Adams & Lizano 1987).

* **Turbulent Core** (TC, McKee & Tan 2002, 2003)—Inside-out collapse of a turbulent core:

$$\dot{m} = \dot{m}_{\rm TC} \left(\frac{m}{m_f}\right)^{1/2} m_f^{3/4}, \quad \text{with } \dot{m}_{\rm TC} \propto \Sigma^{3/4}. \tag{4.2}$$

The accretion rate increases with both the protostellar mass, m, and the final mass, m_f. This model was developed for high-mass star formation, and is equivalent to the accretion rate in the Hennebelle & Chabrier (2008) theory of the IMF.

* **Two-Component Turbulent Core** (2CTC)—Begins as isothermal accretion and evolves to turbulent accretion (similar to the TNT model of Myers & Fuller 1992):

$$\dot{m} = \left[\dot{m}_{\rm IS}^2 + \dot{m}_{\rm TC}^2 \left(\frac{m}{m_f}\right) m_f^{3/2}\right]^{1/2}. \tag{4.3}$$

* **Competitive Accretion** (CA, Zinnecker 1982, Bonnell *et al.* 1997)—Stars form in a common gas reservoir, usually accreting at the tidally limited Bondi rate, $\dot{m} \propto m^{2/3}$, and all having the same formation time:

$$\dot{m} = \dot{m}_{\rm CA} \left(\frac{m}{m_f}\right)^{2/3} m_f, \quad \text{with } \dot{m}_{\rm CA} \propto 1/(\text{free-fall time}). \tag{4.4}$$

4.2. *Comparing the PMF and PLF with Observation*

We adopt a truncated Chabrier (2005) IMF:

$$\psi(m_f) \propto \exp - \left[\frac{\log^2(m_f/0.2)}{2 \times 0.55^2}\right] \qquad (m_f \leqslant 1\ M_\odot) \tag{4.5}$$

$$\propto m_f^{-1.35} \qquad (1\ M_\odot < m_f \leqslant m_u) \tag{4.6}$$

The Evans *et al.* (2009) sample of Class II YSOs has about 400 objects. In this sample, 9 YSOs are expected with masses exceeding $3M_\odot$ if $m_u \gg 3M_\odot$ (i.e., stars with masses much greater than $3M_\odot$ are possible). Since none are seen, we adopt an upper cutoff $m_u = 3M_\odot$ for the IMF.

Figure 1a illustrates the PMF (eq. 3.2) estimated with each of the accretion histories. As discussed by McKee & Offner (2010), the models predict very different mass distributions. For example, in the isothermal sphere case more massive protostars spend a

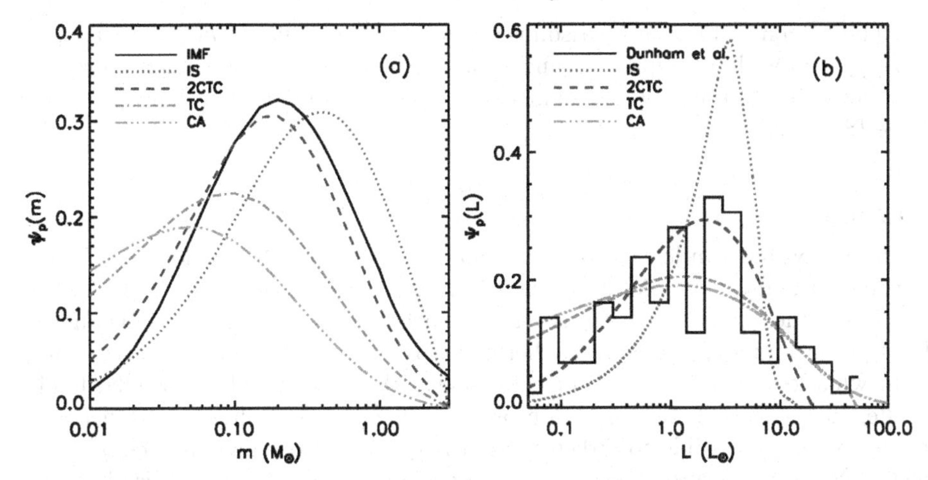

Figure 1. Left: The PMF, $\psi_p(m)$, for the four accretion models and the Chabrier IMF. Right: The PLF, $\Psi_p(L)$, for the same models assuming no tapering of the accretion rate, with $F_{\rm high} = 0.25$ and $\langle t_f \rangle = 0.56$ Myr with the data from Dunham et al. (2010). Both panels adopt a cluster upper mass $m_u = 3\ M_\odot$.

longer time accreting and consequently weight the distribution towards higher masses. In contrast, for competitive accretion, where all the protostars share the same protostellar lifetime, the significantly larger number of low-mass protostars shifts the PMF towards lower masses. Unfortunately, since we can't directly measure the protostellar masses, Figure 1a is not sufficient to observationally discriminate between the models.

For a more direct comparison, Offner & McKee (2010) calculate the PLF using the predicted mass functions in combination with a stellar evolution model (see Offner *et al.* 2009 for details). Figure 1b shows the PLF for each accretion model plotted with the extinction corrected luminosities from Dunham *et al.* (2010). The model curves assume that 25% of the mass is accreted during unseen bursts ($F_{\rm high} = 0.25$) and $\langle t_f \rangle = 0.56$ Myr. This is equivalent to applying both the slow accretion and variable accretion solutions and thus likely represents a lower bound on the predicted luminosities. For comparison with Dunham *et al.* 2010, we adopt an upper bolometric luminosity uncertainty of 50% and use the uncorrected bolometric luminosities to set a lower error bound.

The distributions can be characterized by the mean, median, and standard deviation. For $m_u = 3\ M_\odot$ and $L_{\rm min} = 0.05\ L_\odot$, the fiducial models have means in the range 2.5 $L_\odot$ (2CTC) - 3.6 $L_\odot$ (CA), all a factor of ~1.5-2 below the observed value: $5.3\ ^{+2.6}_{-1.9}\ L_\odot$. We find that the mean luminosities fall in a narrower range, 2.6 $L_\odot$ (2CTC)- 3.4 $L_\odot$ (IS), when the accretion rates taper off towards the end of the protostellar lifetime:

$$\dot{m} = \dot{m}_0(m, m_f)\left[1 - \left(\frac{t}{t_f}\right)\right], \tag{4.7}$$

where $\dot{m}_0(m, m_f)$ is the untapered accretion rate. (Foster & Chevalier 1993 found that the accretion rate tapered off at late times in their 1D calculations, and Myers *et al.* 1998 included an exponential tapering in their models.) Only the non-tapered CA and tapered IS models fall within the uncertainty, suggesting that the adopted lifetime may be too high by as much as a factor of 2. In contrast, the fiducial medians, which range from 0.8 $L_\odot$ (CA) to 2.5 $L_\odot$ (IS), are in better agreement with the observed median of $1.5\ ^{+0.7}_{-0.4}\ L_\odot$. The median of the TC (0.9 $L_\odot$) and 2CTC (1.4 $L_\odot$) models are within error and remain so even for a lifetime reduced by a factor up to 2.4 and 1.6, respectively. The observed standard deviation of $\log L$, $0.7^{+0.2}_{-0.1}$, is consistent with the 2CTC (0.6), TC (0.7) and CA

(0.8) cases. The outcome of the comparison is sensitive to the values of $F_{\rm high}$ and $f_{\rm acc}$, in addition to $\langle t_f \rangle$, and all have significant uncertainty. The parameter dependence may be reduced by comparing to the ratio of the median to the mean luminosity, $0.3\,^{+0.2}_{-0.1}$. This rules out the IS model.

5. Conclusions

The luminosity problem in low-mass star formation can be resolved if low-mass stars form slowly, over a period ~ 0.5 Myr, as suggested by the results of Evans *et al.* (2009). Indeed, if some of the accretion energy is released mechanically ($1 - f_{\rm acc} \simeq 1/4$) and in unseen FU Ori outbursts ($F_{\rm high} \simeq 1/4$), then for most models the star formation time must be somewhat less than 0.5 Myr to be consistent with the observed protostellar luminosities. Different theories of star formation predict different prototellar mass functions, which are currently inaccessible to direct observation, and protostellar luminosity functions, which have been observed. The latter serves as an important metric to discriminate between theories of star formation.

Acknowledgements

We thank Michael Dunham, Melissa Enoch, and Neal Evans for discussions of their work, and Philippe Andre, Lee Hartmann, Jean-Francois Lestrade, and Mark Krumholz for useful comments. This research has been supported by the NSF under grants AST-0908553 (CFM) and AST-0901055 (SSRO). CFM also acknowledges the support of the Groupement d'Intérêt Scientifique (GIS) "Physique des deux infinis (P2I)."

References

Andersen, M., Meyer, M. R., Greissl, J., & Aversa, A. 2008, *ApJL*, 683, L183
Bonnell, I. A., Bate, M. R., Clarke, C. J., & Pringle, J. E. 1997, *MNRAS*, 285, 201
Bonnell, I. A., Bate, M. R., Clarke, C. J., & Pringle, J. E. 2001, *MNRAS*, 323, 785
Bontemps, S., Andre, P., Terebey, S., & Cabrit, S. 1996, *A&A*, 311, 858
Foster, P. N. & Chevalier, R. A. 1993, *ApJ*, 416, 303
Hartmann, L., Cassen, P., & Kenyon, S. J. 1997, *ApJ*, 475, 770
Hennebelle, P. & Chabrier, G. 2008, *ApJ*, 684, 395
Henriksen, R., Andre, P., & Bontemps, S. 1997, *A&A*, 323, 549
Larson, R. B. 1969, *MNRAS*, 145, 271
McKee, C. F. & Ostriker, E. C. 2007, *ARAA*, 45, 565
McKee, C. F. & Tan, J. C. 2002, *Nature*, 416, 59
McKee, C. F. & Tan, J. C. 2003, *ApJ*, 585, 850
Myers, P. C., Adams, F. C., Chen, H. & Schaff, E. 1998, *ApJ*, 492, 703
Myers, P. C. & Fuller, G. A. 1992, *ApJ*, 396, 631
Offner, S. S. R., Klein, R. I., McKee, C. F., & Krumholz, M. R. 2009, 703, 131.
Ostriker, E. C. & Shu, F. H. 1995, *ApJ*, 447, 813
Penston, M. V. 1969, *MNRAS*, 144, 425
Schmeja, S. & Klessen, R. S. 2004, *A&A*, 419, 405
Shibata, K. & Uchida, Y. 1985, *PASJ*, 37, 31
Shu, F. H. 1977, *ApJ*, 214, 488
Stahler, S. W. 1988, *ApJ*, 332, 804
Tan, J. C. & McKee, C. F. 2004, *ApJ*, 603, 383
Zinnecker, H. 1982, Annals of the New York Academy of Sciences, 395, 226

Computational Star Formation
Proceedings IAU Symposium No. 270, 2011
J. Alves, B.G. Elmegreen, J. M. Girart & V. Trimble, eds.

doi:10.1017/S1743921311000214

Internal Structure of Stellar Clusters: Geometry of Star Formation

Emilio J. Alfaro & Néstor Sánchez

Instituto de Astrofísica de Andalucía, CSIC, Apdo. 3004, E-18080, Granada, Spain

Abstract. The study of the internal structure of star clusters provides important clues concerning their formation mechanism and dynamical evolution. There are both observational and numerical evidences indicating that open clusters evolve from an initial clumpy structure, presumably a direct consequence of the formation in a fractal medium, toward a centrally condensed state. This simple picture has, however, several drawbacks. There can be very young clusters exhibiting radial patterns maybe reflecting the early effect of gravity on primordial gas. There can be also very evolved clusters showing fractal patterns that either have survived through time or have been generated subsequently by some (unknown) mechanism. Additionally, the fractal structure of some open clusters is much clumpier than the average structure of the interstellar medium in the Milky Way, although in principle a very similar structure should be expected. Here we summarize and discuss observational and numerical results concerning this subject.

Keywords. ISM: structure, open clusters and associations: general, stars: formation

1. Introduction

Most of visible matter in the universe is condensed into stars, with densities more than 30 orders of magnitude higher than the average density of the universe and more than 20 orders of magnitude higher than the densities of the interstellar clouds in which they form (Larson 2007). Thus, the fundamental question is not how baryons end up as stars, but how some of them form stars and others remain as hot, low-density interstellar gas. This enigma lies at the core of a predictive theory of star formation, one of the main goals of modern astronomy. We are still far away from a global solution to this complex problem, whose answer depends very much on the existence of a well-structured and complete set of empirical data, as well as on the building of reliable and precise simulation tools.

Nowadays, it is widely accepted that stars form in highly hierarchical stellar systems that mimic, in some way, the stepped structure of the interstellar medium (ISM) or, at least, the morphology of the densest regions. This hierarchical pattern, both spatial and temporal, presents singular condensations, the stellar clusters, whose main physical characteristics make them reliable tracers of the star forming processes in galaxies. Hierarchical structure extends from star complexes (or large portions of spiral arms in flocculent galaxies) through embedded clusters to individual young stars inside those embedded clusters. The cluster scale is the best metric to measure and analyze the whole spatial range in the formation of stellar systems.

The study of star forming regions in the infrared range led to the conclusion that most stars, if not all, are born grouped in clusters (Lada & Lada 2003). However, after ten million years, the fraction of stars in clusters is reduced to 10% and this figure tends to decrease with age. This simple temporal pattern suggests that a few million years after their birth star clusters suffer a high mortality rate. The survivors evolve then under different destruction processes and are gradually eroded until diluted into the galactic field.

Table 1. Summary of perimeter dimensions in molecular clouds in the Galaxy

Ref.	D_{per}	Region/Map
(1)	1.40	Extinction maps of dark clouds
(2)	1.12 − 1.40	Dust emission maps of cirrus clouds
(3)	1.17 − 1.30	Infrared intensity and column density maps of several molecular clouds
(4)	1.36	Molecular emission maps (Taurus complex)
(5)	1.38 − 1.52	Visual extinction maps (Chamaeleon complex)
(6)	1.51	Molecular emission map (Taurus complex)
(7)	1.23 − 1.54	HI maps of high-velocity clouds and infrared maps of cirrus clouds
(8)	1.34 − 1.40	Molecular emission maps of clouds in the antigalactic center
(9)	1.31 − 1.35	Molecular emission maps (Ophiuchus, Perseus, and Orion clouds)
(10)	1.50 − 1.53	Molecular emission maps of clouds in the outer Galaxy

Reference index: (1) Beech (1987); (2) Bazell & Desert (1988); (3) Dickman *et al.* (1990); (4) Falgarone *et al.* (1991); (5) Hetem & Lepine (1993); (6) Stutzki (1993); (7) Vogelaar & Wakker (1994); (8) Lee (2004); (9) Sánchez *et al.* (2007a); (10) Lee *et al.* (2008).

The initial conditions, that is, the properties of the cold and dark clouds that eventually form stars are poorly known (Bergin and Tafalla 2007). A few years ago, we started a project aimed to characterize the geometry of the ISM. This information would provide important clues on the physical processes developing and maintaining the internal structure of the clouds. Since the pioneering work by Larson (1981), turbulence is considered the best candidate to do this job. It appears that the distribution of gas and dust in these clouds determines the initial conditions of a newborn cluster because star formation follows the patterns defined by the densest regions (Bonnell *et al.* 2003). Thus, the fractal (self-similar) distribution of the gas in molecular cloud complexes may account for the hierarchical structure observed in some open clusters (Elmegreen 2010). However, observations show that the morphologies of young clusters show a wide variety, from hierarchical to centrally condensed ones, often being elongated or surrounded by a low-density stellar halo (see, for example, Maíz-Apellániz 2001; Hartmann 2002; and Caballero 2008). The reasons for this heterogeneity of shapes are still poorly understood.

Here we review and discuss the formation of star clusters and their evolution from a geometric point of view. In particular, our framework is determined by several questions: What physical mechanisms control and shape the internal structure of fertile gas clouds? What is the influence of the geometric structure of a star-forming cloud on the internal spatial structure of the stellar population formed from it? How does it evolve with time? Because of the hierarchical and self-similar nature involved, fractal geometry appears to be a good descriptor for these physical structures. Thus, the design and development of mathematical tools for determining the fractal dimension of 3D gas clouds and point-like object distributions are also discussed and evaluated in this work.

2. Fractal dimension of the interstellar medium

Gas and dust in the Galaxy are organized into irregular structures in a hierarchical and approximately self-similar manner. This means that interstellar clouds can be well described or characterized as fractal structures (Mandelbrot 1983). Many tools can be used to characterize the complexity of these structures (see Elmegreen & Scalo 2004) but to measure the fractal dimension seems particularly appropriate when dealing with nearly fractal systems. The measurement of the dimension of the projected boundaries D_{per} is the most used method to characterize the fractal properties of interstellar clouds, but there is a wide variation in the estimated values. A summary of results that can be accessed via NASA's ADS service is shown in Table 1.

In general, observed values are spread over the range $1.1 \lesssim D_{per} \lesssim 1.5$. It is not clear, however, whether the different values seen in Table 1 represent real variations or they are consequence of different data quality and/or analysis techniques. For example, it is well known that the obtained results may be affected by factors such as image resolution and/or signal-to-noise ratio (Dickman *et al.* 1990; Vogelaar & Wakker 1994; Lee 2004; Sánchez *et al.* 2005, 2007a). Note, for example, that for CO emission maps of the same region in the Taurus molecular complex Falgarone *et al.* (1991) obtained $D_{per} = 1.36$ whereas Stutzki (1993) found $D_{per} = 1.51$ on a different set of data. Despite those results, the general "belief" is that the fractal dimension of the projected boundaries of interstellar clouds is roughly a constant throughout the Galaxy, with $D_{per} \simeq 1.3 - 1.4$ (Bergin & Tafalla 2007). This constancy in D_{per} would be a natural consequence of a universal picture in which interstellar turbulence is driven by the same physical mechanisms everywhere (Elmegreen & Scalo 2004). In order to get reliable clues about the ISM structure, it is important that any analysis technique is applied systematically on homogeneous data sets. Sánchez *et al.* (2007a) used several maps of different regions (Ophiuchus, Perseus, and Orion molecular clouds) in different emission lines and calculated D_{per} by using an algorithm previously calibrated on simulated fractals (Sánchez *et al.* 2005). In this case the range of obtained values decreased notoriously to $1.31 \lesssim D_{per} \lesssim 1.35$ (reference 9 in Table 1).

But what is the corresponding value of the fractal dimension of interstellar clouds in the three-dimensional space, D_f? It has been traditionally assumed that $D_f = D_{per} + 1 \simeq 2.3 - 2.4$ (Beech 1992). Sánchez *et al.* (2005) used simulated fractal clouds to study the relationship between D_{per} and D_f and showed this assumption is not correct. Their main result (see their Figure 8) indicate that if the perimeter dimension is around $D_{per} \simeq 1.31 - 1.35$ (Sánchez *et al.* 2007a) then the 3D fractal dimension should be in the range $D_f \sim 2.6 - 2.8$. This dimension is clearly higher than the value $D_f \sim 2.3$ that is usually assumed in the literature for interstellar clouds in the Galaxy (Bergin & Tafalla 2007).

3. Fractal dimension of young stellar clusters

The distribution of stars and star-forming regions also exhibits a spatial hierarchy from large star complexes to individual clusters (Efremov 1995; de la Fuente Marcos & de la Fuente Marcos 2006, 2009; Elias *et al.* 2009; Elmegreen 2010). This hierarchical structure is presumably a direct consequence of the fact that stars are formed in a medium with an underlying fractal structure (previous Section). If this were the case, then it is reasonable to assume that the fractal dimension of the distribution of new-born stars should be nearly the same as that of the molecular clouds from which they are formed.

The fractal dimension of a distribution of stars can be measured by using the correlation integral $C(r)$. For a fractal set it holds that $C(r) \sim r^{D_c}$, being D_c the so-called correlation dimension. Calculating the mean surface density of companions (MSDC) per star $\Sigma(\theta)$ as a function of angular separation θ is another widely used way to measure the degree of clustering of stars. For fractals $\Sigma(\theta) \sim \theta^{\gamma}$, and the exponent is related to the fractal dimension through $D_c = 2 + \gamma$. MSDC technique has been used by various authors to study the clustering of protostars, pre-main sequence stars, or young stars in different star-forming regions. Most results seem to indicate that there are two different ranges of spatial scales, the regime of binary and multiple systems on smaller scales and a regime of fractal clustering on the largest scales. The idea prevalent among astronomers is that self-similar clustering above the binary regime is due to, or arises from, the fractal features of the parent clouds. However, such as in the case of gas distribution in the ISM, if one checks the references a wide variety of different values can be found. Nakajima *et al.* (1998) found significantly variations among different star-forming regions with $1.2 \lesssim D_c \lesssim 1.9$. There can be large differences even in the same regions if analyzed by

Table 2. Summary of correlation dimensions for the distribution of stars in clusters

Ref.	N_{dat}	D_c	Cluster
(1)	> 121	1.38	Taurus-Auriga
(2)	80	1.36 ± 0.19	Taurus
	51	1.50 ± 0.19	Ophiuchus
	355	1.80 ± 0.21	Trapezium
(3)	361	1.85 ± 0.02	Orion OB
	488	1.77 ± 0.02	Orion A
	226	1.31 ± 0.01	Orion B
	96	1.64 ± 0.06	Ophiuchus
	103	1.43 ± 0.04	Chamaeleon I
	94	1.45 ± 0.03	Chamaeleon
	278	1.39 ± 0.02	Vela
	65	1.18 ± 0.13	Lupus
(4)	744	1.98 ± 0.01	Trapezium
(5)	137	1.72 ± 0.06	Chamaeleon I
	216	1.13 ± 0.01	Taurus
(6)	204	1.02 ± 0.04	Taurus
(7)	272	1.049 ± 0.007	Taurus-Auriga

Reference index: (1) Larson (1995); (2) Simon (1997); (3) Nakajima *et al.* (1998); (4) Bate *et al.* (1998), their first data set; (5) Gladwin *et al.* (1999); (6) Hartmann (2002); (7) Kraus & Hillenbrand (2008).

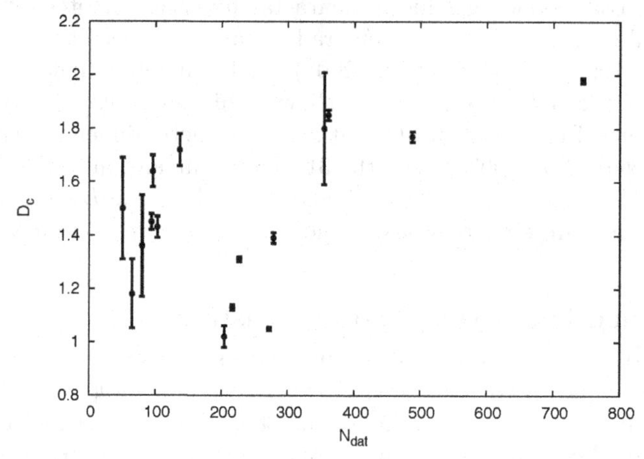

Figure 1. Fractal (correlation) dimension for the distribution of young stars and pre-main-sequence stars in different star-forming regions as a function of the number of data points (Table 2).

different authors and data sets. For example, in the Taurus region both Larson (1995) and Simon (1997) analyzed young stars and their results are in perfect agreement with $D_c \simeq 1.4$, whereas Hartmann (2002) and Kraus & Hillenbrand (2008) both agree in $D_c \simeq 1.0$ for the same region. Table 2 summarizes the wide range of D_c values estimated by using the MSDC technique. Obviously, there can be different results depending on data sources, object selection criteria, and details of the specific calculation procedures. But additionally, it has been shown that if boundary and/or small data-set effects are not taken into account the final results can be seriously biased, given fractal dimension values smaller than the true ones (Sánchez *et al.* 2007b; see also Sánchez & Alfaro 2008). In Figure 1 we have plotted D_c values from Table 2 as a function of the number of data points N_{dat}. The observed behavior seems to be biased (at least in part) in the sense that D_c decreases as N_{dat} decreases (compare Figure 1 here with Figures 2 and 4 in Sánchez & Alfaro 2008). If this kind of effect is not corrected then any real variation in D_c could be hidden or misunderstood.

4. Evolutionary effects

Even in the case of an extremely young cluster, we always are seeing a snapshot of the cluster at a particular age resulting from certain initial conditions (ISM structure) and early dynamical evolution. As a cluster evolves, its initial distribution of stars may be erased, or at least modified. Then, part of the variations observed in Figure 1 could be due to evolutionary effects. The early evolution of the cluster will depend, among other things, on how much gas is removed after the formation process (Gieles 2010). In gravitationally unbound clusters, the separation of the stars increases with age until the cluster dissolves into the field. In principle, the initial clumpy structure disappears after this process of expansion although some simulations suggest that it is possible to keep the initial substructure for a long time in unbound clusters (Goodwin & Whitworth 2004). Gravitationally bound clusters, instead, have to evolve toward a new equilibrium state. Simulations show that this dynamical evolution can be a very complex process (e.g., Moeckel & Bate 2010). It seems that the general trend is to evolve from the initially substructured distribution of stars toward centrally peaked distributions, that is, radial star density profiles. The evidence for this kind of evolution comes from both observations and from numerical simulations (Schmeja & Klessen 2006; Schmeja *et al.* 2008; Sánchez & Alfaro 2009; Allison *et al.* 2009, 2010; Moeckel & Bate 2010).

Roughly speaking, the time interval necessary to erase any initial structure will depend on the crossing time T_{cross}. It should take at least several crossing times to reach an equilibrium state and/or to eliminate the original distribution, although some simulations indicate that the evolution from clumpy to radial distribution may occur on time scales as short as ~ 1 Myr (Allison *et al.* 2009, 2010). In order to address these questions, it is necessary to characterize the internal structure of young clusters and also to get some idea about the evolutionary stage.

5. Minimum spanning tree

For radially concentrated clusters, star distribution ca be characterized by fitting the density profile to some given predefined function. From the fitting procedure it is possible to get parameters such as the central density of stars, the steepness of the density profile, and cluster radius. Obviously, this kind of analysis does not work in clumpy clusters because a smooth function cannot be well fitted to an irregular distribution.

Cartwright & Whitworth (2004) proposed a method to quantify the internal structure of star clusters. Their technique is becoming very widespread and useful for analyzing both observational and simulated data because it is able to distinguish between centrally concentrated and fractal-like distributions. The technique is based on the construction of the minimum spanning tree (MST). The MST is the set of straight lines (called branches or edges) connecting a given set of points without closed loops, such that the total edge length is minimum (see Figure 2). From the MST an adimensional structure parameter Q can be easily calculated (Cartwright & Whitworth 2004; see also Schmeja & Klessen 2006). For an homogeneous distribution of stars $Q \simeq 0.8$. The behavior of Q is such that $Q > 0.8$ for radial clustering whereas $Q < 0.8$ for fractal clustering (see the examples in Figure 2). Moreover, Q increases as the steepness of the profile increases for radial clustering and Q decreases as the fractal dimension decreases for fractal-type clustering (see Figure 5 in Cartwright & Whitworth 2004; and Figure 7 in Sánchez & Alfaro 2009). Thus, Q is able to disentangle between radial and fractal clustering but it is also the strength of clustering.

It is expected that the internal structure of a star cluster evolves with time from initial fractal clustering ($Q < 0.8$) to either homogeneous distribution ($Q \simeq 0.8$) if the cluster is

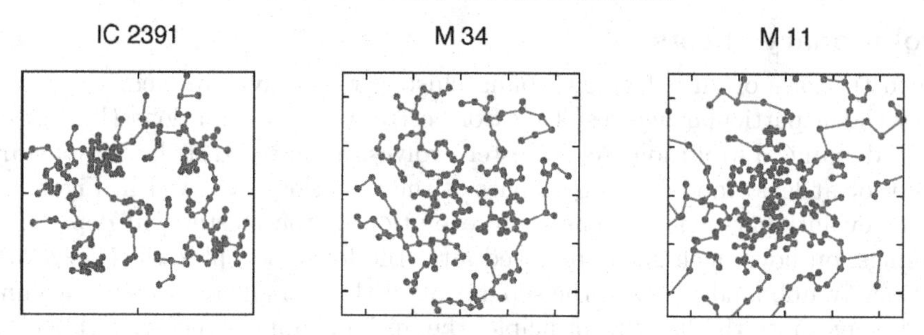

Figure 2. Minimum spanning trees for three open clusters, from which the parameter Q can be calculated (see text). Star positions are indicated with blue circles and red lines represent the trees. The value of Q quantifies the spatial distribution of stars. For IC 2391 the stars are distributed following an irregular, fractal pattern ($Q = 0.66 < 0.8$), for M 34 the stars are distributed roughly homogeneously ($Q = 0.8$), and for M 11 the stars follow a radial density profile ($Q = 1.02 > 0.8$).

dispersing its stars or centrally concentrated distribution ($Q > 0.8$) if it is a bound cluster. Cartwright & Whitworth (2004) calculated Q for several star clusters. They obtained $Q = 0.47$ for stars in Taurus, a value consistent with its observed clumpy structure and with its relatively young evolutionary stage (they estimated an age in crossing time units of $T/T_{cross} \simeq 0.1$). However, they also found some apparent contradictory results. IC 348, a slightly evolved cluster with $T/T_{cross} \simeq 1$ yielded $Q = 0.98$ according to its steep radial density profile. Instead, the highly evolved cluster IC 2391 ($T/T_{cross} \simeq 20$) still exhibits fractal clustering with $Q = 0.66$. Schmeja *et al.* (2008) applied this technique to embedded clusters in the Perseus, Serpens and Ophiuchus molecular clouds, and found that older Class 2/3 objects are more centrally condensed than the younger Class 0/1 protostars. Sánchez & Alfaro (2009) measured Q in a sample of 16 open clusters in the Milky Way spanning a wide range of ages. They found that there can exist clusters as old as ~ 100 Myr exhibiting fractal structure. This means that either the initial clumpiness may last for a long time or other mechanisms may develop some kind of substructure starting from an initially more homogeneous state. Sánchez & Alfaro (2009) obtained a statistically significant correlation between Q and T/T_{cross} in their sample of open clusters. Figure 3a shows this tendency where the crossing times were calculated by assuming a constant velocity dispersion of 2 km s^{-1}. As we can see, the general trend is that young clusters (meaning that dynamically less evolved clusters) tend to distribute their stars following fractal patterns whereas older clusters tend to exhibit centrally concentrated structures. This result supports the idea that stars in newly born clusters likely follow the fractal patterns of their parent molecular clouds, and that they eventually evolve towards more centrally concentrated structures. However, we know that this is only an overall trend. Some very young clusters may exhibit radial density gradients, as for instance σ Orionis for which $Q \simeq 0.9$ (Caballero 2008).

Given the wide variety of physical processes involved in the origin and early evolution of star clusters, it is somewhat surprising that a correlation like that seen in Figure 3a can be observed. Very recent simulations by Allison *et al.* (2009, 2010) and Moeckel & Bate (2010) show that the transition from fractal clustering to central clustering may occur on very short timescales ($\lesssim$1 Myr). Simulations by Maschberger *et al.* (2010) suggest a more complex variety of possibilities. Bound systems may start fractal and evolve towards a centrally concentrated stage whereas unbound systems may stay fractal in time. But this is the evolution for the whole systems. Star clusters in each system may evolve in totally different ways. In fact, the time evolution of the Q parameter of clusters fluctuates dramatically depending on episodes of relaxation or merging (see

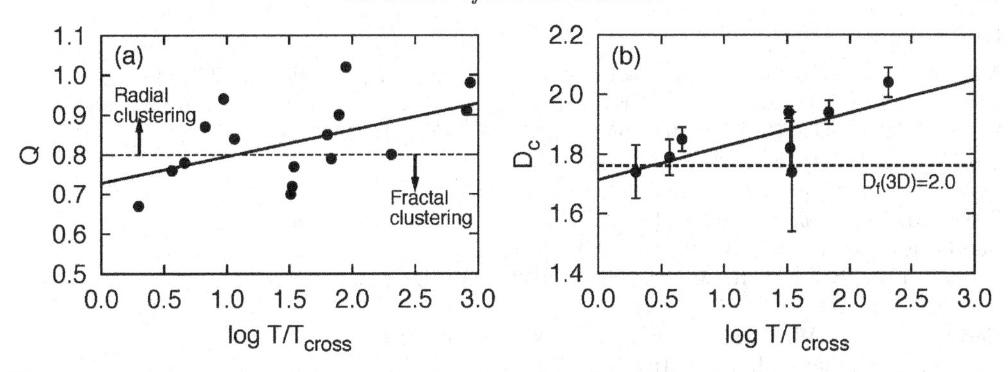

Figure 3. (a) Structure parameter as a function of the logarithm of age in crossing time units. Dashed line separates radial from fractal clustering and solid line is the best linear fit. (b) Correlation dimension as a function of age in crossing-time units. The best linear fit is represented by a solid line. For reference, a horizontal dashed line indicates the value corresponding to three-dimensional distributions with fractal dimensions of $D_f = 2.0$.

Figure 8 in Maschberger *et al.* 2010). It is difficult to argue that, despite all this complex formation history (occurring in $\sim$ 0.5 Myr), we should still observe some correlation between internal structure and age.

6. Initial fractal dimension of star clusters

For those open clusters with internal substructure, the fractal dimension also shows a significant correlation with the age in crossing time units (Sánchez & Alfaro 2009), as it can be seen in Figure 3b. The degree of clumpiness is smaller for more evolved clusters. The horizontal dashed line in Figure 3b shows a reference D_f value estimated from previous papers (Sánchez & Alfaro 2008). It is interesting to note that open clusters with the smallest correlation dimensions $D_c = 1.74$ would have 3D fractal dimensions around $D_f \sim 2$. This value is considerably smaller than the average value estimated for Galactic molecular clouds (see Section 2), which is $D_f \simeq 2.6 - 2.8$.

This result creates an apparent problem to be addressed, because as mentioned before a group of stars born from the same cloud at almost the same place and time is expected to have a fractal dimension similar to that of the parent cloud. If the fractal dimension of the interstellar medium has a nearly universal value around $2.6 - 2.8$, then how can some clusters exhibit such small fractal dimensions? This is still an open question. Several possibilities should be investigated in future studies. First, some simulations demonstrate that it is possible to increase the clumpiness (to decrease D_f) with time (Goodwin & Whitworth 2004). Second, maybe this difference is a consequence of a more clustered distribution of the densest gas from which stars form on the smallest spatial scales in the molecular cloud complexes, according to a multifractal scenario (Chappell & Scalo 2001). Third, perhaps the star formation process itself modifies in some (unknown) way the underlying geometry generating distributions of stars that can be very different from the distribution of gas in the star-forming cloud. A fourth possibility is that the fractal dimension of the interstellar medium in the Galaxy does not have a universal value (i.e., that D_f is different from region to region depending on the main physical processes driving the turbulence). Therefore some clusters could show smaller initial fractal dimensions because they formed in more clustered regions. The possibility of a non-universal fractal dimension for the ISM should not, in principle, be ruled out. However, in this last case, overall correlations as those shown in Figures 3a and 3b should not, in principle, be observed.

References

Allison, R. J., Goodwin, S. P., & Parker, R. J. *et al.* 2009, *ApJ* (Letters), 700, L99
Allison, R. J., Goodwin, S. P., & Parker, R. J. *et al.* 2010, *MNRAS*, in press (arXiv:1004.5244).
Bate, M. R., Clarke, C. J., & McCaughrean, M. J. 1998, *MNRAS*, 297, 1163
Bazell, D. & Desert, F.X. 1988, *ApJ*, 333, 353
Beech, M. 1987, *Ap&SS*, 133, 193
Beech, M. 1992, *Ap&SS*, 192, 103
Bergin, E. A. & Tafalla, M. 2007, *ARAA*, 45, 339
Bonnell, I. A., Bate, M. R., & Vine, S. G. 2003, *MNRAS*, 343, 413
Caballero, J. A. 2008, *MNRAS 383*, 375
Cartwright, A. & Whitworth, A.P. 2004, *MNRAS*, 348, 589
Chappell, D. & Scalo, J. 2001, *ApJ*, 551, 712
de la Fuente Marcos, R. & de la Fuente Marcos, C. 2006, *A&A*, 452, 163
de la Fuente Marcos, R. & de la Fuente Marcos, C. 2009, *ApJ*, 700, 436
Dickman, R. L., Horvath, M. A., & Margulis, M. 1990, *ApJ*, 365, 586
Efremov, Y.N. 1995, *AJ*, 110, 2757
Elias, F., Alfaro, E. J., & Cabrera-Caño, J. 2009, *MNRAS*, 397, 2
Elmegreen, B. G., 2010, *IAU Symposium*, 266, 3
Elmegreen, B. G. & Hunter, D.A. 2010, *ApJ*, 712, 604
Elmegreen, B. G. & Scalo, J. 2004, *ARAA*, 42, 211
Falgarone, E., Phillips, T.G., & Walker, C.K. 1991, *ApJ*, 378, 186
Gieles, M. 2010, *IAU Symposium*, 266, 6
Gladwin, P. P., Kitsionas, S., & Boffin, H.M.J. *et al.* 1999, *MNRAS*, 302, 305
Goodwin, S. P. & Whitworth, A.P. 2004, *A&A*, 413, 929
Hartmann, L. 2002, *ApJ*, 578, 914
Hetem, A. Jr. & Lepine, J.R.D. 1993, *A&A*, 270, 451
Kraus, A. L. & Hillenbrand, L.A. 2008, *ApJ* (Letters), 686, L111
Lada, C. J. & Lada, E.A. 2003, *ARAA*, 41, 57
Larson, R. B. 1981, *MNRAS*, 194, 809
Larson, R. B. 1995, *MNRAS*, 272, 213
Larson, R. B. 2007, *Reports on Progress in Physics*, 70, 337
Lee, Y. 2004, *Journal of Korean Astronomical Society*, 37, 137
Lee, Y., Kang, M., & Kim, B. K. *et al.* 2008, *Journal of Korean Astronomical Society*, 41, 157
Maíz-Apellániz, J. 2001, *ApJ*, 563, 151
Mandelbrot, B. B. 1983, in: *The Fractal Geometry of Nature* (New York: Freeman)
Maschberger, T., Clarke, C. J., & Bonnell, I. A. *et al.* 2010, *MNRAS*, 404, 1061
Moeckel, N. & Bate, M.R. 2010, *MNRAS*, 404, 721
Nakajima, Y., Tachihara, K., & Hanawa, T. *et al.* 1998, *ApJ*, 497, 721
Sánchez, N., Alfaro, E. J., & Pérez, E. 2005, *ApJ*, 625, 849
Sánchez, N., Alfaro, E. J., & Pérez, E. 2007a, *ApJ*, 656, 222
Sánchez, N., Alfaro, E. J., & Elias, F. *et al.* 2007b, *ApJ*, 667, 213
Sánchez, N. & Alfaro, E. J. 2008, *ApJS*, 178, 1
Sánchez, N. & Alfaro, E. J. 2009, *ApJ*, 696, 2086
Schmeja, S. & Klessen, R. S. 2006, *A&A*, 449, 151
Schmeja, S., Kumar, M. S. N., & Ferreira, B. 2008, *MNRAS*, 389, 1209
Simon, M. 1997, *ApJ* (Letters), 482, L81
Stutzki, R. 1993, *Reviews in Modern Astronomy*, 6, 209
Vogelaar, M. G. R. & Wakker, B. P. 1994, *A&A*, 291, 557

Computational Star Formation
Proceedings IAU Symposium No. 270, 2011
J. Alves, B.G. Elmegreen, J.M. Girart, V. Trimble, eds.

doi:10.1017/S1743921311000226

Observations of the IMF in clusters

Joana Ascenso[1]

[1] Centro de Astrofísica da Universidade do Porto, Rua das Estrelas, 4150-762 Porto, Portugal
email: `jascenso@astro.up.pt`

Abstract. Stars form from molecular clouds, mostly in clusters with tens to tens of thousands of members, and the mass distribution within these clusters, or the Initial Mass Function, seems to be invariable against many parameters and over a wide range of masses. However, masses are a very difficult quantity to assess, and the precision of our determinations of the IMF is systematically lower than usually quoted. I will discuss the process of determining masses from observations and the type of uncertainties associated with this process.

Keywords. techniques: photometric, stars: fundamental parameters, stars: luminosity function, mass function, (Galaxy:) open clusters and associations: general

---

1. Introduction

The Initial Mass Function (IMF) describes the distribution of stellar masses at birth for a single star formation event. The significance of understanding the IMF is transversal to most fields of modern astronomy: not only does it help to constrain the process of star formation, but it is also all important as an input parameter for studies of galactic populations and evolution, chemistry of the interstellar (and intergalactic) medium, and cosmology (e.g., the STARBURST99 code, by Leitherer *et al.* (1999), has been used extensively in studies based on assumptions of the IMF).

The IMF was first introduced by Salpeter in 1955 when characterizing the field population of the Galaxy, and it was then defined as a power-law of index -1.35, since known as the Salpeter slope. In the decades that followed, the IMF was measured in all possible environments, including stellar clusters and associations featuring a wide range of ages, age ranges, masses, stellar densities, morphologies, massive star content and metallicities, both in our Galaxy and in other galaxies, some of which with intense starburst activity (e.g. Bastian *et al.* 2010, and references therein). These studies found that the Salpeter power-law breaks around a characteristic mass of $0.3 - 0.8$ $M_{\odot}$, and the IMF becomes flatter, although still rising, until it starts to decline in the substellar regime (Kroupa 2001; Chabrier 2003). Although the exact form of the subsolar IMF is still a matter of active debate, all observational evidence suggests the IMF is invariable over a wide range of densities, environments, and metallicities, at least in the high-mass range. This is becoming more and more clear as new studies using more sophisticated technology and methods reveal "universal" IMFs even in regions previously said to have peculiar mass functions (e.g. Espinoza *et al.* 2009).

Given the predominance of non-observers in this meeting, I chose to focus on the actual process of determining the IMF from observations rather than reviewing the actual results. The reader is referred to the recent excellent reviews of the "observed" IMF by, e.g., Bastian *et al.* (2010); Elmegreen (2009); Clarke (2009); Elmegreen (2005); Chabrier (2005); Luhman (2004); Lada & Lada (2003).

This contribution is not meant to describe all possible ways to solve the main steps toward determining masses from observations. Instead, I will outline the process in moderate detail describing the most commonly used techniques, while emphasizing the associated difficulties and challenges, and conclude with the argument that, in face of so many, often unquantifiable, sources of uncertainty, it is only expected that the IMF shows scatter from region to region, even if it *is* universal.

1.1. *The IMF in clusters*

Stellar clusters are widely accepted as units of star formation. They are the immediate outcome of the process of star formation from molecular clouds, and as such are the ideal end-product probes of the initial conditions for star formation in the current era of near-infrared telescopes and surveys (Lada & Lada 2003). The main advantages of studying the IMF in young clusters follow from their group nature: they are statistically significant samples of coeval, equidistant, and volume-limited stars that share the properties of a common parent molecular cloud, such as metallicity, and initial conditions (pressure, temperature, ambient radiation, magnetic fields, etc.), and are thus ideally suited for the study of a group property such as the IMF. Although the IMF was studied thoroughly in all types of clusters – globular clusters (a few Gyr), open clusters (a few $\times 10$ Myr), and embedded clusters (a few Myr) –, the latter, embedded clusters, present the most favorable conditions for the characterization of the IMF. Being the youngest, their present-day mass function is closest to their initial mass function, as they have not yet had time to lose members to stellar or dynamical evolution, and so no additional corrections, often sources of uncertainty, are needed. Also, their young stars are still bright in the near-infrared, and the stellar mass spectrum presents a much smaller dynamic range in brightness when compared to equivalent observations in the optical, which favors the completeness of the observed samples. Finally, their compactness and embedded state provide a natural shield against background contamination by unrelated sources, thus minimizing a significant potential bias when studying the properties of the cluster.

For these reasons, I will focus on observations of the IMF in young clusters, rather than reviewing the measurements done in all other environments. Many of the methods and challenges are, nonetheless, common to all determinations of mass.

1.2. *Measuring the IMF*

The determination of the IMF of any given population depends on our ability to determine masses. Unfortunately, mass is not a direct observable, hence the difficulty of deriving accurate and consistent IMFs for different regions. The determination of stellar masses relies heavily on assumptions, on models, and on the knowledge of other properties like distance and age that are, by themselves, very challenging to ascertain.

The most accurate way to derive stellar masses is through high-resolution spectroscopy, preferably associated to photometric data, covering a wide range of wavelengths. Spectra allow for the determination of spectral type and evolutionary status of individual stars. Single-band photometry, when combined with the spectral information, helps to assess membership and to determine the distance, whereas the multi-wavelength coverage allows for the construction of a spectral energy distribution, leading to the characterization of the circumstellar material around the star. The spectroscopic approach, although reliable, assuming our knowledge of pre-main-sequence spectral features is accurate, is extremely inefficient and expensive in telescope time. It requires each cluster member to be observed individually, in many bands, and using different instruments and setups, preferentially simultaneously to prevent errors associated with the ubiquitous variability of young stars. Also, spectroscopy of the faint members requires long integration times, even in the largest telescopes, all translating into impractical prerequisites for clusters with more than a few members and more distant than just a few kpc.

The best alternative is to use broad-band photometry in several wavelengths and infer the properties of the individual stars based on their brightness and color. This approach is relatively inexpensive, as it provides useful information about all cluster members in just a few images with the same telescope time investment, meaning less telescope time for more stars. Photometry, by definition, will also reach fainter stars with shorter integration times when compared to spectroscopy. The drawback is that the information given by broad-band photometry is limited, and more needs to be inferred. Using the brightness and color of each cluster member, one can construct a luminosity function that can then be transformed into a mass function using the appropriate mass-luminosity relation. Often, it pays to have spectra of at least the few brightest members to help constrain the distance and age of the cluster, assuming coevality.

In the following sections I will describe the difficulties associated with deriving the IMF from broad-band, multi-wavelength photometry of young, embedded clusters.

2. Observational challenges

2.1. *Membership assessment*

Most molecular clouds and young clusters of our Galaxy are located in the galactic disc. Stars from the field, that are physically unrelated with the cluster, are therefore likely to be present in any image of a cluster, and to be included in the sample as part of the cluster population. But including these stars in the sample is equivalent to observing a mass function that is the convolution of the field present-day mass function with the cluster IMF, worsened by the fact that we will most likely detect a disproportionate amount of background giants, for example, if one is observing in the infrared. Discerning which stars in an image are actually physically associated with the cluster is therefore determinant to an accurate analysis of the cluster properties, and in particular of the IMF.

There are several ways to disentangle the cluster stars from the foreground/background contaminants, each requiring specific observations and analysis. One of the most accurate methods of assessing membership relies on proper motion studies: by observing the cluster at several epochs so that its stars can be seen moving against the background, one can measure the projected velocity of the individual stars. Assuming that the cluster members have similar velocities it is possible to establish the membership of each star individually. Although accurate, this method often requires prohibitively large time baselines, especially if the region of interest is distant, and multiple very high-resolution observations of the same region, which is not easy to obtain given the constraints in telescope time.

The previous method is most frequently used for nearby open clusters (e.g., Caballero 2010; Krone-Martins *et al.* 2010), where membership assessment is hardest. In young, embedded clusters, although it would still be possible to use the dynamical method described above (e.g. Stolte *et al.* 2008) , there are other, less expensive options that can simultaneously be applied to clusters farther away from the Sun. If a cluster is embedded, then the foreground objects will be at "zero" extinction, making them relatively easy to identify. Similarly, the few background stars that will still be detected through the cloud will be behind an extinction "wall", allowing for their identification, in principle, based on extinction arguments. However, all cluster members do not present one single value of extinction, but rather a continuum of extinction values around a mean deriving from the patchy nature of the cloud, making it difficult to determine at which value of extinction a star is no longer a member of the cluster. Also, if the cloud is only thick enough to embed the cluster, then the shielding of background sources becomes less effective. Using ages – most cluster stars are still on the pre-main-sequence, making their identification less challenging using multi-band photometry alone and/or X-ray data – and the related presence of circumstellar disks, one can furthermore separate out the young from the old, main-sequence field objects in the image, and attribute membership to all objects that match a reasonable age for the cluster. Moreover, not all stars will have circumstellar material, nor will they be all in the pre-main-sequence, making the age assessment more difficult.

For these reasons, most studies use statistical methods instead to subtract the contamination from field stars, by observing a nearby control field at the same galactic latitude with the exact same setup. Once properly reddened by the same amount of (mean) extinction or, if available, by the extinction profile of the cloud, this sample becomes a good statistical representation of the foreground/background population toward the science field. Since the IMF is most commonly derived from a single-band luminosity function, one can statistically subtract a reddened control-field population from the science sample, thus correcting for contamination. Although it does not provide individual memberships, this method does provide a reasonable correction from contaminants for IMF purposes.

2.2. *Completeness*

Our ability to detect all stars in a cluster determines the accuracy of our IMF. Stars can go undetected for several reasons, both observational and characteristic of the star forming region itself. In this section I will describe the various parameters that can affect the completeness of the observations, and ultimately bias the IMF.

2.2.1. *Sensitivity*

The most obvious consequence of having time-limited observations is that we cannot detect objects fainter than some brightness limit. Due to exposure time and to the intrinsic characteristics of the detector, we will only be sensitive to signal that falls above some brightness threshold, compromising our ability to sample the IMF to the lowest mass range. To a first approximation, knowing the completeness limit of a sample will delimit the mass range over which one feels confident that the IMF is fully sampled, but many authors use this information further to correct their sample of incompleteness by dividing the observed mass function by the completeness profile (e.g. Stolte *et al.* 2002, but see also discussion in Ascenso *et al.* (2009)).

The best way to estimate the completeness of a given sample is to perform artificial stars experiments. These consist in adding artificial stars of different brightnesses to the observed cluster image and trying to recover them with the correct brightness using the exact same method used for the detection and photometry of the science stars (Ascenso *et al.* 2009). This is done in most studies of clusters and the IMF, the exact details of the method varying from study to study.

2.2.2. *Resolution*

Despite the great advances in the past decades, resolution is still a limiting factor in modern observations. The farthest and the denser the star forming region, the more difficult it is to have a comprehensive view of the full stellar population, even above the sensitivity limit. The first consequence of not being able to resolve (visual) multiples is that the luminosities of "individual" objects, that actually comprise two or more stars, are overestimated, counting as more massive stars. Although it is not a big increase in brightness, and therefore in mass, for a, say, 20 $M_\odot$ star to be blended with a low-mass object, it will make a significant difference for a low-mass object to be blended with an equal mass object, especially if one considers that the mass bins of the IMF are usually logarithmic. The other consequence is one of demographics: blending stars means we lose cluster members. This incompleteness effect is mass-dependent, and thus may bias the IMF.

The most advanced observatories and space telescopes are less affected by this problem, the first due to adaptive optics technology, and the latter to being above the atmosphere, being almost only limited by diffraction. As before, the best solution to blending would be to analyse all stars spectroscopically, looking for signatures of unresolved neighbors. Alternatively, by performing artificial star experiments, one can correct the sample globally (c.f. 2.2.1).

2.2.3. *Variable extinction*

Perhaps the major downside of having a cluster embedded in a molecular cloud is, since the cloud is often patchy, that each star is likely subject to a different amount of extinction. This causes problems of variable completeness in the field, as the extinction dims the brightness of the stars, inevitably causing some to fall below the detection limit. This effect is more important for the youngest clusters, where the cloud is still well within the cluster, permeating the intracluster medium. As the cluster evolves and the stars clear most of the gaseous and dusty material this problem becomes less significant.

The effect of variable extinction in a cluster is somehow a function of mass: whereas the brightest, more massive stars will still be detected as reddened objects, the faint, low-mass stars will most likely be dimmed away from detection, causing the IMF to be incomplete toward the lower-mass end. This effect is very difficult to quantify without an extinction map of the region, which is usually itself very challenging to obtain for regions of star formation since there is not a clear view of the background stars. Since it is possible to estimate the extinction to each individual detected object, one can at most use that information to estimate which mass bins would most likely be affected by the variable extinction. This would always be an incomplete assessment, as one can only measure the extinction caused by the material in front of any specific star, and not the full length of extinction any cluster star is potentially subject to.

2.2.4. *Variable nebula*

Similar to the effect of variable extinction is the effect of having a bright nebula of varying intensity. This is seen frequently in embedded clusters with one or several bright, massive stars that ionise the gas in the intracluster medium, making it bright. The nebula then acts as an extended emission that lowers our ability to detect faint stars by reducing the contrast between the background and the object.

Again, this affects mostly the low-mass end of the mass spectrum, but, unlike the incompleteness due to variable extinction, this can be quantified, namely using the artificial-star completeness tests mentioned above. However, given the usually patchy nature of the nebula, and the fact that the brightness may vary significantly from region to region, the statistical approach of the artificial stars experiments may not be effective as a correction tool: if the areas affected by the nebula are locally small, then it is likely that they are hiding only a small number of stars, making the correction too susceptible to small number statistics.

2.3. *Interstellar Extinction and Circumstellar Material*

Estimating the mass of any embedded source requires, among other things, that the extinction is known or can be derived. Using multi-band photometric observations, this can be done by comparing the color of an object with that of a similar, unreddened object. This implies that we know two things: (1) the underlying, intrinsic nature of the object, and (2) the way extinction acts on brightness and color, *i.e.*, the extinction law.

Determining the intrinsic color of a young object is always a challenge, both because of extinction, and because, depending on the exact age of the cluster, low-mass objects will have circumstellar discs and/or envelopes that increase the object's brightness and further redden its colors. The reddening from the circumstellar material, however, acts on a different direction in a color-color diagram than does the interstellar reddening, so it is possible, at least to some degree, to disentangle the two. Determining the amount of excess emission from an object is crucial to estimating its mass: if a star appears brighter than it actually is due to emission from circumstellar material, its mass will be overestimated, potentially biasing the IMF. To this end, spectroscopy and/or the analysis of the spectral energy distribution (SED) of each object helps to determine the fraction of light that is emitted by the circumstellar material, although building an SED also requires having observations covering a large spectral range, while spectroscopy of many objects is prohibitively time-consuming. The solution is often to de-excess and de-redden each object by, first, roughly estimating the emission from circumstellar material using empirical models (Meyer *et al.* 1997), and then assume (pre-)main-sequence photosphere colors as comparison standards to estimate the extinction.

This last step requires the knowledge of the extinction law. For a long time believed to be universal, it has recently been found that the law may be density-dependent (e.g, Román-Zúñiga *et al.* 2007). The relevant departures from the "universal" law have mostly been found in the mid-infrared regime, and have been proposed to originate in different grain composition and size: in denser areas, the grains are expected to grow and coalesce thus producing different signatures in the extinction properties. The *near-infrared* extinction law appears not to be significantly affected by density. Ideally, one would derive the extinction law for each specific region of star formation, but this would require access to a pristine background population in the direction of the cluster rather than the typical zoo of young objects, ionizing nebula, *and* reddened objects typically found toward a star forming region. The "universal" law (Rieke & Lebofsky 1985) is still used in most studies of embedded clusters.

2.4. *Mass-luminosity relation*

The last step toward a catalog of masses is to derive the actual masses from the observed luminosities. The difficulty here is to find suitable models and have a good estimate of the age of the stars. Massive and low-mass stars evolve on different timescales, so, even in the (unlikely) case of a perfectly coeval population, there will still be massive main-sequence stars, and lower-mass pre-main-sequence objects, the exact fraction of which depending on the age. We must therefore use a combination of both main-sequence (e.g., Lejeune & Schaerer 2001) and pre-main-sequence models (e.g., Palla & Stahler 1993; Baraffe *et al.* 1998), adjusted to whatever age and distance we derive for the cluster. Although the current models for pre-main-sequence objects still have many uncertainties, age and distance estimates are by far the greatest sources of error in the determination of the mass-luminosity relation, and therefore, in the determination of mass.

2.4.1. *Age*

Age estimates rely on placing the cluster members on HR diagrams, and comparing their position with theoretical evolutionary tracks. This is usually done using color information and/or spectra of individual stars, with the caveat that, as mentioned before, the presence of circumstellar material can produce deceiving colors and lead to erroneous age estimates, not to mention the effect on the mass determination itself. The pre-main-sequence models are continuously being improved, and already seem to capture the essence of pre-main-sequence objects, but factors like accretion have been shown to have a high impact on the models (Hartmann *et al.* 1997; Baraffe *et al.* 2009).

The presence of short-lived, massive pre-main-sequence stars in the cluster can also constrain the age if one assumes coevality. The expected lifetime of a massive star in the main-sequence can, in this case, be used as an upper limit for the cluster age. Other youth indicators may also help constrain the approximate age of the cluster. It is expected that the embedded phase lasts 3 to 5 Myr (Lada & Lada 2003), during which time the stars will effectively clear the remains of the molecular cloud. The existence of class I sources, the presence of jets, or phenomena, such as proplyds, believed to be short-lived, also testify to youth. A linear relation of disc frequency with age has been proposed by Haisch *et al.* (2001), but it is not yet solid enough to serve as a reliable constraint (Mayne *et al.* 2007). In any case, one can never expect to be accurate to more than 1 or 2 Myr, at best.

Once the age is known to whatever accuracy, it is possible to estimate which fraction of stars are likely main-sequence, and which are pre-main-sequence, but even then, there is still some degeneracy, in that two pre-main-sequence stars may have the same brightness even though their mass is (slightly) different (see discussion in Ascenso *et al.* 2007). This degeneracy could, in principle, be lifted by the colors, but the differences will most often be too subtle to prevail over observational errors, extinction, and circumstellar material. Finally, the mass-luminosity relation for subsolar-mass objects is still very poorly characterized, which would generate further uncertainty in this mass range even if the age was accurately known.

2.4.2. *Distance*

Distance, on its turn, can be constrained using either trigonometric parallax, for nearby regions, or, most frequently, spectroscopic parallax. The latter consists in comparing the observed brightness with that expected for a star, with its spectroscopically-derived spectral type. Alternatively, the distance can be constrained using dynamical arguments and models of the Galaxy to derive "kinematic distances" (Roman-Duval *et al.* 2009), or even using the density of foreground stars to a molecular cloud or cluster (Lombardi *et al.* 2010).

2.5. *What is a cluster?*

If one defines the IMF as the initial mass function of a single event of star formation, then one would need to observe all stars from one cluster or entire cloud in order to have a good representation of the IMF, depending on the definition of "single event of star formation" they choose. However, a cluster is neither believed to be coeval, nor does it have defined boundaries. This somewhat more philosophical question goes back to the understanding of the very process of star formation, either in clusters or in (relative) isolation throughout the cloud. Since clusters are (still?) believed to be the basic units of star formation, and given the constraints on telescope time, most studies of young populations have analysed single clusters rather than entire clouds, although the paradigm seems to be changing with the improvement of large surveys (e.g., Jørgensen *et al.* 2006). The question of cluster boundaries is pertinent even if one simplifies the analysis by considering single clusters: some numerical simulations of star forming clouds suggest that, along with the identifiable cluster, a significant population of stars formed in the same event are scattered over large areas (Bonnell *et al.* 2010). Rapid dynamical phenomena have also been theoretically described by which a cluster can rapidly disperse a fraction of its members (Binney &

Tremaine 1987). The gas removal, around 3 to 5 Myr into the cluster's age, will also unbound the cluster, leading to a rapid and effective scattering of cluster members over large areas around the identifiable cluster, or to the disruption of the cluster altogether (Bastian & Goodwin 2006). All these possibilities are somehow mass-specific, introducing potential systematic errors when typically observing and characterizing the IMF of single clusters. This problem will be more significant the more effective these processes are in scattering the cluster population spatially, but will be less important the denser and more massive the cluster is, as most of its mass and members will, in principle, reside in the identifiable and gravitationally more bound cluster core.

Large multi-wavelength surveys are and will continue to be helpful in identifying cluster members residing outside the core, thus contributing immensely to our understanding of the star formation process.

3. Technical challenges

Finally, after having a trusted sample of masses, one needs to characterize the functional form of the mass distribution. The two most widely used IMFs are that of Chabrier (2003), consisting of a (Salpeter) power-law above $\sim 1\ M_\odot$ and a log-normal for lower masses; and that of Kroupa (2001), consisting of a three-segment power-law. Both are quasi-equivalent in shape and parameters.

These functions are usually fitted to a histogram of masses, obtained, traditionally, by binning the masses in equal-width dex bins, but this approach has been shown to be biased and should therefore be avoided (Maíz Apellániz & Úbeda 2005). Instead, and according to these authors, the IMF must be built using bins with equal number of stars, therefore dissipating the bias entirely. Alternatively, one can use the modified maximum likelihood method proposed recently by Maschberger & Kroupa (2009).

The fitting technique itself would be a matter of a whole different paper entirely and will not be addressed here.

4. Summary

This contribution describes the difficulties in deriving masses and mass functions from observational data, with special emphasis on the problematics of young clusters. Considering all these sources of uncertainty, it is probably not surprising that the initial mass function has turned out to be universal, even when measured in such different environments. On the other hand, even considering the uncertainties, it is clear that it also cannot be completely different from region to region. Although we are not (yet) in position to say that the IMF is strictly universal, observations continue to show a consistent overall shape of the IMF, as well as a consistent characteristic mass, hinting at some degree of universality reflecting the process of star formation.

It is clear from the literature that we are still struggling with methodology and instrument limitation issues, as one sometimes sees the same region of star formation showing different results with every new publication, but convergence seems to be closer than ever, especially for the high-mass end. Although benefitting from the lessons learned for the high-masses over the last six decades, the subsolar mass function is still taking its first serious steps (e.g., Luhman *et al.* 2009). Although naturally more prone to error and uncertainties, new, deep surveys of entire clouds are revealing the lower-mass end of star formation with unprecedented completeness and detail. The first results already hint at some consistency also in this regime, but the jury is still out on whether the low-mass end of the IMF is as universal as the high-mass end appears to be.

References

Ascenso, J., Alves, J., Beletsky, Y., & Lago, M. T. V. T. 2007, *A&A*, 466, 137
Ascenso, J., Alves, J., & Lago, M. T. V. T. 2009, *A&A*, 495, 147
Baraffe, I., Chabrier, G., Allard, F., & Hauschildt, P. H. 1998, *A&A*, 337, 403
Baraffe, I., Chabrier, G., & Gallardo, J. 2009, *ApJ*, 702, L27
Bastian, N., Covey, K. R., & Meyer, M. R. 2010, *ARA&A*, 48, 339
Bastian, N. & Goodwin, S. P. 2006, *MNRAS*, 369, L9
Binney, J. & Tremaine, S. 1987, Galactic dynamics (Princeton, NJ, Princeton University Press, 1987, 747 p.)
Bonnell, I. A., Smith, R. J., Clark, P. C., & Bate, M. R. 2010, ArXiv e-prints
Caballero, J. A. 2010, *A&A*, 514, A18+
Chabrier, G. 2003, *PASP*, 115, 763
Chabrier, G. 2005, in Astrophysics and Space Science Library, Vol. 327, The Initial Mass Function 50 Years Later, ed. E. Corbelli, F. Palla, & H. Zinnecker, 41–+
Clarke, C. J. 2009, *Ap&SS*, 324, 121
Elmegreen, B. G. 2005, in Astrophysics and Space Science Library, Vol. 329, Starbursts: From 30 Doradus to Lyman Break Galaxies, ed. R. de Grijs & R. M. González Delgado, 57–+
Elmegreen, B. G. 2009, in The Evolving ISM in the Milky Way and Nearby Galaxies
Espinoza, P., Selman, F. J., & Melnick, J. 2009, *A&A*, 501, 563
Haisch, Jr., K. E., Lada, E. A., & Lada, C. J. 2001, *ApJ*, 553, L153
Hartmann, L., Cassen, P., & Kenyon, S. J. 1997, *ApJ*, 475, 770
Jørgensen, J. K., Harvey, P. M., Evans, II, N. J., *et al.* 2006, *ApJ*, 645, 1246
Krone-Martins, A., Soubiran, C., Ducourant, C., Teixeira, R., & Le Campion, J. F. 2010, *A&A*, 516, A3+
Kroupa, P. 2001, *MNRAS*, 322, 231
Lada, C. J. & Lada, E. A. 2003, *ARA&A*, 41, 57
Leitherer, C., Schaerer, D., Goldader, J. D., *et al.* 1999, *ApJs*, 123, 3
Lejeune, T. & Schaerer, D. 2001, *A&A*, 366, 538
Lombardi, M., Lada, C. J., & Alves, J. 2010, *A&A*, 512, A67+
Luhman, K. L. 2004, in IAU Symposium, Vol. 221, Star Formation at High Angular Resolution, ed. M. G. Burton, R. Jayawardhana, & T. L. Bourke, 237–+
Luhman, K. L., Mamajek, E. E., Allen, P. R., & Cruz, K. L. 2009, *ApJ*, 703, 399
Maíz Apellániz, J. & Úbeda, L. 2005, *ApJ*, 629, 873
Maschberger, T. & Kroupa, P. 2009, *MNRAS*, 395, 931
Mayne, N. J., Naylor, T., Littlefair, S. P., Saunders, E. S., & Jeffries, R. D. 2007, *MNRAS*, 375, 1220
Meyer, M. R., Calvet, N., & Hillenbrand, L. A. 1997, *AJ*, 114, 288
Palla, F. & Stahler, S. W. 1993, *ApJ*, 418, 414
Rieke, G. H. & Lebofsky, M. J. 1985, *ApJ*, 288, 618
Roman-Duval, J., Jackson, J. M., Heyer, M., *et al.* 2009, *ApJ*, 699, 1153
Román-Zúñiga, C. G., Lada, C. J., Muench, A., & Alves, J. F. 2007, *ApJ*, 664, 357
Salpeter, E. E. 1955, *ApJ*, 121, 161
Stolte, A., Ghez, A. M., Morris, M., *et al.* 2008, *ApJ*, 675, 1278
Stolte, A., Grebel, E. K., Brandner, W., & Figer, D. F. 2002, *A&A*, 394, 459

Lise Deharveng and Joana Ascenso

Computational Star Formation
Proceedings IAU Symposium No. 270, 2011
J. Alves, B.G. Elmegreen, J. M. Girart & V. Trimble, eds.

doi:10.1017/S1743921311000238

Insights on molecular cloud structure

João Alves[1], Marco Lombardi[2,3], Charles Lada[4]

[1]University of Vienna, Türkenschanzstrasse 17, 1180 Vienna, Austria
[2]ESO, Karl-Schwarzschild-Strasse 2, D-85748 Garching bei München, Germany
[3]University of Milan, Department of Physics, via Celoria 16, I-20133 Milan, Italy (on leave)
[4]Harvard-Smithsonian Center for Astrophysics, MS42, 60 Garden St., Cambridge, MA 02138
email: joao.alves@univie.ac.at

Abstract. Stars form in the densest regions of clouds of cold molecular hydrogen. Measuring structure in these clouds is far from trivial as 99% of the mass of a molecular cloud is inaccessible to direct observation. Over the last decade we have been developing an alternative, more robust density tracer technique based on dust extinction measurements towards background starlight. The new technique does not suffer from the complications plaguing the more conventional molecular line and dust emission techniques, and when used with these can provide unique views on cloud chemistry and dust grain properties in molecular clouds. In this brief communication we summarize the main results achieved so far using this technique.

Keywords. ISM: clouds, dust, extinction, structure, stars: formation

1. Introduction

Stars are the outcome of the collapse of cold molecular clouds and yet we know much more about stellar structure than we know cloud structure. This is not surprising giving that clouds are not only the coldest known objects in the Universe but are made of molecular hydrogen (H_2) and helium both inaccessible to direct observation at the typical temperatures in these clouds (~10 K). Instead, observers have been deriving the basic physical properties of these objects from observations of H_2 trace molecules such as CO and NH_3, as well as dust emission. Unfortunately, decoding observations of trace molecules or dust emission is not a straightforward task and several complications plague the derivation of basic cloud physical properties. An alternative way to derive cloud properties, and in particular cloud density structure, is to map the dust column density in these clouds, as dust is a robust surrogate of the molecular hydrogen. We have been developing a technique that makes use of reddened starlight seen through these clouds†. This technique produces column density maps that are more reliable and have a larger dynamic range than either dust emission or molecular line emission maps. Because dust extinction is such a robust tracer of mass (Goodman *et al.* 2009), the correlation between dust extinction and molecular line/dust emission maps provides unique views on the chemistry and dust grain properties in these clouds.

2. Large Scale Extinction Maps

With the availability of large near-infrared surveys, in particular the all-sky 2MASS survey, it became possible to apply the NICEST technique to essentially the entire sky. We took a systematic approach to produce the highest resolution maps possible of the most nearby cloud complexes, where confusion along the line-of-sight is minimal and

† The technique, first introduced by Lada *et al.* (1994), has had three improved versions: NICE (Alves *et al.* 1998), NICER (Lombardi & Alves 2001), and NICEST (Lombardi 2009)

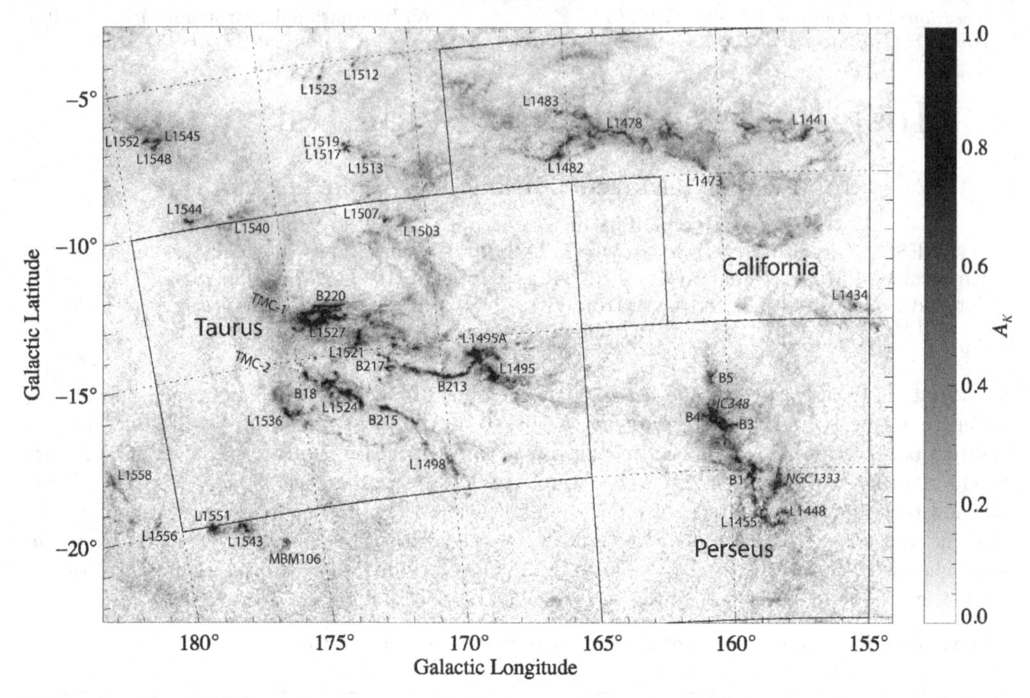

Figure 1. The NICER extinction map of the Taurus, Perseus, California complexes made from millions of background 2MASS stars (Lombardi *et al.* 2010b). The spatial resolution of the map is 2.5 arcmin.

distances are fairly well known (essentially objects belonging to the Gould's belt, e.g., Pipe Nebula, Ophiuchus, Lupus, Taurus, Perseus, California, and Orion). These maps (see Figure 1 for an example) provide an extraordinary view on these objects, from their largest, tens of parsec scales down to scales of about 2000 AU (e.g. Román-Zúñiga *et al.* 2010), and with unprecedented dynamic ranges of $10^{21} < \mathrm{N} < 10^{23}$ (cm^{-2}). One of the main results from these large-scale maps of nearby complexes is that the same cloud description seems to be emerging repeatedly: all the nearby complexes are (large) filaments, with aspect ratios of the order of 10-100:1. At close inspection these filaments seem to be made of entanglements of narrower filaments (the streamers of Ophiuchus being the best example) loosely suggesting a self-similar organization. Apart a mass scale, the largest complexes (e.g., Orion) seem to have the same structure as the smaller complexes (e.g., Pipe). Another important cloud structure result is that about 90% of the mass of the nearby clouds lies at column densities below $A_V \sim 5$ mag, having little chance of ending up in the young stars forming in these clouds.

3. The Universality of Molecular Cloud Structure

An important result arising from the large-scale extinction maps is that different clouds have almost identical average column densities above a given extinction threshold (this statement holds regardless of the extinction threshold), indicating that molecular clouds are characterized by a universal structure (Lombardi *et al.* 2010a), a stringent confirmation of Larson's 3rd law. An apparently independent result that has become obvious from dust extinction maps is that the column density distribution in these clouds can be well fitted by log-normal distributions, e.g., Lombardi *et al.* (2008); Kainulainen *et al.* (2009); Froebrich & Rowles (2010). While it is not clear yet what is the ultimate relevance

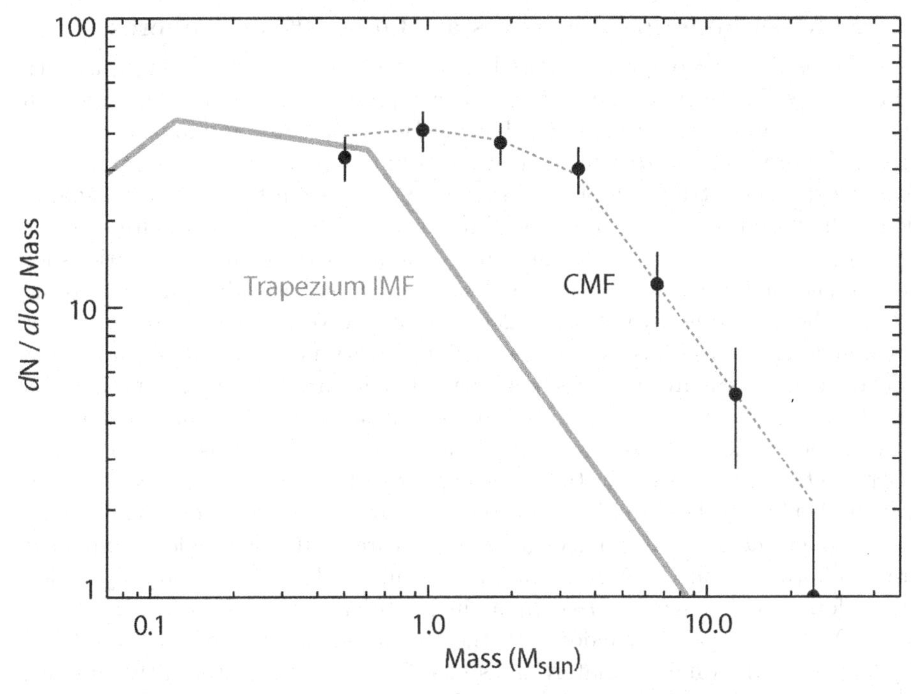

Figure 2. Mass function of dense molecular cores for the Pipe Nebula plotted as filled circles with error bars (Alves *et al.* 2007). The grey line is the stellar IMF for the Trapezium cluster (Muench *et al.* 2002). The dashed grey line represents the stellar IMF in binned form matching the resolution of the data and shifted to higher masses by about a factor of 4. The dense core mass function is similar in shape to the stellar IMF function, apart from a (uniform) star formation efficiency factor ($\sim$30%).

of this result (Goodman *et al.* 2009; Tassis *et al.* 2010), we showed that the universal structure of these clouds can be linked to the log-normal nature of cloud column density distributions.

4. Mass Distribution and the Star Formation Rate (SFR)

Despite having the same overall structure, the fraction of the mass of a cloud at high column density varies immensely from cloud to cloud. For example, the California Nebula molecular cloud (Lada *et al.* 2009), while rivaling the Orion A molecular cloud as the largest and most massive GMC in the solar neighborhood, displays significantly less star formation activity with more than an order of magnitude fewer young stellar objects than found in the OMC. Analysis of extinction maps of both clouds shows that the California cloud contains only 10% the amount of high extinction ($A_K > 1.0$ mag) material as is found in the OMC, suggesting that the level of star formation activity and perhaps the star formation rate in these two clouds may be directly proportional to the total amount of high extinction material. Recently, Lada *et al.* (2010) found that the star formation rate (SFR) in molecular clouds is linearly proportional to the cloud mass ($M_{0.8}$) above an extinction threshold of $A_K \approx 0.8$ magnitudes, corresponding to a gas surface density threshold of $\Sigma_{gas} \approx 116\ M_\odot\ pc^2$ corresponding to a volume density threshold of $n(H_2) \approx 10^4\ cm^3$ [SFR ($M_\odot yr^1$) $= 4.6 \pm 2.6 \times 10^8\ M_{0.8}\ (M_\odot)$]. This relation between the rate of star formation and the amount of dense gas in molecular clouds appears to be in excellent agreement with previous observations of both galactic and extragalactic star forming activity (e.g. Wu *et al.* 2005).

5. The Core Mass Function and cores as Bonnor-Ebert spheres

Moving to the small scales one finds a population of well defined extinction peaks with high column density. These column density peaks compare well in size and density with the dense ammonia cores first discussed in Myers & Benson (1983), and appear to have, at least in the Pipe Nebula, a spatial distribution similar to the stars in Taurus, suggesting that the spatial distribution of stars evolves directly from the primordial spatial distribution of high density peaks (Román-Zúñiga *et al.* 2010). The structure of these dense cores, in particular the higher density ones, is well described by the equations for a pressure-confined, self-gravitating isothermal sphere that is critically stable according to the Bonnor-Ebert criteria (Alves *et al.* 2001). Using wavelet techniques to extract these cores we constructed a mass spectrum of high density peaks, or dense cores, and found evidence for a departure from a single power-law form in the mass function of a population of cores and find that this mass function is surprisingly similar in shape to the stellar IMF but scaled to a higher mass by a factor of about 3 as can be seen in Figure 2 (Alves *et al.* 2007). Most of the cores appear to be pressure confined, gravitationally unbound entities whose fundamental physical properties are determined by only a few factors, which include self-gravity, gas temperature, and the simple requirement of pressure equilibrium with the surrounding environment. The entire core population in the Pipe is found to be characterized by a single critical Bonnor-Ebert mass of approximately 2 $M_\odot$. This mass coincides with the characteristic mass of the Pipe CMF suggesting that the CMF (and ultimately the stellar IMF) has its origin in the physical process of thermal fragmentation in a pressurized medium.

References

Alves, J., Lada, C. J., Lada, E. A., Kenyon, S. J. & Phelps, R. 1998, ApJ, 506, 292
Alves, J., Lombardi, M. & Lada, C. J. 2007, A&A, 462, L17
Alves, J. F., Lada, C. J. & Lada, E. A. 2001, Nat, 409, 159
Froebrich, D. & Rowles, J. 2010, MNRAS, 406, 1350
Goodman, A. A., Pineda, J. E. & Schnee, S. L. 2009, ApJ, 692, 91
Kainulainen, J., Beuther, H., Henning, T. & Plume, R. 2009, A&A, 508, L35
Lada, C. J., Lada, E. A., Clemens, D. P. & Bally, J. 1994, ApJ, 429, 694
Lada, C. J., Lombardi, M. & Alves, J. F. 2009, ApJ, 703, 52
—. 2010, ArXiv e-prints
Lombardi, M. 2009, A&A, 493, 735
Lombardi, M. & Alves, J. 2001, A&A, 377, 1023
Lombardi, M., Alves, J. & Lada, C. J. 2010a, A&A, 519, L7
Lombardi, M., Lada, C. J. & Alves, J. 2008, A&A, 489, 143
—. 2010b, A&A, 512, 67
Muench, A. A., Lada, E. A., Lada, C. J. & Alves, J. 2002, ApJ, 573, 366
Myers, P. C. & Benson, P. J. 1983, ApJ, 266, 309
Román-Zúñiga, C. G., Alves, J. F., Lada, C. J. & Lombardi, M. 2010, ArXiv e-prints
Tassis, K., Christie, D. A., Urban, A., Pineda, J. L., Mouschovias, T. C., Yorke, H. W. & Martel, H. 2010, MNRAS, 408, 1089
Wu, J., Evans, II, N. J., Gao, Y., Solomon, P. M., Shirley, Y. L. & Vanden Bout, P. A. 2005, ApJL, 635, L173

Computational Star Formation
Proceedings IAU Symposium No. 270, 2011
J. Alves, B. G. Elmegreen, J. M. Girart, V. Trimble

doi:10.1017/S174392131100024X

Submillimeter Array Observations of Magnetic Fields in Star Forming Regions

R. Rao[1], J.-M. Girart[2], and D. P. Marrone[3]

[1]Academia Sinica Institute of Astronomy & Astrophysics, Submillimeter Array, 645 N. Aohoku Place, Hilo, HI 96720, USA
email: `rrao@sma.hawaii.edu`

[2]Institut de Ciències de l' Espai (CSIC-IEEC), Campus UAB – Facultat de Ciències, Torre C5 - parell 2, 08193 Bellaterra, Catalunya, Spain

[3]National Radio Astronomy Observatory & Kavli Institute for Cosmological Physics, University of Chicago, 5640 South Ellis Avenue, Chicago, Illinois, 60637, USA

Abstract. There have been a number of theoretical and computational models which state that magnetic fields play an important role in the process of star formation. Competing theories instead postulate that it is turbulence which is dominant and magnetic fields are weak. The recent installation of a polarimetry system at the Submillimeter Array (SMA) has enabled us to conduct observations that could potentially distinguish between the two theories. Some of the nearby low mass star forming regions show hour-glass shaped magnetic field structures that are consistent with theoretical models in which the magnetic field plays a dominant role. However, there are other similar regions where no significant polarization is detected. Future polarimetry observations made by the Submillimeter Array should be able to increase the sample of observed regions. These measurements will allow us to address observationally the important question of the role of magnetic fields and/or turbulence in the process of star formation.

Keywords. stars: formation, ISM: magnetic fields,i ISM: dust, techniques: polarimetric

---

1. Introduction

In the "classical picture" of star formation, magnetic fields are believed to strongly influence star formation activity in molecular clouds. They provide support to a cloud against gravitational collapse and thus explain the low efficiency of the star formation process (See review by Mouschovias 2001). The process of ambipolar diffusion, in which magnetic flux is redistributed in the cloud, leads to the formation of a core that can no longer be magnetically supported and gravitational collapse sets in. In addition, the process of magnetic braking can help to remove angular momentum and slow down the rotation of the cloud as it collapses. In contrast, there have been a number of alternate theories which postulate that magnetic fields are relatively weak and supersonic magnetohydrodynamic turbulence is the dominant process (Mac Low & Klessen 2004). Turbulence controls the evolutions of clouds, and cores form at the intersection of supersonic turbulent flows. Only a fraction of such cores become supercritical and collapse begins to occur on a gravitational free-fall timescale.

It is therefore imperative to be able to measure the structure and morphology of the magnetic field in regions of active star formation in order to distinguish between the two competing theories. One of the ways of detecting magnetic fields is through observations of polarized dust emission from spinning dust grains (See review by Hildebrand 1988). In the presence of magnetic fields, spinning dust grains in the interstellar medium become aligned such that their short axis becomes parallel to the direction of the magnetic field. The exact nature of the alignment process is still being studied by a number of theorists

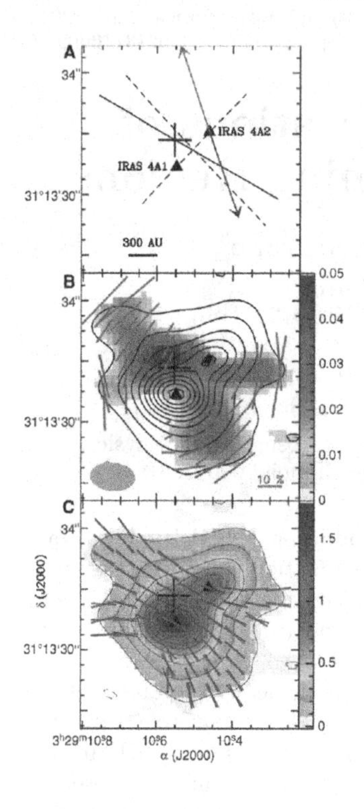

Figure 1. Polarization in NGC 1333 IRAS 4A: (A) Sketch of the axis directions: red/blue arrows show the direction of the redshifted/blueshifted lobes of the molecular outflow, solid lines show the main axis of the magnetic field, and dashed lines show the envelope axes. The solid triangles show the position of IRAS 4A1 and 4A2. The small cross shows the centre of the magnetic field symmetry. (B) Contour map of the 877 μm dust emission (Stokes I) superposed with the color image of the polarized flux intensity. Red vectors: Length is proportional to fractional polarization and the direction is position angle of linear polarization. The synthesized beam is shown in the bottom left corner. (C) Contour and image map of the dust emission. Red bars show the measured magnetic field vectors. Grey bars correspond to the best fit parabolic magnetic field model. Using the residuals from this fit, the strength of the magnetic field can be estimated with the Chandrasekhar-Fermi method.

(See review by Lazarian 2007). Previous observations of the polarization conducted at a number of observatories have either been hampered by sensitivity or angular resolution. The Submillimeter Array (SMA) is the ideal instrument to conduct such observations as it is not limited by the factors mentioned and is currently the only interferometer array that can do so.

2. Observations with the SMA

Figure 1 shows the first SMA observations at 345 GHz of the polarized dust continuum emission observations of a young stellar object (YSO) NGC 1333 IRAS 4A reported by Girart, Rao, & Marrone (2006). For the first time, it was shown that in a low-mass star forming region, the observed properties of the magnetic field are in agreement with the standard theoretical models of isolated star formation in magnetized molecular clouds: the magnetic field traces a clear "hour glass" morphology. On larger scales the polarization direction is quite uniform at a position angle of $\sim 145°$ and is in excellent agreement with earlier lower resolution observations. However, on small scales ($\sim$ 200 AU) the field is significantly distorted or "pinched" and the morphology resembles the "hour glass" shape that is predicted by theory. Using the intrinsic dispersion of the polarization angles (Chandrasekhar & Fermi 1953), it is possible to infer the strength of the magnetic field to be 4 mG. The analysis reveals that the magnetic field is substantially more important than turbulence in the evolution of the NGC 1333 IRAS4A circumbinary envelope. A similar magnetic field topology is also seen in IRAS 16293 which is another well studied region of low mass star formation in Ophiuchus (Rao *et al.* 2009).

In contrast to NGC 1333 IRAS 4A which is a low mass YSO, G31.41+0.31 is a region where high mass stars are forming. It harbors a massive ($\sim$ 500-1500 $M_\odot$) rotating hot molecular core. Embedded inside this core are young stellar objects which are in a very

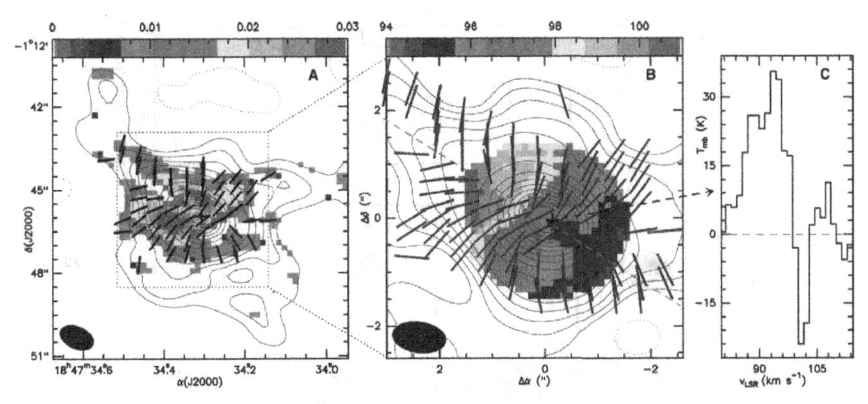

Figure 2. Polarization in G31.41: (A) Contour map of the 879 μm dust emission superposed on the color image of the polarized flux intensity. Black thick bars indicate the position angle of the magnetic field. The full width half maximum synthesized beam is $1.34'' \times 0.83''$ with a position angle of 67° (shown in the bottom left corner). (B) Contours of the 879 μm dust emission superposed on the color image of the flux weighted velocity map of the CH3OH 147-156 A. Black thick bars indicate the direction of the magnetic field. (C) Spectrum of the C34S 7-6 line at the position of the dust emission peak.

early stage of their evolution. The mass of the toroid containing the protostellar objects is much larger than the dynamical mass required for equilibrium, which suggests that the toroid may be gravitationally unstable and undergoing collapse. The dust continuum polarimetry observations (in addition to spectral line maps) carried out with the SMA in the compact and extended configuration at 345 GHz have enabled mapping of this massive hot core at angular resolution slightly below one arcsecond, tracing scales of several thousand AU (Girart *et al.* 2009). The maps (Figure 2) clearly show that the magnetic field lines threading the hot core are pinched along the major axis of the core, where the velocity gradient due to rotation is observed, acquiring a "hour glass" morphology. Furthermore, mapping the spectral line emission from the $C^{34}S$ molecule shows an inverse P-Cygni profile, indicative of infall motions. In addition, a comparison of the velocity gradient along the major axis for different methanol lines show a smaller rotation velocity in the more spatially compact lines (typically the higher excitation ones). This implies that the angular momentum is not conserved during the collapse process. The analysis of the SMA data show that the magnetic field dominates energetically (with respect to centrifugal and turbulence forces) the dynamics of the collapse, and that there is evidence of magnetic braking.

The SMA observations of the polarized emission can also be used to provide information on the nature and efficiency of the dust grain alignment mechanisms. Previous observations (Tamura *et al.*, 1999) and theoretical predictions (Cho & Lazarian 2007) have suggested that a polarization fraction of 2-3% should be commonly observed for protoplanetary disks around young stars. SMA observations of the disk around HD 163296 (Figure 3; Hughes *et al.* 2009), do not detect any polarized emission from either disk. The observations set the most stringent limits to date on the millimeter wavelength polarization from protoplanetary disks, and rule out the fiducial Cho & Lazarian (2007) model at the 10-sigma level. By comparing the SMA observations to the model predictions of Cho & Lazarian (2007), it is determined that the factors most likely contributing to the suppression of polarized emission relative the fiducial model are the roundness of large grains, inefficient alignment of grains with the magnetic field, and a random "tangled" component to the magnetic field lines.

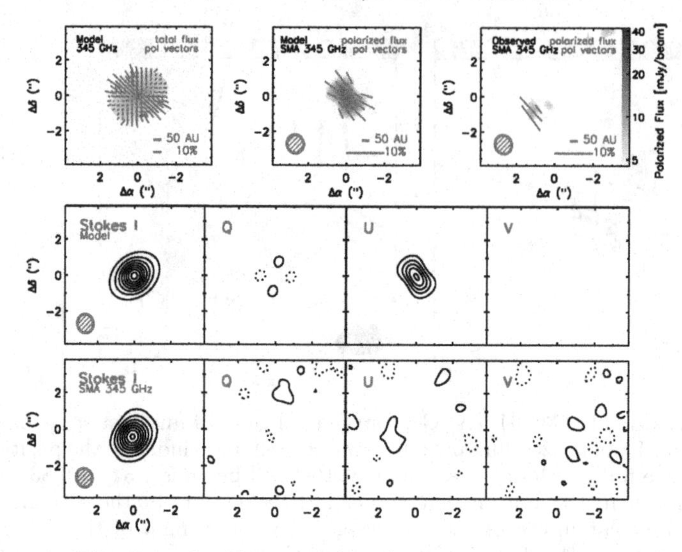

Figure 3. Comparison between the Cho *et al.*, (2007) model and the SMA 340 GHz observations of HD 163296. The top row shows the prediction for the model at full resolution (left), a simulated observation of the model with the SMA (center), and the 2008 SMA observations (right). The grayscale shows either the total flux (left) or the polarized flux (center, right), and the blue vectors indicate the percentage and direction of polarized flux at half-beam intervals. The center and bottom rows compare the model prediction (center) with the observed SMA data (bottom) in each of the four Stokes parameters (I, Q, U, V, from left to right).

3. Summary and Conclusions

These are a few of the results obtained at the SMA which have made a significant contribution towards understanding the nature of the star formation process. Polarimetric observations of a number of other active star forming regions also appears to indicate that the effects of magnetic fields are quite strong and possibly even dominate the star formation process in such regions. These observations are among the most challenging ones conducted at the SMA as they require high sensitivity as well as precise instrumental calibration. The future use of dual polarization receivers at 345 GHz will vastly improve the sensitivity and simplify the operation. This will enable us to make such observations towards an even larger statistical sample of YSOs enabling us to better understand the role of magnetic fields and turbulence in the star formation process.

References

Chandrasekhar, S. & Fermi, E. 1953, *ApJ*, 118, 113

Cho, J., & Lazarian, A. 2007, *ApJ*, 669, 1085

Girart, J. M., Rao, R., & Marrone, D. P. 2006, *Science*, 313, 812

Girart, J. M., Beltrán, M. T., Zhang, Q., Rao, R., & Estalella, R. 2009, *Science*, 324, 1408

Hildebrand, R. H. 1988, *QJRAS*, 29, 327

Hughes, A. M., Wilner, D. J., Cho, J., Marrone, D. P., Lazarian, A., Andrews, S. M., & Rao, R. 2009, *ApJ*, 704, 1204

Lazarian, A. 2007, *Journal of Quantitative Spectroscopy and Radiative Transfer*, 106, 225

Mac Low, M.-M. & Klessen, R. S. 2004, *Rev. of Modern Physics*, 76, 125

Mouschovias, T. 2001, Magnetic Fields Across the Hertzsprung-Russell Diagram, 248, 515

Rao, R., Girart, J. M., Marrone, D. P., Lai, S.-P., & Schnee, S. 2009, *ApJ*, 707, 921

Tamura, M., Hough, J. H., Greaves, J. S., Morino, J.-I., Chrysostomou, A., Holland, W. S., & Momose, M. 1999, *ApJ*, 525, 832

Computational Star Formation
Proceedings IAU Symposium No. 270, 2011
J. Alves, B.G. Elmegreen, J. M. Girart & V. Trimble, eds.

doi:10.1017/S1743921311000251

Modeling High-Mass Star Formation and Ultracompact H II Regions

Ralf S. Klessen[1], Thomas Peters[1], Robi Banerjee[1], Mordecai-Mark Mac Low[2], Roberto Galván-Madrid[3,4] & Eric R. Keto[3]

[1]Zentrum für Astronomie der Universität Heidelberg, Institut für Theoretische Astrophysik, Albert-Ueberle-Str. 2, D-69120 Heidelberg, Germany

[2]Department of Astrophysics, American Museum of Natural History, 79th Street at Central Park West, New York, New York 10024-5192, USA

[3] Harvard-Smithsonian Center for Astrophysics, 60 Garden Street, Cambridge, MA 02138, USA

[4] Centro de Radioastronomía y Astrofísica, UNAM, A.P. 3-72 Xangari, Morelia 58089, Mexico

Abstract. Massive stars influence the surrounding universe far out of proportion to their numbers through ionizing radiation, supernova explosions, and heavy element production. Their formation requires the collapse of massive interstellar gas clouds with very high accretion rates. We discuss results from the first three-dimensional simulations of the gravitational collapse of a massive, rotating molecular cloud core that include heating by both non-ionizing and ionizing radiation. Local gravitational instabilities in the accretion flow lead to the build-up of a small cluster of stars. These lower-mass companions subsequently compete with the high-mass star for the same common gas reservoir and limit its overall mass growth. This process is called fragmentation-induced starvation, and explains why massive stars are usually found as members of high-order stellar systems. These simulations also show that the H II regions forming around massive stars are initially trapped by the infalling gas, but soon begin to fluctuate rapidly. Over time, the same ultracompact H II region can expand anisotropically, contract again, and take on any of the observed morphological classes. The total lifetime of H II regions is given by the global accretion timescale, rather than their short internal sound-crossing time. This solves the so-called lifetime problem of ultracompact H II region. We conclude that the the most significant differences between the formation of low-mass and high-mass stars are all explained as the result of rapid accretion within a dense, gravitationally unstable flow.

Keywords. stars: formation, stars: massive, ISM: H II regions, ISM: kinematics and dynamics

1. Introduction

High-mass stars form in denser and more massive cloud cores (Motte *et al.* 2008) than their low-mass counterparts (Myers *et al.* 1986). High densities result in the large accretion rates, exceeding 10^{-4} $M_\odot$ yr^{-1}, required for massive stars to reach their final mass before exhausting their nuclear fuel (Keto & Wood 2006). High densities also result in local gravitational instabilities in the accretion flow, resulting in the formation of multiple additional stars (Klessen & Burkert 2000, 2001; Klessen 2001; Kratter & Matzner 2006). Young massive stars almost always have companions (Ho & Haschick 1981), and the number of their companions significantly exceeds those of low-mass stars (Zinnecker & Yorke 2007). Such companions influence subsequent accretion onto the initial star (Krumholz *et al.* 2009). Observations show an upper mass limit of about $100\,M_\odot$. It remains unclear whether limits on internal stability or termination of accretion by stellar feedback determines the value of the upper mass limit (Zinnecker & Yorke 2007).

H II regions form around accreting protostars once they exceed $\sim 10\,\mathrm{M}_\odot$, equivalent to a spectral type of early B. Thus, accretion and ionization must occur together in the formation of massive stars. The pressure of the 10^4 K ionized gas far exceeds that in the 10^2 K accreting molecular gas, creating unique feedback effects such as ionized outflows (Keto 2002, 2003, 2007).

Around the most luminous stars the outward radiation pressure force can equal the inward gravitational attraction. A spherically symmetric calculation of radiation pressure on dust yields equality at just under 10 $\mathrm{M}_\odot$ (Wolfire & Cassinelli 1987). However, the dust opacity is wavelength dependent, the accretion is non-spherical, the mass-luminosity ratio is different for multiple companions than for a single star, and to stop accretion more than static force balance is required. The momentum of any part of the accretion flow must be reversed (Larson & Starrfield 1971; Kahn 1974; Yorke & Krügel 1977; Nakano *et al.* 1995). Observations by Keto & Wood (2006) provide evidence for the presence of all these mitigating factors, and numerical experiments combining some of these effects (Yorke & Sonnhalter 2002; Krumholz *et al.* 2007) confirm their effectiveness, showing that radiation pressure is not dynamically significant below the Eddington limit.

The most significant differences between massive star formation and low-mass star formation seem to be the clustered nature of star formation in dense accretion flows and the ionization of these flows. We present results from three-dimensional simulations by Peters *et al.* (2010a,b,c) of the collapse of molecular cloud cores to form a cluster of massive stars that include ionization feedback. These calculations are the first ones that allow us to study these effects simultaneously.

2. Modeling High-Mass Star Formation

Our discussion is based on a series of recent numerical simulations by Peters *et al.* (2010a,b,c). They are based on a modified version of the adaptive-mesh code FLASH (Fryxell *et al.* 2000) that has been extended to include sink particles representing protostars (Federrath *et al.* 2010). The protostars evolve following a prestellar model that determines their stellar and accretion luminosities as function of protostellar mass and accretion rate. We set the stellar luminosity with the zero-age main sequence model by Paxton (2004) and the accretion luminosity by using the tables by Hosokawa & Omukai (2009). The ionizing and non-ionizing radiation from the protostars is propagated through the gas using an improved version of the hybrid characteristics raytracing method on the adaptive mesh developed by Rijkhorst *et al.* (2006). In some calculations, secondary sink formation is suppressed with a density-dependent temperature floor to prevent runaway collapse of dense blobs of gas. For further details, consult Peters *et al.* (2010a).

The simulations start with a $1000\,M_\odot$ molecular cloud. The cloud has a constant density core of $\rho = 1.27 \times 10^{-20}\ \mathrm{g\,cm^{-3}}$ within a radius of $r = 0.5\,\mathrm{pc}$ and then falls off as $r^{-3/2}$ until $r = 1.6\,\mathrm{pc}$. The initial temperature of the cloud is $T = 30\,\mathrm{K}$. The whole cloud is set up in solid body rotation with an angular velocity $\omega = 1.5 \times 10^{-14}\,\mathrm{s^{-1}}$ corresponding to a ratio of rotational to gravitational energy $\beta = 0.05$ and a mean specific angular momentum of $j = 1.27 \times 10^{23}\,\mathrm{cm^2 s^{-1}}$. Peters *et al.* (2010a,b,c) follow the gravitational collapse of the molecular cloud with the adaptive mesh until they reach a cell size of 98 AU. Then sink particles are created at a cut-off density of $\rho_{\mathrm{crit}} = 7 \times 10^{-16}\,\mathrm{g\,cm^{-3}}$. All gas within the accretion radius of $r_{\mathrm{sink}} = 590\,\mathrm{AU}$ above ρ_{crit} is accreted to the sink particle if it is gravitationally bound to it. The Jeans mass on the highest refinement level is $M_{\mathrm{jeans}} = 0.13\,\mathrm{M}_\odot$. A summary of the three simulations discussed in this proceedings article is provided in Table 1.

Table 1. Overview of collapse simulations.

Name	Resolution	Radiative Feedback	Multiple Sinks	M_{sinks} ($M_\odot$)	N_{sinks}	M_{max} ($M_\odot$)
Run A	98 AU	yes	no	72.13	1	72.13
Run B	98 AU	yes	yes	125.56	25	23.39
Run D	98 AU	no	yes	151.43	37	14.64

3. Fragmentation-Induced Starvation

In this section we compare the protostellar mass growth rates from three different calculations. Run A only allows for the formation of a single sink particle, Run B has multiple sinks and radiative heating, while Run D has multiple sinks but no radiative heating. As already discussed by Peters *et al.* (2010a), when only the central sink particle is allowed to form (Run A), nothing stops the accretion flow to the center. Figure 1 shows that the central protostar grows at a rate $\dot{M} \approx 5.9 \times 10^{-4}\,M_\odot\,yr^{-1}$ until we stop the calculation when the star has reached $72\,M_\odot$. The growing star ionizes the surrounding gas, raising it to high pressure. However this hot bubble soon breaks out above and below the disk plane, without affecting the gas flow in the disk midplane much. In particular, it cannot halt the accretion onto the central star. Similar findings have also been reported from simulations focussing on the effects of non-ionizing radiation acting on smaller scales (Yorke & Sonnhalter 2002; Krumholz *et al.* 2007, 2009; Sigalotti *et al.* 2009). Radiation pressure cannot stop accretion onto massive stars and is dynamically unimportant, except maybe in the centers of dense star clusters near the Galactic center.

The situation is different when the disk can fragment and form multiple sink particles. Initially the mass growth of the central protostar in Runs B and D is comparable to the one in Run A. However, as soon as further protostars form in the gravitationally unstable disk, they begin to compete with the central object for accretion of disk material. Unlike in the classical competitive accretion picture (Bonnell *et al.* 2001a, 2004), it is not the most massive object that dominates and grows disproportionately fast. On the contrary, it is the successive formation of a number of low-mass objects in the disk at increasing radii that limits subsequent growth of the more massive objects in the inner disk. Material that moves inwards through the disk due to viscous and gravitational torques accretes preferentially onto the sinks at larger radii.

Similar behavior is found in models of low-mass protobinary disks, where again the secondary accretes at a higher rate than the primary. Its orbit around the common center of gravity scans larger radii and hence it encounters material that moves inwards through the disk before the primary star. This drives the system towards equal masses and circular orbits (Bate 2000). In our simulations, after a certain transition period hardly any gas makes it all the way to the center and the accretion rate of the first sink particle drops to almost zero. This is the essence of the fragmentation-induced starvation process. In Run B, it prevents any star from reaching a mass larger than $25\,M_\odot$. The Jeans mass in Run D is smaller than in Run B because of the lack of accretion heating, and consequently the highest mass star in Run D grows to less than $15\,M_\odot$.

Inspection of Figure 1 reveals additional aspects of the process. We see that the total mass of the sink particle system increases at a faster rate in the multiple sink simulations, Runs B and D, than in the single sink case, Run A. This is understandable, because as more and more gas falls onto the disk it becomes more and more unstable to fragmentation, so as time goes by additional sink particles form at larger and larger radii. Star formation occurs in a larger volume of the disk, and mass growth is not limited by the

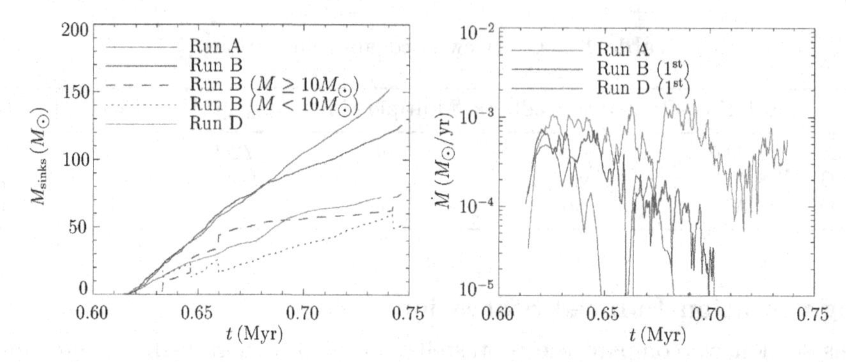

Figure 1. *left* Total accretion history of all sink particles combined forming in Runs A, B, and D. *right* Instantaneous accretion rate as function of time of the first sink particle to form in the three runs.

disk's ability to transport matter to its center by gravitational or viscous torques. As a result the overall star-formation rate is larger than in Run A.

Since the accretion heating raises the Jeans mass and length in Run B, the total number of sink particles is higher in Run D than in B, and the stars in Run D generally reach a lower mass than in Run B. These two effects cancel out to lead to the same overall star formation rate for some time. Eventually, however, the total accretion rate of Run B drops below that of Run D. At time $t \approx 0.68\,\mathrm{Myr}$ the accretion flow around the most massive star has attenuated below the value required to trap the H II region. It is able to break out and affect a significant fraction of the disk area. A comparison with the mass growth of Run D clearly shows that there is still enough gas available to continue constant cluster growth for another 50 kyr or longer, but the gas can no longer collapse in Run B. Instead, it is swept up in a shell surrounding the expanding H II region. The figure furthermore demonstrates that, although the accretion rates of the most massive stars ($M \geqslant 10\,\mathrm{M}_\odot$) steadily decrease, the low-mass stars ($M < 10\,\mathrm{M}_\odot$), which do not produce any significant H II regions, keep accreting at the same rate.

4. Properties of the H II Regions

In all calculations by Peters *et al.* (2010a,c), the H II regions are gravitationally trapped in the disk plane but drive a bipolar outflow perpendicular to the disk. The highly variable rate of accretion onto protostars as they pass through dense filaments causes fast ionization and recombination of large parts of the interior of the perpendicular outflow. The H II regions around the massive protostars do not uniformly expand, but instead rapidly fluctuate in size, shape and luminosity.

We can directly compare these numerical models with radio observations of free-free continuum, hydrogen recombination lines, and $NH_3(3,3)$ rotational lines by generating synthetic maps (Peters *et al.* 2010a,b). The simulated observations of radio continuum emission reproduce the morphologies reported in surveys of ultracompact H II regions (Wood & Churchwell 1989; Kurtz *et al.* 1994). Figure 2 shows typical images from Run B to illustrate this point. It is important to note that even the correct relative numbers of the different morphological types are obtained. Table 2 shows the morphology statistics of UC H II regions in the surveys of Wood & Churchwell (1989) and Kurtz *et al.* (1994) as well as from a random evolutionary sample from Runs A and B of 500 images for each simulation. While the statistics of the cluster simulation B agrees with the observational

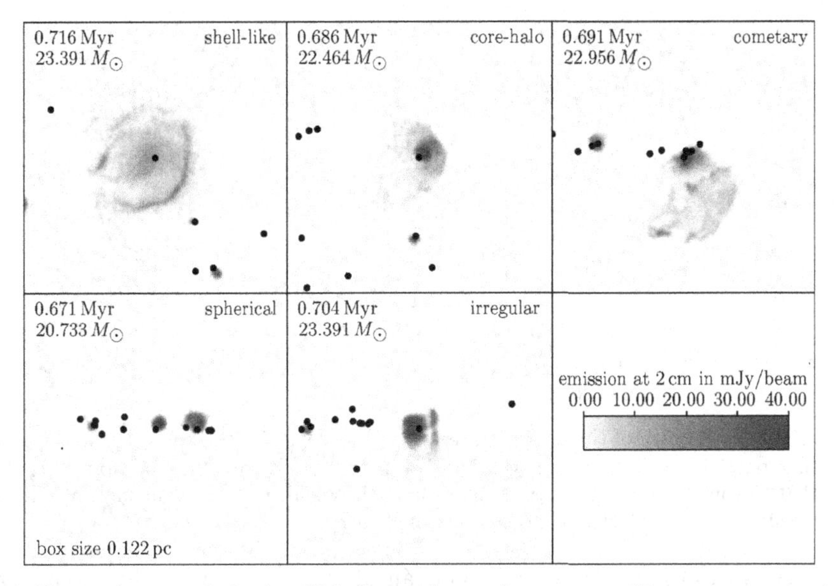

Figure 2. H II region morphologies. This figure shows ultracompact H II regions around massive protostars in Run B at different time steps and from different viewpoints. The cluster is assumed to be 2.65 kpc away, the full width at half maximum of the beam is 0.″14 and the noise level is 10^{-3} Jy. This corresponds to typical VLA parameters at a wavelength of 2 cm. The protostellar mass of the central star which powers the H II region is given in the images. The H II region morphology is highly variable in time and shape, taking the form of any observed type (Wood & Churchwell 1989; Kurtz *et al.* 1994) during the cluster evolution.

Table 2. Percentage Frequency Distribution of Morphologies

Type	Wood & Churchwell (1989)	Kurtz *et al.* (1994)	Run A	Run B
Spherical/Unresolved	43	55	19	60 ± 5
Cometary	20	16	7	10 ± 5
Core-halo	16	9	15	4 ± 2
Shell-like	4	1	3	5 ± 1
Irregular	17	19	57	21 ± 5

data, this is not the case for Run A, in which only one massive star forms. For further discussion, see Peters *et al.* (2010b).

The H II regions in the model fluctuate rapidly between different shapes while accretion onto the protostar continues. When the gas reservoir around the two most massive stars is exhausted, their H II regions merge into a compact H II region, the type that generally accompanies observed ultracompact H II regions (Kim & Koo 2001). These results suggest that the lifetime problem of ultracompact H II regions (Wood & Churchwell 1989) is only apparent. Since H II regions embedded in accretion flows are continuously fed, and since they flicker with variations in the flow rate, their size does not depend on their age until late in their lifetimes.

To compare more directly to observations of the time variability of H II regions, Peters *et al.* (2010b) analyze a few time intervals of interest at a resolution of 10 years. They find that when the accretion rate to the star powering the H II region has a large, sudden increase, the ionized region shrinks, and then slowly re-expands. This agrees with the contraction, changes in shape, or anisotropic expansion observed in radio continuum observations of ultracompact H II regions over intervals of $\sim$ 10 yr (Franco-Hernández &

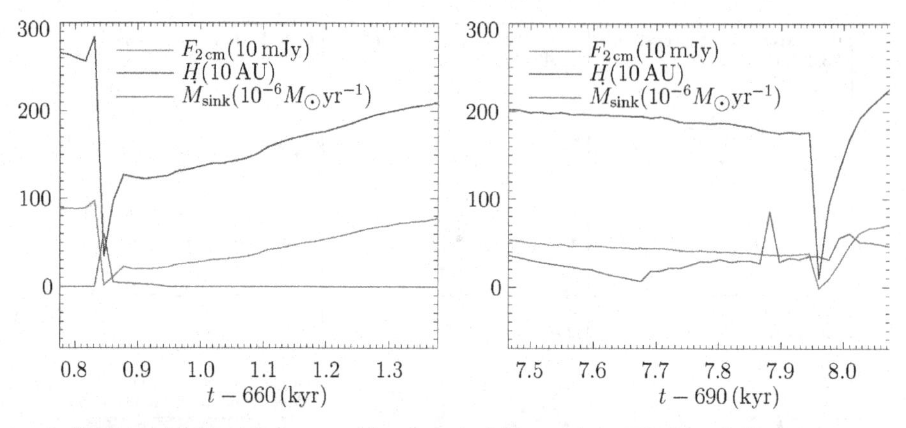

Figure 3. Time variability of the accretion flow and H II region. The image shows accretion rate onto the sink particle $\dot{M}_{\rm sink}$ (sink particle, in units of 10^{-6} M$_\odot$ yr^{-1}), corresponding diameter H of the H II region (in units of 10 AU), as well as the resulting 2-cm continuum flux $F_{\rm 2cm}$ (in units of 10 mJy) for two selected time periods in Run B.

Rodríguez 2004; Rodríguez *et al.* 2007; Galván-Madrid *et al.* 2008). Figure 3 shows the 2-cm continuum flux, the characteristic size of the H II region, and the rate of accretion onto the star. In the left panel, the H II region is initially relatively large, and accretion is almost shut off. A sudden strong accretion event causes the H II region to shrink and decrease in flux. The star at this moment has a mass of 19.8 M$_\odot$. In the right panel, the star has a larger mass (23.3 M$_\odot$), the H II region is initially smaller, and the star is constantly accreting gas. The ionizing-photon flux appears to be able to ionize the infalling gas stably, until a peak in the accretion rate by a factor of 3 and the subsequent continuos accretion of gas makes the H II region to shrink and decrease in flux. The H II region does not shrink immediately after the accretion peak because the increase is relatively mild and the geometry of the infalling gas permitted ionizing photons to escape in one direction. These results show that observations of large, fast changes in ultracompact H II regions (Franco-Hernández & Rodríguez 2004; Rodríguez *et al.* 2007; Galván-Madrid *et al.* 2008) are controlled by the accretion process.

5. Comparison with W51e2

The Peters *et al.* (2010a) model can be compared with the well-studied ultracompact H II region W51e2 (Zhang *et al.* 1998; Keto & Klaassen 2008). In Figure 4 we show simulated and observed maps of the $NH_3(3,3)$ emission, the 1.3 cm thermal continuum emission, and the H53α radio recombination line. The simulated maps were made at a time when the first star in Run B has reached a mass of 20 M$_\odot$.

The brightest $NH_3(3,3)$ emission reveals the dense accretion disk surrounding the most massive star in the model, one of several within the larger-scale rotationally flattened flow. The disk shows the signature of rotation, a gradient from redshifted to blue-shifted velocities across the star. A rotating accretion flow is identifiable in the observations, oriented from the SE (red velocities) to the NW (blue velocities) at a projection angle of 135° east of north (counterclockwise).

The 1.3 cm radio continuum traces the ionized gas, which in the model expands perpendicularly to the accretion disk down the steepest density gradient. As a result, the simulated map shows the brightest radio continuum emission just off the mid-plane of the accretion disk, offset from the central star rather than surrounding it spherically. Continuum emission in the observations is indeed offset from the accretion disk traced in

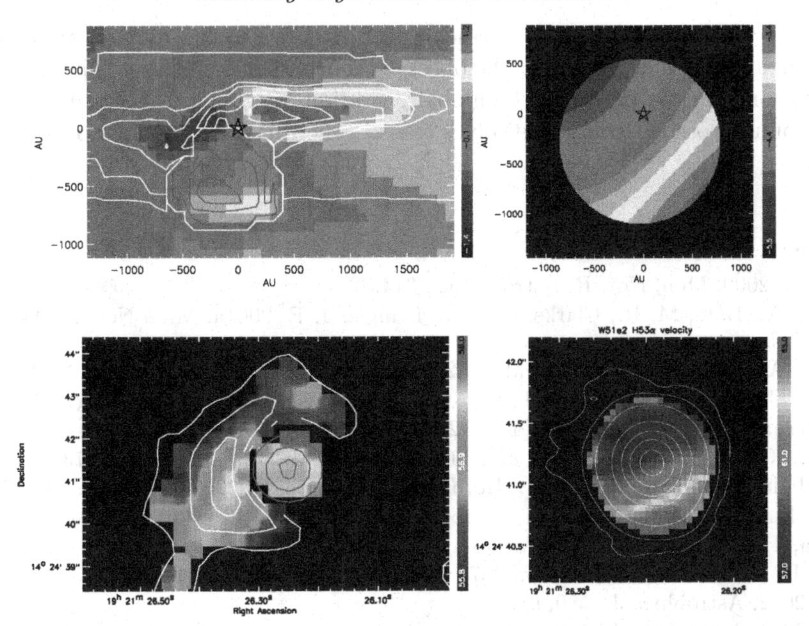

Figure 4. Comparison of line and continuum emission simulated from the model (upper panels) and actually observed from the W51e2 region (lower panels). The left panels show the NH_3(3,3) line emission strength in white contours, the molecular line velocities as the background color, and the 1.3 cm free-free continuum from ionized gas in red contours. The right panels show the H53α recombination line velocities from the ionized gas. The molecular line observations are from Zhang *et al.* (1998) and the H53α observations from Keto & Klaassen (2008). For further details see Peters *et al.* (2010a).

ammonia. The NH_3(3,3) in front of the H II region is seen in absorption and red-shifted by its inward flow toward the protostar. The density gradient in the ionized flow determines the apparent size of the H II region. Therefore the accretion time scale determines the age of the H II region rather than the much shorter sound-crossing time.

Photoevaporation of the actively accreting disk supplies the ionized flow. Therefore, the ionized gas rotates as it flows outward, tracing a spiral. An observation that only partially resolves the spatial structure of the ionized flow sees a velocity gradient oriented in a direction between that of rotation and of outflow, as shown in the simulated observation. The observed H53α recombination line (Keto & Klaassen 2008) in Figure 4 indeed shows a velocity gradient oriented between the directions of rotation and the outflow.

6. Conclusions

Numerical simulations of high-mass star formation regions are now able to resolve the collapse of massive molecular cloud cores and the accretion flow onto the central group of protostars while at the same time treating the transport of ionizing and non-ionizing radiation. This allows us to consistently follow the dynamical evolution of the H II regions that ubiquitously accompany the birth of massive stars. We find that the accretion flow becomes gravitationally unstable and fragments. Secondary star formation sets in and consumes material that would otherwise be accreted by the massive star in the center. We call this process fragmentation-induced starvation. It determines the upper mass limit of the stars in the system and explains why massive stars are usually found as members of larger clusters. These simulations furthermore show that the H II regions forming around massive stars are initially trapped by the infalling gas. But soon, they begin to fluctuate. Over time, the same ultracompact H II region can expand anisotropically, contract again, and take on any of the observed morphological classes. The total lifetime of H II regions thus is given by the global accretion timescale, rather than their short internal sound-crossing time. This solves the so-called lifetime problem of ultracompact H II regions.

References

Bate, M. R. 2000, Mon. Not. R. Astron. Soc., 314, 33
Bonnell, I. A., Bate, M. R., Clarke, C. J., & Pringle, J. E. 2001a, Mon. Not. R. Astron. Soc., 323, 785
Bonnell, I. A., Vine, S. G., & Bate, M. R. 2004, Mon. Not. R. Astron. Soc., 349, 735
Federrath, C., Banerjee, R., Clark, P. C., & Klessen, R. S. 2010, Astrophys. J., 713, 269
Franco-Hernández, R. & Rodríguez, L. F. 2004, Astrophys. J., 604, L105
Fryxell, B., Olson, K., Ricker, P., et al. 2000, Astrophys. J. Suppl. Ser., 131, 273
Galván-Madrid, R., Rodríguez, L. F., Ho, P. T. P., & Keto, E. 2008, Astrophys. J., 674, L33
Ho, P. T. P. & Haschick, A. D. 1981, Astrophys. J., 248, 622
Hosokawa, T. & Omukai, K. 2009, Astrophys. J., 691, 823
Kahn, F. D. 1974, Astron. Astrophys., 37, 149
Keto, E. 2002, Astrophys. J., 580, 980
Keto, E. 2003, Astrophys. J., 599, 1196
Keto, E. 2007, Astrophys. J., 666, 976
Keto, E. & Klaassen, P. 2008, Astrophys. J., 678, L109
Keto, E. & Wood, K. 2006, Astrophys. J., 637, 850
Kim, K.-T. & Koo, B.-C. 2001, Astrophys. J., 549, 979
Klessen, R. S. 2001, Astrophys. J., 556, 837
Klessen, R. S. & Burkert, A. 2000, Astrophys. J. Suppl. Ser., 128, 287
Klessen, R. S. & Burkert, A. 2001, Astrophys. J., 549, 386
Kratter, K. M. & Matzner, C. D. 2006, Mon. Not. R. Astron. Soc., 373, 1563
Krumholz, M. R., Klein, R. I., & McKee, C. F. 2007, Astrophys. J., 656, 959
Krumholz, M. R., Klein, R. I., McKee, C. F., Offner, S. S. R., & Cunningham, A. J. 2009, Science, 323, 754
Kurtz, S., Churchwell, E., & Wood, D. O. S. 1994, Astrophys. J. Suppl. Ser., 91, 659
Larson, R. B. & Starrfield, S. 1971, Astron. Astrophys., 13, 190
Motte, F., Bontemps, S., Schneider, N., Schilke, P., & Menten, K. M. 2008, in Astronomical Society of the Pacific Conference Series, Vol. 387, Massive Star Formation: Observations Confront Theory, ed. H. Beuther, H. Linz, & T. Henning, 22–29
Myers, P. C., Dame, T. M., Thaddeus, P., et al. 1986, Astrophys. J., 301, 398
Nakano, T., Hasegawa, T., & Norman, C. 1995, Astrophys. J., 450, 183
Paxton, B. 2004, Publ. Astron. Soc. Pac., 116, 699
Peters, T., Banerjee, R., Klessen, R. S., Mac Low, M.-M., Galván-Madrid, R., & Keto, E. R. 2010a, Astrophys. J., 711, 1017
Peters, T., Mac Low, M.-M., Banerjee, R., Klessen, R. S., & Dullemond, C. P. 2010b, Astrophys. J., in press
Peters, T., Klessen, R. S., Banerjee, R., & Mac Low, M.-M. 2010c, Astrophys. J., submitted
Rijkhorst, E.-J., Plewa, T., Dubey, A., & Mellema, G. 2006, Astron. Astrophys., 452, 907
Rodríguez, L. F., Gómez, Y., & Tafoya, D. 2007, Astrophys. J., 663, 1083
Sigalotti, L. D. G., de Felice, F., & Daza-Montero, J. 2009, Astrophys. J., 707, 1438
Wolfire, M. G. & Cassinelli, J. P. 1987, Astrophys. J., 319, 850
Wood, D. O. S. & Churchwell, E. 1989, Astrophys. J. Suppl. Ser., 69, 831
Yorke, H. W. & Krügel, E. 1977, Astron. Astrophys., 54, 183
Yorke, H. W. & Sonnhalter, C. 2002, Astrophys. J., 569, 846
Zhang, Q., Ho, P. T. P., & Ohashi, N. 1998, Astrophys. J., 494, 636
Zinnecker, H. & Yorke, H. W. 2007, Ann. Rev. Astron. Astrophys., 45, 481

Computational Star Formation
Proceedings IAU Symposium No. 270, 2011
J. Alves, B. G. Elmegreen, J. M. Girart, & V. Trimble, eds.

doi:10.1017/S1743921311000263

Theory of Cluster Formation: Effects of Magnetic Fields

Fumitaka Nakamura[1,2] and Zhi-Yun Li[3]

[1]Division of Theoretical Astrophysics, National Astronomical Observatory of Japan
[2]Institute of Space and Astronautical Science, Japan Aerospace Exploration Agency, 3-1-1 Yoshinodai, Sagamihara, Kanagawa 229-8510, Japan
email: fumitaka.nakamura@nao.ac.jp

[3]Astronomy Department, University of Virginia, P. O. Box 400325, Charlottesville, VA 22904
email: zl4h@virginia.edu

Abstract. Stars form predominantly in clusters inside dense clumps of molecular clouds that are both turbulent and magnetized. The typical size and mass of the cluster-forming clumps are ~ 1 pc and $\sim 10^2 - 10^3$ $M_\odot$, respectively. Here, we discuss some recent progress on numerical simulations of clustered star formation in such parsec-scale dense clumps with emphasis on the role of magnetic fields. The simulations have shown that magnetic fields tend to slow down global gravitational collapse and thus star formation, especially in the presence of protostellar outflow feedback. Even a relatively weak magnetic field can retard star formation significantly, because the field is amplified by supersonic turbulence to an equipartition strength. However, in such a case, the distorted field component dominates the uniform one. In contrast, if the field is moderately-strong, the uniform component remains dominant. Such a difference in the magnetic structure is observed in simulated polarization maps of dust thermal emission. Recent polarization measurements show that the field lines in nearby cluster-forming clumps are spatially well-ordered, indicative of a rather strong field. In such strongly-magnetized clumps, star formation should proceed relatively slowly; it continues for at least several global free-fall times of the parent dense clump ($t_{ff} \sim$ a few $\times 10^5$ yr).

Keywords. ISM: clouds, ISM: jets and outflows, ISM: magnetic fields, MHD, polarization, stars: formation, turbulence

---

1. Introduction

It is now widely accepted that most stars form in clusters (Lada & Lada 2003). Observations have revealed that clustered star formation occurs in dense compact clumps of molecular clouds that are highly turbulent and magnetized. In addition, almost all massive stars, which greatly impact the interstellar environments and thus galaxy evolution, are believed to be produced in clusters. Thus, understanding how star clusters form is one of the central issues in star formation studies. Although our current knowledge of star cluster formation still remains limited, great efforts have been paid to clarify the relative importance between magnetic fields and turbulence in clustered star formation, on the basis of numerical simulations. In the present paper, we discuss some recent progress on such numerical simulations with emphasis on the role of magnetic fields.

The dynamical stability of a magnetized cloud is determined by the ratio of the mass of a cloud to its magnetic flux. When the mass-to-flux ratio is larger than the critical value $1/(2\pi G^{1/2})$, the magnetic field alone cannot support the whole cloud against gravitational collapse. Such a *magnetically-supercritical* cloud collapses dynamically. On the other hand, when the mass-to-flux ratio is smaller than the critical value, the magnetic field can support the whole cloud. Such a *magnetically-subcritical* cloud cannot collapse without losing magnetic flux. Since molecular gas is almost neutral, ambipolar diffusion can play a role in reducing the magnetic flux from dense cores and clumps, leading to the global gravitational collapse. Thus, it is of great importance to measure the mass-to-flux ratios of molecular clouds and their substructures such as clumps and cores, in order to assess their dynamical stability, although it is difficult to do.

The only technique for directly measuring magnetic field strength in molecular clouds is through the Zeeman effect of molecular lines, which yields the line-of-sight, rather than the total, field strength. Cloud mass determination is also uncertain, due to uncertainties in the abundances of the observed molecules and distances to the clouds. These difficulties make it difficult to determine the mass-to-flux ratio accurately. Available Zeeman measurements indicate that the median mass-to-flux ratios for dense cores of dark clouds (Troland & Crutcher 2008) and massive star-forming dense clumps (Falgarone *et al.* 2008) are within a factor of a few of the critical value (after geometric corrections). Too weak a magnetic field may contradict the observed polarization maps of star forming regions, which often indicate spatially well-ordered magnetic field lines. In Section 3.4, we present the polarization maps of the dust thermal emission from our simulation data. Our polarization maps indicate that in the presence of a weak magnetic field, the polarization vectors have large fluctuations and weak polarization degrees, whereas in the presence of a strong magnetic field, the vectors are spatially well-ordered.

2. Setup of Cluster Formation Simulations

Recent observations have revealed that active cluster-forming regions are not distributed uniformly in parent molecular clouds, but localized and embedded in dense clumps. A good example is the nearby well-studied star-forming region, the Perseus molecular cloud. This cloud has two active cluster-forming regions: IC 348 and NGC 1333, which contain about 80 % of the young stars associated with this cloud (Carpenter 2000). The mass fraction of molecular gas occupied by these two regions is only less than a few tens %. The typical size and mass of such dense clumps is about 1 pc and $10^2 - 10^3 M_\odot$, respectively (e.g., Ridge *et al.* 2003). Here, we choose such a parsec-scale dense clump as the initial condition of our simulations.

The initial cloud is a centrally-condensed spherical clump with an initial density profile of $\rho(r) = \rho_c/[1 + (r/r_c)^2]$, where $r_c = L/6$ is the radius of the central plateau region and $L = 2$ pc is the length of the simulation box. We adopt a central density of 5.0×10^{-20} g cm^{-3}, corresponding to the central free-fall time $t_{\rm ff,c} = 0.30$ Myr. It yields a total clump mass of $M_{\rm tot} = 884 M_\odot$. The average clump density is $\bar{\rho} = 7.5 \times 10^{-21}$ g cm^{-3}, corresponding to the global free-fall time $t_{\rm ff} = 0.77$ Myr. We assume the isothermal equation of state with a sound speed of $c_s = 0.23$ km s^{-1} for a mean molecular weight of $\mu = 2.33$ and the gas temperature of $T = 15$ K. The periodic boundary condition is applied to each side of a cubic simulation box.

At the beginning of the simulation, we impose on the cloud a uniform magnetic field along the x-axis. The field strength is specified by the plasma β, the ratio of thermal pressure to magnetic pressure at the clump center, through $B_0 = 25.8\beta^{-1/2}\mu$G. In units of the critical value, the mass-to-flux ratio in the central flux tube is given by $\Gamma_0 = 8.3\beta^{1/2}$. The mass-to-flux ratio for the initial clump as a whole is $\Gamma = 3.0\beta^{1/2}$. In the present paper, we concentrate on the simulation results of three models with different magnetic field strength, to discuss the role of magnetic fields in cluster formation: (1) $\Gamma = 3.0 \times 10^3$ ($\beta = 10^6$), (2) $\Gamma = 4.3$ ($\beta = 2$), and (3) $\Gamma = 1.4$ ($\beta = 0.2$). Following the standard procedure, we stir the initial clump at the beginning of the simulation with a turbulent velocity field of power spectrum $v_k^2 \propto k^{-4}$ and rms Mach number $\mathcal{M} = 5$. Our simulation

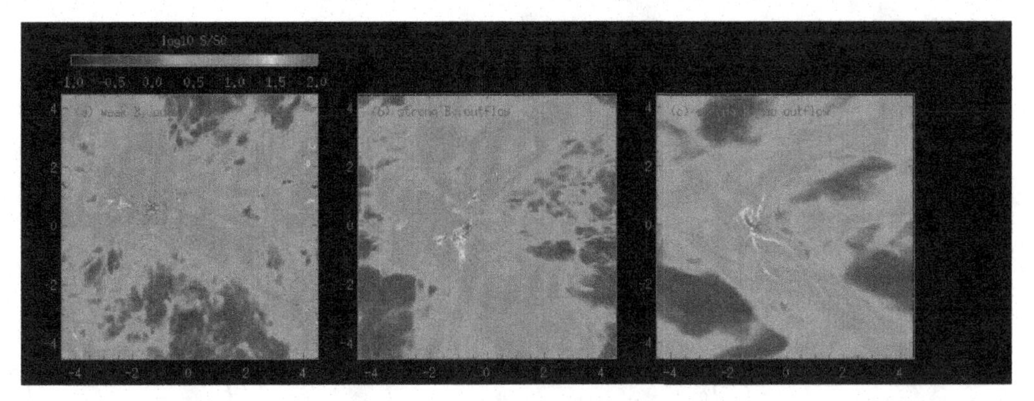

Figure 1. Snapshots of the column density distribution for (a) the weakly-magnetized model with the outflow feedback [$\Gamma = 4.3$ ($\beta = 2.0$) and $t = 2.8t_{\rm ff}$], (b) strongly-magnetized model with the outflow feedback, [$\Gamma = 1.4$ ($\beta = 0.2$) and $t = 4.0t_{\rm ff}$] and (c) strongly-magnetized model without the outflow feedback [$\Gamma = 1.4$ ($\beta = 0.2$) and $t = 1.3t_{\rm ff}$], at the stage when the star formation efficiency has reached 15 %. The initial magnetic field direction is parallel to the horizontal axis. The small red dots indicate the positions of stars. The units of length is the Jeans length of the initial cloud $L_J = 0.22$ pc and the global free-fall time is $t_{\rm ff} = 0.77$ Myr.

has a relatively modest resolution of 256^3. When the density in a cell crosses the threshold $\rho_{\rm th} = 400\rho_0$, we create a Lagrangian particle at the position of the maximum density. The particle is assumed to move with the mass-weighted mean velocity calculated from the extracted gas. The particle is also allowed to accrete the surrounding gas. To mimic the effect of protostellar outflow feedback, the particle injects into the ambient gas a momentum that is proportional to the particle mass M_*. The outflow momentum is scaled with the dimensionless outflow parameter f as $P = f(M_*/M_\odot)(V_w/100{\rm km\ s^{-1}})$, where we adopt $V_w = 100$ km s^{-1} and $f = 0.5$. Each outflow has bipolar and spherical components. The ratio of the bipolar to spherical components is 0.75. The direction of the bipolar component is set to be parallel to the local magnetic field direction.

3. Numerical Results

3.1. *Effects of Magnetic Fields*

In Figs. 1a and 1b, we compare the snapshots of the weakly-magnetized and strongly-magnetized models at the stage when the star formation efficiency (hereafter SFE) has reached 15 %. For the models in panels (a) and (b), the outflow feedback is taken into account. For comparison, we present the snapshot of the strongly-magnetized model without the outflow feedback at the same stage of SFE = 15 %. For all the panels, the initial magnetic field direction is parallel to the horizontal axis. For the weakly-magnetized model, the overall column density distribution appears insensitive to the initial magnetic field direction, implying that the cloud dynamics is controlled by supersonic turbulence. On the other hand, in the presence of the strong magnetic field, the overall density distribution tends to be elongated perpendicular to the initial magnetic field direction. In the presence of outflow feedback, the column density distribution shows many cavities that are created in part by the protostellar outflows. On the other hand, in the absence of the outflow feedback, the cavities are less prominent. This is in part because in the absence of the outflow feedback, the substantial amount of turbulence has decayed and the turbulence is too weak to create many cavities. Another reason is that the protostellar outflow-driven turbulence have more compressible mode that tends to create more cavities (see also Carroll *et al.* 2010).

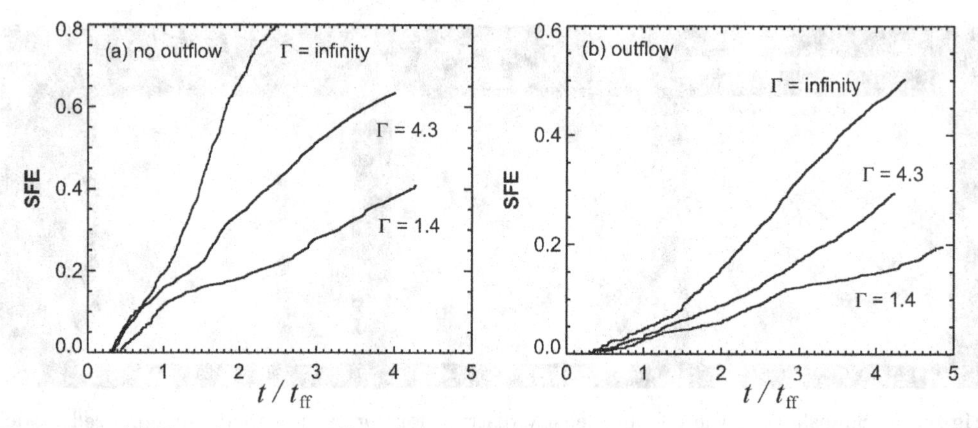

Figure 2. (a) Star formation efficiencies for three models with no outflow feedback against evolution time that is normalized to the global clump free-fall time, (1) $\Gamma = 3.0 \times 10^3$, (2) $\Gamma = 4.3$ (weak magnetic field), and (3) $\Gamma = 1.4$ (moderately-strong magnetic field). Protostellar outflow feedback is not taken into account for these models. (b) Same as panel (a) but for models with outflow feedback.

To illustrate how the initial magnetic field influences star formation in the clump, we present in Fig. 2a the star formation efficiencies against the evolution time for the three models with different magnetic field strengths, turning the outflow feedback off. Clearly, the magnetic field tends to retard star formation. In the period at which the star formation efficiency has reached 15 % from the formation of the first star, the star formation rate per one global free-fall time is estimated to be 29 %, 27 %, and 21 %, for $\Gamma = 3.0 \times 10^3$, 4.3, and 1.4, respectively. The reduction of the star formation rate due to the magnetic field is more significant for later stages, although it is not enough to reproduce the observed level of a few %. This is in good agreement with the results of the recent MHD SPH simulations by Price & Bate (2008) and Price & Bate (2009) who followed the evolution of 50 $M_\odot$ clumps with $\infty \geqslant \Gamma \geqslant 3$ until $t \lesssim 1.5t_{\rm ff}$,where $\Gamma = \infty$ corresponds to a nonmagnetized model. Although their initial clump mass is too small for a cluster-forming clump, their star formation rates per one free-fall time are estimated to be over 10 % in the range of $\infty \geqslant \Gamma \geqslant 3$. These simulations imply that other factors are needed to significantly retard star formation in a pc-scale dense clump.

3.2. *A Significant Role of Protostellar Outflows in Cluster Formation*

There are two ways to slow down the star formation rate more significantly. One way is the feedback from forming stars. In a cluster-forming clump, we usually see stars at different stages of formation, all in close proximity to each other, with one generation of stars potentially affecting the formation of the next. The effects of star formation and its feedback on cluster formation have been studied by two different groups. Price & Bate (2009) included radiative feedback from forming stars in their MHD SPH simulations, and found that the radiative feedback significantly suppress the small-scale fragmentation by increasing the temperature in the high-density material near the protostars. However, it does not much change the global star formation efficiency in parent clumps. Li & Nakamura (2006), Nakamura & Li (2007), and Wang *et al.* (2010) considered the effects of protostellar outflow feedback on cluster formation. They found that the protostellar outflow feedback can significantly reduce the star formation efficiency, although the moderately-strong magnetic fields are necessary to reproduce the observed level of low SFEs. In Fig. 2b, we present the star formation efficiencies against the evolution time for the same three models as in Fig. 2a but including the protostellar outflow feedback. It is

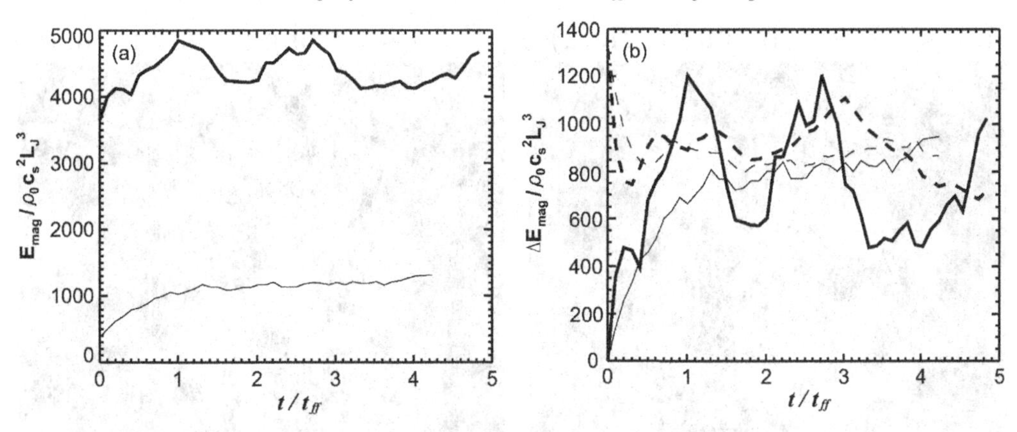

Figure 3. (a) Time evolution of total magnetic energy for two magnetized models. Thick and thin lines are for the models with moderately-strong ($\Gamma = 1.4$) and weak magnetic fields ($\Gamma = 4.3$), respectively. (b) Time evolution of magnetic energy of amplified component (*solid lines*) and kinetic energy of turbulence (*dashed lines*). The magnetic energy of the amplified component tends to approaches the kinetic energy for each model.

clear that star formation rate is greatly reduced by the inclusion of the outflow feedback. In fact, the star formation rate per one free-fall time is estimated to be 7.6 %, 5.6 %, and 4.6 % for the models with $\Gamma = 3.0 \times 10^3$, 4.3, and 1.4, respectively. It is interesting to note that even in the presence of the weak magnetic field, star formation is greatly retarded when the outflow feedback is included. The reason of the great reduction of SFR can be seen in Fig. 3, where the total magnetic energy is plotted against the evolution time.

As shown in Fig. 3a, for the strong magnetic field, the total magnetic energy is dominated by the background uniform field that does not contribute to the force balance at the initial cloud. In Fig. 3b, we illustrate the time evolution of the magnetic energy stored in the distorted component that was amplified by supersonic turbulence. Here, we computed the magnetic energy stored in the distorted component by subtracting the initial magnetic energy from the total magnetic energy. For both the models, the amplified component increases with time and then begins oscillations about a level value, after a free-fall time. Furthermore, the magnetic energy of the amplified component becomes comparable to the kinetic energy of the dense gas for each model, indicating that the energies have reached an equipartition level. For the weaker magnetic field, the amplified component is more important than the initial uniform field, resulting in a significantly-distorted magnetic field structure (left panel of Fig. 4). In contrast, the global field is well-ordered for the stronger initial field (right panel of Fig. 4).

3.3. *Cluster Formation in Initially Magnetically-Subcritical Clumps*

Another way to slow down star formation is to have a stronger magnetic field. Nakamura & Li (2008) considered a magnetically-subcritical cloud, as a model for dispersed, rather than clustered, star formation. They performed 3D MHD simulations including ambipolar diffusion, which reduces the magnetic flux from the dense regions. In Fig. 5, we present the results of the initially magnetically-subcritical model with $\Gamma = 0.8$ and $\mathcal{M} = 10$. In this model, the star formation rate is as small as 0.5 % in the early phase of star formation, but it increases with time and reaches about 1 % by the end of the computation when the SFE has reached about 5 %. In other words, the star formation is accelerated. An unique characteristic of this model is the diffuse filamentary structure seen in the low-density envelope. Such a filamentary structure is an important characteristic of MHD turbulence in the presence of the strong magnetic field. Such a feature has indeed been observed in a

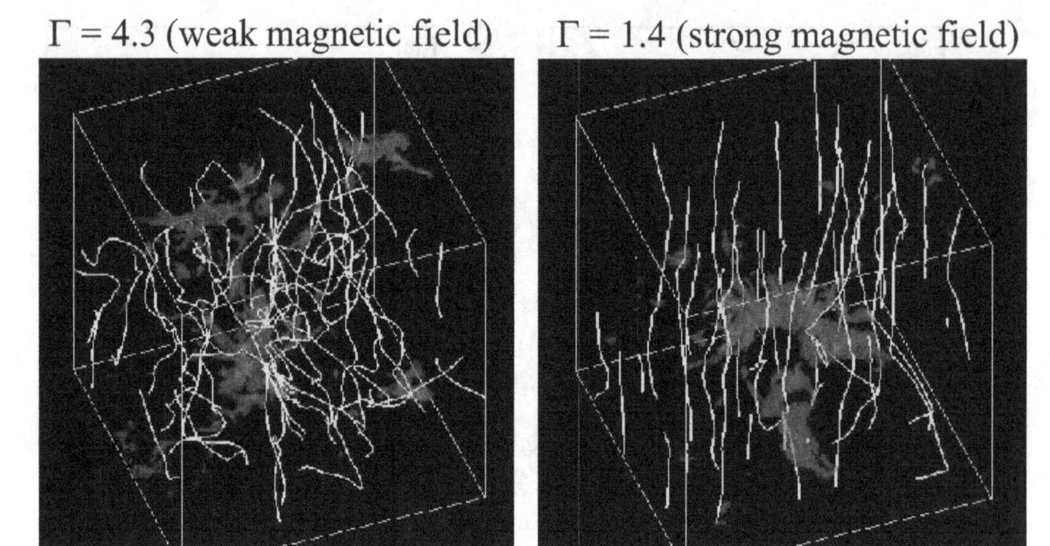

Figure 4. 3D view of the density and magnetic field distributions for the weakly-magnetized model with $\Gamma = 4.3$ (*left*) and the strongly-magnetized model with $\Gamma = 1.4$ (*right*).

nearby low-mass star-forming region, the Taurus molecular cloud. If the cluster-forming clumps are created out of a magnetically-subcritical parent molecular cloud, the low star formation rate can be easily achieved. One possibility to form the subcritical clumps is the external large-scale flows induced by turbulence and/or supernovae. In fact, infrared dark clouds, which are thought to be regions in the early stages of cluster formation, tend to be very filamentary (Gutermuth *et al.* 2008). In the presence of strong magnetic fields, the filamentary clumps are almost perpendicular to the global magnetic fields, and diffuse filaments that are parallel to the magnetic field lines are likely to be observed in the low-density envelope. Very recently, Sugitani *et al.* (2010, in preparation) performed the polarization observations toward Serpens South, the nearby infrared dark cloud, and found that the filament is almost perpendicular to the global magnetic field, implying that the cloud dynamics is controlled by the strong magnetic field. To clarify the possibility of the formation of cluster-forming clumps from the subcritical media, future observations of magnetic fields associated with cluster-forming regions will be needed.

3.4. *Polarization Maps*

Polarization maps of submillimeter thermal dust emission have recently been obtained for nearby star forming regions. Here, we present the polarization maps derived from the simulation data for the two magnetically-supercritical models with different initial magnetic field strengths (the same models as described in Section). We computed the polarized thermal dust emission from the MHD model following Padoan *et al.* (2001). We neglect the effect of self-absorption and scattering because we are interested in the thermal dust emission at submillimeter wavelengths. We further assume that the grain properties are constant and the temperature is uniform. The polarization degree is set such that the maximum is equal to 10 %. Figure 6 shows the dust polarization maps calculated from the two magnetized models with $\Gamma = 4.3$ and 1.4. Only a small portion of the computation box is shown in each panel of Fig. 6. As expected from Fig. 4, in the presence of the weak magnetic field, the spatial distribution of the polarization vectors has relatively large fluctuations. The polarization degree tends to be smaller in less dense parts where the magnetic fields are strongly distorted. The column density

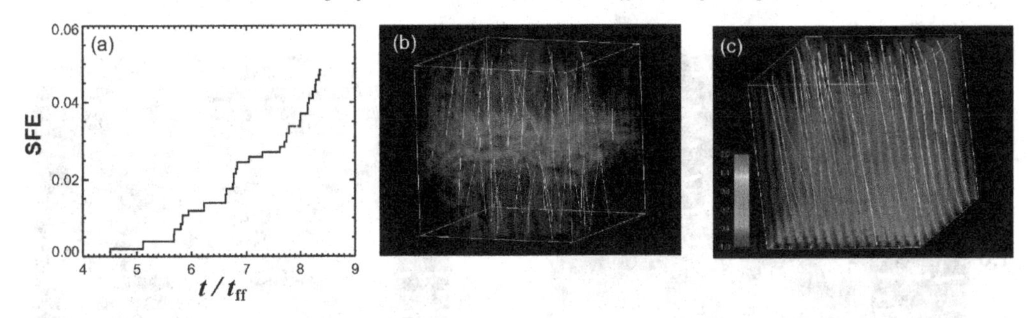

Figure 5. (a) Star formation efficiency for the initially-magnetically-subcritical cloud with $\Gamma = 0.8$ and the initial turbulent Mach number $\mathcal{M} = 10$ against the evolution time that is normalized to the free-fall time of the gas sheet (see Nakamura Li 2008). (b) 3D view of density and magnetic field distribution of the model presented in panel (a). The white lines indicate the magnetic field lines. (c) Blow-up of the low density envelope of the model presented in panel (b). Many non-self-gravitating filaments parallel to the local magnetic field lines are seen.

distribution doesn't show clear filamentary structure. In contrast, in the presence of the strong magnetic field, the filamentary structure is prominent and the filament axes tend to be perpendicular to the polarization vectors that are almost parallel to the initial magnetic field direction. Our polarization maps indicate that the polarization observations can reflect the magnetic field strengths of the cluster-forming clumps.

4. Summary

We have discussed some recent progress on numerical simulations of cluster formation with emphasis on the role of magnetic fields. The numerical simulations indicate that the magnetic field tends to slow down star formation significantly. However, it seems difficult to slow down star formation at the observed level of a few % by the magnetic field alone if the initial field is supercritical ($\Gamma \gtrsim$ a few). We considered two possibilities to retard star formation to the observed level. The first possibility is that the cluster-forming clumps are created out of the parent magnetically-subcritical cloud. If the initial field is magnetically subcritical, the cloud support by the strong magnetic field leads to the great retardation of the global gravitational collapse. The resultant star formation rate decreases significantly. Another possibility is to slow down the star formation by the energy injection due to the protostellar outflows. In the presence of the outflow feedback, the star formation is greatly reduced even for the relatively-weak magnetic fields. This is because the supersonic turbulence amplifies the distorted component of the magnetic fields significantly for the weak magnetic field. However, in this case, the magnetic field structure appears to be random because of the dominant distorted component. According to recent polarization observations of cluster-forming regions, the global magnetic field lines are more or less spatially well-ordered. For example, Sugitani *et al.* (2010) found that the Serpens cloud core is penetrated by a hour-glass shaped well-ordered magnetic field and is elongated in the cross-field direction. Very recently, Sugitani *et al.* (2010, in preparation) also found that the Serpens South filamentary cloud discovered by Gutermuth *et al.* (2008) appears to be penetrated by more or less straight global magnetic field. These observations imply that the magnetic fields associated with the nearby cluster-forming regions are likely to be moderately strong.

Moderately-strong magnetic fields are needed to slow down the internal motions of dense cores where stars form. Observations of nearby cluster-forming clumps have suggested that the internal motions of dense cores tend to be subsonic or at least transonic

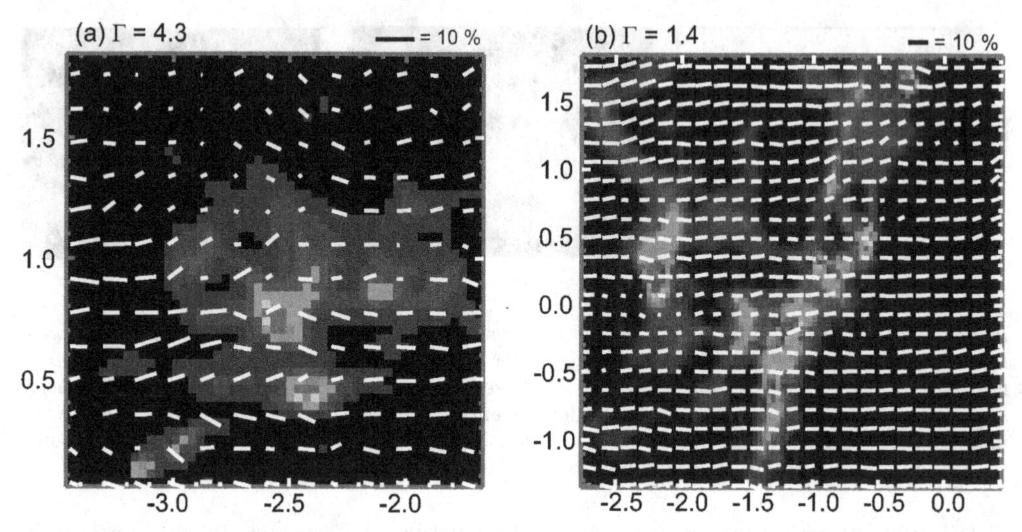

Figure 6. (a) Polarization maps of the models with $\Gamma = 4.3$ (weak magnetic field) and (b) $\Gamma = 1.4$ (moderately-strong magnetic field). The length of the polarization vectors is proportional to the degree of polarization, with the longest vector corresponding to $P = 10\%$. Only one polarization vector is plotted for every four computational cells. The color contour shows the column density distribution. The initial magnetic field lines are parallel to the horizontal line.

[e.g., Maruta *et al.* (2010) for ρ Oph, Saito *et al.* (2008)]. In our numerical models, the internal motions of the dense cores tend to be subsonic for the magnetized models, whereas they are extremely supersonic for the non-magnetized models (Nakamura & Li 2010, in preparation). The moderately-strong magnetic fields (and outflow feedback) also tend to lower the characteristic mass of the stellar IMF (Li *et al.* 2010).

The numerical calculations were carried out mainly on NEC SX8 at YITP in Kyoto University, and on NEC SX9 at CfCA in National Astronomical Observatory of Japan.

References

Carpenter, J. M. 2000, *AJ*, 120, 3139
Carroll, J. J., Frank, A., & Blackman, E. G. 2010, submitted to *ApJ* (arXiv:1005.1098)
Falgarone, E., Troland, T. H., Crutcher, R. M., & Paubert, G. 2008, *A&A*, 487, 247
Lada, C. J. & Lada, E. A. 2003, *ARA&A*, 41, 57
Gutermuth *et al.* 2008, *ApJ* (Letters), 673, L151
Li, Z.-Y., & Nakamura, F. 2006, *ApJ* (Letters), 640, L187
Li, Z.-Y., Wang, P. Abel, T., & Nakamura, F. 2010, *ApJ* (Letters), 720, L26
Maruta *et al.* 2010, *ApJ*, 714, 680
Nakamura, F. & Li, Z.-Y. 2007, *ApJ*, 662, 395
Nakamura, F. & Li, Z.-Y. 2008, *ApJ*, 687, 354
Padoan, P., *et al.* 2001, *ApJ*, 559, 1005
Price, D. J. & Bate, M. R. 2008, *MNRAS*, 385, 1820
Price, D. J. & Bate, M. R. 2009, *MNRAS*, 398, 33
Ridge, N. A., Wilson, T. L., Megeath, S. T., Allen, L. E., & Myers, P. C. 2003, *AJ*, 126, 286
Saito, H., Saito, M., Yonekura, Y., & Nakamura, F. 2008, *ApJS*, 178, 302
Sugitani *et al.* 2010, *ApJ*, 716, 299
Troland, T. H. & Crutcher, R. M. 2008, *ApJ*, 680, 457
Wang, P., Li, Z.-Y., Abel, T., & Nakamura, F. 2010, *ApJ*, 709, 27

Computational Star Formation
Proceedings IAU Symposium No. 270, 2011
J. Alves, B. G. Elmegreen, J. M. Girart, & V. Trimble, eds.

doi:10.1017/S1743921311000275

Magnetic Diffusion in Star Formation

Shantanu Basu and Wolf B. Dapp

Department of Physics and Astronomy, The University of Western Ontario,
London, Ontario N6A 3K7, Canada
email: basu@uwo.ca; wdapp@uwo.ca

Abstract. Magnetic diffusion plays a vital role in star formation. We trace its influence from interstellar cloud scales down to star-disk scales. On both scales, we find that magnetic diffusion can be significantly enhanced by the buildup of strong gradients in magnetic field structure. Large scale nonlinear flows can create compressed cloud layers within which ambipolar diffusion occurs rapidly. However, in the flux-freezing limit that may be applicable to photoionized molecular cloud envelopes, supersonic motions can persist for long times if driven by an externally generated magnetic field that corresponds to a subcritical mass-to-flux ratio. In the case of protostellar accretion, rapid magnetic diffusion (through Ohmic dissipation with additional support from ambipolar diffusion) near the protostar causes dramatic magnetic flux loss. By doing so, it also allows the formation of a centrifugal disk, thereby avoiding the magnetic braking catastrophe.

Keywords. MHD, turbulence, waves, stars: formation, ISM: clouds, ISM: magnetic fields

---

1. Clouds to Cores

Magnetic energy dominates self-gravitational energy in the H I clouds that occupy the interstellar medium (ISM) of our Galaxy (Heiles & Troland 2005). In other words, their mass-to-flux ratio M/Φ is significantly less than the critical value required for gravitational collapse and fragmentation. On the other hand, molecular clouds, the birthplaces of stars, have mass-to-flux ratios that are very close to the critical value (Crutcher 2004). Mass-to-flux ratios in molecular clouds (or cloud fragments) that are significantly greater than the ambient ISM value can be achieved by two distinct but not mutually exclusive mechanisms. One is the accumulation of matter *along* the ambient magnetic field direction in the local spiral arm. However, this places severe constraints on the accumulation length of molecular clouds, and on either the associated formation timescale or the magnitude of streaming motions that can create the clouds (Mestel 1999). The second and rather attractive possibility is that the formation process of the molecular cloud, in a medium pervaded by turbulence or nonlinear flows, will lead to rapid ambipolar diffusion in at least the compressed (filamentary or sheet-like) regions. This scenario of turbulent ambipolar diffusion has been explored in several recent works, e.g., Li & Nakamura (2004), Kudoh & Basu (2008), and Basu *et al.* (2009).

Here, we report the results of Basu & Dapp (2010). Solution of the thin-sheet MHD equations, including ambipolar diffusion, shows that the rapid accumulation of subcritical gas is expected to lead to islands of higher mass-to-flux ratio. These regions undergo enhanced ambipolar diffusion during the compression, and if they are still subcritical, they then undergo ambipolar-diffusion-driven contraction more rapidly than their surroundings due to the elevated density in those regions. The qualitative result is a handful of cores that are formed within elongated ridges. These cores undergo runaway collapse with subsequent near-flux-trapping as soon as gravity overwhelms magnetic and thermal pressure forces within a supercritical region. However, most of the gas in the cloud

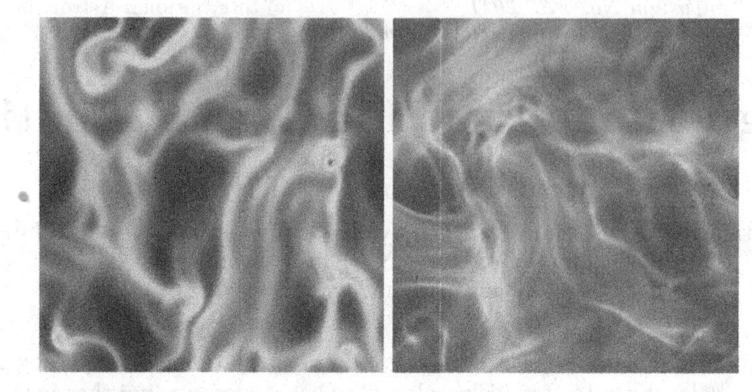

Figure 1. Images of gas column density for initially turbulent models with ambipolar diffusion (left) and flux freezing (right), shown in identical logarithmic color schemes. Both models have initially the same subcritical mass-to-flux ratio and turbulent power spectrum and amplitude, that is allowed to decay. Both models are shown at the same physical time, with the model on the left undergoing runaway collapse in isolated cores and the model on the right in the midst of indefinite supersonic motions. From Basu & Dapp (2010).

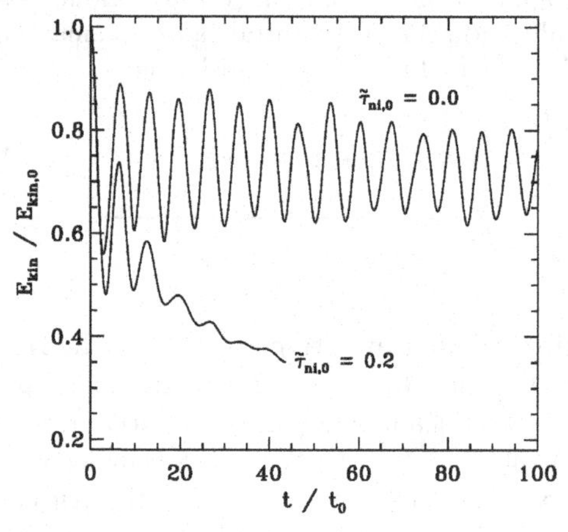

Figure 2. Decay of kinetic energy. The time evolution of total kinetic energy, normalized to its initial value, for a model with flux freezing ($\tilde{\tau}_{\rm ni,0} = 0$) and another with ambipolar diffusion corresponding to a canonical ionization fraction for molecular clouds ($\tilde{\tau}_{\rm ni,0} = 0.2$). From Basu & Dapp (2010).

remains subcritical, does not form stars, and maintains a higher velocity dispersion. This mode of star formation is illustrated in the left panel of Fig. 1. The dark (red) dots represent high density regions of cores that are undergoing runaway collapse. This occurs at a time $43.5\,t_0$ in this model, where $t_0 = c_{\rm s}/(2\pi G\Sigma_0)$, $c_{\rm s}$ is the isothermal sound speed, and Σ_0 is the mean column density of the sheet. The collapse time is a factor ≈ 6 shorter than if starting with small-amplitude initial perturbations. The right panel shows the state of evolution at the same physical time and for statistically identical initial conditions but with ambipolar diffusion turned off, i.e., pure flux freezing. In this case, a startling result is that supersonic motions persist indefinitely. This follows an initial phase in which some but not all of the initial kinetic energy is lost due to shocks. A full explanation and theory is given by Basu & Dapp (2010). Fully three-dimensional simulations by Kudoh & Basu (2008) confirm the rapid core formation found in thin-sheet models such as this and those of Li & Nakamura (2004) and Basu *et al.* (2009). The thin-sheet flux-freezing result of long-lived supersonic motions (Basu & Dapp 2010) remains to be tested fully in three dimensions since the result is dependent upon the existence of a low density

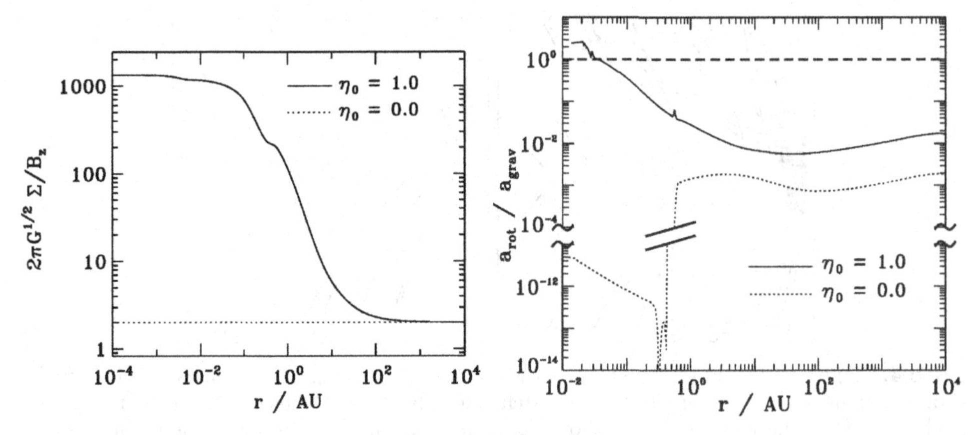

Figure 3. Spatial profiles of mass-to-flux ratio (left) and centrifugal support level (right). Both profiles are shown after the formation of the second core, with extent $\sim R_\odot$. The plot on the right is shown at a time shortly after the introduction of a central sink cell of radius $3\,R_\odot$ that masks the newly-formed second core. From Dapp & Basu (2010).

external medium in which the magnetic field can quickly adjust to a current-free configuration. The latter is built in to the thin-sheet models and may be approximately true in a three-dimensional model of a cold magnetized cloud embedded in a hot tenuous medium.

2. Cores to Star-Disk Systems

Core collapse inevitably begins when the mass-to-flux ratio is a factor ≈ 2 above the critical value (e.g., Basu & Mouschovias 1994). Rapid collapse on a dynamical timescale is able to effectively trap the remaining magnetic flux during the prestellar runaway collapse phase. If this flux trapping continued indefinitely, there would remain a big magnetic flux problem for the final star. A cloud core with twice the critical mass-to-flux ratio would still contain 10^8 times as much magnetic flux per mass as threads the solar surface, and $10^3 - 10^5$ times as much as a magnetic Ap star or a T Tauri star.

Fortunately, a resolution to the magnetic flux problem is facilitated in the post-stellar-core formation epoch, also known as the accretion phase or "$t > 0$", where $t = 0$ is the pivotal moment at which a central protostar is formed. In the spherically symmetric model of Shu (1977), $t = 0$ is pivotal in that an expansion wave subsequently moves outward from the center and envelopes a region of near-free-fall that is dominated by the potential of the central point mass. For the purpose of the magnetic flux problem, $t = 0$ is also pivotal. The subsequent collapse leads to sharp magnetic field gradients in the innermost regions such that the diffusion terms dominate the advection term in the magnetic induction equation, as shown by Li & McKee (1996) and Contopoulos *et al.* (1998). Expressed more loosely, the extreme dragging-in of magnetic field lines in the flux-freezing limit leads to a split-monopole configuration that would in reality be prone to rapid magnetic diffusion due its sharp magnetic field gradient. It is essentially a self-regulation by magnetic diffusion that prevents a split-monopole from forming in the accretion phase.

Another classical problem of star formation is due to angular momentum. Even a rotation rate of $\sim 10^{-14}$ rad s^{-1} as observed in molecular clouds (Goodman *et al.* 1993), while not dynamically important on cloud scales, contains enough angular momentum to prevent nearly all the matter from falling in to a central region of stellar dimensions. However, collapse in the prestellar phase is never sufficient to raise the level of centrifugal

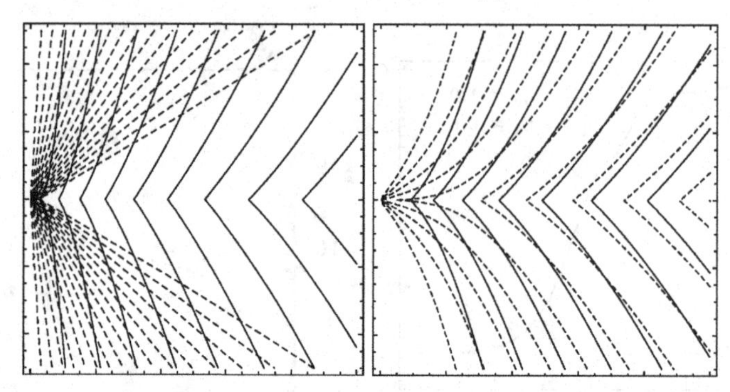

Figure 4. Magnetic field lines. The box on the left has dimensions 10 AU on each side, while the box on the right has dimensions 100 AU on each side. The dashed lines represent the flux-freezing model ($\eta_0 = 0$) while the solid lines show the same field lines for $\eta_0 = 1$. The second core has just formed and is on the left axis midplane.

support relative to gravity, a result decisively shown by the simulations of Norman *et al.* (1980) and explained analytically as a property of self-similar collapse profiles by Basu & Mouschovias (1995). Magnetic braking further weakens the level of centrifugal support, primarily in the core formation epoch, because the subsequent runaway collapse of a prestellar core is generally too rapid for magnetic braking to remain active (Basu & Mouschovias 1994). Here again, $t = 0$ provides a pivot point, after which a centrifugal barrier *does* exist for infalling mass shells, as they fall inward under the gravitational domination of a central protostar. While this implies that a disk will be formed (only) in the $t > 0$ phase, Allen *et al.* (2003) in fact found that magnetic braking gets revitalized in this phase under the assumption of flux freezing. For $t > 0$, the extreme flaring of the magnetic field due to a monopole-like configuration is able to couple near-stellar regions to regions of far greater moment-of-inertia, leading to very efficient magnetic braking. Also, even if a centrifugal disk begins to form during the $t > 0$ phase, the radial velocity is slowed down enough that magnetic braking has time to act and prevent the ultimate formation of the disk.

We have solved the MHD equations for a rotating thin sheet, including the effect of Ohmic dissipation, in axisymmetric geometry. Our results published in Dapp & Basu (2010) show that significant magnetic flux loss occurs within the first core (radius $\sim$ few AU), and effectively shuts down magnetic braking. Ohmic dissipation is added according to the prescription of Nakano *et al.* (2002), similar to the implementation of Machida *et al.* (2007), but accounting for spatial gradients in the resistivity. A dimensionless scaling parameter η_0 characterizes the resistivity, in which the uncertainties hinge largely on the grain size distribution. A canonical value is $\eta_0 = 1$. Additional significant magnetic diffusion can occur at high densities due to ambipolar diffusion, as studied by Kunz & Mouschovias (2010), but is not included in our present study. Ohmic dissipation and its facilitation of disk formation has also been modeled by Krasnopolsky *et al.* (2010) but only on larger scales (>10 AU).

The eventual transformation of a first core to a second core of radius $\sim R_\odot$, due to H_2 dissociation, then occurs with near angular momentum conservation. This is due to rapid Ohmic dissipation having rendered magnetic braking ineffective in this region. The result is that a centrifugal disk does indeed form in the near-environment of the newly-formed second core with mass only $\sim 10^{-3}\, M_\odot$. The rapid flux loss (as seen by a dramatically increased mass-to-flux ratio) is shown in the left panel of Fig. 3. The right panel illustrates the achievement of centrifugal balance in the late stages of the resistive model ($\eta_0 = 1$), and the *magnetic braking catastrophe* in the flux-freezing model ($\eta_0 = 0$). In the latter case, centrifugal support is decimated in the innermost regions by magnetic braking, and a disk cannot form.

Fig. 4 shows the dramatic difference in magnetic field line structure above and below the sheet, calculated using the current-free approximation from a scalar magnetic potential. The split monopole of the $\eta_0 = 0$ model is replaced by a much more relaxed field line structure. The extreme flaring of field lines in the $\eta_0 = 0$ model is a fundamental cause of the magnetic braking catastrophe. More details are in Dapp & Basu (2010). Similar results on magnetic field structure, using a simplified model of Ohmic dissipation, can be found in Galli *et al.* (2009).

References

Allen, A., Li, Z.-Y., & Shu, F. H. 2003, *ApJ*, 599, 363
Basu, S., Ciolek, G. E., Dapp, W. B., & Wurster, J. 2009, *New Astron.*, 14, 483
Basu, S. & Dapp, W. B. 2010, *ApJ*, 716, 427
Basu, S. & Mouschovias, T. Ch. 1994, *ApJ*, 432, 720
Basu, S., & Mouschovias, T. Ch. 1995, *ApJ*, 452, 386
Contopoulos, I., Ciolek, G. E., & Königl, A. 1998, *ApJ*, 504, 247
Crutcher, R. M. 2004, *ApSS*, 292, 225
Dapp, W. B. & Basu, S. 2010, *A&A*, 521, L56
Galli, D., Cai, M., Lizano, S., & Shu, F. H. 2009, *RMxAC*, 36, 143
Goodman, A. A., Benson, P. J., Fuller, G. A., & Myers, P. C. 1993, *ApJ*, 406, 528
Heiles, C. & Troland, T. H. 2005, *ApJ*, 624, 773
Krasnopolsky, R., Li, Z.-Y., & Shang, H. 2010, *ApJ*, 716, 1541
Kudoh, T. & Basu, S. 2008, *ApJ*, 679, L97
Kunz, M. W. & Mouschovias, T. Ch. 2010, 408, 322
Li, Z.-Y. & McKee, C. F. 1996, *ApJ*, 464, 373
Li, Z.-Y. & Nakamura, F. 2004, *ApJ*, 609, L83
Machida, M. N., Inutsuka, S.-i., & Matsumoto, T. 2007, *ApJ*, 670, 1198
Mestel, L. 1999, Stellar Magnetism (Oxford: Oxford Univ. Press)
Nakano, T., Nishi, R., & Umebayashi, T. 2002, *ApJ*, 573, 199
Norman, M. L., Wilson, J. R., & Barton, R. T. 1980, *ApJ*, 239, 968
Shu, F. H. 1977, *ApJ*, 214, 488

Matthew Bate

Computational Star Formation
Proceedings IAU Symposium No. 270, 2011
J. Alves, B.G. Elmegreen, J. M. Girart & V. Trimble, eds.

doi:10.1017/S1743921311000287

Analogues of Cores and Stars in Simulated Molecular Clouds

James Wadsley[1], Michael Reid[2], Farid Qamar[1], Alison Sills[1] and Nicholas Petitclerc[1]

[1]Dept. Physics & Astronomy, McMaster University, 1280 Main St. W., Hamilton, ON, L8S 4M1, Canada
corresponding author email: wadsley@mcmaster.ca

[2]Dept. Astronomy & Astrophysics, University of Toronto, ON, M5S 3H4, Canada

Abstract. We use images derived from collapsing, turbulent molecular cloud simulations without sinks to explore the effects of finite image angular resolution and noise on the derived clump mass function. These effects randomly perturb the clump masses, producing a lognormal clump mass function with a Salpeter-like high mass end. We show that the characteristic break mass of the simulated clump mass functions changes with the angular resolution of the images in a way that is entirely consistent with the observations. We also present some cautionary tales regarding sink particles and highlight the need to ensure that sinks actually correspond to distinct collapsing objects. We test several popular numerical sink criteria in the literature and compare to converged, non-sink results.

Keywords. ISM: structure, submillimeter, methods: data analysis, stars: mass function, stars: formation, methods: numerical

1. Clump Mass Functions

The properties of molecular cloud clumps are increasingly being used as a tool to test theories of star formation. Recent measurements of the shape of the clump mass function (CMF) in nearby low-mass star-forming regions appear to agree with the shape of the stellar initial mass function (Motte *et al.* 2001; Johnstone *et al.* 2001; Tothill *et al.* 2002). The different peak masses are taken to be indicative of the star formation efficiency. However, Reid & Wilson (2006) have shown that the interpretation of observational clump mass functions is biased by the effects of small-number statistics and certain fitting techniques.

Maps of the dust continuum emission in star-forming regions are sectioned into clumps either by eye or using one of a variety of algorithms, such as clfind2d (Williams, de Geus, & Blitz 1994). Clfind2d has been criticized as inaccurate (Pineda, Rosolowsky, & Goodman 2009;Curtis & Richer 2010) but it remains popular and the effects investigated here affect all methods. We used clfind2d to identify clumps in the simulated image in the left panel of fig. 1. The cloud was simulated with the GASOLINE (Wadsley, Stadel & Quinn, 2004) SPH code and contains 5000 solar masses (see Reid *et al.* 2010, Petitclerc 2009). We applied different beams, equivalent to placing the cloud at distances of 160 pc (ρ-Ophiuchus), 450 pc (Orion), 1 kpc and 2 kpc. We also added noise approximating a 10'x10' SCUBA 850 μm scan map over 10 hours in grade 2 weather (roughly 0.03 Jy beam^{-1}) and progressively worse noise by factors of two. The number of objects found changes at all masses. Thus there is no equivalent of a completeness limit as for point sources. The perturbing effects of changing beam and noise generate a log-normal CMF in all cases (c.f. Larson 1973). Furthermore, the high mass end is well-fit by Salpeter-like

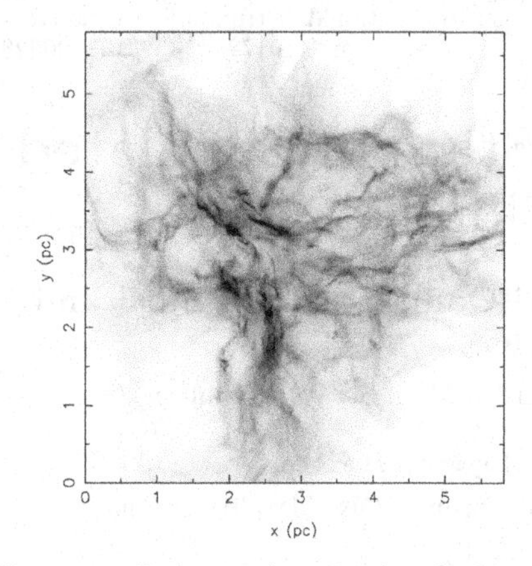

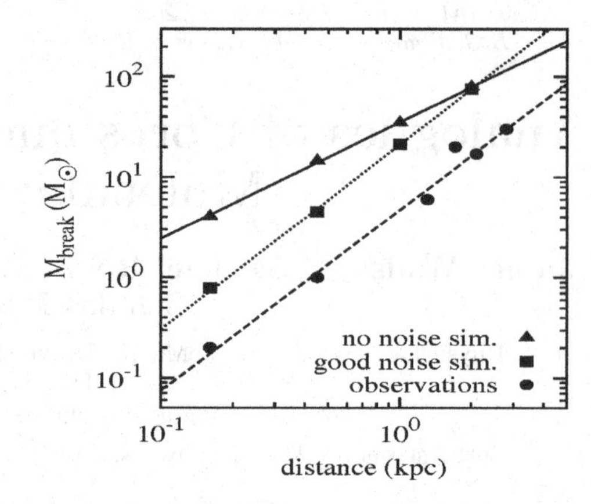

Figure 1. Left panel – simulated observation (noise-free). Right Panel – Break mass versus distance to various observed and simulated star-forming regions. Break masses from the observational data are shown with filled circles, while those from the simulations with noise-free and good noise are shown with filled triangles and squares, respectively. The lines of best fit shown are power laws in distance with exponents of 1.2 for the noise-free simulations (solid line) and 1.8 for both the simulations with good noise (dotted line) and the observations (dashed line).

power-law. Simulated images, mass functions and fits for all cases are presented in Reid *et al.* (2010).

The high mass end for each case was fit with a double power-law intersecting at a characteristic break mass. The break moves to higher masses for more distant objects with a fixed beam (i.e. poorer angular resolution). This behavior tracks that seen for observations of progressively more distant star forming regions, as shown in the right panel of fig. 1.

We conclude that CMF measurements to date have not provided unambiguous measurements of either the intrinsic shape or the characteristic mass scales of the clump mass function due to the effects of noise and resolution. This situation may be about to change, thanks to upcoming observations to be made with instruments such as the Submillimetre Common-User Bolometer Array 2 (SCUBA2) on the James Clerk Maxwell Telescope (JCMT) as well as the Spectral and Photographic Imaging Receiver (SPIRE) and the Photodetector Array Camera and Spectrometer (PACS) on the Herschel Space Observatory.

2. Sink Particles

Sink particles are a numerical technique that allows simulators to model individual star formation over the time-scale of an entire molecular cloud collapse: to go from the CMF to the IMF. The basic premise behind sink particles is that strongly bound regions arise in self-gravitating flows where material becomes permanently locked up. In the case of star formation, neglecting stellar outflows (which may be treated separately), this is probably a reasonable assumption near a proto-star. If inflows are supersonic there is little hydrodynamic back reaction from the accreted material to the surroundings. A region including each protostar is replaced with a sink with the same mass and a reasonable boundary condition for the gas dynamics. Sinks have been used in star formation simulations for some time (e.g. Bate, Bonnell & Price 1995) and have been implemented in many different codes and forms.

As sink formation is irreversible, it is important to introduce sinks only where individual collapse is inevitable. In all sink approaches, spherical candidate sink regions are

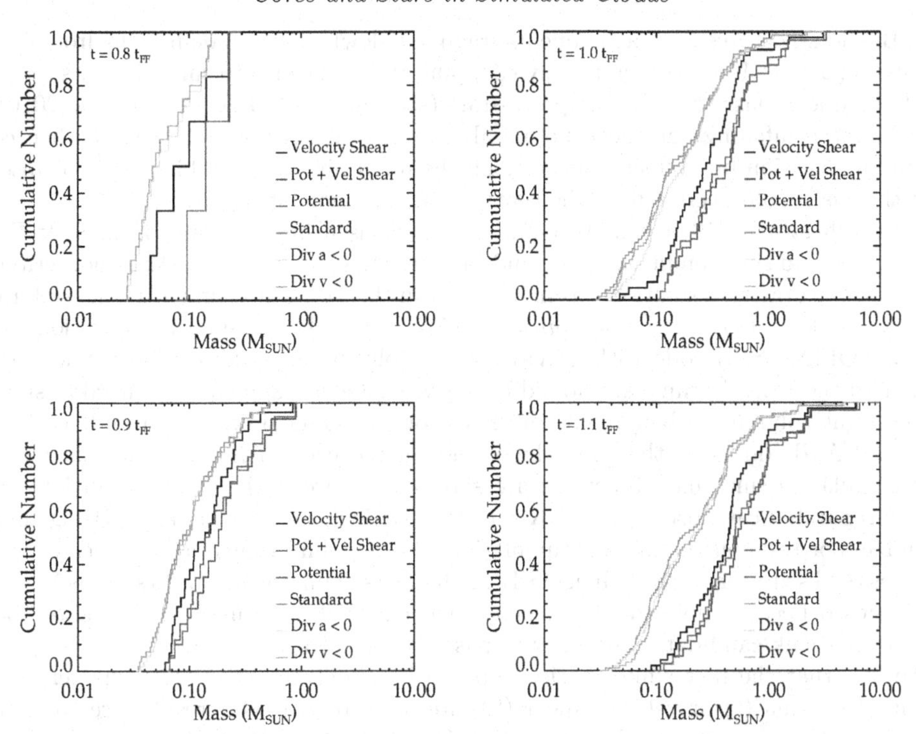

Figure 2. Cumulative mass function of sink particles with different sink formation criteria. Potential and/or velocity shear based criteria are necessary to avoid spurious low mass sinks.

identified when the local gas density exceeds either a fixed value (e.g. Bate 1995) or a value tied to the local Jeans criterion (e.g. Krumholz *et al.* 2004). Additional criteria have been introduced to ensure that the region is collapsing individually, e.g.:

(*a*) Bound: $E_{Total} < 0$
(*b*) Rotationally Bound: $E_{rotational} + E_{Thermal} + E_{Gravity} < 0$
(*c*) Virialized (Thermally): $E_{Thermal} < \frac{1}{2}|E_{Gravity}|$
(*d*) Flow is accelerating: $\nabla \cdot \mathbf{a} < 0$, i.e. $d^2 \log \rho/dt^2 > 0$
(*e*) Density (ρ) maximum
(*f*) Potential (ϕ) minimum (NEW)
(*g*) Flow is converging: $\nabla \cdot \mathbf{v} < 0$, i.e. $d \log \rho/dt > 0$
(*h*) Flow is converging in all directions: eigenvalues of dv_i/dx_j negative (NEW)

Bate, Bonnell & Price (1995) (SPH) implemented criteria (a-d). (a) implies (b) so test (b) will not be considered further. Test (e) tends to occur naturally as sink candidates first reach the threshold density. Krumholz *et al.* 2004 (AMR) performed no explicit tests but the use of a Jeans criterion for the threshold density is equivalent to test (a) for boundness. They also aggressively merge sinks to mop up excess sink creation though no merging is considered here. Attwood *et al.* 2009 (SPH) used test (a) and flow convergence (g). Federrath *et al.* 2010 tested various criteria on a 100 solar mass turbulent cloud collapse, using the FLASH AMR code (Fryxell *et al.* 2000), trying standard: bound density maxima (a,c,e) and new approaches: bound potential minima with flows converging in all directions (a,c,f,h). A key result was that standard approaches create extra sinks in filaments. There are clear circumstances when bound gas cannot be assumed to be forming new collapsed object:

1) Bound structures such as filaments fragment slowly via well defined hydrodynamic processes (Larson 1985). Preemptively chopping the filament into sinks imposes an artificial numerical mass scale for the proto-stars (see Figure 14 of Federrath *et al.* 2010).

2) Material infalling onto deep potentials well may be intrinsically bound when considered in isolation but velocity shear along the infall direction inhibits local collapse so that the gas should accrete onto the centre without fragmenting.

Federrath *et al.* 2010 re-ran their initial conditions using the Bate, Bonnell & Price (1995) code and sink conditions (a-d) and saw similar sink numbers to their new criteria. However, the gravity resolution was much lower in the SPH runs, similar to the sink radii ($r_{SINK} = 500AU$) (*private communication*). We re-ran the same initial condition using the GASOLINE SPH code with gravitational resolution an order of magnitude better, similar to the FLASH runs, and found excessive filament fragmentation in SPH similar to the right column of their figure 14. Thus for the same sink criteria and resolution SPH and AMR do agree that boundness and related criteria (i.e. a-c) are insufficient. Our cumulative sink mass functions are shown in figure 2. The results fall into two, clearly separated categories (e.g. by K-S tests), particularly at later times. Using simple boundness or tests on the mass accumulation (d,g), we find that roughly twice as many low mass sinks are generated, particularly in filaments. Numerically, tests such as $\nabla \cdot \mathbf{v} < 0$ are ill-posed because noise in the velocity field can make the measure fluctuate about zero so that sink candidates are likely to pass if they are tested a few times.

We find that the new sink criteria proposed by Federrath *et al.* 2010, namely using potential minima (f) or velocity shear (h), are more robust. Our results are consistent with theirs, as presented in their figure 16. However, the authors did not demonstrate that this is the correct result. To do so, we ran the collapse without sinks to much higher densities at the same mass resolution and again with 8 times higher mass resolution and confirmed that filaments form bound collapsing objects only within $\sim$ 500 AU of the sinks allowed by the new criteria.

References

Attwood, R., Goodwin, S., Stamatellos, D. & Whitworth, A. 2009, *A&A*, 495, 201
Bate, M., Bonnell, I. & Price, N. 1995, *MNRAS*, 277, 362
Curtis, E. I. & Richer, J. S. 2010, *MNRAS*, 402, 603
Federrath, C., Banerjee, R., Clark, P. & Klessen, R. 2010, *ApJ*, 713, 269
Fryxell *et al.* 2000, *ApJS*, 131, 273
Johnston, D. *et al.* 2001, *ApJ*, 559, 307
Kratter, K., Matzner, C., Krumholz, M., & Klein, R. 2010, *ApJ*, 708, .1585
Krumholz, M., McKee, C., Klein, R. 2004, *ApJ*, 611, 399
Larson, R. B. 1973, *MNRAS*, 161, 133
Larson, R. B., 1985, *MNRAS*, 214, 379
Motte, F., *et al.* 2001, *A&A*, 372, L41
Petitclerc, N. 2009, Ph.D. Thesis, McMaster University
Pineda, J. E., Rosolowsky, E. W., & Goodman, A. A. 2009, *ApJ*, 699, L134
Reid, M., Wadsley, J., Petitclerc, N., & Sills, A. 2010, *ApJ*, accepted
Reid, M. A. & Wilson, C. D. 2006, *ApJ*, 650, 970
Tothill *et al.* 2002, *ApJ*, 580, 285
Wadsley, J. W., Stadel, J., & Quinn, T. 2004, *New Astronomy*, 9, 137
Williams, J. P., de Geus, E. J., & Blitz, L. 1994, *ApJ*, 428, 693

Computational Star Formation
Proceedings IAU Symposium No. 270, 2011
J. Alves, B.G. Elmegreen, J. M. Girart & V. Trimble, eds.

doi:10.1017/S1743921311000299

Dependence of star formation on initial conditions and molecular cloud structure

Matthew R. Bate

School of Physics, University of Exeter,
Stocker Road, Exeter EX4 4QL, United Kingdom
email: mbate@astro.ex.ac.uk

Abstract. I review what has been learnt so far regarding the origin of stellar properties from numerical simulations of the formation of groups and clusters of stars. In agreement with observations, stellar properties are found to be relatively robust to variations of initial conditions in terms of molecular cloud structure and kinetics, as long as extreme initial conditions (e.g. strong central condensation, weak or no turbulence) and small-scale driving are avoided, but properties may differ between bound and unbound clouds. Radiative feedback appears crucial for setting stellar masses, even for low-mass stars, while magnetic fields can provide low star formation rates.

Keywords. gravitation; hydrodynamics; magnetic fields, MHD; radiative transfer; stellar dynamics; methods: numerical; stars: formation; stars: low-mass, brown dwarfs; stars: luminosity function, mass function

1. Introduction

The star formation process is complex, involving a huge range of spatial scales (roughly 10 orders of magnitude from $\sim$ 10 pc to $\sim$R$_\odot$) and temporal scales (roughly 12 orders of magnitude, from several million years to a few minutes, the latter being the time required for a sound wave to propagate across the Sun). It also involves a large number of physical processes, including gravity, supersonic fluid dynamics, radiative transfer, magnetic fields, and chemistry. No numerical simulation can include all of this complexity currently.

To simulate the formation of a star, the bare minimum of physics required even to get started is self-gravitating compressible fluid dynamics. The first calculations of star formation were performed more than 40 years ago by Larson (1969). These one-dimensional calculations also included radiative transfer. Soon after, hydrodynamical calculations were extended to two dimensions (e.g. Larson 1972) and three dimensions (e.g. Boss & Bodenheimer 1979). However, until the late 1990s, most calculations were limited to investigating the formation of single stars or small multiple systems (e.g. binary or triple systems) within isolated molecular cloud cores (e.g. Boss 1986; Bonnell *et al.* 1991).

However, the three-dimensional calculations of even these relatively simple systems could not be followed very far in time because as the collapse to form the first star occurred, the timesteps required to evolve the calculation decreased by orders of magnitude. Thus, each calculation would grind to a halt as the first star forms. Order to get past this point, an approximation must be made. The standard procedure now used in both particle-based (Bate, Bonnell & Price 1995) and grid-based numerical methods (Krumholz, McKee & Klein 2004) is to replace collapsing protostars by sink particles. Here the dense gas within a specified radius of the centre of the protostar, the accretion radius, is combined into a single point mass with the same total mass and momentum as the gas it replaces. Gas that subsequently falls within this radius is accreted by the sink particle. This method allows the very short timesteps that would be required to evolve the gas deeper inside the protostar to be avoided, so hydrodynamical calculations can be evolved well beyond the collapse to form the first protostar. The sink particle method opened the way for large-scale hydrodynamical calculations of stellar groups and clusters to be performed, since the evolution of a molecular cloud could be followed over its dynamical time (typically hundreds of thousands to millions of years) at the same time as resolving small scales (down to the sink particle accretion radii).

The first hydrodynamical calculations of the formation of groups or clusters of stars began with those of Chapman *et al.* (1992), who studied star formation due to colliding molecular clouds (without sink particles), and Klessen, Burkert & Bate (1998) who studied the collapse and fragmentation of a clumpy molecular cloud to form 55 protostars (modelled by sink particles) and obtained an approximately log-normal mass function. At the same time, Bonnell *et al.* (1997) used sink particles to model the growth of protostellar 'seeds' as they accreted from a molecular cloud and found that the stellar mass distribution resulted from 'competitive accretion' between the protostars. Subsequent calculations showed that the peak of the mass function obtained from such calculations was located near the mean Jeans mass of the initial clouds (Klessen & Burkert 2000, 2001; Bate, Bonnell & Bromm 2003; Bate & Bonnell 2005).

The ability, over the past decade, to begin simulating the formation of clusters of stars directly using self-gravitating hydrodynamical calculations has opened up the possibility of trying to understand the origin of the statistical properties of stellar systems by conducting 'numerical experiments'. In these experiments, the initial conditions for star formation and/or physical processes included are varied and the effect of these changes on the outputs of the star formation process (i.e. the statistical properties of the stellar systems) are measured. In this way, we can hope to learn what conditions and processes are important for determining stellar properties, and which have only second-order effects.

2. Hydrodynamical simulations

Despite the use of sink particles, many hydrodynamical calculations of star formation published to date only resolve the collapsing protostars to scales of several hundreds of AU (i.e. the sink particles had accretion radii of hundreds of AU). Thus, they do not capture the opacity limit for fragmentation that occurs when collapsing gas becomes optically thick to its own radiation and so do not capture all of the expected fragmentation. This leads to incomplete mass functions, with the lowest mass objects (e.g. brown dwarfs) missing and, presumably, affects the higher mass protostars as well because there are fewer objects accreting competitively. Most stellar multiple systems and discs are also unresolved. This severely limits the comparisons that can be made with observations.

The first cluster formation calculation that resolved the opacity limit for fragmentation, thus capturing all fragmentation and resolving even the lowest mass brown dwarfs, was that of Bate, Bonnell & Bromm (2002a,b, 2003). The calculation began with a uniform sphere of molecular gas at a temperature of 10 K. Structure was generated in the gas by imposing an initial solenoidal random Gaussian velocity field on the gas to mimic the turbulence that is observed to be present in molecular clouds. These 'turbulent' motions were imposed initially and then allowed to decay during the calculation; the turbulence was not 'driven'. The opacity limit for fragmentation was modelled using a barotropic equation of state in which the gas temperature increased with the gas density once the gas once the gas began to trap its own radiation. The calculation also resolved binaries with separations as close as 1 AU and discs with radii down to ≈ 10 AU. The calculation demonstrated that star formation in clusters could be highly chaotic and dynamic, with discs being truncated by dynamical encounters, fragmention to form multiple systems, and stars and brown dwarfs ejected from unstable multiple systems and escaping the cluster. The calculation produced a modest 50 stars and brown dwarfs, enough for a crude comparison with observations. Properties such as stellar multiplicity, and the fraction of close binaries, were in agreement with observations while the initial mass function (IMF) was in qualitative agreement, though over produced brown dwarfs. As with the earlier, more poorly resolved, simulations, the stellar masses were found to originate through a process of competitive accretion with the characteristic stellar mass being roughly the Jeans mass. All objects began with low masses and accreted from the cloud, typically to the mean thermal Jeans mass, but with a few stars in the centres of sub-clusters, reaching higher masses and forming the high-mass end of the IMF. For those objects with final masses much less than the typical Jeans mass (i.e. brown dwarfs), dynamical interactions and ejections from multiple systems were the key to their low masses (Bate *et al.* 2002a) as proposed by Reipurth & Clarke (2001). Brown dwarfs began with low masses (as did those objects that ended up with stellar masses), but their accretion was terminated when they were involved in dynamical interactions in small groups which increased their velocities (but typically only to a few km/s) and ejected them from the dense cores in which they began forming. Thus, they were unable to accrete to the typical Jeans mass and ended up with substellar masses. Approximately 3/4 of the brown dwarfs were found to originate from the fragmentation of massive circumstellar discs, while the remainder formed in dense filaments, fell into existing multiple systems and then were ejected.

2.1. *The dependence of star formation on the Jeans mass*

This first hydrodynamical calculation of star cluster formation that resolved the opacity limit for fragmentation was followed by three similar calculations that investigated the dependence of stellar properties on the initial conditions in the molecular clouds and variations in the opacity limit for fragmentation. Bate & Bonnell (2005) performed an identical calculation to that of Bate *et al.* (2003), but for a cloud with a smaller radius and nine times higher density. Thus, the mean thermal Jeans mass in the cloud was 1/3 of that in the original calculation. The calculation produced a median stellar mass a factor of 3 lower than the original calculation, exactly matching the change in the mean Jeans mass. This confirmed the results of the earlier, more poorly resolved, calculations mentioned above which indicated that the typical stellar mass was similar to the mean thermal Jeans mass. More recently, Jappsen *et al.* (2005) and Bonnell, Clarke & Bate (2006) using calculations with non-isothermal equations of state at very low molecular densities showed that the transition from atomic line to dust cooling could set an appropriate Jeans mass which in turn could produce the characteristic mass of the IMF.

2.2. *The dependence on the opacity limit for fragmentation*

Bate (2005) performed a third opacity limited calculation, identical to that of Bate *et al.* (2003) except that the opacity limit for fragmentation was moved to a lower density by a factor of nine (i.e. the transition from an isothermal collapse to the gas becoming optically thick to its own radiation was assumed to occur at earlier in the collapse). This increased the minimum mass by a factor of 3 to ≈ 9 Jupiter masses. Such a change may occur, for example, in lower metallicity gas which cools less effectively. Bate found that apart from increasing the minimum mass of a brown dwarf, this change to the equation of state produced no large change in the rest of the IMF or the other stellar properties.

2.3. *The dependence on turbulent motions and molecular cloud structure*

Bate (2009c) performed a fourth opacity limited calculation, identical to Bate *et al.* (2003) except the power spectrum of the initial turbulent velocity field was $P(k) \propto$

k^{-6} to give more power on large scales, rather than $P(k) \propto k^{-4}$ (which was chosen to match the observed Larson (1981) scaling relations). The structures produced in the cloud during the evolution were very different to those found in the original calculation with large shocks and little small-scale structure. However, despite this difference, the stellar properties obtained were indistinguishable from those of the original calculation. In particular, the IMFs were almost identical.

Other studies have also investigated the dependence of the star formation on the properties of the turbulence, though these calculations have not resolved the opacity limit for fragmentation. For example, another approach is to use a periodic box and drive turbulence (in Fourier space), only turning on gravity once the turbulence attains a quasi-steady state. Neither this nor the above approach is physically consistent and it is unclear which is more realistic. Klessen (2001) investigated clouds with driven turbulence and found that the resulting stellar mass function was broader and flatter if the turbulence was driven on small scales than for large-scale driving. More recently, Offner *et al.* (2008) showed that the outcomes of the star formation process does not depend greatly on whether decaying turbulence or large-scale driving are used.

There is also the question of the magnitude of the turbulence. The above calculations assumed that the energy of the turbulent motions is approximately equal to the gravitational potential energy of the clouds. Of course, in the extreme case of no motions, such uniform spherical initial conditions would collapse to a single massive protostar. Similarly, if the initial conditions are strongly centrally condensed, this favours the formation of a single massive object rather than fragmentation into many objects (e.g. Dobbs, Bonnell, Clark 2005; Krumholz, Klein, McKee 2007; Girichidis *et al.* 2010). On the other hand, there is the question of how strongly centrally-condensed molecular clouds could arise in nature without collapse and fragmentation occurring while the object was being assembled. Furthermore, in order for an unusual IMF to be obtained, the object would need to be formed and collapse in relative isolation so that other low-mass stars did not form nearby and result in a more normal mass function overall. Except in strongly-centrally condensed initial conditions, hydrodynamical calculations usually find that star formation proceeds through the formation of multiple groups or sub-clusters. If the global cloud is bound, these sub-clusters typically fall together and merge into larger clusters. Thus, large stellar clusters are thought to form hierarchically (Bonnell, Bate & Vine 2003).

If the magnitude of the turbulence is such that the cloud is globally unbound, however, the star-forming groups and sub-clusters will not merge into a large bound cluster. Stars will still form from dense filaments and other structures formed by colliding flows, but the efficiency with which the gas is converted to stars is decreased as the level of turbulence increases (Clark & Bonnell 2004; Clark *et al.* 2005). Recently, Bonnell *et al.* (2010) has also shown that the IMF may also vary between global bound and globally unbound star-forming regions with star formation in unbound regions resulting in fewer brown dwarfs and high-mass stars relative to solar-type stars than are found in bound regions.

2.4. *Comparison of hydrodynamical calculations with observed stellar properties*

The above calculations either produced only a few dozen objects, or did not resolve the opacity limit for fragmentation, brown dwarfs, or many multiple systems. Thus, detailed comparison with the observed properties of stars and brown dwarfs was not possible.

Recently, Bate (2009a) performed a hydrodynamical simulation of star cluster formation that resolved the opacity limit for fragmentation *and* produced well over 1000 stars and brown dwarfs. The calculation was identical to that of Bate *et al.* (2003), except that the cloud was an order of magnitude more massive (500 $M_{\odot}$). With so many objects, the statistical properties of low-mass stars and brown dwarfs are well determined and detailed comparison with observed stellar properties is possible. Bate found that many stellar properties were in good agreement with observations. For example, stellar multiplicity was found to be a strongly increasing function of primary mass with the multiplicity of very-low-mass objects ($0.03 - 0.10$ $M_\odot$), M-dwarfs, and solar-type stars increasing from ≈ 20 to 60% as observed. The trends for low-mass binaries to have smaller separations and equal-mass components were also reproduced, and even the distribution of relative orientations of the orbital planes of triple systems were found to be in agreement with the observed distribution. The two main areas of disagreement with observed stellar properties were that the calculation produced a much higher ratio of brown dwarfs to stars than is observed, and there was a deficit of unequal-mass solar-type binaries.

2.5. *Conclusions based on hydrodynamical simulations*

The fact that numerical calculations only including gravity and fluid dynamics (without the more complicated physics of radiative transfer, magnetic fields, and chemistry) can reproduce many of the observed statistical properties of stellar systems implies the origin of these stellar properties is primarily due to dissipative gravitational dynamics and may not be significantly altered by additional processes.

The hydrodynamical calculations also indicate that the resulting stellar properties do not vary greatly with variations in the initial conditions such as the metallicity (as long as dust is still the primary coolant at intermediate densities), the power spectrum of the initial velocity field, and or whether or not decaying turbulence or large-scale driving is used. This is because, so long as there is sufficient structure in the gas and the equation of state allows fragmentation to produce many objects that interact dynamically, the processes of competitive accretion (Bonnell *et al.* 1997) and dynamical interactions and ejections (Bate *et al.* 2003) do not depend significantly on the initial structure and turbulent motions in the molecular cloud. On the other hand, extreme initial conditions such as highly centrally-condensed molecular clouds, a very weak (or no) velocity field, or small-scale turbulent driving can lead to statistically different stellar mass distributions. However, it is difficult to see how such initial conditions would arise in nature.

The relative invariance of the IMF to many variations in initial conditions is consistent with the lack of variation in the observed IMF (e.g. Bastian, Covey & Meyer 2010). However, the hydrodynamical calculations do leave us with one serious difficulty: the characteristic mass of the IMF obtained from hydrodynamical calculations is found to scale with the typical Jeans mass of the molecular cloud. One potential solution to this problem is that the typical Jeans mass in molecular clouds does not vary greatly with different environmental conditions as argued by Elmegreen, Klessen & Wilson (2008). But there is also the question of what impact extending the calculations beyond simple self-gravitating hydrodynamics will have.

3. Radiation hydrodynamical simulations

Thus, the question moves on to what the role of additional physical processes is on the star formation process. Boss *et al.* (2000) first pointed out that a barotropic equation of state is a poor approximation to including radiative transfer. Bate (2009b) recently repeated the two cluster formation calculations of Bate *et al.* (2003) and Bate & Bonnell (2005), this time including a realistic equation of state and radiative transfer in the flux-limited diffusion approximation. The calculations modelled gas to within 0.5 AU of each protostar and the energy released down to such scales, but did not include the radiative feedback coming from the stars themselves or discs within 0.5 AU. Despite this, the inclusion of radiative feedback from the forming protostars back into the cloud had a huge effect on the fragmentation. Massive circumstellar discs, in particular, were found to be much hotter and more resistant to fragmentation than in the barotropic calculations. The result was than in both calculations, the numbers of objects formed were reduced by a factor of $\approx 4-5$ from the original calculations. This in turn led to fewer dynamical interactions between objects and fewer ejections. Since objects ejected soon after they form end up as low-mass stars and brown dwarfs, the ratio of brown dwarfs to stars was reduced by the inclusion of radiative feedback, bringing the IMFs produced by the calculations into much better agreement with observations.

As mentioned in the previous section, using a barotropic equation of state the stellar IMF was found to scale with the mean Jeans mass of the clouds. However, with radiative transfer, the IMFs produced from the two clouds were found to be indistinguishable despite their differing mean Jeans masses. When each protostar forms, it heats the gas surrounding it, inhibiting fragmentation nearby (i.e. locally the 'effective' Jeans length and Jeans mass are increased). For a denser cloud, where the initial Jeans length and mass are smaller than for a less dense cloud, the local heating increases the 'effective' Jeans mass by a proportionally greater amount than in a lower-density cloud, largely erasing the differences in the 'effective' Jeans mass between the two clouds. Thus, Bate (2009b) proposed that radiative feedback from protostars self-regulates star formation and erases, or severely weakens, the dependence of the IMF on the mean Jeans mass of the progenitor cloud. This may help to explain the observed invariance of the IMF.

Following these calculations, Offner *et al.* (2009) also performed radiation hydrodynamical calculations low-mass star formation. Their calculations were performed using adaptive mesh refinement (AMR) whereas Bate used smoothed particle hydrodynamics (SPH). Unlike Bate, Offner *et al.* included radiative feedback from the stellar objects themselves and their accretion luminosity, but they used much larger accretion radii of $\approx$ 130 AU. Urban, Martel & Evans (2010) also investigated the effects of protostellar heating on the star formation process, using a simplified method to calculate dust temperatures near protostars and, again, large sink particles ($\approx$ 150 AU). Despite the differences between the calculations, each of these studies found that radiative feedback dramatically reduced the number of objects formed relative to calculations performed using a barotropic equation of state and, therefore, that even in the case of low-mass star formation, it is crucial to include the effects of radiative feedback.

4. Magnetohydrodynamical simulations

Magnetic fields have long been recognised as a potentially important physical process in star formation. However, while their role in some aspects of star formation is clearly crucial (e.g. protostellar jets), their role in star cluster formation is less clear. Recent ideal magnetohydrodynamical (MHD) calculations have confirmed that magnetic fields can play a crucial role in protostellar disc formation (Price & Bate 2007; Hennebelle & Fromang 2008) and the fragmentation of isolated molecular cloud cores to form binary and multiple systems (Price & Bate 2007; Hennebelle & Teyssier 2008). However, their role in cluster formation has only begun to be investigated numerically.

Price & Bate (2008) recently repeated the calculation of Bate *et al.* (2003), this time including magnetic fields of varying strengths (mass-to-flux ratios from 20 to 3) using the ideal MHD approximation. For magnetic fields where the ratio of gas to magnetic pressure was less than unity (i.e. plasma $\beta < 1$) the results were substantially different to the hydrodynamic case. Anisotropic turbulent motions and column density striations aligned with the magnetic field lines were found, both of which have been observed in the Taurus molecular cloud (Goldsmith *et al.* 2008). Large-scale magnetically supported voids were also produced. The additional large-scale support provided by the magnetic field strongly suppressed collapse in the clouds, reducing the star formation efficiency and leading to a more quiescent mode of star formation. Price & Bate found an indication that the relative formation efficiency of brown dwarfs was lower in the strongly magnetized runs due to a reduction in the importance of protostellar ejections.

Conversely, Li *et al.* (2010) recently argued from AMR calculations that included magentic fields and a prescription for outflows that magnetic fields and outflows lower the characteristic stellar mass which they defined as the location of the break in a power-law fit to the high-mass end of the IMF. However, as with Price & Bate (2008), the statistics from these calculations are still relatively poor and the median stellar masses of Li *et al.*'s hydrodynamical and magnetohydrodynamical calculations are similar.

5. Radiation magnetohydrodynamical simulations

Most recently, Price & Bate (2009) combined their treatments of magnetic fields and radiation hydrodynamics to perform the first radiation magnetohydrodynamical calculations of star cluster formation. Again, the calculations were of small 50-$M_\odot$ molecular clouds as modelled by Bate *et al.* (2003), Bate (2009b), and Price & Bate (2008).

The effects of radiative feedback and magnetic fields, found separately in the earlier studies, were combined in these calculations. The main effect of radiative feedback was to inhibit fragmentation on small scales, while the main effect of magnetic fields was to provide support to the low-density gas on large scales, decreasing the star formation rate. With strong magnetic fields and radiative feedback the net result was an inefficient star formation process with a star formation rate of $\approx 10\%$ per free-fall time. This is much less than the rates found without magnetic fields (typically $\approx 50\%$ per free-fall time) and approaches the observed rate of $\sim 3 - 6\%$ (Evans *et al.* 2009). However, it is also important to note that the star formation efficiency also depends on other parameters such as the boundedness of the molecular cloud (Clark & Bonnell 2004) and whether and how the turbulence is driven (Klessen 2001; Offner *et al.* 2009).

6. Conclusions

Calculations of star cluster formation that only take into account gravity and fluid dynamics can reproduce many observed properties of stellar systems, including various binary properties. Many properties do not appear to depend sensitively on the properties of the molecular cloud structure and turbulence as long as extreme initial conditions are avoided (e.g. no motions or strongly centrally-condensed clouds) and the turbulence is decaying or driven on large scales (as seems to be observed; Brunt, Heyer & Mac-Low 2009). This is due to the nature of competitive accretion and dynamical interactions between protostars which determine the spectrum of stellar masses in such calculations. These are local processes that have little memory of the large-scale initial conditions.

An exception is that, with only self-gravitating hydrodynamics, the characteristic mass of the IMF is found to scale linearly with the typical Jeans mass in the molecular cloud. The IMF may also differ in unbound clouds. Radiative feedback, even from low-mass protostars, appears to be crucial to obtain quantitative agreement with the observed IMF and may help to weaken the dependence of the IMF on the typical Jeans mass. Dynamically important magnetic fields also seem to be required to explain the low rate of star formation and many of the structures observed in molecular clouds.

References

Bastian, N., Covey, K. R., & Meyer, M. R. 2010, *arXiv*, 1001.2965
Bate, M. R. 2005, *MNRAS*, 363, 363
Bate, M. R. 2009a, *MNRAS*, 392, 590
Bate, M. R. 2009b, *MNRAS*, 392, 1363
Bate, M. R. 2009c, *MNRAS*, 397, 232
Bate, M. R. & Bonnell, I. A., 2005, *MNRAS*, 356, 1201
Bate, M. R., Bonnell, I. A., & Bromm, V. 2002a, *MNRAS*, 332, L65
Bate, M. R., Bonnell, I. A., & Bromm, V. 2002b, *MNRAS*, 336, 705
Bate, M. R., Bonnell, I. A., & Bromm, V. 2003, *MNRAS*, 339, 577
Bonnell, I. A., Bate, M. R., Clarke, C. J., Pringle, J. E. 1997, *MNRAS*, 285, 201
Bonnell, I. A., Bate, M. R., & Vine, S. G. 2003, *MNRAS*, 343, 413
Bonnell, I. A., Clarke, C. J., & Bate, M. R. 2006, *MNRAS*, 368, 1296
Bonnell, I., Martel, H., Bastien, P., Arcoragi, J.-P., Benz, W. 1991, *ApJ*, 377, 553
Bonnell, I. A., Smith, R. J., Clark, P. C., Bate, M. R. 2010, *arXiv*, 1009.1152
Boss, A. P., 1986, *ApJS*, 62, 519
Boss, A. P. & Bodenheimer, P., 1979, *ApJ*, 234, 289
Boss, A. P., Fisher, R. T., Klein, R. I., McKee, C. F., 2000, *ApJ*, 528, 325
Brunt, C. M., Heyer, M. H., & Mac Low, M.-M. 2009, *MNRAS*, 504, 883
Chapman, S., Pongracic, H., Disney, M., Nelson, A., Turner, J, Whitworth, A. 1992, *Nature*, 359, 207
Clark, P. C. & Bonnell, I. A., 2004, *MNRAS*, 347, 36
Clark, P. C., Bonnell, I. A., Zinnecker, H., Bate, M. R. 2005, *MNRAS*, 359, 809
Dobbs, C. L., Bonnell, I. A., & Clark, P. C. 2005, *MNRAS*, 360, 2
Elmegreen, B.G., Klessen, R.S., & Wilson, C.D. 2008, *ApJ*, 681, 365
Evans, N. J., Dunham, M. M., Jørgensen, J. K., Enoch, M. L., Merín, B., van Dishoeck, E. F., Alcalá, J. M., Myers, P. C., Stapelfeldt, K. R., Huard, T. L., Allen, L. E., Harvey, P. M., van Kempen, T., Blake, G. A., Koerner, D. W., Mundy, L. G., Padgett, D. L., Sargent, A. I. 2009, *ApJS*, 181, 321
Girichidis, P., Federrath, C., Banerjee, R., Klessen, R. S. 2010, *arXiv*, 1008.5255
Goldsmith, P. F., Heyer, M., Narayanan, G., Snell, R., Li, D., Brunt, C. 2008, *ApJ*, 680, 428
Hennebelle, P. & Fromang, S. 2008, *A&A*, 477, 9
Hennebelle, P. & Teyssier, R. 2008, *A&A*, 477, 25
Jappsen, A.-K., Klessen, R. S., Larson, R. B., Li, Y., Mac Low, M.-M. 2005, *A&A*, 435, 611
Klessen, R. S. 2001, *ApJ*, 550, 77
Klessen, R. S.& Burkert, A. 2000, *ApJS*, 128, 287
Klessen, R. S. & Burkert, A. 2001, *ApJ*, 549, 386
Klessen, R. S., Burkert, A., & Bate, M. R. 1998, *ApJ*, 501, L205
Krumholz, M. R., Klein, R. I., & McKee, C. F. 2007, *ApJ*, 656, 959
Krumholz, M. R., McKee, C. F., & Klein, R. I. 2004, *ApJ*, 611, 399
Larson, R. B. 1969, *MNRAS*, 145, 271
Larson, R. B. 1972, *MNRAS*, 156, 437
Larson, R. B. 1981, *MNRAS*, 194, 809
Li, Z.-Y., Wang, P., Abel, T., Nakamura, F. 2010, *ApJ*, 720, L26
Offner, S. S. R., Klein, R. I., & McKee, C. F. 2008, *ApJ*, 686, 1174
Offner, S. S. R., Klein, R. I., McKee, C. F., Krumholz, M. R. 2009, *ApJ*, 703, 131
Price, D. J. & Bate, M. R. 2007, *MNRAS*, 377, 77
Price, D. J. & Bate, M. R. 2008, *MNRAS*, 385, 1820
Price, D. J. & Bate, M. R. 2009, *MNRAS*, 398, 33
Reipurth, B. & Clarke, C. 2001, *MNRAS*, 122, 432
Urban, A., Martel, H., & Evans, N. J. 2010, *ApJ*, 710, 1343

Computational Star Formation
Proceedings IAU Symposium No. 270, 2011
J. Alves, B.G. Elmegreen, J. M. Girart & V. Trimble, eds.

doi:10.1017/S1743921311000305

The universality hypothesis: binary and stellar populations in star clusters and galaxies

Pavel Kroupa

Argelander-Institut für Astronomie, Universität Bonn, Auf dem Hügel 71, D-53121 Bonn, Germany
email: pavel@astro.uni-bonn.de

Abstract. It is possible to extract, from the observations, distribution functions of the birth dynamical properties of a stellar population, and to also infer that these are quite invariant to the physical conditions of star formation. The most famous example is the stellar IMF, and the initial binary population (IBP) seems to follow suit. A compact mathematical formulation of the IBP can be derived from the data. It has three broad parts: the IBP of the dominant stellar population ($0.08 - 2\,M_\odot$), the IBP of the more-massive stars and the IBP of brown dwarfs. These three mass regimes correspond to different physical regimes of star formation but not to structure in the IMF. With this formulation of the IBP it becomes possible to synthesize the stellar-population of whole galaxies.

Keywords. stars: formation; stars: low-mass, brown dwarfs; stars: early-type; stars: pre–main-sequence; stars: luminosity function, mass function; binaries: general; open clusters and associations: general; galaxies: star clusters; galaxies: stellar content; methods: n-body simulations,

---

1. Introduction

The fundamental dynamical properties of stellar populations are the masses of the stars and their correlation in multiple stellar systems. The distribution of stellar masses at birth, the IMF, is rather well constrained and has (surprisingly) been found to be invariant despite theoretical models predicting systematic variation for example of the mean stellar mass or even of the minimum mass with the physical conditions of star formation. This problematical issue has been discussed at some length by Kroupa (2008), where the IMF UNIVERSALITY HYPOTHESIS is stated.

Equivalently, the question may be raised whether the other distribution functions characterizing a stellar population, namely the distribution functions of binary systems, are just as invariant. If this were the case then it would have important bearings on the theory of star formation as the fragmentation length-scale may then not depend much on the physical conditions of the molecular cloud core. A change in the properties of the binary-star distribution functions with mass scale, if found, would yield important clues to the fragmentation and angular momentum re-distribution processes during star formation.

The three important distribution functions describing the initial binary population (IBP) are the distribution of periods (P, here always in days), or equivalently of semi-major axes (a, in AU), the distribution of mass-ratios ($q = m_2/m_1 \leqslant 1$) and the distribution of orbital eccentricities (e). These are related by Kepler's third law: $a^3/P_{\rm yr}^2 = m_1 + m_2$, where $P_{\rm yr}$ is the orbital period in years ($P = 365.25\,P_{\rm yr}$) and m_1, m_2 are the primary- and secondary-star masses in $M_\odot$. Because the periods of binary stars range over many orders of magnitude the shorthand $lP \equiv \log_{10} P$ is used throughout this text.

2. Star formation and the initial binary population (IBP)

Observations have shown that the star-formation process is intrinsically linked to the production of binary stars. Indeed, binary stars must be the dominant formation channel because the observed multiplicity fraction (the number of multiple systems divided by the number of sources in the survey) is indistinguishably high among old metal-poor (Carney *et al.* (2005)) and among thin-disk main-sequence stars (Duquennoy & Mayor (1991)), and is near unity for pre-main sequence stars and proto-stars (Duchêne (1999), Connelley *et al.*(2008)). If, on the other hand, higher-order multiple systems were a major outcome of late-type star formation, then the dynamical decay of these on a time-scale $< 10^5$ yr would pollute the pre-main-sequence stellar population with single stars which are not observed in large numbers. Indeed, this is evidently the case for massive stars ($m \gtrsim$ few $M_\odot$) which appear to form preferentially in binary-rich dense cores of populous embedded clusters which rapidly decay dynamically by ejecting massive stars (Clarke & Pringle (1992), Pflamm-Altenburg & Kroupa (2010)). Thus, according to the BINARY-STAR CONJECTURE OR THEOREM (Kroupa (2008)) the vast number of stars form as binaries, while non-hierarchical higher order multiple systems cannot be a significant outcome of late-type star formation.

The formation of binary systems remains an essentially unsolved problem theoretically. Fisher (2004) shows analytically that isolated turbulent cloud cores can produce an unquantifiable fraction of binary systems with the very wide range of orbital periods as observed. But direct cloud collapse calculations are very limited in predicting binary-star properties owing to the severe computational difficulties of treating the magneto-hydrodynamics together with correct radiation transfer and evolving atomic and molecular opacities during collapse.

The currently most advanced hydrodynamical simulations have been reported by Moeckel & Bate (2010). They allow a turbulent SPH cloud to collapse forming a substantial cluster of 1253 stars and brown dwarfs amounting to 191 $M_\odot$. The cluster has a half-mass radius of about 0.05 pc and contains a very substantial binary and higher-order multiple stellar population with a large spread of semi-major axes but peaking at a few AU. After dynamical evolution with or without expulsion of the residual gas the distribution of orbits ends up being quite strongly peaked at a few AU with a significant deficit of orbits with $a > 10$ AU, and with a deficit of systems with a mass ratio $q < 0.8$, when compared to the main-sequence population (their fig. 11). This state-of-the art computation therewith confirms the above stated issue that it remains a significant challenge for star-formation theory to account for the Gaussian-type distribution of a spanning $10^{-1} - 10^5$ AU as for Galactic-field binaries. One essential aspect which is still missing from such computations is stellar feedback which starts heating the cloud as soon as the first proto stars appear. These heating sources are likely to counter the gravitational collapse such that in reality the extreme densities are not achieved allowing a much larger fraction of wide binaries to survive.

More general theoretical considerations suggest that star-formation in dense clusters ought to have a tendency towards a *lower* binary proportion in warmer molecular clouds (i.e. in cluster-forming cores) because of the reduction of available phase-space for binary-star formation with increasing temperature (Durisen & Sterzik (1994), hereinafter DS). On the other hand, an *enhanced* binary proportion for orbital periods $lP \lesssim 5.6$ may be expected in dense clusters due to the stimulation of binary formation through tidal shear (Horton, Bate & Bonnell (2001)), thus possibly compensating the DS effect. The initial period distribution function (IPF) may thus appear similar in dense and sparse clusters, apart from deviations at long periods due to encounters and the cluster tidal field.

The multiple-star population in the Galactic field is build-up by star-forming events in star-clusters or groups containing from a dozen to possibly millions of stars. Indeed, a certain but presently not well known fraction of stars form in small–N systems that typically have a size ≈ 100 AU, and the dynamical decay of these is likely to affect the final distribution of $P \lesssim 10^5$ binaries (Sterzik & Durisen (1998)), giving rise to non-uniform jet activity (Reipurth (2000)). But again, quantification of their properties is next-to-impossible given the neglect of the hydrodynamical component. But, by the BINARY-STAR CONJECTURE above, the binary formation channel must be vastly dominating over the formation of non-hierarchical higher-order multiples.

3. The IBP universality hypothesis

It is hoped that star-forming simulations will allow essential insights into reproducing the stellar population stemming from an individual modest star-forming event. But given the computational complexity, synthesizing for example the binary population in a massive star cluster or of a whole galaxy or even parts thereof such as the solar neighborhood are not possible. Therewith it becomes rather apparent that current theory has no predictive power concerning the binary properties of stellar populations in different environments.

However, by proposing that initial distribution functions of the binary-star properties (see end of § 1) exist, that is, that there exists an outcome of the isolated (i.e. low-density, such as in Taurus-Auriga) star-formation process that can be quantified in terms of an IBP, we would be put into the situation of being able to synthesize populations. This would become feasible if it is understood how this IBP is affected by physical processes that are inherent to a binary system and that are due to stellar-dynamical encounters in denser star-forming regions.

In fact, the invariance of the IMF must be implying an insensitivity of the star-formation outcome to physical conditions. The IMF being invariant constitutes a statistical statement on one of the birth dynamical properties of stars (namely their distribution of masses). So, since both the IMF and the IBP are the result of the same (star-formation) process, and since the IMF is a result of this process "one level deeper down" than the IBP, it is quite natural to suggest that the formal mathematical distribution function of all of the birth dynamical properties of stars are invariant. Thus the following hypothesis follows:

THE STAR-FORMATION UNIVERSALITY HYPOTHESIS:
IMF universality $\Longleftrightarrow$ IBP universality.

4. Inferring the IBP for $m < 2\,M_\odot$ stars

The challenge of inferring the period-, mass-ratio- and eccentricity-distribution functions characterizing the IBP can be formulated as follows: assuming the STAR-FORMATION UNIVERSALITY HYPOTHESIS to be valid and using observational constraints on the pre-main sequence binary population, the observed main-sequence Galactic-field binary-star distribution function must be corrected for the dynamical processes acting in the birth-groups or birth-clusters of stars as these emerge into the Galactic field. Fig. 1 visualises this idea.

These dynamical processes are well understood and are detailed in Kroupa (2008): Energy arguments imply the HEGGIE-HILLS LAW according to which the wide-binaries are disrupted preferentially compared to short-period binaries and tight binaries typically become tighter. The boundary between the short and long-period binaries depends on the velocity dispersion in the birth population. At the same time, binaries with a small mass ratio (small q) are also preferentially disrupted. The tightening of tight binaries implies that such systems and individual stars interacting with them can be ejected from the cluster. Pre-main-sequence EIGENEVOLUTION evolves the short-period binaries within a time-scale of few 10^5 yr due to system-internal dissipative processes such as tidal-circularisation, primary-star-disk–secondary-star interactions and disk–disk interactions.

The dynamical disruption of binaries induced through the birth cluster and characterized by the stellar-dynamical operator $\Omega^{M_{ecl},R_{0.5}}$, and the mass-ratio, eccentricity and period evolution induced through pre-main sequence eigenevolution, can both be calculated (see Kroupa (2008) for details). M_{ecl} is the stellar mass of the embedded cluster with half-mass radius $R_{0.5}$. Note that this "embedded cluster" does not need to be a cluster which survives the first few 10 Myr as a bound star cluster. It can readily be taken to include sparse star-forming aggregates of stars as observed e.g. in the about 0.5 pc radius sub-groups of a dozen pre-main sequence stars in Taurus-Auriga.

The above two-stage transformation of the formal mathematical birth distribution functions (f_P, f_q, f_e) to the final distribution function (e.g. of the dispersed cluster) can be written as

$$\mathcal{D}_{\rm outcome}(lP, e, q : m_1) = \Omega^{M_{ecl},R_{0.5}} \left[\Omega_{\rm eigenevol} \left[\mathcal{D}_{\rm birth}(lP, e, q : m_1)\right]\right]. \tag{4.1}$$

$\mathcal{D}_{\rm birth} = f_P(lP)\, f_q(q)\, f_e(e)$ is the birth distribution function taking the birth period, mass-ratio and eccentricity distribution functions to be separable, that is, the birth parameters P_b, q_b, e_b are not correlated.

A simple theoretical treatment of eigenevolution which is based on pre-main sequence tidal-circularisation theory has been shown to quite nicely account for the correlations between eccentricity, mass-ratio and period for short-period ($P \lesssim 10^3$ d) binaries. The operator $\Omega_{\rm eigenevol}$ generates the correlations between period, eccentricity and mass ratio.

The stellar-dynamical operator, $\Omega^{M_{ecl},R_{0.5}}$, is given by an Nbody star-cluster, and can be envisioned as a transformation of the number of binaries in different binding-energy bins such that the reduction is largest in the most weakly-bound binaries. The tightening of binaries can increase the fraction of short-period binaries in a population.

Putting this together, the birth binary population is defined by the following formal mathematical rules:

THE BIRTH BINARY POPULATION (BBP):

- random pairing from the canonical IMF for $0.08 \lesssim m/M_\odot \lesssim 2$;
- thermal eccentricity distribution of eccentricities, $f_e(e) = 2\,e$;
- the period distribution function

$$f_{P,birth} = \eta \, \frac{lP - lP_{\rm min}}{\delta + (lP - lP_{\rm min})^2}, \tag{4.2}$$

where $\eta = 2.5, \delta = 45, lP_{\rm min} = 1$ and $\int_{lP_{\rm min}}^{lP_{\rm max}} f_{P,birth}\, dlP = 1$ such that the birth binary fraction is unity ($lP_{\rm max} = 8.43$).

Details are provided in Kroupa (2008). Note that customarily the "initial binary population" (IBP) derives from the BBP after pre-main sequence eigenevolution. Given that pre-main sequence eigenevolution occurs mostly for short-period binaries on a time-scale of 10^5 yr within the system, it is expected to be approximately universal. The STAR FORMATION UNIVERSALITY HYPOTHESIS thus remains valid.

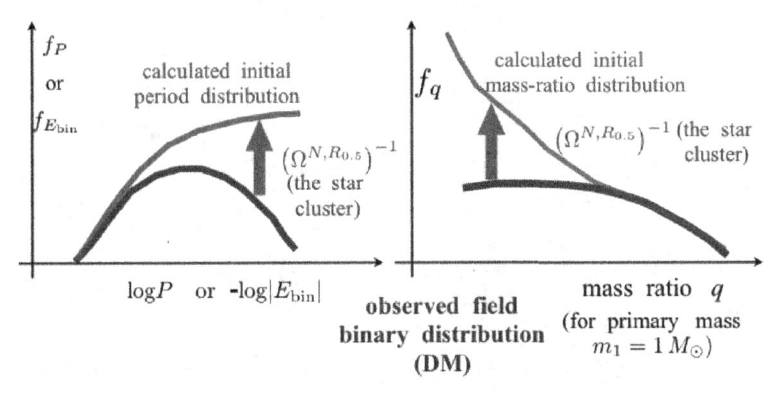

Figure 1. Schematic of how the observed Galactic-field main-sequence (thick black curves, Duquennoy & Mayor (1991)) period or binding energy distribution (**left panel**) and the main-sequence mass-ratio distribution (**right panel**) are back-computed (upwards pointing arrows) with the stellar-dynamical operator $\Omega^{M_{\rm ecl},R_{0.5}}$ to yield an estimate of the initial (thin red curves) period and mass-ratio distributions. Note that the physical reality of $\Omega^{M_{\rm ecl},R_{0.5}}$ is established if both f_P and f_q become consistent with the pre-main sequence data for one and the same $\Omega^{M_{\rm ecl},R_{0.5}}$. The eccentricity distribution is not affected by $\Omega^{M_{\rm ecl},R_{0.5}}$.

Passing the above easily generated *birth* distributions through eigenevolution and then letting the stellar-dynamical operator $\Omega^{M_{\rm ecl},R_{0.5}}$ act on this resulting *initial* distribution leads to the Galactic field population as observed. This is shown in Figs 2–3 for the case $M_{\rm ecl} = 130\,M_\odot, R_{0.5} = 0.8\,{\rm pc}$, where the arrows indicate the action of $\Omega^{M_{\rm ecl},R_{0.5}}$. These results are valid for long-lived cluster models without gas. Short-lived embedded clusters would have a smaller $R_{0.5}$ to give a *dynamically-equivalent* $\Omega^{M_{\rm ecl},R_{0.5}}$ (Kroupa (1995a)). Note that in Fig. 2 the normalisation of the distribution functions follows the custom

$$f = \frac{N_{\rm orbits}}{N_{\rm sys}}, \tag{4.3}$$

where $N_{\rm orbits}$ is the number of binary-star orbits found in a sample of $N_{\rm sys} = N_{\rm orbits} + N_{\rm singles}$ systems or sources, and the $N_{\rm orbits}$ may be the total number of binaries (for the total binary fraction in a population), or just the number of orbits in an lP interval yielding f_P.

5. The IBP for massive stars ($m > 2\,M_\odot$) and for brown dwarfs

A seminal argument suggesting that massive stars form in regions of high density void of low-mass stars but preferentially in tight binaries with similar component masses has been provided by Clarke & Pringle (1992). They reach this conclusion on the basis of the distribution of OB runaway stars finding these conditions to be necessary. In Bonn, Seungkyung Oh is building on these semi-analytical results by performing direct Nbody calculations of realistic young clusters and testing different pairing rules for massive stars in initially mass-segregated and unsegregated clusters. The general result is that massive stars indeed need high densities and tight binaries with mass-ratios nearby unity in order to account for their spatial distribution around star forming regions.

Brown dwarfs (BDs) form an altogether distinct population from stars as is deduced by Thies & Kroupa (2008) on the basis of the observed different pairing properties of brown dwarfs and very-low-mass stars on the one hand side, and stars on the other hand side. The most famous evidence for this comes from the brown-dwarf desert: while stars pair up irrespectively of their masses (e.g. G-dwarfs typically have M-dwarf companions), BDs

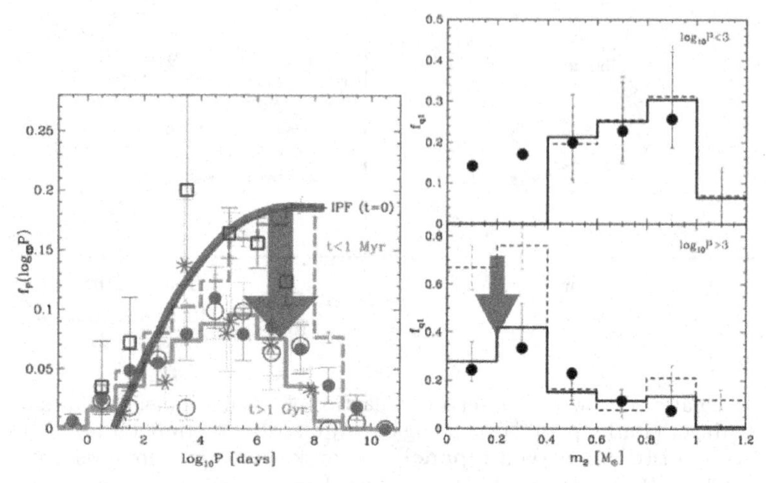

Figure 2. Left panel: The transformation of the birth period distribution function (BPF, Eq. 4.2, thick red curve with $\eta = 2.5, \delta = 45, lP_{\min} = 1$, Kroupa (1995b)=K2) first to the eigenevolved "initial" period distribution function (IPF, dashed green histogram which can be described by Eq. 4.2 with $\eta \approx 3.5, \delta \approx 100, lP_{\min} = 0$, Kroupa (1995a)=K1). The green solid histogram is the final PF after $\Omega^{130\,M_\odot, 0.8\,\mathrm{pc}}$ acts on the IBP. The solid dots, open circles and stars are G-, K-, and M-dwarf binaries, and the open squares are pre-main sequence systems (see Kroupa (2008) for references). The normalisation of this plot is such that each period bin contains the fraction of binary orbits in the whole sample of stars plus binaries. **Right panel:** The transformation of the *initial* mass-ratio distribution for primary masses $m_1 \approx 1\,M_\odot$ (dashed histogram) to the final mass-ratio distribution after the population emerges from its star cluster (solid histogram, from K2). The upper panel is for short-period binaries which are not affected by $\Omega^{130\,M_\odot, 0.8\,\mathrm{pc}}$, while the bottom panel shows the long-period binaries. Note that the *birth* mass-ratio distribution, which results from random pairing from the IMF, is given by the dashed histogram in the lower panel, while eigenevolution transforms this distribution to the dashed histogram evident in the upper panel. That pre-main sequence, i.e. dynamically unevolved, mass-ratios are consistent with random paring from the IMF has been found by Woitas *et al.* (2001).

are exempt from participating (there are exceptions, but such exceptions are a natural outcome of stellar-dynamical processes). BDs also have a very different semi-major axis distribution, by being limited to $a \lesssim 20\,\mathrm{AU}$, while for stars a ranges over five orders of magnitude. Also, BDs have a small binary fraction $f \approx 0.15$ with a bias towards similar companion masses. Computing the effects of the pairing properties of BDs on the observed mass function, Thies & Kroupa (2008) deduce that the IMF must be discontinuous near $0.08\,M_\odot$.

Massive stars and BDs thus cannot be included in the BBP formalism above, but require their own mathematical rules. In terms of physics this means that the formation of massive stars and brown dwarfs occur in a physical regime different to that of the typical star.

This is naturally understandable by noting that massive stars can only form in the densest regions of their parent embedded cluster. BDs, on the other hand, typically form in the outer regions of extended circum-stellar discs of the average star, either by disc instability (Goodwin & Whitworth (2007)) or by tidally-induced gravitational instability caused by passing stars in the parent cluster (Thies *et al.* (2010)), or are ejected embryos (Reipurth & Clarke (2001)).

In terms of numbers, massive stars and brown dwarfs are much rarer than the typical star: on average one BD forms per five late-type stars, while one massive star forms per few hundred late-type stars.

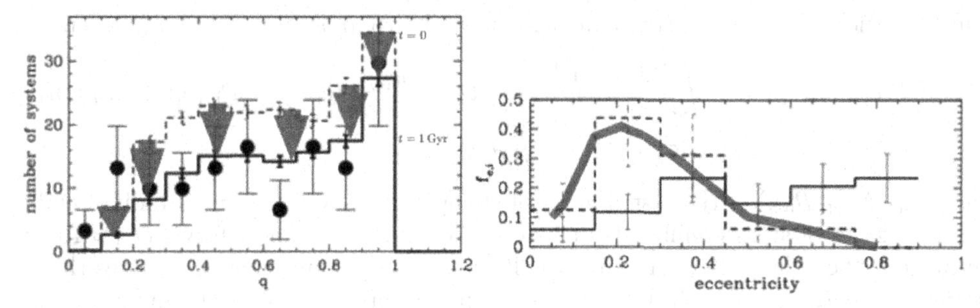

Figure 3. Left panel: The transformation of the overall *initial* mass-ratio distribution for $0.8 < m_1/M_\odot < 1$ (dashed histogram) to the final mass-ratio distribution after the population emerges from its star cluster (solid histogram, from Kroupa (2008)). The $q = 1$ peak results from pre-main sequence eigenevolution. The solid circles are observational data by Reid & Gizis (1997). **Right panel:** The eccentricity distribution for short-period ($lP < 3$, dashed histogram) and long period ($lP > 3$, solid histogram) binaries. The thick red solid curve visualises the bell-shaped distribution of short-period orbits which results from the thermal distribution (the solid histogram) after eigenevolution. Stellar-dynamical encounters, i.e. $\Omega^{M_{\rm ecl},R_{0.5}}$ have no effect on the thermal eccentricity distribution (it is an invariant to $\Omega^{M_{\rm ecl},R_{0.5}}$).

6. IBP mass regimes and IMF structure

A possibly interesting issue thus emerges: While the BD/star schism is clearly evident in the different pairing rules and the discontinuity in the IMF near $0.08\,M_\odot$, the change of pairing rules between late-type and early-type stars (roughly at a few $M_\odot$) does not seem to be evident in any corresponding structure in the IMF. In fact, the flattening of the IMF near $0.5\,M_\odot$ does not seem to correspond to a change in birth binary-star properties. While this fact does not contradict the above statements, it does provide an additional constraint on star-formation theories.

7. Dynamical Population Synthesis (DyPoS)

Having thus obtained a formal mathematical description of the invariant birth binary population it now becomes possible to perform DYNAMICAL POPULATION SYNTHESIS (DyPoS). DyPoS rests on the same ansatz as has already been applied to model the vertical structure of the Milky Way disk (Kroupa (2002)) and to model the stellar initial mass function of a whole galaxy (Kroupa & Weidner (2003)) by adding up the contributions by each embedded cluster. This LEGO PRINCIPLE comes about from the realisation that stars that ultimately end up populating the field form in groups embedded in gas (e.g. Lada & Lada (2003)).

If the birth stellar population in the groups *and* the transformations that occur within the groups before the population hatches into the field are understood and are known, then an entire galaxy can be synthesized. This principle is easy to implement for the integrated initial mass function (IGIMF). It is a little bit more involved when implementing it for an integrated description of the vertical structure of the Milky Way disk because the stellar velocity field emerging from an embedded stellar group needs additional assumptions on its time-evolving properties.

Along the same lines, the binary population in a galaxy follows from adding up all the binary populations born in the embedded groups taking into account the dynamical transformation that acts on the population in each group as it evolves and disperses,

$$\mathcal{D}_{\rm field}(lP, e, q : m_1) = \int_{M_{\rm ecl,min}}^{M_{\rm ecl,max}} \int_{R_{0.5,min}}^{R_{0.5,max}} \Omega^{M_{\rm ecl},R_{0.5}} \left[\Omega_{\rm eigenevol}\left[\mathcal{D}_{\rm birth}(lP, e, q : m_1)\right]\right] \times \xi_{\rm ecl}(M_{\rm ecl}, R_{0.5})\, dR_{0.5}\, dM_{\rm ecl}, \quad (7.1)$$

where $\xi_{\rm ecl}(M_{\rm ecl}, R_{0.5})$ is the distribution of stellar masses and half-mass radii of embedded aggregates of stars ("embedded clusters") forming in a time interval δt ($\approx 10\,$Myr) throughout the galaxy. Eq. 7.1 sums up all the evolved populations of binaries that are formed in $\xi_{\rm ecl}\, dR_{0.5}\, dM_{\rm ecl}$ clusters. $R_{0.5} \approx 0.4\,$pc for embedded clusters and $\xi_{\rm ecl} \propto M_{\rm ecl}^{-\beta}$ is the power-law mass function ($\beta = 2$ is the usually found index from observational surveys).

Two examples of DyPoS have already been computed by Michael Marks at Bonn University. Eq. 7.1 is solved on a grid of $R_{0.5}$ and power-law index, β, of the embedded group or cluster initial mass function, $\xi_{\rm ecl}(M_{\rm ecl})$, where the maximal star-cluster mass, $M_{\rm ecl,max}$ follows from the star-formation-rate (SFR) versus $M_{\rm ecl,max}$ relation of Weidner *et al.* (2004). In an elliptical galaxy which formed in a burst with $SFR = 10^4\,M_\odot$/yr the most-massive star-cluster weighs about $10^8\,M_\odot$ corresponding to the mass-scale of ultra-compact dwarf galaxies. Binary systems are disrupted efficiently in the massive clusters. Using Eq. 7.1 and assuming the typical embedded star cluster has a radius of 0.4 pc, the binary fraction of the late-type stellar population is computed to be $f_{\rm bin} = 0.45$. In a dIrr galaxy with a $SFR = 10^{-4}\,M_\odot$/yr the most massive cluster that can form has a mass of about $M_{\rm ecl,max} = 100\,M_\odot$. For $R_{0.5} = 0.4\,$pc DyPoS yields $f_{\rm bin} = 0.85$, because many more wider binaries survive the on average less-massive embedded clusters in the dIrr galaxy, compared to the above star-burst E galaxy.

A full mathematical treatment is found in Marks, Oh & Kroupa (2010).

8. Conclusions

It appears that the star-formation outcome in terms of stellar masses and multiple systems can be formulated by the STAR FORMATION UNIVERSALITY HYPOTHESIS (§ 3). For stars with $m \lesssim 2\,M_\odot$ the BIRTH BINARY POPULATION (§ 4) can be defined. This is the outcome of star formation in low to intermediate density ($\rho \lesssim 10^4\,M_\odot/{\rm pc}^3$) cloud regions (e.g. of an embedded cluster). For $m \gtrsim 2\,M_\odot$ stars the pairing rules change (§ 5) perhaps reflecting the outcome of star formation in dense regions such as in the cores of embedded clusters ($\rho \gtrsim 10^5\,M_\odot{\rm pc}^3$). Brown dwarfs follow entirely separate rules (§ 5) being an accompanying but distinct population to stars. It remains to be understood why these changing IBP properties do not correspond to the structure evident in the IMF (§ 6).

Acknowledgments: I thank the organizers for a stimulating and memorable conference in Barcelona. My warmest gratitude I wish to express to Sverre Aarseth for his brilliant work on Nbody codes which are freely available and without which this work would not have been possible, and for his unwavering support. This text was written while being a Visitor at ESO/Garching, and I am thankful for the kind hospitality of my colleagues there.

References

Carney, B. W., Aguilar, L. A., Latham, D. W., & Laird, J. B. 2005, *AJ*, 129, 1886
Clarke, C. J. & Pringle, J. E. 1992, *M*NRAS, 255, 423
Connelley, M. S., Reipurth, B., & Tokunaga, A. T. 2008, *AJ*, 135, 2526

Duchêne, G. 1999, *A*&A, 341, 547
Duquennoy A. & Mayor M., 1991, *A*&A, 248, 485 (DM)
Durisen R.H. & Sterzik M.F., 1994, *A*&A, 286, 84
Fisher, R. T. 2004, *A*pJ, 600, 769
Goodwin, S. P. & Whitworth, A. 2007, *A*&A, 466, 943
Horton A.J., Bate M.R., & Bonnell I.A., 2001, *M*NRAS, 321, 585
Kroupa, P. 1995a, *M*NRAS, 277, 1491 (K1)
Kroupa, P. 1995b, *M*NRAS, 277, 1507 (K2)
Kroupa, P. 2002, *M*NRAS, 330, 707
Kroupa, P. 2008, The Cambridge N-Body Lectures, *L*ecture Notes in Physics, 760, 181
Kroupa, P. & Weidner, C. 2003, *A*pJ, 598, 1076
Lada, C. J. & Lada, E. A. 2003, *A*RAA, 41, 57
Marks, M., Oh, S., & Kroupa, P., 2010, submitted
Moeckel, N. & Bate, M. R. 2010, *M*NRAS, 404, 721
Pflamm-Altenburg, J. & Kroupa, P. 2010, *M*NRAS, 404, 1564
Reid, I. N. & Gizis, J. E. 1997, *A*J, 113, 2246
Reipurth B., 2000, *A*J, 120, 3177
Reipurth, B. & Clarke, C. 2001, *A*J, 122, 432
Sterzik M.F. & Durisen R.H., 1998, *A*&A, 339, 95
Thies, I. & Kroupa, P. 2008, *M*NRAS, 390, 1200
Thies, I., Kroupa, P., Goodwin, S. P., Stamatellos, D., & Whitworth, A. P. 2010, *A*pJ, 717, 577
Weidner, C., Kroupa, P., & Larsen, S. S. 2004, *M*NRAS, 350, 1503
Woitas, J., Leinert, C., Köhler, R. 2001, *A*&A, 376, 982

Carsten Weidner, Jan Pflamm-Altenburg, and Pavel Kroupa at the banquet

Computational Star Formation
Proceedings IAU Symposium No. 270, 2011
J. Alves, B.G. Elmegreen, J. Miquel, & V. Trimble, eds.

doi:10.1017/S1743921311000317

Simulations of the IMF in Clusters

Ralph E. Pudritz

Origins Institute, McMaster University, ABB 241,
Hamilton, Ontario L8S 4M1, Canada
email: pudritz@mcmaster.ca

Abstract. We review computational approaches to understanding the origin of the Initial Mass Function (IMF) during the formation of star clusters. We examine the role of turbulence, gravity and accretion, equations of state, and magnetic fields in producing the distribution of core masses - the Core Mass Function (CMF). Observations show that the CMF is similar in form to the IMF. We focus on feedback processes such as stellar dynamics, radiation, and outflows can reduce the accreted mass to give rise to the IMF. Numerical work suggests that filamentary accretion may play a key role in the origin of the IMF.

Keywords. initial mass function, turbulence, accretion, filaments, magnetic fields, feedback

1. Introduction

The Initial Mass Function (IMF) plays a central role in astrophysics because it encapsulates the complex physics of star formation. Observations suggest that the IMF can be variously described as a piece-wise power-law (Kroupa 2002), a lognormal (Miller & Scalo 1979), or a lognormal distribution with a high-mass power-law tail (Chabrier 2003). The high mass behaviour of the IMF Salpeter (1953) is a power-law, $dN \propto m^{-2.3}dm$ for stellar masses $m \geqslant 0.5M_{odot}$. The peak mass for isolated stars in the galactic disk is $0.1M_{\odot}$ and $0.2-0.3M_{\odot}$ (Chabrier 2003) for the bulge. The form of the IMF is similar in many different galactic and extragalactic environments such as globular clusters, wherein one has a large range of metallicities and concentrations (Paresce & De Marchi 2000). The current evidence therefore tends to support the notion that the IMF is universal.

What physical processes produce the IMF? Observations show that it emerges during the early stages of the formation of star clusters (Meyer *et al.* 2000; Zinnecker *et al.* 1993). Young stars are formed within gravitationally bound subunits of a cluster-forming environment known as "cores" whose mass distribution - the core mass function (CMF) - strongly resembles the IMF. Cores are closely associated with filaments, as recent observations using the Spitzer and Herschel observatories clearly show (André *et al.* 2010). Competing physical processes such as turbulence, gravity, cooling and thermodynamics, as well as magnetic fields play significant roles in building the CMF, as well as filaments, within molecular clouds. Feedback processes such as radiation from massive stars, jets and outflows, as well as stellar dynamics serve to truncate the accretion of material onto stars and their natal disks. These may lead to the emergence of the IMF from the CMF.

This review focuses on the critical role that computation is playing in exploring the origin of the IMF in clusters. We focus first on the processes leading to the CMF, and then discuss feedback processes that may convert the CMF into the IMF. Recent reviews of the computational aspects of the IMF may be found in Mac Low & Klessen (2004), Bonnell *et al.* (2007), Larson (2007), and Klessen *et al.* (2009). The theory of the IMF is covered by Hennebelle (this volume), and McKee & Ostriker (2007).

2. Structure formation and the IMF

Structure formation in the diffuse ISM as well as the dense molecular medium is a shocking affair. Diffuse atomic hydrogen near the midplane of our Milky Way is observed to be organized as a plethora of filamentary structure, bubbles, supernova remnants, HII regions (Taylor *et al.* 2003). Filamentary structure also characterizes giant molecular clouds - as seen in the large scale extinction maps of the Orion and Monocerus clouds (Cambrésy 1999). On smaller scales, a 850 micron continuum map of a 10pc region in Orion shows that Bonner-Ebert like cores are associated with filaments (Johnstone & Bally 2006). Herschel observations, such as those of the IRDC filament known as the "snake", show massive stars and a star cluster in formation (Henning *et al.* 2010).

How can cloud structure be characterized? One fruitful approach is to measure the probability distribution functions (PDFs) of the cloud column density of all of the gas in a given molecular cloud which is readily measured directly from the extinction data. This PDF for clouds without star formation (such as Lupus V and the Coal Sack clouds) turns out to be lognormal, whereas that of star forming clouds is a lognormal plus high mass, power-law tail (Kainulainen *et al.* 2009). This is interpreted as arising from the effects of gravity which drives collapse.

Filaments have various origins and arise in supersonic turbulent media due to the intersection of shocks waves, by gravitational break-up of self-gravitating sheets, or by thermal instabilities of various kinds. A physically plausible picture for the origin of the CMF is that filaments with sufficiently high values of their mass per unit length produce cores by gravitational instability. Gravitational fragmentation of filaments is well studied theoretically (Nagasawa 1987; Fiege & Pudritz 2000) but has received renewed emphasis with the Herschel observations.

The similarity in the functional form of the CMF and the IMF has been observed in many clouds starting with the study of Motte *et al.* (1998) in ρ Oph. The CMF in the Pipe Nebula, as another example, can be shifted into the IMF by converting a fraction $\epsilon \simeq 1/3$ of its mass to stars Alves *et al.* 2007.

3. From clouds to the CMF

3.1. *Turbulence*

Supersonic turbulence rapidly compresses gas into a hierarchy of sheets and filaments wherein the denser gas undergoes gravitational collapse to form stars (Porter *et al.* 1994; Vazquez-Semadeni *et al.* 1995; Ostriker *et al.* 1999; Klessen & Burkert 2001; Padoan *et al.* 2001; Bonnell *et al.* 2003; Tilley & Pudritz 2004; Krumholz *et al.* 2007). Supersonic gas is also highly dissipative and without constant replenishment, damps within a crossing time. There are many sources of turbulent motions that can affect molecular clouds, such as galactic spiral shocks in which most giant molecular clouds form, supernovae, expanding HII regions, radiation pressure, cosmic ray streaming, Kelvin-Helmholtz and Rayleigh-Taylor instabilities, gravitational instabilities, and bipolar outflows from regions of star formation (Elmegreen & Scalo 2004).

Supernova driven bubbles and turbulence in the galactic disk have been simulated by several groups. The work of de Avillez & Breitschwerdt (2004) simulates the high resolution (down t0 1.5 pc scales) global structure of the ISM as a function of the supernova rate. Densities range over six orders of magnitude, $10^{-4} \leqslant n \leqslant 10^{2}$ cm^{-3} and multiphase density PDFs are given. Simulations of the global structure of a supernova lashed, multi-component, instellar medium (Tasker & Bryan 2008; Wada & Norman 2007) find that the density PDF follows a lognormal distribution.

Supersonic turbulence produces hierarchical structure that can be described by a lognormal distribution (Vazquez-Semadeni 1994). A lognormal arises whenever the probability density of each new step in density increment in the turbulence is independent of the previous one. As an example, consider a medium that undergoes a series of random shocks whose strengths are uniformly distributed (Kevlahan & Pudritz 2009). The density at any point is the product of the shock-induced, density jumps. Taking the log of this relation, and then the limit of a large number of shocks, the central limit theorem then shows that the log of the gas density should be normally distributed - hence the lognormal. It can also be shown mathematically that the convergence of the distribution for a finite number of shocks is very rapid - just 3 or 4 shocks will give a distribution that is very highly converged to a lognormal.

Lognormal behaviour for the CMF has been found in a wide variety of simulations using various types of codes (eg. SPH, AMR) and setups (driven or not driven, periodic boxes, initial uniform spheres, etc.). Early results showed that lognormal behaviour in periodic box simulations is independent of details on how the turbulence is driven (Klessen 2001).

It is interesting that lognormal distributions appear across science (eg. physics, biology, medicine, etc.), not just in fluid mechanics. The key difference between normal and lognormal distributions in general is that the former arises for additive processes, whereas the latter arise in multiplicative ones (Limpert *et al.* 2001). As a concrete example, consider a simple dice game where one first adds the values on the faces of two thrown dice - the distribution of results (ranging from 2 to 12) is a normal distribution whose mean is 7. If one multiplies the two values however, the resulting distribution of numbers is highly skewed (ranging between 1 and 36) and is described by a lognormal.

Thus, lognormals characterize structure in the diffuse ISM as well as in denser molecular gas because shocks are the dominant process for configuring the gas. This is independent of exactly how the shocks are produced. Of physical significance are the mass of the peak of this distribution, and its width σ_o. The latter depends on both the thermal state of the gas as well as the rms Mach number of the turbulence. The standard deviation σ_o is found from the simulations, and takes the form; $\sigma_o^2 = ln(1+b^2M^2)$ where M is the rms Mach number of the turbulence and $b \simeq 0.5$ is a fitting parameter (Padoan *et al.* 1997).

3.2. *Gravity and accretion*

Adding gravity to turbulence changes this distribution - a power-law tail appears at the high mass end of the simulated CMF (Li *et al.* 2003; Tilley & Pudritz 2004). In the semi-analytic treatment of Padoan & Nordlund (2002), the power-law arises because of the turbulence spectrum of the turbulence. In Hennebelle & Chabrier (2008), a Press-Schecter formalism is adopted to argue that a lognormal plus power-law behaviour is the consequence of imposing a star formation threshold (eg. Jeans' criterion) and gravity at high mass.

Much of the debate concerning the origin of the CMF and IMF has focused on whether stars form by competitive accretion (Bonnell *et al.* 2001) or by the gravitational collapse of discrete cores (Krumholz *et al.* 2005). Cluster formation simulations often use initial top-hat density profiles (uniform spheres) that are chosen to mimic the observed initial conditions of cluster forming clumps (eg. a hundred solar masses of material, at temperature of 10K, size of half a pc, and mean density of at least 10^5 cm^{-3}). Such simulated clumps start to undergo global gravitational collapse in less than a free-fall time (eg. Bonnell *et al.* (2001), Tilley & Pudritz (2004)) as the turbulent energy is dissipated. The collapsing background ramps up the density of the gas including those in fluctuations such as the filaments. This drives up the accretion of gas into the filaments pushing some of them towards gravitational fragmentation. Sink-particles (taken as proxies for cores) first appear within the filaments. Cores will collapse more quickly than the collapsing background clump because they are denser and therefore have shorter free-fall times. The the collapse carries the collection of sink particles and filaments into the ever deepening potential well of the clump.

Cores are not isolated objects within filaments. Rather, they continue to undergo considerable filamentary accretion. The first objects to appear within most simulations generally become the most massive cores. In Banerjee *et al.* (2006), the rapid growth of the first, and most massive star by filamentary accretion was followed with an AMR (FLASH) code. These simulations showed that the filament was formed at the intersection of two sheets- shocks.

3.3. *Equations of state*

The thermal state of the bulk gas is a major factor in determining the local Jeans mass from point to point in molecular clouds. Both theory and simulations show that equation of state plays a very important role in controling the gravitational fragmentation of the gas. The local Jeans mass scales with the local temperature and density as $M_J \propto T^{3/2}\rho^{1/2}$. For simple polytropic equations of state $P \propto \rho^{\gamma}$, the Jean's mass can be written purely as a power law of the density; $M_J \propto \rho^{3/2(\gamma-(4/3))}$. This scaling suggests that for $\gamma > 4/3$, the Jeans mass increases with density, which puts an end to fragmentation. Indeed, simulations show that strong fragmentation prevails for $\gamma \leqslant 1$, gets progressively weaker for $\gamma > 1$, and stops altogether for $\gamma > 1.4$ (Li *et al.* 2003).

Local energy sources can raise the fragmentation mass by changing the temperature of the region. Thus, the accretion luminosity released by massive stars in particular, must reduce the degree of fragmentation within a localized region around such a heat source, as has been demonstrated by several groups (Krumholz *et al.* 2007). For massive stars, this region is limited in extent - roughly 1000 AU or so (see §4).

3.4. *Magnetic fields*

Magnetic fields play several different roles in star formation. As has long been known, they can control the gravitational stability of a gas if their energy density exceeds that of gravity. This is formalized by mass to flux ratio; $\Gamma = 2\pi\sqrt{G}\Sigma/B = 1.4\beta^{1/2}n_J^{1/3}$ where n_J is the number of Jeans masses and β is the ratio of gas to magnetic pressure. The fragmentation of a uniform cloud is highly suppressed for subcritical clouds ($\Gamma < 1$). Simulations of slightly supercritical clumps in uniform magnetic fields show that the field channels the collapse into large sheets that are perpendicular to the direction of the field. Slightly more supercritical clouds however, break up into more substructure including filaments (Tilley & Pudritz 2007). Turbulence creates a very broad range of magnetizations of cores. This is because shocks sweep material along field lines where it accumulates in filaments - increasing the mass to flux ratio in those regions, and greatly reducing the mass to flux ratio in the more diffuse zones left behind. Simulations show that initially supercritical magnetized clouds results in cores that range from critical to strongly supercritical (Padoan & Nordlund 2002; Tilley & Pudritz 2007), in agreement with the observations (Crutcher 2007). This may also account for the fact that magnetic fields are more dominant in the diffuse gas than in molecular gas.

When turbulence is added to a subcricital cloud, the column density PDF of a cloud is a lognormal but with a very small standard deviation. When ambipolar diffusion is added, the lognormal broadens considerably (Nakamura & Li 2008)). Supercritical subregions can form within the subcritical cloud, and it is within these regions that star formation can proceed, albeit at a very heavily reduced rate.

The second major aspect of magnetic fields is that twisted fields exert torques on gas and can therefore transport angular momentum away from spinning bodies. The origin of angular momentum in supersonic turbulence is that oblique shocks produce spinning cores (Jappsen & Klessen 2004; Tilley & Pudritz 2007)). Turbulence simulations produce a broad distribution of angular momenta of cores ranging over nearly two orders of magnitude. This implies that a broad distribution of disk sizes should result from the collapse of these systems - from very small to very large disks. The distribution of angular momentum vectors is also quite varied and is not particularly aligned with the filament principal axes.

Magnetic fields participate in several kinds of braking. On the level of the cores, torsional Alfvén waves have long been known to be able to extract significant amounts of angular momentum Basu & Mouschovias (1994). The collapse of rotating cores produces magnetized outflows as demonstrated in a variety of initial core models; cylinders (Tomisaka 2002), Bonner-Ebert spheres (Banerjee & Pudritz 2006), uniform spheres (Hennebelle & Fromang 2008), and singular isothermal spheres (Mellon & Li 2008). Early outflows could sweep up significant amounts of material and have been implicated as the basic physics in the CMF to IMF efficiency factor ϵ (see §4.3).

Finally, the combination of radiative and MHD effects (RMHD) limits the fragmentation of cluster-forming gas. Attempts to model RMHD on cluster scales have been made by Price & Bate (2009) who used Euler potentials to approximate the MHD in SPH. This work shows that the MHD in their code strongly supresses fragmentation even for clouds that are fairly supercritical (eg. $\Gamma = 3$). On smaller scales, grid-based RMHD methods applied to collapse and outflows show that radiative heating (in flux limited diffusion limit) makes substantial changes to the extent of outflows (Commerçon *et al.* 2010; Tomida *et al.* 2010).

4. Feedback: from the CMF to the IMF

4.1. *Stellar dynamics and filamentary accretion*

The collapse of $10^2 M_\odot$ clumps pulls both the material in the filaments and the sink particles deeper into the central potential well. The sink-sink interactions, which are modeled to be N-body gravity then come into play and are responsible for the creation of a small stellar cluster. The competitive accretion scenario predicts that these objects compete for gas in the dense centre of the gravitational potential well, resulting in an IMF (Bonnell *et al.* 2001). For larger clumps ($10^3 - 10^4 M_\odot$) the evolution of a cluster forming region is a hierarchical process in which subclusters form and ultimately merge (Bonnell *et al.* 2003). The resulting stellar interactions are frequent and close enough to truncate protostellar disks.

Sink particles as implemented by Federrath *et al.* (2010) have the additional desirable feature that they form in local potential minima giving them a more hydrodynamic character. This has an important consequence. As opposed to competitive accretion onto particles in a general potential well, filamentary accretion largely ends when the N-body interactions become strong enough to kick the sink particles out of their feeding filaments Evidence for filamentary accretion is seen in the simulations of Duffin *et al.* (2010a), and shown in Figure 1 (featuring the collapse of a $100 M_\odot \simeq 100 M_J$ initial tophat clump). The first frame, shows that sink particles form in filaments. The second shows that dynamical interactions have started between them as the collapse of the clump proceeds, and the third shows the dynamical end state. The final frame shows that the accretion histories of each particle shuts off when when dynamical interactions become important.

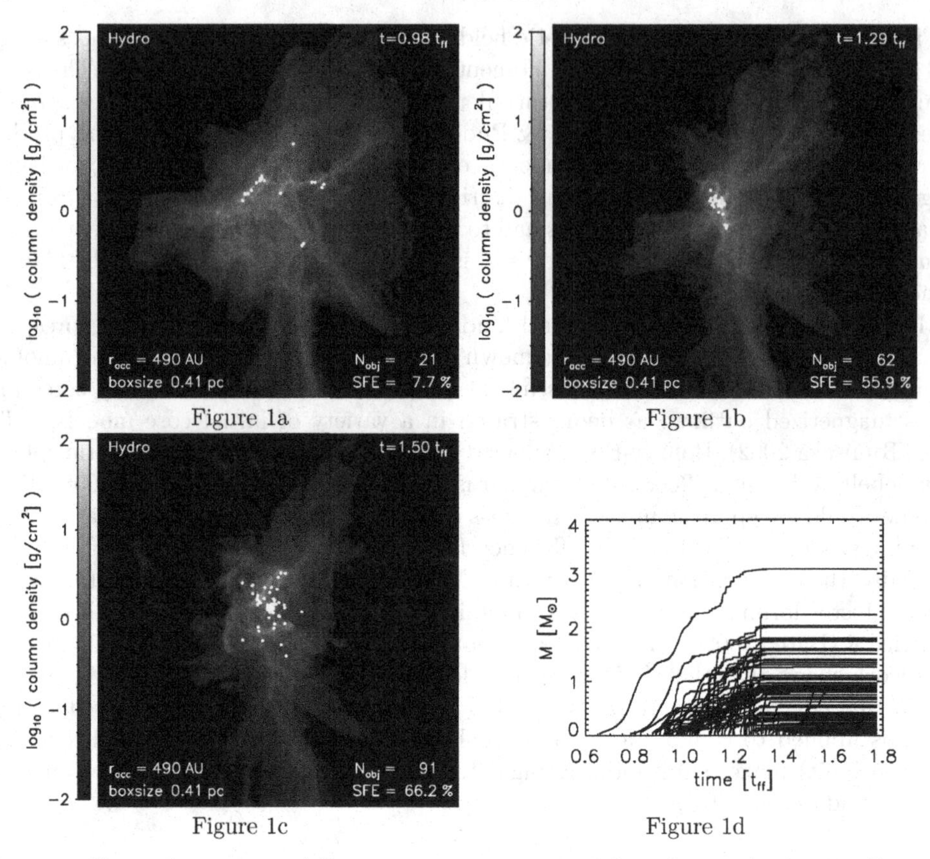

Figure 1. Cluster formation and filamentary accretion. Top left: sink particles form in filaments. Top right: sink particles begin dynamical interaction. Bottom left: cluster dynamics, accretion has ceased. Bottom right: mass accretion history for each sink particle - accretion ends when dynamical interactions begin (Duffin *et al.* 2010a).

4.2. *Radiative feedback*

Comprehensive 2D simulations of massive star formation including full frequency, radiative feedback effects and dust are presented by Yorke & Sonnhalter (2002). For cluster formation, a 3D treatment of radiative transfer in a highly inhomogenous medium is essential. Since Krumholz *et al.* (2007), it has become widely appreciated that radiative heating of the gas is needed to prevent excessive gravitational fragmentation. Radiative heating of the gas prevents the filaments from fragmenting as much as they might. Their gas drains primarily into the central massive forming disk and star, and fragmentation out to 1000 AU scales is prevented. These calculations still invoke grey, flux limited diffusion so there is still a need to examine the role of different frequency regimes in this process. Simulations by Bate (2009) confirm this picture using SPH techniques - wherein suppression of brown dwarfs by a factor of 4 is often observed compared to simulations without radiative feedback. The ionizing radiation from massive young stars becomes important during cluster formation as HII regions drive hot ioinized flows in the cluster environment. Results show that companions to massive stars limit the growth of the latter by a starvation effect (Peters *et al.* 2010).

4.3. *Outflows*

Outflows carry substantial amounts of mechanical energy which, if coupled into clump dynamics, could affect the IMF. Two aspects of feedback are important outflows in a cluster region (Klessen *et al.* 2009). The first is at the core level, in which an outflow removes the collapsing gas at some efficiency. If this efficiency is quite high (so that $\epsilon \simeq 1/3$), then one may resolve the hypothesized conversion of the CMF to IMF (Matzner & McKee 2000). Simulations show that the efficiency may be much lower (Duffin *et al.* 2010b). The second is at the level of the clump and the question as to whether or not the collection of outflows can continue to excite turbulence and thereby regulate cluster formation Norman & Silk (1980).

The situation at the local level is still unclear. Simulations by Banerjee *et al.* (2007) show that supersonic turbulence is not driven by magnetized jets. At the level of clumps, Nakamura & Li (2008) included ambipolar diffusion and outflows in a cloud that is nearly critical ($\Gamma = 1.1$ initially), and found, not unexpectedly, that the magnetic fields regulated a rather slow rate of star formation. On the other hand, Wang *et al.* (2010) used a mass to flux of $\Gamma = 1.4$ in a code without AD, and observed long time regulation of the cluster formation by outflows which maintained turbulence. These results may all depend upon the limited modeling of the the full dynamics of outflows that have been incorporated into the simulations.

5. Synthesis: the IMF of clusters

Numerical simulations have become the primary tool with which to investigate the origin of the IMF in clusters. Turbulence, the equation of state, and gravity play the key roles of filamenting the gas and breaking it into accreting cores describable by a lognormal distribution with high mass power-law tail. Radiative feedback controls fragmentation rates at the low mass end of the IMF whereas feedback by outflows may not be as efficient as previously claimed in converting the CMF to the IMF. Filamentary structure from clouds to clumps may turn out to play a key role in the entire process.

References

Alves, J., Lombardi, M., & Lada, C. J. 2007, *A&A*, 462, L17
André, P., *et al.* 2010, *A&A*, 518, L102+
Banerjee, R., Klessen, R. S., & Fendt, C. 2007, *ApJ*, 668, 1028
Banerjee, R. & Pudritz, R. E. 2006, *ApJ*, 641, 949
Banerjee, R., Pudritz, R. E., & Anderson, D. W. 2006, *MNRAS*, 373, 1091
Basu, S. & Mouschovias, T. C. 1994, *ApJ*, 432, 720
Bate, M. R. 2009, *MNRAS*, 392, 1363
Bonnell, I. A., Bate, M. R., Clarke, C. J., & Pringle, J. E. 2001, *MNRAS*, 323, 785
Bonnell, I. A., Bate, M. R., & Vine, S. G. 2003, *MNRAS*, 343, 413
Bonnell, I. A., Larson, R. B., & Zinnecker, H. 2007, in Protostars and Planets V, ed. B. Reipurth, D. Jewitt, & K. Keil, 149–164
Cambrésy, L. 1999, *A&A*, 345, 965
Chabrier, G. 2003, *PASP*, 115, 763
Commerçon, B., Hennebelle, P., Audit, E., Chabrier, G., & Teyssier, R. 2010, *A&A*, 510, L3+
Crutcher, R. 2007, in Star Formation, Then and Now
de Avillez, M. A. & Breitschwerdt, D. 2004, *A&A*, 425, 899
Duffin, D. F., Federrath, C., Pudritz, R. E., Banerjee, R., & Klessen, R. 2010a, in preparation
Duffin, D. F., Pudritz, R. E., Banerjee, R., & Seifried, D. 2010b, in preparation
Elmegreen, B. & Scalo, J. 2004, Ann. Rev. Astron. Astrophys., 42, 211

Federrath, C., Banerjee, R., Clark, P. C., & Klessen, R. S. 2010, *ApJ*, 713, 269
Fiege, J. D. & Pudritz, R. E. 2000, *MNRAS*, 311, 105
Hennebelle, P. & Chabrier, G. 2008, *ApJ*, 684, 395
Hennebelle, P. & Fromang, S. 2008, *A&A*, 477, 9
Henning, T., Linz, H., Krause, O., Ragan, S., Beuther, H., Launhardt, R., Nielbock, M., & Vasyunina, T. 2010, *A&A*, 518, L95+
Jappsen, A. & Klessen, R. S. 2004, *A&A*, 423, 1
Johnstone, D. & Bally, J. 2006, *ApJ*, 653, 383
Kainulainen, J., Beuther, H., Henning, T., & Plume, R. 2009, *A&A*, 508, L35
Kevlahan, N. & Pudritz, R. E. 2009, *ApJ*, 702, 39
Klessen, R. S. 2001, *ApJ*, 556, 837
Klessen, R. S. & Burkert, A. 2001, *ApJ*, 549, 386
Klessen, R. S., Krumholz, M. R., & Heitsch, F. 2009, ArXiv e-prints
Kroupa, P. 2002, Science, 295, 82
Krumholz, M. R., Klein, R. I., & McKee, C. F. 2007, *ApJ*, 656, 959
Krumholz, M. R., McKee, C. F., & Klein, R. I. 2005, *Nat*, 438, 332
Larson, R. B. 2007, Reports on Progress in Physics, 70, 337
Li, Y., Klessen, R. S., & Mac Low, M. 2003, *ApJ*, 592, 975
Limpert, E., Stahel, W. A., & Abbt, M. 2001, Bioscience, 51, 341
Mac Low, M.-M. & Klessen, R. S. 2004, Reviews of Modern Physics, 76, 125
Matzner, C. D. & McKee, C. F. 2000, *ApJ*, 545, 364
McKee, C. F. & Ostriker, E. C. 2007, *ARA&A*, 45, 565
Mellon, R. R. & Li, Z. 2008, *ApJ*, 681, 1356
Meyer, M. R., Adams, F. C., Hillenbrand, L. A., Carpenter, J. M., & Larson, R. B. 2000, Protostars and Planets IV, 121
Miller, G. E. & Scalo, J. M. 1979, *ApJS*, 41, 513
Motte, F., Andre, P., & Neri, R. 1998, *A&A*, 336, 150
Nagasawa, M. 1987, Progress of Theoretical Physics, 77, 635
Nakamura, F. & Li, Z. 2008, *ApJ*, 687, 354
Norman, C. & Silk, J. 1980, *ApJ*, 238, 158
Ostriker, E. C., Gammie, C. F., & Stone, J. M. 1999, *ApJ*, 513, 259
Padoan, P., Juvela, M., Goodman, A. A., & Nordlund, Å. 2001, *ApJ*, 553, 227
Padoan, P. & Nordlund, A. 2002, ApJ, 576, 870
Padoan, P., Nordlund, A., & Jones, B. J. T. 1997, *MNRAS*, 288, 145
Paresce, F. & De Marchi, G. 2000, *ApJ*, 534, 870
Peters, T., Banerjee, R., Klessen, R. S., Mac Low, M., Galván-Madrid, R., & Keto, E. R. 2010, *ApJ*, 711, 1017
Porter, D., Pouquet, A., & Woodward, P. 1994, Phys. Fluids, 6, 2133
Price, D. J. & Bate, M. R. 2009, *MNRAS*, 398, 33
Salpeter, E. E. 1953, Annual Review of Nuclear and Particle Science, 2, 41
Tasker, E. J. & Bryan, G. L. 2008, *ApJ*, 673, 810
Taylor, A. R., *et al.* 2003, *AJ*, 125, 3145
Tilley, D. A. & Pudritz, R. E. 2004, *MNRAS*, 353, 769
—. 2007, *MNRAS*, 382, 73
Tomida, K., Tomisaka, K., Matsumoto, T., Ohsuga, K., Machida, M. N., & Saigo, K. 2010, *ApJL*, 714, L58
Tomisaka, K. 2002, *ApJ*, 575, 306
Vazquez-Semadeni, E. 1994, *ApJ*, 423, 681
Vazquez-Semadeni, E., Passot, T., & Pouquet, A. 1995, *ApJ*, 441, 702
Wada, K. & Norman, C. A. 2007, *ApJ*, 660, 276
Wang, P., Li, Z., Abel, T., & Nakamura, F. 2010, *ApJ*, 709, 27
Yorke, H. W. & Sonnhalter, C. 2002, *ApJ*, 569, 846
Zinnecker, H., McCaughrean, M. J., & Wilking, B. A. 1993, in Protostars and Planets III, ed. E. H. Levy & J. I. Lunine, 429–495

Computational Star Formation
Proceedings IAU Symposium No. 270, 2011
J. Alves, B.G. Elmegreen, J. M. Girart & V. Trimble, eds.

doi:10.1017/S1743921311000329

Theories of the initial mass function

Patrick Hennebelle[1] **and Gilles Chabrier**[2]

[1]Laboratoire de radioastronomie, Ecole normale supérieure and Observatoire de Paris, UMR CNRS 8112 24 rue Lhomond, 75231 Paris Cedex 05, France
email: `patrick.hennebelle@ens.fr`

[2]CRAL, Ecole normale supérieure de Lyon, UMR CNRS 5574
Univesrité de Lyon, 69364 Lyon Cedex 07, France
email: `chabrier@ens-lyon.fr`

Abstract. We review the various theories which have been proposed along the years to explain the origin of the stellar initial mass function. We pay particular attention to four models, namely the competitive accretion and the theories based respectively on stopped accretion, MHD shocks and turbulent dispersion. In each case, we derive the main assumptions and calculations that support each theory and stress their respective successes and failures or difficulties.

Keywords. STARS: formation, ISM: clouds, gravitation, turbulence

1. Introduction

Stars are the building blocks of our universe and understanding their formation and evolution is one of the most important problems of astrophysics. Of particular importance is the problem of the initial mass function of stars (Salpeter 1955, Scalo 1986, Kroupa 2002, Chabrier 2003) largely because the star properties, evolution and influence on the surrounding interstellar medium strongly depend on their masses. It is generally found that the number of stars per logarithmic bin of masses, $dN/d\log M$, can be described by a lognormal distribution below 1 $M_\odot$, peaking at about $\simeq$0.3 $M_\odot$, and a power-law of slope -1.3 for masses between 1 and 10 $M_\odot$† (e.g. Chabrier 2003). It should be stressed that the IMF of more massive stars is extremely poorly known.

Several theories have been proposed to explain the origin of the IMF, invoking various physical processes that we tentatively classify in four categories: theories based on recursive fragmentation or pure gravity (e.g. Larson 1973), theories based on pure statistical argument, invoking the central limit theorem (e.g. Zinnecker 1984, Elmegreen 1997, Adams & Fatuzzo 1996), theories based on accretion and, finally, theories invoking the initial Jeans mass in a fluctuating environment. We will focus on the two latter ones, which appear to be favored in the modern context of star formation.

2. Theories based on accretion

2.1. *Competitive accretion*

The theory of competitive accretion has been originally proposed by Zinnecker (1982) and Bonnell *et al.* (2001). It has then been used to interpret the series of numerical simulations similar to the ones performed by Bate *et al.* (2003).

† It should be kept in mind that the *initial* mass function is measured only up to about 8 $M_\odot$, in young stellar clusters, and is inferred only indirectly for larger masses (see e.g. Kroupa 2002)

The underlying main idea is that the accretion onto the star is directly linked to its mass in such a way that massive stars tend to accrete more efficiently and thus become disproportionally more massive than the low mass stars. The accretion rate is written as:

$$\dot{M}_* = \pi \rho V_{rel} R_{acc}^2, \tag{2.1}$$

where ρ is the gas density, V_{rel} is the relative velocity between the star and the gas while R_{acc} is the accretion radius. Bonnell *et al.* (2001) consider two situations, namely the cases where the gravitational potential is dominated by the gas or by the stars.

2.1.1. *Gas dominated potential*

Let R be the spherical radius, ρ the gas density and n_* the number density of stars. In the gas dominated potential case, Bonnell *et al.* (2001) assume, following Shu (1977), that the gas density profile is proportional to R^{-2}. They further assume that $n_* \propto R^{-2}$. The accretion radius, R_{acc} is assumed to be equal to the tidal radius given by

$$R_{tidal} \simeq 0.5 \left(\frac{M_*}{M_{enc}} \right)^{1/3} R, \tag{2.2}$$

where M_{enc} is the mass enclosed within radius R. This choice is motivated by the fact that a fluid particle located at a distance from a star smaller than R_{tidal} is more sensitive to the star than to the cluster potential and will thus be accreted onto this star.

The mass of gas within a radius R, $M(R)$, is proportional to R (since $\rho \propto R^{-2}$). The infall speed is about $V_{in} \simeq \sqrt{GM(R)/R}$, and, assuming that the stars are virialized, one gets $V_{rel} \simeq V_{in}$. The number of stars, dN_*, located between R and $R+dR$ is given by the relation $dN_* = n_*(R) \times 4\pi R^2 dR \propto dR$. Thus eqn 2.1 combined with the expression of V_{rel} and R_{acc} leads to the relation $\dot{M}_* \propto (M_*/R)^{2/3}$ and, after integration, $M_* \propto R^{-2}$, $R \propto M_*^{-1/2}$. Consequently, we obtain:

$$dN \propto M_*^{-3/2} dM_*. \tag{2.3}$$

Even though the mass spectrum is too shallow compared to the fiducial IMF, $dN/dM \propto M^{-2.3}$, it is interesting to see that this power law behaviour can be obtained from such a simple model. Note, however, that the model implies that stars of a given mass are all located in the same radius, which points to a difficulty of the model.

2.1.2. *Star dominated potential*

When the potential is dominated by the stars located in the centre of the cloud, the density is given by $\rho \propto R^{-3/2}$, which corresponds to the expected density distribution after the rarefaction wave has propagated away (Shu 1977). The velocity is still assumed to be $V_{rel} \propto R^{-1/2}$. The accretion radius is now supposed to correspond to the Bondi-Hoyle radius as the gas and the star velocities are no longer correlated. This leads to $\dot{M}_* = \pi \rho V_{rel} R_{BH}^2$ where $R_{BH} \propto M_*/V_{rel}^2$. It follows

$$\dot{M}_* \propto M_*^{-2}. \tag{2.4}$$

One can then show (Zinnecker 1982, Bonnell *et al.* 2001) that under reasonable assumptions, $dN \propto M_*^2 dM_*$. This estimate is in better agreement with the Salpeter exponent, although still slightly too shallow. These trends seem to be confirmed by the simulations performed by Bonnell *et al.* (2001) which consists in distributing 100 sink particles in a cloud of total mass about 10 times the total mass of the sinks initially (their Fig. 3).

2.1.3. *Difficulties of the competitive accretion scenario*

As obvious from the previous analytical derivations, finding an explanation for the Salpeter exponent with the competitive accretion scenario appears to be difficult, even though numerical simulations (as the ones presented in Bonnell *et al.* 2001) seem to successfully achieve this task. However, although the IMF exponent is close to the Salpeter one in the star dominated potential case, this scenario entails by construction the Bondi-Hoyle accretion, which is at least a factor 3 lower than the mass infall rate resulting from gravitational collapse at the class O and I stages and leads to too long accretion times compared with observations (André *et al.* 2007, 2009). Another difficulty of this scenario is that it does not explain the peak of the IMF which might be related to the Jeans mass (see § 3). Finally, it is not clear that competitive accretion can work in the case of non-clustered star formation, for which the gas density is much too small. As no evidence for substantial IMF variation among different regions has yet been reported, this constitutes a difficulty for this model as a general model for star formation. Perhaps this scenario applies well to the formation of massive stars.

2.2. *Stopped accretion*

The principle of this type of models is to assume that the accretion of gas onto the stars or the dense cores is a non-steady process, stopped because of either the finite reservoir of mass or the influence of an outflow which sweeps up the remaining gas within the vicinity of the accreting protostar.

The first studies were performed by Silk (1995) and Adams & Fatuzzo (1996). They first relate the mass of the stars to the physical parameters of the cloud such as sound speed and rotation and then assume that an outflow whose properties are related to the accretion luminosity stops the cloud collapse. Using the Larson (1981) relations, they can link all these parameters to the clump masses. Since the mass spectrum of these latter is known (e.g. Heithausen *et al.* 1998), they infer the IMF.

A statistical approach has been carried out by Basu & Jones (2004). These authors assume that the dense core distribution is initially lognormal, justifying it by the large number of processes that control their formation (and invoking the central limit theorem). Then, they argue that the cores grow by accretion and postulate that the accretion rate is simply proportional to their mass, $\dot{M} = \gamma M \rightarrow M(t) = M_0 \exp(\gamma t)$, leading to $\log M = \mu = \mu_0 + \gamma t$. Finally, they assume that accretion is lasting over a finite period of time given by $f(t) = \delta \exp(-\delta t)$. The star mass distribution is thus obtained by summing over the accretion time distribution.

$$
\begin{aligned}
f(M) &= \int_0^\infty \frac{\delta \exp(-\delta t)}{\sqrt{2\pi}\sigma_0 M} \exp\left(-\frac{(\ln M - \mu_0 - \gamma t)^2}{2\sigma_0^2}\right) dt \\
&= \frac{\alpha}{2} \exp(\alpha\mu_0 + \alpha^2\sigma_0^2/2) M^{-1-\alpha} \mathrm{erf}\left(\frac{1}{\sqrt{2}}(\alpha\sigma_0 - \frac{\ln M - \mu_0}{\sigma_0})\right),
\end{aligned} \tag{2.5}
$$

where $\alpha = \delta/\gamma$ and σ_0 characterizes the width of the initial dense core distribution. As δ and α are controlled by the same types of processes, their ratio is expected to be of the order of unity and thus $f(M)$ exhibits a powerlaw behaviour close to the fiducial IMF.

In a recent study, Myers (2009) develops similar ideas in more details, taking into account the accretion coming from the surrounding background. Adjusting two parameters, he reproduces quite nicely the observed IMF (his figure 5).

A related model has also been developed by Bate & Bonnell (2005) based on an idea proposed by Price & Podsiadlowski (1995). They consider objects that form by fragmentation within a small cluster and are ejected by gravitational interaction with the other fragments, which stops the accretion process. Assuming a lognormal accretion rate and an exponential probability of being ejected, these authors construct a mass distribution that can fit the IMF for some choices of parameters.

In summary, the stopped accretion scenario presents interesting ideas and, providing (typically 2 or more) adequate adjustable parameters, can reproduce reasonably well the IMF. However, the very presence of such parameters, which characterizes our inability to precisely determine the processes that halt accretion, illustrates the obvious difficulties of this class of models, and their lack of predictive power and accuracy.

3. Gravo-turbulent theories

While in the accretion models, turbulence is not determinant, it is one of the essential physical processes for the two theories presented in this section, although the role it plays differs in both models, as shown below. The theories proposed along this line seemingly identify cores or *pre-cores* and are motivated by the strong similarity between the observed CMF and the IMF (e.g. André *et al.* 2010).

The first theory which combined turbulence and gravity was proposed by Padoan *et al.* (1997). In this paper, the authors consider a lognormal density distribution - density PDF computed from numerical simulations (e.g. Vázquez-Semadeni 1994, Kritsuk *et al.* 2007, Schmidt *et al.* 2009, Federrath *et al.* 2010) are indeed nearly lognormal - and select the regions of the flow which are Jeans unstable. By doing so, they get too stiff an IMF (typically $dN/dM \propto M^{-3}$) but nevertheless find a lognormal behaviour at small masses, a direct consequence of the lognormal density distribution, and a powerlaw one at large masses.

3.1. *Formation of cores by MHD shocks*

The idea developed by Padoan & Nordlund (2002) is slightly different. These authors consider a compressed layer formed by ram pressure in a weakly magnetized medium. They assume that the magnetic field is parallel to the layer and thus perpendicular to the incoming velocity field. The postshock density, ρ_1, the thickness of the layer, λ, and the postshock magnetic field, B_1, can be related to the Alfvénic Mach numbers, $\mathcal{M}_a = v/v_a$ (v is the velocity and v_a the Alfvén speed), and preshocked quantities, ρ_0 and B_0 according to the shock conditions:

$$\rho_1/\rho_0 \simeq \mathcal{M}_a \ , \ \lambda/L \simeq \mathcal{M}_a^{-1} \ , \ B_1/B_0 \simeq \mathcal{M}_a \ , \tag{3.1}$$

where L is the scale of the turbulent fluctuation. Note that for classical hydrodynamical isothermal shocks, the jump condition is typically $\propto \mathcal{M}^2$. The dependence on $\mathcal{M}_a$ instead of $\mathcal{M}^2$ stems from the magnetic pressure which is quadratic in B. As we will see, this is a central assumption of this model.

The typical mass of this perturbation is expected to be

$$m \simeq \rho_1 \lambda^3 \simeq \rho_0 \mathcal{M}_a \left(\frac{L}{\mathcal{M}_a} \right)^3 \simeq \rho_0 L^3 \mathcal{M}_a^{-2}. \tag{3.2}$$

As the flow is turbulent, the velocity distribution depends on the scale and $v \simeq L^\alpha$, with $\alpha = (n-3)/2$, $E(k) \propto k^{-n}$ being the velocity powerspectrum†. Combining these expressions with eqn (3.2), they infer

$$m \simeq \frac{\rho_0 L_0^3}{\mathcal{M}_{a,0}} \left(\frac{L}{L_0} \right)^{6-n}, \tag{3.3}$$

† n is denoted β in Padoan & Nordlund (2002), more precisely $n - 2 = \beta$

where L_0 is the largest or integral scale of the system and $\mathcal{M}_{a,0}$ the corresponding Mach number. To get a mass spectrum, it is further assumed that the number of cores, $N(L)$, formed by a velocity fluctuation of scale L, is proportional to L^{-3}. Combining this last relation with eqn (3.3) leads to

$$N(m) d\log m \simeq m^{-3/(6-n)} d\log m. \tag{3.4}$$

For a value of $n = 3.74$ (close to what is inferred from 3D numerical simulations), one gets $N(m) \simeq m^{-1.33}$, very close to the Salpeter exponent.

So far, gravity has not been playing any role in this derivation and the mass spectrum that is inferred is valid for arbitrarily small masses. In a second step, these authors consider a distribution of Jeans masses within the clumps induced by turbulence. As the density in turbulent flows presents a lognormal distribution, they *assume* that this implies a lognormal distribution of Jeans lengths and they multiply the mass spectrum (3.4) by a distribution of Jeans masses, which leads to

$$N(m) d\log m \simeq m^{-3/(6-n)} \left(\int_0^m p(m_J) dm_J \right) d\log m. \tag{3.5}$$

The shape of the mass spectrum stated by eqn 3.5 is very similar to the observed IMF (see for example the figure 1 of Padoan & Nordlund 2002).

Note, however, that difficulties with this theory have been pointed out by McKee & Ostriker (2007) and Hennebelle & Chabrier (2008). Eqn (3.1), in particular, implies that in the densest regions where dense cores form, the magnetic field is proportional to the density, in strong contrast with what is observed both in simulations (Padoan & Nordlund 1999, Hennebelle *et al.* 2008) and in observations (e.g. Troland & Heiles 1986). This is a consequence of the assumption that the magnetic field and the velocity field are perpendicular, which again is not the trend observed in numerical simulations. In both cases, it is found that at densities lower than about 10^3 cm^{-3}, B depends only weakly on n while at higher densities, $B \propto \sqrt{\rho}$. This constitutes a problem for this theory, as the index of the power law slope is a direct consequence of eqn (3.1). Assuming a different relation between B and ρ, as the aforementioned observed one, would lead to a slope stiffer than the Salpeter value. Furthermore, the Salpeter IMF is recovered in various purely hydrodynamical simulations (e.g. Bate *et al.* 2003), while the Padoan & Nordlund theory predicts a stiffer distribution ($dN/dM \propto M^{-3}$) in the hydrodynamical case. Another important shortcoming of this theory is that it predicts that turbulence, by producing overdense, gravitationally unstable areas, *always* promotes star formation, while it is well established from numerical simulations that the net effect of turbulence is to reduce the star formation efficiency (e.g. MacLow & Klessen 2004).

3.2. *Turbulent dispersion*

Recently, Hennebelle & Chabrier (2008, 2009, HC08, HC09) proposed a different theory which consists in counting the mass of the fluid regions within which gravity dominates over the sum of all supports, thermal, turbulent and magnetic, according to the Virial condition. In this approach, the role of turbulence is dual: on one hand it promotes star formation by locally compressing the gas but on the other hand, it also quenches star formation because of the turbulent dispersion of the flow, which is taken into account in the selection of the pieces of fluid that collapse.

The theory is formulated by deriving an extension of the Press & Schechter (1974, PS) statistical formalism, developed in cosmology. The two major differences are (i) the underlying density field, characterized by small and Gaussian fluctuations in the cosmological case while lognormal in the star formation case, and (ii) the selection criterion, a simple scale-free density threshold in cosmology while scale-dependent, based on the Virial theorem in the second case. That is, fluid particles which satisfy the criterion (see HC08)

$$\langle V_{\rm rms}^2 \rangle + 3\,(C_s^{eff})^2 < -E_{\rm pot}/M \tag{3.6}$$

are assumed to collapse and form a prestellar bound core. The turbulent rms velocity obeys a power-law correlation with the size of the region, the observed so-called Larson relation, $\langle V_{\rm rms}^2 \rangle = V_0^2 \times \left(\frac{R}{1{\rm pc}}\right)^{2\eta}$, with $V_0 \simeq 1\,{\rm km\,s^{-1}}$ and $\eta \simeq 0.4$-0.5 (Larson 1981).

The principle of the method is the following. First, the density field is smoothed at a scale R, using a window function. Then, the total mass contained in areas which, at scale R, have a density contrast larger than the specified density criterion δ_R^c, is obtained by integrating accordingly the density PDF. This mass, on the other hand, is also equal to the total mass located in structures of mass larger than a scale dependent critical mass M_R^c, which will end up forming structures of mass *smaller than or equal* to this critical mass for collapse (see HC08 §5.1).

$$\int_{\delta_R^c}^{\infty} \bar{\rho} \exp(\delta) \mathcal{P}_R(\delta) d\delta = \int_0^{M_R^c} M' \mathcal{N}(M')\, P(R, M')\, dM'. \tag{3.7}$$

In this expression, M_R^c is the mass which at scale R is gravitationally unstable, $\delta_R^c = log(\rho_R^c/\bar{\rho})$ and $\rho_R^c = M_R^c/(C_m R^3)$, C_m being a dimensionless coefficient of order unity. $\mathcal{P}_R$ is the (turbulent) density PDF, assumed to be lognormal, while $P(R, M')$ is the conditional probability to find a gravitationally unstable mass, M' embedded into M_R^c at scale R, assumed to be equal to 1 (see HC08 §5.1.2 and App. D). Note that this expression is explicitly solving the cloud in cloud problem as the mass which is unstable at scale R is spread over the structures of masses *smaller* than M_R^c (right hand side integral). Therefore, by construction, all the gravitationally unstable regions in the parent clump that will eventually collapse to form individual prestellar cores are properly accounted for in this theory (see HC08 §5.1).

Taking the derivative of eqn (3.7) with respect to R, we obtain the mass spectrum

$$\mathcal{N}(M_R^c) = \frac{\bar{\rho}}{M_R^c} \frac{dR}{dM_R^c} \left(-\frac{d\delta_R^c}{dR} \exp(\delta_R^c) \mathcal{P}_R(\delta_R^c) + \int_{\delta_R^c}^{\infty} \exp(\delta) \frac{d\mathcal{P}_R}{dR} d\delta \right). \tag{3.8}$$

While the second term is important to explain the mass spectrum of unbound clumps defined by a uniform density threshold (as the CO clumps), it plays a minor role for (Virial defined) bound cores and can generally be dropped (see HC08 for details).

After soma algebra and proper normalisation, one gets

$$\mathcal{N}(\widetilde{M}) = 2\mathcal{N}_0 \frac{1}{\widetilde{R}^6} \frac{1 + (1-\eta)\mathcal{M}_*^2 \widetilde{R}^{2\eta}}{[1 + (2\eta+1)\mathcal{M}_*^2 \widetilde{R}^{2\eta}]} \times \left(\frac{\widetilde{M}}{\widetilde{R}^3}\right)^{-\frac{3}{2} - \frac{1}{2\sigma^2} \ln(\widetilde{M}/\widetilde{R}^3)} \times \frac{\exp(-\sigma^2/8)}{\sqrt{2\pi}\sigma} \tag{3.9}$$

where $\widetilde{R} = R/\lambda_J^0$, $\delta_R^c = \ln\left\{(1 + \mathcal{M}_*^2 \widetilde{R}^{2\eta})/\widetilde{R}^2\right\}$, $\mathcal{N}_0 = \bar{\rho}/M_J^0$ and M_J^0, λ_J^0 denote the usual thermal Jeans mass and Jeans length, respectively, and

$$\widetilde{M}(R) = M/M_J^0 = \widetilde{R}\,(1 + \mathcal{M}_*^2 \widetilde{R}^{2\eta}) \tag{3.10}$$

denotes the unstable mass at scale R in the turbulent medium. The theory is controlled by two Mach numbers, namely

$$\mathcal{M}_* = \frac{1}{\sqrt{3}} \frac{V_0}{C_s} \left(\frac{\lambda_J^0}{1\text{pc}} \right)^\eta \approx (0.8 - 1.0) \left(\frac{\lambda_J^0}{0.1\,\text{pc}} \right)^\eta \left(\frac{C_s}{0.2\text{km}\,\text{s}^{-1}} \right)^{-1}, \tag{3.11}$$

defined as the non-thermal velocity to sound speed ratio at the mean Jeans scale λ_J^0 (and not at the local Jeans length), and the usual Mach number, $\mathcal{M}$, which represents the same quantity at the scale of the turbulence injection scale, L_i, assumed to be the characteristic size of the system, $\mathcal{M} = \frac{\langle V^2 \rangle^{1/2}}{C_s}$.

The global Mach number, $\mathcal{M}$, broadens the density PDF, as $\sigma^2 = \ln(1 + b^2\mathcal{M}^2)$, illustrating the trend of supersonic turbulence to promote star formation by creating new overdense collapsing seeds.

The effect described by $\mathcal{M}_*$ is the additional non thermal support induced by the turbulent dispersion. In particular, at large scales the net effect of turbulence is to stabilize pieces of fluid that would be gravitationally unstable if only the thermal support was considered. This is illustrated by eqn (3.10) which reduces to $\widetilde{M} = \widetilde{R}$ when $\mathcal{M}_* = 0$. In particular, for a finite cloud size, the gas whose associated turbulent Jeans length is larger than the cloud size is not going to collapse.

When $\mathcal{M}_* \ll 1$, i.e. the turbulent support is small compared to the thermal one, eqn (3.9) shows that the CMF at large masses is identical to the Padoan *et al.* (1997) result, i.e. $dN/d\log M \propto M^{-2}$. On the other hand, when $\mathcal{M}_* \simeq 1$, $dN/d\log M \propto M^{-(n+1)/(2n-4)}$, where the index of the velocity powerspectrum n is related to η by the relation $\eta = (n-3)/2$ (see HC08). As $n \simeq 3.8 - 3.9$ in supersonic turbulence simulations (e.g. Kritsuk *et al.* 2007), turbulent dispersion leads to the correct Salpeter slope and Larson velocity-size relation.

Comparisons with the Chabrier (2003) IMF have been performed for a series of cloud parameters (density, size, velocity dispersion) and good agreement has been found (HC09) for clouds typically 3 to 5 times denser than the mean density inferred from Larson (1981) density-size relation. Comparisons with numerical simulations have also been performed. In particular, Schmidt *et al.* (2010), performing supersonic isothermal simulations with various forcing, have computed the mass spectrum of cores supported either by pure thermal support or by turbulent plus thermal support. Their converged simulations show very good *quantitative* agreement with the present theory, confirming that turbulent support is needed to yield the Salpeter index. Note that Schmidt *et al.* (2010) use for the density PDF the one they measure in their simulations which is nearly, but not exactly lognormal. Comparisons with the results of SPH simulations (Jappsen *et al.* 2005) including self-gravity and thermal properties of the gas have also been found to be quite successful (HC09).

3.3. *Difficulties of the gravo-turbulent theories*

One natural question about any IMF theory is to which extent it varies with physical conditions. Indeed, there is strong observational support for a nearly invariant form and peak location of the IMF in various environments under Milky Way like conditions (see e.g. Bastian *et al.* 2010). Jeans length based theories could have difficulty with the universality of the peak position, since it is linked to the Jeans mass which varies with the gas density. Various propositions have been made to alleviate this problem. Elmegreen *et al.* (2008) and Bate (2009) propose that the gas temperature may indeed increase with density, resulting in a Jeans mass which weakly depends on the density, while HC08 propose that for clumps following Larson relations, there is a compensation between the density dependence of the Jeans mass and the Mach number dependence of the density PDF, resulting in a peak position that is insensitive to the clump size.

A related problem is the fact that massive stars are often observed to be located in the densest regions, where the Jeans mass is smaller. Indeed, $M_J \propto \rho^{-1/2}$ when a purely thermal support is considered, whereas $M_J \propto \rho^{-2}$ when turbulence is taken into account (assuming that $V \propto L^{0.5}$) (see HC09). This constitutes a difficulty for theories based on Jeans mass although, as seen above, the issue is much less severe when turbulent support is considered as massive stars can be formed at densities only few times smaller than the densities at which low mass stars form. Another possibility is that dynamical interactions between young protostars may lead to the migration of massive stars in the center of the gravitational well.

Furthermore, the dependence of the freefall time on the Jeans mass should also modify the link between the CMF and the IMF, as pointed out by Clark *et al.* (2007). This is particularly true for theories which invoke only thermal support. When turbulent support is included, the free-fall time is found to depend only weakly on the mass, with $t_{ff} \propto M^{1/4}$ (see McKee & Tan 2003 and HC09 App. C), resolving this collapsing time problem. We stress that this time represents the time needed for the whole turbulent Jeans mass to be accreted. It is certainly true that, within this turbulent Jeans mass, small structures induced by turbulent compression will form rapidly. Their total mass, however, is expected to represent only a fraction of the total turbulent Jeans mass because the net effect of turbulence is to decrease the star formation efficiency (see HC08).

Generally speaking, the fragmentation that occurs during the collapse could constitute a problem for theories invoking Jeans masses. Although this problem is far from being settled, it should be stressed that such a fragmentation process is not incompatible with the calculations performed by HC08. As shown by eqn 3.7, the presence of small self-gravitational condensations induced by turbulence at the early stages of star formation and embedded into larger ones is self-consistently taken into account in the theory. Moreover, the SPH simulations performed by Smith *et al.* (2008) show a clear correlation between the initial masses within the gravitational well and the final sink masses up to a few local freefall times (see Chabrier & Hennebelle 2010 for a quantitative analysis), suggesting that the initial prestellar cores do not fragment into many objects. As time goes on, the correlation becomes weaker but seems to persist up to the end of their run. Massive stars, on the other hand, are weakly correlated with the mass of the potential well in which they form. Whether their mass was contained into a larger more massive well with which the final sink mass would be well correlated remains an open issue, which needs to be further investigated. At last, both the magnetic field (e.g. Machida *et al.* 2005, Hennebelle & Teyssier 2008) and the radiative feedback (Bate 2009, Offner *et al.* 2009) will reduce the fragmentation, suggesting that the core-sink correlation found in Smith *et al.* (2008) should improve if such processes were included. Clearly, these questions require careful investigations.

4. Conclusion

We have reviewed the most recent theories which have been proposed to explain the origin of the IMF. Due to limited space, it was not possible to cover all of them (e.g. Kunz & Mouschovias 2009). Two main categories received particular attention: the theories based on accretion and the ones based explicitly on turbulence. It should be stressed that these theories are not all exclusive from each other and may apply in different ranges of mass. For instance, the turbulent dispersion theory calculates the distribution of the initial mass accretion reservoirs; it is not incompatible with the stopped accretion theories and with the competitive accretion as long as mass redistribution/competition occurs within one parent core reservoir. The question as to whether one of these mechanisms is dominant is yet unsettled. Detailed comparisons between systematic sets of simulations, as done in Schmidt *et al.* (2010), or observations, and the various analytical predictions is clearly mandatory to make further progress.

References

Adams, F. & Fatuzzo, M. 1996, *ApJ*, 464, 256

André, Ph., Belloche, A., Motte, F., & Peretto, N., 2007, *A&A*, 472, 519

André, Ph., Basu, S., & Inutsuka, S-I., 2009, in *Structure Formation in Astrophysics*, G. Chabrier Ed., Cambridge University Press

André, P., Men'shchikov, A., & Bontemps, S. *et al.* *A&A*, in press, arXiv1005.2618

Bastian, N., Covey, K., & Meyer, M., 2010, *ARA&A*, 48

Basu, S. & Jones, C., 2004, *MNRAS*, 347, L47

Bate, M., Bonnell, I., & Bromm, V., 2003, *MNRAS*, 339, 577

Bate, M., 2009, *MNRAS*, 392, 1363

Bate, M. & Bonnell, I., 2005, *MNRAS*, 356, 1201

Bonnell I., Bate M., Clarke C., & Pringle J., 2001, *MNRAS* 323, 785

Chabrier, G., 2003, *PASP*, 115, 763

Chabrier, G. & Hennebelle, P., 2010, *ApJL in press*, arXiv1011.1185

Clark, P., Klessen, R., & Bonell, I., 2007, *MNRAS* 379, 57

Elmegreen B.G., 1997, *ApJ* 486, 944

Elmegreen, B., Klessen, R., & Wilson, C., 2008, *ApJ*, 681, 365

Federrath, C., Roman-Duval, J., Klessen, R., Schmidt, W., & Mac Low, M.-M., 2010, *A&A*, 512, 81

Heithausen, A., Bensch, F., Stutzki, J., Falgarone, F., & Panis, J.-F., 1998, *A&A*, 331, L65

Hennebelle, P. & Teyssier, R., 2008, *A&A*, 477, 25

Hennebelle, P., Banerjee, R., Vázquez-Semadeni, E., Klessen, R. & Audit, E., 2008, *A&A*, 446, 43

Hennebelle, P. & Chabrier, G., 2008, *ApJ*, 684, 395 (HC08)

Hennebelle, P. & Chabrier, G., 2009, *ApJ*, 702, 1428 (HC09)

Jappsen, A., Klessen, R., Larson, R., Li, Y., & Mac Low,M.-M., 2005, A&A, 435, 611

Kritsuk, A. G., Norman, M. L., Padoan, P., & Wagner, R. 2007, *ApJ*, 665, 416

Kroupa P., 2002, *Sci.* 295, 82

Kunz M. & Mouschovias T., 2009, *MNRAS* 399L, 94

Larson R., 1973, *MNRAS*, 161, 133

Larson R., 1981, *MNRAS*, 194, 809

Machida, M., Matsumoto, T., Hanawa, T., & Tomisaka, K., 2005, *MNRAS*, 362, 382

McKee C.F. & Ostriker J.P., 2007, *ApJ* 218, 448

McKee C.F. & Tan T., 2003, *ApJ* 585, 850

MacLow, M.-M. & Klessen, R., 2004, *Rev. Mod. Phys.*, 76, 125

Myers P.C., 2009, *ApJ*, 706, 1341

Offner, S., Klein, R., McKee, C., & Krumholz, M., 2009, *ApJ*, 703, 131

Padoan, P., Nordlund, A., & Jones, B., 1997, *MNRAS*, 288, 145

Padoan, P. & Nordlund, 1999, *ApJ*, 526, 279

Padoan, P. & Nordlund, 2002, *ApJ*, 576, 870

Press, W. & Schechter, P., 1974, *ApJ*, 187, 425

Price, N. & Podsiadlowski, P., 1995, MNRAS, 275, 1041

Salpeter, E., 1955, *ApJ*, 121, 161

Scalo J., 1986, *FCPh*, 11, 1

Silk, J., 1995, *ApJ*, 438, L41

Shu F., 1977, *ApJ*, 214, 488

Schmidt, W., Federrath, C., Hupp, M., Kern, S., & Niemeyer, J., 2009, *A&A*, 494, 127

Schmidt, W., Kern, S., Federrath, C., & Klessen, R., 2010, *A&A*, 516, 25
Smith, R., Clark, P., & Bonnell, I., 2008, *MNRAS*, 391, 1091
Troland T. & Heiles C., 1986, *ApJ*, 301, 339
Vázquez-Semadeni, E., 1994, *ApJ*, 423, 681
Zinnecker H., 1982, in Glassgold A. E. *et al.*, eds, Symposium on the Orion Nebula to Honour Henry Draper. New York Academy of Sciences, New York, p. 226
Zinnecker H., 1984, *MNRAS*, 210, 43

Patrick Hennebelle

Computational Star Formation
Proceedings IAU Symposium No. 270, 2011
J. Alves, B.G. Elmegreen, J. M. Girart & V. Trimble, eds.

doi:10.1017/S1743921311000330

Magnetic fields and Turbulence in Star Formation using Smoothed Particle Hydrodynamics

Daniel J. Price

Centre for Stellar and Planetary Astrophysics,
Monash University, Clayton Vic 3800, Australia.
email: daniel.price@monash.edu

Abstract. Firstly, we give a historical overview of attempts to incorporate magnetic fields into the Smoothed Particle Hydrodynamics method by solving the equations of Magnetohydrodynamics (MHD), leading an honest assessment of the current state-of-the-art in terms of the limitations to performing realistic calculations of the star formation process. Secondly, we discuss the results of a recent comparison we have performed on simulations of driven, supersonic turbulence with SPH and Eulerian techniques. Finally we present some new results on the relationship between the density variance and the Mach number in supersonic turbulent flows, finding $\sigma^2_{\ln\rho} = \ln(1 + b^2\mathcal{M}^2)$ with $b = 0.33$ up to Mach 20, consistent with other numerical results at lower Mach number (Lemaster & Stone 2008) but inconsistent with observational constraints on σ_ρ and $\mathcal{M}$ in Taurus and IC5146.

Keywords. magnetic fields, magnetohdyrodynamics (MHD), turbulence, methods: numerical, ISM: clouds

---

1. Introduction

Magnetic fields and turbulence are thought to be two of the most important ingredients in the star formation process, so it is critical that we are able to model the effect of both in star formation simulations. Smoothed Particle Hydrodynamics (SPH, for recent reviews see Monaghan 2005; Price 2004), since it is a Lagrangian method where resolution follows mass, is a very natural method to use in order to model star formation. However the introduction of magnetic fields in SPH has a somewhat troubled history, and there are perceptions that the explicit use of artificial viscosity terms in order to treat shocks is too crude to enable accurate modelling of supersonic turbulence. We will discuss both of these aspects in this talk.

2. Magnetic fields in SPH

2.1. *A prehistory of MHD in SPH*

Magnetic fields were introduced in one of the founding SPH papers by Gingold & Monaghan (1977), though a detailed investigation did not follow until around a decade later with the publication of Graham Phillips' PhD work (Phillips & Monaghan 1985) (PM85), with successful application of the method to star formation problems involving the collapse of isothermal, non-rotating, magnetised clouds (e.g. Phillips 1986)†. Indeed, it was Phillips & Monaghan (1985) who first discovered that the equations of motion for MHD

† Graham Phillips now pursues a rather different career as presenter of the science show 'Catalyst' on Australian television (I suspect this was easier than getting MHD in SPH to work).

– when expressed in a momentum-conserving form – were unstable to a clumping or 'tensile' instability whereby particles attract each other unstoppably along field lines, that occurs in the regime where the magnetic pressure exceeds the gas pressure (i.e., $\beta < 1$).

Whilst Phillips & Monaghan proposed a way of achieving stability, another decade followed before the matter was investigated in any more detail, by another PhD student of Joe Monaghan's, Joe Morris. Morris' PhD thesis (Morris 1996) contained detailed stability analysis of the equations of the now-named Smoothed Particle Magnetohydrodynamics (SPMHD) and proposed a way of avoiding the instability by using more accurate derivatives for the anisotropic part of the magnetic force in place of exact conservation of momentum. Around the same time Meglicki (1995) applied a non-conservative SPMHD formulation to magnetic fields in the Galaxy. The period after these works saw a few but limited applications of SPMHD to 'real' astrophysical problems, including notably an application to magnetic fields in galaxy clusters by Dolag, Bartelmann & Lesch (1999)† and at least one foray into star formation by Byleveld & Pongracic (1996). However it would be fair to say that there was no real consensus on a standard approach and that many problems remained.

2.2. *21st century Smoothed Particle Magnetohydrodynamics*

Børve, Omang & Trulsen (2001) suggested an alternative to Morris' stabilisation based on explicitly subtracting the non-zero source term ($-\mathbf{B}\nabla\cdot\mathbf{B}$) that is the cause of the stability problems in the momentum-conserving formulation. They also proposed a regularisation scheme for the SPH particle distribution which, though complicated, was found to give excellent results on MHD shock problems by the focussing of resolution where it is needed. Price & Monaghan (2004a) adopted (but later abandoned) a stability fix proposed by Monaghan (2000) and showed how the dissipative terms could be formulated for treating MHD shocks and other discontinuities in a standard SPMHD approach (that is, without regularisation). Paper II (Price & Monaghan 2004b) showed how the terms necessary for strict conservation in the presence of a spatially variable smoothing length could be included by deriving the SPMHD equations of motion from a variational principle. At the same time Børve *et al.* (2004) undertook a detailed stability analysis showing that indeed their source-term approach was indeed stable in more than one dimension.

With these improvements meant that SPMHD could be successfully run on many of the test problems used to test grid-based MHD codes (Paper III, Price & Monaghan 2005). However, accuracy in realistic problems was found to depend on the accuracy of the divergence-free condition on the magnetic field – perhaps the key problem for any numerical MHD scheme – since the $\nabla\cdot\mathbf{B} = 0$ constraint enters the ideal MHD equations only as an initial condition:

$$\frac{\partial \mathbf{B}}{\partial t} = \nabla \times (\mathbf{v} \times \mathbf{B}); \qquad \frac{\partial}{\partial t}(\nabla \cdot \mathbf{B}) = 0. \tag{2.1}$$

As aptly summarised by Mike Norman (this proceedings), approaches to the divergence constraint can be roughly categorised into "ignore", "clean" or "prevent". Many test problems for MHD can be run quite happily (in SPH or otherwise) with the "ignore" approach: simply evolve the induction equation (2.1) and cross your fingers (I refer to this as the "hope and pray" method). For realistic applications this is almost never a good idea and in SPMHD simulations of star formation manifests as 'exploding stars' due to large errors in the magnetic force (see Price & Federrath 2009 for a figure). Price & Monaghan (2005) examined a range of cleaning methods for the divergence constraint, including most promisingly the hyperbolic/parabolic cleaning scheme proposed by Dedner *et al.*

† Harald Lesch is now also a science show presenter, on German television.

(2002). However this was not found to be sufficient to prevent divergence errors generated in realistic flows, least of all during protostellar collapse.

2.3. *The Euler Potentials*

Thus, Price & Bate (2007) and Rosswog & Price (2007) adopted a preventative approach, based on writing the magnetic field in terms of the 'Euler Potentials' (Stern 1976), variously known as 'Clebsch' or 'Sakurai' variables:

$$\mathbf{B} = \nabla\alpha \times \nabla\beta, \tag{2.2}$$

with which the divergence constraint is satisfied by construction. Furthermore the ideal MHD induction equation takes the particularly simple form

$$\frac{d\alpha}{dt} = 0; \qquad \frac{d\beta}{dt} = 0, \tag{2.3}$$

corresponding physically to the advection of magnetic field lines by Lagrangian particles. This is then combined with a standard SPMHD approach by computing $\mathbf{B}$ according to Eq. 2.2 and using the standard equation of motion with either the Morris or Børve *et al.* (2001) force stabilisation (thus it is usually trivial to revert to a $\mathbf{B}$-based formulation – that is, solving Eq. 2.1 instead of Eqs. 2.2–2.3 – for comparison purposes.

Adoption of this approach meant that we were able to study realistic star formation problems in 3D, including the magnetised collapse of a rotating core to form single and binary stars (Price & Bate 2007) and the effect of magnetic fields on star cluster formation from turbulent molecular clouds (Price & Bate 2008), most recently including both radiation and magnetic fields (Price & Bate 2009). Around this time magnetic fields were also implemented in the widely used SPH code GADGET (Dolag & Stasyszyn 2009).

From these we have learnt that magnetic fields can significantly change star formation even with weak (supercritical) fields, significantly affecting the formation of circumstellar discs and binary stars (Price & Bate 2007, similar to the findings of many other authors, e.g. Hennebelle & Teyssier 2008; Hennebelle & Ciardi 2009; Mellon & Li 2008; Machida *et al.* 2008) and having a profound effect on the star formation rate (Price & Bate 2008, 2009).

However there are important limitations to the Euler Potentials (EPs) formulation. The most important are that:

(*a*) They can only be used to represent topologically trivial fields,

(*b*) As a consequence of a), the wind-up of a magnetic field in a rotating or turbulent flow cannot be followed indefinitely,

(*c*) It is difficult to incorporate single fluid non-ideal MHD effects such as resistivity.

The topological restrictions are a consequence of the fact that the helicity $\mathbf{A}\cdot\mathbf{B}$ is identically zero in the EP representation. The lack of winding processes is easily understood as a consequence of Eq. 2.3: Since the potentials are simply advected by the particles, the reconstruction of the magnetic field via Eq. 2.2 relies on a one-to-one mapping between the initial and final particle positions. Thus, for example, a complete rotation of an inner ring of particles will return the magnetic field to its original configuration, whereas physically the field should 'remember' the winding [in the talk this was illustrated by a dance with J. Monaghan which is difficult to insert in the proceedings].

In practice this means that in using the EPs one is limited to the study of problems where the magnetic field is initially simple (e.g. imposed as a constant field) and at some scale the field winding will be lost (underestimated) in the calculations. The difficulty in representing physical resistive processes is also an issue given the importance of non-ideal MHD effects for star formation.

2.4. *Why don't you just use the Vector Potential?*

At one of the last major computational star formation meetings, held at the Kavli Institute for Theoretical Physics during 2007, Axel Brandenburg suggested that the use of the vector potential, with the gauge set to give Galilean invariance, may provide similar advantages in terms of the divergence constraint without the topological restrictions of the Euler potentials. Three years and 25 pages of pain later (Price 2010) we had derived the Ultimate Vector Potential Formulation^{TM} for SPMHD. It was beautiful, derived elegantly from a variational principle, the method was exactly conservative, with a brand new force equation that in principle might not suffer from instabilities since the divergence constraint was built-in... could it be that we had uncovered the ultimate method for MHD in SPH? Unfortunately the vector potential approach failed spectacularly in practice due to two problems: 1) The same old force-related instability in 2D and 3D, which could, however, be cured by reverting to the stable force formulations discussed above, though with more difficulty than in the standard SPMHD case, and 2) An instability peculiar to the 3D vector potential related to the unconstrained growth of non-physical components of **A**, occurring, for example, when a 2D test problem like the Orszag-Tang vortex is run in 3D. Further investigation suggests the latter is related to not enforcing the gauge condition $\nabla \cdot \mathbf{A} = 0$ on the vector potential – that is, simply pushing the divergence problem one level up. Thus, we must unfortunately report that the answer to the vector potential question is that it not a good approach for SPMHD.

2.5. *Alternatives to the Euler Potentials*

The most promising approach to lifting the restrictions of the Euler Potentials whilst preserving the benefits involve moving to a generalised formulation, where the most general involves three sets of potentials, in the form

$$\mathbf{B} = \nabla\alpha_1 \times \nabla\beta_1 + \nabla\alpha_2 \times \nabla\beta_2 + \nabla\alpha_3 \times \nabla\beta_3, \tag{2.4}$$

essentially taking advantage of the fact that, whilst sets of potentials cannot be added together, the fields can. Thus one can represent arbitrary field geometries. The reason for the three sets is because one of the potentials β is essentially preserving one of the initial spatial coordinates so that a Jacobian map can be constructed – for example 2D fields are represented by $\alpha = \alpha(x, y)$ and $\beta = z_0$. Thus in the most general case all three coordinates of the particles $(\beta_1, \beta_2, \beta_3) = (x_0, y_0, z_0)$ are used to compute the mapping. Better still, adopting this more general formulation means that the potentials can be re-mapped to a new set whenever the gradients break down, meaning that physical field winding can be followed indefinitely, and with explicit control on the amount of numerical dissipation. Watch this space.

An alternative is to return to the 'clean' approach, despite the only limited success of the formulations considered by Price & Monaghan (2005), at least for star formation problems. Use of cleaning methods has found more success in cosmological applications (Dolag & Stasyszyn 2009) and there are further possibilities for improving the accuracy of the projection methods discussed in Price & Monaghan (2005) that offer some promise.

3. Supersonic turbulence with SPH

Progress on solving the issues with magnetic fields in SPH generally slows due to the need to periodically address concerns over the core SPH method. Most recently this has involved a rather inflamed and unhelpful discussion of Kelvin-Helmholtz instabilities

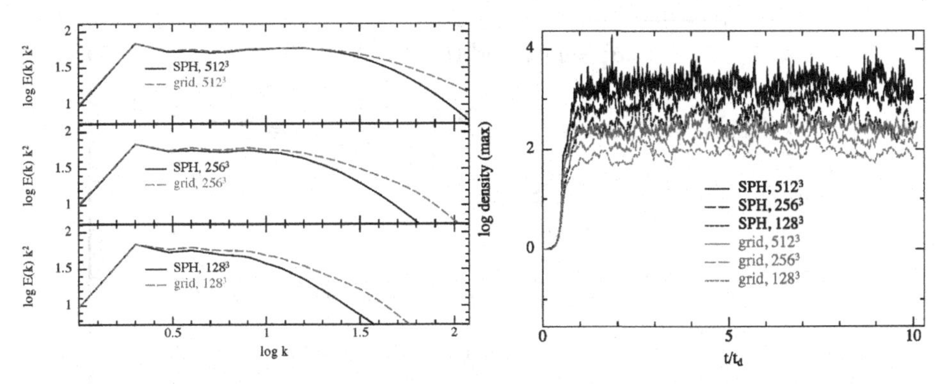

Figure 1. Time-averaged kinetic-energy power spectra (left), and the time evolution of the maximum density (right) at three different resolutions in the SPH vs. grid driven turbulence comparison by Price & Federrath (2010). Comparable results on volume-weighted quantities such as the kinetic energy power spectrum requires roughly $n_{cells} \approx n_{particles}$ (left), but SPH resolves maximum densities at 128^3 similar to a grid at 512^3 (right).

in SPH†, and which many authors are now using as the primary justification for the often fruitless pursuit of all manner of alternative particle methods. Computational star formation is no stranger to such debates, though thankfully at least the heated discussions surrounding 'artificial fragmentation' (Klein *et al.* 2004) went quietly away, leading to a more helpful consensus on the implications of the physical approximations (e.g. use of a barotropic equation of state vs. radiation hydrodynamics) rather than arguments over numerics. However there remain sharp disagreements over numerics in some areas, for example the following in a discussion of turbulence simulations by Ballesteros-Paredes *et al.* (2006) given in Padoan *et al.* (2007):

> "SPH simulations of large scale star formation fields to date fail in all three fronts: numerical diffusivity, numerical resolution and presence of magnetic fields. This should case serious doubts on the value of comparing predictions based on SPH simulations with observational data (see also Agertz *et al.* 2007)",

(note the completely irrelevant reference to Agertz *et al.* 2007). Such concerns over turbulence helped motivate at least two major comparison projects – the 'Potsdam' comparison (Kitsionas *et al.* 2009) and the 2007 KITP comparison project, both studying the decay of supersonic turbulence from pre-evolved initial conditions.

Given that comparisons of decaying turbulence are necessarily limited to a few snapshots and that there are issues surrounding the setup of initial conditions between codes, we additionally undertook a side-by-side comparison of driven turbulence, starting from simple initial conditions and over many crossing times, using just two codes – D. Price's SPH code PHANTOM and the grid-based FLASH code (Price & Federrath 2010). The devastatingly obvious conclusion from such efforts was that to achieve comparable results between codes required – wait for it – comparable resolutions. Actually the point was rather more subtle, because "comparable resolution" depended on what quantity one was interested in. Thus, for example to obtain comparable power spectra in a volume-weighted quantity such as the kinetic energy required roughly equal numbers of SPH particles to Eulerian grid cells (and was thus about an order of magnitude more costly

† Following the publication of Agertz *et al.* (2007). The issue itself has nothing to do with Kelvin-Helmholtz instabilities but rather to the (non-)treatment of contact discontinuities in standard SPH schemes, in this case manifested when a KH instability problem is run across a contact discontinuity. Adding a small amount of artificial conductivity is a simple fix (see Price 2008).

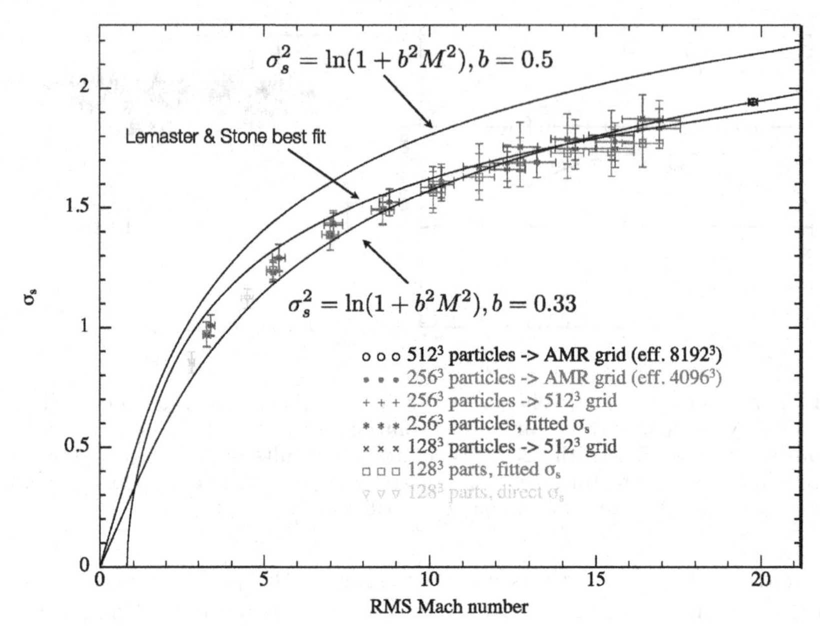

Figure 2. The measured relationship (points) between the standard deviation in the log density, σ_s and the RMS Mach number, $\mathcal{M}$, from a series of calculations of solenoidally-driven, supersonic turbulence at various Mach numbers and at different resolutions in both SPH particles and on the grid to which the density field is interpolated. Overplotted is the standard relationship, Eq. 4.2, with $b = 0.33$ (lower) and $b = 0.5$ (above). Also shown is the best fit obtained by Lemaster & Stone (2008) at lower Mach number, Eq. 4.3, which differs only slightly from the standard relation with $b = 0.33$. Error bars show the 1σ deviations from the time-averaged values.

for SPH in terms of cpu time). On the other hand, the maximum density achieved in the SPH calculations at 128^3 particles was roughly equivalent to that reached in the grid-based calculations at 512^3 (thus being about an order of magnitude more costly for FLASH). Quantities involving a mix of the density and velocity fields – such as the spectra of the mass weighted kinetic energy $\rho^{1/3}v$ – showed intermediate results. Other quantities such as structure function slopes were not well converged in either code at 512^3 computational elements. We refer the reader to Price & Federrath (2010) for more details.

4. The density variance–Mach number relation in supersonic turbulence

One of the questions to arise from the turbulence comparison project was the question of exactly how the width of the density Probability Distribution Function (PDF) in driven, supersonic turbulence depends on the Mach number. Early calculations by Padoan *et al.* (1997) found/assumed a relationship between the linear density variance σ_ρ^2 and the Mach number $\mathcal{M}$ of the form

$$\sigma_\rho^2 = b^2\mathcal{M}^2, \tag{4.1}$$

such that in the log-normal distribution the standard deviation is given by

$$\sigma_s^2 = \ln(1 + b^2\mathcal{M}^2), \tag{4.2}$$

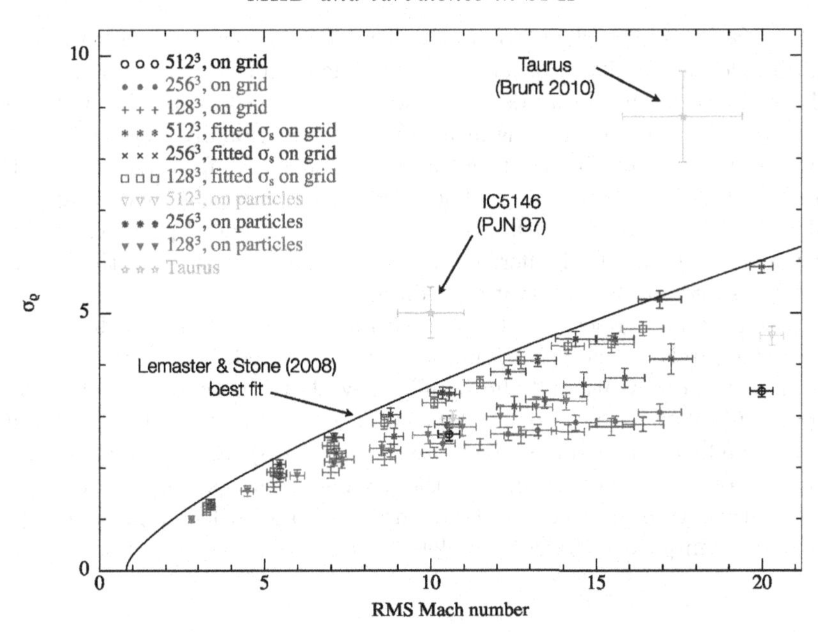

Figure 3. The directly measured (lower points) and inferred (assuming a log-normal distribution: $\sigma_\rho^2 = \exp(\sigma_s^2) - 1$ higher points) relationship between the standard deviation in linear density, σ_ρ, and the RMS Mach number, $\mathcal{M}$, in supersonic turbulence simulations, together with the best-fitting relationship from Lemaster & Stone (2008) (solid line, similar to $b = 0.33$ in Eq. 4.1) and the observational results from IC5146 by Padoan *et al.* (1997) and in Taurus from Brunt (2010).

where $\sigma_s \equiv \sigma_{\ln\rho}$ is the standard deviation in the *log* of the density and b is a parameter that was determined empirically to be $b \sim 0.5$ (Padoan, Nordlund & Jones 1997). Later authors have found considerable variation in the measured value of b based on individual calculations performed at a fixed Mach number. Recently Federrath *et al.* (2010) have also shown that the value of b also depends strongly on particulars of the fourier-space driving, specifically the balance between energy in solenoidal and compressible modes (with $b \approx 0.33$ for purely solenoidal and $b \approx 1$ for purely compressible driving), which they relate physically to whether or not turbulence is driven by compressions or shear flows.

However, relatively few studies have attempted to constrain the relation over a range in Mach number, though the early empirical (but unpublished) work by Padoan *et al.* was calibrated by such calculations. Lemaster & Stone (2008) have recently done so, though only in a limited range of Mach numbers (up to $\mathcal{M} \sim 6$), deriving a best-fitting relationship

$$\sigma_s^2 = -0.72 \ln\left(1 + 0.5\mathcal{M}^2\right) + 0.20. \tag{4.3}$$

However we have recently developed a technique for inferring the true 3D variance σ_ρ from observational data (Brunt *et al.* 2010) that, together with the measured Mach number, indicate $b \approx 0.5 \pm 0.05$ in the Taurus Molecular Cloud at Mach 20 (Brunt 2010), similar to an earlier result derived in IC5146 at Mach 10 by Padoan, Jones & Nordlund (1997). In order to make comparisons with observations it is therefore necessary to pin down the relationship up to these Mach numbers.

We have thus performed a series of calculations with RMS Mach numbers in the range $1 - 20$, the results of which are shown in Fig. 2 in terms of σ_s, measured directly as the (volume-weighted) standard deviation in the (gridded) log density field with no assumptions regarding log-normality or otherwise. The difficulty in comparing with observations is that the variance in the *linear* density field is severely truncated by finite resolution effects, but it is this quantity that is measurable observationally (though it will also be underestimated). One way of estimating the true value is to simply assume that the PDF is log-normal, i.e., $\sigma_\rho^2 = \exp(\sigma_s^2) - 1$. The calculations were observed to approach this relationship as the resolution – either in terms of the number of particles or the grid to which they are interpolated – is increased. The standard deviation in the linear density is what is measured in the observational data, a comparison with which is shown in Fig. 3. Whilst we find a good fit to Eq. 4.2 with $b = 0.33$ (and also to the LS08 best fit, Eq. 4.3) for the purely solenoidal driving we have employed, it is clear that the density variances measured observationally lie significantly above those in the calculations, with $b \approx 0.5$ (and this noting that our assumption of log-normality will *overestimate* σ_ρ, if anything). In the heuristic model by Federrath *et al.* (2008) this can be explained if there is a significant compressive component to the driving mechanism, or alternatively due to the fact that star-forming molecular clouds are quite obviously self-gravitating which is not accounted for in pure turbulence models.

5. Conclusions

In summary:

- Being a television presenter is easier than getting MHD in SPH to work
- Development of MHD in SPH would proceed faster in the absence of arguments over basic misunderstandings of SPH which require lengthy comparison projects to resolve.
- Simulations with supercritical field strengths already tell us that magnetic fields can significantly change star formation even with weak fields.
- SPH and grid codes agree very well on the statistics of supersonic turbulence provided comparable resolutions are used (for power spectra this means $n_{cells} \approx n_{particles}$, but SPH was found to be much better at resolving dense structures).
- We have constrained the density variance – Mach number relation in supersonic turbulence to $\sigma_s^2 = \ln(1 + b^2\mathcal{M}^2)$, with $b \approx 0.33$ up to Mach 20, but observed density variances suggest $b \approx 0.5$, higher than can be produced with purely solenoidally-driven turbulence – rather, some form of compressive driving or additional physics such as gravity is required to explain the observations.

References

Agertz O., Moore B., Stadel J., Potter D., Miniati F., Read J., Mayer L., Gawryszczak A., Kravtsov A., Nordlund Å., Pearce F., Quilis V., Rudd D., Springel V., Stone J., Tasker E., Teyssier R., Wadsley J., & Walder R., 2007, MNRAS, 380, 963

Ballesteros-Paredes J., Gazol A., Kim J., Klessen R. S., Jappsen A.-K., & Tejero E., 2006, ApJ, 637, 384

Børve S., Omang M., & Trulsen J., 2001, ApJ, 561, 82

Børve S., Omang M., & Trulsen J., 2004, ApJS, 153, 447

Brunt C. M., 2010, A&A, 513, A67

Brunt C. M., Federrath C., & Price D. J., 2010, MNRAS, 403, 1507

Byleveld S. E. & Pongracic H., 1996, PASA, 13, 71

Dedner A., Kemm F., Kröner D., Munz C.-D., Schnitzer T., & Wesenberg M., 2002, J. Comp. Phys., 175, 645

Dolag K., Bartelmann M., & Lesch H., 1999, A&A, 348, 351

Dolag K. & Stasyszyn F., 2009, MNRAS, 398, 1678

Federrath C., Klessen R. S., & Schmidt W., 2008, ApJL, 688, L79
Federrath C., Roman-Duval J., Klessen R. S., Schmidt W., & Mac Low M., 2010, A&A, 512, A81
Gingold R. A. & Monaghan J. J., 1977, MNRAS, 181, 375
Hennebelle P. & Ciardi A., 2009, A&A, 506, L29
Hennebelle P. & Teyssier R., 2008, A&A, 477, 25
Kitsionas S., Federrath C., Klessen R. S., Schmidt W., Price D. J., Dursi L. J., Gritschneder M., Walch S., Piontek R., Kim J., Jappsen A., Ciecielag P., & Mac Low M., 2009, A&A, 508, 541
Klein R. I., Fisher R., & McKee C. F., 2004, in G. Garcia-Segura, G. Tenorio-Tagle, J. Franco, & H. W. York ed., RMAA Conf. Ser. Vol. 22 of RMAA Conf. Ser., pp 3–7
Lemaster M. N. & Stone J. M., 2008, ApJL, 682, L97
Machida M. N., Tomisaka K., Matsumoto T., & Inutsuka S., 2008, ApJ, 677, 327
Meglicki Z., 1995, PhD thesis, Australian National University
Mellon R. R. & Li Z., 2008, ApJ, 681, 1356
Monaghan J. J., 2000, J. Comp. Phys., 159, 290
Monaghan J. J., 2005, Rep. Prog. Phys., 68, 1703
Morris J. P., 1996, PhD thesis, Monash University, Melbourne, Australia
Padoan P., Jones B. J. T., & Nordlund A. P., 1997, ApJ, 474, 730
Padoan P., Nordlund A., & Jones B. J. T., 1997, MNRAS, 288, 145
Padoan P., Nordlund Å., Kritsuk A. G., Norman M. L., & Li P. S., 2007, ApJ, 661, 972
Phillips G. J., 1986, MNRAS, 221, 571
Phillips G. J. & Monaghan J. J., 1985, MNRAS, 216, 883
Price D. J., 2004, PhD thesis, University of Cambridge, Cambridge, UK. astro-ph/0507472
Price D. J., 2008, J. Comp. Phys., 227, 10040
Price D. J., 2010, MNRAS, 401, 1475
Price D. J. & Bate M. R., 2007, MNRAS, 377, 77
Price D. J. & Bate M. R., 2008, MNRAS, 385, 1820
Price D. J. & Bate M. R., 2009, MNRAS, 398, 33
Price D. J. & Federrath C., 2009, arXiv:0910.0285
Price D. J. & Federrath C., 2010, MNRAS, p. 960
Price D. J. & Monaghan J. J., 2004a, MNRAS, 348, 123
Price D. J. & Monaghan J. J., 2004b, MNRAS, 348, 139
Price D. J. & Monaghan J. J., 2005, MNRAS, 364, 384
Rosswog S. & Price D., 2007, MNRAS, 379, 915
Stern D. P., 1976, Rev. Geophys. & Space Phys., 14, 199

Daniel Price and Joe Monaghan

Computational Star Formation
Proceedings IAU Symposium No. 270, 2011
J. Alves, B.G. Elmegreen, J. M. Girart & V. Trimble, eds.

doi:10.1017/S1743921311000342

Interstellar Turbulence and Star Formation

Alexei G. Kritsuk,[1] Sergey D. Ustyugov,[2] and Michael L. Norman[1]

[1]UC San Diego, 9500 Gilman Drive MC 0424, La Jolla CA 92093-0424, USA
email: akritsuk@ucsd.edu, mlnorman@ucsd.edu

[2]Keldysh Institute of Applied Mathematics, Miusskaya Sq. 4, 125047, Moscow, Russia
email: ustyugs@keldysh.ru

Abstract. We provide a brief overview of recent advances and outstanding issues in simulations of interstellar turbulence, including isothermal models for interior structure of molecular clouds and larger-scale multiphase models designed to simulate the formation of molecular clouds. We show how self-organization in highly compressible magnetized turbulence in the multiphase ISM can be exploited in simple numerical models to generate realistic initial conditions for star formation.

Keywords. ISM: structure, ISM: clouds, ISM: magnetic fields, turbulence, methods: numerical

1. Introduction

Since most of the ISM is characterized by very large Reynolds numbers, turbulent motions control the structure of nearly all temperature and density regimes in the interstellar gas (Elmegreen & Scalo 2004). Because of that, turbulence is often viewed as an organizing agent forming and shaping hierarchical cloudy structures in the diffuse ISM and ultimately in star-forming molecular clouds (e.g., Vázquez-Semadeni & Passot 1999). Nonlinear advection dominating the dynamics of such highly compressible magnetized multi-scale self-gravitating flows makes computer simulations practically the only tool to study fundamental aspects of interstellar turbulence, even though effective Reynolds numbers in numerical models are always limited by the available computational resource (e.g., Kritsuk *et al.* 2006).

Over the last five years three-dimensional numerical simulations fostered the development of theoretical concepts concerning the interstellar medium undergoing nonlinear self-interaction and self-organization in galactic disks. One can conventionally divide these models designed to tackle various aspects of interstellar turbulence into three different classes depending on the range of resolved scales and physics included: (i) *mesoscale* models that cover evolution of multiphase ISM in volumes with linear size of a few-to-ten kpc and resolve the flow structure down to a fraction of 1 pc (e.g., galactic fountain models developed by de Avillez & Breitschwerdt 2005-07); (ii) *sub-mesoscale* models resolving the scale-height of the diffuse HI ($\sim$ 100 pc) and usually limited to only warm-to-cold neutral phases (WNM and CNM) (e.g., Kissmann *et al.* 2008; Gazol *et al.* 2009; Gazol & Kim 2010; Seifried *et al.* 2010); and (iii) *microscale* models for molecular cloud (MC) turbulence that assume an isothermal equation of state and deal with $<$ 10 pc-sized subvolumes within MCs. *Global* galactic disk models (e.g., Tasker & Bryan 2006-08, Wada 2008 and references therein) which represent the future of direct ISM turbulence modeling, are currently resolving scales down to $\sim$ 10 pc, i.e. insufficient to properly follow the thermal structure of self-gravitating multiphase ISM.

Mesoscale models of supernova-powered (SNe) galactic fountain have demonstrated the important role of dynamic pressure in the ISM that keeps large fractions of the gas mass out of thermal equilibrium and elevates gas pressures of GMCs to the observed levels even without direct action of self-gravity (Korpi *et al.* 1999; Mac Low *et al.* 2005; de Avilles & Breitschwerdt 2005-07; Joung *et al.* 2006-09). They also show that the effective integral scale of the SN-driven turbulence (~ 75 pc) is about half the scale height of the HI gas in the inner Galaxy [100 – 150 pc (Malhotra 1995)] and outline a general picture of probability distributions for the mass density, magnetic field strength, and thermal pressure in the turbulent ISM in disk-like galaxies.

Supersonic isothermal turbulence simulations in periodic boxes representative of the microscale models provided many important insights into the physics of interstellar turbulence and helped to guide the interpretation of observations. These numerical experiments highlighted the importance of nonlinear advection as a major feature of compressible turbulence (Pouquet *et al.* 1991). At high Mach numbers, turbulent flows are dominated by shocks; therefore the velocity spectra are steeper than the Kolmogorov slope of $-5/3$ and closely resemble the Burgers -2 scaling (Kritsuk *et al.* 2006b). The physics of three-dimensional supersonic turbulence is, however, quite different from burgulence (Frisch & Bec 2001) as the solenoidal velocity component always remains dominant (Pouquet *et al.* 1991; Pavlovski *et al.* 2006; Kritsuk *et al.* 2007; Pan *et al.* 2009; Schmidt *et al.* 2009; Kritsuk *et al.* 2010). At sonic Mach numbers $M_s > 3$, strong shock interactions and associated nonlinear instabilities create sophisticated multi-scale pattern of nested U-shaped structures in dynamically active regions morphologically similar to what is observed in molecular clouds (Kritsuk *et al.* 2006a). Scaling of the first-order velocity structure functions $S_1(\delta\mathbf{u}) \sim \ell^{0.54}$ (where $\delta\mathbf{u}(\ell) = \mathbf{u}(\mathbf{x}) - \mathbf{u}(\mathbf{x} + \hat{\mathbf{e}}\ell)$, $S_p(\delta\mathbf{u}) = \langle[\delta\mathbf{u}(\ell)]^p\rangle$ and $\langle\ldots\rangle$ indicates averaging over an ensemble of random point pairs separated by the lag ℓ) obtained in simulations (Kritsuk *et al.* 2007) is similar to the velocity scaling observed in molecular clouds $S_1(\delta\mathbf{u}) \sim \ell^{0.56}$ (Heyer & Brunt 2004). Simulations also support the concept of (lossy) energy cascade in compressible turbulence (e.g., Vázquez-Semadeni *et al.* 2003), suggesting that the kinetic energy directly lost in shocks constitutes a small fraction of the total energy dissipation. The fact that the Richardson-Kolmogorov cascade picture does approximately hold for supersonic turbulence follows from the linear scaling of the third-order structure function of the mass-weighted velocity, $S_3(\delta\sqrt[3]{\rho}\mathbf{u}) \sim \ell$, indicating constant turbulent energy transfer rate across the hierarchy of scales (Kritsuk *et al.* 2007; Kowal & Lazarian 2007; Schwarz *et al.* 2010). The power spectra of $\sqrt[3]{\rho}\mathbf{u}$, accordingly, demonstrate the Kolmogorov scaling independent of the Mach number (Kritsuk *et al.* 2007; Schmidt *et al.* 2008; Kritsuk *et al.* 2009; Federrath *et al.* 2010; Price & Federrath 2010). It seems that this result can be also extended to supersonic MHD turbulence, where the incompressible 4/3-law of Politano & Pouquet (1998) also approximately holds in its scaling part, $S^{\pm}_{\|,3} \equiv \left\langle \delta\mathbf{Z}^{\mp}_{\|}(\ell)[\delta\mathbf{Z}^{\pm}_{i}(\ell)]^2 \right\rangle \sim \ell$ (here $\delta\mathbf{Z}_{\|}(\ell) \equiv [\mathbf{Z}(\mathbf{x}+\hat{\mathbf{e}}\ell) - \mathbf{Z}(\mathbf{x})]\cdot\hat{\mathbf{e}}$, and $\hat{\mathbf{e}}\ell$ is the displacement vector), if reformulated in terms of the mass-weighted Elsässer fields $\mathbf{Z}^{\pm} \equiv \rho^{1/3}(\mathbf{u}\pm\mathbf{B}/\sqrt{4\pi\rho})$ (Kritsuk *et al.* 2009). The presence of magnetic field effectively reduces compressibility of the gas making the velocity spectra more shallow with slopes approaching the Iroshnikov-Kraichnan index of -1.5 in trans-Alfvénic flows. The observed spectral slope of about -1.8 for the Perseus molecular cloud is thus consistent with the super-Alfvénic turbulence regime dominant in that cloud (Padoan *et al.* 2006).

Recent results from numerical experiments on highly compressible turbulence stimulated theorists to reconsider the steady-state statistics of turbulence in the inertial interval. Falkovich *et al.* (2010) have shown that the Kolmogorov 4/5-law is a particular case of the general relation on the current-density correlation function. They derived an analog of the flux relation for compressible turbulence that can be used as a test for direct numerical simulations and as a guide for the development of subgrid scale models for astrophysical turbulence (Schmidt & Federrath 2010).

Most of the recent sub-mesoscale models belong to a class of so-called converging (or colliding) flows of diffuse HI originally developed to study thermal, dynamic, and gravitational instabilities in shock-bounded slabs (Hunter *et al.* 1986; Vishniac 1994; Walder & Folini 1998; Folini *et al.* 2010). These models remain popular as a framework to directly simulate star formation in molecular clouds (Vázquez-Semadeni *et al.* 2006; Hennebelle & Inutsuka 2006; Vázquez-Semadeni *et al.* 2007; Hennebelle *et al.* 2008; Inoue & Inutsuka 2008-09; Heitsch *et al.* 2008-09; Banerjee *et al.* 2009; Niklaus *et al.* 2009; Audit & Hennebelle 2010; Rosas-Guevara *et al.* 2010). Recent numerical experiments with converging flows have demonstrated strong sensitivity of results to adopted initial and boundary conditions as well as to model parameters that control the density of colliding gas streams, mean thermal pressure, orientation and strength of the mean magnetic field, levels and character of "turbulence" at infinity, etc. All these parameters live their unique imprints in the statistics of derived stellar populations and any comprehensive parameter study based on computational modeling in this framework would be prohibitively expensive.

One way to circumvent these difficulties is to exploit Prigogine's concept of self-organization in non-equilibrium nonlinear dissipative systems (Nicolis & Prigogine 1977) in application to the ISM (e.g., Biglari & Diamond 1989). With this approach, one can use interstellar turbulence as an agent that imposes "order" in the form of coherent structures and correlations between various flow fields emerging in a simple periodic box simulation when a *statistical* steady state develops. In this case, the initial conditions are no longer important, instead the steady state would provide the "correct" turbulent initial conditions for star formation when self-gravity is turned on. While this idea is not new,† it remained largely undeveloped so far. In the following sections we will discuss this concept in more detail and report first results from a series of MHD simulations of turbulent multiphase ISM with the piecewise parabolic method on a local stencil (PPML; Ustyugov *et al.* 2009).

2. Self-organization in the magnetized multiphase ISM

In out numerical experiments, we treat the ISM as a turbulent, driven system, with kinetic energy being injected at the largest scales by supernova explosions, shear associated with differential rotation of the galactic disk, gas accretion onto the disk, etc. (Mac Low & Klessen 2004; Klessen & Hennebelle 2010). This kinetic energy is then being transferred from large to small scales in a cascade-like fashion. As our models include a mean magnetic field, B_0, some part of this kinetic energy gets stored in the turbulent magnetic field component, b, generated by stretching, twisting, and folding of magnetic field lines. The ISM is also exposed to the far-ultraviolet (FUV) background radiation due to OB associations of quickly evolving massive stars that form in molecular clouds. This FUV radiation is the main source of energy input for the neutral gas phases and this volumetric thermal energy source is in turn balanced by radiative cooling (Wolfire *et al.* 2003). The ISM is thus exposed to various energy fluxes, and self-organization arises as a result of the relaxation through nonlinear interactions of different physical constituents of the system subject to usual MHD constraints in the form of conservation laws. In this picture, molecular clouds with their hierarchical internal structure form as dissipative structures that represent active regions of highly intermittent turbulent cascade that drain the kinetic energy supplied by the driving forces.

† See, for instance, summary of the panel discussion on Phases of the ISM during the 1986 Grand Teton Symposium in Wyoming (Shull 1987).

Table 1. Model parameters.

Model	N^3	n_0 cm^{-3}	$u_{\rm rms,0}$ km/s	B_0 μG	$\beta_{\rm th,0}$	$\beta_{\rm turb,0}$	$M_{A,0}$
A	512^3	5	16	9.54	0.2	3.3	1.3
B	512^3	5	16	3.02	2	33	4.0
C	512^3	5	16	0.95	20	330	13
D	256^3	2	16	3.02	2	13.2	2.6
E	256^3	5	7	3.02	2	8.3	2.0

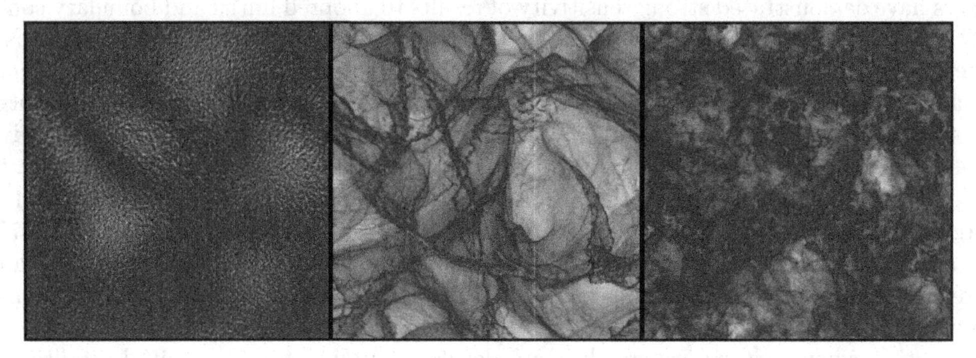

Figure 1. Three snapshots of projected gas density in model A taken at $t = 2$, 3, and 4 Myr. The white-blue-yellow colors correspond to low-intermediate-high projected density values.

3. Modeling the formation of molecular clouds

To illustrate these ideas, we consider a set of simple periodic box models, which ignore gas stratification and differential rotation in the disk and employ an artificial large-scale solenoidal force to mimic the supply of kinetic energy from various galactic sources. This naturally leads to an upper bound on the box size, L, which determines our choice of $L = 200$ pc. Our models are, thus, fully defined by the following three parameters: the mean gas density in the box, n_0; the rms velocity, $u_{\rm rms,0}$; and the mean magnetic field strength, B_0. All three would ultimately depend on L; Table 1 provides the summary of parameters for models A, B, C, D, and E assuming $L = 200$ pc. The table also gives the grid resolution, N, the initial values for plasma beta, $\beta_{\rm th,0} \equiv 8\pi p_0/B_0^2$, turbulent beta, $\beta_{\rm turb,0} \equiv 8\pi\rho_0 u_{\rm rms,0}^2/B_0^2$, and Alfvénic Mach number, $M_{A,0} = (4\pi\rho_0)^{1/2}u_{\rm rms,0}/B_0$, where p_0 is the initial thermal pressure of the gas (see the phase diagram in Fig. 2 for more detail).

We initiate our numerical experiments with a uniform gas distribution in the computational domain. An addition of small random isobaric density perturbations at $t = 0$ triggers a phase transition in the thermally bi-stable gas that quickly turns $\sim 25 - 65\%$ of the gas mass into the stable cold phase (CNM with temperature below $T = 184$ K), while the rest of the mass is shared between the unstable and stable warm gas (WNM). In models A, B, and C, CNM and WNM each contain roughly $\sim 50\%$ of the total H I mass in agreement with observations (Heiles & Crutcher 2005). We then turn on the forcing and after a few large-eddy turn-over times the simulation approaches a statistical steady state. If we replace this two-stage initiation process with a one-stage procedure by turning the driving on at $t = 0$, the properties of the steady state remain unchanged. Figure 1 illustrates this evolutionary sequence for the two-stage case with three snapshots of projected gas density for model A. The left panel shows two-phase medium at $t = 2$ Myr right before we turn on the forcing; the panel in the middle illustrates an early stage of turbulization with transient "colliding flows" at $t = 3$ Myr. The right panel

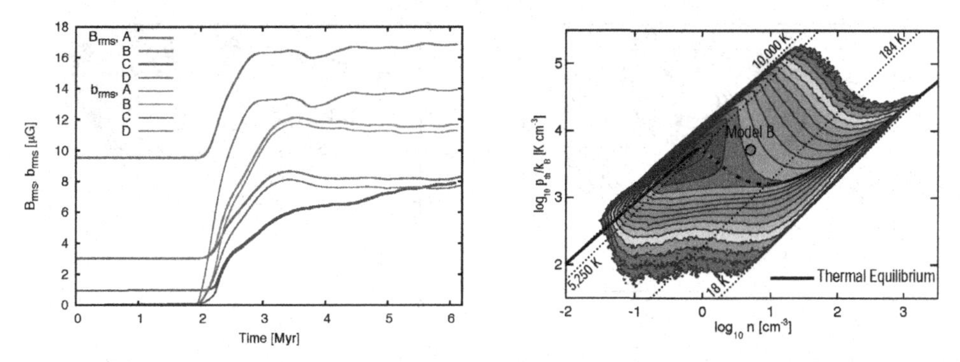

Figure 2. Time evolution of the rms magnetic field strength, B_{rms}, and its turbulent component, b_{rms}, for models A, B, C, and D (*left* panel). Phase diagram (thermal pressure versus gas density) for model B at $t = 5$ Myr (*right* panel).

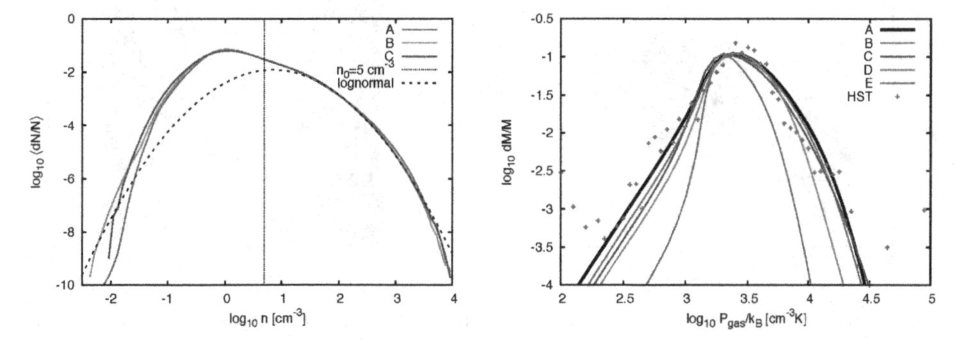

Figure 3. Time-average density distributions for fully developed turbulence in models A, B, and C (*left* panel) and time-average mass-weighted thermal pressure distributions for models A, B, C, and D (*right* panel); see text for more detail.

shows the projected density at $t = 4$ Myr for a statistically developed turbulent state. Molecular clouds can be seen in the right panel as filamentary brown-to-yellow structures (note that these are morphologically quite different from the transient dense structures in the middle panel). The rms magnetic field is amplified by the forcing and saturates when the relaxation in the system results in a steady state, see Fig. 2. The level of saturation depends on B_0 and on the rate of kinetic energy injection by the large-scale force, which is in turn determined by u_{rms} and n_0. This level can be easily controlled with the model parameters. In the saturated regime, models A and B tend to establish energy equipartition ($E_{\mathrm{K}} \sim E_{\mathrm{M}}$), while the saturation level of magnetic energy in model C is a factor of ~ 3 lower than the equipartition level. The mean thermal energy also gets a slight boost due to forcing, but remains subdominant in all the models. A typical phase diagram is shown in Fig. 2. The contours indicate constant levels of volume fraction for different regimes of the thermal pressure, p_{th}, and density, n, separated by factors of 2. About 23% of the domain volume is filled with the stable warm phase at $T > 5250$ K, the stable cold phase ($T < 184$ K) occupies $\sim 7\%$, and $\sim 70\%$ of the volume resides in the thermally unstable regime at intermediate temperatures. The big orange dot at the center indicates the (forgotten) initial conditions for models A, B, C, and E. The phase diagram indicates that turbulence supports an enormously wide range of thermal pressures and also that p_{th} in the molecular gas ($n > 100$ cm^{-3}) is higher than that in the diffuse ISM, even though self-gravity is ignored in the model.

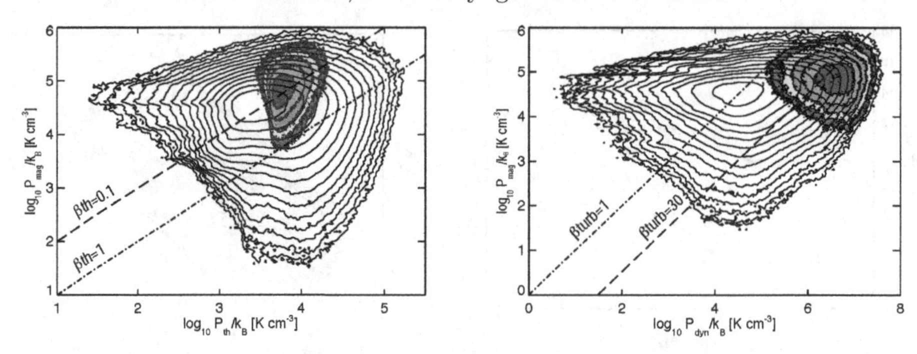

Figure 4. Distributions of magnetic pressure vs. thermal (*left*) and dynamic (*right*) pressure for a snapshot from model B at $t = 5$ Myr.

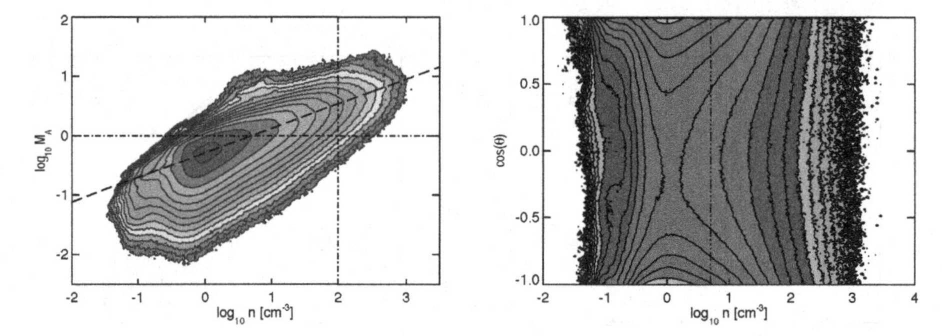

Figure 5. Distributions of the Alfvénic Mach number and the cosine of the alignment angle vs. density for a snapshot from model A at $t = 5$ Myr.

Figure 3 (*left*) shows the time-average density PDFs for models A, B, and C in the steady state. The effect of magnetic field on the density PDF is apparently very weak on average, and the high-density part of the PDF can be well approximated by a lognormal as in the isothermal case. The signature of the two stable thermal phases in the PDF is smeared by the (relatively) high level of turbulence, $u_{\rm rms}(100\ {\rm pc}) = 16$ km/s (Brunt & Heyer 2004), but the overall shape of the distribution is not lognormal. The distribution of thermal pressure for models A, B, and C spans about 6 dex leaving no room for the old pressure-supported cloud picture in the violent ISM. All distributions match the characteristic pressure typical for the Milky Way disk at the solar radius and show only weak dependence on B_0, while the width of the distribution remains sensitive to $u_{\rm rms}$ and n_0. Figure 3 (*right*) shows how the mass-weighted pressure distributions obtained in our models compare with the distribution reconstructed from high-resolution UV spectra of hot stars in the HST archive (Jenkins & Tripp 2010). It seems that models with $u_{\rm rms} = 16$ km/s reproduce both the shape and the width of the observed distribution quite nicely, while a lower turbulence level in model E makes that model distribution too narrow.

These numerical experiments allow us to probe the levels of magnetic field strength in molecular clouds that form self-consistently in the magnetized turbulent diffuse ISM. In the *left* panel of Fig. 4, we show a scatter plot of magnetic vs. thermal pressure for model B at $t = 5$ Myr. The black contour lines show the distribution for the whole domain, which is centered at $\beta_{\rm th} \approx 0.1$. The subset of cells representing the molecular gas ($T < 100$ K and $n > 100$ cm^{-3}, color contour plot) shows very similar mean values of $\beta_{\rm th}$. This indicates a target plasma beta for realistic isothermal molecular cloud turbulence simulations. The *right* panel of the same Figure shows a scatter plot of magnetic vs. dynamic pressure for the same snapshot from the same model. The distribution for the whole domain is centered at $\beta_{\rm turb} \approx 1$ because kinetic and magnetic energy levels are close to equipartition on average. At the same time, the subset of cells representing the molecular gas (color contours) shows a distribution centered at $\beta_{\rm turb} \approx 30$ meaning that turbulence in the molecular gas is super-Alfvénic. Figure 5, *left* panel, shows the distribution of Alfvénic Mach number, $M_{\rm A}$, as a function of density for the strongly magnetized model A that further supports this result. There is a clear positive correlation, $M_{\rm A} \sim n^{0.4}$, indicated by the dashed line and most of the dense material ($n > 100$ cm^{-3}) clearly falls into the super-Alfvénic part of the distribution.

A key to understanding the origin of this super-Alfvénic regime in the cold and dense molecular gas lies in the process of self-organization in magnetized ISM turbulence that we briefly introduced in Section 2. The statistical steady state that our models attain on a time-scale of a few million years is characterized by a certain degree of alignment between the velocity and magnetic field lines. Figure 5, *right* panel, shows the distribution of the cosine of the alignment angle, $\cos\theta \equiv \mathbf{B}\cdot\mathbf{u}/(Bu)$, for model A at $t = 5$ Myr. The contours indicate a saddle-like structure of this probability distribution with a strong alignment regime ($\cos\theta = \pm 1$) in the WNM at and around $n \sim 1$ cm^{-3}. This means that compressions in the WNM gas, which is on average trans-Alfvénic (e.g., $M_{\rm A} \in [0.6, 0.9]$ in model A), occur preferentially along the field lines. If molecular clouds form in the turbulent ISM via large-scale compression of the diffuse HI, then turbulence in such molecular clouds can only be super-Alfvénic, see also Padoan *et al.* (2010).

4. Conclusion

Rapid development of computational astrophysics in the recent years enabled progress in understanding the basics of interstellar turbulence. These new advances will help us to move forward with direct star formation simulations from turbulent initial conditions.

Acknowledgements. This research was supported in part by the National Science Foundation through grants AST-0607675, AST-0808184, and AST-0908740, as well as through TeraGrid resources provided by NICS and SDSC (MCA07S014) and through DOE Office of Science INCITE-2009 and DD-2010 awards allocated at NCCS (ast015/ast021).

References

Audit, E. & Hennebelle, P. 2010, *A&A*, 511, A76
de Avillez, M. A. & Breitschwerdt, D. 2007, *ApJL*, 665, L35
de Avillez, M. A. & Breitschwerdt, D. 2005, *A&A*, 436, 585
Banerjee, R., Vázquez-Semadeni, E., Hennebelle, P., & Klessen, R. S. 2009, *MNRAS*, 398, 1082
Biglari, H. & Diamond, P. H. 1989, Physica D Nonlinear Phenomena, 37, 206
Elmegreen, B. G. & Scalo, J. 2004, *ARA&A*, 42, 211
Falkovich, G., Fouxon, I., & Oz, Y. 2010, Journal of Fluid Mechanics, 644, 465
Federrath, C., Roman-Duval, J., Klessen, R. S., Schmidt, W., & Mac Low, M.-M. 2010, *A&A*, 512, A81
Folini, D., Walder, R., & Favre, J. M. 2010, ASP Conf. Ser., 429, 9
Frisch, U. & Bec, J. 2001, New Trends in Turbulence, 341
Gazol, A. & Kim, J. 2010, *ApJ*, 723, 482
Gazol, A., Luis, L., & Kim, J. 2009, *ApJ*, 693, 656
Heiles, C. & Crutcher, R. 2005, *Cosmic Magnetic Fields*, Lect. Notes Phys., 664, 137

Heitsch, F., Hartmann, L., Slyz, A., Devriendt, J., & Burkert, A. 2008, *ApJ*, 674, 316
Heitsch, F., Stone, J. M., & Hartmann, L. W. 2009, *ApJ*, 695, 248
Hennebelle, P., Banerjee, R., Vázquez-Semadeni, E., Klessen, R. S., & Audit, E. 2008, *A&A*, 486, L43
Hennebelle, P. & Inutsuka, S.-i. 2006, *ApJ*, 647, 404
Heyer, M. H. & Brunt, C. M. 2004, *ApJL*, 615, L45
Hunter, J. H., Jr., Sandford, M. T., II, Whitaker, R. W., & Klein, R. I. 1986, *ApJ*, 305, 309
Inoue, T. & Inutsuka, S.-i. 2009, *ApJ*, 704, 161
Inoue, T. & Inutsuka, S.-i. 2008, *ApJ*, 687, 303
Jenkins, E. B. & Tripp, T. M. 2010, to appear in *ApJ*
Joung, M. K. R. & Mac Low, M.-M. 2006, *ApJ*, 653, 1266
Joung, M. R., Mac Low, M.-M., & Bryan, G. L. 2009, *ApJ*, 704, 137
Kissmann, R., Kleimann, J., Fichtner, H., & Grauer, R. 2008, *MNRAS*, 391, 1577
Klessen, R. S. & Hennebelle, P. 2010, *A&A*, 520, A17
Korpi, M., Brandenburg, A., Shukurov, A., Tuominen, I., & Nordlund, Å. 1999, *ApJL*, 514, L99
Kowal, G. & Lazarian, A. 2007, *ApJL*, 666, L69
Kritsuk, A. G., Ustyugov, S. D., Norman, M. L., & Padoan, P. 2010, ASP Conf. Ser., 429, 15
Kritsuk, A. G., Ustyugov, S. D., Norman, M. L., & Padoan, P. 2009, ASP Conf. Ser., 406, 15
Kritsuk, A. G., Norman, M. L., Padoan, P., & Wagner, R. 2007, *ApJ*, 665, 416
Kritsuk, A. G., Norman, M. L., & Padoan, P. 2006a, *ApJL*, 638, L25
Kritsuk, A. G., Wagner, R., Norman, M. L., & Padoan, P. 2006b, ASP Conf. Ser., 359, 84
Mac Low, M.-M., Balsara, D. S., Kim, J., & de Avillez, M. A. 2005, *ApJ*, 626, 864
Mac Low, M.-M. & Klessen, R. S. 2004, Rev. Mod. Phys., 76, 125
Malhotra, S. 1995, *ApJ*, 448, 138
Nicolis, G. & Prigogine, I., Self-Organization in Nonequilibrium Systems, Wiley, New York, 1977
Niklaus, M., Schmidt, W., & Niemeyer, J. C. 2009, *A&A*, 506, 1065
Padoan, P., *et al.* 2010, AIP Conf. Proc., 1242, 219
Padoan, P., Juvela, M., Kritsuk, A., & Norman, M. L. 2006, *ApJL*, 653, L125
Pan, L., Padoan, P., & Kritsuk, A. G. 2009, Phys. Rev. Lett., 102, 034501
Pavlovski, G., Smith, M. D., & Mac Low, M.-M. 2006, *MNRAS*, 368, 943
Politano, H. & Pouquet, A. 1998, Geophys. Res. Lett., 25, 273
Pouquet, A., Passot, T., & Leorat, J. 1991, Proc. IAU Symp. 147: Fragmentation of Molecular Clouds and Star Formation, p. 101
Price, D. J. & Federrath, C. 2010, *MNRAS*, 406, 1659
Rosas-Guevara, Y., Vázquez-Semadeni, E., Gómez, G., & Jappsen, A. 2010, *MNRAS*, 406, 1875
Schmidt, W. & Federrath, C. 2010, arXiv:1010.4492
Schmidt, W., Federrath, C., Hupp, M., Kern, S., & Niemeyer, J. C. 2009, *A&A*, 494, 127
Schmidt, W., Federrath, C., & Klessen, R. 2008, Phys. Rev. Lett., 101, 194505
Schwarz, C., Beetz, C., Dreher, J., & Grauer, R. 2010, Physics Letters A, 374, 1039
Seifried, D., Schmidt, W., & Niemeyer, J. C. 2010, arXiv:1009.2871
Shull, J. M. 1987, Interstellar Processes, 134, 225
Solomon, P. M., Rivolo, A. R., Barrett, J., & Yahil, A. 1987, *ApJ*, 319, 730
Tasker, E. J. & Bryan, G. L. 2006, *ApJ*, 641, 878
Tasker, E. J. & Bryan, G. L. 2008, *ApJ*, 673, 810
Ustyugov, S., Popov, M., Kritsuk, A., & Norman, M. 2009, J. Comp. Phys., 228, 7614
Vázquez-Semadeni, E., *et al.* 2007, *ApJ*, 657, 870
Vázquez-Semadeni, E., Ryu, D., Passot, T., González, R. F., & Gazol, A. 2006, *ApJ*, 643, 245
Vázquez-Semadeni, E., Ballesteros-Paredes, J., & Klessen, R. 2003, ASP Conf. Ser., 287, 81
Vázquez-Semadeni, E. & Passot, T. 1999, Interstellar Turbulence, Proc. 2nd Guillermo Haro Conference, Eds. J. Franco and A. Carraminana. Cambridge University Press, p. 223
Vishniac, E. T. 1994, *ApJ*, 428, 186
Wada, K. 2008, *ApJ*, 675, 188
Walder, R. & Folini, D. 1998, *A&A*, 330, L21
Wolfire, M. G., McKee, C. F., Hollenbach, D., & Tielens, A. G. G. M. 2003, *ApJ*, 587, 278

Computational Star Formation
Proceedings IAU Symposium No. 270, 2011
J. Alves, B.G. Elmegreen, J. M. Girart & V. Trimble, eds.

doi:10.1017/S1743921311000354

Star Formation with Adaptive Mesh Refinement Radiation Hydrodynamics

Mark R. Krumholz[1]

[1]Dept. of Astronomy & Astrophysics, University of California, Santa Cruz, Interdisciplinary Sciences Building, Santa Cruz, CA 95064, USA
email: krumholz@ucolick.org

Abstract. I provide a pedagogic review of adaptive mesh refinement (AMR) radiation hydrodynamics (RHD) methods and codes used in simulations of star formation, at a level suitable for researchers who are not computational experts. I begin with a brief overview of the types of RHD processes that are most important to star formation, and then I formally introduce the equations of RHD and the approximations one uses to render them computationally tractable. I discuss strategies for solving these approximate equations on adaptive grids, with particular emphasis on identifying the main advantages and disadvantages of various approximations and numerical approaches. Finally, I conclude by discussing areas ripe for improvement.

Keywords. hydrodynamics, methods: numerical, radiative transfer, stars: formation

1. Introduction

While gravity is the dominant force in star formation, radiative processes are crucial as well. Most basically, radiation removes energy, allowing a collapsing cloud to maintain a nearly constant, low temperature as its density and gravitational binding energy rise by many orders of magnitude. Radiative cooling is what makes star formation a dynamic rather than a quasi-static process, a point understood quite early (see the review by Hayashi 1966). At densities below $\sim 10^{4-5}$ H atoms cm^{-3}, cooling occurs primarily via collisionally-excited atomic or molecular line emission, while at higher densities dust and gas become thermally well-coupled via collisions, and thermal radiation by dust grains is the dominant cooling process (e.g. Goldsmith 2001). At yet higher densities the coupled dust-gas fluid becomes optically thick to its own cooling radiation, causing the temperature to rise in a quasi-adiabatic fashion (Larson 1969).

While one-dimensional simulations of this collapse can include quite sophisticated treatments of radiative transfer between gas parcels (e.g. Masunaga *et al.* 1998), the computational difficulty of handling RHD in three dimensions led to a long period in which most 3D simulations of star formation simply dispensed with radiation entirely, choosing instead to represent the cooling processes either by a cooling function that removes energy at a rate dependent only on local gas properties, or by an even simpler equation of state that prescribes the gas temperature as a function of density. However, this approach has a major flaw: it does not allow energy exchange between parcels of gas, or between stars and diffuse gas. The last few years have seen a number of studies pointing out that this energy exchange is crucial to the dynamics of star formation, and that it can be dealt with only by RHD simulations.

The most important radiative transfer processes can be broken into three rough categories: thermal feedback, in which collapsing gas and stars heat the gas and thereby change its pressure; force feedback, in which radiation exerts forces on the gas that alter its motion; and chemical feedback, in which radiation changes the chemical state of the gas (e.g. by ionizing it), and this chemical change somehow affects the dynamics. Each of these processes has been the object of intense study in the last few years.

With regard to thermal feedback, Larson (2005) pointed out that how gas fragments, and thus the stellar mass distribution that results from fragmentation, is extremely sensitive to how the temperature varies with density. Krumholz (2006) and Krumholz & McKee (2008) showed that one consequence of this sensitivity is that radiation produced by the first few stars to form in a given region can heat the gas around them, reducing the ability of that gas to fragment, favoring monolithic collapse to massive stars and suppressing formation of low mass objects. Subsequent numerical simulations (Krumholz *et al.* 2007a, 2010; Bate 2009; Offner *et al.* 2009; Urban *et al.* 2010) using a variety of methods have confirmed this effect.

Force feedback becomes important in the context of massive stars, which produce such high luminosities that the radiative force they exert on dusty interstellar gas can exceed their gravitational force. A number of authors have argued based on one-dimensional models that this process sets an upper limit on stellar masses (Kahn 1974; Yorke & Kruegel 1977; Wolfire & Cassinelli 1987). More recent work in two (Nakano 1989; Nakano *et al.* 1995; Jijina & Adams 1996; Yorke & Bodenheimer 1999; Yorke & Sonnhalter 2002; also see the contribution by Kuiper *et al.* in these proceedings) and three (Krumholz *et al.* 2007a, 2009) dimensions instead suggests that force feedback does not limit stellar masses. Deciding the question requires radiation-hydrodynamic simulations.

The most important type of chemical feedback is photoionization, which raises the gas from ~ 10 K to $\sim 10^4$ K. This causes a number of important dynamic effects, including ejecting mass from star-forming clouds (Dale *et al.* 2005; Peters *et al.* 2010), altering the magnetic field structure (Krumholz *et al.* 2007c), and driving turbulent motions (Gritschneder *et al.* 2009). Radiation also drives other chemical processes, but these are usually not important for dynamics, and thus may be handled by post-processing simulations. In this review I limit myself to radiative effects that are dynamically important and must therefore be simulated in tandem with the hydrodynamic evolution of the system.

2. The Equations of Radiation Hydrodynamics

The equations of radiation hydrodynamics (RHD) in conservation form read (Mihalas & Weibel-Mihalas 1999)

$$\frac{\partial}{\partial t}\begin{pmatrix} \rho \\ \rho\mathbf{v} \\ \rho e \end{pmatrix} + \nabla\cdot\begin{pmatrix} \rho\mathbf{v} \\ \rho\mathbf{v}:\mathbf{v}+P \\ (\rho e+P)\mathbf{v} \end{pmatrix} = \begin{pmatrix} 0 \\ \mathbf{G} \\ cG^0 \end{pmatrix}, \tag{2.1}$$

where ρ, $\mathbf{v}$, e, and P are the gas density, velocity, specific energy, and pressure. The source term on the right hand side represents the rate at which radiation transfers momentum and energy to the gas. For simplicity I have omitted terms describing gravity and magnetic fields, which would appear as additional sources on the right hand side.

The rate at which radiation transfers energy and momentum to the gas is described by the radiation four-force vector $(G^0, \mathbf{G})$,

$$c\begin{pmatrix} G^0 \\ \mathbf{G} \end{pmatrix} = \int d\nu \int d\Omega\, [\kappa_\nu(\mathbf{n})I_\nu - \eta_\nu(\mathbf{n})]\begin{pmatrix} 1 \\ \mathbf{n} \end{pmatrix} \tag{2.2}$$

where $\mathbf{n}$ is a unit vector, I_ν is the radiation intensity at frequency ν in direction $\mathbf{n}$, and κ_ν and η_ν are the extinction coefficient and emissivity of the gas as a function of frequency

ν and direction $\mathbf{n}$. Finally, the intensity is governed by the transfer equation,

$$\frac{1}{c}\frac{\partial}{\partial t}I_\nu + \mathbf{n}\cdot\nabla I_\nu = \eta_\nu(\mathbf{n}) - \kappa_\nu(\mathbf{n})I_\nu. \tag{2.3}$$

For simplicity I have omitted terms describing scattering, which is generally not important for the radiation-hydrodynamics of star formation.

The equations of RHD are, like the ordinary equations of fluid dynamics, characterized by dimensionless numbers. The two most important ones for radiation-hydrodynmics are $\beta \sim v/c$, where v is the characteristic value of $\mathbf{v}$, and $\tau \sim L/\lambda_p$, where L is the size-scale of the system and λ_p is the photon mean free path. The first ratio β characterizes how relativistic the system is. In writing equation (2.1) in non-relativistic form, we have already assumed that $\beta \ll 1$. The second, τ, characterizes how optically thick it is. Systems with $\tau \ll 1$ are described as being in the streaming limit, and are characterized by weak matter-radiation interaction. Those with $\tau \gg 1$ are in the diffusion limit and have strong-matter radiation coupling. Both limits appears in star formation problems.

3. Approximations and Solution Methods

The full system of equations (2.1) and (2.3) is seven-dimensional, with quantities varying in time, space (3 dimensions), direction (2 dimensions), and frequency. Unfortunately, this makes full numerical solution prohibitively expensive even on modern supercomputers. We are therefore reduced to approximations. There are two broad classes of approximation in common use in star formation simulations: moment methods and characteristic methods. A third category, Monte Carlo methods, has been used extensively in post-processing radiative transfer calculations, but has not been used extensively in radiation-hydrodynamic simulations. I will not discuss it in detail.

3.1. *Codes*

In the discussion that follows I introduce the most common methods used in AMR RHD methods, and in Table 1 I summarize which method(s) are used in each code. The codes I include in the table are as follows: **Orion** is the oldest and probably best-tested AMR RHD code, but is not publicly available, and currently lacks a magnetohydrodynamic (MHD) capability (Klein 1999; Howell & Greenough 2003; Krumholz *et al.* 2007b). **Ramses** is an AMR MHD code that was recently upgraded with a radiation solver, although the latter is not (as of this writing) publicly available (Fromang *et al.* 2006; Commerçon *et al.* 2010). **Flash** is an open source AMR MHD code with extensive chemistry capabilities and several add-on modules that handle radiation in different ways (Fryxell *et al.* 2000; Rijkhorst *et al.* 2006; Peters *et al.* 2010). **Enzo** is an open source AMR HD code (an MHD version exists but is not public) with cosmology capabilities and two different radiation methods (Abel & Wandelt 2002; Norman *et al.* 2008; Reynolds *et al.* 2009). **Pluto** is an open source AMR MHD code with a newly-developed radiation solver (Mignone *et al.* 2007; Kuiper *et al.* 2010). There are also a number of fixed, nested grid codes in use, but I will not mention these by name.

3.2. *Moment Methods*

3.2.1. *Basic Theory of Moment Methods*

The basic idea of a moment method is to take moments of the transfer equation, in exact analogy to the Chapman-Enskog procedure used to derive the equations of fluid dynamics from the kinetic theory of gases. To take the zeroth moment we integrate the transfer equation over all directions; to obtain the first moment we multiply both sides

Moment methods			Characteristic methods		
Code	**ν resolution**	**Frame**	**Code**	**ν resolution**	**Ray scheme**
Orion	MG, 2T	Mixed	Orion	1F	HEALpix
Ramses	2T	Comoving	Flash	MG	Hybrid characteristics
Enzo	2T	Comoving	Enzo	1F	HEALpix
Pluto	1T	Comoving	Pluto	MG	Sphere

Table 1. Summary of Codes. See text for explanation of fields. All moment codes use first order closure (flux-limited diffusion). Enzo and Orion appear in both lists because they can operate in characteristic or moment mode. Pluto appears ion both lists because it uses a hybrid characteristic-moment method (see text for details). Orion has both a MG and a 2T moment method.

by the unit vector $\mathbf{n}$ and then integrate; for the second moment we multiply by the rank two tensor $\mathbf{n} : \mathbf{n}$ and integrate, and so forth. As in the analogous fluid case, the procedure yields an exact solution if carried out to infinitely many orders, but one instead makes an approximation by truncating the procedure after finitely many orders. Usually "finitely many" here means one or two, and the first two moments of equation (2.3) are

$$\frac{\partial}{\partial t}\begin{pmatrix} E \\ \mathbf{F}/c^2 \end{pmatrix} + \nabla \cdot \begin{pmatrix} \mathbf{F} \\ \mathbf{\Pi} \end{pmatrix} = -\begin{pmatrix} cG^0 \\ \mathbf{G} \end{pmatrix}, \tag{3.1}$$

where

$$E = \frac{1}{c}\int d\nu \int d\Omega\, I_\nu \tag{3.2}$$

$$\mathbf{F} = \int d\nu \int d\Omega\, I_\nu \mathbf{n} \tag{3.3}$$

$$\mathbf{\Pi} = \frac{1}{c}\int d\nu \int d\Omega\, I_\nu \mathbf{n} : \mathbf{n} \tag{3.4}$$

are the first three moments of the radiation intensity: the radiation energy density, radiation flux, and radiation pressure tensor. In these equations the direction-dependence in the transfer equation is removed and the dimensionality of the problem is lowered by two, but at the price of introducing the radiation pressure tensor, for which we do not have an equation (since we have not expanded the transfer equation to the next order) and must instead make an approximation.

Because moment methods necessarily involve smoothing the angular dependence of the radiation field, they are best suited to representing diffuse, smooth radiation fields. This makes them ideal for handling thermal and force feedback, which tend to be dominated by diffuse infrared light re-radiated by dust grains. They are less suited for chemical feedback, which is usually dominated by direct, highly beamed radiation from a handful of stellar sources. A further advantage of moment methods is that their computational cost is independent of the number of sources, and usually scales only as N or $N \log N$, where N is the number of cells. They also parallelize fairly easily.

3.2.2. *Types of Moment Methods*

Moment methods can be classified based on several design choices that are made in their construction, all of which involve some tradeoff between computational expense and accuracy. See Table 1 for a list of the design choices made in each of the popular codes.

Closure order. The first choice is whether to retain both of the first two moment equations, or to retain only the zeroth moment and adopt a closure approximation for the radiation flux; all the moment methods currently used in practical star formation simulations make the latter choice, using a closure known as flux-limited diffusion (FLD). In a medium where the photon mean free path is small compared to the size of the system, Eddington showed that the radiation flux in the fluid rest frame approaches

$$\mathbf{F} = -\frac{c}{3\kappa_{\mathrm{R}}}\nabla E, \tag{3.5}$$

where $\kappa_R = \int d\nu\,(\partial B_\nu/\partial T)/\int d\nu\,\kappa^{-1}(\partial B_\nu/\partial T)$ is the Rosseland mean opacity, $B_\nu(T)$ is the Planck function. This is known as the diffusion approximation. While it is quite good in optically thick media, it breaks down in optically thin regions because as $\kappa_{\mathrm{R}} \to 0$ the signal speed approaches infinity rather than properly limiting to c. The FLD approximation is an attempt to fix this problem by instead setting the radiation flux to

$$\mathbf{F}_0 = -\frac{c\lambda}{\kappa_{\mathrm{R}}}\nabla E_0, \tag{3.6}$$

where λ is a function that has the property that $\lambda \to 1/3$ in optically thick regions and $\lambda \to \kappa_{\mathrm{R}} E_0/\nabla E_0$ in optically thin regions, so that $|\mathbf{F}_0| \to cE_0$, its correct limiting value. Many choices for the function λ are possible; the most common one in astrophysical applications is the Levermore & Pomraning (1981) limiter, $\lambda(R) = R^{-1}(\coth R - R^{-1})$, where $R = |\nabla E_0|/(\kappa_{\mathrm{R}} E_0)$. However, all of these limiters are of unknown accuracy in the intermediate optical depth regime, and all FLD methods suffer from the problem that they discard information about the directionality and momentum content of the radiation field. Higher order moment methods exist and can solve some of these problems (e.g. variable tensor Eddington factor methods, Hayes & Norman 2003, and the M1 closure method, González *et al.* 2007), but none have yet proven cheap and robust enough for use in practical AMR star formation simulations.

Frequency resolution. The second choice in setting up an RHD moment method is the level of resolution in frequency. Most accurate is the multigroup (MG) method, in which one discretizes equation (3.1) in frequency, solving one version of the equation for each frequency bin. All the bins are coupled via the matter temperature, which affects the extinction and emissivity and thus the radiation four-force vector (equation 2.2). Next most accurate is the 2T method, in which one takes the radiation spectrum to be a Planck function characterized by a single temperature T_{rad} (thus removing a dimension from the problem), but allows T_{rad} to be different than the gas temperature T_{gas}. Simplest of all is the 1T method, in which one assumes $T_{\mathrm{rad}} = T_{\mathrm{gas}}$, allowing a further simplification of the equations. As one might expect, this choice involves a tradeoff between accuracy and cost; the 1T method is cheapest but badly misestimates both the radiation spectrum and T_{gas} in the streaming regime. In comparison 2T methods still produce incorrect radiation spectra in the streaming regime, but make much smaller errors in the matter temperature. They are intermediate in cost. MG provides a good representation of both the spectrum and the matter temperature but is most expensive.

Choice of frame. The third design choice is whether to formulate the equations in the comoving frame or using mixed frames. Extinction κ and emissivity η are simple, isotropic functions in the frame comoving with the fluid, but in other frames they contain complex directional dependence as a result of relativistic boosting. One might think that this effect is small in non-relativistic flows, but ignoring it is equivalent to neglecting the work done by radiation on the gas (Mihalas & Klein 1982), which is obviously unacceptable if radiation forces are non-negligible. To avoid complex velocity dependence in η and κ, it is highly desirable to write them in the comoving frame. The simplest choice after that is to write the radiation quantities (E, $\mathbf{F}$) in the comoving frame as well (e.g. see Mihalas & Klein 1982), but this carries a significant cost. Since the comoving frame is non-inertial in any system with non-constant fluid velocity, the comoving radiation energy is not a conserved quantity, and thus comoving-frame codes cannot be exactly conservative. Moreover, every time the resolution changes and the velocity field is refined, the reference frame change as well, introducing errors in conservation that are likely to accumulate with increasing refinement. The alternative is a mixed-frame formulation in which one writes the radiation quantities in the inertial lab frame. Deriving this formulation is tricky, since one must carefully account for the Lorentz transformations between frames, but the resulting equations are explicitly conservative, and can be discretized in a manner that conserves energy to machine precision (Krumholz *et al.* 2007b).

3.3. *Characteristic Methods*

In characteristic methods one solves the transfer equation (2.3) directly, but only along selected rays. Given this solution, one can compute the radiation four-force vector (equation 2.2) directly. This provides much greater accuracy in computing the radiation intensity along those rays, but at the cost of neglecting all other rays. For this reason it is best used for chemical feedback, which tends to be dominated by the rays coming from a small number of sources. It is not well-suited to handling diffuse radiation fields, and simulations based on this technique are generally no better than non-radiative simulations when it comes to handling effects like fragmentation being altered by a diffuse IR radiation field. The cost of these methods scales as the number of cells times the number of sources, but with a coefficient that is generally smaller than for moment methods. Thus these methods are cheaper than moment methods when the number of sources and rays is small, but become impractically expensive if the number of sources is even a small fraction of the number of cells. As with moment methods, there are several design choices to be made in setting up a characteristic method.

Frequency resolution. As in the moment method case, one can choose either to retain the frequency-dependence in the transfer equation by using multiple frequency groups (the MG method), or one can integrate over frequency. In this case one does not generally assign a temperature to the radiation; instead, since for chemical feedback one cares about photons only above a certain energy, the usual approximation is to adopt an average photon energy. This is the single-frequency (1F) approximation.

Ray-drawing scheme. There are also a number of schemes for drawing rays from the point source(s). The simplest is the spherical ray technique: one uses spherical coordinates, requires that the source be at the origin, and simply draws rays aligned with the computational grid. This parallelizes almost perfectly and is very simple to code, but is obviously unsuited to any problem with multiple stellar sources or where the locations at which stars form is not known *a priori*. The other two schemes in wide use are HEALPix-based adaptive trees (Abel & Wandelt 2002) and the hybrid characteristics (Rijkhorst *et al.* 2006), both of which operate on top of an underlying Cartesian grid. The HEALPix scheme is based on the Hierarchical Equal Area isoLatitude Pixelization of the sphere, which divides the sphere into a equal area pixels that can be subdivided indefinitely, allowing the angular resolution to adapt to match that of the underlying grid. The hybrid characteristics scheme works by combining a method of short characteristics within individual AMR grids with a method of long characteristics between grids. Neither suffer from the limitations of the spherical ray method.

3.4. *Hybrid Methods*

Hybrid methods attempt to combine the best features of characteristic and moment methods. Characteristics work well for the highly beamed radiation coming directly from a star or some other point source. However, once this radiation is absorbed, following its re-emission with characteristics becomes unreasonably expensive, and a moment method is a far better choice. The underlying idea of a hybrid method, first used in multidimensional simulations by Murray *et al.* (1994) and recently implemented in the Pluto code (Kuiper *et al.* 2010), is to use characteristics for the "first absorption", then switch to a moment method. One does this by performing a characteristic trace from the star or stars to determine the rate of energy and momentum input into each cell by direct radiation. One adds this as a source term on the right hand side of the moment equation (3.1), then solves as one would in a pure moment method. The computationally-cheap characteristic step can use MG, while the more expensive moment solve uses the 1T or 2T approximation. Since the first absorption is generally the part of the problem where the frequency dependence is most important, this method achieves some of the accuracy of a fully frequency-dependent calculation at significantly lower cost.

4. Future Directions

I close this review with a brief discussion of possible future directions for AMR RHD in the star formation context. One such likely direction comes from combining realistic treatments of thermal, force, and chemical feedback into a single simulation, rather than treating only one or two of them at a time as in current simulations. Since the first two effects are most easily handled by moment methods and the third by characteristic methods, hybrid techniques are the natural solution. To be competitive in handling thermal and force feedback, however, these methods will have to be linked to more advanced moment methods than has been attempted before. Nonetheless, a hybrid method combining frequency-dependent characteristic tracing with a 2T, mixed frame FLD method is a straightforward extension of current techniques, and seems a logical place to start.

Further in the future, second order moment methods are likely to become important. Developing them to the point where they can run reliably in a parallel environment on adaptive grids will be a major algorithmic undertaking, one that is likely to require the assistance of computer scientists and applied mathematicians in addition to astronomers. If successful, these methods will be able to handle both beamed and diffuse radiation fields within a single framework, and will remove many of the limitations of the FLD approximation. Monte Carlo methods are another candidate to replace the current dominance of FLD and characteristic methods. They are very computationally costly, since many photons are required to suppress Poisson noise, but they have the advantage of near perfect parallelization. Since the future of supercomputing seems to be heading toward the development of ever-larger numbers of processors, each of which is no faster than the processors of the previous generation, this may prove to be a decisive advantage.

Finally, a caveat: for all the limitations of our current RHD methods, it is not entirely obvious that our numerical techniques are the limiting factor in the accuracy of our simulations. The main agent for coupling radiation and gas at the high densities where stars form is dust. Our knowledge of dust grain properties, such as sublimation temperatures and grain size distributions in dense environments, is quite limited, and appears likely to advance less quickly than either computer power or algorithmic technique. It may be the case in the future that our knowledge of dust becomes the limiting factor in the accuracy of RHD simulations.

Acknowledgements

I thank Richard Klein, Chris McKee, and Loius Howell for helpful discussions. I acknowledge support from: an Alfred P. Sloan Fellowship; NASA through ATFP grant

NNX09AK31G and through the Spitzer Theoretical Research Program, through a contract issued by the JPL; and the NSF through grant AST-0807739.

References

Abel, T. & Wandelt, B. D. 2002, *MNRAS*, 330, L53
Bate, M. R. 2009, *MNRAS*, 392, 1363
Commerçon, B., Hennebelle, P., Audit, E., Chabrier, G., & Teyssier, R. 2010, *A&A*, 510, L3+
Dale, J. E., Bonnell, I. A., Clarke, C. J., & Bate, M. R. 2005, *MNRAS*, 358, 291
Fromang, S., Hennebelle, P., & Teyssier, R. 2006, *A&A*, 457, 371
Fryxell, B., *et al.* 2000, *ApJS*, 131, 273
Goldsmith, P. F. 2001, *ApJ*, 557, 736
González, M., Audit, E., & Huynh, P. 2007, *A&A*, 464, 429
Gritschneder, M., Naab, T., Walch, S., Burkert, A., & Heitsch, F. 2009, *ApJL*, 694, L26
Hayashi, C. 1966, *ARA&A*, 4, 171
Hayes, J. C. & Norman, M. L. 2003, *ApJS*, 147, 197
Howell, L. H. & Greenough, J. A. 2003, JCP, 184, 53
Jijina, J. & Adams, F. C. 1996, *ApJ*, 462, 874
Kahn, F. D. 1974, *A&A*, 37, 149
Klein, R. I. 1999, *J. Comp. App. Math.*, 109, 123
Krumholz, M. R. 2006, *ApJL*, 641, L45
Krumholz, M. R., Cunningham, A. J., Klein, R. I., & McKee, C. F. 2010, *ApJ*, 713, 1120
Krumholz, M. R., Klein, R. I., & McKee, C. F. 2007a, *ApJ*, 656, 959
Krumholz, M. R., Klein, R. I., McKee, C. F., & Bolstad, J. 2007b, *ApJ*, 667, 626
Krumholz, M. R., Klein, R. I., McKee, C. F., Offner, S. S. R., & Cunningham, A. J. 2009, Science, 323, 754
Krumholz, M. R. & McKee, C. F. 2008, *Nature*, 451, 1082
Krumholz, M. R., Stone, J. M., & Gardiner, T. A. 2007c, *ApJ*, 671, 518
Kuiper, R., Klahr, H., Dullemond, C., Kley, W., & Henning, T. 2010, *A&A*, 511, A81+
Larson, R. B. 1969, *MNRAS*, 145, 271
—. 2005, *MNRAS*, 359, 211
Levermore, C. D. & Pomraning, G. C. 1981, *ApJ*, 248, 321
Masunaga, H., Miyama, S. M., & Inutsuka, S. 1998, *ApJ*, 495, 346
Mignone, A., Bodo, G., Massaglia, S., Matsakos, T., Tesileanu, O., Zanni, C., & Ferrari, A. 2007, *ApJS*, 170, 228
Mihalas, D. & Klein, R. I. 1982, JCP, 46, 97
Mihalas, D. & Weibel-Mihalas, B. 1999, Foundations of Radiation Hydrodynamics (Mineola, New York: Dover)
Murray, S. D., Castor, J. I., Klein, R. I., & McKee, C. F. 1994, *ApJ*, 435, 631
Nakano, T. 1989, *ApJ*, 345, 464
Nakano, T., Hasegawa, T., & Norman, C. 1995, *ApJ*, 450, 183
Norman, M. L., Bryan, G. L., Harkness, R., Bordner, J., Reynolds, D., O'Shea, B., & Wagner, R. 2008, in Petascale Computing: Algorithms and Applications, ed. D. A. Bader, Computational Science Series (Chapman & Hall/CRC), 83–101, arXiv:0705.1556
Offner, S. S. R., Klein, R. I., McKee, C. F., & Krumholz, M. R. 2009, *ApJ*, 703, 131
Peters, T., Banerjee, R., Klessen, R. S., Mac Low, M., Galván-Madrid, R., & Keto, E. R. 2010, *ApJ*, 711, 1017
Reynolds, D. R., Hayes, J. C., Paschos, P., & Norman, M. L. 2009, Journal of Computational Physics, 228, 6833
Rijkhorst, E., Plewa, T., Dubey, A., & Mellema, G. 2006, *A&A*, 452, 907
Urban, A., Martel, H., & Evans, N. J. 2010, *ApJ*, 710, 1343
Wolfire, M. G. & Cassinelli, J. P. 1987, *ApJ*, 319, 850
Yorke, H. W. & Bodenheimer, P. 1999, *ApJ*, 525, 330
Yorke, H. W. & Kruegel, E. 1977, *A&A*, 54, 183
Yorke, H. W. & Sonnhalter, C. 2002, *ApJ*, 569, 846

Computational Star Formation
Proceedings IAU Symposium No. 270, 2011
J. Alves, B.G. Elmegreen, J. M. Girart & V. Trimble, eds.

doi:10.1017/S1743921311000366

SPH Radiative Hydrodynamics Methods

Hajime Susa

Departiment of Physics, Konan University, Kobe, Japan
email: susa@konan-u.ac.jp

Abstract. In this paper, we review the radiative hydrodynamics methods based upon Smoothed Particle Hydrodynamics(SPH). There are already various implementations so far, which can be categorized into three types: moment equation solvers, Monte Carlo methods, and ray-tracing schemes. These codes have been applied to various astrophysical problems including dynamics of dense proto-stellar cores, photoionization feedback of massive stars on molecular clouds, radiative feedback in the early universe, etc. Among these different methods, we focus on the ray-tracing schemes. We also describe one particular ray-tracing code RSPH in some details.

Keywords. methods: numerical, radiative transfer, hydrodynamics

1. Various astrophysical problems and different codes

Smoothed Particle Hydrodynamics(SPH) is the most widely used Lagrangian scheme for the studies of star formation and galaxy formation. Thanks to its Lagrange nature, it has an advantage over the Eulerian grid codes in resolving the collapsing regions which are always found in numerical simulations of star/galaxy formation. On the other hand, radiative processes including radiative cooling/heating, radiation pressure, photo-chemical reactions, play central roles in star/galaxy formation. Thus, implementing radiation physics in SPH is quite important and a natural pathway to tackle such astrophysical problems.

A number of authors have already developed radiative hydrodynamics codes with SPH. Most of the codes use approximations for radiation transfer calculations depending on the problem which they try to solve. Studies of the formation of proto-stellar cores have the longest history of this type of numerical simulations. Because of the relatively large optical depth in the dense collapsing core, the diffusion/flux limited diffusion approximation has been used in these studies (Lucy 1977, Brookshaw 1994, Whitehouse *et al.* 2005, Viau *et al.* 2006, Mayer *et al.* 2007, Price & Bate 2009). Flux limited diffusion is also used in the studies of supernova explosions(Herant & Woosley 1994, Fryer *et al.* 2006), where the system is also very dense, so optically thick. The advantage of the diffusion/flux limited diffusion approximation is that it is much less computationally costly than direct methods like Monte Carlo schemes. It utilizes the moment equations of the radiation transfer equation, with a simple closure relation between the mean intensity and pressure tensor of the radiation field. Thus, the five dimensional radiation transfer equation is reduced to equations for three dimensional variables like energy density or radiation flux. As a result, the cost of the computation is reduced dramatically. Another type of moment equation solver is called Optically Thin Variable Eddington Tensor (OTVLT) scheme developed by Gnedin & Abel (2001). In this scheme, the variable Eddington tensor is calculated under the assumption that the system is optically thin, whereas the diffusion approximation is basically applicable in the optically thick region. OTVLT has been coupled with SPH (Gadget) by Petkova & Springel (2009), and was applied to cosmic reionization problem.

Apart from these studies with moment equation solvers, there is another direction for the development of radiative hydrodynamics with SPH. In case we want to obtain realistic images or the spectra of proto-stellar cores from theoretical calculations, rough approximations with moment equation solvers are not good enough to be compared with observations. In these cases, Monte Carlo schemes are often used, which have also been implemented in SPH density fields (Oxley & Woolfson 2003, Stamatellos & Whitworth 2005, Forgan & Rice 2010). If we could use enough photon packets in a simulation, the outcome of the calculation approaches the precise solution, while the computational cost becomes very large. So, most of the Monte Carlo solvers at present are used for post-processing the results of hydrodynamical calculations. Monte Carlo schemes with SPH are also widely used in studies of cosmic reionization (Semelin *et al.* 2007, Pawlik & Schaye 2008, Altay *et al.* 2008, Maselli *et al.* 2009), in which large scale ionization pattern is not significantly changed by the coupling of radiation transfer and hydrodynamics.

On the other hand, if we are interested in much smaller scales $\lesssim 10$ kpc ($\simeq$ 10km/s $\times$ 10Gyr), radiative hydrodynamical effects come into play. For instance, if we consider the radiative feedback effects on the formation of the first galaxies, or the first/second generation of stars, intense radiation flux from neighboring sources can ionize and heat up the gas (e.g. Susa & Umemura 2004, 2006). Another important example is the positive/negative radiative feedback by massive stars in local star formation processes (e.g. Bisbas *et al.* 2009, Gritschneder *et al.* 2009). In order to tackle these problems, a careful treatment of photoionization as well as a radiation transfer solver fully coupled with hydrodynamics is required. Because of the too heavy cost of Monte Carlo radiation transfer coupled with hydrodynamics, several authors have developed a different type of radiation transfer solver, those are categorized as ray-tracing codes. In these codes, the transfer of diffuse photons is not solved. These photons are assumed to be absorbed "on-the-spot" and never scattered or re-emitted to a spatially distant position. Thus, we only have to solve the transfer of photons directly from the source star under such approximation. As a result, the computational cost of the radiation transfer calculation is greatly reduced. In this paper, we focus on the description of such ray-tracing schemes. We also describe in some detail one particular ray-tracing scheme RSPH developed by ourselves.

2. Ray-Tracing schemes coupled with SPH

2.1. *Ray-Tracing*

The core part of the ray-tracing scheme is the optical depth integrator. As for the method to calculate the optical depth, the codes developed so far are basically divided into two categories. the first contains the "neighbor connecting" schemes (Kessel-Deynet & Burkert 2000, Susa & Umemura 2004, Susa 2006, Miao *et al.* 2006, Dale *et al.* 2007, Yoshida *et al.* 2007, Gritschneder *et al.* 2009). In these schemes the optical depth from the point source is assessed utilizing the neighbor lists of SPH particles. Although there are some variations in the usage of the neighbor lists, all codes basically follow the neighbor lists along the light rays from the sources to all SPH particles to create the evaluation points of the gas density and optical depth on the rays. Summing up all the contributions of SPH particles on the evaluation points of the light rays, we can evaluate the optical depth from the source stars to each SPH particle.

On the other hand, a few codes (Alvarez *et al.* 2006, Bisbas *et al.* 2009) utilize a public domain code called HEALPix (Górski *et al.* 2005). HEALPix is a software which produces a subdivision of a spherical surface in which each pixel covers the same surface area as every other pixel. In these codes, light rays are generated corresponding to the small solid angles provided by HEALPix. Then, the SPH density field is mapped to the grids along these rays to calculate the optical depth. Another advantage of HEALPix is its hierarchical structure. The light ray from the source can be split as the distance from the source becomes large, based on the tree structure of solid angles in HEALPix(Abel & Wandelt 2002). Such a ray-splitting procedure reduces the computational cost dramatically.

2.2. *Photoionization solver*

Since most of the ray-tracing schemes are designed to investigate the radiative feedback from the photoionization process, it is crucial to implement a photoionization solver in the code.

The photoionization rate of HI and the photoheating rate for each SPH particle labeled as i are given by

$$k_{\rm ion}^{(1)}(i) = n_{\rm HI}(i) \int_{\nu_L}^{\infty}\int \frac{I_\nu(i)}{h\nu}\sigma_\nu d\Omega d\nu, \tag{2.1}$$

$$\Gamma_{\rm ion}^{(1)}(i) = n_{\rm HI}(i) \int_{\nu_L}^{\infty}\int \frac{I_\nu(i)}{h\nu}\sigma_\nu (h\nu - h\nu_L) d\Omega d\nu. \tag{2.2}$$

Here $n_{\rm HI}(i)$ represents the number density of neutral hydrogen of the i-th particle, and σ_ν is the photoionization cross section. The frequency at the Lyman limit is denoted by $\nu_{\rm L}$, and Ω is the solid angle. $I_\nu(i)$ is the intensity of the ultraviolet radiation that irradiates the i-th particle, which is obtained by the ray-tracing discussed in the previous subsection.

In case the optical depth for a single SPH particle is less than $\sim O(1)$, equations (2.1) and (2.2) are valid. If the optical depth becomes much larger than unity, however, those expressions could lead to essentially zero ionization and heating rates because the equations do not conserve the number of photons numerically.

There are two directions to avoid this difficulty. The first one is the "photon conserving method" (Kessel-Deynet & Burkert 2000, Susa 2006, Miao *et al.* 2006, Gritschneder *et al.* 2009) similar to the scheme developed in grid codes (e.g. Abel, Norman & Madau 1999).

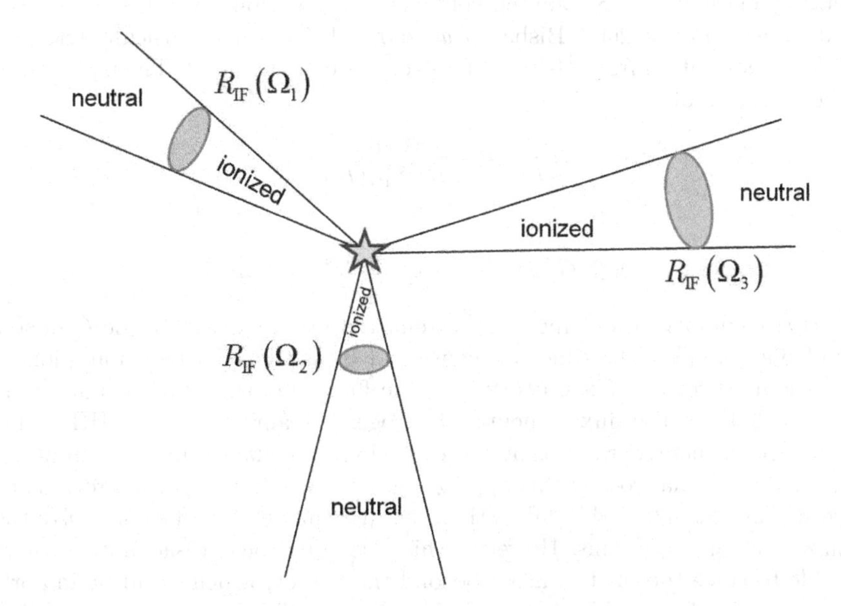

Figure 1. Schematic description of I-front tracking schemes. Here the positions of the ionization front along particular three directions corresponding to $\Omega_1, \Omega_2, \Omega_3$ are shown.

We combine equations (2.1),(2.2) without the suffix i and radiation transfer equation to rewrite the ionization rate and the photoheating rate as follows:

$$k_{\rm ion} = -\frac{1}{4\pi r^2}\frac{d}{dr}\int_{\nu_L}^{\infty}\frac{L_\nu^* \exp(-\tau_\nu)}{h\nu}d\nu, \tag{2.3}$$

$$\Gamma_{\rm ion} = -\frac{1}{4\pi r^2}\frac{d}{dr}\int_{\nu_L}^{\infty}\frac{L_\nu^* \exp(-\tau_\nu)}{h\nu}(h\nu - h\nu_L)d\nu, \tag{2.4}$$

where L_ν^* denotes the intrinsic luminosity of the source and r is the distance from the source. Then we integrate these rates in a small volume within $(r_i - \Delta r_i/2, r_i + \Delta r_i/2)$ to obtain the "volume averaged" rates:

$$k_{\rm ion}^{(2)}(i) \equiv \overline{k_{\rm ion}} = \frac{3}{\Delta r_i}\frac{\Phi_1(r_i - \Delta r_i/2) - \Phi_1(r_i + \Delta r_i/2)}{3r_i^2 + \Delta r_i^2/4}, \tag{2.5}$$

$$\Gamma_{\rm ion}^{(2)}(i) \equiv \overline{\Gamma_{\rm ion}} = \frac{3}{\Delta r_i}\frac{\Phi_2(r_i - \Delta r_i/2) - \Phi_2(r_i + \Delta r_i/2)}{3r_i^2 + \Delta r_i^2/4}, \tag{2.6}$$

where

$$\Phi_1(r) = \int_{\nu_L}^{\infty}\frac{L_\nu^*}{4\pi}\frac{\exp(-\tau_\nu)}{h\nu}d\nu, \tag{2.7}$$

$$\Phi_2(r) = \int_{\nu_L}^{\infty}\frac{L_\nu^*}{4\pi}\frac{\exp(-\tau_\nu)}{h\nu}(h\nu - h\nu_L)d\nu. \tag{2.8}$$

Here r_i is the distance between the source and i-th particle, Δr_i denotes the spatial step of the ray-tracing integration. Using these volume averaged photoionization rates, we can solve the time dependent rate equations of chemical species by ordinary implicit time integration. The method described above has the important advantage that the propagation of the ionization front is properly traced even for a large particle separation with optical depth greater than unity.

Another approach to overcome the difficulty of the ionizing photon transfer is called "I-front tracking method" or "Strömgren volume approximation"(Alvarez *et al.* 2006, Dale *et al.* 2007, Yoshida *et al.* 2007, Bisbas *et al.* 2009). This method basically tries to follow the position of ionization front (I-front) for every direction (Fig.1). The equation used to follow the evolution of I-front is

$$\frac{dR_{\rm IF}}{dt} = \frac{Q(R_{\rm IF}, t)}{4\pi R_{\rm IF}^2 n_{\rm HI}(R_{\rm IF}, t)} \tag{2.9}$$

where

$$Q(R,t) = Q_* - 4\pi\alpha_B\int_0^R r^2 n^2(r,t)dr \tag{2.10}$$

Here Q_* is the number of ionizing photons emitted by the source, while Q denotes the number of photons per unit time not consumed by the case B recombination process within the ionized region. Thus, $Q/4\pi R^2$ is the flux of ionizing photons at the I-front, which should balance the flux of neutral hydrogen streaming into the HII region over the I-front. Consequently, we obtain the evolutionary equation of the I-front position $R_{\rm IF}$. The computational cost of this approximation is smaller than the previous method discussed in the first half of this subsection, because this scheme does not solve the local photoionization rate equations. However, this also could be a disadvantage, since it is not possible to trace the photoheating beyond the I-front, which could be important in case the spectrum of the radiation source is very hard like for first stars or QSOs (e.g. Susa & Umemura 2006).

3. RSPH

3.1. Brief description of the code

In this section, we briefly describe the ray-tracing code RSPH which is developed by ourselves. The code was originally designed to investigate the formation and evolution of the first generation of objects at $z \gtrsim 10$, where the radiative feedback from various sources plays important roles. The code can compute the fraction of chemical species e, H^+, H, H^-, H_2, and H_2^+ by fully implicit time integration. It also can deal with multiple sources of ionizing radiation as well as radiation at the Lyman-Werner band. We use the version of SPH by Umemura(1993) with modifications according to Steinmetz & Müller(1993), and also adopt the particle resizing formalism by Thacker *et al.* (2000).

The non-equilibrium chemistry and radiative cooling for primordial gas are calculated by the code developed by Susa & Kitayama(2000), where the H_2 cooling and reaction rates are mostly taken from Galli & Palla (1998).

As for the photoionization process, we employ the on-the-spot approximation (Spitzer 1978) which has already been discussed in section 1. We solve the transfer of ionizing photons directly from the source, whereas we do not solve the transfer of diffuse photons. Instead, it is assumed that the recombination photons are absorbed in the neighborhood of the spatial position where they are emitted. Because of the absence of the source term in this approximation, the radiation transfer equation becomes very simple. The method to solve the transfer equation reduces to the simple problem of assessing the optical depth from the source to every SPH particle.

The optical depth is integrated utilizing the neighbor lists of SPH particles. In our scheme, we do not create many grid points on the light ray. Instead, we just create one grid point per SPH particle in its neighborhood. In Fig.2, the scheme is schematically shown for a particular case. In case we try to assess the optical depth at the particle labeled as P0, we find the 'upstream' particle for P0 on its line of sight to the source. In this case, the 'upstream' particle is P2. The selection criterion of the 'upstream' particle is that it has the smallest angle θ (see Fig.2) among the particles in the neighbor list of P0. Then the optical depth from the source to P0 is obtained by summing up the optical depth at P2 and the differential optical depth between P0 and P2. Remark that the the differential optical depth $d\tau_{\rm P0}$ is measured from P0 to the point on the light ray

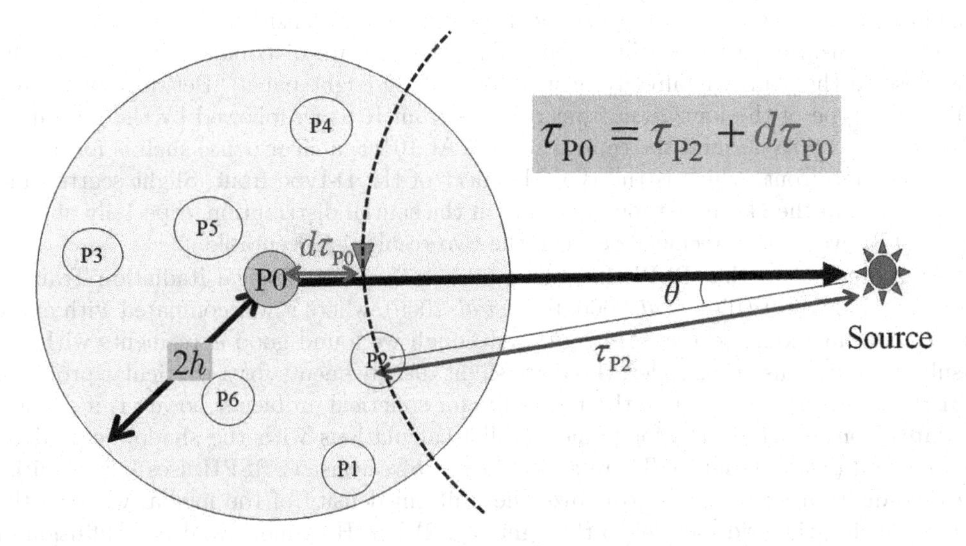

Figure 2. Schematic description of the ray-tracing procedure of RSPH.

marked by the wedge in Fig.2. The distance between the wedge and the source is set to equals the interval between P2 and the source. Thus, the geometrical prolongation of path length due to the deviation from the straight line is avoided. It is also worth noting that this procedure naturally introduces ray-splitting. The number of light rays in the neighborhood of the source should be similar to the number of neighbor particles of the source, while each SPH particle has its own ray to the source. Thus, the rays from the source are split to reach many SPH particles.

The code is already parallelized with the MPI library. The computational domain is divided by the so called Orthogonal Recursive Bisection method. The parallelization method for the radiation transfer part is similar to the Multiple Wave Front method developed by Nakamoto, Umemura, & Susa (2001) and Heinemann *et al.* (2006), but it is changed to fit to the SPH code. The details are described in Susa (2006). The code is also able to handle self-gravity with a Barns-Hut tree, which is also parallelized. We also remark that the code has already been applied to various astrophysical problems including radiative feedback during star formation / galaxy formation in the early universe (e.g. Susa & Umemura 2006, Susa 2008).

3.2. *Test simulations*

We have performed various tests with the code. Among several test calculations, here we show the results of most simple and fundamental tests. The test results are compared with the numerical solutions from one dimensional radiation hydrodynamic simulations. In this test calculation, we use 8 nodes (16 Xeon processors with gigabit ethernet) and 1048576 SPH particles. The clock time of this particular run is about 3 hours.

We put a single source at the center of an uniform gas cloud and trace the propagation of the ionization front. The initial number density of the gas is $n_{\rm H} = 0.01{\rm cm}^{-3}$, and the temperature is $T = 3 \times 10^2{\rm K}$. The ionizing photon luminosity of the source is $S = 1.33 \times 10^{50}{\rm s}^{-1}$ and the spectrum is black body with $T_* = 9.92 \times 10^4{\rm K}$, which is typical for POPIII stars. This parameter set is chosen so that the we can trace the well known transition of the ionization front from R-type to D-type in uniform media.

Figure 3 shows the spatial distributions of density and temperature at different times ($t = 10^7{\rm yr}, 10^8{\rm yr}, 10^9{\rm yr}$). The results are compared with highly accurate one dimensional simulation. At 10^7yr, no hydrodynamical change of gas density is found (left panel), since the ionization front is still R-type. The temperature distribution shows that the gas close to the source is already heated to $\sim 10^4$K (right panel). Between 10^7yr and 10^8yr, the type of the ionization front changes from R to D, followed by the formation of less dense cavity around the central source. At 10^9yr, a clear dense shell is formed at the ionization front, which is the typical aspect of the D-type front. Slight scatter and deviation from the 1D simulation are found in the spatial distribution, especially at later epoch. However, the agreement between the two results is acceptable.

We also point out that RSPH has taken part in the Cosmological Radiation Transfer Comparison Project (Iliev *et al.* 2006, Iliev *et al.* 2009), where it was compared with many other codes including several grid codes. Although we found good agreements with the results from various other codes, there are slight disagreements for a particular problem. RSPH is basically very close to the 1D results for spherical problems, however, it slightly deviates from other results for plane parallel calculations with the shadow casted by dense cloud (Tests 3 and 7). There are mainly two reasons: 1) RSPH uses 'glass' initial particle distribution in order to realize the uniform density of the media, whereas the grid codes use the grid parallel to the light ray. 2) RSPH cannot avoid the 'diffusion of optical depth' because of its ray-tracing scheme. Thus, it is difficult to obtain very sharp

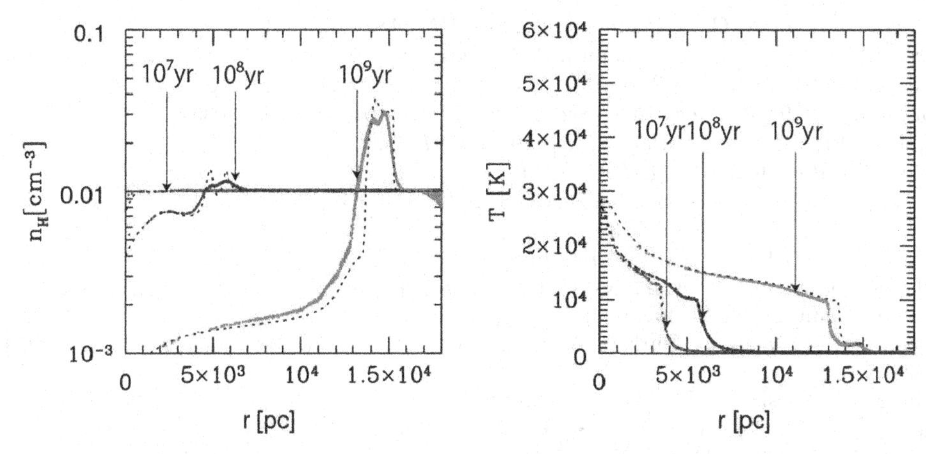

Figure 3. Spatial distribution of physical quantities at three times ($t = 10^7, 10^8, 10^9$yr) are shown for the test run. The two panels show the density and temperature as functions of the distance from the source. The dotted lines denote the results from the 1D simulation.

boundaries between the shadow and the ionized region. We have to keep in mind this disadvantage of RSPH scheme.

4. Summary

In this paper, we reviewed the radiative hydrodynamics methods coupled with SPH. Previous implementations can be categorized into three types: moment equation solvers, Monte Carlo methods, and ray-tracing schemes.

The core part of the ray-tracing schemes consists of an optical depth integrator and a photoionization solver. As for the optical depth integration, most of the codes employ "neighbor connecting" schemes, whereas a few utilize HEALPix to generate hierarchical light rays. There are two basic types of photoionization solvers: the simple photon conserving scheme and I-front tracking method. Both of them are useful to trace the photoionization, however, in case we have to consider photoheating beyond the I-front, the I-front tracking method is not a good approximation.

We also described the ray-tracing code RSPH in some detail. The results of the test calculations agree well with the results from 1D simulations and various other radiative hydrodynamics codes including grid codes.

Acknowledgements

HS thanks M. Umemura for insightful comments on the construction of RSPH code. HS also thanks all the collaborators in the Cosmological Radiation Transfer Code Comparison Project for fruitful discussions during the three workshops in CITA, Leiden and Austin. The analysis has been made with computational facilities at the Center for Computational Science at the University of Tsukuba and Rikkyo University and Konan University. This work was supported in part by the Ministry of Education, Culture, Sports, Science, and Technology (MEXT), Grants-in-Aid, Specially Promoted Research 16002003, Scientific Research (C), 22540295.

References

Abel T., Norman M. L., & Madau P., 1999, *ApJ*, 523, 66 +
Abel, T. & Wandelt, B. D. 2002, *MNRAS*, 330, L53

Altay, G., Croft, R. A. C., & Pelupessy, I. 2008, *MNRAS*, 386, 1931
Alvarez, M. A., Bromm, V., & Shapiro, P. R. 2006, *ApJ*, 639, 621
Bisbas, T. G., Wünsch, R., Whitworth, A. P., & Hubber, D. A. 2009, *A&A*, 497, 649
Brookshaw, L. 1994, *Memorie della Societa Astronomica Italiana*, 65, 1033
Dale, J. E., Ercolano, B., & Clarke, C. J. 2007, *MNRAS*, 382, 1759
Forgan, D. & Rice, K. 2010, *MNRAS*, 1049
Fryer, C. L., Rockefeller, G., & Warren, M. S. 2006, *ApJ*, 643, 292
Galli, D. & Palla, F. 1998, *A&A*, 335, 403
Gnedin, N. Y. & Abel, T. 2001, *New Astronomy*, 6, 437
Górski, K. M., Hivon, E., Banday, A. J., Wandelt, B. D., Hansen, F. K., Reinecke, M., & Bartelmann, M. 2005, *ApJ*, 622, 759
Gritschneder, M., Naab, T., Burkert, A., Walch, S., Heitsch, F., & Wetzstein, M. 2009, *MNRAS*, 393, 21
Herant, M. & Woosley, S. E. 1994, *ApJ*, 425, 814
Heinemann, T., Dobler, W., Nordlund, Å., & Brandenburg, A. 2006, *A&A*, 448, 731
Iliev, I. T., *et al.* 2006, *MNRAS*, 371, 1057
Iliev, I. T., *et al.* 2009, *MNRAS*, 400, 1283
Kessel-Deynet, O. & Burkert, A. 2000, *MNRAS*, 315, 713
Lucy, L. B. 1977, *AJ*, 82, 1013
Maselli, A., Ciardi, B., & Kanekar, A. 2009, *MNRAS*, 393, 171
Mayer, L., Lufkin, G., Quinn, T., & Wadsley, J. 2007, *ApJL*, 661, L77
Miao, J., White, G. J., Nelson, R., Thompson, M., & Morgan, L. 2006, *MNRAS*, 369, 143
Nakamoto, T., Umemura, M., & Susa, H. 2001, *MNRAS*, 321, 593
Pawlik, A. H. & Schaye, J. 2008, *MNRAS*, 389, 651
Petkova, M. & Springel, V. 2009, *MNRAS*, 396, 1383
Price, D. J. & Bate, M. R. 2009, *MNRAS*, 398, 33
Oxley, S. & Woolfson, M. M. 2003, *MNRAS*, 343, 900
Semelin, B., Combes, F., & Baek, S. 2007, *A&A*, 474, 365
Spitzer, L. Jr. 1978, in Physical Processes in the Interstellar Medium (John Wiley & Sons, Inc. 1978)
Stamatellos, D., & Whitworth, A. P. 2005, *A&A*, 439, 153
Steinmetz, M. & Müller, E. 1993, *A&A*, 268, 391
Susa, H. 2006, *PASJ*, 58, 445
Susa, H. & Kitayama, T. 2000, *MNRAS*, 317, 175
Susa, H. & Umemura, M. 2004, *ApJ*, 600, 1
Susa, H. & Umemura, M. 2006, *ApJL*, 645, 93
Susa, H. 2008, *ApJ*, 684, 226
Thacker, J., Tittley, R., Pearce, R., Couchman, P. & Thomas, A. 2000, *MNRAS* 319, 619
Umemura, M. 1993, *ApJ*, 406, 36
Viau, S., Bastien, P., & Cha, S.-H. 2006, *ApJ*, 639, 559
Whitehouse, S. C., Bate, M. R., & Monaghan, J. J. 2005, *MNRAS*, 364, 1367
Yoshida, N., Oh, S. P., Kitayama, T., & Hernquist, L. 2007, *ApJ*, 663, 68

Computational Star Formation
Proceedings IAU Symposium No. 270, 2011
J. Alves, B.G. Elmegreen, J. M. Girart & V. Trimble, eds.

doi:10.1017/S1743921311000378

Moving-mesh hydrodynamics with the AREPO code

Volker Springel[1,2]

[1]Heidelberg Institute for Theoretical Studies, Schloss-Wolfsbrunnenweg 35, 69118 Heidelberg, Germany
[2]Zentrum für Astronomie, Universität Heidelberg, Mönchhofstr. 12-14, 69120 Heidelberg, Germany
email: `volker.springel@h-its.org`

Abstract. At present, hydrodynamical simulations in computational star formation are either carried out with Eulerian mesh-based approaches or with the Lagrangian smoothed particle hydrodynamics (SPH) technique. Both methods differ in their strengths and weaknesses, as well as in their error properties. It would be highly desirable to find an intermediate discretization scheme that combines the accuracy advantage of mesh-based methods with the automatic adaptivity and Galilean invariance of SPH. Here we briefly describe the novel AREPO code which achieves these goals based on a moving unstructured mesh defined by the Voronoi tessellation of a set of discrete points. The mesh is used to solve the hyperbolic conservation laws of ideal hydrodynamics with a finite volume approach, based on a second-order unsplit Godunov scheme with an exact Riemann solver. A particularly powerful feature is that the mesh-generating points can in principle be moved arbitrarily. If they are given the velocity of the local flow, an accurate Lagrangian formulation of continuum hydrodynamics is obtained that features a very low numerical diffusivity and is free of mesh distortion problems. If the points are kept fixed, the scheme is equivalent to a Eulerian code on a structured mesh. The new AREPO code appears especially well suited for problems such as gravitational fragmentation or compressible turbulence.

Keywords. hydrodynamics, methods: numerical, radiative transfer

1. Introduction

In astrophysics, a variety of fundamentally quite different numerical methods for hydrodynamic simulations are in use, the most prominent ones are SPH (e.g. Monaghan 1992) and Eulerian mesh-based hydrodynamics (e.g. Stone *et al.* 2008) with (optional) adaptive mesh refinement (AMR). However, it has become clear over recent years that both SPH and AMR suffer from fundamental problems that make them inaccurate in certain regimes. Indeed, these methods sometimes yield conflicting results even for basic calculations that only consider non-radiative hydrodynamics (e.g. Frenk *et al.* 1999, Agertz *et al.* 2007, Tasker *et al.* 2008, Mitchell et al. 2009). SPH codes have comparatively poor shock resolution, and offer only low-order accuracy for the treatment of contact discontinuities. Worse, they appear to suppress fluid instabilities under certain conditions, as a result of a spurious surface tension and inaccurate gradient estimates across density jumps. On the other hand, Eulerian codes are not free of problems either. They do not produce Galilean-invariant results, which can make the results sensitive to the presence of large bulk velocities. Another concern lies in the comparatively strong mixing inherent in multi-dimensional Eulerian hydrodynamics. This provides for an implicit source of entropy, with unclear consequences (e.g. Wadsley, Veeravalli & Couchman 2008).

There is hence substantial motivation to search for new hydrodynamical methods that improve on these weaknesses. We would like to retain the accuracy of mesh-based hydrodynamical methods while at the same time outfit them with the Galilean-invariance, geometric flexibility, and automatic adaptivity that is characteristic of SPH. The principal idea for achieving such a synthesis is to allow the mesh to move with the flow itself. This is an obvious and old idea, but one fraught with many practical difficulties that have so far prevented widespread use of any of the few past attempts to introduce moving-mesh methods in astrophysics (e.g. Gnedin 1995, Pen 1998).

2. The AREPO code

In our new AREPO code (see Springel 2010 for a detailed description), we introduce a new formulation of continuum hydrodynamics based on an unstructured mesh. The mesh is defined as the Voronoi tessellation of a set of discrete mesh-generating points, which are in principle allowed to move freely. For the given set of points, the Voronoi tessellation of space consists of non-overlapping cells around each of the sites such that each cell contains the region of space closer to it than any of the other sites. In practice, we construct the Voronoi mesh in terms of the Delaunay tessellation, which is the topological dual of the Voronoi diagram and can be more easily calculated with fast geometric algorithms.

The Voronoi cells can be used as control volumes for a finite-volume formulation of hydrodynamics, using the same principal ideas for reconstruction, evolution and averaging (REA) steps that are commonly employed in many Eulerian techniques. However, it is possible to consistently include the mesh motion in the formulation of the numerical steps, allowing the REA-scheme to become Galilean-invariant. Even more importantly, due to the mathematical properties of the Voronoi tessellation, the mesh continuously deforms and changes its topology as a result of the point motion, without ever leading to the dreaded mesh-tangling effects that are the curse of traditional moving mesh methods.

In our approach, each computational cell carries as conserved fluid variables the mass, momentum and total energy, while the primitive variables are determined through the Voronoi volume of a cell. We then estimate spatial gradients for the unstructured mesh, apply slope limiting techniques to them, and finally define a piece-wise linear reconstruction for all primitive fluid variables. Appropriate spatial and temporal extrapolation steps are applied to estimate the fluid states at both sides of each mesh face. These are fed to a Riemann solver, yielding numerical fluxes that update in a detailed balance the conserved quantities of the cells. The overall method is a second-order accurate scheme in space and time, and closely corresponds to a MUSCL approach.

Self-gravity is calculated in the code with a hierarchical multipole expansion (a tree algorithm), which we have ported from the GADGET-3 code. Also, a TreePM extension for a more efficient long-range gravity calculation is available. The AREPO code is fully parallelized for distributed memory machines, allows for individual and adaptive timesteps, and can also treat additional collisionless particle components.

3. Example calculations

In Springel (2010), a number of test problems are discussed that demonstrate that the new scheme achieves high accuracy in the treatment of shocks, shear waves, and fluid instabilities, making it an attractive alternative to currently employed SPH and AMR codes. To illustrate a few principal features of the code, we here briefly discuss the Rayleigh-Taylor (RT) instability as a representative example for the behavior of the method in multi-dimensional flow.

The RT instability can arise in stratified layers of gas in an external gravitational field. If higher density gas lies on top of low-density gas, the stratification is unstable to

Figure 1. Rayleigh-Taylor instability calculated at low resolution with the moving-mesh approach. A denser fluid lies above a less dense fluid in an external gravitational field. The hydrostatic equilibrium of the initial state is buoyantly instable. The three frames show the time evolution of the density field of the system at times $t = 5.0$, 10.0, and 15.0, after a single mode has been perturbed to trigger the stability, as described in the text.

buoyancy forces, and characteristic finger-like perturbations grow that will mix the fluids with time. To illustrate the motion of the mesh in our new code, we show in Figure 1 the evolution of a single Rayleigh-Taylor mode, calculated at the low resolution of 12×36 cells. The simulation domain is two-dimensional (extension $[0.5, 1.5]$), and has periodic boundaries on the left and right and solid walls at the bottom and top. There is an external gravitational field with acceleration $g = -0.1$, and the bottom and top halves of the box are filled with gas of density $\rho = 1$ and $\rho = 2$, respectively. The initial hydrostatic pressure profile is of the form $P(y) = P_0 + (y - 0.75)\, g\, \rho(y)$ with $P_0 = 2.5$ and $\gamma = 1.4$. To seed the perturbation, one mode is excited with a small velocity perturbation of the form $v_y(x, y) = w_0[1 - \cos(4\pi x)][1 - \cos(4\pi y/3)]$, where $w_0 = 0.0025$.

As can be clearly seen in the time evolution shown in Figure 1, the Rayleigh-Taylor instability is captured well by the moving-mesh method even at this low resolution. Note that the sharp boundary between the phases can be maintained for relatively long time during the early evolution of the instability, simply because the contact discontinuity is not smeared out as it bends, thanks to the mesh's ability to follow this motion in an approximately Lagrangian fashion. A stationary mesh on the other hand would automatically wash out the boundary due to advection errors.

This important improvement becomes clearer in Figure 2, where we compare a high-resolution version of the RT instability between the moving-mesh approach and the same calculation carried out with a stationary Cartesian mesh. Here 1024×1024 cells have been used in the unit domain, and the instability was triggered by adding small random noise to the velocity field everywhere. While the instability shows a similar overall growth rate in both cases, there are also striking differences. The calculation with the fixed mesh produces a lot of intermediate density values due to the strong mixing of the phases

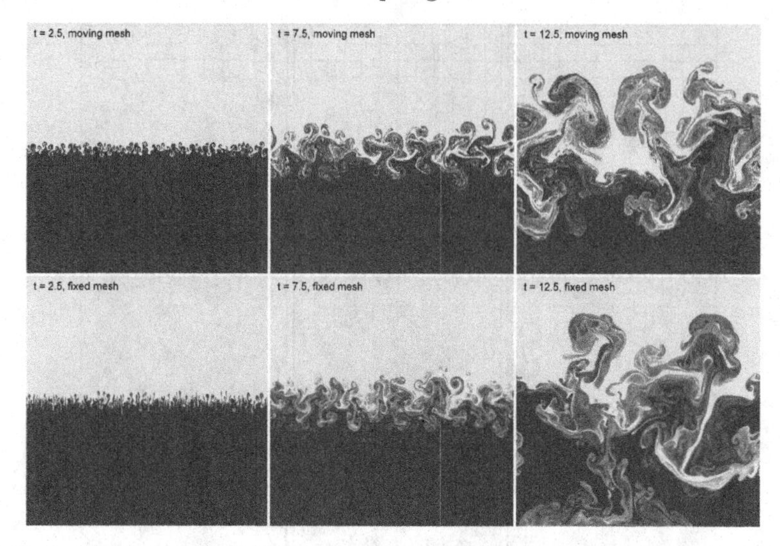

Figure 2. Rayleigh-Taylor instability calculated at high resolution with 1024×1024 points in the unit domain. The instability is here seeded by small random noise added to the velocity field. The top and bottom rows compare the time evolution for calculations with a moving and a stationary mesh, respectively.

on small scales, whereas the moving mesh approach maintains finely stratified regions where different layers of the fluid phases have been folded over each other. The contact discontinuities between these layers can be kept sharp by the code even when they are moving relative to the rest-frame of the box. In the early phase of the growth, it also appears as if small-scale RT fingers grow somewhat too quickly in the Eulerian case as a result of grid alignment effects.

4. Outlook

With the addition of an extensive chemical cooling network, the AREPO code is presently already employed to study formation of the first stars (Greif *et al.*, in preparation). The new code should be especially well suited for problems of gravitational fragmentation and for hydrodynamical turbulence. We presently also work on a radiative transfer solver that is directly operating on the Voronoi mesh (Petkova *et al.*, in preparation), and think about a potential implementation of magnetohydrodynamics in AREPO. With these additions, the code should become even more attractive for problems of computational star formation.

References

Agertz, O., Moore, B., Stadel, J., Potter, D., & Miniati, F., 2007, *MNRAS*, 380, 963
Frenk, C. S., *et al.*, 1999, *ApJ*, 525, 554
Gnedin, N. Y., 1995, *ApJS*, 97, 231
Mitchell, N. L., *et al.*, 2009, *MNRAS*, 395, 180
Monaghan, J. J., 1992, *ARAA*, 30, 543
Pen, U. L., 1998, *ApJS*, 115, 19
Springel, V., 2010, *MNRAS*, 401, 791
Stone, J. M., Gardiner, T. A., Teuben, P., Hawley, J. F., & Simon J. B., 2008, *ApJS*, 178, 137
Tasker, E. J., etal., 2008, *MNRAS*, 390, 1267
Wadsley, J. W., Veeravalli, G., & Couchman, H. M. P., 2008, *MNRAS*, 387, 427

Computational Star Formation
Proceedings IAU Symposium No. 270, 2011
J. Alves, B.G. Elmegreen, J. M. Girart & V. Trimble, eds.

doi:10.1017/S174392131100038X

Ray Casting and Flux Limited Diffusion

Åke Nordlund[1]

[1]Centre for Star and Planet Formation, and
Niels Bohr Institute, University of Copenhagen
Blegdamsvej 17, 2100 Copenhagen, Denmark
email: aake@nbi.dk

Abstract. Solving radiative transfer problems with ray casting methods is compared with the commonly used 'Flux Limited Diffusion' approximation. Whereas ray casting produces solutions that converge to the exact one as the number of rays is increased, flux-limited-diffusion is fundamentally a 'look-alike' method, which produces solutions that are reminiscent of the correct solution but which cannot be made to converge to it.

---

1. Introduction

Solving the 'radiative transfer problem' in general means computing specific radiation intensities $I_\nu(\mathbf{r}, \mathbf{\Omega})$ at relevant frequencies ν in a given physical situation. With a sufficient set of frequencies and space angle directions $\mathbf{\Omega}$, and given also the absorption and scattering cross sections κ_ν and σ_ν, one can then compute the exchange of energy and momentum between the radiation field and the medium the radiation is propagating through.

A number of sub-classes of the 'radiative transfer (RT) problem' are of relevance in different astrophysical situations: In some cases the time-of-flight of the radiation is important, in others not. In some cases the momentum and energy transfer are both important, while in other cases only the energy transfer is of significance. In some cases the populations of atomic states, which again determine the absorption cross sections, are strongly influenced by the radiation field, and computing the population numbers becomes in practice an integral part of the 'radiative transfer problem'. Scattering leads to a similar situation, where the 'source function' in the RT problem contains a term directly proportional to the radiation intensity, averaged over space angle with some (isotropic or non-isotropic) angular re-distribution function.

Common to all these cases is the fact that the specific intensity $I_\nu(\mathbf{r}, \mathbf{\Omega})$ depends on both frequency ν, direction $\mathbf{\Omega}$, and position $\mathbf{r}$; i.e., the radiation field has six degrees of freedom. This is often, but as we shall see incorrectly, considered an insurmountable problem with respect to obtaining direct numerical solutions of the 'radiative transfer problem'. It is argued that it would be too expensive to obtain sufficiently accurate solutions in a 6-dimensional space, and that therefore one needs to use approximate methods that reduce the dimensionality.

Flux limited diffusion (FLD; Minerbo 1978; Levermore & Pomraning 1981) has been, and still is, a popular method for obtaining approximate solutions of radiative transfer problems (Turner & Stone 2001; Krumholtz *et al.* 2007). The flux limited diffusion approximation reduces the dimensionality from six to four, by assuming that the radiation field can be reasonably represented by the mean radiation intensity,

$$J_\nu(\mathbf{r}) = \frac{1}{4\pi} \int_\Omega I_\nu(\mathbf{r}, \mathbf{\Omega}) d\mathbf{\Omega}. \tag{1.1}$$

Approximate expressions, which fulfill certain physically motivated asymptotic conditions, are then used to compute estimates of the radiative energy flux,

$$\mathbf{F}_\nu(\mathbf{r}) = \frac{1}{4\pi}\int_\Omega I_\nu(\mathbf{r},\mathbf{\Omega})\,\mathbf{\Omega}\,d\mathbf{\Omega}, \tag{1.2}$$

based upon the mean intensity (as well as gradients thereof) and the local opacity. Note that the flux-limited diffusion approximation is only concerned with approximating the space angle dependence of the radiation field; the frequency dependence of the radiation field is a different affair.

A reduction of the dimensionality from six to four would appear to be a significant advantage and could be seen as such a dramatic gain that it would motivate the use of an approximation, if the results obtained were still reasonable. However, a proper cost comparison must also factor in the relative complexity of the different methods; i.e., what is, in the end, the cost per $(\mathbf{r},\nu,\mathbf{\Omega})$-point for a direct solution, relative to the cost per $(\mathbf{r},\nu)$-point for the approximate solution?

Because the FLD leads to an elliptic equation for the mean intensity, while the direct solution only involves a small number of floating point operations per degree of freedom, the cost balance actually comes out in favor of the direct method, even for rather large numbers of rays. To put this differently: The key circumstance that renders direct solutions of the 6-dimensional RT problem not only tractable but even advantageous is that, even though the computation effort scales with six degrees of freedom,

$$t_{\rm update} = c_{\rm update} N_{\rm r} N_\nu N_\Omega, \tag{1.3}$$

the scaling constant c_{update} *is very small*, of the order of just a few nanoseconds per update per point with current CPU-cores. Since the time required to update the physical state of an MHD-model is typically several microseconds per point on the same CPU-cores, one can afford of the order of a thousand frequency points times space-angle directions, without increasing the computing time per time step with more than a factor of two.

Therefore, in many common astrophysical situations it is indeed possible, and actually advantageous, to solve radiative transfer directly, using ray tracing.

The main advantage with ray-tracing methods is that solutions converge to the correct ones as the number of rays are increased, and that they therefore lend themselves to traditional types of convergence studies and validation. However, as discussed below, ray tracing methods in general also have several other advantages relative to flux-limit-diffusion: They are simpler to implement, often use less computing time, and they can be made to have near-perfect parallelization properties.

2. Radiative Transfer: Formal Solutions

The central task that needs to be performed in all ray-tracing based methods is computing the radiation intensity $I_\nu(\mathbf{r},\mathbf{\Omega})$, given a source function $S_\nu(\mathbf{r},\mathbf{\Omega})$, from the radiative transfer equation

$$\frac{dI_\nu(\mathbf{r},\mathbf{\Omega})}{d\tau} = S_\nu(\mathbf{r},\mathbf{\Omega}) - I_\nu(\mathbf{r},\mathbf{\Omega}), \tag{2.1}$$

where $d\tau$ is the optical depth increment in the direction $\mathbf{\Omega}$, defined as

$$d\tau = (\kappa_\nu + \sigma_\nu)ds. \tag{2.2}$$

This is often referred to as the 'formal problem', and the task is thus to compute the solution to the formal problem as rapidly and accurately as possible.

Note that solving the 'formal problem' is part of (while still totally independent of) any requirement of consistency between $I_\nu(\mathbf{r}, \mathbf{\Omega})$ and $S_\nu(\mathbf{r}, \mathbf{\Omega})$, e.g. due to scattering or atomic level population balance. A number of well-known and thoroughly tested iterative techniques exist for solving for source-intensity consistency (e.g. Hubeny 2003), but these need not be discussed here, since they all have similar requirements with respect to solving the formal problem—several solutions of the formal problem are typically needed to evaluate residual imbalance between the radiation field and the source function. Thus, for the purpose of this discussion we can focus on considering the cost of obtaining the radiation field from a given source function.

As a test case where the results of using direct ray casting on the one hand and the flux-limited-diffusion approximation on the hand can be easily compared, I consider below the computation of the rate of energy transfer between the radiation and the gas, in a case where velocities are small with respect to the speed of light, so the radiation field can be considered to be determined instantaneously from the source function.

2.1. *Ray Tracing / Ray Casting*

Direct solutions of the radiative transfer problem generally rely on solving the radiative transfer equation (2.1) along a number of rays through the volume under consideration. In general one distinguishes between *short characteristics* methods and *long characteristics* methods.

Short characteristics solutions are obtained incrementally, going typically from one grid point to the next cell boundary, where the solution then needs to be interpolated to the nearest grid point in order to continue.

Long characteristics methods rely on solutions obtained along rays traversing the entire volume. For each direction $\mathbf{\Omega}$ parallel inclined rays are passed through all grid points in a central plane. One thus obtains solutions near all grid point in one sweep. Rays generally then do not pass through grid points, but with a density of rays similar to the density of grid points across rays, values at grid points can be found by interpolation between rays. With this method all information that is obtained along each single ray is being utilized, and the total cost thus scales as $N_{\text{update}} \sim N_{\mathrm{r}} N_{\Omega} N_\nu$.

The denomination 'long characteristics' is sometimes used to refer to a theoretical case where each point in three dimensions is connected by a specific ray to every other point. Such a method can of course never be used in practice, since it would scale as the square of the number of points; $N_{\text{update}} \sim N_{\mathrm{r}}^2 N_\nu$. The method corresponds to setting $N_\Omega = N_{\mathrm{r}}$, and would thus correspond to a huge over-resolution in space angle. Not even discrete sources needs to be handled in this way; these can instead be handled by adaptive schemes (Wang *et al.* 2004; Razoumov & Cardall 2005).

A common misconception is that direct, ray tracing solutions are only suitable for discrete sources. In fact, it is easy to use a combination of a set of rays emanating from point sources with sets of parallel rays that take care of diffuse sources. For diffuse sources, the number of space-angles needed is generally quite small, as illustrated by the test below. However, since the requirements depend on the detailed properties (e.g. smoothness) of the source field, no general rule can be given.

In analogy with the concept of 'multi-casting' on the internet (where many clients can receive information transmitted by a single source), one could argue for referring to the efficient long characteristics method as 'ray multi-casting', or just 'ray casting'.

2.2. *Flux Limited Diffusion*

In the flux-limited-diffusion approximation one finds the mean radiation intensity as follows: One assumes that the radiative flux can be estimated by an expression of the

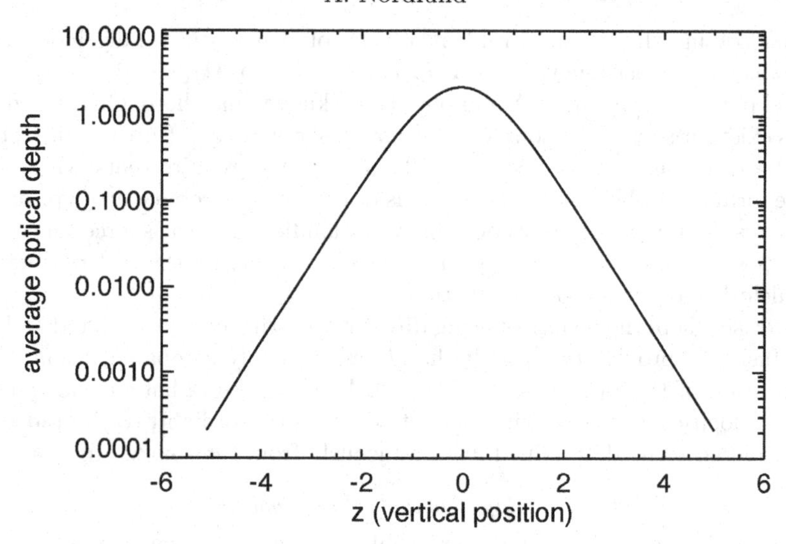

Figure 1. Horizontally averaged optical depth in the test case.

form

$$\mathbf{F}_{\mathrm{rad},\nu} = -\frac{\lambda}{\kappa_\nu}\nabla J, \tag{2.3}$$

where $\lambda = \lambda(|\nabla J|, J, \kappa_\nu)$ is a 'flux limiter' function (e.g. Levermore & Pomraning 1981). One then requires that the divergence of the radiative flux equals the rate of energy transfer between the radiation field and the gas, so

$$\nabla \cdot \mathbf{F}_{\mathrm{rad},\nu} = \rho\kappa_\nu(S - J), \tag{2.4}$$

where S and J are, respectively, the source function and the mean radiation intensity (Turner & Stone 2001).

The result of combining Eqs. 2.3 and 2.4 is a non-linear elliptic equation in J. In the test cases reported on below this problem was solved using successive over-relaxation, starting from the 'exact' solution for J provided by the ray casting method. To monitor convergence the residual between the left and right hand sides of Eq. 2.4 was used.

Tests with three different flux limiters showed that the Levermore & Pomraning (1981) limiter produced slightly better results than the Minerbo (1978) and Krumholtz *et al.* (2007) limiters in the current case.

It should be noted that, even though the radiative flux is fully specified by Eq. 2.4, the actual boundary value of the mean intensity needs to be determined independently. To ensure that this choice did not disfavor the FLD solution the mean intensity of the most accurate (reference) ray casting case was adopted as a boundary condition on J.

3. Results

The test case represents a 'fragmented disk' — a proxy of a turbulent, low temperature stellar accretion disk. The opacity is assumed to be due to pure absorption, with a constant absorption κ_ν per unit mass. The opacity $\rho\kappa_\nu$ thus varies in proportion to the mass density. The source function is chosen to vary in unison with the density, mimicking temperature fluctuations at near-constant entropy. The horizontal variation of the mass density is chosen to be a pattern generated by smoothing an image of white noise with a

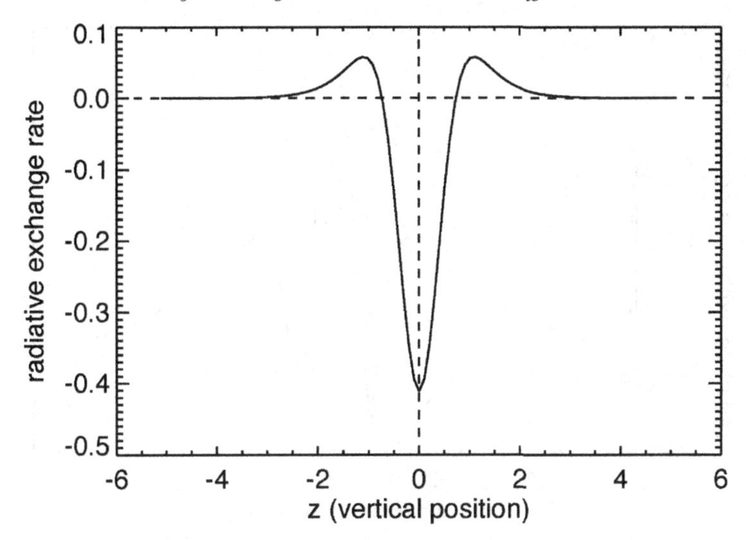

Figure 2. Average heating / cooling rate per unit volume for the reference solution, as a function of the vertical coordinate.

Gaussian filter, leaving approximately 16 degrees of freedom in each horizontal direction (cf. Fig 5a).

The opacity scaling has been chosen so the maximum optical depth across the disk (from bottom to top) is about 5, while in the least dense spots on the disk the total optical depth is about 0.5. Above and below the mid plane the mass density (and hence the optical depth) decreases exponentially. The optical depth is of the order of 10^{-4} at the upper / lower boundaries of the domain. Figure 1 shows the horizontally averaged optical depth in the disk, as a function of the vertical position. The horizontal variation is illustrated in Fig. 5a.

Figure 2 shows the average heating / cooling rate for the reference solution as a function of the vertical coordinate. Generally, there is cooling (energy loss) in a region near the disk mid plane, because the finite optical depth allows energy to leak out. However, sideways heating can occur in spots, as radiation 'leaks' from hotter regions into cooler ones. There is a general re-heating above the optical surface, as the mean intensity is larger than the local source function. Because of the exponential drop in density away from the mid plane the heating / cooling rate drops to very small values there.

Figure 3 shows the horizontal root-mean-square variation of the heating / cooling rate as a function of the vertical coordinate. The root mean square variation is largest in the mid plane, then drops, but is again fairly large in the re-heating layer just outside of optical depth unity.

The FLD solutions are initialized with the 'exact' mean intensity from the reference solution, and the elliptic problem is then iterated to near consistency (note that, since the iterations drive the solution *away* from the correct solution any residual iteration error generally tends to *reduce* the difference with respect to the ray tracing reference solution).

Figure 4 shows minimum and maximum errors of the FLD heating / cooling rate, relative to the global root-mean-square variation of the reference heating / cooling rate. Note that the largest error of the FLD solution is more than twice as large as the overall root-mean-square variation of the heating / cooling rate.

The horizontal variation of the error is illustrated in Fig. 5, where the right hand side panel shows an image of the variation of the error in the horizontal plane.

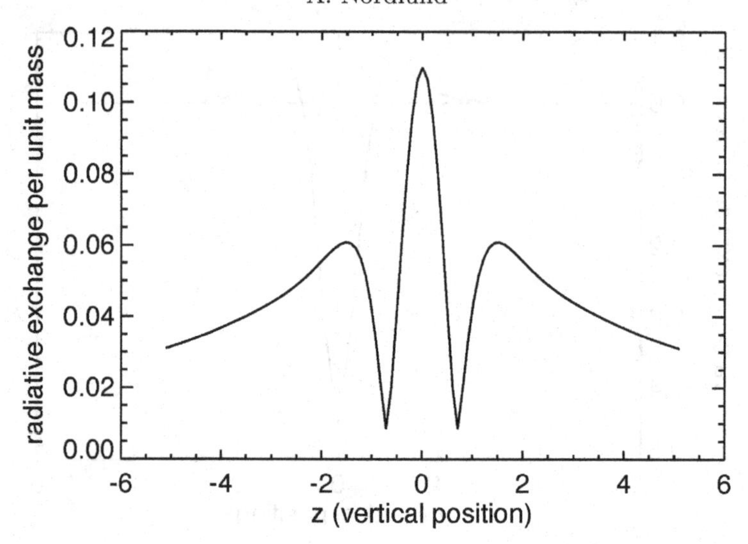

Figure 3. Horizontal root-mean-square variation of the heating / cooling rate, as a function of vertical position

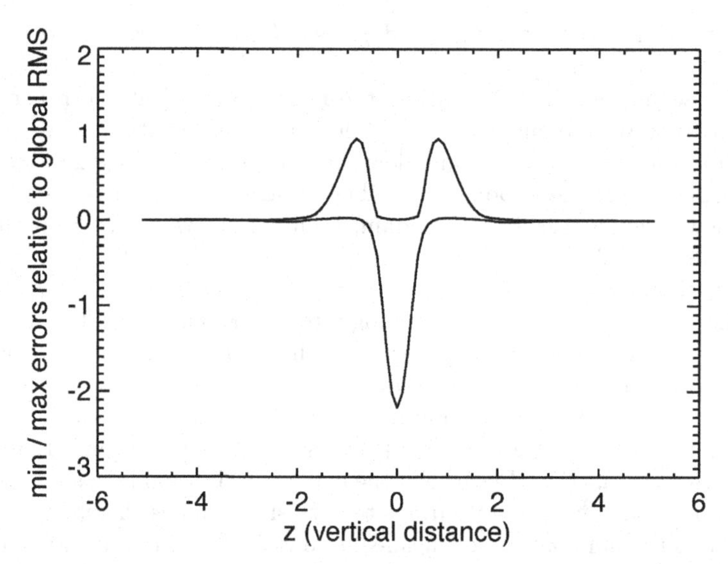

Figure 4. Minimum and maximum errors of the FLD heating / cooling rate, relative to the global root-mean-square variation of the reference heating / cooling rate

Ray tracing solutions were obtained by using Radau integration over inclination angle (Radau integration differs from Gaussian integration in that the vertical direction is always included). For inclined rays integration in the azimuth direction is performed with evenly distributed ray directions.

Figure 6 shows a comparison of the horizontal root-mean-square errors of the heating / cooling rates for the FLD solution and ray tracing solutions with varying number of ray directions (note that there are two rays for each ray direction). As a reference case we use a ray-casting solution with 96 ray directions; this is sufficient for a very accurate solution, given the source function and opacity profiles of this test case. As illustrated by Fig. 6, even the ray tracing solution with only four ray direction is superior to the

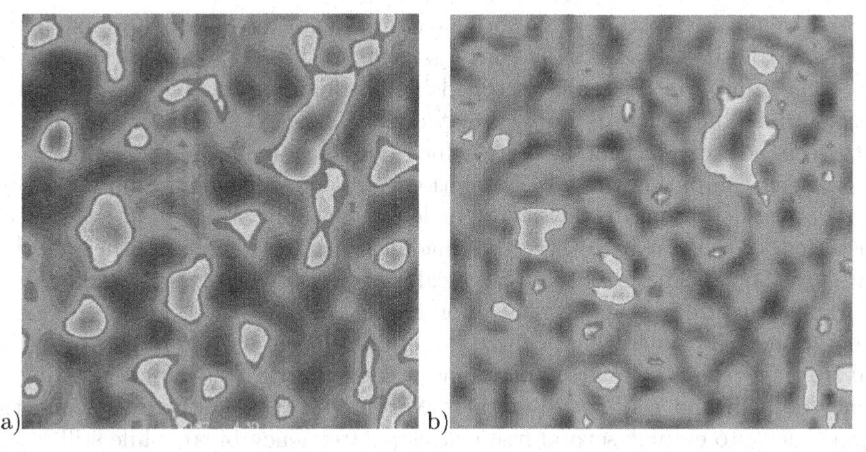

Figure 5. Images of the horizontal variation of the optical depth in the mid plane (left) and of the error in the heating / cooling rate (right). Positive and negative values are shown in distinct colors / shades of gray.

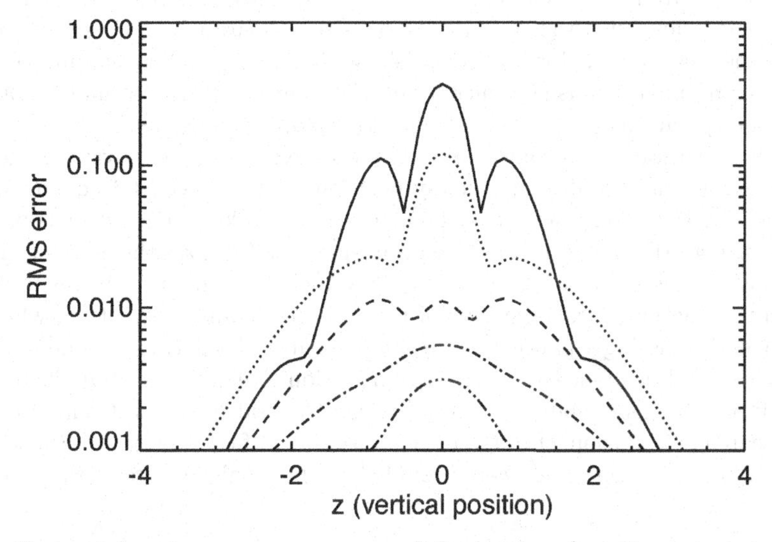

Figure 6. Horizontal root-mean-square errors of the heating / cooling rate, relative to the global root-mean-square variation of the reference solution. The full drawn curve is for the FLD solution. The other curves are for ray tracing with 4 (dotted), 7 (dashed), 16 (dashed-dotted), and 26 (dash-dot-dot-dot) ray directions.

FLD solution at the levels where the heat exchange is strongest, and a solution with 7 ray directions is good to within 1% of the reference solution.

4. Conclusions and concluding remarks

Ray-tracing radiative transfer methods have a challenging ('6-dimensional') scaling of the required work. However, because the scaling constant is very small (about 1-5 nanoseconds per mesh point) a fair number of angle-frequency pairs is still affordable. In practice, the solution of the radiative transfer equation is so fast that similar amounts of time is needed for table lookup of opacities and source functions, and for interpolations of variables to and from the inclined rays.

Ray-tracing (or 'ray-casting') solutions have the big advantage over approximate methods such as FLD that they actually converge towards the exact solution. FLD methods, on the other hand, are 'look-alike methods'; the solutions behave more or less as the correct solutions, but there is no way to force the solution to converge to the correct one—there is no 'cost parameter' with which one can buy higher accuracy at the expense of longer computing time. FLD solutions converge to the diffusion approximation at large optical depth, and behave more or less reasonably at small optical depths, but as illustrated by the test here the errors made can be quite large in a neighborhood of optical depth unity. These layers are typically of central importance when calculating the thermal structure of stars and accretion disks.

Whether one uses ray-tracing or flux-limited-diffusion it is often important to include several frequency points (or frequency bins)—representing for example heating in one frequency domain and cooling in another (Nordlund 1982). With ray-tracing, there is enough capacity to employ several frequencies (or frequency bins), while still being able to afford a reasonable space-angle resolution.

An important requirement for ray tracing to be affordable is that the solution along all rays should be fully utilized; the solution along each ray should contribute to the knowledge of the radiation field in all cells that it passes through. This is trivial in Cartesian geometries, where rays can be chosen to be parallel. However, in more complicated geometries, such as for example disks with near-cylindrical symmetry and huge ratios of outer to inner radius (Pascucci *et al.* 2004), choosing an optimal set of rays and organizing interpolations of results back and forth can be non-trivial.

Ray tracing methods with long characteristics have the great advantage to be easy to parallelize. One can parallelize over ray positions, ray directions, frequencies and frequency bins, but also along each ray (Heinemann *et al.* 2006). The latter is particularly useful when using MPI-domain decomposition, since it minimizes the amount of information that must be passed from cell to cell, and parallelizes without problems to thousands of cores. In this method local solutions of the radiative transfer equation, which can be obtained in 'embarrassingly parallel mode', can be patched together by sending boundary data up-stream and down-stream in the ray direction (Heinemann *et al.* 2006).

The work of ÅN was supported by the Danish Natural Research Council and by the Danish Research Foundation, through its establishment of the Centre for Star and Planet Formation. Computing resources were provided by the Danish Center for Scientific Computing.

References

Heinemann T., Dobler W., Nordlund Å., & Brandenburg A. 2006, *AA* 448, 731

Hubeny, I. 2003, ASP Conference Proceedings, Vol. 288, 17

Krumholz M. R., Klein R. I., McKee C. F., & Bolstad J. 2007, *ApJ* 667, 626

Levermore C. D. & Pomraning G. C. 1981, *ApJ* 248, 321

Minerbo G. N. 1978, *JQSRT* 31, 149

Nordlund, A. 1982, *AA* 107, 1

Pascucci I., Wolf S., Steinacker J., Dullemond C. P., Henning T., Niccolini G., Woitke P., & Lopez B. 2004, *AA* 417, 793

Razoumov A. O. & Cardall C. Y. 2005, *MNRAS* 362, 1413

Turner N. J. & Stone J. M. 2001, *ApJS* 135, 95

Wang P., Abel T., & Zhang W. 2008, *ApJS* 176, 467

Computational Star Formation
Proceedings IAU Symposium No. 270, 2011
J. Alves, B.G. Elmegreen, J. M. Girart & V. Trimble, eds.

doi:10.1017/S1743921311000391

The role of accretion disks in the formation of massive stars

R. Kuiper[1,2], H. Klahr[2], H. Beuther[2] & Th. Henning[2]

[1] Argelander-Institut für Astronomie, Rheinische Friedrich-Wilhelms-Universität Bonn, Auf dem Hügel 71, D-53121 Bonn, Germany
email: kuiper@astro.uni-bonn.de

[2] Max-Planck-Institut für Astronomie, Königstuhl 17, D-68169 Heidelberg, Germany

Abstract. We present radiation hydrodynamics simulations of the collapse of massive pre-stellar cores. We treat frequency dependent radiative feedback from stellar evolution and accretion luminosity at a numerical resolution down to 1.27 AU. In the 2D approximation of axially symmetric simulations, it is possible for the first time to simulate the whole accretion phase of several 10^5 yr for the forming massive star and to perform a comprehensive scan of the parameter space. Our simulation series show evidently the necessity to incorporate the dust sublimation front to preserve the high shielding property of massive accretion disks. Our disk accretion models show a persistent high anisotropy of the corresponding thermal radiation field, yielding to the growth of the highest-mass stars ever formed in multi-dimensional radiation hydrodynamics simulations. Non-axially symmetric effects are not necessary to sustain accretion. The radiation pressure launches a stable bipolar outflow, which grows in angle with time as presumed from observations. For an initial mass of the pre-stellar host core of 60, 120, 240, and 480$\odot$ the masses of the final stars formed in our simulations add up to 28.2, 56.5, 92.6, and at least 137.2$\odot$ respectively.

Keywords. stars: formation, accretion, accretion disks, stars: masses, stars: outflows, dust, hydrodynamics, radiative transfer, methods: numerical

---

1. Introduction

The understanding of massive stars suffers from the lack of a generally accepted formation scenario. If the formation of high-mass stars is treated as a scaled-up version of low-mass star formation, a special feature of these high-mass proto-stars is the interaction of the accretion flow with the strong irradiation emitted by the newborn stars due to their short Kelvin-Helmholtz contraction timescale (Shu *et al.* 1987). Early one-dimensional studies (e.g. Larson & Starrfield 1971; Kahn 1974; Yorke & Krügel 1977) agree on the fact that the growing radiation pressure potentially stops and reverts the accretion flow onto a massive star. But this radiative impact strongly depends on the geometry of the stellar environment (Nakano 1989). The possibility was suggested to overcome this radiation pressure barrier via the formation of a long-living massive circumstellar disk, which forces the generation of a strong anisotropic feature of the thermal radiation field. Earlier investigations by Yorke & Sonnhalter (2002) tried to identify such an anisotropy, but their simulations show an early end of the disk accretion phase shortly after its formation due to strong radiation pressure feedback.

2. Method

2.1. *The aim*

Aim of this study is to reveal the details of the radiation-dust interaction in the vicinity of the most massive star formed during the collapse of a pre-stellar core (gravitationally unstable fragment of a molecular cloud). Due to this strong focus onto the core center, the physics in the outer core region have to be simplified, e.g. the usage of quiescent initial conditions suppresses further fragmentation of the pre-stellar core.

2.2. *The code*

For this purpose, we use our newly developed self-gravity radiation hydrodynamics code. The evolution of the gas density, velocity, pressure, and total energy density is computed using the magneto-hydrodynamics code Pluto3 (Mignone *et al.* 2007), including full tensor viscosity. The derivation and numerical details of the newly developed frequency dependent hybrid radiation transport method are summarized by Kuiper *et al.* (2010a). Our implementation of Poisson's equation as well as the description of the dust and stellar evolution model are given in Kuiper *et al.* (2010b).

The simulations are performed on a time independent grid in spherical coordinates. The radially inner and outer boundary of the computational domain are semi-permeable walls, i.e. the gas can leave but not enter the domain. The resolution of the non-uniform grid is chosen to be $(\Delta r \text{ x } r\ \Delta\theta)_{\text{min}} = 1.27$ AU x 1.04 AU around the forming massive star and decreases logarithmically in the radial outward direction. The accurate size of the inner sink cell is determined in a parameter scan presented in Sect. 3.1.

2.3. *Initial Conditions*

Our basic initial condition is very similar to the one used by Yorke & Sonnhalter (2002). We start from a cold ($T_0 = 20$ K) pre-stellar core of gas and dust. The initial dust to gas mass ratio is chosen to be $M_{\text{dust}}/M_{\text{gas}} = 1\%$. The model describes a so-called quiescent collapse scenario without turbulent motion ($\vec{u}_r = \vec{u}_\theta = 0$). The core is initially in slow rigid rotation $\left(|\vec{u}_\phi|/R = \Omega_0 = 5 * 10^{-13} \text{ Hz}\right)$. The initial density slope drops with r^{-2} and the outer radius of the cores is fixed to $r_{\text{max}} = 0.1$ pc. The total mass M_{core} varies in the simulations from 60 up to 480 $\text{M}_\odot$.

3. Results

3.1. *The dust sublimation front*

In the following, we check the dependency of the stellar accretion rate on the radius r_{min} of the inner sink cell in four simulations with $r_{\text{min}} = 1$, 5, 10, and 80 AU. We follow the long-term evolution of the runs for at least 10^5 yrs. The results are displayed in Fig. 1.

The chosen location r_{min} of the inner boundary of the computational domain influences the resulting accretion rate in two distinguishable effects: First, a smaller sink cell leads to a shorter free fall epoch, i.e. to an earlier onset of the disk formation phase. This effect of the artificial inner cut-off of the gas disk results on one hand in an overestimation of the final mass of the central star by approximately $1\odot$ or below (left panel of Fig. 1), but on the other hand does not influence the disk accretion epoch (right panel of Fig. 1).

The second effect is related to a potential cut-off of the inner dust disk. The region in the vicinity of the forming massive star will be heated up to temperatures beyond the dust sublimation temperature and a gap is formed between the central star and the dust disk. For an inner sink cell radius r_{min} larger than the dust sublimation radius r_{subl}, the region of radiative feedback is artificially shifted to higher radii including a strong

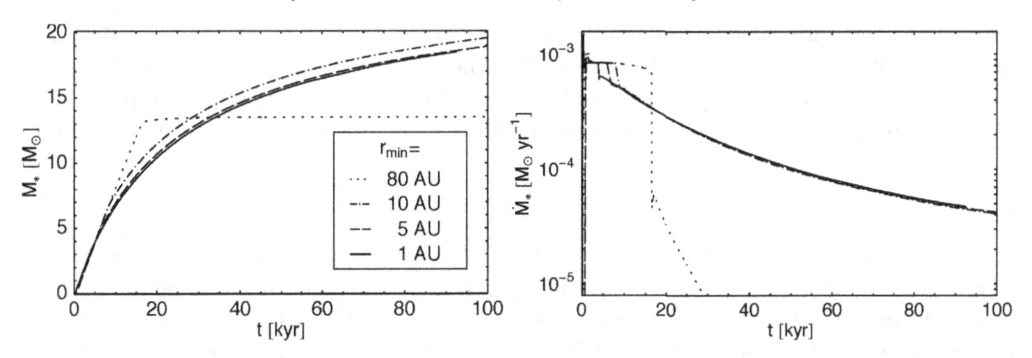

Figure 1: Stellar mass M_* (left panel) and accretion rate $\dot{M}_*$ (right panel) as a function of time t for different radii $r_{\rm min}$ of the central sink cell during a 60$\odot$ core collapse.

decrease in density, opacity, and gravity. As a result, the dust disk looses its shielding property, the thermal radiation field retains in major parts its isotropic character and the radiation pressure therefore stops the emerging disk accretion phase (cp. the case of $r_{\rm min} = 80$ AU in Fig. 1). This dependency of the radiation pressure on the radius of the sink cell explains also the abrupt end of the accretion phase in the simulations by Yorke & Sonnhalter (2002), who presented simulations of collapsing pre-stellar cores of $M_{\rm core} = 30\odot$, 60 M$_\odot$ and 120 M$_\odot$ with a radius of the inner sink cell chosen to be 40, 80, and 160 AU respectively. Our subsequent simulations meet this concern by using an adequate central sink cell radius of $r_{\rm min} = 10$ AU.

3.2. *The radiation pressure barrier*

The spherically symmetric accretion flow simulations yield a maximum stellar mass of less than 40 M$_\odot$ independent of the initial core mass $M_{\rm core} \geqslant 60\odot$ due to radiative feedback (Kuiper *et al.* 2010b). We attack this radiation pressure barrier in axially and midplane symmetric circumstellar disk geometry now. Resulting accretion histories as a function of the actual stellar mass for different initial core masses of $M_{\rm core} = 60$ M$_\odot$, 120 M$_\odot$, 240 M$_\odot$, and 480 M$_\odot$ are displayed in Fig. 2. In face of the additional centrifugal forces, the disk accretion easily breaks through the upper mass limit of the final star of $M_*^{\rm 1D} < 40$ M$_\odot$ found in the spherically symmetric accretion models!

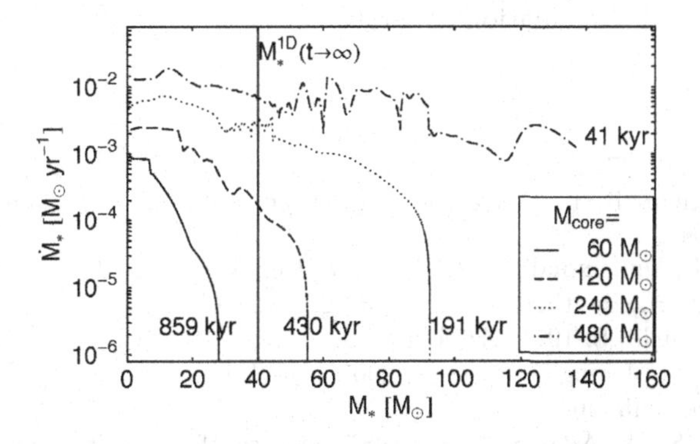

Figure 2: Accretion rate $\dot{M}_*$ as a function of the actual stellar mass M_* for four different initial core masses. Also the periods of accretion are mentioned for each run. The vertical line marks the upper mass limit found in the spherically symmetric accretion models.

No upper mass limit of the final star is detected so far, but the star formation efficiency declines for higher mass cores. The depletion of the envelope by radiative forces decreases the large-scale accretion onto the midplane. The disk looses its shielding property and the radiation pressure starts to accelerate the remnant material in the outward direction.

4. Summary

We performed high-resolution radiation hydrodynamics simulations of monolithic pre-stellar core collapses including frequency dependent radiative feedback. The dust sublimation front of the forming star could be resolved down to 1.27 AU. The whole accretion phase of several 10^5 yrs was computed. The frequency dependent ray-tracing of our newly developed radiation module denotes the most realistic radiation transport method used in multi-dimensional hydrodynamic simulations of massive star formation by now.

The broad parameter studies, especially regarding the size of the sink cell and the initial core mass, reveal new insights of the radiative feedback onto the accretion flow during the formation of a massive star: In the case of disk accretion, the thermal radiation field generates a strong anisotropic feature. We found that it is strict necessary to include the dust sublimation front in the computational domain to reveal the persistent anisotropy during the long-term evolution of the accretion disk. The short accretion phases of the disks in the simulations by Yorke & Sonnhalter (2002) are a result of the fact that they did not include the dust sublimation front in their simulations, as clearly shown in our result of the parameter scan of the size of the central sink cell (see Sect. 3.1). Additional feeding of the disk by unstable outflow regions due to the so-called "3D radiative Rayleigh-Taylor instability", as proposed in Krumholz *et al.* (2009), would enhance this anisotropy but is not necessary. In fact, preliminary analyses of our ongoing three-dimensional collapse simulations identify evolving gravitational torques in the massive circumstellar disk as well as the launching of non-axially symmetric outflows. The accretion rate, driven by the angular momentum transport of the gravitational torques, is more episodically compared to the axially-symmetric runs, but results in a similar mean accretion rate. The radiation pressure driven ouflow remains stable.

Finally, the central stars in our simulations of the disk accretion scenario grow far beyond the upper mass limit found in the case of spherically symmetric accretion flows. For an initial mass of the pre-stellar host core of 60, 120, 240, and $480\odot$ the masses of the final stars formed add up to 28.2, 56.5, 92.6, and at least $137.2\odot$ respectively. Indeed, the final massive stars are the most massive stars ever formed in a multi-dimensional radiation hydrodynamics simulation so far.

References

Kahn, F. D. 1974, *A&A*, 37, 149

Krumholz, M. R., Klein, R. I., McKee, C. F., Offner, S. S. R., & Cunningham, A. J. 2009, *Science*, 323, 754

Kuiper, R., Klahr, H., Dullemond, C., Kley, W., & Henning, T. 2010a, *A&A*, 511, 81

Kuiper, R., Z ZKlahr, H., Beuther, H., & Henning, T. 2010b, *ApJ*, in press

Larson, R. B. & Starrfield, S. 1971, *A&A*, 13, 190

Mignone, A., Bodo, G., Massaglia, S., et al. 2007, *ApJS*, 170, 228

Nakano, T. 1989, *ApJ*, 345, 464

Shu, F. H., Lizano, S., & Adams, F. C. 1987, in: Star forming regions, ed. M. Peimbert, *J. Jugaku*, 115, 417

Yorke, H. W. & Krügel, E. 1977, *A&A*, 54, 183

Yorke, H. W. & Sonnhalter, C. 2002, *ApJ*, 569, 846

Computational Star Formation
Proceedings IAU Symposium No. 270, 2011
J. Alves, B.G. Elmegreen, J. M. Girart & V. Trimble, eds.

doi:10.1017/S1743921311000408

Embedded disks around low-mass protostars

Eduard I. Vorobyov[1,2] and Shantanu Basu[3]

[1] The Institute for Computational Astrophysics, Saint Mary's University, Halifax, Canada
email: vorobyov@ap.smu.ca

[2] Research Institute of Physics, Southern Federal University, Rostov-on-Don, Russia

[3] The University of Western Ontario, London, Canada

Abstract. The time evolution of protostellar disks in the embedded phase of star formation (EPSF) is reviewed based on numerical hydrodynamics simulations of the gravitational collapse of two cloud cores with distinct initial masses. Special emphasis is given to disk, stellar, and envelope masses and also mass accretion rates onto the star. It is shown that accretion is highly variable in the EPSF, in agreement with recent theoretical and observational expectations. Protostellar disks quickly accumulate mass upon formation and may reach a sizeable fraction of the envelope mass ($\sim 35\%$) by the end of the Class 0 phase. Systems with disk-to-star mass ratio $\xi \approx 0.5$ are common but systems with $\xi \geqslant 1.0$ are rare because the latter quickly evolve into binary or multiple systems. Embedded disks are characterized by radial pulsations, the amplitude of which increases with growing core mass.

Keywords. accretion, accretion disks, (stars:) circumstellar matter, stars: formation.

---

1. Introduction

The embedded phase of the evolution of a protostellar disk, starting from its formation and ending with the clearing of a parent cloud core, set the course along which a young stellar object (YSO) will evolve later in the T Tauri phase. Despite its pivotal role, the embedded phase of star formation (EPSF) phase is poorly understood owing to difficulties with both observations and modeling. Directly observing disks in this phase is difficult as they are hidden within dense, extincting protostellar envelopes and only a handful of attempted studies exist (e.g. Jorgensen *et al.* 2009).

Self-consistent numerical simulations of protostellar disks in the EPSF are no less difficult than observations due to vastly changing spatial and temporal scales involved. The matter is that it is not sufficient to just consider an *isolated* system with some presumed disk-to-star mass ratio. A self-consistent treatment of the interaction of the star/disk system with the natal cloud core is of considerable importance for the disk physics and this inevitably requires solving for a much larger spatial volume than in isolated systems. Such multidimensional numerical simulations only recently have started to emerge (e.g. Vorobyov & Basu 2006, Vorobyov 2009a, and Attwood *et al.* 2009).

In this article, we make use of numerical hydrodynamics simulations in the thin-disk approximation to compute the gravitational collapse of rotating, gravitationally unstable cloud cores with an accurate treatment of disk thermodynamics. This allows us to realistically model the formation and long-term evolution of protostellar disks. The basic equations, initial conditions, and details of the code can be found in Vorobyov & Basu (2010).

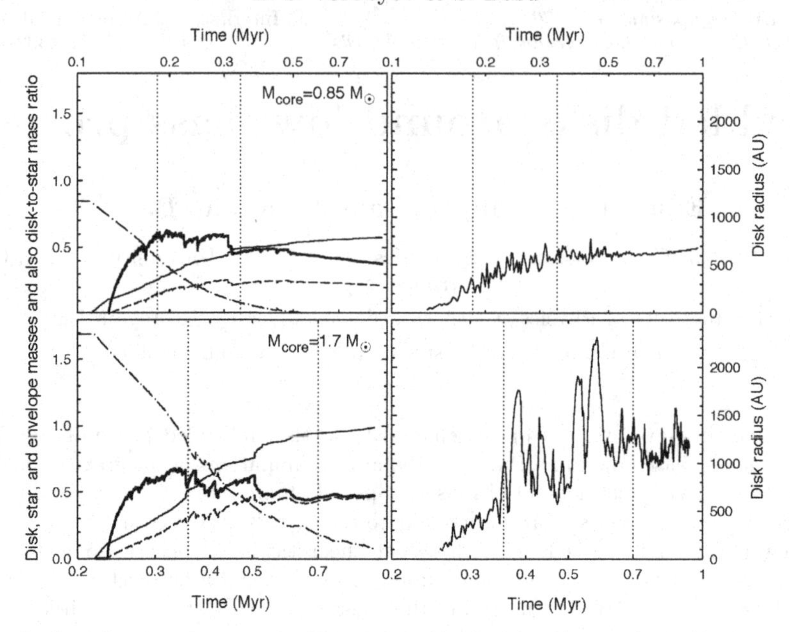

Figure 1. Left column. Time evolution of disk (dashed lines), envelope (dash-dotted lines), and stellar (thin solid lines) masses and also disk-to-star mass ratio (thick solid lines) in model 1 (top) and model 2 (bottom). **Right column.** Time evolution of disk outer radii in model 1 (top) and model 2 (bottom). The vertical dotted lines mark the onset of Class I (left) and Class II (right) phases of stellar evolution.

2. Time evolution of disk, stellar, and envelope masses

We start our numerical integration in the pre-stellar phase, which is characterized by a collapsing *starless* cloud core, continue into the embedded phase of star formation, during which a star, disk, and envelope are formed, and terminate our simulations in the T Tauri phase, when most of the envelope has accreted onto the forming star/disk system. In this section, we present results for two model cores with initial masses $M_{\rm core} = 0.85\ M_{\odot}$ (model 1) and $M_{\rm core} = 1.7\ M_{\odot}$ (model 2). The ratio of rotational to gravitational energy is $\beta = 5.6 \times 10^{-3}$. The left column in Figure 1 shows the time evolution of the stellar mass M_* (thin solid line), disk mass $M_{\rm d}$ (dashed line), envelope mass $M_{\rm env}$ (dash-dotted line), and disk-to-star mass ratio ξ (thick solid line) in model 1 (upper-left panel) and model 2 (lower-left panel). The separation of the burgeoning disk from the infalling envelope is done using a method described in detail in Vorobyov (2010), which is based on the typical transitional density $\Sigma_{\rm d2e} = 0.1$ g cm^{-2} and the radial velocity field. The time is counted since the onset of core collapse. The vertical dotted lines mark the onset of the Class I (left) and Class II (right) phases of star formation as inferred from the total mass remaining in the envelope (André *et al.* 1993).

It is evident that by the end of the Class 0 phase the disk mass reaches a significant fraction of the envelope and stellar masses. For instance, in model 1 the disk-to-star mass ratio is $\xi = 0.59$ and the disk-to-envelope mass ratio is $\zeta = 0.37$, while in model 2 the corresponding values are $\xi = 0.56$ and $\zeta = 0.35$. Our simulations suggest that quite massive disks can be present as early as in the Class 0 phase, a conclusion that finds recent observational support (Enoch *et al.* 2009a).

Another interesting feature of Figure 1 is that the disk mass never exceeds that of the star, though ξ may become a substantial fraction of unity in the EPSF. It turns out that systems with $\xi \geqslant 1$ are unstable and quickly evolve via disk fragmentation into a binary/multiple system. We therefore argue that such massive disks are feasible but they must be statistically rare.

The right column in Figure 1 depicts the time evolution of the disk outer radius in model 1 (upper right panel) and model 2 (lower-right panel). A remarkable feature of embedded disks is ongoing radial pulsations, the amplitude of which increases with growing core mass. These radial pulsations are caused by disk fragmentation and inward migration of massive fragments. The chain of events is as follows. Fragments form near corotation (most favourable place for fragmentation due to small sheer) via fragmentation of dense spiral arms. These fragments interact gravitationally with the parent spiral arm and, as a result of this interaction, lose angular momentum and migrate radially inward. The disk expands in response to this migration to conserve the total angular momentum of the system. As a consequence, the gas surface density drops and the disk stabilizes temporarily until it contracts and accumulates enough mass from the infalling envelope for a next round of fragmentation to commence. Such radial pulsations continue as long as there is enough mass reservoir in the envelope (Class 0 and I phases), as Figure 1 suggests. In this sense, mass loading from the envelope can be regarded as the driving force of the radial pulsations but the actual disk expansion is caused by inward migration of the fragments.

3. Mass accretion history

The mass accretion history provides information on how stars accumulate mass and this may affect the internal structure and position of a star on the HR diagram (Baraffe *et al.* 2009). There is growing observational evidence that accretion onto a star is highly variable in the EPSF (e.g. Enoch *et al.* 2009b), in contrast to the theoretical predictions of the inside-out collapse (Shu 1977). Recent numerical simulations of polytropic protostellar disks by Vorobyov (2009b) have shown that the mass accretion rates $\dot{M}$ in the Class I stage are expected to have a lognormal distribution, with its shape controlled by disk viscosity and disk temperature.

Our numerical modeling confirms that the accretion process onto the star is highly variable throughout much of the embedded phase. The top row in Figure 2 presents the mass accretion rates (in $M_\odot$ yr^{-1}) in model 1 (left panel) and model 2 (right panel) as a function of time. The vertical dotted lines mark the onset of the Class I (left) and Class II (right) phases of star formation. It is seen that $\dot{M}$ in the EPSF exhibits short-term variations of several orders in magnitude, with the highest rates exceeding 10^{-4} $M_\odot$ yr^{-1} and lowest rates dropping below 10^{-9} $M_\odot$ yr^{-1}. The bursts of high-rate accretion are caused by disk fragmentation. It turns out that most of the fragments spiral down onto the star due to the loss of angular momentum via gravitational interaction with spiral arms (Vorobyov & Basu 2006). On the other hand, an order of magnitude variability seen throughout much of the EPSF is caused by gravitational instability and non-linear interaction of low-order spiral modes. As shown by Vorobyov (2009b), this accretion flickering greatly reduces when disk self-gravity is artificially turned off, confirming that the disk gravitational destabilization lies behind this effect.

The highly variable accretion makes the star sporadically increase its *total* luminosity, as illustrated in the bottom panel of Figure 2. The solid lines present the accretion luminosity $L_{\rm accr}$, while the red dashed lines depict the photospheric luminosity $L_{\rm ph}$. Several clear-cut luminosity outbursts with $L_{\rm accr}$ as high as 100 $L_\odot$ and many more weaker bursts (solid line) are evident against the background of a near-constant photospheric luminosity with $L_{\rm ph} \sim 1.0$ $L_\odot$ (dashed line). The stronger bursts may represent FU Orionis-like eruptions, while weaker ones may manifest EX Lupi-like eruptions (EXors).

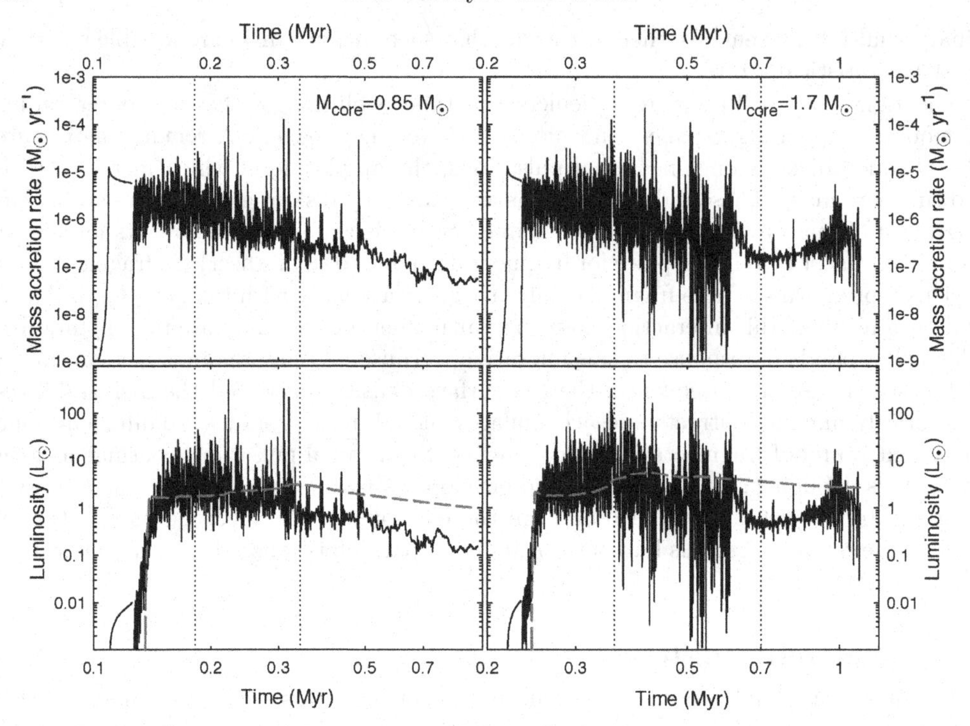

Figure 2. Top row. Mass accretion rate onto the star as a function of time in model 1 (left) and model 2 (right). **Bottom row**. Photospheric (red dashed lines) and accretion (solid lines) luminosity in model 1 (left) and model 2 (right). Vertical dotted lines mark the onset of Class I (left) and Class II (right) phases.

Acknowledgements

The authors thank the anonymous referee for useful comments. E.I.V. gratefully acknowledges present support from an ACEnet Fellowship, RFBR grant 10-02-00278, and Ministry of Education grant RNP 2.1.1/1937. Numerical simulations were done on the Atlantic Computational Excellence Network (ACEnet), on the Shared Hierarchical Academic Research Computing Network (SHARCNET), and at the Center of Collective Supercomputer Resources, Taganrog Technological Institute. S. B. was supported by a Discovery Grant from the Natural Sciences and Engineering Research Council of Canada.

References

André, P., Ward-Thompson, D., & Barsony, M. 1993, *ApJ*, 406, 122

Attwood, R. E., Goodwin, S. P., Stamatellos, D., Whitworth, A. P. 2009, *A&A*, 495, 201

Baraffe, I., Chabrier, G., & Gallardo, J. 2009, *ApJ* (Letters), 702, 27

Enoch, M. L., Corder, S., Dunham, M. M., & Duchéne, G. 2009, *ApJ*, 707, 103

Enoch, M. L., Evans, N. J., II, Sargent, A. I., & Glenn, J. 2009, *ApJ*, 692, 973

Jørgensen, J. K., van Dishoeck, E. F., Visser, R., Bourke, T. L., Wilner, D. J., Lommen, D., Hogerheijde, M. R., & Myers, P. C. 2009, *A&A*, 507, 861

Shu, F. H. 1977, ApJ, 214, 488

Vorobyov, E. I. & Basu, S. 2006, *ApJ*, 650, 956

Vorobyov, E. I., 2009a, *ApJ*, 692, 1609

Vorobyov, E. I. 2009b, *ApJ*, 704, 715

Vorobyov, E. I. 2010, *ApJ*, 713, 1

Vorobyov, E. I. & Basu, S. 2010, *ApJ*, 719, 1896

Computational Star Formation
Proceedings IAU Symposium No. 270, 2011
J. Alves, B.G. Elmegreen, J. M. Girart & V. Trimble, eds.

doi:10.1017/S174392131100041X

The formation of brown dwarfs in discs: Physics, numerics, and observations

Dimitris Stamatellos & Anthony Whitworth

School of Physics & Astronomy, Cardiff University, 5 The Parade, Cardiff, CF24 3AA, UK

Abstract. A large fraction of brown dwarfs and low-mass stars may form by gravitational fragmentation of relatively massive (a few 0.1 $M_{\odot}$) and extended (a few hundred AU) discs around Sun-like stars. We present an ensemble of radiative hydrodynamic simulations that examine the conditions for disc fragmentation. We demonstrate that this model can explain the low-mass IMF, the brown dwarf desert, and the binary properties of low-mass stars and brown dwarfs. Observing discs that are undergoing fragmentation is possible but very improbable, as the process of disc fragmentation is short lived (discs fragment within a few thousand years).

Keywords. Stars: formation – Stars: low-mass, brown dwarfs – accretion, accretion disks – Methods: Numerical, Radiative transfer, Hydrodynamics

---

1. Introduction

The formation of brown dwarfs (BDs) and low-mass stars is not well understood (Whitworth *et al.* 2007). Low-mass objects are difficult to form by gravitational fragmentation of unstable gas, as for masses in the BD regime ($\lesssim$ 80 M_J, where M_J is the mass of Jupiter), a high density ($\gtrsim 10^{-16}$g cm^{-3}) is required for the gas to be Jeans unstable. Padoan & Nordland (2004) and Hennebelle & Chabrier (2008) suggest that these high density cores can be formed by colliding flows in a turbulent magnetic medium. However, this model requires a large amount of turbulence, and has difficulty in explaining the binary properties of BDs. Additionally, the large number of brown-dwarf mass cores that the theory predicts have not been observed. Another way to reach the high densities required for the formation of BDs and low-mass stars is in gravitationally unstable discs (Whitworth & Stamatellos 2006; Stamatellos *et al.* 2007b; Stamatellos & Whitworth 2008, 2009a,b). These discs form around newly born stars and grow quickly in mass by accreting material from the infalling envelope. They become unstable if the mass accreted onto them cannot efficiently redistribute its angular momentum outwards in order to accrete onto the central star (Attwood *et al.* 2009). BDs are also thought to form as ejected embryos from star forming regions, i.e. as a by-product of the star formation process (Reipurth & Clarke 2001). In this model BDs form the same way as low-mass stars, i.e. in collapsing molecular cores, but shortly after their formation they are ejected from their parental core and stop accreting any further material. Hence, they do not realise their potential to become hydrogen-burning stars. In this paper we focus on the mechanism of BD formation by fragmentation of gravitationally unstable discs.

2. The fragmentation of gravitationally unstable discs

Discs can fragment if (i) they are massive enough so that gravity overcomes thermal and local centrifugal support (Toomre 1964), and (ii) they can cool fast enough so that the energy provided by the collapse of a proto-fragment is radiated away and the growth of the proto-fragment continues (Gammie 2001; Rice *et al.* 2005). Analytical and numerical

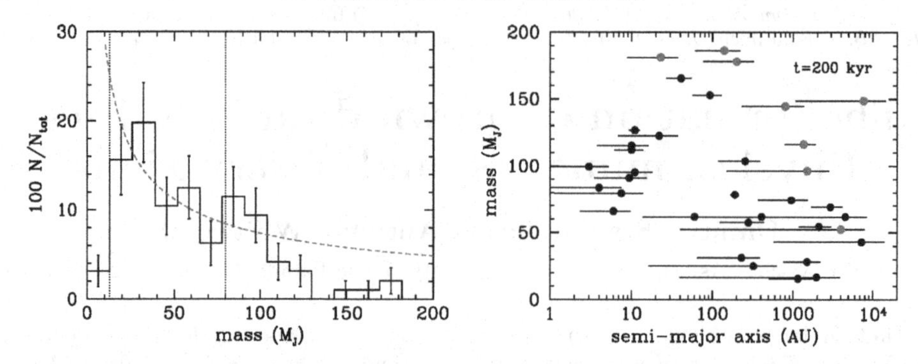

Figure 1. Left: The mass spectrum of the objects produced by disc fragmentation. Most of these are BDs (70%); the rest are low-mass stars. The vertical dotted lines correspond to the D-burning limit ($\sim$ 13 M_J) and the H-burning limit ($\sim$ 80 M_J). The red dashed line refers to a low-mass IMF with $\Delta N/\Delta M \propto M^{-0.6}$. Right: The BD desert. The semi-major axes of the objects (at t = 200 kyr) formed in the disc and remained bound to the central star plotted against their mass. The bars indicate the minimum and maximum extent of the orbit. There is a lack of BD companions close to central star, but there is a population of BDs loosely bound to the central star. The red dots correspond to low-mass binary systems formed in the disc.

studies have shown that the cooling time must be on the order of the dynamical time which happens to be similar to the orbital time.

Massive (a few times 0.1 $M_\odot$), extended ($>$100 AU) discs exist (e.g. Enoch *et al.* 2009; Jogersen *et al.* 2005; 2009; Greaves *et al.* 2008; Andrews *et al.* 2009; Duchene this volume). We have performed an ensemble of radiative hydrodynamical simulations of such discs. We assume a star-disc system in which the central star has initial mass $M_1 = 0.7\,M_\odot$. Initially the disc has mass $M_D = 0.7\,M_\odot$, inner radius $R_{IN} = 40\,$AU, outer radius $R_{OUT} = 400\,$AU, surface density $\Sigma(R) \propto R^{-7/4}$, temperature $T(R) \propto R^{-1/2}$, and hence approximately uniform initial Toomre parameter $Q \sim 0.9$. Thus, the disc is at the outset marginally gravitationally unstable. The only parameter that is different between different runs is the random noise from which the gravitational instabilities grow in the disc. The evolution of the disc is initially followed using SPH, until $\sim$ 70% of the disc mass has been accreted, either onto the stars condensing out of the disc, or onto the central star; this typically happens within 20 kyr. Then the residual gas is ignored and the long term dynamical evolution of the system is followed up to 200 kyr, using an N-body code. The energy equation and associated radiative transfer are treated with the method of Stamatellos *et al.* (2007a). The intrinsic radiation of the central star is taken into account but its accretion luminosity is ignored (Stamatellos & Whitworth 2009a). The radius of the sink representing the star is set to 1 AU. The gravitational instabilities develop quickly and the disc fragments within a few thousand years. Typically 5-10 objects form in each disc, most of them BDs. The final status of these objects is determined by subsequent accretion of material onto them and by their mutual interactions.

3. The statistical properties of objects formed by disc fragmentation

The mass spectrum. Most of the objects ($\sim$ 70%) formed in the discs are BDs, including a few planetary-mass BDs. The rest are low-mass hydrogen-burning stars. The typical mass of an object produced is $\sim$ 20 − 30 M_J. The shape of the initial mass function (Fig. 1, left) is similar to what observations suggest, i.e. it is consistent with $\Delta N/\Delta M \propto M^{-\alpha}$, where $\alpha \approx 0.6$ (e.g. in Pleiades $\alpha = 0.6 \pm 0.11$ – Moraux *et al.* 2003 – and in σ Orionis $\alpha = 0.6 \pm 0.1$ – Lodieu *et al.* 2009).

The brown dwarf desert. Most of the fragments form in the discs with initial masses as low as 3 M_J (Stamatellos & Whitworth 2009c). After a fragment forms it accretes mass from the disc, and interacts with the disc through drag forces, and dynamically with other fragments. As a result some fragments migrate close to the central star. This region is rich in gas, hence these accreting proto-fragments eventually become low-mass hydrogen-burning stars. Fragments that form farther out also accrete material from the disc but not quite as much; these become BDs. If any of the BDs happen to migrate in the region close to the central star they tend to be ejected back into the outer region through 3-body interactions. Hence, the disc fragmentation model produces a lack of BD close companions to Sun-like stars, i.e the BD desert (Fig. 1, right). The BD desert may extend out to 300 AU but it is less dry outside 100 AU. We predict a population of BDs at distances from 20 to 5000 AU from the central star.

Free-floating planetary-mass objects. The disc fragmentation model provides an explanation of the existence of free-floating planetary-mass objects (Lucas & Roche 2000; Zapatero Osorio *et al.* 2000). In our model these objects form in the disc by gravitational fragmentation and quickly after their formation they are ejected from the system due to 3-body interactions. Hence, they stop accreting and their mass remains low.

Low-mass binary statistics. Close and wide BD-BD binaries are common outcomes of disc fragmentation. The components of the binary form independently in the disc and then pair up. The simulations produce all kinds of low-mass binaries: star-star, star-BD, BD-BD, and BD-planetary mass object binaries. Our model predicts a low-mass binary fraction of 16%. This is comparable with the low-mass binary fraction observed in star-forming regions (e.g. in Taurus : $> 20\%$, Kraus *et al.* 2006, in the field $15 \pm 5\%$, Gizis *et al.* 2003). We also predict that close low-mass binaries should outnumber wide ones, and this seems to be what is observed (Burgasser *et al.* 2007). Most of the low-mass binaries (55%) have components with similar masses ($q > 0.7$), in agreement with the observed properties of low-mass binaries (e.g. Burgasser *et al.* 2007). The model also predicts that BDs that are companions to Sun-like stars are more likely to be in binaries (binary frequency 25%) than BDs in the field (frequency $5 - 8\%$). This trend is consistent with what is observed (Burgasser *et al.* 2005; Faherty *et al.* 2009).

4. Numerical Issues

We use 1.5×10^5 particles to represent each disc, which means that the minimum resolvable mass (corresponding to the number of neighbours used, i.e. 50 SPH particles) is $\approx 0.25\,M_J$. We have also performed simulations using 2.5×10^5 and 4×10^5 particles. In these simulations the growth of gravitational instabilities, and the properties of the proto-fragments formed as a result of these instabilities, follow the same patterns as in the simulation with lower resolution, but the details of the final outcomes (i.e. the number of objects formed and their exact formation positions) are different.

The minimum Jeans mass ($M_J = (4\pi^{5/2}/24)(c_s^3/(G^3\rho)^{1/2})$) in a typical simulation using 1.5×10^5 particles is $M_{\rm JEANS,MIN} \approx 2$ M_J. Thus, according to the Bate & Burkert (1997) condition, this mass is adequately resolved as it corresponds to $\sim 8 \times N_{\rm NEIGH}$ (a minimum factor of $2 \times N_{\rm NEIGH}$ is recommended). The Toomre mass ($M_T = \pi c_s^4/G^2\Sigma$) is also adequately resolved. The minimum Toomre mass in our simulation is $M_{\rm TOOMRE,MIN} \approx$ 2.5 M_J). This mass corresponds to $\sim 10 \times N_{\rm NEIGH}$ (a factor of $6 \times N_{\rm NEIGH}$ is recommended; Nelson 2006). Finally, the vertical disc structure is resolved by at least ~ 5 smoothing lengths, satisfying the Nelson (2006) condition. This is important as the disc mainly cools in this direction.

5. Observing fragmenting discs

Typically a few BDs form in each fragmenting disc around a Sun-like star. Hence, it may be that only $\sim$ 20% of Sun-like stars need to have large and massive unstable discs to produce a large fraction of the observed BDs. Assuming that the lifetime of the Class 0 phase is 10^5 yr and that in the disc fragmentation scenario the disc fragments and therefore dissipates within 10^4 yr, then the probability of observing a fragmenting disc around a Class 0 object is only 10%. Then, considering that only 20% of Sun-like stars may have such unstable discs, the probability of observing such discs is only 2%. Hence, fragmenting discs should be very difficult to discover (Maury *et al.* 2009).

6. Conclusions

Discs can fragment at distances $\gtrsim$ 100 AU (Stamatellos & Whiworth 2009a) from the central star to form predominately brown dwarfs, but also low-mass hydrogen burning stars and planetary-mass objects. Despite the fact that the model does not include magnetic fields, radiative feedback due to the accretion luminosity from newly formed protostars, and mechanical feedback (i.e. jets), it can reasonably reproduce (i) the shape of the low-mass IMF, (ii) the brown dwarf desert, (iii) the binary properties of low-mass objects and (iv) the formation of free-floating planetary mass objects.

References

Andrews, S. M., Wilner, D. J., Hughes, A. M., Qi, C., & Dullemond, C. P. 2009, *ApJ*, 700, 1502
Attwood, R. E., Goodwin, S. P., Stamatellos, D., & Whitworth, A. P. 2009, *A&A*, 495, 201
Burgasser, A. J., Kirkpatrick, J. D., & Lowrance, P. J. 2005, *AJ*, 129, 2849
Burgasser, A. J., Reid, I. N., Siegler, N., Close, L., Allen, P., Lowrance, P., & Gizis, J. 2007, Protostars and Planets V, 427
Faherty, J. K., Burgasser, A. J., West, A. A., Bochanski, J. J., Cruz, K. L., Shara, M. M., & Walter, F. M. 2009, arXiv:0911.1363
Gammie, C. F. 2001, *ApJ*, 553, 174
Gizis, J. E., Reid, I. N., Knapp, G. R., Liebert, J., Kirkpatrick, J. D., Koerner, D. W., & Burgasser, A. J. 2003, *AJ*, 125, 3302
Hennebelle, P. & Chabrier, G. 2008, *ApJ*, 684, 395
Kraus, A. L., White, R. J., & Hillenbrand, L. A. 2006, *ApJ*, 649, 306
Lodieu, N., Zapatero Osorio, M. R., Rebolo, R., Martín, E. L., & Hambly, N. C. 2009, *A&A*, 505, 1115
Lucas, P. W. & Roche, P. F. 2000, *MNRAS*, 314, 858
Maury, A. J., *et al.* 2010, *A&A*, 512, A40
Moraux, E., Bouvier, J., Stauffer, J. R., & Cuillandre, J.-C. 2003, *A&A*, 400, 891
Padoan, P., & Nordlund, Å. 2004, *ApJ*, 617, 559
Reipurth, B. & Clarke, C. 2001, *AJ*, 122, 432
Rice, W. K. M., Lodato, G., & Armitage, P. J. 2005, *MNRAS*, 364, L56
Stamatellos, D., Whitworth, A. P., Bisbas, T., & Goodwin, S. 2007a, *A&A*, 475, 37
Stamatellos, D., Hubber, D. A., & Whitworth, A. P. 2007b, *MNRAS*, 382, L30
Stamatellos, D. & Whitworth, A. P. 2008, *A&A*, 480, 879
Stamatellos, D. & Whitworth, A. P. 2009a, *MNRAS*, 392, 413
Stamatellos, D. & Whitworth, A. P. 2009b, *MNRAS*, 1548
Toomre, A. 1964, *ApJ*, 139, 1217
Whitworth, A. P. & Stamatellos, D. 2006, *A&A*, 458, 817
Whitworth, A., Bate, M. R., Nordlund, Å., Reipurth, B., & Zinnecker, H. 2007, Protostars and Planets V, 459
Zapatero Osorio, M. R., Béjar, V. J. S., Martín, E. L., Rebolo, R., y Navascués, D. B., Bailer-Jones, C. A. L., & Mundt, R. 2000, *Science*, 290, 103

Computational Star Formation
Proceedings IAU Symposium No. 270, 2011
J. Alves, B.G. Elmegreen, J. M. Girart & V. Trimble, eds.

doi:10.1017/S1743921311000421

Radiative, magnetic and numerical feedbacks on small-scale fragmentation

Benoît Commerçon[1], **Patrick Hennebelle**[2], **Edouard Audit**[3], **Gilles Chabrier**[4] **and Romain Teyssier**[3,5]

[1]Max PLanck Institut für Astronomie
Königstuhl 17, D-69117 Heidelberg, Germany
email: benoit@mpia-hd.mpg.de

[2]Laboratoire de radioastronomie, École Normale Supérieure et Observatoire de Paris,
24 rue Lhomond, F-75231 Paris Cedex 05, France

[3]Laboratoire AIM, CEA/DSM - CNRS - Université Paris Diderot,
IRFU/SAp, F-91191 Gif sur Yvette, France

[4]École Normale Supérieure de Lyon, Centre de recherche Astrophysique de Lyon,
46 allée d'Italie, F-69364 Lyon Cedex 07, France

[5]Universität Zürich, Institute für Theoretische Physik,
Winterthurerstrasse 190, CH-8057 Zürich, Switzerland

Abstract. Radiative feedback and magnetic field are understood to have a strong impact on the protostellar collapse. We present high resolution numerical calculations of the collapse of a 1 $M_\odot$ dense core in solid body rotation, including both radiative transfer and magnetic field. Using typical parameters for low-mass cores, we study thoroughly the effect of radiative transfer and magnetic field on the first core formation and fragmentation. We show that including the two aforementioned physical processes does not correspond to the simple picture of adding them separately. The interplay between the two is extremely strong, via the magnetic braking and the radiation from the accretion shock.

1. Introduction

The protostellar collapse of low mass dense cores follows a well-defined sequence of different stages, down to the formation of a protostar. In the first collapse phase (Larson 1969), the compressed gas cools efficiently thanks to the coupling between gas and dust. At higher densities ($\rho > 10^{-13}$ g cm^{-3}), the radiation is trapped and the first hydrostatic core (the first Larson core) is formed. At this stage, the grain opacities and the radiation transport play a major role. In the recent past few years, a lot of progress in the computational star formation field has be done. For instance, a lot of radiation hydrodynamics (RHD) methods have been developed for grid based codes (e.g., Krumholz *et al.* 2007, Kuiper *et al.* 2010) and for smoothed particles hydrodynamics (SPH) codes (Whitehouse & Bate 2006, Stamatellos *et al.* 2007). Applying these methods to star formation, it turns out that a barotropic EOS cannot account for realistic cooling and heating of the gas (e.g., Commerçon *et al.* 2010). On larger scales, radiative transfer has been found to efficiently reduce the fragmentation thanks to radiative feedback due to the accretion and the protostellar evolution (Bate 2009, Offner *et al.* 2009). Regarding magnetic field in the star formation context, a gradually improved expertise has been developed for magnetohydrodynamical (MHD) flows (e.g., Hennebelle & Teyssier 2008, Machida *et al.* 2008). All these studies showed that magnetic fields reduce efficiently the fragmentation of prestellar cores. Recently, it has been shown that both radiative transfer and ideal MHD are important and cannot be neglected in the collapse and fragmentation of protostellar cores (Price & Bate 2009, Commerçon *et al.* 2010, Tomida *et al.* 2010).

In this study, we present radiation magnetohydrodynamics (RMHD) calculations of prestellar dense core collapse, using the adaptive mesh refinement (AMR) code RAMSES (Teyssier 2002). The paper is organized as follows: in section 2, we briefly introduce the RMHD solver we designed and present our initial conditions. In section 3, we present our results of dense core collapse calculations using the RMHD solver with various numerical resolutions. Finally, we draw our conclusion in section 4.

2. Numerical method and initial conditions

We use the RAMSES code (Teyssier 2002), in which we have implemented a RHD solver using the grey flux-limited diffusion (FLD, e.g. Minerbo 1978) approximation. We use the comoving frame, which is valid to leading order in v/c in the static diffusion and streaming limits. The solver we designed is coupled to the ideal magneto hydrodynamics one developed by Fromang *et al.* (2006). The coupled RMHD equations read

$$\left\{\begin{array}{llll} \partial_t \rho & + & \nabla\left[\rho \mathbf{u}\right] & = 0 \\ \partial_t \rho \mathbf{u} & + & \nabla\left[\rho \mathbf{u}\otimes\mathbf{u} - \mathbf{B}\otimes\mathbf{B} + P\mathbb{I}\right] & = -\rho\nabla\Phi - \lambda\nabla E_{\mathrm{r}} \\ \partial_t E_{\mathrm{T}} & + & \nabla\left[\mathbf{u}\left(E_{\mathrm{T}} + P_{\mathrm{tot}}\right) - \mathbf{B}(\mathbf{B}.\mathbf{u})\right] & = -\rho\mathbf{u}\cdot\nabla\Phi - \mathbb{P}_{\mathrm{r}}\nabla:\mathbf{u} - \lambda\mathbf{u}\nabla E_{\mathrm{r}} \\ & & & \quad +\nabla\cdot\left(\frac{c\lambda}{\rho\kappa_{\mathrm{R}}}\nabla E_{\mathrm{r}}\right) \\ \partial_t E_{\mathrm{r}} & + & \nabla\left[\mathbf{u}E_{\mathrm{r}}\right] & = -\mathbb{P}_{\mathrm{r}}\nabla:\mathbf{u} + \nabla\cdot\left(\frac{c\lambda}{\rho\kappa_{\mathrm{R}}}\nabla E_{\mathrm{r}}\right) \\ & & & \quad +\kappa_{\mathrm{P}}\rho c(a_{\mathrm{R}}T^4 - E_{\mathrm{r}}) \\ \partial_t \mathbf{B} & - & \nabla\times(\mathbf{u}\times\mathbf{B}) & = 0, \end{array}\right. \tag{2.1}$$

where ρ is the density, $\mathbf{u}$ is the velocity vector, $\mathbf{B}$ is the magnetic field vector, P_{tot} is the total pressure, sum of the thermal and magnetic pressures, $P_{\mathrm{tot}} = P + \frac{\mathbf{B}.\mathbf{B}}{2}$, E_{r} is the radiative energy, λ is the flux limiter (Minerbo 1978), E_{T} is the total fluid energy per unit volume, $E_{\mathrm{T}} = \rho(\epsilon + \frac{\mathbf{u}.\mathbf{u}}{2}) + E_{\mathrm{r}} + \frac{\mathbf{B}.\mathbf{B}}{2}$, Φ is the gravitational potential, $\mathbb{P}_{\mathrm{r}}$ is the radiative pressure, and κ_{R} and κ_{P} are the Rosseland and Planck mean opacities. This system of equations is closed by the perfect gas equation of state $P/\rho = (\gamma - 1)\epsilon$, where $\gamma = 5/3$ is the adiabatic exponent. An additional constraint comes from the divergence of the magnetic field, which has to vanish everywhere at all times ($\nabla.\mathbf{B} = 0$). Note that the radiation diffusion and coupling terms are integrated implicitly in time, since it involves very short timescale processes, compared to the hydrodynamical evolution.

- *Initial conditions:* We consider a uniform-density sphere of molecular gas, rotating about the z-axis with a uniform angular velocity. The prestellar core mass is fixed at $M_0 = 1\ \mathrm{M}_\odot$ and the temperature at 11 K, which corresponds to an isothermal sound speed $c_{s0} \sim 0.19$ km s^{-1}. To promote fragmentation, we add an m = 2 azimuthal density perturbation with an amplitude of 10%. The magnetic field is initially uniform and parallel to the rotation axis. The strength of the magnetic field is expressed in terms of the mass-to-flux to critical mass-to-flux ratio $\mu = 20 = (M_0/\Phi)/(M_0/\Phi)_c$. The initial ratio of the thermal to gravitational energies is $\alpha = 0.37$, and the initial ratio of the rotational to gravitational energies is $\beta = 0.045$.

Calculations were performed using either the rather diffusive Lax Friedrich (LF) Riemann solver or the more accurate HLLD Riemann solver (Miyoshi & Kusano 2005). Following up on former studies (Commerçon *et al.* 2008), we impose at least 15 cells per Jeans length as a grid refinement criterion (parameter N_{J}). We also use one additional parameter, N_{exp}, which indicates the number of cells refined in each direction around a

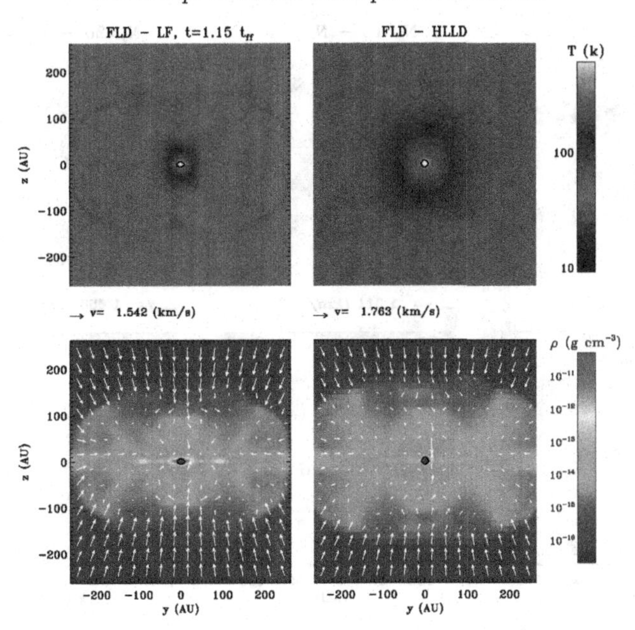

Figure 1. Temperature and density maps in the yz-plane at time $t = 1.15t_{\rm ff}$ for two calculations with the FLD and the LF (left) and the HLLD (right) Riemann solvers.

cell violating the Jeans length criterion. The initial resolution of the grid contains 64^3 cells. We use the low temperature grey opacities of Semenov *et al.* (2003).

3. Results

- *Effect of the solver:* Figure 1 shows temperature and density maps in the yz-plane for two calculations with the FLD and the LF and the HLLD Riemann solvers. Each calculation has been performed using $N_{\rm J} = 15$ and $N_{\rm exp} = 4$. The FLD-LF case leads to spurious fragmentation as it is shown in Commerçon *et al.* (2010). The bubble (dense region, $\rho > 10^{-15}$ g cm^{-3}), driven by magnetic pressure due to the magnetic field line wrapping, is less extended in the FLD-LF case, and the disc is thus more massive and more prone to fragmentation. We identify the accretion shock on the first Larson core as a *supercritical radiative shock*, i.e. all the infalling kinetic energy is radiated away. Consequently, the radiative feedback due to the accretion on the first Larson core is much larger in the FLD-HLLD case, since the magnetic braking and thus the infall velocity are larger thanks to the less diffusive HLLD Riemann solver (Commerçon *et al.* 2010).

- *Effect of the numerical resolution:* Figure 2 shows temperature and density maps in the xy-plane for three calculations using the FLD and the LF Riemann solver, with various resolutions: ($N_{\rm J} = 15$; $N_{\rm exp} = 2$), ($N_{\rm J} = 15$; $N_{\rm exp} = 4$) and ($N_{\rm J} = 20$; $N_{\rm exp} = 4$). We clearly see that increasing the resolution, from left to right, leads to a decrease in the number of fragments produced. As resolution increases, the diffusivity of the LF solver is reduced, the disk is then less massive, and the magnetic braking more efficient. Using the HLLD solver and barotropic calculations, Commerçon *et al.* (2010) show that the correct behavior is the case without fragmentation.

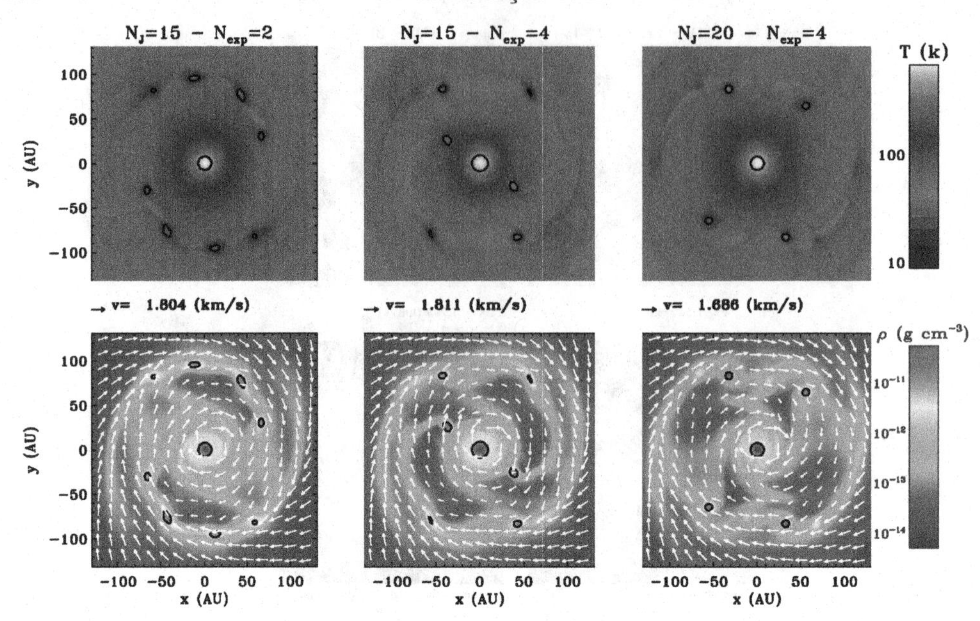

Figure 2. Temperature and density maps in the xy-plane at time $t = 1.15t_{\rm ff}$ for three calculations using the FLD and the LF Riemann solver, with various resolutions.

4. Conclusion

We show that taking into account both radiative transfer and magnetic field is not a straightforward linear process. We show that the magnetic braking and the magnetic bubble extent influence: i) the radiative feedback via the infall velocity and ii) the fragmentation via the disc mass and the rotational velocity. Last but not least, the results are extremely sensitive to the numerical resolution and to the numerical diffusivity of the code used, which readers and authors should be aware of.

References

Bate, M. R. 2009, *MNRAS*, 392, 1363

Commerçon, B., Hennebelle, P., Audit, E. and Chabrier, G., & Teyssier, R. 2008, *A&A*, 483, 371

Commerçon, B., Hennebelle, P., Audit, E. and Chabrier, G., & Teyssier, R. 2010, *A&A*, 510, L3

Fromang, S., Hennebelle, P., & Teyssier, R. 2006, *A&A*, 457, 371

Hennebelle, P. & Teyssier, R. 2008, *A&A*, 477, 25

Krumholz, M. R., Klein, R. I., & McKee, C. F. 2007, *ApJ*, 656, 959

Kuiper, R., Klahr, H., Dullemond, C., Kley, W., & Henning, T. 2010, *A&A*, 511, A81

Larson, R. B. 1969, *MNRAS*, 145, 271

Machida, M. N., Tomisaka, K., Matsumoto, T., & Inutsuka, S.-i. 2008, *ApJ*, 677, 327

Minerbo, G. N. 1978, *JQSRT*, 20, 541

Miyoshi, T. & Kusano, K. 2005, *JCP*, 208, 315

Offner, S. S. R., Klein, R. I., McKee, C. F., & Krumholz, M. R. 2009, *ApJ*, 703, 131

Price, D. J. & Bate, M. R. 2009, *MNRAS*, 398, 33

Semenov, D., Henning, T., Helling, C., Ilgner, M., & Sedlmayr, E. 2003, *A&A*, 410, 611

Stamatellos, D., Whitworth, A. P., Bisbas, T., & Goodwin, S. 2007, *A&A*, 475, 37

Teyssier, R. 2002, *A&A*, 385, 337

Tomida, K., Tomisaka, K., Matsumoto, T., Ohsuga, *et al.* 2010,*ApJ*, 714, L58

Whitehouse, S. C. & Bate, M. R. 2006, *MNRAS*, 367, 32

Computational Star Formation
Proceedings IAU Symposium No. 270, 2011
J. Alves, B.G. Elmegreen, J. M. Girart & V. Trimble, eds.

doi:10.1017/S1743921311000433

The Effects of Radiation Feedback on Early Fragmentation and Stellar Multiplicity

Stella S. R. Offner[1]

[1]Harvard-Smithsonian Center for Astrophysics, 60 Garden St, Cambridge MA 02138, USA
email: soffner@cfa.harvard.edu

Abstract. Forming stars emit a significant amount of radiation into their natal environment. While the importance of radiation feedback from high-mass stars is widely accepted, radiation has generally been ignored in simulations of low-mass star formation. I use ORION, an adaptive mesh refinement (AMR) three-dimensional gravito-radiation-hydrodynamics code, to model low-mass star formation in a turbulent molecular cloud. I demonstrate that including radiation feedback has a profound effect on fragmentation and protostellar multiplicity. Although heating is mainly confined within the core envelope, it is sufficient to suppress disk fragmentation that would otherwise result in low-mass companions or brown dwarfs. As a consequence, turbulent fragmentation, not disk fragmentation, is likely the origin of low-mass binaries.

Keywords. stars:formation, binaries, hydrodynamics, turbulence, radiative transfer

---

1. Introduction

The origin of stellar multiplicity remains an unsolved problem in star formation. The dense conditions in molecular cloud cores and dim luminosities of young protostars make estimations of the initial multiplicity distribution challenging (Duchêne et al. 2007). However, the present-day multiplicity can be observed among field stars, where the likelihood of companions is strongly correlated with the primary stellar mass. Nearly all O and B stars are found in binaries or multiple systems, while only ~20% of M stars have companions (Lada 2006). Successful simulations and theories of star formation must be able to predict the multiplicity fraction and explain why it depends so strongly on stellar mass.

A number of mechanisms have been proposed of which two appear to have the most potential for producing the observed number of multiple systems (see Tohline 2002 for a review). First, gravitational instability within a protostellar accretion disk may produce companions within a few 100 AU (Adams et al. 1989; Bonnell *et al.* 1994). Repeated fragmentation over the disk lifetime may generate numerous companions. Second, perturbations within a turbulent core may seed additional fragmentation on scales of ~ 0.001-0.1 pc (Fisher 2004; Goodwin *et al.* 2004). This must occur within the first ~ 0.5 Myr of collapse, when the core still contains at least $0.1 M_\odot$.

In this paper, we discuss the effect of radiation feedback on early fragmentation and stellar multiplicity using 3D adaptive mesh refinement (AMR) simulations of turbulent molecular clouds. We compare the cases with and without radiation from forming stars. We describe the simulations in §2, present results in §3, and conclude in §4.

2. Simulations

The ORION code solves the equations of compressible gas dynamics, Poisson equation, and radiation energy equation in the flux-limited diffusion approximation (Krumholz *et al.* 2007):

$$\frac{\partial \rho}{\partial t} + \nabla \cdot (\rho \mathbf{v}) = 0, \quad (2.1)$$

$$\frac{\partial (\rho \mathbf{v})}{\partial t} + \nabla \cdot (\rho \mathbf{v}\mathbf{v}) = -\nabla P - \rho \nabla \phi, \quad (2.2)$$

$$\frac{\partial (\rho e)}{\partial t} + \nabla \cdot [(\rho e + P)\mathbf{v}] = \rho \mathbf{v} \nabla \phi - \kappa_R \rho (4\pi B - cE), \quad (2.3)$$

$$\frac{\partial E}{\partial t} - \nabla \cdot (\frac{c\lambda}{\kappa_{\rm R} \rho} \nabla E) = \kappa_{\rm P} \rho (4\pi B - cE) + \sum_n L_n \delta(\mathbf{x} - \mathbf{x}_n), \quad (2.4)$$

$$\nabla^2 \phi = 4\pi G[\rho + \sum_n m_n \delta(\mathbf{x} - \mathbf{x}_n)], \quad (2.5)$$

where ρ, P, $\mathbf{v}$ and e are the fluid density, pressure, velocity, and specific kinetic plus internal energy of the gas, ϕ is the gravitational potential, m_n, $\mathbf{x}_n$, and L_n are the mass, position, and luminosity of the n$^{\rm th}$ star, E is the radiation energy density, and $\kappa_{\rm R}$ and $\kappa_{\rm P}$ are the Rosseland and Planck dust opacities. For comparison, a second calculation closes the equations with a barotropic equation of state (EOS) in lieu of equation 2.4:

$$P = \rho c_{\rm s}^2 + \left(\frac{\rho}{\rho_{\rm c}}\right)^{\gamma} \rho_{\rm c} c_{\rm s}^2, \quad (2.6)$$

where $c_{\rm s} = (k_{\rm B}T/\mu)^{1/2}$ is the sound speed, $\gamma = 5/3$, the average molecular weight $\mu = 2.33 m_{\rm H}$, and the critical density, $\rho_{\rm c} = 2 \times 10^{-13}$ g cm^{-3}.

The calculations insert Lagrangian sink particles in regions exceeding the Jeans density on the maximum AMR level (Krumholz *et al.* 2004). In the simulation with radiation, the particles are endowed with a sub-grid model based upon McKee & Tan (2003) that includes the accretion energy, Kelvin-Helmholtz contraction, and nuclear burning (Offner *et al.* 2009).

The calculations have a Mach number of $\mathcal{M}_{\rm 3D}$=6.6, domain size $L = 0.65$ pc, and mass M=185 $M_\odot$, which correspond to an approximately virialized cloud. We adopt periodic boundary conditions for the gas and Marshak boundary conditions for the radiation field, which allows the cloud to cool. We use a 256^3 base grid with 4 levels of grid refinement, where $\Delta x_4 = 32$ AU. High-resolution convergence tests are discussed in detail in Offner *et al.* (2009). We drive the boxes for three crossing times using random velocity perturbations with wavenumbers $1 \leqslant k \leqslant 2$ after which self-gravity is turned on. The initial gas temperature is 10 K. Since the gas cools efficiently during the driving phase, the radiation calculation remains nearly isothermal.

3. Results

3.1. *Fragmentation*

At the end of a freefall time, the two calculations have very different temperature distributions. Temperatures in the EOS calculation do not exceed 15 K, while temperatures in the radiation calculation reach $\sim$ 100 K. In the latter, because the heated cores are turbulent, a range of densities is heated to various temperatures such that gas temperature is not a single valued function of gas density. The heating also varies with the number of stars and their instantaneous accretion rate, such that any EOS fit to a mean temperature-density curve would be a generally poor approximation over the course of the simulation.

As a consequence of the feedback, the calculation with radiation has a fundamentally different mass distribution (Figure 1A). The protostellar heating is sufficient to raise the temperature of the gas significantly within a few hundred AU and suppress fragmentation

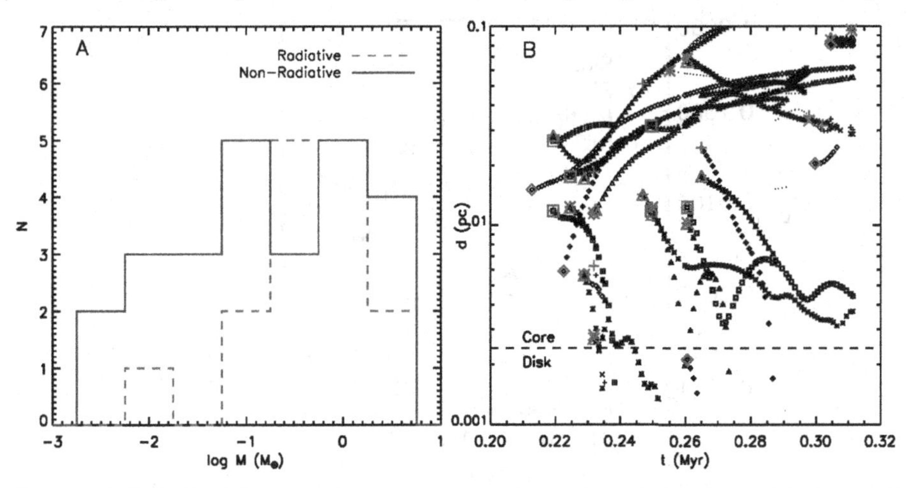

Figure 1. Left: Distribution of star masses produced from simulations with (dashed) and without (solid) radiative feedback. Right: Pair separation as a function of time in 1 kyr bins for all particle pairs from Offner *et al.* (2010). The dashed line at 500 AU indicates a rough boundary between the disk and core scales. The large majority of pairs have separations > 0.1 pc and are not shown. The large (red) symbols indicate the first time bin.

in accretion disks that would otherwise be unstable. As a result, the radiative calculation has far fewer brown dwarfs and, despite small number statistics, its mass distribution more closely resembles the stellar initial mass function. Bate (2009) has previously demonstrated similar results using smoothed-particle hydrodynamics simulations.

Although the heating is efficient at small scales, it is limited to the parent core and does not inhibit fragmentation of other cores or even wide fragmentation within the same core. Figure 1B shows the separations of all star pairs as a function of time in the calculation with radiation feedback. The plot is restricted to separations of 0.1 pc or less to highlight stars forming within the same core that may comprise a binary. The plot shows that fragmentation on scales of $\sim$ 2000 AU is not suppressed by heating. The one instance of fragmentation with $d < 500$ AU is actually filament rather than disk fragmentation. This suggests that turbulent core fragmentation remains a viable mechanism for binary formation of low-mass stars, while disk fragmentation is much less likely.

3.2. *Disk Analysis*

Although the disks in the simulations are not well resolved, it is still possible to use their mean properties to draw robust conclusions about the protostellar multiplicity. Kratter *et al.* (2010) define a two-dimensional parameter space for characterizing accretion and disk stability:

$$\xi = \frac{\dot{M}_{\rm in} G}{c_{\rm s,d}^3}, \qquad \Gamma = \frac{\dot{M}_{\rm in}}{M_{*\rm d}\Omega_{\rm k,in}} = \frac{\dot{M}_{\rm in}\langle j\rangle_{\rm in}^3}{G^2 M_{*\rm d}^3}, \tag{3.1}$$

The thermal parameter, ξ, compares the core sound speed to the disk sound speed, $c_{\rm s,d}$, where $\dot{M}_{\rm in}$ is the infall mass accretion rate. A collapsing isothermal sphere has $\xi \simeq 1$ (Shu 1977). For $\xi > 1$, a disk will be unable to efficiently process accreting material and will eventually fragment. The rotational parameter, Γ, compares the disk orbital time to the gas infall time, where $M_{*\rm d}$ is the total mass in the star-disk system, $\Omega_{\rm k,in}$ is the Keplerian angular velocity at the circularization radius of the infall, and $\langle j\rangle_{\rm in}$ is the specific angular momentum. For large Γ ($\Gamma \sim 0.1$) the disk mass changes quickly over an orbital time.

In Figure 2 we estimate these parameters for the disks in each calculation. Our analysis confirms that disks in the radiative calculation are stable, where the binaries to the left

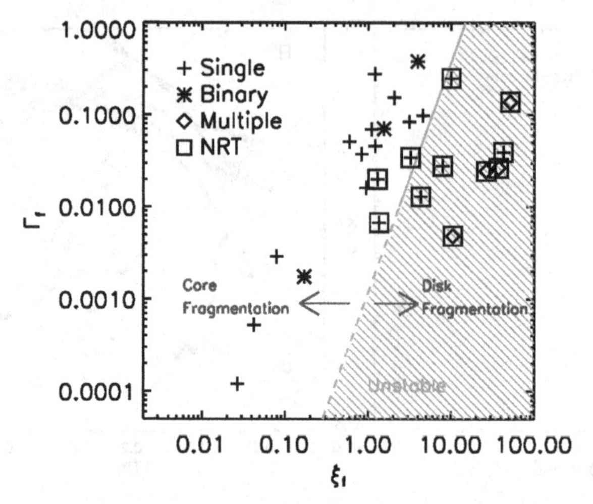

Figure 2. The values of Γ and ξ at 1 freefall time for the protostellar disks in each simulation, where the Non-Radiative Transfer (NRT) cases are denoted by boxes. The diagonal line indicates the boundary between stable and unstable disks found by Kratter *et al.* (2010).

of the line are in fact products of turbulent rather than disk fragmentation. Without radiation feedback large unstable disks yield high multiplicity systems, which fall to the right of the line as expected. Although some single systems exist within the unstable regime these tend to have low Q values and often have previously fragmented.

4. Conclusions

Radiation feedback from low-mass stars is important in shaping the stellar mass distribution. Heating works to stabilize protostellar disks and suppress fragmentation that would otherwise over-produce brown dwarfs. However, fragmentation of the parent core may still occur on thousand AU scales, suggesting that turbulent core fragmentation, not disk fragmentation, is the most likely origin of low-mass binaries.

References

Adams, F. C., Ruden, S. P., & Shu, F. H. 1989, *ApJ*, 347, 959
Bate, M. R. 2009, *MNRAS*, 392, 1363
Bonnell, I. A. & Bate, M. R. 1994, *MNRAS*, 269, L45
Duchêne, G., Delgado-Donate, E., Haisch, Jr., K. E., Loinard, L., & Rodríguez, L. F. 2007, Protostars and Planets V, 379
Fisher, R. T. 2004, *ApJ*, 600, 769
Goodwin, S. P., Whitworth, A. P., & Ward-Thompson, D. 2004, *A&A*, 414, 633
Kratter, K. M., Matzner, C. D., Krumholz, M. R., & Klein, R. I. 2010, *ApJ*, 708, 1585
Krumholz, M. R., McKee, C. F., & Klein, R. I. 2004, *ApJ*, 611, 399
Krumholz, M. R., Klein, R. I., McKee, C. F., & Bolstad, J. 2007, *ApJ*, 667, 626
Lada, C. J. 2006, *ApJL*, 640, L63
McKee, C. F., & Tan, J. C. 2003, *ApJ*, 585, 850
Offner, S. S. R., Klein, R. I., McKee, C. F., & Krumholz, M. R. 2009, *ApJ*, 703, 131
Offner, S. S. R., Kratter, K. M., Matzner, C. D., Krumholz, M. R., & Klein, R. I. 2010, in prep.
Shu, F. H. 1977, *ApJ*, 214, 488
Tohline, J. E. 2002, *ARAA*, 40, 349

Computational Star Formation
Proceedings IAU Symposium No. 270, 2011
J. Alves, B.G. Elmegreen, J. M. Girart & V. Trimble, eds.

doi:10.1017/S1743921311000445

Simulations of Massive Star Cluster Formation and Feedback in Turbulent Giant Molecular Clouds

Elizabeth Harper-Clark[1,2] and Norman Murray[1]

[1]Canadian Institute for Theoretical Astrophysics, Toronto, Ontario, Canada
[2] email: h-clark@cita.utoronto.ca

Abstract. Using the AMR code ENZO we are simulating the formation of massive star clusters within turbulent Giant Molecular Clouds (GMCs). Here we discuss the simulations from the first stages of building realistic turbulent GMCs, to accurate star formation, and ultimately comprehensive feedback. These simulations aim to build a better understanding of how stars affect GMCs, helping to answer the questions of how long GMCs live and why only a small fraction of the GMC gas becomes stars.

Keywords. stars: formation, galaxies: ISM, galaxies: star clusters, ISM: clouds, ISM: bubbles, ISM: structure, ISM: jets and outflows, turbulence, radiative transfer, and MHD

---

1. Introduction

Within the Milky Way there are $\sim$ 300 Giant Molecular Clouds (GMCs) with masses of the order a million solar masses. For these GMCs, the average star formation rate per free-fall time (SFR_{ff}) is 2% (Kennicutt 1998). However, the majority of the star formation is contained within fewer than 50 of these GMCs (Murray & Rahman 2010 and Murray 2010). The small fraction of GMCs with high SFR_{ff} suggests that the SFR_{ff} of a GMC changes over its lifetime. We propose that the star formation rate increases over a cloud's lifetime, as shown in cartoon form in Fig. 1. The GMCs with highest

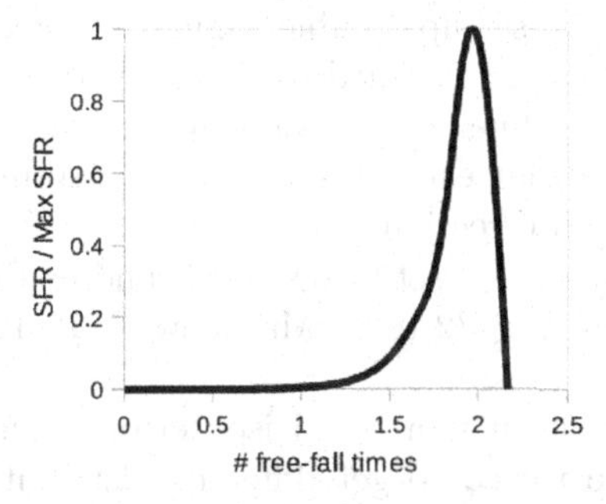

Figure 1. A cartoon of our propose star formation rate per free-fall time over an individual clouds lifetime ending with its destruction at $\sim$ 2 free fall times.

SFR_{ff} are seen to be in the process of disruption, Fig. 2. Once a large star cluster forms it will blow apart its parent GMC (Harper-Clark and Murray 2009 and Fig. 2), thus the last period of star formation is likely the GMCs most rapid.

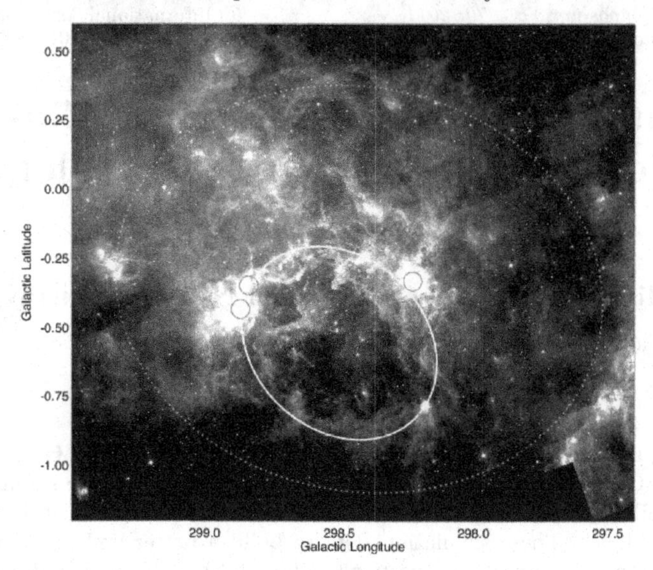

Figure 2. A Spitzer Glimpse image of G298-0.34, a GMC in the process of being blow apart by an invisible central massive star cluster (Murray & Rahman 2010). The small red dashed ellipses show compact HII regions from triggered star formation. The solid oval is the bubble wall and the large dotted oval is the extent of the WMAP source, suggesting the hot gas is leaking from the bubble (Harper-Clark and Murray 2009).

2. Code

The high-resolution Eulerian AMR code ENZO (Bryan & Norman 1997 and O'Shea *et al.* 2004 etc.) contains all necessary physics for accurate testing of turbulent GMC gravitational collapse. We are using the developer's version which includes MHD, protostellar jets, radiative transfer, and an alternative hydro solver using a total variation diminishing second order Runge-Kutter scheme for time-integration (see Wang and Abel 2009 and Wise and Abel 2008 for details).

3. Simulation Set-up

To model an entire GMC we set up a cubic simulation box 128pc wide. The top grid has dimensions 256^3 (0.5pc) and is refined by 2 for up to 8 levels (effective resolution of 65 536, 0.002pc, 400AU). Refinement is based upon Jeans' length or the presence of stars. Subgrids are set up so that each star is always surrounded by at least 0.03pc of maximally refined region in all directions.

The initial velocities of the gas are set by a seeded random distribution fitting a Larson turbulent spectrum with $2 \leqslant k \leqslant 32$ (e.g., Mac Low *et al.* 1998) at Mach 9 and decay appropriately.

The initial density distribution is a cored isothermal sphere with radius of 44.8pc. The size of the core can be varied to get different distributions whilst maintaining a mass of a million solar masses of gas within the cloud. Although a smooth spherically symmetric density distribution is not realistic for a GMC the random velocities cause filamentary structures to form rapidly and before any stars form the densities look much more realistic (Fig. 3).

Stars are formed according to the Truelove criterion (Truelove *et al.* 1997). Once the Jeans' length (l_J) is less than 4 maximum refinement level cell widths only the mass

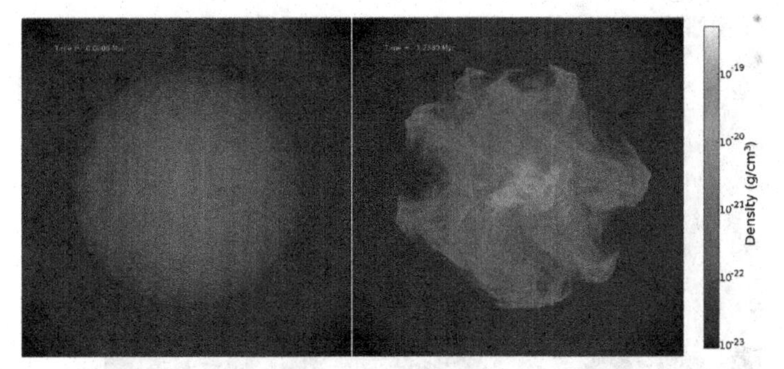

Figure 3. Initial density distributions. Left: time = 0 the cloud is a spherically symmetric cored isothermal sphere. Right: after 1.238 million years the turbulence has formed a filamentary structure and the first stars are about to form.

needed to bring the Jeans length back to 4 cell widths is taken out of the cell and forms a sink particle conserving momentum and mass.

4. Preliminary Results

While the simulations have only run as far as forming the first star cluster to date, there are some interesting results from our simulations worth discussing.

4.1. *Local conditions for star formation*

By construction the stars in our simulation form where the Jeans length becomes less than 0.008pc. Upon close inspection this typically occurs where three or more filaments of inflow meet or when two filaments collide, as shown in Fig. 4.

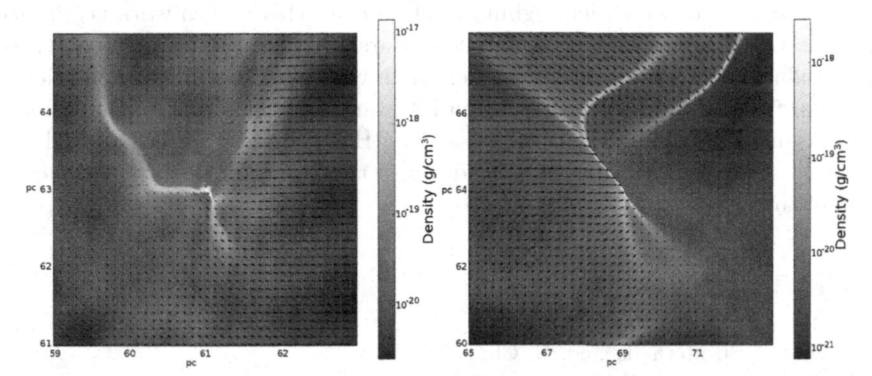

Figure 4. Density slices of a core about to form a star from hydrodynamic simulations showing the velocities of the surrounding gas (black arrows). Left: three or more filaments of inflow meeting. Right: just after two filaments have collided

4.2. *Location of first stars within the GMC*

Precisely where within a GMC the first stars form depends strongly upon the central concentration of the density. For example Fig. 5 shows two different GMCs with identical initial velocity distribution, the same mass but different radial density profiles. Both clouds are cored isothermal spheres, one with density varying as: $\rho(r) = 4.25\rho_c/(1 + (9r/r_{cl})^2)$ so half mass radius = 0.57 (left in Fig. 5) and the other: $\rho(r) = 1.05\rho_c/(1 + (4r/r_{cl})^2)$ so half mass radius = 0.63 (right in Fig. 5) where ρ_c is a set density and r_{cl} is the radius of the cloud.

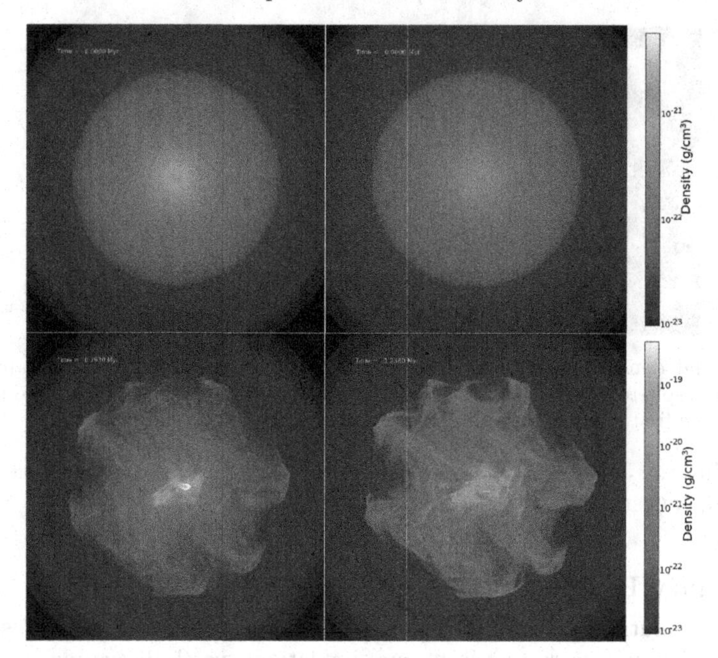

Figure 5. Two clouds with different central concentrations but identical initial velocity distribution and total mass. The top two images show the initial density when spherically symmetric and the bottom images are the corresponding clouds just before the first stars form in each. Please note the different colourbars between the top and bottom images (colourbars on right).

5. Future Work

These results provide an enticing glimpse into a field where much work is still needed. It is important to investigate further where stars first form with different turbulent seeds and different density profiles to understand what effect the initial conditions have upon results. Also, how does star formation rate vary with time within simulations with no feedback and with some or all feedbacks (jets, HII, radiation pressure, winds, supernovae). Additionally, How do magnetic fields affect the formation of stars and subsequent evolution of the clouds?

References

Bryan, G. L. and Norman, M. L. Mar. 1997, In "Workshop on Structured Adaptive Mesh Refinement Grid Methods", ed. N. Chrisochoides

Harper-Clark, E. & Murray, N. 2009, *ApJ*, 693, 1696

Kennicutt, R. C. 1998, *ApJ*, 498, 541

Mac Low, M.-M., Klessen, R.S., Burkert, A., & Smith, M.D. 1998, Phys. Rev. Lett., 80, 2754

Murray, N. 2010, *submitted*

Murray, N. & Rahman, M. 2010, *ApJ*, 709, 424

O'Shear, B.W. *et al.* 2004, In "Adaptive Mesh Refinement - Theory and Applications," Eds. T. Plewa, T. Linde & V. G. Weirs, Springer Lecture Notes in Computational Science and Engineering.

Truelove, J. K. et al 2009, *ApJL*, 489, 179

Wang, P. & Abel, T. 2010, *ApJ*, 696, 96

Wise, J. & Abel, T. 2008, *ApJ*, 685, 40

Computational Star Formation
Proceedings IAU Symposium No. 270, 2011
J. Alves, B.G. Elmegreen, J. M. Girart & V. Trimble, eds.

doi:10.1017/S1743921311000457

Observations of star formation triggered by H II regions

Lise Deharveng[1] **and Annie Zavagno**[1]

[1]Laboratoire d'Astrophysique de Marseille (UMR 6110, CNRS & Université de Provence)
38 rue F. Joliot-Curie, 13388 Marseille Cedex 13, France
email: lise.deharveng@oamp.fr

Abstract. Observations show that expanding H II regions may trigger star formation. We discuss several aspects of this type of star formation, and try to estimate its prevalence. We show how LMC H II regions may help us to understand what we see in our Galaxy.

Keywords. stars: formation, ISM: H II regions, ISM: bubbles, infrared: ISM, infrared: stars

1. Introduction

Massive stars are often presented as having a negative feedback on their surroundings in terms of star formation. Via ionization of the gas, and via stellar winds and supernovae explosions at the end of their life, they may disperse their parental molecular cloud, thereby preventing the formation of subsequent stars. Numerous surveys are now available, at all wavelengths, allowing a better understanding of the vicinity of H II regions, with improved detection of young stellar objects (YSOs). These observations show that massive stars may also provide positive feedback affecting star formation, via the accumulation of neutral material at the periphery of H II regions, or via the compression of pre-existing dense condensations as they are reached by ionization fronts. In the following, we will give examples of the accumulation of neutral material around expanding H II regions. We will discuss the various processes able to trigger star formation, how they can be identified, and what kind of stars they form.

2. The accumulation of neutral material around expanding H II regions

This point concerns the accumulation of neutral material around an H II region, in a layer where the conditions for star formation are fulfilled.

Massive OB2 stars ionize the surrounding material, forming H II regions. Due to the high pressure of the warm ionized gas with respect to that of the cold surrounding neutral material, H II regions expand (Dyson & Williams 1997). During their expansion, neutral material accumulates between the ionization front (IF) and the shock front (SF) which precedes it on the neutral side. These layers of accumulated material are dense, and with time they may become massive. We can use a very rough model of expansion in a homogeneous medium to get an order of magnitude. For example an O6V (B0V) star evolving in a medium of 10^3 cm^{-3} (10^2 cm^{-3}) will form, after one megayear, an H II region of radius 5 pc (same size) and of mass 900 $M_\odot$(90 $M_\odot$), surrounded by a neutral shell of ~ 17000 $M_\odot$(~ 1700 $M_\odot$).

How can we observe these shells of collected material?

• We can observe the CO emission of the shells. Knowledge of the velocity allows us to consider only the molecular material associated with the central H II regions. But the CO emission is not sensitive to high densities; it may be optically thick and, also, in cold dense condensations the CO molecules condense onto the dust grains so that CO may be heavily depleted (Caselli *et al.* 1999; Bacmann *et al.* 2002). All these effects result in underestimation of the masses of the shells.

An illustration is given by the Sh 104 H II region (Deharveng *et al.* 2003); excited by an O6 star, it has a radius ~4 pc and a mass ~450 $M_{\odot}$. It is completely surrounded by a shell of molecular material with a mass of ~6000 $M_{\odot}$. This shell contains four dense fragments (mapped in CS) elongated along the IF. A deeply embedded and probably young cluster, exciting a compact H II region, is observed in the direction of the most massive fragment.

Many other examples can be found in Beaumont & Williams (2010). They mapped the CO(3-2) emission in the direction of 43 bubbles from Churchwell *et al.*'s catalogue (2006; hereafter CHU06). These bubbles detected at 8 μm enclose classical H II regions. Most of these bubbles are surrounded by shells of CO emitting material.

• We can observe the (sub)millimeter thermal emission of the cold dust associated with dense molecular material. But we do not have the velocity information and thus the association may be uncertain.

This is illustrated by the RCW 79 and RCW 120 H II regions. Zavagno *et al.* (2006) have shown that RCW 79 is surrounded by a shell of cold dust observed at 1.2-mm; the mass of this shell is estimated to be >3600 $M_{\odot}$. RCW 120 has been observed at 1.2-mm (Zavagno *et al.* 2007) and at 870 μm (Deharveng *et al.* 2009). It is surrounded by a shell of cold dust, with a mass of ~2000 $M_{\odot}$. In both objects massive fragments are elongated along the IF, and contain young stellar objects.

Many other examples can be found in Deharveng *et al.* (2010; hereafter DEH10) who studied the distribution of the cold dust in the direction of 102 bubbles from CHU06, using the ATLASGAL survey of the Galaxy at 870 μm (Schuller *et al.* 2009). This survey

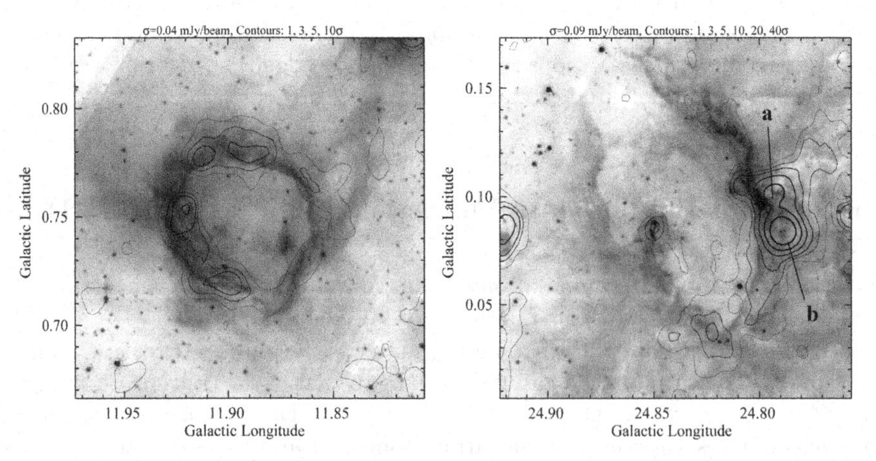

Figure 1. N4 (left) and N36 (right) bubbles from the CHU06 catalogue. The 870 μm cold dust emission is presented as contours superimposed on the *Spitzer* image at 8 μm; these two bubbles enclose classical H II regions (see DEH10 for details). N4 is surrounded by a shell of collected material, with a mass of ~ 1150 $M_{\odot}$. N36 also shows a faint 870 μm emission shell following the bubble, probably formed of collected material, but the massive a and b condensations were probably pre-existing condensations reached by the IF. Condensation a contains a compact H II region and condensation b contains two UC H II regions and various masers.

covers the inner Galactic plane, $l = 300°$ to $60°$, $|b| \leqslant 1.°5$, with an angular resolution of $19.''2$ and an rms noise ~0.06 Jy/beam. DEH10 find that at least 86% of these bubbles, detected at 8 μm, enclose H II regions detected thanks to their radio-continuum emission. Considering only the 65 bubbles, all enclosing H II regions, for which the angular resolution is enough to study the distribution of the associated cold dust, they find that 38% of these bubbles are surrounded by a shell of collected material; 31% more bubbles are surrounded by numerous dust condensations, but DEH10 are unable to estimate if they were pre-existing and reached by the ionization fronts or if they formed out of collected material. Fig. 1 shows the N4 and N36 bubbles and their associated cold dust condensations. The association is based on velocity measurements (see details in DEH10).

Are the collected neutral shells expanding?

It is hard to answer this question. So far there are few observations concerning the expansion of the shells, and they give contradictory results. Examples of expansion can be found in Patel *et al.* (1995; the IC1396 H II region presents a ring of CO material, of radius 12 pc, expanding with a velocity of 5 km s^{-1}), in Dent *et al.* (2009; in the Rosette Nebula, a CO ring is also identified, expanding with a velocity $\sim$ 15 km s^{-1}), in Kang *et al.* (2009; a CO shell, expanding with a velocity $\sim$ 3–4 km s^{-1}, surrounds the N102 bubble).

On the other hand, Beaumont & Williams (2010) do not detect CO emission in the direction of the centers of their bubbles. If the bubbles were surrounded by spherical expanding shells faint blue- and redshifted emissions would be expected in the direction of the central regions. As this is not the case, they conclude that the parental molecular clouds are somewhat flat, and that the bubbles are not three dimensional spherical structures but two dimensional rings.

3. Different processes for triggering star formation around H II regions

Different processes can trigger star formation around H II regions. We will discuss here the two processes most often mentioned. A sketch of the various processes and more details about them can be found in DEH10.

The radiation driven compression of pre-existing condensations.

The H II region expands into a medium where pre-existing dense condensations are present. What happens when one of them is reached by the IF? If the pressure of the ionized gas in the H II region is higher than that inside the molecular condensation an IF progresses inside the condensation, preceded by an SF. A dense ionized boundary layer forms at the periphery of the condensation, and a layer of compressed neutral material accumulates between the two fronts. This layer may fragment and collapse. Alternatively, the whole condensation may implode; thus the name of this process: "radiation-driven implosion of a globule". Star formation follows.

What are the morphological signatures of such a process? First, the condensations are dense. If they are situated in front of an optically visible H II region they are seen in absorption. Also they are bordered by a bright emission zone (Hα or radio-continuum emission of the dense ionized gas at their border, or 8 μm emission of polycyclic aromatic hydrocarbons (PAHs) in the photodissociation region); thus the name "bright rim clouds" (BRCs). Generally, their velocity differs from that of the IF, and they protrude inside the ionized region.

A sample of 89 BRCs has been selected on the basis of a high extinction zone bordered by a bright rim at Hα, and the presence in this direction of an IRAS source with the colors of an YSO (Sugitani *et al.* 1991, 1994). Most of these BRCs have been observed at several wavelengths (for example, Thompson *et al.* 2004, Morgan *et al.* 2004 for radio-continuum observations; Sugitani *et al.* 2000 for millimeter observations; Morgan *et al.* 2008 for sub-millimeter and IR observations; De Vries *et al.* 2002, Morgan *et al.* 2009 for CO observations; Urquhart *et al.* 2009 for CO and IR observations). These observations allow to compare the pressure in the ionized gas and in the BRC. The conclusion is that about 55% of the BRCs of the sample are good candidates for triggered star formation (Morgan *et al.* 2009, Urquhart *et al.* 2009). Most of the second-generation sources have low or intermediate masses (depending on their luminosity); however a few BRCs harbor a massive second-generation source (see Fig. 3).

Small-scale sequential star formation is a characteristic of star formation in the vicinity of many BRCs. BCR37, on the border of the IC 1396 H II region, is a good illustration of this characteristic. As one gets away from the first-generation star exciting the large H II region, he finds: i) on the ionized side of the BRC, thus outside the molecular core, some Hα-emission stars (Ikeda *et al.* 2008); these are believed to be T Tauri stars, thus YSOs of about one megayear; ii) inside the molecular core, two sources with a near-IR excess, one corresponding to an IRAS source. *Spitzer*-IRAC images show that these two sources are bright at longer wavelengths. One of them is associated with H_2 jets and a CO outflow of dynamic age 0.3 Myr (Duvert *et al.* 1990), the other is associated with a water maser (Valdettaro *et al.* 2008). Thus, in the vicinity of this BRC we see a gradient of age among the second-generation sources. Star formation has progressed in time and space from the first generation massive star to the present location of the molecular core, thus confirming its triggering by the expanding H II region. Also it shows that star formation has not occurred during an implosion (an unique event) but more probably that stars have formed in the compressed layer progressing inside the cloud, as the cloud got ionized.

The collect and collapse process of star formation.

Another mechanism of star formation is the gravitational fragmentation and subsequent collapse of the neutral collected shell surrounding an H II region.

What are the signatures of such a process? First, the fragments are massive and elongated along the IF. Since the collected shell has the same velocity as the IF, the fragments also have the same velocity and do not protrude inside the ionized region. The young sources formed in these fragments share the velocity of the gas from which they form, and they are observed later on in the direction of the collected layer.

The RCW 120 H II region, at the center of an almost perfect bubble (see details in Zavagno *et al.* 2007 and Deharveng et al. 2009), illustrates such a process. This H II region is surrounded by a shell of collected material of about 2000 $M_{\odot}$, which is fragmented; it probably expands in a medium with a density gradient. Two massive fragments, which are elongated along the IF, are present south of RCW 120; they do not protrude inside the ionized region. The more massive one ($\sim$ 1000 $M_{\odot}$; Fig. 2) contains several YSOs of various masses and in various evolutionary stages: i) a chain of about 10 class I–class II sources, parallel to the IF, regularly spaced (separation $\sim$0.1 pc), with an uncertain mass, possibly $\sim$ 1 $M_{\odot}$. The formation of these YSOs probably results from small-scale gravitational instabilities in the collected layer; ii) a massive core, which appears as a resolved source at 100 μm (*Herschel* image, Zavagno *et al.* 2010a). It contains a class 0 source, of relatively high mass (8–10 $M_{\odot}$), with a high accretion rate ($\gtrsim 10^{-3}$ $M_{\odot}$ yr^{-1}). This massive class 0 YSO is a very good candidate for massive-star formation by collect &

collapse. Many other YSOs of low or intermediate mass are observed all around RCW 120, in other condensations, possibly formed by different processes.

4. What is the prevalence of massive-star formation triggered by H II regions?

In this section we consider only the formation of *massive* stars.

It is difficult to answer this question because of the complex morphology of most H II regions. A beginning of an answer can be obtained if we consider very simple H II regions, as is done by DEH10. Considering only the 65 bubbles, all enclosing H II regions, for which the angular resolution is enough to study the distribution of the associated cold dust, they find that: i) seven bubbles show associated UC H II regions and nearby 6.7 GHz methanol masers (considered as a signature of massive-star formation at work) in dust condensations adjacent to their IFs; ii) another six show only UC H II regions; iii) another five show only methanol masers in similar condensations. This result suggest that more than a quarter of the bubbles have triggered the formation of massive objects (either via collect and collapse, or via compression of pre-existing clouds). Therefore, star formation triggered by H II regions may be a significant process, especially for *massive*-star formation.

The N36 bubble illustrates such a configuration (Fig. 1). The massive condensation adjacent to N36's IF (probably a pre-existing condensation) contains several H II regions and various masers. The N49 bubble is another candidate for triggered massive-star formation. A shell of neutral material, of $\sim$4000 $M_\odot$, surrounds the central H II region. A compact H II region lies in the direction of the shell. And a massive fragment ($\sim$2000 $M_\odot$) contains two YSOs of intermediate mass along with various masers (see DEH10 and Zavagno *et al.* 2010b for details).

The masses of the second-generation massive-stars

Fig. 3 compares the spectral types of the first- and second-generation massive stars, in regions candidates for triggered star formation. It shows that the second-generation stars

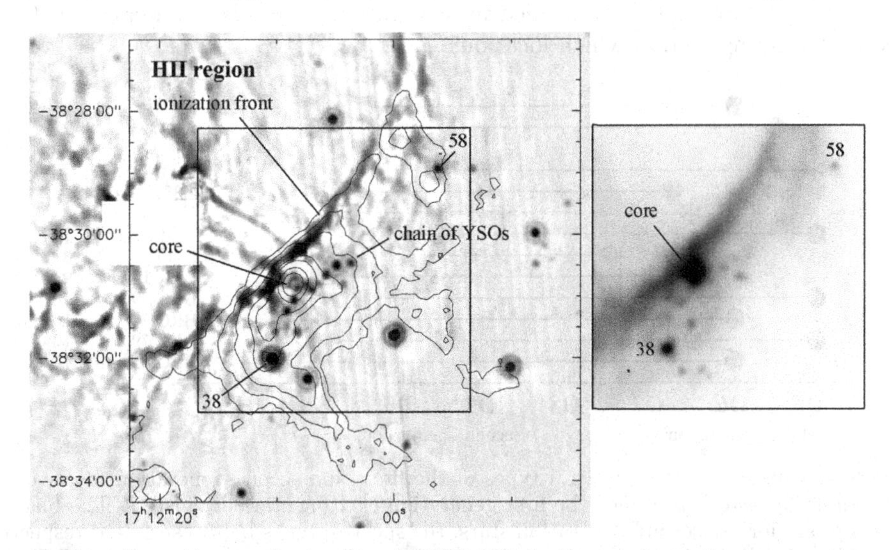

Figure 2. Star formation on the border of RCW 120, in the most massive condensation. Left: the contours show the 870 μm emission of the cold dust (APEX-LABOCA observations) superimposed on an unsharp mask image at 24 μm (*Spitzer*-MIPSGAL observations). Right: the same region observed by *Herschel*-PACS at 100 μm.

are less massive than the first-generation ones; this is not a prediction of the collect & collapse model of star formation, as a late type O star can accumulate with time a very massive layer of neutral material, which by collapse can form massive fragments and thus massive objects. Also, the second-generation stars are less massive than an O7.5 star.

These points are very important: if it is true that the second-generation stars are always less massive than those of the first-generation, star formation triggered by H II regions cannot be an important process of massive-star formation. And it cannot explain the formation of early O stars.

The spectral types of the second generation sources are rather uncertain. They are often estimated from the radio continuum flux of the associated H II regions, assuming a main exciting star, no ionizing photons absorbed by dust, and an optically thin H II region; if it is not the case the spectral types of the second-generation sources are underestimated. Also, it is possible that some of these massive YSOs are still accreting material, and have not yet reached their final masses.

Thus it is very important to identify, if they exist, any good candidate second-generation H II regions excited by early O stars. Where can we find them? Possibly at the border of older and larger structures evolving under the influence of stellar winds and supernova explosions, as are present in the Large Magellanic Cloud (see Sect. 6). They are very difficult to detect in our crowded Galactic plane.

5. Conclusions 1

Thus, we have seen that:

- The accumulation of neutral material around H II regions is a normal process.
- Stars of all masses form on the borders of H II regions by different processes, either by the compression of pre-existing dense condensations, or by the fragmentation and collapse of the collected layer.
- The formation of *massive* stars may be triggered by H II regions.

But a few questions are still without a clear answer. Are the bubbles spherical expanding structures, or are they rings formed in flat molecular clouds? Can early O stars be formed by triggering, and in which locations?

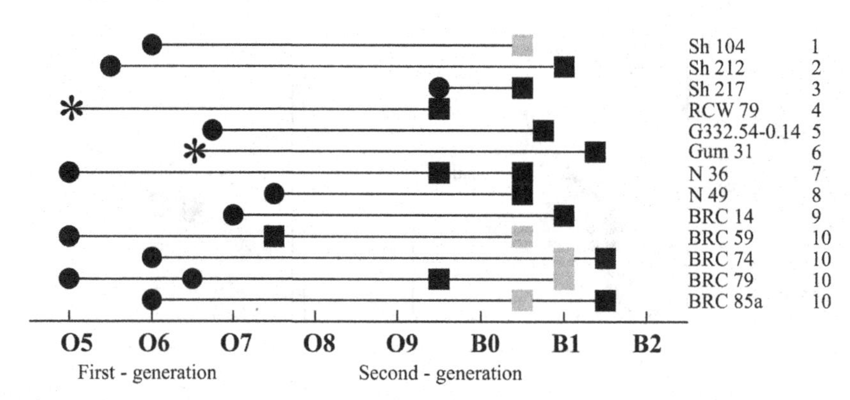

Figure 3. Comparison of the spectral types of the first- and second-generation massive stars. The full circles or asterisks are for the first-generation exciting stars or clusters. The black and grey squares are for the second-generation stars, the spectral types being estimated respectively via the radio flux or via the luminosity of the associated H II regions. *References:* 1 – Deharveng *et al.* 2003; 2 – Deharveng *et al.* 2008; 3 – Brand *et al.* 2010; 4 – Zavagno *et al.* 2006; 5 – Deharveng, in preparation; 6 – Cappa *et al.* 2008; 7 – DEH10; 8 – Kantharia *et al.* 2007; 9 – Morgan *et al.* 2004; 10 – Thompson *et al.* 2004.

6. What can we learn from LMC H II regions?

The Large Magellanic Cloud presents us with a unique opportunity to study the interaction of massive stars with their interstellar environment. All the sources are at the same distance. The plane of the LMC is inclined by only 27° to the plane of the sky, so that the three-dimensional structure of the bubbles can be mapped without confusion. And the reddening is relatively low everywhere. We will consider one example, that of the N44 H II region (α(2000)= $05^{\rm h}22^{\rm m}25.0^{\rm s}$, δ(2000)= $-67°57'00''$).

The brightest part of N44 is a superbubble ($\sim$ 70 pc$\times$50 pc). Many bright H II regions lie at its periphery. This superbubble is expanding with a velocity of 40 km s^{-1}. It is a three-dimensional bubble as the two sides of the expanding bubble are observed (Meaburn & Laspias 1991) . The stellar population of this region is well known (OB association LH47/48). The age of the stars inside the superbubble has been estimated to be $\sim$ 10 Myr; the stars situated outside, at the periphery of the superbubble, are younger, with an age of $\sim$ 5 Myr (Oey & Massey 1995). Extended X-ray emission has been detected in this region, from a nearby classical supernova remnant, inside the superbubble, and leaking towards the south-east through a hole in the central bubble. The X-ray emission of the central superbubble is well explained by off-center supernovas exploding in a bubble already formed by the stellar winds of the central cluster (Chu *et al.* 1993; Magnier *et al.* 1996). Molecular material, in three massive clouds, is associated with N44 (Yamaghchi *et al.* 2001; Kim *et al.* 2004).

Star formation has been studied in this region, based on *Spitzer*-GLIMPSE and -MIPSGAL observations (Whitney *et al.* 2008; Chen *et al.* 2009). Sixty YSOs of high or intermediate mass have been identified in the vicinity of N44 (Chen *et al.* 2009). Eighty percent of the most massive YSOs (types O and B) lie in the direction of H II regions or in adjacent directions. The highest concentration of such massive YSOs is observed in the direction of the central molecular cloud, where the brightest H II regions are also found. Thus this central region clearly shows three generations of massive stars, with ages of ten, five, and less than one Myr. All the young stellar sources lie in the direction of H II regions or of their PDRs, suggesting that their formation has been triggered.

A number of YSOs, some characteristic of massive stars, lie in the outskirts of the H II region and very far from the central superbubble. However, deep Hα images show that the ionized gas emission, mainly in the form of filaments, is very extended. This is confirmed by the emission at 8 μm, mapped by *Spitzer*, of the polycyclic aromatic hydrocarbons located in the PDR which covers about the same area. We see here the long-distance influence, in terms of star-formation, of the central ionized region.

7. Conclusions 2

A few new directions, in the field of star formation triggered by H II regions, seem interesting to follow:

- Look for the long-distance influence of H II regions in terms of star formation, which may explain the formation of apparently isolated YSOs.
- Search for second-generation early O stars. If there are any, they may be found on the border of large structures, like superbubbles, formed via the combined actions of ionization, stellar winds, and supernovae explosions.
- Use *Herschel*, and especially the Hi-GAL survey (Molinari et al. 2010), to estimate the extent of star formation triggered by H II regions.

Acknowledgments: We thank all our collaborators, and especially L. Anderson, J. Brand, J. Caplan, M. Cunningham, P. Jones, S. Kurtz, B. Lefloch, F. Massi, M. Pomarès, D. Russeil, and F. Schuller

References

Bacmann, A., Lefloch, B., Ceccarelli, C. *et al.* 2002 *A&A*, 389, 6
Beaumont, C.N., Williams, J.P. 2010, *ApJ*, 709, 791
Brand, J., Massi, F., Zavagno, A., Deharveng, L. 2010, *A&A*, submitted
Cappa, C., Niemela, V.S., Amorin, R., Vasquez, J. 2008, *A&A*, 477, 173
Caselli, P., Walmsley, C.M., Tafalla, M. *et al.* 1999, *ApJ*, 523,165
Chen, C.-H.R., Chu, Y.-H., Gruendl, R.A. *et al.* 2009, *ApJ*, 695, 511
Chu, Y.-H., Mac Low, M.-M., Garcia-Segura, G. *et al.* 1993, *ApJ*, 414, 213
Churchwell, E., Povich, M.S., Allen, D. *et al.* 2006, *ApJ*, 649, 759 (CHU06)
Deharveng, L., Lefloch, B., Zavagno, A. *et al.* 2003, *A&A*, 408, L25
Deharveng, L., Lefloch, B., Kurtz, S. *et al.* 2008, *A&A*, 482, 585
Deharveng, L., Zavagno, A., Schuller *et al.* 2009, *A&A*, 496, 177
Deharveng, L., Schuller, F., Anderson, L.D. *et al.* 2010, *A&A*, accepted, (DEH10)
Dent, W.R.F., Hovey, G.J., Dewdney, P.E. *et al.* 2009, *MNRAS*, 395, 1805
De Vries, C.H., Narayanan, G., Snell, R.L. 2002, *ApJ*, 577, 798
Duvert, G., Cernicharo, J., Bachiller, R., Gomez-Gonzalez, J. 1990 *A&A*, 233, 190
Dyson, J.E., Williams, D.A. 1997 *The physics of the interstellar medium*
Ikeda, H., Sugitani, K., Watanabe, M. *et al.* 2008 *AJ*, 135, 2323
Kang, M., Bieging, J.H., Kulesa, C.A., Lee, Y. 2009, *ApJ*, 701, 454
Kantharia, N.G., Goss, W.M., Roshi, D. *et al.* 2007, *JApJ*, 28, 41
Kim, S., Walsh, W., Xiao, K. 2004, *ApJ*, 616, 865
Magnier, A.E., Chu, Y.-H., Points, S.D. *et al.* 1996, *ApJ*, 464, 829
Meaburn, J., Laspias, V.N. 1991, *A&A*, 245, 635
Molinari, S., Swinyard, B., Bally, J. *et al.* 2010, *A&A*, accepted, arXiv:1001.2106
Morgan, L.K., Thompson, M.A., Urquhart, J.S. *et al.* *A&A*, 426, 535
Morgan, L.K., Thompson, M.A., Urquhart, J.S., White, G.J. 2008, *A&A*, 477, 557
Morgan, L.K., Urquhart, J.S., Thompson, M.A. 2009, *MNRAS*, 400, 1726
Oey, M.S., Massey, P. 1995, *ApJ*, 452, 210
Patel, N.A., Goldsmith, P.F., Snell, R.L. *et al.* 1995, *ApJ*, 447, 721
Schuller, F., Menten, K.M., Contreras, Y. *et al.* 2009, *A&A*, 504, 415
Sugitani, K., Fukui, Y., Ogura, K. 1991, *ApJS*, 77, 59
Sugitani, K., Ogura, K. 1994, *ApJS*, 92, 163
Sugitani, K., Matsuo, H., Nakano, M. *et al.* *AJ*,119, 323
Thompson, M.A., Urquhart, J.S., White, G.J. 2004, *A&A*, 415, 627
Urquhart, J.S., Morgan, L.K., Thompson, M.A. 2009, *A&A*, 497, 789
Valdetarro, R., Migenes, V., Trinidad, M.A. *et al.* 2008, *ApJ*, 675, 1352
Whitney, B.A., Sewilo, M., Indebetouw, R. *et al.* 2008, *AJ*, 136, 18
Yamaguchi, R., Mizuno, N., Mizuno, A. *et al.* 2001, *PASJ*, 53, 985
Zavagno, A., Deharveng, L., Comeron, F. *et al.* 2006, *A&A*, 446, 171
Zavagno, A., Pomarès, M., Deharveng, L. *et al.* 2007, *A&A*, 472, 835
Zavagno, A., Russeil, D., Motte, F. *et al.* 2010a, *A&A*, accepted, arXiv:1005,1615
Zavagno, A., Anderson, L.D., Russeil, D. *et al.* 2010b, *A&A*, accepted, arXiv:1005,1591

Computational Star Formation
Proceedings IAU Symposium No. 270, 2011
J. Alves, B.G. Elmegreen, J. M. Girart & V. Trimble, eds.

doi:10.1017/S1743921311000469

Observations of Winds, Jets, and Turbulence Generation in GMCs

John Bally[1]

[1]Astrophysical and Planetary Sciences Department,
Center for Astrophysics and Space Astronomy,
UBC 389, University of Colorado at Boulder, Boulder CO 80309, USA
email: john.bally@colorado.edu

Abstract. Protostellar outflows can inject sufficient mass, momentum, and kinetic energy into their parent star-forming clumps to dramatically alter their structure, generate turbulence, and even to disrupt them. Outflows represent the lowest rung on a 'feedback ladder' consisting of increasingly powerful mechanisms which kick-in if star formation escalates towards the production of more massive stars, higher efficiency, and larger clusters. Outflow feedback may dominate turbulence generation and cloud disruption on the scale of cluster-forming clumps having dimensions up to a few parsecs. Outflows inject energy and momentum on a wide-range of length-scales from less than 0.01 pc to over 30 pc. However, they fail by several orders of magnitude to inject sufficient momentum and kinetic energy to drive turbulent motions on the size and mass-scales of GMCs. Injection from higher rungs on the feedback ladder or momentum injected by Galactic-scale processes are needed to power the observed turbulence on the 10 to 100 pc scales of GMCs.

Keywords. stars: formation, ISM: jets and outflows, ISM: Herbig-Haro objects

1. Introduction

Protostellar jets and outflows produce the visual-wavelength shocks known as Herbig-Haro (HH) objects (Reipurth & Bally 2001). As jets entrain and accelerate surrounding molecules, they become visible as molecular outflows. When they break out into the atomic or ionized interstellar medium, they may become visible as externally irradiated flows. Because most forming and young stars produce bipolar jets and outflows, they are abundant with hundreds examples located within 1 kpc of the Sun.

Protostellar outflows have profound impacts on the star formation environment. Their terminal shocks probe the ambient medium. The most distant shocks from a source provide information about the density, velocity structure, ionization state, and chemical composition of the impacted region. In the absence of massive stars, momentum and energy injection by jets can be the dominant source of turbulence generation and cloud disruption. Thus, outflows in low- to intermediate-mass star forming regions may dominate star formation feedback and self-regulation.

2. The Feedback Ladder

Protostellar outflows represent the lowest rung in the *Feedback Ladder* of self-regulating star formation. In molecular cloud clumps and cores where only low-mass stars are forming, outflows may be the dominant agent of energy injection and feedback. Low-mass stars have long pre main-sequence contraction time-scales ranging from one to many tens of Myr during which they have cool photospheres and K and M spectral types. Thus, while their X-rays may penetrate deep into their parent clouds, the bulk of their

Table 1. The Proto-Stellar Feedback Ladder

Type of Feedback	Mechanism	Most Massive Star
Outflows	Momentum Injection	$<$ few $M_\odot$)
X-rays	Ionization & Heating	All stellar masses
FUV ($>$ 912A)	Cloud surface heating	$>\sim$ 3 $M_\odot$)
EUV ($<$ 912A)	Ionization of H	$>\sim$ 8 $M_\odot$)
Stellar Winds	Expansion of Wind-bubble	$>\sim$ 20 $M_\odot$)
Supernovae	Impact of Blast Waves	$>\sim$ 8 $M_\odot$)
Radiation Pressure	$P \sim L/c$	Massive clusters

luminosity is an inefficient heater of gas more than a few hundred AU from the star. The luminosity produced by accretion and nuclear energy sources will only impact the immediate environment of low-mass proto-stars. On the other-hand, outflows from even the lowest mass young stellar objects (YSOs) can propagate over distances of more than a parsec.

Observations show that outflows tend to be collimated into bipolar structures. Outflows in isolated globules forming one or only a few stars tend to have long-lasting stable orientations. While they may push-away portions of the cloud along the poles of their accretion disks, they are unlikely to have much impact in the plane of the disk where the outflow is blocked. Examples of relatively isolated YSOs with steady orientations forming in small cores include B335, HH34, and HH 46/47.

The situation in regions forming clusters of low-mass YSOs may be different. Observations show that in clusters, the orientations of disks and outflows are random. Thus, most portions of the cloud will be impacted by outflows. Examples of such regions include the NGC 1333 region in the Perseus Molecular Cloud (Bally *et al.* 1996) and the 'Gulf of Mexico' region in the North America and Pelican Nebula complex (W80 = NGC 7000; see Figure 1).

If outflow kinetic energy and momentum injection rates are comparable to or greater than the dissipation rate of turbulence and consequent accretion onto YSOs (or their disks and envelopes), outflows can regulate or even terminate star formation by blowing-out the gas. On the other hand, if outflow feedback is less efficient, then dissipation in the clump will allow accretion onto proto-stellar seeds to continue. However, as stars grow in mass, the power of their outflows (momentum injection rates) tend to increase.

If outflows fail to stop star-fomation in a given region, star formation will continue until a more powerful feedback mechanism intervenes. The next rung of the ladder is non-ionizing UV (FUV) radiation. As one or more stars reach masses of a few $M_\odot$, their Kelvin-Helholtz contraction time-scales ($\approx$ pre main-sequence time-scale $\sim GM^2/R_*$) become short compared to their accretion time-scales ($\sim M/\dot{M}$). Such stars quickly develop hot photospheres with spectral types ranging from early F, A, to late B that radiate much of their luminosity in the far-UV between 912 and about 2,000 Angstroms. FUV radiation is an efficient heater of cloud surfaces by means of the grain photo-electric emission and molecular hydrogen dissociation followed by re-formation. Depending on the incident fluxes, such *photon dominated regions* (PDRs) can reach temperatures of 10^2 to nearly 10^4 K which correspond to sound-speeds of 2 to 10 km s^{-1}. The expansion of the heated gas in a PDR is a potent mechanism for injecting momentum and kinetic energy into a cloud by means of the FUV-dominated rocket effect. As moderate-mass (2 to 8 $M_\odot$) stars form, the momentum injected by FUV radiation can dominate feedback. Examples of such regions include NGC 2023 in Orion, NGC 7023 in Cepheus, and the infrared reflection nebulae surrounding late B stars in the ρ-Ophiuchus dark clouds.

Low-mass young stars are highly variable X-ray sources with luminosities ranging from $L_x \sim 10^{27}$ to 10^{31} ergs in the 0.2 to 10 keV range (Feigelson & Montmerle 1999). Hard radiation can penetrate most of the column density of typical molecular clouds. Although an insignificant global heat source, the penetrating power of X-rays results in ionization throughout the volume of a star-forming GMC. Such radiation can alter the chemistry and heat the local environments of YSOs and is a form of feedback. X-ray luminosity of young star tends to increase with the stellar mass.

If outflows and FUV fail to halt star formation, it will continue until stars having masses greater than about 8 $M_\odot$ (spectral type earlier than B3) form that produce radiation which can ionize hydrogen. Such stars produce HII regions with temperatures of about 10^4 K and effective sound speeds of 10 to 15 km s^{-1}. When the sound speed is greater than the gravitational escape speed at the ionization front, the plasma expands. Thus, under most conditions, photo-ionization will halt star formation in the region by first dissociating molecules in a PDR, then ionizing the resulting atoms. While expanding HII regions certainly halt star formation in the immediate vicinity of UV sources, there is evidence that many parsecs from the UV source, ionization fronts may trigger star formation. Examples of ionization-driven blow-out, and possible triggering of star formation in adjacent clouds include the Orion Nebula and NGC 1977 in Orion A, NGC 2024 in Orion B, the M17 complex in Sagittarius, and the W3, W4, and W5 in Cassiopeia.

Main-sequence stellar winds increase in power with stellar mass, and can contribute to feedback by generating mega-Kelvin bubbles that contribute to HII region expansion and feedback.

In high-mass and high density clumps where the escape speed is comparable to or greater than the sound speed in photo-ionized gas, the expansion of HII regions and their tendency to stop star formation may be curtailed. In extremely massive and dense clouds such as in GMCs near the Galactic center and in starburst galaxies such as M82 and the Antennae, ionization may not halt star formation. Stars can continue to form

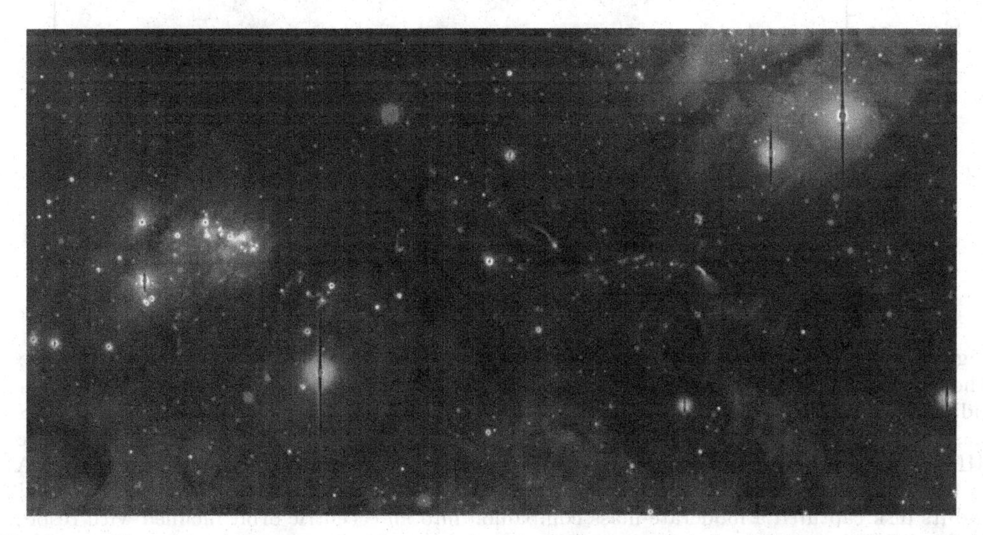

Figure 1. A color image showing dozens of protostellar outflows bursting out of the "Gulf of Mexico" region located directly in front of the W80 (NGC 7000) HII region in Cygnus. The image shows λ 2.12 μm H_2 emission (red) superimposed on λ 0.6563 μm Hα (blue), and $\lambda\lambda$ 0.6716/0.6731 μm [SII] (green). The λ 2.12 μm image was obtained by J. Bally and G. Stringfellow in November 2009 with the NEWFIRM wide-field infrared camera on the Mayall 4 meter telescope on Kitt Peak. The Hα and [SII] images were obtained by Bo Reipurth using the SUPRIME camera at the prime-focus of the Subaru 8.4 meter telescope on Mauna Kea.

and grow despite the presence of massive stars whose ionization fronts remain trapped by gravity (Keto 2002; Keto 2003). As Lyman continuum luminosity increases, HII region may break-out and halt star formation. But, if HII regions remain trapped, the next rung on the feedback ladder will be reached when stars with M >100 $M_{\odot}$ explode as supernovae about 3 Myr after formation.

In super-star-cluster forming clouds birthing thousands of O stars, even supernovae may not halt continued star birth. In such environments, most of the gas can be converted into stars, raising the star formation efficiency to near 100%. As the gas depletes by forming stars, the left-overs of the cloud may eventually be blown out by ionization, stellar winds, supernova explosions, and radiation pressure (Krumholz & Matzner 2009).

3. Jets, Herbig-Haro Objects, and Molecular Outflows

Protostellar outflows are traced by radiative shock waves at near-UV, visual, and infrared wavelengths, or by high-velocity line-wings on sub-mm to mm wavelength emission lines. Shocks form where faster ejecta slams into slower material with supersonic speeds. Collision velocities less than about 60 km s^{-1} excite visual-wavelength forbidden transitions such as [OI] and [SII], and the 1.26 and 1.64 μm transitions of [FeII] if the medium is weakly ionized, and the 2.12 μm and mid-IR lines of H_2 if the medium is molecular. Velocities >60 km s^{-1} dissociate and ionize hydrogen. Charge exchange and collisional

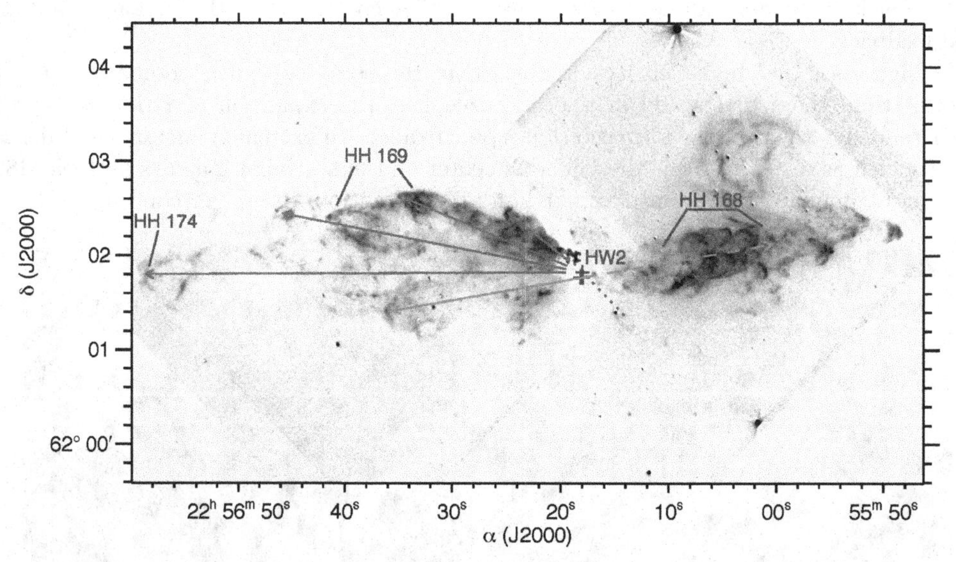

Figure 2. Continuum-subtracted 2.12 μm H_2 emission in the 10^4 $L_{\odot}$ Cep A outflow complex. The successive orientations of a suspected precessing jet from the $\sim$ 15 $M_{\odot}$ protostar HW2 are indicated by red arrows; the position of HW2 is marked with the upper blue cross. The oldest (eastward) ejection powers HH 174 to the east. The next two ejections may be responsible for HH 169. The current orientation of the radio jet emerging from HW2 is at position angle (PA) = 45°, indicated by the dashed magenta line. Cunningham *et al.* (2009) proposed that HW2 and its disk captured a moderate-mass companion into an eccentric orbit inclined with respect to the disk. The companion passes through the disk twice during each orbit, triggering a mass accretion and ejection event. The inclined orbit causes the disk to precess. These two effects combine to produce a pulsed, precessing jet. HH 168, and the HH bows at far right (dashed green line), trace an outflow from HW3c or HW3d (marked with the lower blue cross) along the green axis. A faint H_2 bow marks the eastern lobe of this flow. A collision between the two flows may be responsible for the bright HH 168 complex. Taken from Cunningham, Moeckel, & Bally (2009).

excitation at the shock form thin zones that radiate only in hydrogen recombination lines, producing "Balmer filaments". In fast shocks, the thin (about one mean-free-path) Balmer filament is followed by a thicker layer of fully ionized hydrogen and highly ionized trace elements. Most near-UV, visual, and near-IR radiation produced by a shock emerges from the zone where recombining hydrogen and trace elements radiate away the heat generated by the shock. This layer tends to have and extended tail with temperatures of order 10^4 K set by the $\sim 1 - 3$ eV energy gaps of the visual and near-IR wavelength forbidden transitions of the most common species such as [OI], [OIII], [NII], [SII], and [FeII]. In this region, at most only one Hα photon can be produced by each recombining H atom because collisions do not have sufficient energy to excite the n=2, 3, or higher energy levels of H. Once hydrogen recombines, the low electron density insures that trace ions have long lifetimes. Thus, the few eV forbidden transitions of common ions can be excited thousands of times by collisions before they recombine. These forbidden emission lines can be as bright or brighter than Hα. Shocks speeds higher than about 300 km s^{-1} can sometimes be detected in X-rays and in the non-thermal radio continuum.

The structure, velocity field, and symmetries of outflows provide powerful diagnostics of protostellar accretion and interactions between members of multiple star systems and star clusters. The structure and kinematics of protostellar outflows point to large variations in jet ejection velocities and mass-loss rates of the source YSOs. The giant, parsec-scale outflows trace mass-loss histories over time-scales comparable to YSO accretion times. Protostellar jets are variable in mass-loss rate, ejection velocity, degree of collimation, and orientation over time-scales ranging form years to millennia. The close-connection between accretion and mass-loss implies that accretion onto YSOs is episodic.

Jets and winds transfer momentum and entrain their surroundings by means of shocks. Species such as CO probe the mass and radial velocity of swept-up, entrained gas, but only where the flow interacts with the molecular cloud. Most molecular transitions (and 21 cm emission from HI) are excited by collisions at low ($\sim$ 10 K) temperatures and do not require shocks to be observable. These transitions trace the total amount of entrained mass and the momentum injected into the parent cloud by an outflow over its lifetime.

As jets blow-out of their parent clouds, they entrain atomic gas or ionized plasmas from their surroundings. Because densities tend to be lower than the parent cloud, these swept-up layers are more difficult to observe. They can occasionally be seen in the 21 cm transition of HI, but diffuse Galactic HI emission produces a strong and structured background, making outflow-entrained HI difficult to detect (e.g. Russell *et al.* 1992). When outflows propagate in UV-rich environments, the external radiation field can render the outflow visible. Such 'irradiated jets' and 'irradiated outflows' are especially common in OB associations such as Orion and in HII regions such as the Orion Nebula (Bally & Reipurth 2001; Bally *et al.* 2006.)

4. The Impacts of Outflows

While HH objects and near-IR [FeII] and H_2 emission trace currently active shocks in great detail, the determination of the overall impacts of outflows are better determined from sub-mm and mm molecular line observations. The statistical properties of about 400 molecular outflows were reviewed by Wu *et al.*(2004) and Wu *et al.*(2005). They found the following correlations: The outflow force (momentum injection rate) scales as $log\ \dot{P} = (-4.92 \pm 0.15) + (0.65 \pm 0.043) log\ L_{bol}$ with a correlation coefficient of 0.72 where L_{bol} is the bolometric luminosity of the source in Solar units and $\dot{P}$ is in units of $M_{\odot}$ km s^{-1} yr^{-1}. The outflow force is between 3 to 5 orders of magnitude greater than radiation pressure at L_{bol} =1 $L_\odot$, but decreases to 1 to 3 orders of magnitude greater than L_{bol}/c at L_{bol} =10^6 $L_\odot$. Thus, winds and jets can't be powered by radiation. The outflow mechanical luminosity (in Solar units) scales as $log\ L_{mech} = (-1.98 \pm 0.14) + (0.62 \pm 0.04) log\ L_{bol}$ with a correlation coefficient of 0.69. While the mechanical luminosity is nearly comparable to the bolometric luminosity at L_{bol} =1 $L_\odot$, it is 2 to 4 orders of magnitude lower at L_{bol} =10^6 $L_\odot$. The outflow mass-loss rate (in $M_\odot$ yr^{-1}) scales as $log\dot{M} = (-5.57 \pm 0.096) + (0.50 \pm 0.03) log L_{bol}$ with a correlation coefficient of 0.73. The outflow mass (in $M_\odot$ units) scales as $log\ M = (-1.04 \pm 0.08) + (0.56 \pm 0.02) log\ L_{bol}$ with a correlation coefficient of 0.78. Outflow masses range from $\sim 0.001 - 1$ $M_\odot$ at L_{bol} =1 $L_\odot$, they increase to $10 - 10^3$ $M_\odot$ at L_{bol} =10^6 $L_\odot$. There is also a trend toward poor collimation at high luminosity. The 'dynamical ages' range from 10^3 to 10^6 years with a mean at about 5×10^4 years. Although the 'noise' in these correlations is 1 to 2 orders of magnitude, the range in luminosity is 6 orders of magnitude.

The dynamic ages of outflows from massive stars may be shorter than outflows from low-mass stars because of their rapid evolution. Once ionizing radiation breaks out of their gravitation well, UV radiation will enter the outflow cavity and erase the molecular outflow signatures. Multiplying the momentum injection rate by a typical dynamic age of 5×10^4 years implies that YSOs with L_{bol} = 1, 100, 10^4, and 10^6 $L_\odot$ inject about

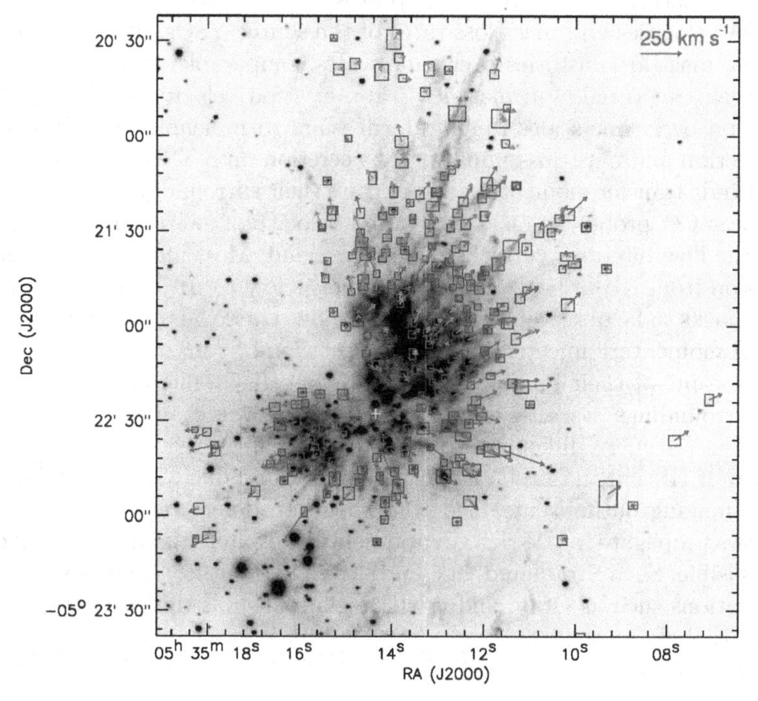

Figure 3. The explosive Orion BN/KL outflow imaged in 2.12 μm H_2 emission in December 2004 (Cunningham 2006). This outflow has dynamic age of about 500 to 1,000 years, has a kinetic energy of order 10^{47} to 10^{48} ergs, and a momentum of about 190 $M_\odot$ km s^{-1}. It may have been triggered by the dynamic ejection of at least three massive stars; the Becklin-Neugebauer (BN) object, radio source I, and infrared source n (Gomez 2008; Zapata 2009). The Trapezium stars which ionize the Orion Nebula are located in the lower-left corner of the image. Red and green arrows denote proper motion measurements based on comparison of the 2004 data with two different prior-epoch images, from September 1992 (Allen & Burton 1993 and January 1999 (Kaifu 2000), respectively. Blue boxes indicate features with proper motion measurements. Arrow lengths indicate velocities according to the scale at upper right, and are equivalent to 122-year motions. The yellow cross marks apparent point of origin of the ejected massive stars.

0.6, 12, 240, and 4,700 $M_{\odot}$ km s^{-1} of momentum respectively. Taking into account a possible decreasing lifetime for massive-star outflows, the momenta scale as L_{bol}^{α} with $\alpha = 0.5 \pm 0.2$.

Ages and sizes deduced from mm/sub-mm observations are lower bounds because CO (and similar) emission is confined to the extent of the parent clouds. In most cases, optical outflows extend far beyond associated CO flows. For example, the HH 111 outflow only is a few arcminutes long in CO; but at visual wavelengths, it can be traced for nearly a degree to HH 113 and HH 311 (Reipurth, Bally, & Devine 1997). Other examples include HH 1/2, HH 34, HH 46/47. The most extreme example is the nearly 30 pc long outflow marked by HH 131 in Orion (Wang *et al.* 2005) powered from L1641N more than 2.5 degrees (18 pc) away.

Observations of YSOs reveal evolutionary trends (Reipurth & Bally 1991; Wu *et al.* 2004, 2005; Wang *et al.* 2005). The youngest, most embedded protostars (Class 0 or young Class I objects) drive slower (10 to 100 km s^{-1}) flows predominantly traced by CO, SiO, and shocked H_2. These flows tend to be dense with n(H_2) $\sim 10^4$ to over 10^7 cm^{-3}, have mass-loss rates of order 10^{-6} to more than 10^{-5} $M_{\odot}$ yr^{-1}, and high mechanical luminosities. Weak masers in species such as H_2O are occasionally seen. Bright maser emission is generally associated only with high-mass protostars. More evolved Class I YSOs tend to drive faster jets dominated by HI and low-ionization potential metals rendered visible by their forbidden lines, have lower densities around 10^2 to 10^4 cm^{-3}, and higher speeds in the range 100 to 400 km s^{-1}. Class II YSOs (classical T-Tauri stars) tend to have much fainter and lower mass-loss rate jets. Thus, as YSOs age, their jets become faster, but have lower densities, mass-loss rates, and momenta.

Outflows in forming clusters may regulate the evolution of their parent clump. NGC 1333 region in the Perseus Molecular Cloud contains approximately 150 YOSs formed within the last Myr that drive hundreds of individual shocks and dozens of outflows; many are bursting out of the parent clump (Walawender *et al.* 2008). Long-since faded, older outflows may be responsible for dozens of cavities in the cloud (Quillen *et al.* 2005). Bally *et al.* (1996) estimated that the mean time between the passage of a supersonic shock in a typical location in NGC 1333 is 10^4 to 10^5 years. Outflow feedback may be responsible for sculpting the cloud and may inject sufficient energy and momentum to maintain a quasi-balance between star formation and turbulent energy dissipation. Surveys of HH objects (Walawender *et al.* 2005) and outflows (Arce *et al.* 2010) over the entire Perseus cloud show that while outflows may self-regulate star formation on parsec scales, they would require 10 to 100 collapse time-scales to supply the observed motions on the 30 pc scale of the entire cloud.

The two nearest massive star forming complexes are in Orion and Cepheus. Figure 2 shows the $L \approx 2 \times 10^4$ $L_{\odot}$ Cepheus A (Cep A) region in the near-infrared (Cunningham, Moeckel, & Bally 2009). The most luminous object HW2 appears to be a 15 $M_{\odot}$ protostar which has trapped its ionizing radiation. It appears to power a pulsed, precessing jet, possibly caused by a moderate-mass companion captured into a non-coplanar (with the HW2 disk), eccentric orbit around HW2.

Figure 3 shows the spectacular BN/KL outflow emerging from OMC1 behind the Orion Nebula, the closest (414 $\pm$ 7 pc) site of on-going massive star formation with $L_{bol} \approx 10^5$ $L_{\odot}$. The explosive outflow may be a consequence of the dynamic decay of a system of massive stars 500 years ago (Zapata *et al.* 2009). This flow has a momentum of 190 $M_{\odot}$ km s^{-1} and kinetic energy of order 10^{47} ergs (uncertain by at least an order of magnitude). Over its dynamic age of $\sim 10^3$ years, it has impacted a region about 0.1 pc in extent.

The most powerful and massive outflows are found in massive star and cluster forming regions. The HII region and associated cluster, DR21 in Cygnus, powers a giant outflow with a mass of about 3,000 $M_{\odot}$ (Russell *et al.* 1992). It may be powered by ionizing radiation which ablates gas from adjacent dense clouds. The plasma flows through a recombination front to produce a powerful neutral flow detected by means of its 21 cm HI emission.

Outflows sculpt their parent clouds by creating cavities and accelerating the displaced gas. Combined effects of variations in jet/outflow mass-loss-rates, ejection velocities, degree of collimation and outflow orientation, and clustering result in generation of chaotic cloud structures and velocity fields. Outflows inject energy and momentum on a range of length-scales form less than 0.01 to 30 pc. Numerical simulations of the impacts of outflows on cloud structure, kinematics, and star formation have been conducted by Carroll *et al.* (2010), Nakamura & Li (2007), and Wang *et al.* (2010). These studies do not incorporate the full range of length, energy, and momentum scales of observed outflows. The absence of bumps in observed power-spectra of cloud turbulence (Padoan *et al.* 2009) in star forming clouds may indicate that outflow injection occurs on length-scales extending over many orders of magnitude.

References

Allen, D. A. & Burton, M. G. 1993, *Nature*, 363, 54
Arce, H. G., Borkin, M. A., Goodman, A. A., Pineda, J. E., & Halle, M. W. 2010, *ApJ*, 715, 1170
Bally, J., Devine, D., & Reipurth, B. 1996, *ApJ*, 473, L49
Bally, J. & Reipurth B. 2001, *ApJ*, 546, 299
Bally, J., Licht D., Smith N., & Walawender J. 2006, *AJ*, 131
Carroll, J. J., Frank, A., & Blackman, E. G 2010, arXiv:1005.1098
Cunningham, N. 2006, *PhD Thesis*, University of Colorado at Boulder
Cunningham, N. J., Moeckel, N., & Bally, J. 2009, *ApJ*, 692, 943
Feigelson, E. D. & Montmerle, T. 1999, *Ann.Rev.Astron.&Astrophys.*, 37, 363
Gómez, L., Rodríguez, L.F., Loinard, L., Lizano, S., Allen, C., Poveda, A., & Menten, K.M. 2008, *ApJ*, 685, 333
Kaifu, N., *et al.* 2000, *PASJ*, 52, 1
Keto, E. 2002, *ApJ*, 580, 980
Keto, E. 2003, *ApJ*, 599, 1196
Krumholz, M. R. & Matzner, C. D. 2009, *ApJ*, 703, 1352
Nakamura, F. & Li, Z.-Y. 2007, *ApJ*, 662, 395
Padoan, P., Juvela, M., Kritsuk, A., & Norman, M. L. 2009,*ApJ*, 707, L153
Quillen, A. C., Thorndike, S. L., Cunningham, A., Frank, A., Gutermuth, R. A., Blackman, E. G., Pipher, J. L., & Ridge, N. 2005, *ApJ*, 632, 941
Reipurth, B., Bally, J., & Devine, D. 1997, *AJ*, 114, 2708
Reipurth, B. & Bally, J. 2001, *Ann.Rev.Astron.&Astrophys.*, 39, 403
Russell, A. P. G., Bally, J., Padman, R., & Hills, R. E. 1992, *ApJ*, 387, 219
Walawender, J., Bally, J., Francesco, J. D., Jørgensen, J., & Getman, K. 2008, Handbook of Star Forming Regions, Volume I, 346
Walawender, J., Bally, J., & Reipurth, B. 2005, *AJ*, 129, 2308
Wang, M., Noumaru, J., Wang, H., Yang, J., & Chen, J. 2005, *AJ*, 130, 2745
Wang, P., Li, Z.-Y., Abel, T., & Nakamura, F. 2010, *ApJ*, 709, 27
Wu, Y., Zhang, Q., Chen, H., Yang, C., Wei, Y., & Ho, P. T. P. 2005, *AJ*, 129, 330
Wu, Y., Wei, Y., Zhao, M., Shi, Y., Yu, W., Qin, S., & Huang, M. 2004, *A&A*, 426, 503
Zapata, L. A., Schmid-Burgk, J., Ho, P. T. P., Rodríguez, L. F., & Menten, K. M. 2009, *ApJ*, 704, L45

Computational Star Formation
Proceedings IAU Symposium No. 270, 2011
J. Alves, B.G. Elmegreen, J. M. Girart & V. Trimble, eds.

doi:10.1017/S1743921311000470

Origin of the prestellar core mass function and link to the IMF – *Herschel* first results

Ph. André[1], A. Men'shchikov[1], V. Könyves[1], and D. Arzoumanian[1]

[1]Laboratoire AIM, CEA Saclay, IRFU/Service d'Astrophysique,
F-91191, Gif-sur-Yvette, France
email: pandre@cea.fr

Abstract. We briefly review ground-based (sub)millimeter dust continuum observations of the prestellar core mass function (CMF) and its connection to the stellar initial mass function (IMF). We also summarize the first results obtained on this topic from the *Herschel* Gould Belt survey, one of the largest key projects with the *Herschel* Space Observatory. Our early findings with *Herschel* confirm the existence of a close relationship between the CMF and the IMF. Furthermore, they suggest a scenario according to which the formation of prestellar cores occurs in two main steps: 1) complex networks of long, thin filaments form first, probably as a result of interstellar MHD turbulence; 2) the densest filaments then fragment and develop prestellar cores via gravitational instability.

Keywords. stars: formation – ISM: clouds – ISM: structure – submillimeter

---

1. Prestellar core mass functions from ground-based observations

Wide-field (sub)mm dust continuum mapping is a powerful tool to take a census of prestellar dense cores and young protostars within star-forming clouds. At the end of the 1990s, the advent of bolometer cameras such as MAMBO and SCUBA on ground-based (sub)millimeter radio-telescopes like the IRAM 30m and the JCMT led to the identification of numerous cold, compact condensations (see Fig. 1–left for examples) that do not obey the Larson (1981) self-similar scaling relations and are intermediate in their properties between diffuse CO clumps and infrared young stellar objects (cf. Motte *et al.* 1998, 2001 and André *et al.* 2000, Ward-Thompson *et al.* 2007 for reviews). These (sub)millimeter continuum condensations detected from the ground are ~ 3 orders of magnitude denser than typical CO clumps (e.g. Kramer *et al.* 1998) and feature large ($\gg 50\%$) mean column density contrasts over their parent background clouds, strongly suggesting they are self-gravitating. The latter is directly confirmed by line observations in a number of cases. When available, the virial masses of the condensations indeed agree within a factor of ~ 2 with the masses derived from their (sub)millimeter dust continuum emission (e.g. André *et al.* 2007). A small fraction of these condensations lie at the base of powerful jet-like outflows and correspond to Class 0 protostars. However, the majority of them are starless/jetless and appear to be the immediate *prestellar* progenitors of individual protostars or protostellar systems.

As first pointed out by Motte, André, Neri (1998) in the case of the ρ Ophiuchi (L1688) cloud, the core mass function (CMF) found for these starless dust continuum condensations is very similar in shape to the stellar IMF. While the early determination of the CMF by Motte *et al.* (1998), based on only 57 prestellar condensations, was clearly limited by small-number statistics, similar results were consistently found by a number of independent groups in the past decade and the statistics improved somewhat over the years (e.g. Testi & Sargent 1998; Johnstone *et al.* 2000, 2001; Motte *et al.* 2001; Stanke *et al.*

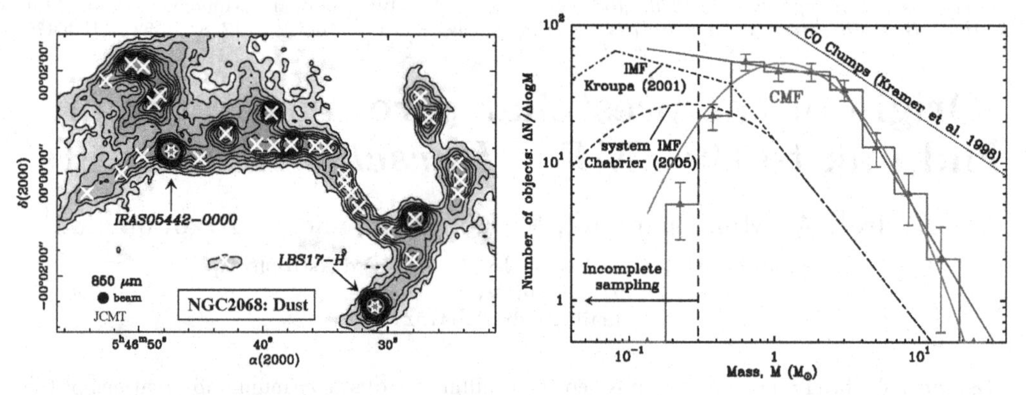

Figure 1. Left: SCUBA 850 μm dust continuum map of the NGC 2068 protocluster extracted from the mosaic of Orion B by Motte *et al.* (2001). A total of 30 compact prestellar condensations (marked by crosses), with masses between $\sim 0.4\, M_{\odot}$ and $\sim 4.5\, M_{\odot}$, are detected in this ~ 1 pc $\times$ 0.7 pc field. **Right:** Differential mass function [dN/dlogM] of the 229 starless dust continuum condensations detected at 850 μm with SCUBA in the Orion A/B cloud complex excluding the crowded OMC1 and NGC 2024 regions (histogram with error bars – from Motte *et al.* 2001, Johnstone *et al.* 2001, and Nutter & Ward-Thompson 2007). A two-segment power-law fit and a log-normal fit are shown for comparison. For reference, the field star IMF (Kroupa 2001), the IMF of multiple systems (Chabrier 2005), and the typical mass spectrum found for CO clumps (Kramer *et al.* 1998) are also shown.

2006; Nutter & Ward-Thompson 2007; Enoch *et al.* 2008). There is thus good evidence in nearby star-forming regions such as Ophiuchus, Serpens, Orion A & B, and Perseus that the shape of the prestellar CMF (e.g. Fig. 1–right) is consistent with the Salpeter power-law IMF at the high-mass end (dN/dlogM $\propto$ M$^{-1.35}$), and significantly steeper than the mass distribution of diffuse CO clumps (dN/dlogM $\propto$ M$^{-0.7}$ – e.g. Kramer *et al.* 1998). The difference presumably arises because CO clumps are primarily structured by supersonic turbulence (e.g. Elmegreen & Falgarone 1996) while prestellar condensations are largely free of supersonic turbulence and clearly shaped by self-gravity (e.g. Motte *et al.* 2001, André *et al.* 2007). The slope of the observed CMF becomes shallower than the Salpeter power law and more similar to the slope of the CO clump mass distribution at the low-mass end. For instance, in the Orion A/B complex which contains the largest sample of starless submm continuum cores identified from the ground (229 objects), the entire prestellar CMF can be fit equally well with either a two-segment broken power law or a log-normal distribution down to the completeness limit of the observations (cf. Fig. 1–right – compare Nutter & Ward-Thompson 2007 and André *et al.* 2009).

The median prestellar core mass observed in regions such as ρ Ophiuchi and Orion ($\sim 0.2 - 1.5\, M_{\odot}$) is only slightly larger than the characteristic $\sim 0.5\, M_{\odot}$ set by the peak of the IMF in dN/dlogM format (cf. Fig. 1–right). Such a close resemblance of the CMF to the IMF in both shape and mass scale is consistent with the view that the prestellar condensations identified in (sub)millimeter dust continuum surveys are about to form stars on a $\sim$ one-to-one basis, with a fixed and high local efficiency, i.e., $M_\star = \epsilon_{\rm core}\, M_{\rm core}$ with $\epsilon_{\rm core} \sim 30 - 100\%$. Taken at face value, this finding suggests that the IMF is at least partly determined by cloud fragmentation at the prestellar core stage (or earlier).

Interestingly, in their near-IR extinction imaging study of the Pipe dark cloud, Alves *et al.* (2007) found a population of 159 starless cores whose mass distribution similarly follows the shape of the IMF. This finding is reminiscent of the CMF results obtained with ground-based (sub)millimeter dust continuum observations, although most of the starless cores in the Pipe Nebula are gravitationally unbound objects confined by external pressure (Lada *et al.* 2008). Hence, a large fraction of them do not qualify as prestellar cores and may never evolve into stars. Assuming nevertheless that most of them will evolve into self-gravitating prestellar cores and form stars, the Alves *et al.* (2007) result suggests that the IMF may be determined even earlier than the prestellar core stage.

Appealing as a direct connection between the prestellar CMF and the IMF might be, several caveats should be kept in mind. First, although core mass estimates based on optically thin (sub)millimeter dust continuum emission are straightforward, they rely on uncertain assumptions about the dust (temperature and emissivity) properties (e.g. Stamatellos *et al.* 2007). Second, existing ground-based determinations of the CMF are limited by small-number statistics at both ends of the mass spectrum. In particular, this has led to concerns that the shape of the CMF may be strongly affected by incompleteness effects at the low-mass end (e.g. Johnstone *et al.* 2000). Third, there are hints that the star formation efficiency at the level of an individual core ($\epsilon_{\rm core}$), corresponding to a global shift in mass scale between the CMF and the IMF, may vary from cloud to cloud, with values ranging from $\epsilon_{\rm core} \sim 15\%$ in Taurus (Onishi *et al.* 2002), to $\epsilon_{\rm core} \sim 30\%$ in the Pipe (Alves *et al.* 2007), to $\epsilon_{\rm core} \sim$ 50–100% in Ophiuchus (Motte *et al.* 1998). It is presently unclear whether this reflects real variations of $\epsilon_{\rm core}$ with environment or whether this results from, e.g., the difficulty of adopting a uniform, precise definition of the boundaries between individual cores and the local background cloud when comparing different core samples.

With an angular resolution at 70–300 μm comparable to, or better than, the largest ground-based millimeter-wave radio-telescopes, the *Herschel* Space Observatory successfully launched by ESA in May 2009 (Pilbratt *et al.* 2010), now makes it possible to address the above issues on the CMF–IMF connection. In particular, *Herschel* can dramatically improve on the statistics of the CMF based on deep, extensive surveys of nearby clouds. *Herschel* can also help to greatly reduce the uncertainties in the core masses through direct measurements of the dust temperatures. In the following sections, we summarize and discuss the first results from the Gould Belt survey, one of the largest key projects with *Herschel*, whose main initial motivation is precisely to clarify the nature of the relationship between the CMF and the IMF (cf. André & Saraceno 2005).

2. First results from *Herschel* on the CMF–IMF connection

The observational objective of the *Herschel* Gould Belt survey is to image the bulk of nearby ($d \lesssim 500$ pc) molecular clouds, mostly located in Gould's Belt (e.g. Guillout 2001), at 6 wavelengths between 70 μm and 500 μm. The total surface area covered by the survey will exceed 160 deg^2 (cf. André *et al.* 2010 and references therein). Two extreme regions among a sample of 15 nearby cloud complexes were selected for imaging during the science demonstration phase of *Herschel* in October 2009: the Polaris Flare and the Aquila Rift. While the Polaris flare field is a high-latitude *translucent* cloud with little to no star formation at $d \sim 150$ pc (e.g., Heithausen *et al.* 2002), the Aquila field is a very active star-forming complex at $d \sim 260$ pc (e.g., Gutermuth *et al.* 2008). Our initial results for these two regions are discussed in 6 papers published in the A&A special issue on *Herschel* (André *et al.* 2010, Könyves *et al.* 2010, Bontemps *et al.* 2010, Miville-Deschênes *et al.* 2010, Men'shchikov *et al.* 2010, Ward-Thompson *et al.* 2010).

Briefly, 302 starless cores but no protostars were detected with *Herschel* in the Polaris field (~ 8 deg^2). The locations of the Polaris starless cores in a mass versus size diagram (Fig. 2–left) shows that they are ~ 2 orders of magnitude less dense than self-gravitating isothermal Bonnor-Ebert spheres and therefore cannot be gravitationally bound. The mass function of these unbound starless cores peaks at an order of magnitude smaller

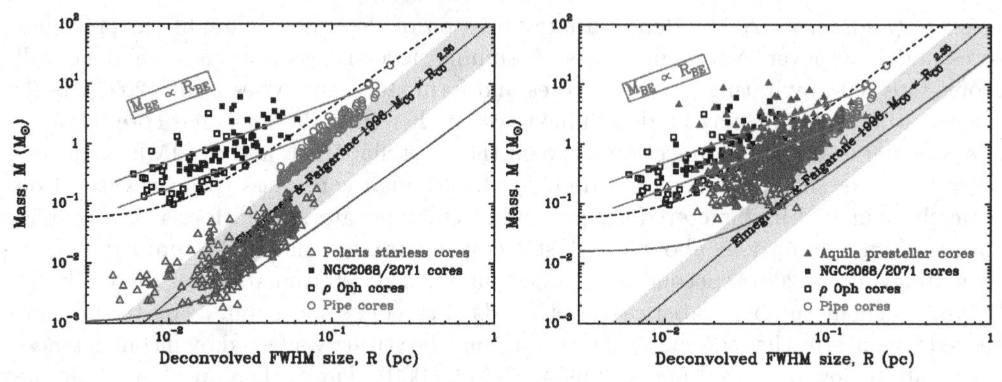

Figure 2. Mass vs. size diagrams for the starless cores detected with *Herschel*-SPIRE/PACS in Polaris (left) and Aquila (right) (triangles, from André *et al.* 2010 and Könyves *et al.* 2010). For comparison, the locations of the (sub)mm continuum prestellar cores identified by Motte *et al.* (1998, 2001) in ρ Oph and NGC2068/2071, respectively, are shown, along with the dust extinction cores of the Pipe nebula (Alves *et al.* 2007, Lada *et al.* 2008), and the correlation observed for diffuse CO clumps (shaded band – cf. Elmegreen & Falgarone 1996). The two solid lines ($M_{\rm BE} \propto R_{\rm BE}$) mark the loci of critically self-gravitating isothermal Bonnor-Ebert spheres at $T = 7$ K and $T = 20$ K, respectively. The 302 starless cores of Polaris lie much below the two Bonnor-Ebert lines in the left panel, suggesting they all are unbound. The 341 cores classified as prestellar by Könyves *et al.* (2010), out of a total of 541 starless cores in Aquila, are shown as filled triangles in the right panel. The (5σ) detection threshold at $d = 150$ pc of existing ground-based (sub)mm (e.g., SCUBA) surveys as a function of size is shown by a dashed curve. The 5σ detection thresholds of the SPIRE 250 μm observations, given the estimated levels of cirrus noise and distances (150 pc for Polaris and 260 pc for Aquila), are shown by solid curves.

mass than the stellar IMF (Fig. 3–left – André *et al.* 2010). In contrast, more than 200 (Class 0 & Class I) protostars can be identified in the *Herschel* images of the whole ($\sim$ 11 deg^2) Aquila field (Bontemps *et al.* 2010), along with a total of 541 starless cores ($\sim$ 0.01–0.1 pc in size). Most (> 60%) of these 541 starless cores lie close to the loci of critical Bonnor-Ebert spheres in a mass versus size diagram (Fig. 2–right), suggesting that they are self-gravitating and prestellar in nature (Könyves *et al.* 2010). The CMF derived for the entire sample of 541 starless cores in Aquila is well fit by a log-normal distribution and closely resembles the IMF (Fig. 3–right – Könyves *et al.* 2010, André *et al.* 2010). The similarity between the Aquila CMF and the Chabrier (2005) system IMF is consistent with a $\sim$ one-to-one correspondence between core mass and stellar system mass ($M_{\star \rm sys} = \epsilon_{\rm core}\, M_{\rm core}$). Comparing the peak of the CMF to the peak of the system IMF suggests that the efficiency $\epsilon_{\rm core}$ of the conversion from core mass to stellar system mass is between $\sim$ 0.25 and $\sim$ 0.4 in Aquila, depending on whether one considers the reduced sample of the best 341 candidate prestellar cores or the entire sample of all 541 starless cores (see discussion in Könyves *et al.* 2010).

The early results of the *Herschel* Gould Belt survey therefore confirm the existence of a close relationship between the prestellar CMF and the stellar IMF, using data with already a factor of $\sim$ 2 to 9 better counting statistics than the ground-based studies discussed in § 1. These results seem difficult to reconcile with models in which competitive accretion plays a key role in shaping the distribution of stellar masses (e.g. Bate & Bonnell 2005) and are in much better agreement with the gravo-turbulent fragmentation picture (e.g. Larson 1985, Klessen & Burkert 2000, Padoan & Nordlund 2002, Hennebelle & Chabrier 2008). In any event, the observations clearly suggest that one of the keys to the problem of the origin of the IMF lies in a good understanding of the formation process of prestellar cores. This is true even if additional processes, such as rotational

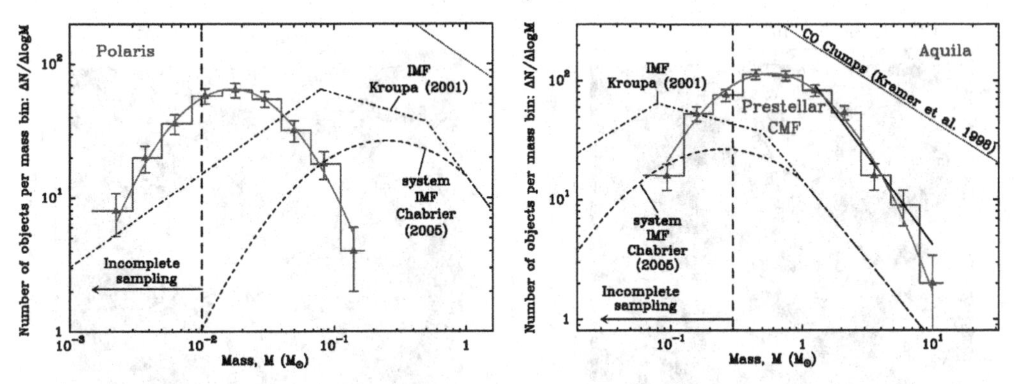

Figure 3. Core mass functions (histograms with error bars) derived from our *Herschel* observations of the Polaris (left) and Aquila (right) regions, which reveal of total of 302 starless cores and 541 candidate prestellar cores, respectively (André *et al.* 2010 and Könyves *et al.* 2010). The IMF of single stars (corrected for binaries – e.g., Kroupa 2001), the IMF of multiple systems (e.g., Chabrier 2005), and the typical mass spectrum of CO clumps (e.g., Kramer *et al.* 1998) are shown for comparison. Log-normal fits to the observed CMFs are superimposed. These fits peak at $\sim 0.02\,M_\odot$ (Polaris) and $\sim 0.6\,M_\odot$ (Aquila), and have standard deviations of ~ 0.41 and ~ 0.43 in $\log_{10}M$, respectively. (For reference, the log-normal part of the Chabrier system IMF peaks at $0.25\,M_\odot$ and has a standard deviation of ~ 0.55 in $\log_{10}M$.)

sub-fragmentation of prestellar cores into binary/multiple systems during collapse, probably also play an important role and may help to populate the low-mass end of the IMF (e.g. Bate *et al.* 2003, Goodwin *et al.* 2008).

3. Spatial distribution of *Herschel* cores vs. large-scale cloud structure

The high quality and dynamic range of the *Herschel* images are such that they provide key information on both dense cores on small (< 0.1 pc) scales *and* the structure of the parent background cloud on large (> 1 pc) scales. In particular, one of the most spectacular early findings made with *Herschel* is the fascinating omnipresence of long ($>$ pc scale) filamentary structures in the cold interstellar medium and the apparently tight connection between the filaments and the formation process of dense cores (e.g. André *et al.* 2010, Men'shchikov *et al.* 2010, Molinari *et al.* 2010). In the Aquila Rift and Polaris flare clouds, for instance, an excellent correspondence is observed between the filaments and the spatial distribution of compact cores (see Men'shchikov *et al.* 2010 and Fig. 4). More precisely, the cores identified with *Herschel* are preferentially found within the *densest* filaments. In Aquila, the distribution of background cloud column densities for the prestellar cores shows a steep rise above $N_{\rm H_2}^{\rm back} \sim 5 \times 10^{21}$ cm^{-2} (cf. Fig. 5–left) and is such that $\sim 90\%$ of the candidate bound cores are found above a background column density $N_{\rm H_2}^{\rm back} \sim 7 \times 10^{21}$ cm^{-2}, corresponding to a background visual extinction $A_V^{\rm back} \sim 7$. The *Herschel* observations of the Aquila Rift complex therefore strongly support the existence of a visual extinction threshold for the formation of prestellar cores (at $A_V^{\rm back} \sim 5$–10), which had been suggested based on earlier ground-based studies of, e.g., Taurus and Ophiuchus (cf. Onishi *et al.* 1998, Johnstone *et al.* 2004). In the Polaris flare, our results are also consistent with such an extinction threshold since the observed background column densities are all below $A_V^{\rm back} \sim 7$ and there are no examples of bound prestellar cores in this cloud. Another visualization of the extinction threshold is obtained by considering the probability density function (PDF) of column densities in the ~ 11 deg^2 field observed with *Herschel* in Aquila (cf. Fig. 5–right). This column density PDF is reasonably well described by a log-normal distribution below $A_V^{\rm back} \sim 7$

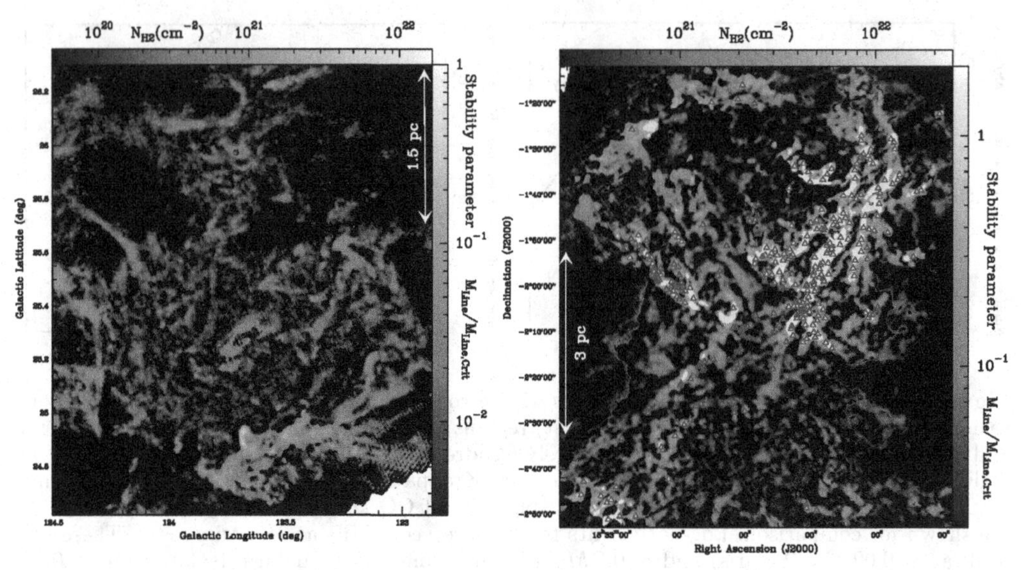

Figure 4. Column density maps of two subfields in Polaris (left) and Aquila (right) derived from our *Herschel* data (André *et al.* 2010). The contrast of the filaments with respect to the non-filamentary background has been enhanced using a curvelet transform (cf. Men'shchikov *et al.* 2010 and Starck *et al.* 2003). Given the typical width $\sim$ 10000–15000 AU of the filaments, these column density maps are equivalent to *maps of the mass per unit length along the filaments.* The scale shown on the right of each panel is given in approximate units of the critical line mass $2\,c_s^2/G$ (cf. Inutsuka & Miyama 1997). The areas where the filaments have a mass per unit length larger than half the critical value and are thus likely gravitationally unstable have been highlighted in white. The bound prestellar cores identified by Könyves *et al.* (2010) in Aquila are shown as small triangles in the right panel; there are no bound cores in Polaris.

and shows a pronounced power-law tail above $A_V^{\rm back} \sim 7$. While similar column density PDFs have been recently reported from near-IR extinction studies (e.g. Kainulainen *et al.* 2009), the power-law tail at high column densities is particularly clear in the *Herschel* data of Aquila. Based on Fig. 5, we believe that this power-law tail reflects the dominant role of gravity above $A_V^{\rm back} \sim 7$ and is intimately related to the formation of prestellar cores (see also § 4 below).

4. Implications for our understanding of the core formation process

Our *Herschel* results in Polaris and Aquila also provide key insight into the core formation issue. They support an emerging picture (see, e.g., Myers 2009) according to which complex networks of long, thin filaments form first within molecular clouds, probably as a result of interstellar MHD turbulence (e.g. Padoan *et al.* 2001), and then the densest filaments fragment into a number of prestellar cores via gravitational instability. That the formation of filaments in the diffuse ISM represents the first step toward core/star formation is suggested by the filaments *already* being omnipresent in a diffuse, non-star-forming cloud such as Polaris (cf. Fig. 4–left, Men'shchikov *et al.* 2010, and Miville-Deschênes *et al.* 2010). The second step appears to be the gravitational fragmentation of a subset of the filaments into self-gravitating cores. Indeed, most (> 60%) of the bound prestellar cores and Class 0 protostars identified in Aquila are concentrated in *gravitationally unstable filaments* for which the mass per unit length exceeds the critical mass per unit length required for hydrostatic equilibrium (cf. Ostriker 1964, Inutsuka & Miyama 1997), $M_{\rm line,crit} = 2\,c_s^2/G \sim 15\,M_\odot/{\rm pc}$, where $c_s \sim 0.2$ km/s is the isothermal sound speed for $T \sim 10$ K (André *et al.* 2010 – cf. Fig. 4–right). Note that the critical

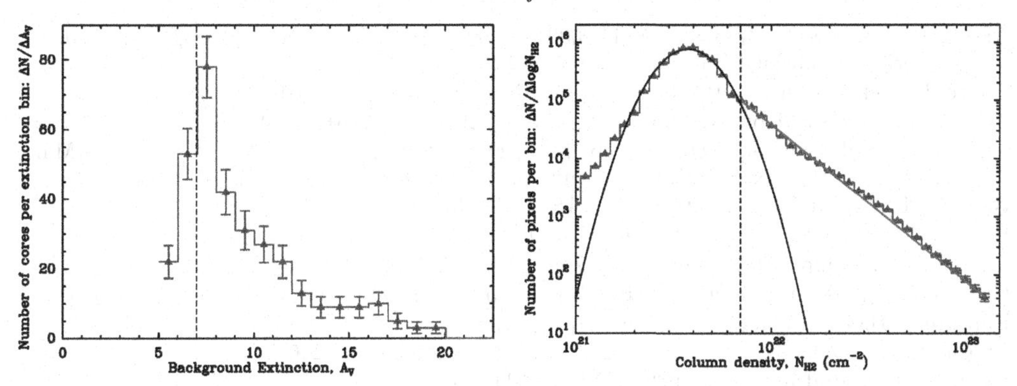

Figure 5. Left: Distribution of background column densities (converted to equivalent extinctions) for the best 341 candidate prestellar cores identified with *Herschel* in the Aquila Rift complex. **Right:** Probability density function of column density in the Aquila complex, based on the column density images derived from our *Herschel* data (cf. Könyves *et al.* 2010). A log-normal fit at low column densities and a power-law fit at high column densities are superimposed. In both panels, the vertical dashed line marks the extinction threshold $A_V^{\rm back} = 7$ (see § 3).

line mass of a filament depends only on gas temperature and is modified by only a factor of order unity for filaments with realistic levels of magnetization (cf. Fiege & Pudritz 2000). In contrast, in the non-star-forming, translucent Polaris cloud (where only unbound starless cores are found but no prestellar cores nor protostars), all of the filaments have subcritical masses per unit length (cf. Fig. 4–left), which is consistent with the view that they are gravitationally stable, hence neither collapsing nor forming stars.

The above scenario provides an *explanation* of the visual extinction threshold discussed in § 3 above (cf. Fig. 5). Given the typical width $\sim$ 10000–15000 AU measured for the filaments, the extinction threshold at $A_V^{\rm back} \sim 7$ corresponds to within a factor of 2 to the critical mass per unit length $M_{\rm line,crit} \sim 15\, M_\odot/{\rm pc}$ at $T \sim 10$ K. Thus, the extinction threshold approximately corresponds to the threshold above which the filaments are gravitationally unstable. Prestellar cores are only observed above this threshold because they form out of a filamentary background and only the supercritical, gravitationally unstable filaments are able to fragment into bound cores.

Interestingly, the median spacing between starless cores in Aquila is $\sim$ 16000 AU which roughly matches the thermal Jeans length at $T \sim 10$ K for a background column density corresponding to the $A_V^{\rm back} \sim 7$ threshold. This is consistent with the idea that gravitational fragmentation is the dominant physical mechanism generating prestellar cores within the filaments. Naively, one would expect gravitational fragmentation to result in a narrow prestellar CMF sharply peaked at the median thermal Jeans mass. However, a broad CMF resembling the IMF (e.g. Fig. 3) can be produced if turbulence has generated a field of initial density fluctuations in the filaments (cf. Inutsuka 2001).

To conclude, our *Herschel* first results support the view that the form of the stellar IMF is largely inherited from the form of the prestellar CMF. They also reveal a tight connection between the large-scale ($>$ 1 pc) filamentary structure of the parent clouds and the formation process of prestellar cores on small ($<$ 0.1 pc) scales in the densest, gravitationally unstable parts of the filaments.

References

Alves, J. F., Lombardi, M., & Lada, C. J. 2007, *A&A*, 462, L17

André, Ph. , & Saraceno, P. 2005, in *The Dusty and Molecular Universe: A Prelude to Herschel and ALMA*, ESA SP-577, p. 179

André, P., Basu, S., & Inutsuka, S.-I. 2009, in *Structure Formation in Astrophysics*, Ed. G. Chabrier, Cambridge University Press, p. 254
André, P., Belloche, A., Motte, F., & Peretto, N. 2007, *A&A*, 472, 519
André, Ph., Men'shchikov, A., Bontemps, S. *et al.* 2010, *A&A*, 518, L102
André, P., Ward-Thompson, D., Barsony, M. 2000, in *Protostars and Planets IV*, Eds. V. Mannings, A. P. Boss, & S. S. Russell (Univ. of Arizona Press, Tucson), p. 59
Bate, M. R. & Bonnell, I. A. 2005, *MNRAS*, 356, 1201
Bate, M. R., Bonnell, I. A., & Bromm, V. 2003, *MNRAS*, 339, 577
Bontemps, S., André, Ph., Könyves, V. *et al.* 2010, *A&A*, 518, L85
Chabrier, G. 2005, in *The Initial Mass Function 50 years later*, Eds. E. Corbelli *et al.*, p.41
Elmegreen, B.G. & Falgarone, E. 1996, *ApJ*, 471, 816
Enoch, M. L., Evans, N. J., Sargent, A. I. *et al.* 2008, *ApJ*, 684, 1240
Fiege, J. D., & Pudritz, R. E. 2000, *MNRAS*, 311, 85
Goodwin, S. P., Nutter, D., Kroupa, P., Ward-Thompson, D., Whitworth, A. P. 2008, *A&A*, 477, 823
Guillout, P. 2001, in *From Darkness to Light*, Eds. T. Montmerle & P. André, ASP Conf. Ser., 243, p. 677
Gutermuth, R. A., Bourke, T. L., Allen, L. E. *et al.* 2008, *ApJ*, 673, L151
Heithausen, A. *et al.* 2002, *ApJ*, 383, 591
Hennebelle, P. & Chabrier, G. 2008, *ApJ*, 684, 395
Inutsuka, S.-I. 2001, *ApJ*, 559, L149
Inutsuka, S.-I. & Miyama, S.M. 1997, *ApJ*, 480, 681
Johnstone, D., Wilson, C. D., Moriarty-Schieven, G., *et al.* 2000, *ApJ*, 545, 327
Johnstone, D., Fich, M., Mitchell, G. F., Moriarty-Schieven, G. 2001, *ApJ*, 559, 307
Johnstone, D., Di Francesco, J., & Kirk, H. 2004, *ApJ*, 611, L45
Kainulainen, J., Beuther, H., Henning, T., & Plume, R. 2009,*A&A*, 508, L35
Klessen, R. S., & Burkert, A. 2000, *ApJS*, 128, 287
Könyves, V., André, Ph., Men'shchikov, A. *et al.* 2010, *A&A*, 518, L106
Kramer, C., Stutzki, J., Rohrig, R., Corneliussen, U. 1998, *A&A*, 329, 249
Kroupa, P. 2001, *MNRAS*, 322, 231
Lada, C. J., Muench, A. A., Rathborne, J. M., Alves, J., & Lombardi, M. 2008, *ApJ*, 672, 410
Larson, R. B., 1981, *MNRAS*, 194, 809
Larson, R. B. 1985, *MNRAS*, 214, 379
Men'shchikov, A., André, Ph., Didelon, P. *et al.* 2010, *A&A*, 518, L103
Miville-Deschênes, M.-A., Martin, P. G., Abergel, A. *et al.* 2010, *A&A*, 518, L104
Molinari, S., Swinyard, B., Bally, J. *et al.* 2010, *A&A*, 518, L100
Motte, F., André, P., Neri, R. 1998, *A&A*, 336, 150
Motte, F., André, P., Ward-Thompson, D., & Bontemps, S. 2001, *A&A*, 372, L41
Myers, P. C. 2009, *ApJ*, 700, 1609
Nutter, D. & Ward-Thompson, D. 2007, *MNRAS*, 374, 1413
Onishi, T., Mizuno, A., Kawamura, A., Ogawa, H., Fukui, Y. 1998, *ApJ*, 502, 296
Onishi, T., Mizuno, A., Kawamura, A., Tachihara, K. & Fukui, Y. 2002, *ApJ*, 575, 950
Ostriker, J. 1964, *ApJ*, 140, 1056
Padoan, P. & Nordlund, A. 2002, *ApJ*, 576, 870
Padoan, P., Juvela, M., Goodman, A. A., & Nordlund, A. 2001, *ApJ*, 553, 227
Pilbratt, G.L., Riedinger, J.R., Passvogel, T. *et al.* 2010, *A&A*, 518, L1
Stamatellos, D., Whitworth, A. P., & Ward-Thompson, D. 2007, *MNRAS*, 379, 1390
Stanke, T, Smith, M. D., Gredel, R., & Khanzadyan, T. 2006, *A&A*, 447, 609
Starck, J. L., Donoho, D. L., Candès, E. J. 2003, *A&A*, 398, 785
Testi, L., Sargent, A. I. 1998, *ApJ*, 508, L91
Ward-Thompson, D., André, Ph., Crutcher, R., Johnstone, D., Onishi, T., & Wilson, C. 2007, in *Protostars & Planets V*, Eds. B. Reipurth *et al.*, Univ. of Arizona Press, p. 33
Ward-Thompson, D., Kirk, J.M., André, P. *et al.* 2010, *A&A*, 518, L92

Computational Star Formation
Proceedings IAU Symposium No. 270, 2011
J. Alves, B.G. Elmegreen, J. M. Girart & V. Trimble, eds.

doi:10.1017/S1743921311000482

Radiation Driven Implosion and Triggered Star Formation

T. G. Bisbas[1], A. P. Whitworth[2], R. Wünsch[1], D. A. Hubber[3], and S. Walch[2]

[1]Astronomical Institute, Academy of Sciences of the Czech Republic, Boční II 1401, 141 31 Prague, Czech Republic. Email: t.bisbas@astro.cf.ac.uk

[2]School of Physics and Astronomy, Cardiff University, Queens Buildings, The Parade, Cardiff, CF24 3AA, United Kingdom

[3]Department of Physics and Astronomy, University of Sheffield, Hicks Building, Hounsfield Road, Sheffield S3 7RH, United Kingdom

Abstract. We present simulations of stable isothermal clouds exposed to ionizing radiation from a discrete external source, and identify the conditions that lead to Radiatively Driven Implosion and Star Formation. We use the Smoothed Particle Hydrodynamics code SEREN (Hubber *et al.* 2010) and the HEALPix-based photoionization algorithm described in Bisbas *et al.* (2009). We find that the incident ionizing flux is the critical parameter determining the evolution; high fluxes disperse the cloud, whereas low fluxes trigger star formation. We find a clear connection between the intensity of the incident flux and the parameters of star formation.

Keywords. hydrodynamics, methods: numerical, stars: formation, (ISM:) HII regions

---

1. Introduction

When an expanding HII region overruns a pre-existing cloud, it compresses it by driving an ionization front and a shock wave into it (Sandford *et al.* 1982; Bertoldi 1989; Lefloch & Lazareff 1994). The inner parts may become gravitationally unstable and collapse to form new stars. This mechanism is known as Radiation Driven Implosion (RDI). Observations (Lefloch & Lazareff 1995; Lefloch *et al.* 1997; Sugitani *et al.* 1999, 2000; Ikeda *et al.* 2008; Morgan *et al.* 2008; Chahuan *et al.* 2009) strongly support a connection between the RDI mechanism and the formation of Young Stellar Objects (YSO). Simulations of the interaction of ultraviolet ionizing radiation with self-gravitating clouds have been presented by various authors (Kessel-Deynet & Burkert 2003; Esquivel & Raga 2007; Gritschneder *et al.* 2009; Miao *et al.* 2009). However, no model can explain *where* star formation takes place (in the core or at the periphery) or *when* (during the maximum compression phase or earlier – Deharveng *et al.* 2005). The aim of this work is to answer questions of whether the ionizing radiation incident upon stable clouds is able to trigger the formation of new stars or not, and how the process and the properties of this star formation are connected with the intensity of the incident flux. In Section 2 we give a brief description of the numerical treatment and the initial conditions we use. In Section 3 we discuss the results of our simulations. We summarize in Section 4.

2. Numerical Treatment and Initial Conditions

We use the Smoothed Particle Hydrodynamics (SPH) code SEREN†, fully described in Hubber *et al.* (2010), with an ionization routine (Bisbas *et al.* 2009) based on the

† http://www.astro.group.shef.ac.uk/seren

HEALPix‡ sphere tesselation code (Górski *et al.* 2005). We use a barotropic equation of state (i.e. Bonnell 1994) to set the temperature of the neutral gas as $T_{\rm N}(\rho) = T_{\rm ISO}\left\{1+(\rho/\rho_{\rm CRIT})^{\gamma-1}\right\}$, where $T_{\rm ISO} = 10\,{\rm K}$, $\rho_{\rm CRIT} = 10^{-13}\,{\rm g\,cm^{-3}}$ and $\gamma = 5/3$ is the ratio of specific heats. The temperature of the ionized gas is set to $T_{\rm i} = 10^4\,{\rm K}$, except in the transition zone between the two extremes, where it changes smoothly from $T_{\rm i}$ to $T_{\rm N}$ (see Bisbas *et al.* 2009). We include sink particles (Bate *et al.* 1995) with radii $R_{\rm SINK} = 2.5\,{\rm AU}$ created if $\rho > \rho_{\rm SINK} = 10^{-11}\,{\rm g\,cm^{-3}}$.

Our clouds are stable Bonnor-Ebert spheres (heareafter 'BES') with dimensionless cutoff radii $\xi_{\rm B} = 4, 5, 6$ (see Bonnor 1956 for its definition) and with masses $M = 2, 5, 10\,{\rm M}_\odot$. The particle resolution we use is 5×10^4 SPH particles per solar mass (cf. Hubber *et al.* 2006). We use a single source emitting Lyman-α photons. We place the BESs at distance $D = 10R$ from the ionizing source, where R is the radius of the cloud (in pc), in order to keep constant the divergence of the incident flux and as parallel as possible. We run simulations with a wide range of emission rates $\dot{\mathcal{N}}_{\rm LyC} = 10^x\,{\rm s^{-1}}$, where $x = 48, 48.5, ...52$.

3. Results

In Fig.1a we present a semi-logarithmic diagram where we correlate the intensity of the incident ionizing flux with the initial mass of each BES. The lines define subsets of parameter space where models either show star formation (left) or not (right), with accuracy 0.25 dex. It can be seen that as the mass of the BES decreases (and as a result $\xi_{\rm B}$ increases) the clouds appear to dissolve in higher fluxes. This is because for a given $\xi_{\rm B}$, the density $\rho_{\rm c}$ at the centre of each BES increases with decreasing M.

We also find that the Strømgren radius at the end of the *R*-type expansion determines whether stars are formed or not; if the ionization front has not overrun the central core of the BES, then the incident flux will trigger star formation during the *D*-type expansion of the HII region. In the opposite case there is not enough material to undergo gravitational collapse and form stars.

The time, $t_{\rm SINK}$, between the beginning of the *D*-type expansion and the first sink creation (beginning of star formation) increases with decreasing ionizing flux. This finding is in agreement with simulations of the RDI performed by Gritschneder *et al.* (2009). Figure 1b is a logarithmic diagram where we plot the values of $t_{\rm SINK}$ versus the incident flux Φ_D. Remarkably, it can be seen that $t_{\rm SINK}$ does not depend on $\xi_{\rm B}$. Results from our simulations can be described with a power law of the form $t_{\rm SINK} = 80 \times \Phi_{\rm D}^{-0.3}$ ($t_{\rm SINK}$ in Myr, $\Phi_{\rm D}$ in ${\rm cm^{-2}s^{-1}}$).

Figure 2 shows column density plots of a BES with $M = 10{\rm M}_\odot$ and with $\xi_{\rm B} = 6$ at $t_{\rm SINK}$ for different fluxes. A common feature in all our simulations is that stars form close to the symmetry axis joining the centre of the cloud to the exciting star. This is probably a consequence of the initial spherical symmetry of the cloud and it is in an agreement with observations by Sugitani *et al.* (1999). The distance $d_{\rm t}$ between the first sink particle and the ionization front is a function of the ionizing flux and the BES parameters (see Fig.1c where we plot $d_{\rm t}/2R$ for all BESs with $\xi_{\rm B} = 6$). We find that for low fluxes stars tend to form away from the periphery, whereas for high fluxes stars tend to form at the periphery of the cloud. Similar results are found also with $\xi_{\rm B} = 4$ and $\xi_{\rm B} = 5$.

Figure 2 shows that the lateral compression, $w_{\rm d}$, of the BESs at the beginning of star formation is connected to the intensity of the incident flux. We see that for low fluxes, $w_{\rm d}$ is quite high and the cloud has a **U**-shape structure, whereas for high fluxes $w_{\rm d}$ is small and the cloud has a V-shape structure. In Fig.1d we plot $w_{\rm d}/2R$ for all BESs with $\xi_{\rm B} = 6$

‡ http://healpix.jpl.nasa.gov

and we find that stars tend to form during maximum compression once the incident flux is increased. Similar results are found also for the rest of the clumps.

4. Conclusions

We present simulations of RDI in stable clouds represented by BESs. We performed 75 simulations with clouds of different masses, different dimensionless radii, and with a wide range of incident fluxes. In general we find a connection between the incident ionizing flux and whether the cloud is induced to form stars or not by the flux. Our results only apply to clouds which have similar density structures.

We introduce a semi-logarithmic diagram (flux-mass diagram) where we correlate the intensity of the incident flux and the initial mass of each BES, and we define zones of Star Formation and no-Star Formation. We find that if the Strømgren radius has not overrun the central core of the BES by the end of the R-type expansion, the ionizing radiation will trigger star formation. The time when star formation occurs increases with decreasing flux, and does not depend on $\xi_{\rm B}$. A power-law of the form $t_{\rm SINK} = 80 \times \Phi_{\rm D}^{-0.3}$ fits very well with the results of our models. Finally, as the incident flux increases, stars tend to form closer to the periphery of the cloud and during its maximum compression phase.

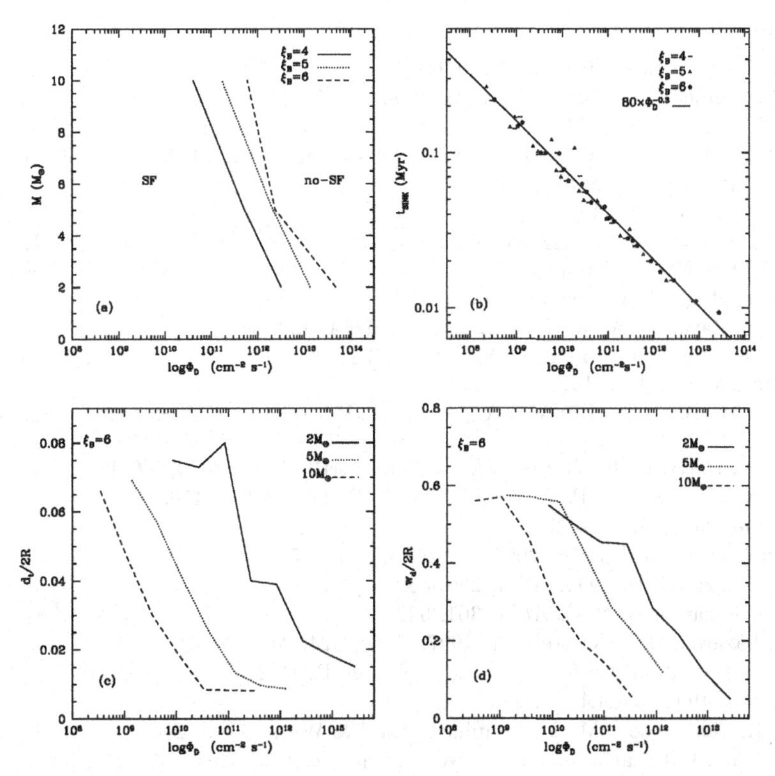

Figure 1. (a) The flux-mass semi-logarithmic diagram where we define areas where stars are formed (SF) and areas where stars are not formed (no-SF) depending on the dimensionless radius $\xi_{\rm B}$ of a BES. (b) Logarithmic diagram of the incident flux versus $t_{\rm SINK}$. The power law we propose (solid line) fits very well with our simulations ($t_{\rm SINK}$ is in Myr and $\Phi_{\rm D}$ is in cm^{-2} s^{-1}). (c) Star formation occurs at the periphery with increasing flux. (d) Star formation occurs during maximum compression with increasing flux.

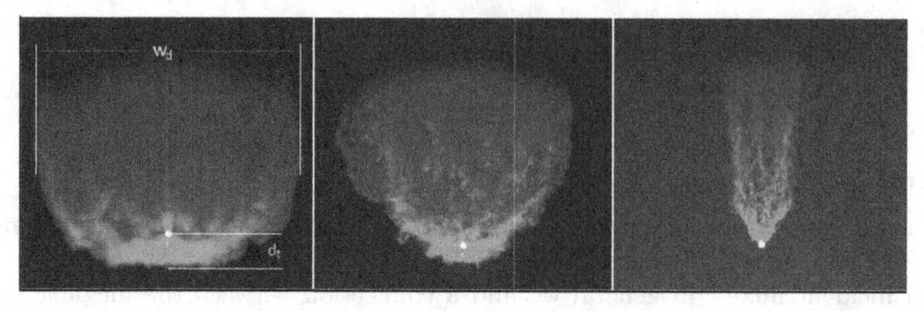

Figure 2. Column density plots of a BES with $M = 10\,M_{\odot}$ and $\xi_B = 6$ at t_{SINK} when it is exposed to three different intensities of flux (flux increases from left to right). The white dots are sink particles. In the left plot we draw the values of d_t and w_d.

Acknowledgments: TGB and RW acknowledge support from the project LC06014-Centre for Theoretical Astrophysics of the Ministry of Education, Youth and Sports of the Czech Republic. APW and SW gratefully acknowledge the support of the Marie Curie Research Training Network CONSTELLATION (Ref. MRTN-CT-2006-035890). DAH is funded by a Leverhulme Trust Research Project Grand (F/00 118/BJ). The computations in this work were carried out on Merlin Supercomputer of Cardiff University. The column density plots were made using the SPLASH visualization code (Price 2007). The authors acknowledge the anonymous referee for the useful comments.

References

Barnes, J. & Hut, P. 1986, *Nature*, 324, 446

Bate, M. R., Bonnell, I. A., & Price, N. M. 1995, *MNRAS*, 277, 362

Bertoldi, F. 1989, *ApJ*, 346, 735

Bisbas, T. G., Wünsch, R., Whitworth, A. P., & Hubber, D. A. 2009, *A&A*, 497, 649

Bonnell, I. A. 1994, *MNRAS*, 269, 837

Bonnor, W. B. 1956, *MNRAS*, 116, 351

Chauhan, N., Pandey, A. K., Ogura, K., Ojha, D. K., Bhatt, B. C., Ghosh, S. K., & Rawat, P. S. 2009, *MNRAS*, 396, 964

Esquivel, A. & Raga, A. C. 2007, *MNRAS*, 377, 383

Deharveng, L., Zavagno, A., & Caplan, J. 2005, *A&A*, 433, 565

Górski, K. M., Hivon, E., Banday, A. J., Wandelt, B. D., Hansen, F. K., Reinecke, M., & Bartelmann, M. 2005, *ApJ*, 622, 759

Gritschneder, M., Naab, T., Burkert, A., Walch, S., Heitsch, F., & Wetzstein, M. 2009, *MNRAS*, 393, 21

Hubber, D. A., Batty, C. P., McLeod, A., & Whitworth, A. P., 2010, *A&A*, submitted

Hubber, D. A., Goodwin, S. P., & Whitworth, A. P. 2006, *A&A*, 450, 881

Ikeda, H., *et al.* 2008, *AJ*, 135, 2323

Kessel-Deynet, O. & Burkert, A. 2003, *MNRAS*, 338, 545

Lefloch, B. & Lazareff, B. 1994, *A&A*, 289, 559

Lefloch, B. & Lazareff, B. 1995, *A&A*, 301, 522

Lefloch, B., Lazareff, B., & Castets, A. 1997, *A&A*, 324, 249

Miao, J., White, G. J., Thompson, M. A., & Nelson, R. P. 2009, *ApJ*, 692, 382

Monaghan, J. J. 1992, *ARAA*, 30, 543

Morgan, L. K., Thompson, M. A., Urquhart, J. S., & White, G. J. 2008, *A&A*, 477, 557

Price, D. J. 2007, Publications of the Astronomical Society of Australia, 24, 159

Sandford, M. T., II, Whitaker, R. W., & Klein, R. I. 1982, *ApJ*, 260, 183

Sugitani, K., Tamura, M., & Ogura, K. 1999, Star Formation 1999, Proceedings of Star Formation 1999, held in Nagoya, Japan, June 21 - 25, 1999, Editor: T. Nakamoto, Nobeyama Radio Observatory, p. 358-364, 358

Sugitani, K., Matsuo, H., Nakano, M., Tamura, M., & Ogura, K. 2000, *AJ*, 119, 323

Computational Star Formation
Proceedings IAU Symposium No. 270, 2011
J. Alves, B.G. Elmegreen, J. M. Girart & V. Trimble, eds.

doi:10.1017/S1743921311000494

Action of Winds Inside and Outside of Star Clusters

Jan Palouš[1], Jim Dale[1], Richard Wünsch[1], and Sergiy Silich[2] & Guillermo Tenorio-Tagle[2] and Anthony Whitworth[3]

[1]Astronomical Institute, ASCR
BočníII 1401, CZ-141 31, Prague 4, Czech Republic
email: palous@ig.cas.cz

[2]Instituto Nacional de Astrofísica Optica y Electrónica,
AP 51, 72000 Puebla, Mexico
email: silich@inaoep.mx

[3]School of Physics and Astronomy, Cardiff University, Queens Building, The Parade, Cardiff, CF24 3AA, UK
email: ant@ac.uk

Abstract. The feedback form pre-main sequence and young stars influences their vicinity. The stars are formed in clusters, which implies that the winds of individual stars collide with each other. Inside of a star cluster, winds thermalize a fraction of their kinetic energy, forming a very hot medium able to escape from the cluster in the form of a large-scale wind. Outside of the cluster, the cluster wind forms a shock front as it interacts with the ambient medium which is accreted onto the expanding shell. A variety of instabilities may develop in such shells, and in some cases they fragment, triggering second generation of star formation. However, if the cluster surpasses a certain mass (depending on the radius and other parameters) the hot medium starts to be thermally unstable even inside of the cluster, forming dense warm clumps. The formation of next generations of stars may start if the clumps are big enough to self-shield against stellar radiation creating cold dense cores.

Keywords. stars: formation, ISM: structure, globular clusters: general, open clusters and associations: general

1. Introduction

The star formation in galaxies transforms ISM to stars, which at the same time return a fraction of their mass to the ISM. The gas recycling redistributes the yield of stellar evolution causing the chemical evolution of galaxies. Star formation is an important process of internal galactic evolution, however, in some cases, it may be triggered by galaxy external agents.

Stars are formed in groups in cold, dense cores of molecular clouds. The feedback of young and massive stars influences their vicinity by radiation, stellar winds and by supernova explosions. In this contribution we shall separately discuss the actions of winds, and of other contributors to the feedback, outside and inside of stellar clusters. The feedback in star forming regions drives turbulence, and it may trigger the next generation of star formation in the compressed shells of the ISM outside of star clusters, or it may initiate star formation in warm clumps inside massive and sufficiently compact young stellar clusters.

2. Outside of a star cluster

The combined feedback of many stars in clusters forms expanding dissociation, ionization and shock fronts creating bubbles and expanding shells in the ISM. The shells are observed in HI in the Milky Way and many nearby galaxies. Radio surveys of the Northern (Hartmann & Burton, 1997) and Southern (McClure-Griffiths *et al.*, 2002) skies discovered hundreds of shells and shell-like structures on the size scales 10 pc - 1 kpc. Ehlerová & Palouš (2005) have used an automated search algorithm to find more than 600 shells in the outer Milky Way. Observation of HI shells in external galaxies are summarized by Walter & Brinks (1999), many more discoveries are expected in THINGS: the HI nearby galaxy survey (Walter *et al.*, 2008). In the infrared, Churchwell *et al.* (2006) and Churchwell *et al.* (2007) found many 0.1 - 10 pc size shells in the GLIMPSE survey covering only a part of the plane of the Milky Way. Deharveng *et al.* (2003), Zavagno *et al.* (2006), Sidorin (2008), and Deharveng *et al.* (2009) identified shells observationally in optical, infrared, radio and other wavebands, some of them appear to have triggered star formation at their borders. Watson *et al.* (2009) analyzed some of the shells from the GLIMPSE survey to determine properties of YSOs related to them.

The observed shells seem to be created by the feedback from young and massive stars in clusters, following the "collect and collapse" scenario proposed by Elmegreen & Lada (1977). It assumes that the massive stars of a star cluster supply energy into the interstellar medium as stellar winds and radiation creating an expanding bubble filled with a high temperature gas. The bubble expands due to its internal overpressure, sweeping the ambient medium into a dense shell. When the shell cools down it becomes gravitationally unstable forming fragments that are seeds of a new stellar generation.

Gravitational fragmentation of expanding shells has been studied under the assumption that their thickness is infinitesimally small. Vishniac (1983) derived the dispersion relation using a decomposition into spherical harmonics. With the linear analysis of hydrodynamical and Poisson equations in 2 dimensions Elmegreen (1994) and Whitworth *et al.* (1994) derived very similar dispersion relations composed of two stretching terms depending on the general expansion velocity of the shell V_{sh} and on its radius R_{sh} plus a term due to shell internal pressure depending on the speed of sound inside of the shell c_{sh}, which are opposed by the gravity term depending on the shell surface density Σ_{sh}.

$$\omega(l) = -\frac{AV_{sh}}{R_{sh}} + \sqrt{\frac{BV_{sh}^2}{R_{sh}^2} - \frac{c_{sh}^2 l^2}{R_{sh}^2} + \frac{2\pi G \Sigma_{sh} l}{R_{sh}}}, \tag{2.1}$$

where G is the gravitational constant and $(A, B) = (\frac{3}{2}, \frac{1}{4})$ or $(3, 1)$ for non-accreting or accreting shells, l is the wave number and ω gives the growth rate of a perturbation. From this dispersion relation it is visible that initially, when the radius of the shell is small and its expansion velocity is large, the first term dominates and the shell is stable; we may say that it is stabilized by expansion. Later, when the shell radius becomes large and its expansion is decelerated, the first two terms are negligible. The time t_b, when it becomes unstable for the first time can be derived provided the functions $R_{sh}(t), V_{sh}(t)$ and $\Sigma_{sh}(t)$ are known. Then, there is a certain range of wave numbers given by the dispersion relation, that are unstable, and we may ask how quickly they grow, and try to construct the mass spectrum of fragments.

A shell expanding due to internal overpressure into a higher density medium is subjected to instabilities: initially, when the expansion velocity is large, the expansion driven by high internal thermal pressure is opposed by ram pressure decelerating the shell. However, the thermal pressure acts isotropically and the ram pressure is always directed against expansion, which creates, in the case of a small deviation from sphericity, a pressure component parallel to the shell surface leading to growth of perturbations, which is described as the Vishniac instability (Vishniac, 1983). Another instability appears when a low-density gas accelerates the high-density gas. This Rayleigh-Taylor instability, which forms long 'fingers' of high-density fluid extending to the low-density medium, is particularly important for large density differences. Later, when Vishniac and Rayleigh-Taylor instabilities combine with the Jeans gravitational instability, the shell breaks into fragments.

In order to see the effect of individual instabilities, in a series of papers (Dale *et al.*, 2009; Wünsch *et al.*, 2010; and Dale *et al.*, 2010) we try to isolate the gravitational instability. Therefore, we analyze the shells expanding into a medium of very low density and set the internal pressure equal to the external one. The lack of effective gravitational/inertial force (the shell evolves ballistically) eliminates the Rayleigh-Taylor instability. At the same time, the negligible ram pressure eliminates the Vishniac instability. We explore with 3 dimensional hydrodynamical simulations the gravitational instability during the ballistic motion of expanding shells.

The value of external and internal pressures is essential, since it influences the shell thickness: at low pressures the shell thickness increases during its ballistic motion, which makes the short wavelengths more stable, compared to dispersion relation 2.1, the low mass fragments do not form in this case. On the other hand, at high pressures, when the shell thickness decreases, even shorter wavelengths become unstable, compared to dispersion relation 2.1, producing an excess of low mass fragments. We also give the thick shell dispersion relation and propose a new Pressure Assisted Gravitational Instability (PAGI), which involves the external pressures and explains the above described behavior. Finally, we follow the evolution of individual fragments: their mass distribution corresponds to the PAGI dispersion relation only at the beginning of fragmentation during its linear phase. Later, they accumulate the remaining mass of the shell in a strongly nonlinear process, which we call oligarchic accretion: the old fragments that were created initially accrete most of the mass and they leave only a small portion for others that were formed later.

As a next step, we shall also include the mass accumulation during its expansion, thus we shall be able to study other instabilities and explore the PAGI dispersion relation and oligarchic accretion in more general situations.

3. Inside of a star cluster

The young and massive stars in clusters deposite energy and mass into the intracluster medium. It happens in the form of radiation, winds and supernovae composing the stellar feedback in star clusters. They range from open clusters (OCl) of $10^4 M_\odot$ to globular clusters (GC) and super star clusters (SSC) of $10^6 M_\odot$. We also consider the compact dwarf galaxies (CDG), which may be the most massive compact stellar groups, of mass up to $10^8 M_\odot$. The energy deposited by individual young stars thermalizes in wind - wind collisions causing a large overpressure inside of the cluster driving the large-scale star cluster wind. In the adiabatic solution of Chevalier & Clegg (1985) the overpressure results in the wind blowing from the center to the periphery of the cluster where it reaches the sound speed c_s. Outside of the cluster, the adiabatic wind accelerates up to a velocity of $2c_s$.

In a series of papers Tenorio-Tagle *et al.* (2007), Wunsch *et al.* (2007), Wunsch *et al.* (2008), Tenorio-Tagle *et al.* (2010), Silich *et al.* (2010), and Hueyotl-Zahuantitla *et al.* (2010) we describe a model where the stellar wind's mechanical energy is thermalized

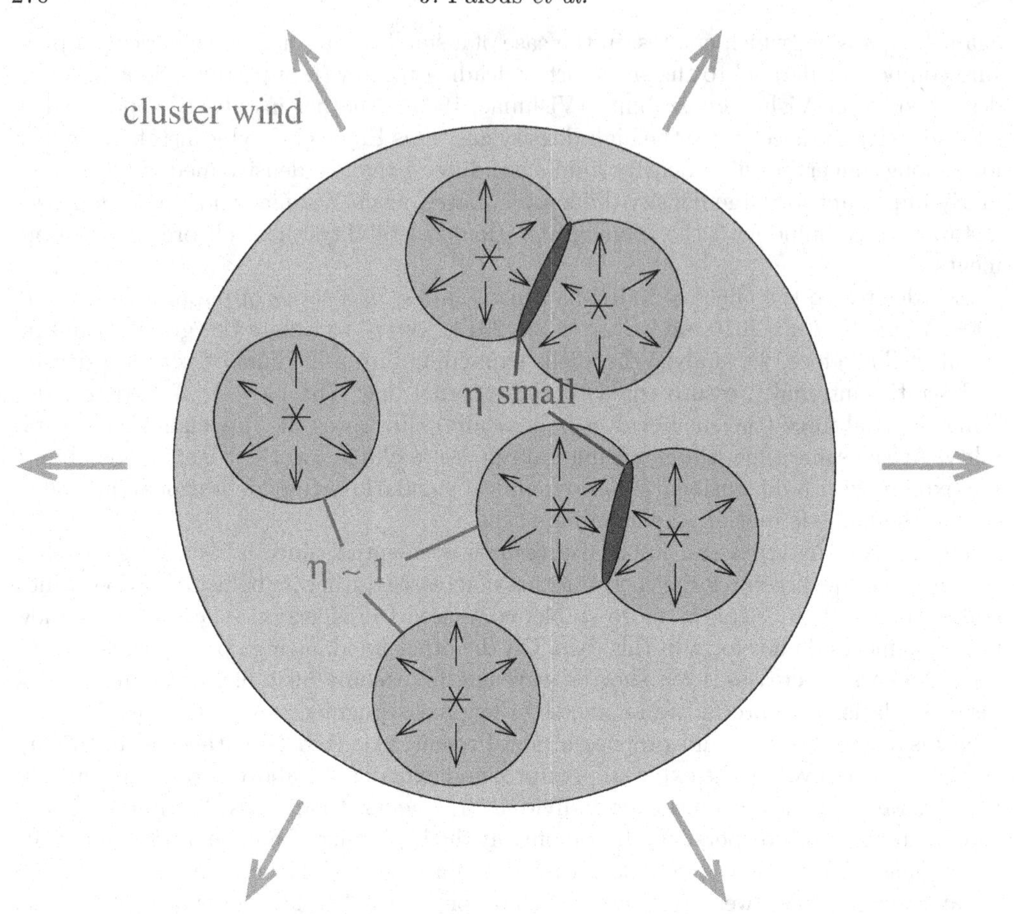

Figure 1. A model of the star cluster. Stars are homogeneously distributed in the cluster volume. The individual stellar winds collide and thermalize a fraction η of their mechanical energy in shocks.

with an efficiency η. The value of η is unknown; however, we try to estimate it from observations and also from numerical simulations. A schematic view of the modelled cluster is shown in Fig. 1.

In our model we give the solution of the hydrodynamical equations (3.1 - 3.3). In some cases the external gravitational force is included:

$$\frac{\partial \rho}{\partial t} + \nabla \cdot (\rho u) = q_m \tag{3.1}$$

$$\frac{\partial u}{\partial t} + (u \cdot \nabla) u + \nabla P / \rho = -\nabla \phi_{\mathrm{BH+NSB}} \tag{3.2}$$

$$\frac{\partial e}{\partial t} + \nabla \cdot (eu) + P \nabla u = q_e - Q \tag{3.3}$$

where q_m and q_e are the mass and energy deposition rates per unit of volume, ρ is the density, u is the velocity, and e is the internal energy of the medium, $Q = n_i n_e \Lambda(T, Z)$ is the cooling rate, n_i and n_e are the ion and electron number densities, and $\Lambda(T, Z)$ is the Raymond and Cox (Plewa, 1995) cooling function, which depends on the thermalized gas temperature, T, and metalicity, Z. The right hand side of the equation (3.2) represents

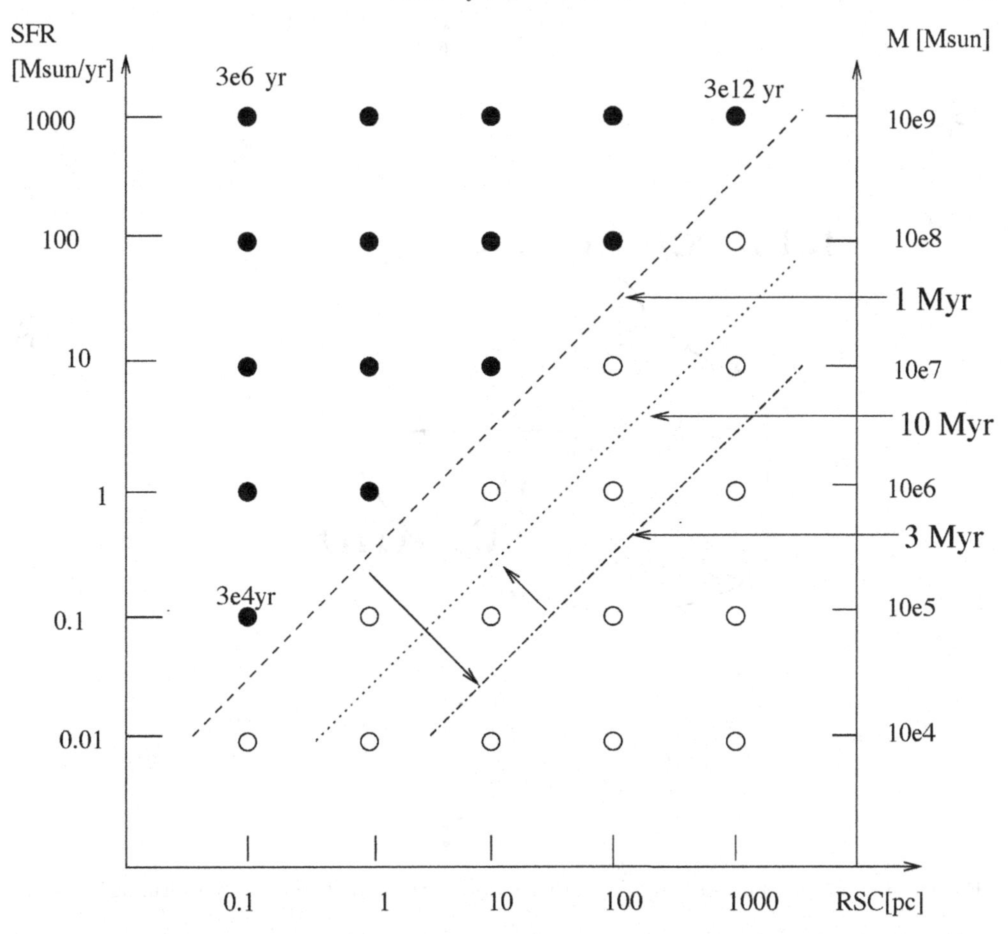

Figure 2. SFR versus R_{SC} plane for $\eta = 0.01$. The positions of clusters at the age of 1 Myr with the pure wind solutions (open circles) and bimodal solutions (filled circles) are given. The dashed line is the critical line dividing the pure wind area from the bimodal solution area. It changes its position with the age of the cluster as shown by arrows. The self-shielding time for a 10 Myr old cluster is given for bimodal solutions at the extreme positions in the SFR - R_{SC} plane.

the external gravitational force of the central black hole and of the gravitating mass of the starburst itself. This external gravitational term has been considered in some cases only.

The adiabatic solution of hydrodynamical equations by Chevalier & Clegg (1985) is valid when for certain cluster radii the mechanical wind energy is low enough. When we increase it above a certain critical value, the wind solution applies to the outer part of the cluster volume only. Inside of a stagnation radius R_{st}, the intracluster medium is thermally unstable, forming warm clumps. The position of the critical line in the cluster mass versus cluster radius plane is very sensitive to the value of η: small η moves the critical line to smaller masses.

In Fig. 2 we show how the position of the critical line, in the case of $\eta = 0.01$, changes its position during the first 10 Myr of the cluster life. We assume that a cluster is formed during the first 1 Myr, and that its final mass is given by the star formation rate (SFR).

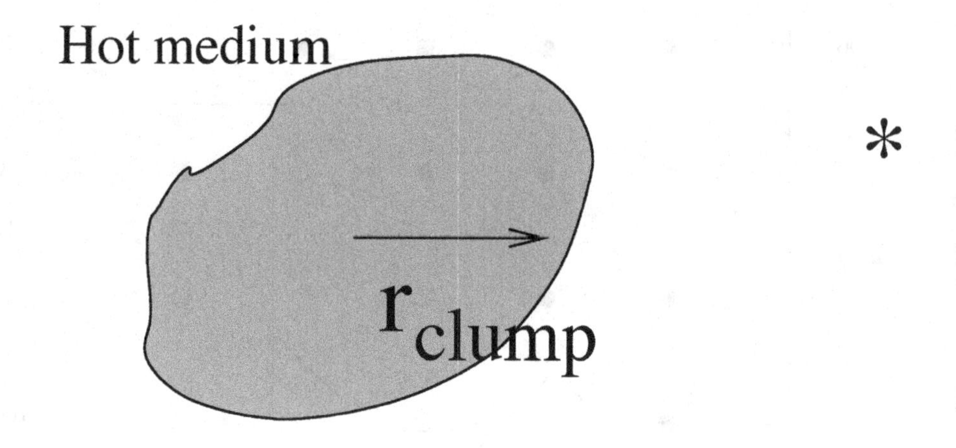

Figure 3. A warm clump inside of the thermally unstable central part of a forming cluster.

However, for larger values of η the critical line shifts to the left upper corner of the SFR versus R plane. When η is close to 1, almost all the clusters are described by the single mode wind solution.

We are interested if such clumps (see Fig. 3) may be Jeans unstable. We assume that a warm clump is in pressure equilibrium in the hot medium, $n_{clump} = \frac{P_{hot}}{kT}$, where $T = 10^4 K$ is the clump temperature. Thus the Jeans mass, $M_{Jeans} = 2.5 \times 10^5 M_{\odot} \left(\frac{n_{clump}}{10^3 cm^{-3}}\right)^{-3/2}$, depends sensitively on the pressure of the hot medium P_{hot}. However, P_{hot} is very sensitive to the thermalization efficiency η. Low values of η produce low values of P_{hot}, which implies large M_{Jeans}. We conclude that if the value of η is small, masses of warm clumps with the temperatures $10^4 K$ have, after 10 Myr of mass accumulation in the thermally unstable region of a young star cluster, masses much smaller than M_{Jeans}, and therefore they are gravitationally stable.

Another question is whether the thermally unstable clumps are able to self-shield against the UV photon field of the young star cluster. The time t_{SS} of the clump growth needed to build clumps able to self-shield may be estimated with the following formula

$$t_{SS} = \frac{3}{4}\mu m_H \left(\frac{kT}{P_{hot}}\right)^5 \left(\frac{3\dot{N}_{UV,SC}}{4\pi R_{SC}\alpha_*}\right)^3 \left(\frac{N_{OB}}{\dot{M}_{SC}}\right), \tag{3.4}$$

where $\alpha_* = 1.58 \times 10^{-13} cm^3 s^{-1}$, $\dot{N}_{UV,SC}$ is the number of UV photons produced by the star cluster, $\dot{M}_{SC}$ is the mass flux inserted by all the stars of the cluster and N_{OB} is the number of OB stars. We assume that N_{OB} also gives the number of warm clumps. A more detailed derivation of this will be given elsewhere (Palouš *et al.*(2010)).

The self-shielding times for 10 Myr old clusters are given in Fig. 2. t_{SS} is less than 1 Myr for small clusters with radius less that 1 pc. There, the clumps are able to self-shield, which may enable the decrease of temperature in their central parts, and trigger secondary star formation there. On the other hand, in star forming regions larger that 10 pc, t_{SS} is larger that 100 Myr, which probably eliminates the warm clumps as seeds of secondary star formation.

4. Conclusions

Outside of star clusters, the common action of the feedback of many stars forms shells, which may become gravitationally unstable, forming second generations of stars. We discuss the analytical and numerical models of expanding shells and analyze the mass spectra of fragments.

Inside of star clusters the bimodal wind solution applies above the critical line in the cluster mass versus radius plane. For small values of the thermalization efficiency, the warm clumps seem to be gravitationally stable, however, for clusters less that 1 pc in radius the self-shielding time is short. The cores of warm clumps may further decrease their temperature, becoming seeds of secondary star formation.

Acknowledgements

The authors thank the referee for careful reading of the paper. The authors gratefully acknowledge the support by the Institutional Research Plan AV0Z10030501 of the Academy of Sciences of the Czech Republic and by the project LC06014 Center for Theoretical Astrophysics of the Ministry of Education, Youth and Sports of the Czech Republic. This study has been supported by CONACYT-México research grant 60333 and 47534-F. JD acknowledges support from a Marie Curie fellowship as part of the European Commission FP6 Research Training Network 'Constellation' under contract MRTN-CT-2006-035890.

References

Chevalier, R. A. & Clegg, A. W. 1985, *Nature*, 317, 44

Dale, J., Wünsch, R., Whitworth, A., & Palouš, J. 2009, *MNRAS*, 398, 1537

Dale, J. Wünsch, R. Smith, R. J., Whitworth, A., & Palouš, J 2010, *MNRAS*, submitted

Deharveng, L., Lefloch, B., Zavagno, A., Caplan, J., Whitworth, A. P., Nadeau, D., & Martín, S. *A&A*, 408, L25

Deharveng, L., Zavagno, A., Schuller, F., Caplan, J., Pomarès, M., & De Breuck, C. 2009, *A&A*, 496, 177

Ehlerová, S., Palouš, J. 2005, *A&A*, 437, 101

Elmegreen, B. G. & Lada, C. J. 1977, *ApJ*, 214, 725

Elmegreen, B. G. 1994 *ApJ*, 427, 384

Hartmann, D. & Burton, W. B. 1997, *Atlas of Galactic Neutral Hydrogen*, Cambridge University Press

Hueyotl-Zahuantitla, F., Tenorio-Tagle, G., Wünsch, R., Silich, S., & Palouš, J. 2010, *ApJ*, 716, 324

McClure-Griffiths, N. M., Dickey, J. M., Gaensler, B.M., & Green, A., J. 2002, *ApJ*, 578, 176

Palouš, J., Wünsch, R., Silich, S., & Tenorio-Tagle, G. 2010 *in preparation*

Plewa, T. 1995, *MNRAS*, 275, 143

Sidorin, V. 2008 *IR, optical and X-ray counterparts of HI shells in the Milky Way*, Diploma Thesis, Charles University, Prague

Silich, S., Tenorio-Tagle, G., Muñoz-Tuñon, C., Hueyotl-Zahuantiutla, F., Wünsch, R., & Palouš, J. 2010, *ApJ*, 711, 25

Tenorio-Tagle, G., Wünsch, R., Silich, S., & Palouš, J. 2007, *ApJ*, 658, 1196

Tenorio-Tagle, G., Wünsch, R., Silich, S., Muñoz-Tuñon, C., & Palouš, J. 2010, *ApJ*, 708, 1621

Vishniac, E. T. 1983 *ApJ*, 274, 152

Walter, F. & Brinks, E. 1999, *AJ*, 118, 273

Walter, F., Brinks, E., de Blok, W. J. G., Bigiel, F., Kennicutt, R. C. Jr., Thornley, M. D., & Leroy, A. 2008, *AJ*, 136, 2563

Watson, C., Corn, T., Churchwell, E. B., Babler, B. L., Povich, M. S., Meade, M. R., & Whitney, B. A. 2009 *ApJ*, 694, 546

Whitworth, A.P., Bhattal, A. S., Chapman, S. J., Disney, M. J., & Turner, J. A. 1994 *MNRAS*, 268, 1341

Wünsch, R., Silich, S., Palouš, J., & Tenorio-Tagle, G. 2007, *A&A*, 471, 579

Wunsch, R., Tenorio-Tagle, G., Palouš, J., & Silich, S. 2008, *ApJ*, 683, 683

Wünsch, R., Dale, J., Palouš, J., & Whitworth, A. P. 2010, *MNRAS*, in press

Zavagno, A., Deharveng, L., Comerón, F., Brand, J., Massi, F., Caplan, J., & Russeil, D. 2008 *A&A*, 446, 171

Computational Star Formation
Proceedings IAU Symposium No. 270, 2011
J. Alves, B.G. Elmegreen, J.M. Girart, and V. Trimble, eds.

doi:10.1017/S1743921311000500

Theory of Feedback in Clusters and Molecular Cloud Turbulence

Enrique Vázquez-Semadeni

Centro de Radioastronomía y Astrofísica, UNAM, Campus Morelia
P.O. Box 3-72 (Xangari), Morelia, Michoacán, México
email: e.vazquez@.crya.unam.mx

Abstract. I review recent numerical and analytical work on the feedback from both low- and high-mass cluster stars into their gaseous environment. The main conclusions are that i) outflow driving appears capable of maintaing the turbulence in parsec-sized clumps and retarding their collapse from the free-fall rate, although there exist regions within molecular clouds, and even some examples of whole clouds, which are not actively forming stars, yet are just as turbulent, so that a more universal turbulence-driving mechanism is needed; ii) outflow-driven turbulence exhibits specific spectral features that can be tested observationally; iii) feedback plays an important role in reducing the SFR; iv) nevertheless, numerical simulations suggest feedback cannot completely prevent a net contracting motion of clouds and clumps. Therefore, an appealing source for driving the turbulence everywhere in GMCs is the accretion from the environment, at all scales. In this case, feedback's most important role may be to prevent a fraction of the gas nearest to newly formed stars from actually reaching them, thus reducing the SFE.

Keywords. ISM: clouds, methods: numerical, stars: winds, outflows; turbulence

1. Introduction

Stars form in dense cores within molecular clouds (MCs), and simultaneously they feed back on their environment through a variety of mechanisms, including ionizing radiation, winds, outflows, and supernova explosions. This feedback crucially affects the stars' environment, in particular its physical conditions, evolution, as well as the subsequent pattern of star formation (SF). In particular, it is believed that stellar feedback may be responsible for the observed non-thermal motions in clouds and clumps, which are generally interpreted as small-scale supersonic turbulence (Zuckerman & Evans 1974). The latter, in turn, is thought to provide support against the clouds' self-gravity, and to be able to maintain a quasi-virial equilibrium state in the clouds, thus preventing global collapse and maintaining a low global SFR. This notion is at the basis of several theoretical models of SF (e.g., McKee 1989; Matzner 2002; Krumholz & McKee 2005; Hennebelle & Chabrier 2008). In this contribution I review recent results on the role of feedback in driving the turbulence in MCs and their substructure, specifically on its coupling with the ambient gas, on the ability of the turbulence to support the clouds, and on its role in regulating the SFR and the SF efficiency (SFE). For a discussion on the role of feedback in the determination of the stellar masses, see the review by Bate in this volume.

2. Background

More than half a century ago, Oort (1954) suggested that an interstellar cycle should exist, in which the HII regions produced by the ionizing radiation of O stars would form a dense shell which, upon fragmenting, would produce cold cloudlets with a significant velocity dispersion. These would then grow by coalescence until they became gravitationally unstable, at which point they would form new stars, and the cycle would repeat. This *Oort model* was later formulated mathematically by Field & Saslaw (1965), who concluded that the SFR should be proportional to the square of the gas density, in agreement with the SF law known at the time (Schmidt 1959), and that the cloud mass spectrum should scale as $M^{-3/2}$, is in agreement with the observed mass spectrum of CO clouds (see, e.g., Blitz 1993 and references therein) and of low-mass clumps (Motte *et al.* 1998). These results exemplify the fundamental implications of feedback on their environment.

Another seminal study was that of Norman & Silk (1980, hereafter NS80), who considered the low-mass analogy of the Oort-Field-Saslaw model, with the driving energy provided in this case by the outflows produced by low-mass T-Tauri stars. In their model, the collision of expanding, wind-driven shells would form dense clumps that would then evolve under the competition of growth by coalescence, and destruction by leak, drag, or very energetic collisions.

Based on the Oort-NS80 scenario, McKee (1989) advanced a model in which MCs form from atomic cloudlets that grow by coalescence, and become molecular (in the sense that most of the carbon is in CO) when the extinction $A_V \sim 1$. He argued that at this point the clouds must be magnetically supercritical, and therefore begin to contract as a whole, although their substructures (clumps and cores) are still subcritical, and are thus magnetically supported. Nevertheless, the contraction speeds up ambipolar diffusion (AD) in the cores, rendering them supercritical and allowing them to collapse and form stars. The latter begin driving turbulence in the cloud, eventually halting the collapse, and allowing the cloud to reach a stable equilibrium. This occurs when the mean extinction has increased to $A_V \sim 4$–8. In this model, the gas depletion time for typical giant MCs (GMCs) was estimated to be ~ 2–4×10^8 yr, and the typical magnetic field strength $B \sim 20$–40μG.

Today, however, we know that clouds do not grow only by coalescence, but that instead a significant, and perhaps dominant, mechanism determining their mass is direct accretion from their warm, diffuse surrounding medium, allowing for much shorter growth timescales (Blitz & Shu 1980; Hennebelle & Pérault 1999; Ballesteros-Paredes *et al.* 1999a,b; Koyama & Inutsuka 2002; Audit & Hennebelle 2005; Hennebelle & Audit 2007; Vázquez-Semadeni *et al.* 2006; Banerjee *et al.* 2009), and that this mechanism can drive strong turbulence in the clouds. Thus, more recent research has focused on the details of the physics that regulates the energy transfer from the stellar sources, especially bipolar outflows from low-mass stars, to their environment, as well as to whether indeed stellar feedback is capable of feeding the clouds' turbulence and maintaining them in rough hydrostatic equilibrium. I now briefly summarize some of the main results in these respects.

3. Feedback from outflows

3.1. *Efficiency of coupling with the environment*

Estimates of the amount of energy injected to the environment by an outflow ($\sim 10^{47}$ erg per $M_\odot$; Shu *et al.* 1988) suggest that outflows may deposit enough kinetic energy in their parent clump as to maintain its turbulence (see Reipurth & Bally 2001 and references therein). However, an important, recurrent question is whether the bipolar outflows couple efficiently to their environment.

In this regard, Quillen *et al.* (2005) investigated the correlation of the molecular gas kinematics in NGC1333 with the distribution of young stellar objects (YSOs) within this cloud, finding that the velocity dispersion does not vary significantly across the cloud and is uncorrelated with the number of nearby young stellar outflows. However, they did find about 20 cavities in the velocity channel maps that they interpreted as remnants of past outflow activity. Those authors concluded that, while outflows may not directly drive the turbulence in the clumps, the cavities ("fossil outflows") may provide an efficient coupling mechanism to the environment. Later numerical work by Cunningham *et al.* (2006b) supported this conclusion.

The interaction of single, or a few, outflows with their environment has been extensively studied. De Colle & Raga (2005) performed magnetohydrodynamic (MHD) simulations of high-density clumps propagating through a high-density medium, in order to represent the interaction of a jet with the medium. They found that $\sim$ 10–30% of the jet's kinetic energy can be transferred to the medium, depending on the magnetic field's strength and orientation relative to the jet. Cunningham *et al.* (2006a) investigated the effect of collisions among outflows, finding that the collisions reduce the efficiency of transfer to the cloud, so that the most efficient drivers are isolated outflows. However, Banerjee *et al.* (2007) noted, using isothermal simulations, that outflows should be inefficient drivers of *supersonic* turbulent motions in their parent clouds because the compressions they produce can only re-expand sonically, although then Cunningham *et al.* (2009) argued that this problem can be circumvented if the cloud is previously turbulent (i.e, the turbulence is *maintained* rather than generated), and adequate care is taken of modeling the cooling.

It can be concluded from this section that, through the mediation of long-lived cavities, magnetic fields, and a pre-existent turbulent velocity field, outflows appear capable of at least maintaining the turbulent motions in a clump.

3.2. *Feedback and nature of the turbulence*

In real MCs, stars are almost always born in clusters, and so the effect of an *ensemble* of outflows on their parent clump is also a crucial issue. In particular, many studies have focused on the nature of the turbulent motions induced in the parent clump by an ensemble of outflows. We now turn to this issue.

Numerical simulations of evolving clumps at the parsec scale suggest that the initial ("interstellar") turbulence is quickly replaced by "outflow" turbulence, at least within the scales modeled by these simulations (Li & Nakamura 2006; Nakamura & Li 2007). This transition consists in a secular variation of the topology of the density and velocity fields in the clump. The density field develops a central concentration, with a power-law profile of the form $\rho \sim r^{-3/2}$, and the velocity field develops a circulation pattern, with gravity driving infall motions that balance the outward motions driven by the outflows. Moreover, the turbulent energy spectrum develops a knee at a characteristic "outflow scale" (Matzner 2007), and a slope steeper than that of isotropic random driving at scales smaller than this (Nakamura & Li 2007; Carroll *et al.* 2009, 2010).

A particularly interesting feature is that the presence of the outflows does not seem to impede the development of coherent streams of infalling gas towards the center of the gravitational potential well (Nakamura & Li 2007). This implies that the most massive stars forming in the clump are fed from a mass supply that extends out to the scale of the whole clump, instead of being restricted to the scale of the very dense core that contains the forming star (Wang *et al.* 2010). This result is in stark contrast with the currently popular notion that the masses of forming stars are determined by the masses of the cores in which they reside (e.g., Padoan & Nordlund 2002, Krumholz *et al.* 2005, Alves *et al.* 2007), and more in agreement with the scenario of competitive accretion for star formation (e.g., Bonnell *et al.* 1997; Bonnell & Bate 2006; Bonnell *et al.* 2007; Smith *et al.* 2009), in which the material reaching a star is collected from distant locations (up to several tenths of a parsec away) within the clump. In fact, the development of a density profile with slope $-3/2$ may also be indicative of a generalized state of collapse in the clumps, as it is the signature of ongoing dynamic collapse (Shu 1977).

In summary, the simulations discussed above suggest that outflow feedback is capable of maintaining the turbulence within parsec-scale clumps that are already forming stars. However, there exist a few observational features of clouds that cannot be explained by this mechanism. First, it is well known that the majority of a cloud's volume is not in the process of forming stars; the star-forming regions within clouds are generally limited to only a few localized, high-column density spots (e.g, Kirk *et al.* 2006) within the clouds. Moreover, there exist clouds with very little or no significant star-forming activity, such as Maddalena's cloud (Maddalena & Thaddeus 1985) or the Pipe Nebula (Onishi *et al.* 1999; see also Lombardi *et al.* 2006) that nevertheless have turbulent properties essentially indistinguishable from those of clouds that are actively forming stars. Outflow feedback clearly cannot explain the origin or maintenance of the turbulence in the non-star-forming regions.

Second, principal component analysis (PCA) of spectroscopical line emission data (Heyer & Brunt 2007; Brunt *et al.* 2009) shows that the turbulent velocity dispersion in the clouds and their substructure appears to be dominated by large-scale, dipolar, velocity gradients spanning the entire structure. Such universal large-scale nature of the turbulent motions in clouds does not seem to be attainable by outflows, due to their small-scale, localized nature. Note, however, that Carroll *et al.* (2010) have recently questioned these results, a claim that requires further investigation. In any case, the problem of driving the turbulence in non-star-forming regions remains, and in general it can be argued that a different source of turbulence is required there.

4. Cloud evolution and control of the SFE

4.1. *Cloud destruction*

The expansion of HII regions from massive stars, especially from those that are sufficiently close to the cloud's periphery to produce a "blister HII region", is probably the dominant feedback mechanism at the scale of GMCs (for a discussion of the contributing feedback mechanisms, see Matzner 2002), and is generally believed to be capable of effectively dispersing the cloud within 10^7 yr after SF starts in the cloud (Blitz & Shu 1980). Whitworth (1979) analytically estimated the eroding effect of O stars producing blister HII regions in MCs, and concluded that the cloud would be completely dispersed after only 4% of its mass had been converted to stars, assuming a standard Salpeter (1955) initial mass function (IMF).

Extending on this result, and considering that a fraction of the massive stars are completely interior to the cloud and do not produce a blister HII region, Franco *et al.* (1994) estimated the maximum number of OB stars that can be hosted by a GMC without it being destroyed, concluding that the resulting SFE should range between 2% and 16%, with an average of 5%. On the observational side, Williams & McKee (1997) compared the Galactic distribution of GMC masses to the distribution of OB association luminosities, in order to statistically estimate the GMC mass that contains at least one O star. They found that the median GMC mass to satisfy this condition is $\sim 10^5\,M_\odot$, and that the average SFE is $\sim 5\%$. Also, they estimated that the typical GMC lifetime as the time for the cloud to be photo-evaporated by the O stars is $\sim 3 \times 10^7$ yr, although a more recent estimate by Matzner (2002) yields a somewhat shorter timescale of $\sim 2 \times 10^7$ yr.

4.2. *Support of isolated clouds and the SFE*

Together with the erosion inflicted on clouds by massive-star feedback, it is also generally believed that the latter can also maintain GMCs close to virial equilibrium for times significantly longer than their free-fall times. Taking this as a working assumption, the SFR and SFE can be derived, similarly to the procedure used in studies of SF self-regulation due to outflows (e.g., Franco & Cox 1983; McKee 1989). For example, Matzner (2002) analytically derived a cloud destruction timescale of $\sim 2 \times 10^7 (M_{\rm cl}/10^6 M_\odot)^{-1/3}$ yr, implying that more massive clouds should be destroyed in shorter times.

A semi-analytic model of the energy balance in GMCs was presented by Krumholz *et al.* (2006). In this model, the fully time-dependent Virial Theorem was written and solved numerically for a spherical cloud under the influence of its self-gravity and the HII-region feedback, with no a-priori assumption of equilibrium. The result was that clouds undergo a few expansion-contraction oscillations, until they are finally dispersed, with lower-mass clouds ($M \sim 2 \times 10^5 M_\odot$) are more quickly dispersed (typically within ~ 1.5 crossing times) than more massive ones ($M \sim 5 \times 10^6 M_\odot$), which last ~ 3 crossing times. Note that this result is opposite to that of Matzner (2002). The SFEs over the clouds' lifetimes were found to be $\sim$ 5–10%.

Full numerical simulations including self-consistent SF prescriptions, aimed at studying the role of stellar feedback on reducing the SFR have been generally carried out at the clump-scale level. There is a general agreement that feedback reduces the SFR and, consequently, the SFE of a cloud, regardless of whether the energy is injected by low-mass outflows (Nakamura & Li 2007; Wang *et al.* 2010), low-mass protostellar luminosity (Bate 2009; Price & Bate 2009; Offner *et al.* 2009), or high-mass-star winds (Dale & Bonnell 2008). Although the details vary depending on the specific implementation, all studies conclude that the SFR is reduced with respect to the case with no feedback. In particular, some of these studies quantify the SFE, reporting values $\lesssim 10\%$ per free-fall time (e.g., Wang *et al.* 2010), in agreement with observations of similar regions (e.g., Evans *et al.* 2009), although Dale & Bonnell (2008) warn that the the SFR may accelerate in time, rendering any conclusions based on the average SFE uncertain.

4.3. *Evolution of cloud-environment systems*

All the numerical studies mentioned in the previous section have been performed at the clump scale, thus omitting the interaction between the clumps and the larger-scale cloud in which they are immersed. This may be of crucial importance, since recent simulations of GMC formation and evolution in the presence of self-gravity (Vázquez-Semadeni *et al.* 2007, 2009; Hartmann & Burkert 2007; Heitsch & Hartmann 2008; Heitsch *et al.* 2008; Hennebelle *et al.* 2008; Banerjee *et al.* 2009; see also the review by Vázquez-Semadeni 2010) as well as observations of massive-star forming regions (Galván-Madrid *et al.* 2009; Csengeri *et al.* 2010) suggest that there is a continuous infalling flow from the large to the small scales, most probably driven by gravity. In this case, the mass reservoir of the local star-forming regions is not limited to its immediate environment, but rather includes material from regions farther away in the cloud than the local clump.

A numerical study incorporating the effect of massive-star ionizing radiation in the context of globally contracting clouds was recently performed by Vázquez-Semadeni *et al.* (2010). These authors considered the evolution of clouds formed by converging flows in the warm neutral atomic medium (WNM), and followed it until the time of active SF. In this type of simulations, the clouds are found to enter a global state of contraction, causing the localized star-forming regions within the clouds to have a continuous inflow of mass from their environment, rather than having a fixed mass. Thus, the SFE, defined as SFE$= M_\star/(M_{\rm cl} + M_\star)$, where $M_\star$ is the total mass in stars and $M_{\rm cl}$ is the gas mass of the cloud, can maintain realistic values, of order of a few percent for GMCs, over extended periods of time, because the gas mass is replenished by the infall while the cloud continues forming stars. These authors also found that the feedback acts on size scales much smaller than the gravitational potential well of the whole GMC, and therefore the global inflow persists, even if locally the gas on route to forming stars is dispersed, reducing the SFE. This effect is stronger for more massive clouds, which have deeper and more extended potential wells. Smaller clouds were found to be more easily destroyed, in agreement with the semi-analytic model of Krumholz *et al.* (2006). However, the results from Vázquez-Semadeni *et al.* (2010) suggest that perhaps termination of the SF activity on the scale of the largest GMCs may require the termination of the inflows, rather than being accomplished by the feedback. Further exploration of parameter space and feedback modeling is needed in order to obtain firmer conclusions in this regard.

The possibility that clouds are in a generalized state of contraction and accreting from WNM has the additional advantage that it may provide the needed universal source of turbulence in GMCs and their substructure, since it is by now well established that the dense layers produced by converging flows naturally develop turbulence, as a consequence of several instabilities acting on them (Hunter *et al.* 1986; Vishniac 1994; Folini & Walder 2006; Heitsch *et al.* 2006; Vázquez-Semadeni *et al.* 2006). Recently, Klessen & Hennebelle (2009) have compared the energy input rate from the accretion to the energy dissipation rate by the turbulence, concluding that the former is sufficient to maintain the turbulence even if only 10% of the accretion energy is converted to turbulence in the dense regions. These authors suggest that this mechanism can operate at all scales from Galactic disks to protostellar disks, passing through GMCs and their substructure, extending the suggestion that a universal mass cascade, driven by gravity, exists at all scales within GMCs (Field *et al.* 2008).

5. Discussion and conclusions

The results from the works discussed in this review imply that feedback from clusters has a complex and strongly nonlinear effect on their parent clouds and clumps, which still remains elusive in some respects. Numerical simulations of outflows from low-mass stars in parsec-sized clumps agree in general that the outflows inject sufficient momentum into the clump to sustain the turbulence in it. In these simulations, the turbulence develops a peculiar form of the turbulent energy spectrum, with a knee at the characteristic outflow scale, and a steep slope (~ -2.5) below that scale. These features should in principle be observationally detectable, and so this prediction is directly testable. However, we pointed out that outflow driving cannot account for the turbulence in non-star-forming regions of clouds, or in clouds that have no significant star-forming activity at all. Thus, we concluded that a more universal source of turbulence is needed.

Analytical and semi-analytical calculations of the effect of feedback from massive stars at the GMC scale suggest that the feedback is also capable of slowing down the SFR to realistic rates, while simultaneously supporting the cloud over a few crossing times, after which the cloud is finally destroyed. However, by their very nature, these models cannot account for the spatial distribution of the stellar sources, an ingredient which is suggested to be crucial by numerical studies of the formation and evolution of a large molecular complex. Such simulations show that the entire cloud complex begins contracting gravitationally even before it becomes predominantly molecular, so that, by the time a GMC is fully formed and begins to form stars, it is already contracting. Moreover, the contraction occurs in a highly non-uniform fashion, due to the turbulence produced at the cloud's formation, so that SF and their feedback occur at a few localized spots in the clouds. This prevents the feedback from reaching the more distant, yet also infalling, regions of the clouds.

This suggests again that the source of the turbulence at the scale of whole GMCs should be a more universal one than the feedback, which is applied very locally. Within this scenario, a natural candidate source is the accretion energy from the environment, since it is well known that the dense layers formed by colliding streams are naturally turbulent due to several dynamical instabilities acting on them. In this picture, the GMCs are the dense "layers" within converging WNM flows, while clumps and cores are the dense "layers" within converging molecular and/or cold atomic convergent flows, all probably driven by gravity rather than by the stellar feedback. The latter is only a byproduct of the gravitational contraction, and acts mainly to prevent a fraction of the infalling gas mass from actually reaching the forming stars, reducing the SFR and the SFE from their free-fall value. More theoretical, numerical and observational work is clearly needed to confirm this picture, and sort out its details.

References

Alves, J., Lombardi, M., & Lada, C. J. 2007, *A&A*, 462, L17
Audit, E. & Hennebelle, P. 2005, A&A, 433, 1
Ballesteros-Paredes, J., Vázquez-Semadeni, E., & Scalo, J. 1999, *ApJ*, 515, 286
Ballesteros-Paredes, J., Hartmann, L., & Vázquez-Semadeni, E. 1999, *ApJ*, 527, 285
Banerjee, R., Klessen, R. S., & Fendt, C. 2007, *ApJ*, 668, 1028
Banerjee, R., Vázquez-Semadeni, E., Hennebelle, P., & Klessen, R. S. 2009, MNRAS, 398, 1082
Bate, M. R. 2009, *MNRAS*, 392, 1363
Blitz, L. 1993, *Protostars and Planets III*, 125
Blitz, L. & Shu, F. H. 1980, ApJ, 238, 148
Bonnell, I. A. & Bate, M. R. 2006, *MNRAS*, 370, 488
Bonnell, I. A., Bate, M. R., Clarke, C. J., & Pringle, J. E. 1997, *MNRAS*, 285, 201
Bonnell, I. A., Larson, R. B., & Zinnecker, H. 2007, Protostars and Planets V, 149
Brunt, C. M., Heyer, M. H., & Mac Low, M.-M. 2009, *A&A*, 504, 883
Carroll, J. J., Frank, A., Blackman, E. G., Cunningham, A. J., & Quillen, A. C. 2009, *ApJ*, 695, 1376
Carroll, J. J., Frank, A., & Blackman, E. G. 2010, *ApJ*, 722, 145
Csengeri, T., Bontemps, S., Schneider, N., Motte, F., & Dib, S. 2010, arXiv:1009.0598
Cunningham, A. J., Frank, A., & Blackman, E. G. 2006a, *ApJ*, 646, 1059
Cunningham, A. J., Frank, A., Quillen, A. C., & Blackman, E. G. 2006b, *ApJ*, 653, 416
Cunningham, A. J., Frank, A., *et al.* 2009, *ApJ*, 692, 816
Dale, J. E. & Bonnell, I. A. 2008, *MNRAS*, 391, 2
De Colle, F. & Raga, A. C. 2005, *MNRAS*, 359, 164
Evans, N. J., *et al.* 2009, ApJS, 181, 321
Field, G. B., Blackman, E. G., & Keto, E. R. 2008, MNRAS, 385, 181
Field, G. B. & Saslaw, W. C. 1965, ApJ, 142, 568
Folini, D. & Walder, R. 2006, *A&A*, 459, 1
Franco, J. & Cox, D. P. 1983, *ApJ*, 273, 243
Franco, J., Shore, S. N., & Tenorio-Tagle, G. 1994, *ApJ*, 436, 795
Galván-Madrid, R., Keto, E., Zhang, Q., Kurtz, S., Rodríguez, L. F., & Ho, P. T. P. 2009, *ApJ*, 706, 1036
Hartmann, L. & Burkert, A. 2007, ApJ, 654, 988
Heitsch, F., Burkert, A., Hartmann, L. W., Slyz, A. D., & Devriendt, J. E. G. 2005, ApJ, 633, L113

Heitsch, F. & Hartmann, L. 2008, *ApJ*, 689, 290
Heitsch, F., Hartmann, L. W., Slyz, A. D., Devriendt, J. E. G., & Burkert, A. 2008, *ApJ*, 674, 316
Heitsch, F., Slyz, A. D., Devriendt, J. E. G., Hartmann, L. W., & Burkert, A. 2006, *ApJ*, 648, 1052
Hennebelle, P. & Audit, E. 2007, A&A, 465, 431
Hennebelle, P., Banerjee, R., Vázquez-Semadeni, E., Klessen, R. S., & Audit, E. 2008, *A&A*, 486, L43
Hennebelle, P. & Chabrier, G. 2008, ApJ, 684, 395
Hennebelle, P. & Pérault, M. 1999, A&A, 351, 309
Heyer, M. H. & Brunt, C. 2007, *IAU Symposium 237*, 9
Hunter, J. H., Jr., Sandford, M. T., II, Whitaker, R. W., Klein, R. I. 1986, ApJ, 305, 309
Kirk, H., Johnstone, D., & Di Francesco, J. 2006, *ApJ*, 646, 1009
Klessen, R. S. & Hennebelle, P. 2009, A&A in press (arXiv:0912.0288)
Koyama, H. & Inutsuka, S.-i. 2002, ApJ, 564, L97
Krumholz, M. R., Matzner, C. D., & McKee, C. F. 2006, *ApJ*, 653, 361
Krumholz, M. R. & McKee, C. F. 2005, ApJ, 630, 250
Krumholz, M. R., McKee, C. F., & Klein, R. I. 2005, *Nature*, 438, 332
Li, Z.-Y. & Nakamura, F. 2006, *ApJ*, 640, L187
Lombardi, M., Alves, J., & Lada, C. J. 2006, *A&A*, 454, 781
Maddalena, R. J. & Thaddeus, P. 1985, *ApJ*, 294, 231
Matzner, C. D. 2002, *ApJ*, 566, 302
Matzner, C. D. 2007, *ApJ*, 659, 1394
McKee, C. F. 1989, ApJ, 345, 782
Motte, F., Andre, P., & Neri, R. 1998, A&A, 336, 150
Nakamura, F. & Li, Z.-Y. 2007, *ApJ*, 662, 395
Norman, C. & Silk, J. 1980, ApJ, 238, 158
Offner, S. S. R., Klein, R. I., McKee, C. F., & Krumholz, M. R. 2009, *ApJ*, 703, 131
Onishi, T., *et al.* 1999, PASJ, 51, 871
Oort, J. H. 1954, Bull. Astron. Inst. Netherlands, 12, 177
Padoan, P. & Nordlund, Å. 2002, ApJ, 576, 870
Price, D. J. & Bate, M. R. 2009, *MNRAS*, 398, 33
Quillen, A. C., Thorndike, S. L., Cunningham, A., Frank, A., Gutermuth, R. A., Blackman, E. G., Pipher, J. L., & Ridge, N. 2005, *ApJ*, 632, 941
Reipurth, B. & Bally, J. 2001, ARAA, 39, 403
Salpeter, E. E. 1955, *ApJ*, 121, 161
Schmidt, M. 1959, ApJ, 129, 243
Shu, F. H. 1977, *ApJ*, 214, 488
Shu, F. H., Lizano, S., Ruden, S. P., & Najita, J. 1988, ApJL, 328, L19
Smith, R. J., Clark, P. C., & Bonnell, I. A. 2009, *MNRAS*, 396, 830
Vazquez-Semadeni, E. 2010, in The Dynamic ISM: A celebration of the Canadian Galactic Plane Survey, ASP Conference Series (arXiv:1009.3962)
Vázquez-Semadeni, E., Colín, P., Gómez, G. C., Ballesteros-Paredes, J., & Watson, A. W. 2010, *ApJ*, 715, 1302
Vázquez-Semadeni, E., Gómez, G. C., Jappsen, A. K., Ballesteros-Paredes, J., González, R. F., & Klessen, R. S. 2007, ApJ, 657, 870
Vázquez-Semadeni, E., Gómez, G. C., Jappsen, A.-K., Ballesteros-Paredes, J., & Klessen, R. S. 2009, ApJ, 707, 1023
Vázquez-Semadeni, E., Ryu, D., Passot, T., González, R. F., & Gazol, A. 2006, ApJ, 643, 245
Vishniac, E. T. 1994, ApJ, 428, 186
Wang, P., Li, Z.-Y., Abel, T., & Nakamura, F. 2010, *ApJ*, 709, 27
Whitworth, A. 1979, *MNRAS*, 186, 59
Williams, J. P. & McKee, C. F. 1997, *ApJ*, 476, 166
Zuckerman, B. & Evans, N. J., II 1974, ApJL, 192, L149

Computational star formation
Proceedings IAU Symposium No. 270, 2011
J. Alves, B.G. Elmegreen, J. M. Girart & V. Trimble, eds.

doi:10.1017/S1743921311000512

The theory of young cluster disruption

Simon P. Goodwin

Dept. of Physics & Astronomy, University of Sheffield, Hounsfield Road, Sheffield, S3 7RH, UK
email: `s.goodwin@sheffield.ac.uk`

Abstract. Most stars seem to form in clusters, but the vast majority of these clusters do not seem to survive much beyond their embedded phase. The most favoured mechanism for the early destruction of star clusters is the effect of the removal of residual gas by feedback which dramatically changes the cluster potential. The effects of feedback depend on the ratio of the masses of stars and gas, and the velocity dispersion of the stars at the onset of gas removal. As gas removal is delayed by a few Myr from star formation these crucial parameters can change significantly from their initial values. In particular, in dynamically cool and clumpy clusters, the stars will collapse to a far denser state and if they decouple from the gas then gas removal may be far less destructive than previously thought. This might well help explain the survival of very massive clusters, such as globular clusters, without the need for extremely high star formation efficiencies or initial masses far greater than their current masses.

Keywords. clusters: general, stars: formation

---

1. Introduction

In our own Galaxy most star formation appears to occur in clusters (see reviews by Lada & Lada 2003; Allen *et al.* 2007; Lada 2010), and in external galaxies most (at least high-mass) star formation is also observed in clusters (de Grijs 2010; Larsen 2010).

Observations of star formation and young clusters are relatively complete out to around 2 kpc and find that clusters appear to form with a power-law initial cluster mass function $N(M) \propto M^{-\alpha}$ where $\alpha = 1.7 - 2$ (Lada 2010). An interesting feature of an initial cluster mass function of this form is that all masses of clusters are equally important for populating the Galactic field. Low-mass clusters are far more numerous than high-mass clusters (e.g. Porras *et al.* 2003), but contribute the same total number of stars to the Galactic field (see e.g. Goodwin 2010).

When the numbers of clusters within a particular age range are examined however, it is found that there are far fewer old clusters than would be expected for a constant cluster formation rate. In particular, there are around ten times more embedded clusters than young (naked) open clusters in the Galaxy (Lada & Lada 2003). This suggests that some form of 'infant mortality' it at work, destroying many clusters at very young (< 10 Myr) ages.

In this contribution I would like to discuss the theory of young cluster disruption: what might cause infant mortality, and what parameters of some young clusters might affect their survival and/or destruction.

2. Gas expulsion

Stars form from molecular gas, and the efficiency with which gas is turned into stars (the star formation efficiency, SFE) is often fairly low. The SFEs observed in cluster-forming clumps in GMCs tend to be around 30% (Lada & Lada 2003). Therefore the mass of embedded clusters is mostly gas and this gas is removed after a few Myr by feedback from the most massive stars in a phase of 'gas removal'.

During gas removal the potential of the cluster will change drastically. Prior to gas removal the stars will be moving in a potential dominated by the gas, but the potential felt by the stars will reduce significantly, and often very rapidly, during gas removal. The effects of gas removal have been studied in detail by many authors both analytically and with the use of simulations (Tutukov 1978; Hills 1980; Mathieu 1983; Elmegreen 1983; Lada *et al.* 1984; Elmegreen & Clemens 1985; Pinto 1987; Verschueren & David 1989; Goodwin 1997a,b; Gyer & Burkert 2001; Boily & Kroupa 2003a,b; Bastian & Goodwin 2006; Goodwin & Bastian 2006; Baumgardt & Kroupa 2007; Parmentier *et al.* 2008; Goodwin 2009; Chen & Ko 2009).

It is often stated that the key parameter that controls the destructiveness of gas removal on a cluster is the SFE, modified by the timescale of gas removal. The lower the SFE the greater the contribution of gas to the potential and the more destructive gas removal will be (typically a figure of an SFE of $\sim 30\%$ is given for a cluster to survive instantaneous gas removal). In addition, the slower the gas removal timescale, the less disruptive it is as the stars have a chance to adapt to the changing potential (lowering the critical SFE from $\sim 30\%$ to $\sim 20\%$).

Observed SFEs in cluster-forming clumps (of around $10^3 M_\odot$) are observed to be of the same level as the theoretical minimum for cluster survival (Lada & Lada 2003), therefore it might be expected that a lucky few clusters with a high-enough SFE can survive gas expulsion. However, the critical SFEs quoted for cluster survival are for the survival of a bound core which can be significantly less massive than the original cluster ('infant weightloss'). If the SFE is close to the critical SFE then 90% of the cluster mass may be lost with only a small bound cluster remaining (see Baumgardt & Kroupa 2007). Therefore in the surviving cluster population (open and globulars) either these clusters were significantly more massive at birth (especially when considering evaporation from very old clusters), or they had extremely high SFEs.

However, there are two critical assumption that are generally made. Firstly, that the cluster is relatively smooth and spherical (most often it is modelled as a Plummer sphere). Secondly, that the cluster is in virial equilibrium†. It is not clear that either of these assumptions are correct.

3. The initial conditions of clusters

Observations strongly suggest that the stars in clusters form highly out-of-equilibrium with both non-virial velocity dispersions and large amounts of substructure (see for example Elmegreen & Elmegreen 2001; Bertout & Genova 2006; Allen *et al.* 2007; Kraus & Hillenbrand 2008; Allison *et al.* 2009; Gutermuth *et al.* 2009; Clarke 2010 and references in all of these papers). This should not be surprising as in the gravoturbulent model of star formation stars will form in dense gas in filaments and clumps in a turbulent environment (see e.g. Elmegreen 2004; McKee & Ostriker 2007; Bergin & Tafalla 2007; Clarke 2010).

If a cluster is born substructured and sub-virial as observations and theory suggest that they often are (see above) then it will collapse in an attempt to reach a new equilibrium. However, initially a cluster will contain both collisionless stars and collisional gas so it is possible that the gas and stars may decouple, the stars collapsing to the centre of the

† A number of authors do consider non-virialised initial conditions such as Lada *et al.* (1984), Elmegreen & Clemens (1985), and Verschueren & David (1989).

local potential well whilst the gas remains supported (although some, or a significant amount of gas may also fall into the potential well, see e.g. Maschberger *et al.* 2010; Moeckel & Bate 2010).

In this process the stellar distribution will be approaching virial equilibrium, but the size of the stellar distribution will decrease (possibly very significantly) and the mass of gas in the same volume of the stars might well change (again, possibly significantly).

The effects of gas expulsion depend on the gas mass associated with the stars, and the velocity dispersion of the stars *at the onset of gas expulsion* – not at the time of formation. Therefore, there are two crucial parameters that evolve with time that determine the destructive effects of gas expulsion. Firstly, the effective SFE (eSFE, see Verschueren & David 1989; Goodwin & Bastian 2006; Goodwin 2009) which is the ratio of the stellar mass to gas mass at the onset of gas expulsion. Secondly, the velocity dispersion of the stars at the onset of gas expulsion (or how close to virial equilibrium the stars are with the combined gas and star potential, see Goodwin 2009).

Without large suites of full hydrodynamical simulations including feedback to correctly and self-consistently expel the gas it is unclear what exactly happens within clusters and how the effects of gas expulsion vary with the initial conditions of star formation (but see Allison *et al.* 2009, 2010; Maschberger *et al.* 2010; Moeckel & Bate 2010).

4. Conclusion

To conclude. To form extremely massive clusters, such as globular clusters, it is not required to have either extremely high SFEs or

masses far greater at birth than now in order to overcome infant mortality and weight-loss. Rather, if they are born in such as way that the effective SFE is low at the onset of gas expulsion then they will survive and lose little of their initial mass.

References

Allen, L., Megeath, S. T., Gutermuth, R., Myers, P. C., Wolk, S., Adams, F. C., Muzerolle, J., Young, E., & Pipher, J. L. 2007, in 'Protostars and Planets V' eds. Reipurth, B., Jewitt, D. & Keil, K. (University of Arizona Press: Tuscon), p 361

Allison, R. J., Goodwin, S. P., Parker, R. J., de Grijs, R., Portegies Zwart, S. F., & Kouwenhoven, M. B. N. 2009, *ApJ*, 700, L99

Allison, R. J., Goodwin, S. P., Parker, R. J., Portegies Zwart, S. F., & de Grijs, R. 2010, *MNRAS*, in press (arXiv:1004.5244)

Bastian, N. & Goodwin, S. P. 2006, *MNRAS*, 369, L9

Baumgardt, H. & Kroupa, P. 2007, *MNRAS*, 380, 1589

Bergin, E. & Tafalla, M. 2007, *ARAA*, 45, 339

Bertout, C. & Genova, F. 2006; *A&A*, 460, 499

Boily, C. M. & Kroupa, P. 2003a, *MNRAS*, 338, 643

Boily, C. M. & Kroupa, P. 2003a, *MNRAS*, 338, 673

Chen, H. & Ko, C. 2009, *ApJ*, 698, 1659

Clarke, C. J. 2010, *RSPTA*, 368, 733

Elmegreen, B. G. 1983, *MNRAS*, 203, 1011

Elmegreen, B. G. & Clemens, C. 1985, *ApJ*, 294, 523

Elmegreen, B. G. & Elmegreen, D. M. 2001, *AJ*, 121, 1507

Elmegreen, B. G. 2004, *ARAA*, 42, 211

de Grijs, R. 2010, *RSPTA*, 368, 693

Goodwin, S. P. 1997a, *MNRAS*, 284, 785

Goodwin, S. P. 1997b, *MNRAS*, 286, 669

Goodwin, S. P. & Bastian, N. 2006, *MNRAS*, 373, 752

Goodwin, S. P. 2009, *Ap&SS*, 324, 259
Goodwin, S. P. 2010, *RSPTA*, 368, 851
Gutermuth, R. A., Megeath, S. T., Myers, P. C., Allen, L. E., Pipher, J. L. & Fazio, G. G. 2009, *ApJS*, 184, 18
Gyer, M. P. & Burkert, A. 2001, *MNRAS*, 323, 988
Hills, J. G. 1980, *ApJ*, 235, 986
Kraus, A. & Hillenbrand, L. 2008, *ApJ*, 686, L111
Lada, C. J., Margulis, M. & Dearborn, D. 1984, *ApJ*, 285, 141
Lada, C. J. & Lada E. A. 2003, *ARAA*, 41, 57
Lada, C. J. 2010, *RSPTA*, 368, 713
Larsen, S. S. 2010, *RSPTA*, 368, 867
McKee, C. & Ostriker, E. 2007, *ARAA*, 45, 565
Maschberger, Th., Clarke, C. J., Bonnell, I. A. & Kroupa, P. 2010, *MNRAS*, 404, 1061
Mathieu, R. D. 1983, *ApJ*, 267, 97
Moeckel, N. & Bate, M. R. 2010, *MNRAS*, 404, 721
Parmentier, G., Goodwin, S. P., Kroupa, P. & Baumgardt, H. 2008, *ApJ*, 678, 347
Pinto, F. 1987, *PASP*, 99, 1161
Porras, A., Christopher, M., Allen, L., Di Francesco, J., Megeath, S. T., & Myers, P. C. 2003, *ApJ*, 126, 1916
Tutukov, A. V. 1978, *A&A*, 70, 57
Verschueren, W. & David, M. 1989, *A&A*, 219, 105

Computational Star Formation
Proceedings IAU Symposium No. 270, 2011
J. Alves, B.G. Elmegreen, J. M. Girart & V. Trimble, eds.

doi:10.1017/S1743921311000524

Outflows and Turbulence in Young Stellar Clusters —An Observer's View

Héctor G. Arce

Department of Astronomy, Yale University, P.O. Box 208101, New Haven CT 06520
email: hector.arce@yale.edu

Abstract. Recent numerical studies have focused their interest on the impact outflows have on the cloud's turbulence. The contradictory results obtained by these studies indicate that it is essential for observers to provide the required data to constrain the models. Here we discuss the impact of outflows on the environment surrounding clusters of young stellar objects, from an observer's point of view. We have conducted several studies of outflows in different active star-forming regions. In all cases it is clear that outflows have the power to sustain the observed turbulence in the gas surrounding protostellar clusters. We investigate whether there is a correlation between outflow strength and star formation efficiency, as predicted by numerical simulations, for six different regions in the Perseus molecular cloud complex. We argue that results of other recent studies that use CO line maps to study the turbulence driving length should not be used to discard outflows as major drivers of turbulence in clusters.

Keywords. stars: formation, stars: pre–main-sequence, ISM: jets and outflows, turbulence

---

1. Introduction

The outflowing supersonic wind from a protostar can accelerate the surrounding molecular gas to velocities significantly greater than those of the quiescent cloud gas thereby producing a molecular outflow that injects momentum and kinetic energy into its surroundings (Arce *et al.* 2007). Recent numerical studies show that outflows can couple strongly to the cloud and are highly efficient at driving turbulent motions (e.g., Carroll *et al.* 2009) and can also regulate the cloud's star formation efficiency (Nakamura & Li 2007; Wang *et al.* 2010). Yet, other numerical studies suggest that protostellar outflows do a poor job at driving cloud turbulence, but can disrupt dense clumps and affect the cloud structure (Banerjee *et al.* 2007). These studies show the increased attention that research on the impact of outflows has obtained, as well as the need for targeted observations required to constrain the models and to reconcile their discrepancies.

Observational studies have shown that outflows, even from low-mass stars, can have an impact on their cloud at different distances from the source, ranging from a few thousand AU to several parsecs. Survey studies of the circumstellar gas within 10^4 AU of low-mass protostars indicate outflows contribute significantly to the mass-loss of the surrounding dense gas (Fuller & Ladd 2002; Arce & Sargent 2006). An outflow's impact on its parent core, at distances of about 0.1 to 0.3 pc from the forming star, is evidenced through the detection of outflow-blown cavities and outflowing dense gas (e.g., Tafalla & Myers 1997). Giant outflows from young stars with sizes exceding 1 pc in length are common (Reipurth *et al.* 1997). These outflows can interact with the surrounding medium and induce changes in the velocity and density distribution of the parent cloud's gas at large distances from the source (Arce & Goodman 2001; 2002). Millimeter studies have shown that many molecular outflows produced by a group or cluster of young stars interact with a substantial volume of the cluster's environment, may sweep up the gas and dust and have the energy required for driving turbulence (e.g., Knee & Sandell 2000). However, most of these studies concentrate on individual regions, making it difficult for a consistent comparison of the outflow impact in different clusters.

2. Survey of Molecular Outflows in Perseus

Recently, Arce *et al.* (2010, hereafter A10) conducted an unbiased outflow survey of the Perseus molecular cloud complex and were able to study how individual and groups of outflows effect the dynamics of the gas in the entire cloud complex and how they interact with their surroundings at different distances from the driving source (from about 0.1 pc to a few parsecs). They discovered numerous new molecular outflow candidates in regions that had been poorly studied before (e.g., outskirts of famous star forming regions and regions with very low star formation activity). The new outflow candidates more than double the amount of outflow mass, momentum, and kinetic energy in the Perseus cloud complex compared to the total obtained from previously known outflows. Their results indicate that outflows have significant impact on the environment immediately surrounding localized regions of active star formation, but lack the energy needed to feed the observed turbulence in the *entire* Perseus complex.

As expected, outflows are concentrated in regions of high star formation activity, and it is in these regions (with sizes of 1 to 4 pc) that outflows have the biggest impact. In their study, A10 defined six regions of active star formation in the Perseus molecular cloud complex where clusters or groups of outflows were found (e.g., L1448, NGC 1333, B1-ridge, B1, IC 348, and B5). Their unbiased outflow search allowed them to study the impact of outflows in each region and to compare the different regions in a consistent manner. One way to assess the importance of outflows in driving the turbulence in their local environment is to compare the total outflow energy input rate into the cloud (i.e., outflow luminosity) with the energy rate needed to maintain the turbulence in the gas. For all six regions, the total outflow energy input rate into the molecular gas is at least 80% of the power needed to maintain the turbulence in the surrounding cluster gas. These results indicate that outflows in an active region of star formation can be a source of non-negligible power for driving turbulence in the molecular gas, and they should be treated as such in numerical simulations of star-forming clouds.

3. Comparing Data with Numerical Simulations

Recent numerical simulations have shown that the turbulence in an isolated cluster-forming region fed by the energy and momentum injected by bipolar outflows can maintain the region close to (but not entirely in) dynamic equilibrium (Nakamura & Li 2007; Wang *et al.* 2010). Simulations show that there is a small net flux of mass towards the bottom of the potential well that results in a slower collapse, compared to simulations without outflows, and only a few percent of the total gas mass is converted into stars within a free-fall time. These studies show that more powerful outflows result in an increase in the turbulence in the gas, which then lead to a delay of the gravitational collapse and, consequently, a lower star formation efficiency per free-fall time. In their study, A10 attempted to investigate whether there is any relationship between the star formation efficiency and the total outflow strength in the different star-forming regions in the Perseus cloud, and Figure 1 summarizes their results. This plot shows the normalized star formation efficiency (SFE_n) as a function of r_L, the ratio of the total outflow mechanical luminosity in the region (L_{flow}) to the energy rate needed to maintain the

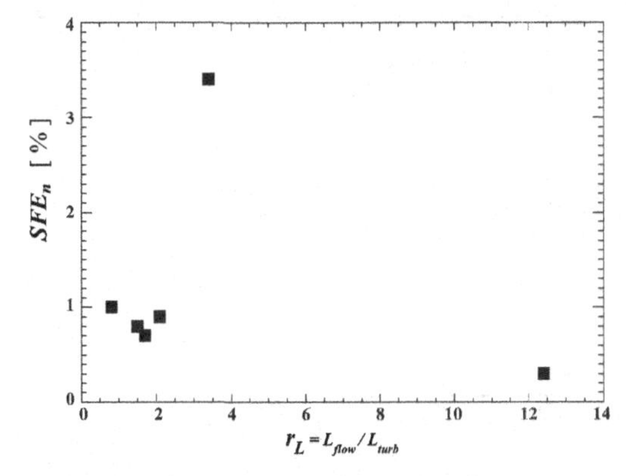

Figure 1. Normalized star formation efficiency as a function of r_L for different regions of star formation in Perseus. No error bars are shown as it is hard to estimate accurate $1-\sigma$ uncertainties for these points. The uncertainties in the values of SFE_n and r_L could easily be a factor of two (or more). Figure from Arce *et al.* (2010)

turbulence in the gas (L_{turb}). Here, SFE_n is the star formation efficiency (SFE) multiplied by the ratio of the free-fall timescale of the region to the age of the region. This normalization is needed in order to take care of the fact that: 1) a cloud with a constant star formation rate will exhibit a higher star formation efficiency as the cloud evolves, since more of the cloud gas will be transformed into stars; and 2) regions with different free-fall times will collapse at different rates and thus will exhibit different values of the current SFE, even for clouds of the same age (see A10 for details). We use r_L as a way to quantify the outflow strength and their impact on the cloud's turbulence. Clouds with $r_L > 1$ harbor outflows that input enough energy into the surrounding gas to maintain the observed turbulence in the cloud. Hence, we would expect clouds with a high r_L to have a low SFE_n.

Figure 1 does not show a clear dependence of SFE_n with r_L. We see that the region with the largest r_L exhibits the lowest value of SFE_n, but otherwise there is no correlation between r_L and SFE_n. It is hard to draw any strong conclusion from these results. Given the large uncertainties in SFE_n, the small size of the sample and the limited range in values of the total outflow momentum and r_L among all regions, it is of no surprise that we do not find a significant correlation between SFE_n and outflow strength, even if such correlation exists. Clearly, similar additional studies are needed to increase the number of star-forming regions and eventually obtain a statistically sound sample of clusters with outflows.

Other recent studies have claimed that even in a region full of outflows like NGC 1333, outflows are not the main source of turbulence in the gas (Brunt *et al.* 2009; Padoan *et al.* 2009). The results of the analysis of the gas kinematics of these two studies, using a ^{13}CO(1-0) map of the region, suggest that turbulence is mostly driven at scales larger than the region of interest. They claim that outflows are not the main drivers of the region's turbulence because if that were the case the velocity power spectrum would show clear evidence that the turbulence is being driven at a scale of approximately 0.3 pc. We argue that in clusters outflow-driven turbulence should not necessarily exhibit a driving scale close to 0.3 pc. One possibility is that outflows drive turbulence at a very small scale (about 0.05 pc), as suggested by Swift & Welch (2008) — a scale that would not have been resolved by the data used by Brunt *et al.* (2009) and Padoan *et al.* (2009). Alternatively, the fact that outflows exhibit a range of sizes from less than 0.1 pc to more than a few parsecs in length may result in outflows driving turbulence at different

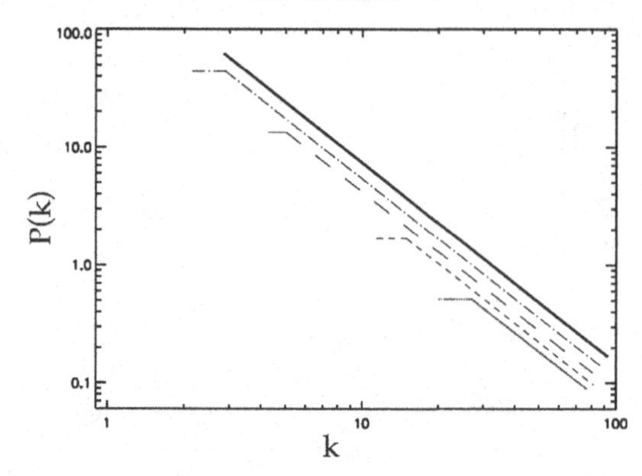

Figure 2. Schematic representation of turbulence power spectrum (P) as a function of wavenumber (k). Different lines represent the expected power spectrum for turbulence driven by outflows at scales of 0.3 pc (dotted), 0.5 pc (small dash), 1.5 pc (long dash), 2.5 pc (dash-dot). The solid line represents the resultant (total) observed power spectrum for a region with outflows that drive turbulence at all the scales listed above.

size scales. As a consequence we should not expect outflow-driven turbulence in a cluster (with many outflows of different sizes and from different sources) to exhibit clear evidence of being driven at a specific scale. In fact, given that outflows may extent to sizes of a few parsecs, they can drive turbulence all the way at scales similar to the size scales of the region, mimicking the expected behavior of the velocity power spectrum (as a function of scale) of turbulence driven by sources external to the cloud. A schematic picture of this is depicted in Figure 2, where the velocity power spectrum as a function of wavenumber (approximately the reciprocal of the size scale) is shown for turbulence driven at different scales. It can be seen that if outflows drive turbulence at different scales, then the shape of the resulting (total) power spectrum will be similar to the power spectrum of turbulence driven at the largest scales. Hence, the results of these recent studies cannot be used to reject the idea that outflows may be a major source of turbulence in clusters.

References

Arce, H. G. & Goodman, A. A. 2001, *ApJ*, 554, 132

Arce, H. G. & Goodman, A. A. 2002, *ApJ*, 575, 911

Arce, H. G. & Sargent, A. I. 2006, *ApJ*, 646, 1070

Arce, H. G., Shepherd, D., Gueth, F., Lee, C.-F., Bachiller, R., Rosen, A., & Beuther, H. 2007, in Protostars and Planets V, eds. B. Reipurth, D. Jewitt, and K. Keil, (University of Arizona Press: Tucson), 245

Arce, H. G.; Borkin, M. A., Goodman, A. A., Pineda, J. E., & Halle, M. 2010, *ApJ*, 715, 1170

Banerjee, R., Klessen, R. S., & Fendt, C. 2007, *ApJ*, 668, 1028

Brunt, C. M., Heyer, M. H., & Mac Low, M.-M. 2009, *A&A*, 504, 883

Carroll, J., Frank, A., Blackman, E., Cunningham, A., & Quillen, A. 2009, *ApJ*, 695, 1376

Fuller, G. A. & Ladd, E. F. 2002, *ApJ*, 573, 699

Knee, L. B. G. & Sandell, G. 2000, *A&A*, 361, 671

Nakamura, F. & Li, Z.-Y. 2007, *ApJ*, 662, 395

Padoan, P., Juvela, M., Kritsuk, A., & Norman, M. L. 2009, *ApJ*, 707, L153

Reipurth, B., Bally, J., & Devine, D. 1997, *AJ*, 114, 2708

Swift, J. J. & Welch, W. J. 2008, *ApJS*, 174, 202

Tafalla, M. & Myers, P. C. 1997, *ApJ*, 491, 653

Wang, P., Li, Z-.Y., Abel, T., & Nakamura, F. 2010, *ApJ*, 709, 27

Computation Star Formation
Proceedings IAU Symposium No. 270, 2011
J. Alves, B.G. Elmegreen, J. M. Girart & V. Trimble, eds.

doi:10.1017/S1743921311000536

Discs, outflows, and feedback in collapsing magnetized cores

Dennis F. Duffin[1] and Ralph E. Pudritz[2]

[1]Department of Physics and Astronomy, McMaster University,
Hamilton ON, L8S 4M1, Canada
email: duffindf@mcmaster.ca

[2]Origins Institue, McMaster University
Hamilton ON, L8S 4M1, Canada
email: pudritz@physics.mcmaster.ca

Abstract. The pre-stellar cores in which low mass stars form are generally well magnetized. Our simulations show that early protostellar discs are massive and experience strong magnetic torques in the form of magnetic braking and protostellar outflows. Simulations of protostellar disk formation suggest that these torques are strong enough to suppress a rotationally supported structure from forming for near critical values of mass-to-flux. We demonstrate through the use of a 3D adaptive mesh refinement code – including cooling, sink particles and magnetic fields – that one produces transient 1000 AU discs while simultaneously generating large outflows which leave the core region, carrying away mass and angular momentum. Early inflow/outflow rates suggest that only a small fraction of the mass is lost in the initial magnetic tower/jet event.

Keywords. stars: formation, stars: winds, outflows, stars: magnetic fields, accretion, accretion disks, (magnetohydrodynamics:) MHD, methods: numerical

1. Introduction

The early evolution of low mass, isolated protostellar cores in the pre-Class 0 stage is now understood as the interplay between gravity, rotation, magnetic fields, and radiative cooling. Strong magnetic fluxes have been observed in molecular clouds (e.g. Crutcher 1999). Recent simulations have focused on the changes endued by a magnetic field in the collapse, through either a 3D SPHR or AMR approach (Basu & Mouschovias 1994; Machida *et al.* 2004; Hosking & Whitworth 2004; Machida *et al.* 2005b,a; Banerjee & Pudritz 2006; Price & Bate 2007; Hennebelle & Fromang 2008; Duffin & Pudritz 2009). It has been shown that magnetic fields slow the collapse timescale and brake the rotation of the initial core and of massive disc-like structures that subsequently form. Magnetic fields have also been shown to suppress fragmentation and the formation of bars and spiral waves. Furthermore, they facilitate the launching of molecular outflows.

Several theoretical problems have arisen from these simulations. First, from axisymmetric simulations outside of 6.7 AU, it is argued that Keplerian discs cannot form in an ideal MHD collapse (Mellon & Li 2008), except in the limit of very weak magnetic flux or high magnetic diffusivity whether numerical (Krasnopolsky *et al.* 2010) or through strong ambipolar diffusion (Mellon & Li 2009). Secondly, magnetic tension and pressure seem to efficiently suppress fragmentation in the early stages of protostellar collapse so that it is difficult to form multi-star systems. With the ability to extend the magnetized collapse further, we can begin to examine questions such as how the Core Mass Function (CMF) is related to the Initial Mass Function of stars (e.g. Matzner & McKee 2000), the nature of the molecular outflow on larger scales and whether fragmentation is suppressed even in later stages. We present our results of the early collapse using 3D ideal and non-ideal magnetohydrodynamic simulations, and the first results of the later stages of the magnetized collapse and outflow. We are able to evolve the simulation an additional $10^4 - 10^5$ yr due to the implementation of the sink particle in the FLASH AMR code (Federrath *et al.* 2010).

2. Numerical Methods and Initial Conditions

We model our a stellar core as in previous papers (e.g. Duffin & Pudritz 2009; Banerjee *et al.* 2004), by embedding a slightly over-critical 1 $M_\odot$ Bonnor-Ebert sphere (Bonnor 1956; Ebert 1955) in a low density environment. We add to this density distribution a 10% over-density to ensure collapse and a 10% $m = 2$ perturbation to break symmetry (see e.g. Duffin & Pudritz 2009; Banerjee *et al.* 2004). The background is an isothermal, low density environment in pressure equilibrium with the sphere (the density is set by choosing a background that is 10 times warmer than the sphere). The box is roughly 10 times the size of the BE radius (0.81 pc in these models). We add to the sphere uniform rotation such that the ratio of rotational and gravitational energy is moderate ($\beta_{\rm rot}$=0.046, similar to the value used in Hennebelle & Fromang (2008)) and a constant $\beta_{\rm plasma} = 2c_s^2/v_{\rm A}^2 = 46.01$. In this model, one can relate the mass to flux ratio μ/μ_0, where μ_0 is the critical mass to flux of $\mu_0 = (2\pi G^{1/2})^{-1}$, by $\mu/\mu_0 \simeq 0.74 c_s/v_{\rm A} = 3.5$.

We use sink radii of 12.7 AU for longer runs, and radii of 3.2 AU to test the effect of sink particle size on the result. The accretion radius of a sink corresponds to 2.5 cells at the highest refinement level, and the critical gas density (beyond which gas can be accreted into particles) corresponds to the Jeans' density at the core temperature (20 K) of these cells. This gives $\rho_{\rm acc} = 3.69 \times 10^{-12}$ g cm^{-3} and $\rho_{\rm acc} = 5.91 \times 10^{-11}$ g cm^{-3} for 12.7 and 3.2 AU sinks respectively. Each Jeans' length is refined by at least 8 cells and de-refined for more than 32 cells. Our customized version of the FLASH AMR code is described in previous work (Fryxell *et al.* 2000; Banerjee *et al.* 2006; Duffin & Pudritz 2008; Federrath *et al.* 2010).

3. The Early Collapse Phase

The advantage of *not* using a sink particle is that we properly resolve the gas in the collapse. The disadvantage is that we are limited to pre-Class 0 times ($\approx 10^5$ yr). We compare the following early collapse models: i) ambipolar diffusion, ii) ideal MHD and iii) hydrodynamics (e.g. $\boldsymbol{B} = 0$). The rotational properties of these early structures are shown in Figure 1a. We compare surface density averaged values of v_ϕ/v_r and $v_\phi/v_{\rm Kepler}$ to measure the degree of rotation in these early collapsed structures. The comparison is done at the maximum common central surface density (Σ_z). This is limited primarily by the ambipolar diffusion model which has the most constrained timestep. More evolved versions of ideal MHD and hydro models are included as thin lines in the graph, hinting at the evolution of the possible structures. The inward velocity has not halted (e.g. the flow is sub-Keplerian) in these early structures, however we are indeed seeing flattened rotationally dominated accretion discs (with associated outflows) being generated with sizes of 6-7 AU, facilitated by the suppression of gravitational instabilities. Bars produced in the hydrodynamic model have prevented a similar structure from forming by efficiently redistributing angular momentum.

To demonstrate the suppression of fragmentation further, we ran the simulation with 10 times more rotational energy in order to study early fragmentation ($\beta_{\rm rot} = 0.74$, corresponding to high values seen in simulations of core formation in a turbulent medium (Tilley & Pudritz 2007)). Indeed our hydrodynamic collapse produces a wide binary with

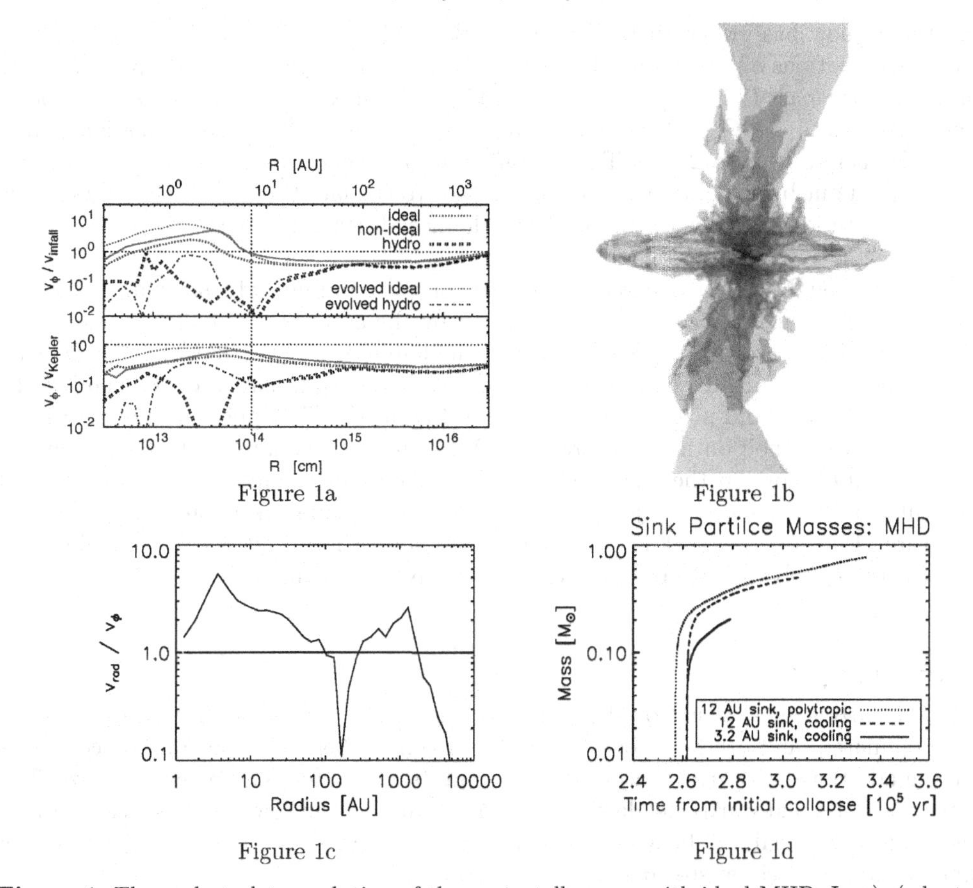

Figure 1c Figure 1d

Figure 1. The early to late evolution of the protostellar core with ideal MHD. In a). (adapted from Duffin & Pudritz 2009), the rotation properties v_ϕ/v_r and $v_\phi/v_{\rm Kepler}$ of the early collapse (no sink particle). In b), contours of density (purple or medium grey, 3.3×10^{-17} g cm^{-3} and black, 1.33×10^{-15} g cm^{-3} at 2000 and 100 AU radii respectively) and v_z (yellow or light grey, $1km/s$) at the end of the simulation (with 3.2 AU sinks). In c), rotational properties of disk at the end of the simulation (with 3.2 AU sink). In d), mass evolution of different sink particle models, from left to right: 12 AU sink particle and a polytropic equation of state, 12 AU sink particle and 3.2 AU sink particle both with molecular cooling.

a separation of about 1000 AU. The ideal MHD model suppresses the fragmentation, as documented in previous studies (Price & Bate 2007; Hennebelle & Teyssier 2008), resulting in a large bar and a central condensed structure. Meanwhile, the ambipolar diffusion model produces an intermediate result, a bar that has fragmented to form a binary. It perhaps through ambipolar diffusion that any "fragmentation crisis" can be solved. These results are discussed further in (Duffin & Pudritz 2009).

4. The Later Evolution: Sink Particles

Using sink particles and ideal MHD, we were able to take the next step and evolve the collapse to much later times. The result is shown in Figure 1b, wherein evolution over an additional 30 kyr is achieved. The sink particle mass evolution is shown in Figure 1d for 12 AU sink particles (with molecular cooling or polytropic equation of state similar to that of Machida *et al.* (2010)) and a smaller 3.2 AU sink particle with molecular cooling (Banerjee *et al.* 2006). By the end of the smaller sink collapse, the sink particle is $0.2M_\odot$

and the outflow has grown to 10^4 AU above the mid-plane of the disk. We are able to run the simulations of larger sinks further, and indeed these particles end up with nearly all of the core mass (80-90%). This is much higher than estimates of theoretical models seeking to relate the Core Mass Function (CMF) to the Initial Mass Function (IMF) (e.g. Matzner & McKee 2000). These results suggest that not much mass has in fact been cleared out by the outflow. This may be due to the fact that most mass must settle into the accretion disk before the outflow is launched, leaving less mass available to be cleared.

The accretion disk is represented by the ratio of v_ϕ/v_r in Figure 1c, showing a 2000 AU accretion disk, which appears to be dissipating by the end of the simulation (as seen in Figure 1d). There are two types of outflows, one lower speed ($\lesssim 1$ km/s) and a central, higher speed centrifugally driven wind. The inner disk and central component of the outflow are warped and precessing, as shown in Figure 1b. The appearance of precessing warped discs and their outflow is extremely interesting and is related to the back reaction of the MHD outflow on the disk (e.g. Lai 2003). Many of the qualitative properties of molecular outflows, including jet precession, clumpiness and an onion layered velocity structure (Pudritz *et al.* 2007) are seen in our simulation, and occur naturally as a consequence of solving the gravito-magnetohydrodynamic equations.

5. Summary

In the early stages, accretion discs are small (<10 AU), massive, flattened, rotationally dominated, and are held together by the magnetic field. As the collapse continues over an additional 10^5 yr, the accretion disc and its associated outflow grow in size. The outflow torques and warps the disk leading to disc and jet outflow and precession respectively. Most material will be accreted by the star raising issues concerning the extent of protostellar feedback on stellar masses.

References

Banerjee, R. & Pudritz, R. E. 2006, *ApJ*, 641, 949
Banerjee, R., Pudritz, R. E., & Anderson, D. W. 2006, *MNRAS*, 373, 1091
Banerjee, R., Pudritz, R. E., & Holmes, L. 2004, *MNRAS*, 355, 248
Basu, S. & Mouschovias, T. C. 1994, *ApJ*, 432, 720
Bonnor, W. B. 1956, *MNRAS*, 116, 351
Crutcher, R. M. 1999, *ApJ*, 520, 706
Duffin, D. F. & Pudritz, R. E. 2008, *MNRAS*, 391, 1659
—. 2009, *ApJL*, 706, L46
Ebert, R. 1955, Zeitschrift fur Astrophysik, 37, 217
Federrath, C., Banerjee, R., Clark, P. C., & Klessen, R. S. 2010, *ApJ*, 713, 269
Fryxell, B., *et al.* 2000, *ApJs*, 131, 273
Hennebelle, P. & Fromang, S. 2008, *A&A*, 477, 9
Hennebelle, P. & Teyssier, R. 2008, *A&A*, 477, 25
Hosking, J. G. & Whitworth, A. P. 2004, *MNRAS*, 347, 1001
Krasnopolsky, R., Li, Z., & Shang, H. 2010, *ApJ*, 716, 1541
Lai, D. 2003, *ApJL*, 591, L119
Machida, M. N., Inutsuka, S., & Matsumoto, T. 2010, ArXiv e-prints
Machida, M. N., Matsumoto, T., Hanawa, T., & Tomisaka, K. 2005a, *MNRAS*, 362, 382
Machida, M. N., Matsumoto, T., Tomisaka, K., & Hanawa, T. 2005b, *MNRAS*, 362, 369
Machida, M. N., Tomisaka, K., & Matsumoto, T. 2004, *MNRAS*, 348, L1
Matzner, C. D. & McKee, C. F. 2000, *ApJ*, 545, 364

Mellon, R. R. & Li, Z.-Y. 2008, *ApJ*, 681, 1356
—. 2009, *ApJ*, 698, 922
Price, D. J. & Bate, M. R. 2007, *MNRAS*, 377, 77
Pudritz, R. E., Ouyed, R., Fendt, C., & Brandenburg, A. 2007, in Protostars and Planets V, ed. B. Reipurth, D. Jewitt, & K. Keil, 277–294
Tilley, D. A. & Pudritz, R. E. 2007, *MNRAS*, 382, 73

Jane Arthur

Computational Star Formation
Proceedings IAU Symposium No. 270, 2011
J. Alves, B.G. Elmegreen, J. M. Girart & V. Trimble, eds.

doi:10.1017/S1743921311000548

Radiation-MHD Simulations of HII Region Expansion in Turbulent Molecular Clouds

S. J. Arthur[1], W. J. Henney[1], G. Mellema[2], F. de Colle[3], E. Vázquez-Semadeni[1]

[1]Centro de Radioastronomía y Astrofísica, UNAM, Campus Morelia, C.P. 58090 Morelia, Michoacán, México

[2]AlbaNova University Center, Stockholm University, SE-106 91 Stockholm, Sweden

[3]Department of Astronomy and Astrophysics, University of California, Santa Cruz

Abstract. We use numerical simulations to investigate how the expansion of an HII region is affected by an ambient magnetic field. First we consider the test problem of expansion in a uniform medium with a unidirectional magnetic field. We then describe the expansion of an HII region in a turbulent medium, taking as our initial conditions the results of and MHD turbulence simulation. We find that although in the uniform medium case the magnetic field does produce interesting effects over long length and timescales, in the turbulent medium case the main effect of the magnetic field is to reduce the efficiency of fragmentation of the molecular gas.

Keywords. HII regions, hydrodynamics, ISM: kinematics and dynamics

1. Introduction

Star formation triggered by the expansion of HII regions is strongly suggested by IR observations of the vicinity of HII regions such as RCW 120 (Deharveng *et al.* 2009). The expanding photoionized gas collects a layer of dense, neutral gas ahead of it, which could then become Jeans unstable, fragment and form new stars. Indeed, a number of Class I low-mass young stellar objects have been found in filaments parallel to the ionization front in RCW 120. Radiation-driven implosion of neutral material within or at the periphery of photoionized regions is another possible route to triggered star formation.

Magnetic fields have been detected in and around molecular clouds with strengths of, for example $10.2\,\mu$G (Ophiucus dark cloud large-scale field; Goodman & Heiles (1994)) to $\sim 500\,\mu$G (in globules of M17, and even higher fields are measured at higher spatial resolution; Brogan & Troland (2001)), indicating that the magnetic field has small-scale structure itself. In this work, we are interested in studying the extent to which the magnetic field on both local and larger scales, influences the evolution of an expanding HII region and what consequences can be expected for triggered star formation.

2. Radiation MHD Code and Test Case

We have developed a radiation-MHD code, which includes a full treatment of the heating and cooling processes in the neutral gas, taking into account the FUV and X-ray radiation field not only from the main ionizing source but also from the associated low-mass stellar cluster, as well as cosmic rays and dust (Henney *et al.* 2009). We have used this code to study the evolution of an HII region in a uniform magnetic field. This problem was first studied by Krumholz *et al.* (2007) and we adopt their parameters in order to aid comparison, that is, an ionizing source of $10^{46.5}$ photons s^{-1}, ambient density $n_0 = 100$ cm^{-2} and temperature $T = 11$ K, and uniform magnetic field of strength

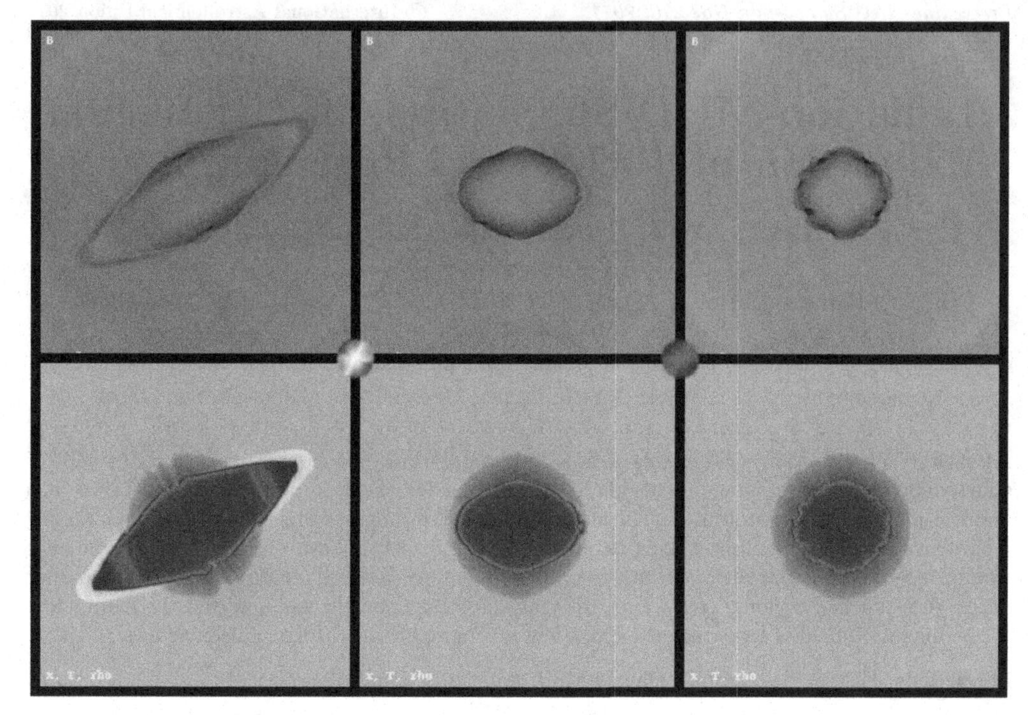

Figure 1. Cuts through the computational cube in the central xy-, xz- and yz-planes for an HII region expanding in a uniform medium with a unidirectional magnetic field after 6.8 Myr. *Top row:* Magnetic field strength and direction (black indicates perpendicular to plane). *Bottom row:* Ionization fraction, temperature and density. Low density, hot, ionized gas fills the center while high density, warm neutral gas piles up at the poles.

$|B| = 14.2\,\mu$G. In order to distinguish between physical instabilities and grid-induced effects we put our magnetic field at 30° to the x-axis in the xy-plane.

As reported by Krumholz *et al.* (2007), the photoionized gas expands, initially pushing the magnetic field aside because the thermal pressure dominates. However, once the thermal pressure in the ionized gas drops to the level of the magnetic pressure in the neutral gas, the magnetic field returns almost to its initial value and direction. The expansion of the HII region is fastest along the direction of the magnetic field, since the slow shock which precedes the ionization front travels fastest along the field lines and cannot travel perpendicular to them. At late times, we see the formation of instabilities in the ionization front where the ionization front is parallel to the magnetic field lines but cannot advance because of this restriction on the slow shock. Clumps form which, if slightly displaced from the equator of the HII region, are subject to the rocket effect and move off along the field lines like beads on a wire. As the magnetic field refills the HII region, it pulls in gas with it. This gas flows in at the equator then is directed along the field lines out of the poles, where it recombines forming dense neutral caps.

3. HII Region evolution in a magnetized turbulent molecular cloud

We use the results of MHD turbulence simulations (Vázquez-Semadeni *et al.* 2005) as the initial conditions for our simulations of HII region evolution in a magnetic, turbulent medium (cf. our previous paper Mellema *et al.* 2006a). The densest clump is taken to harbor the ionizing source and is moved to the center of the grid (the boundary conditions are periodic). The grid is a 256^3 box, 4 pc on a side with a mean number

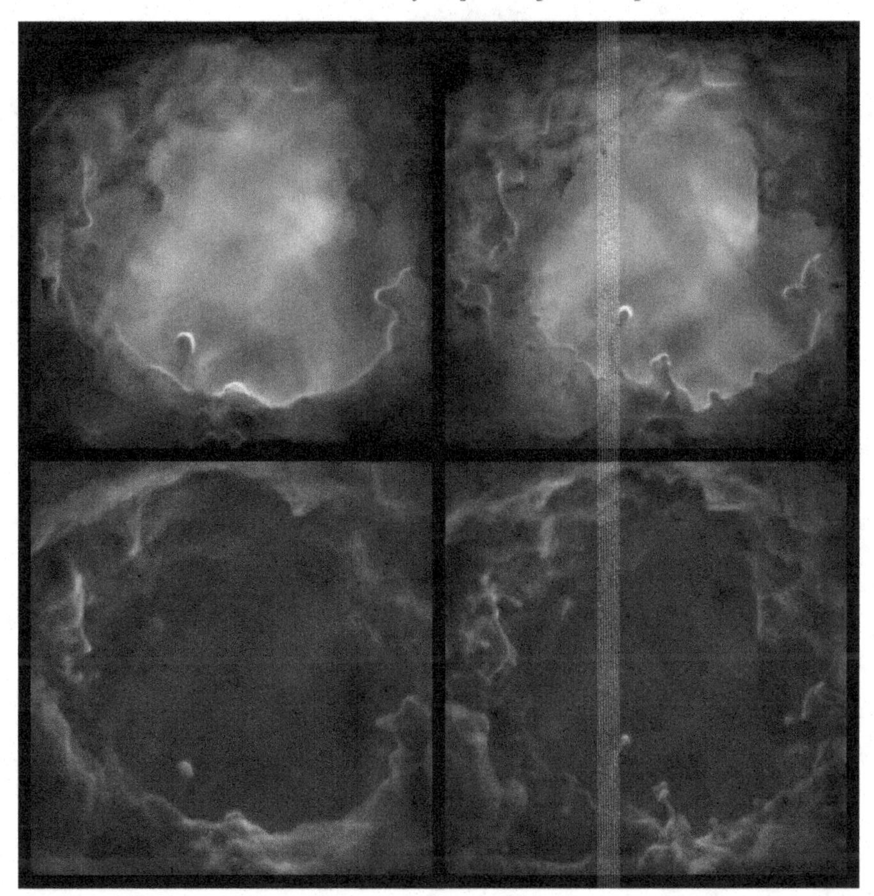

Figure 2. Simulated optical (top) and long-wavelength (bottom) emission maps for the expansion of an HII region due to an O star for MHD (left) and pure hydrodynamic (right) cases after 310,000 yrs of evolution. *Electronic verion only: Colors for the optical emission are: blue—[OIII], green—Hα, red—[NII]. Colors for the long-wavelength emission are: blue—6 cm radio free-free, green—generic PAH emission, red—cold neutral gas column density.*

density of $\langle n \rangle = 1000 \text{ cm}^{-3}$, a neutral gas temperature of 5 K, and a mean magnetic field strength of $\langle B \rangle = 14.5\,\mu$G and r.m.s. magnetic field $B_{\rm rms} = 24.6\,\mu$G.

For these simulations we use ionizing sources corresponding both to a B-star ($Q_{\rm H} = 10^{46.5}$ s^{-1}) and to an O-star ($Q_{\rm H} = 10^{48.5}$ s^{-1}). For the O-star, we do not expect the magnetic field to become important compared to the thermal pressure of the expanding HII region in a global sense within the time it takes for the HII region to expand beyond the confines of the box. Indeed, this is the case, as seen in Figure 2, which shows the optical and long-wavelength emission maps of both MHD and pure hydro simulations after 310,000 yrs. However, locally there are differences between the amount of fragmentation into globules seen in the pure hydro case and the MHD case. Neutral globules are overrun by the ionization front, protruding into the photoionized region until they are swept away by radiation-driven implosion and the rocket effect. In general, the pure hydro case has more fragmentation, since there is no support for the globules from magnetic pressure.

For the B-star simulations, the mean initial conditions suggest that it is possible that the magnetic field could have an effect on the global properties of the HII region, within the time and spatial scales of the simulation. After 10^6 years of evolution, there are a few indications that the magnetic field is having a large-scale effect on the HII region

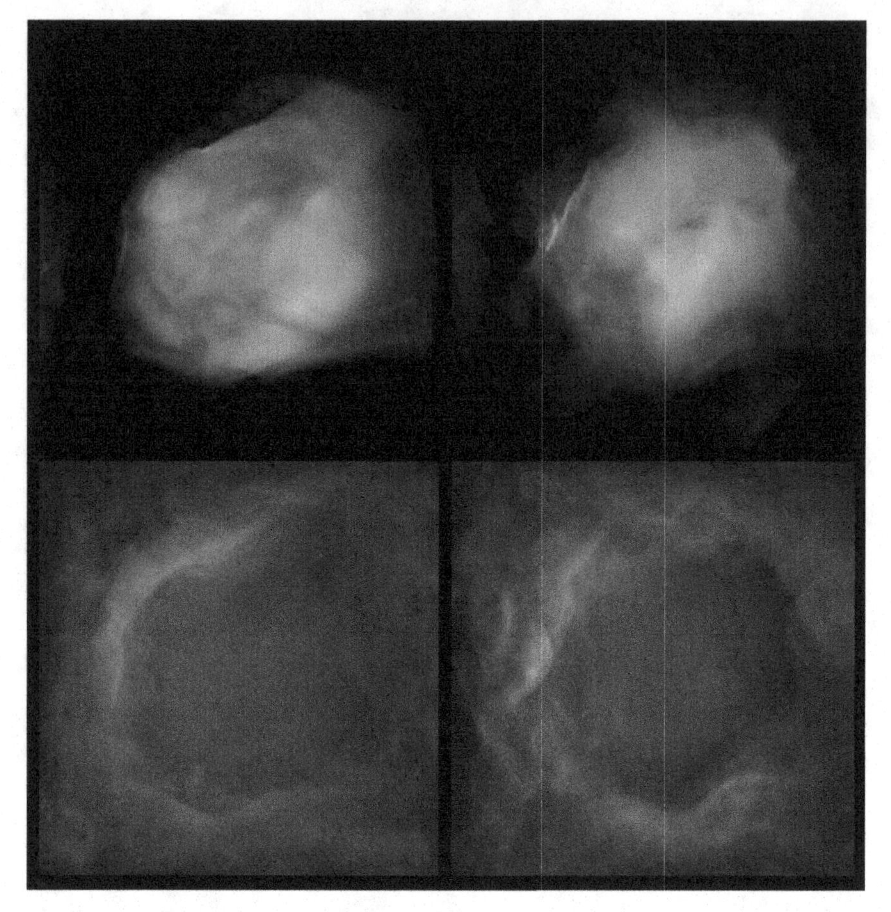

Figure 3. Same as Figure 2 but for a B star after 1 Myr of evolution. *Electronic version only: Colors for optical emission are: blue—[NII], green—Hα, red—[SII]. Colors for the long-wavelength emission are the same as in Figure 2.*

expansion, most notably in the lower part of the HII region as seen in Figure 3, where the magnetic field has become aligned with the ionization front and provides support for the neutral gas against expansion of the ionized gas in this direction. There is much less fragmentation in the B star case than in the O star case. This is because the clumps in the B star simulation are eroded by photoevaporation due to FUV radiation before the ionization front reaches them. The pressure of the warm neutral gas is insufficient to collapse underlying density inhomogeneities into dense globules and filaments.

References

Brogan, C. L. & Troland, T. H. 2001, *ApJ*, 560, 821
Deharveng, L., *et al.* 2009, *A&A*, 496, 177
Goodman, A. A. & Heiles, C. 1994, *ApJ*, 424, 208
Henney, W. J., Arthur, S. J., de Colle, F., & Mellema, G. 2009, *MNRAS*, 398, 157
Krumholz, M. R., Stone, J. M., & Gardiner, T. A. 2007, *ApJ*, 671, 518
Mellema, G., Arthur, S. J., Henney, W. J., Iliev, I. T., & Shapiro, P. R. 2006, *ApJ*, 647, 397
Vázquez-Semadeni, E., Kim, J., Shadmehri, M., & Ballesteros-Paredes, J. 2005, *ApJ*, 618, 344

Computational Star Formation
Proceedings IAU Symposium No. 270, 2011
J. Alves, B.G. Elmegreen, J. M. Girart & V. Trimble, eds.

doi:10.1017/S174392131100055X

Ionisation Feedback in Star and Cluster Formation Simulations

Barbara Ercolano[1] **and Matthias Gritschneder**[2]

[1]School of Physics, University of Exeter, Stocker Road, Exeter, EX4 4QL, UK
email: barbara@astro.ex.ac.uk

[2]KIAA, Peking University, Yi He Yuan Lu 5, Hai Dian Qu Beijing 100871, P. R. China
email: gritschneder@kiaa.pku.edu.cn

Abstract. Feedback from photoionisation may dominate on parsec scales in massive star-forming regions. Such feedback may inhibit or enhance the star formation efficiency and sustain or even drive turbulence in the parent molecular cloud. Photoionisation feedback may also provide a mechanism for the rapid expulsion of gas from young clusters' potentials, often invoked as the main cause of 'infant mortality'. There is currently no agreement, however, with regards to the efficiency of this process and how environment may affect the direction (positive or negative) in which it proceeds. The study of the photoionisation process as part of hydrodynamical simulations is key to understanding these issues, however, due to the computational demand of the problem, crude approximations for the radiation transfer are often employed.

We will briefly review some of the most commonly used approximations and discuss their major drawbacks. We will then present the results of detailed tests carried out using the detailed photoionisation code MOCASSIN and the SPH+ionisation code iVINE code, aimed at understanding the error introduced by the simplified photoionisation algorithms. This is particularly relevant as a number of new codes have recently been developed along those lines.

We will finally propose a new approach that should allow to efficiently and self-consistently treat the photoionisation problem for complex radiation and density fields.

Keywords. stars: formation, HII regions, methods: n-body simulations, radiative transfer

1. Introduction

Ionising radiation from OB stars influences the surrounding interstellar medium (ISM) on parsec scales. As the gas surrounding a high mass star is heated, it expands forming an HII region. The consequence of this expansion is twofold, on the one hand gas is removed from the centre of the potential, preventing further gravitational collapse and perhaps even disrupting the parent molecular cloud. On the other hand gas is swept up and compressed beyond the ionisation front producing high density regions that may be susceptible to gravitation collapse (i.e. the "collect and collapse" model, Elmegreen *et al.* 1995). Furthermore, pre-existing, marginally gravitationally stable clouds may also be driven to collapse by the advancing ionisation front (i.e. "radiation-driven implosion", Bertoldi 1989). Finally, ionisation radiation has also been suggested as a driver for small scale turbulence in a cloud (Gritschneder *et al.* 2009b). Observations (e.g. Deharveng, these proceedings) and theory (e.g. Dale *et al.* 2005, 2007, Gritschneder *et al.* 2009b) often present examples for positive and negative feedback, however, the net effect on the global star formation efficiency is still under debate.

From a theoretical point of view, different groups have performed a number of numerical experiments demonstrating that the efficacy and direction of photoionisation feedback are very sensitive to the specific initial conditions, in particular, to the location of the ionising source(s) and to whether the cloud is initially bound or unbound. This suggests that a parameter space study may be necessary to assess what environmental variables may affect the direction in which feedback proceeds. Several authors in these proceedings discuss the results of recent ionisation feedback simulations (see oral contributions by Arthur, Bisbas, Gritschneder and Walch, and poster contributions by Choudhury, Cornwall, Miao, Motoyama, Rodon and Tremblin).

As the field matures and the codes become more sophisticated it becomes important to assess the accuracy and limitations of the methods currently employed. The computational demand of treating the radiation transfer (RT) and photoionisation (PI) problem within a large scale hydrodynamical simulation has led to the development of approximate algorithms that drastically simplify the physics of RT and PI. In this review we will describe some of the most common approximations employed by current RT+PI implementations, highlighting some potentially important shortcomings. We will then present the result of our ongoing efforts to test current implementations against the 3D Monte Carlo code MOCASSIN (Ercolano *et al.* 2003, 2005, 2008) which includes all the necessary micro physics and solves the ionisation, thermal and statistical equilibrium in detail.

2. Some Common Approximations

The importance of studying the photoionisation process as part of hydrodynamical star formation simulations has long been recognised. Until very recently, however, due to the complexity and the computational demand of the problem, the evolution of ionised gas regions had only been studied in rather idealised systems (e.g. Yorke *et al.* 1989; Garcia-Segura & Franco 1996), with simulations often lacking resolution and dimensions. The situation in the latest years has been rapidly improving, however, with more sophisticated implementations of ionised radiation both in grid-based codes (e.g. Mellema *et al.* 2006; Peters *et al.* 2010) and Smoothed Particle Hydrodynamical (SPH) codes (e.g. Kessel-Deynet & Burkert 2000; Miao *et al.* 2006; Dale *et al.* 2007; Gritschneder *et al.* 2009; Bisbas *et al.* 2009). Klessen *et al.* (2009) and Mac Low *et al.* (2007) present recent reviews of the numerical methods employed.

While the new codes can achieve higher resolutions and can treat more realistic geometries, the treatment of RT and PI is still rather crude in most cases. Even in the current era of parallel computing, an exact solution of the radiative transfer (RT) and photoionisation (PI) problem in three dimensions within SPH calculations is still prohibitive. Some common approximations include the following:

(*a*) Monochromatic radiation field: In order to avoid the burden of frequency resolved RT calculations, monochromatic calculations are often carried out, where all the ionising flux is assumed to be at 13.6 eV (i.e. the H ionisation potential). This approximation is often implicit in the choice of a single value for the gas opacity, and it is of course implicit to Strömgren-type calculations. Implicit or explicit monochromatic fields have the serious drawback that the ionisation and temperature structure of the gas cannot be calculated.

(*b*) Ionisation and thermal balances: Its equations are not solved *or* simple heating/cooling functions are employed *or* the temperature is a simple function of an approximate ionisation fraction. When monochromatic fields are employed it is not possible to calculate the necessary terms to solve the balance equations and idealised temperature distributions must be used.

(*c*) On-the-spot (OTS) approximation (no diffuse field): The OTS approximation is described in detail by Osterbrock & Ferland (2006, page 24). In the OTS approximation the diffuse component of the radiation field is ignored under the assumption that any ionising photon emitted by the gas will be reabsorbed elsewhere, close to where it was emitted, hence not contributing to the net ionisation of the nebula. This is not a bad approximation in the case of reasonably dense homogeneous or smoothly varying density fields, but it is certain to fail in the highly inhomogeneous star-forming gas, where the ionisation and temperature structure of regions that lie behind high density clumps and filaments is often dominated by the diffuse field.

(*d*) Steady-state calculations (instantaneous ionisation): The ionisation structure and the gas temperature of a photoionised region is often obtained by simultaneously solving the *steady state* thermal balance and ionisation equilibrium equations. This approximation is valid when the atomic physics timescales are shorter than the dynamical timescales and the rate of change of the ionising field. In this case, the photoionisation problem is completely decoupled from the dynamics and it can be solved for a given gas density distribution obtained as a snapshot at a given time in the evolution of a cloud. This is a fair assumption for the purpose to study of ionisation feedback on large scales, as most of the gas will be in equilibrium. Non-equilibrium effects, however, should still be kept in mind when interpreting the spectra of regions close to the ionisation front or where shocks are present.

3. How good are the approximations?

In cases where the steady-state calculations are relevant, it is possible to test the effects of approximations a-c from the above list by comparing the temperature distributions obtained by the hydro+ionisation codes against those obtained by a specialised photoionisation code, like the MOCASSIN code, for density snapshots at several times in the hydrodynamics simulations.

MOCASSIN is a fully three-dimensional photoionisation and dust radiative transfer code that employs a Monte Carlo approach to the fully frequency resolved transfer of radiation. The code includes all the microphysical processes that influence the gas ionisation balance and thermal balance as well as those that couple the gas and dust phases. In the case of an HII region ionised by an OB star the dominant heating process for typical gas abundances is H photoionisation, balanced by cooling via collisionally excited line emission (dominant), recombination line emission and free-bound and free-free emission. The atomic database included in MOCASSIN includes opacity data from Verner *et al.* (1996), energy levels, collision strengths and transition probabilities from Version 5.2 of the CHIANTI database (Landi *et al.* 2006, and references therein) and the improved hydrogen and helium free-bound continuous emission data of Ercolano & Storey (2006).

Dale *et al.* (2007, DEC07) performed detailed comparisons against MOCASSIN's solution for the temperature structure of a complex density field ionised by a newly born massive star located at the convergence of high density accretion streams. They found that the two codes were in fair agreement on the ionised mass fractions in high density regions, while low density regions proved problematic for the DEC07 algorithm. The temperature structure, however, was poorly reproduced by the DEC07 algorithm, highlighting the need for more realistic prescriptions. For more details see DEC07.

More recently we have used the MOCASSIN code to calculate the temperature and ionisation structure of the turbulent ISM density fields presented by Gritschneder *et al.* (2009b, hereafter: G09b). The SPH particle fields were obtained with the IVINE code (Gritschneder *et al.* 2009a) and mapped onto a regular 128^3 Cartesian grid. In order to compare with IVINE, which calculates the RT along parallel rays, the stellar field in MOCASSIN was forced to be plane parallel, while the following RT was performed in three dimensions thus allowing for an adequate representation of the diffuse field. The incoming stellar field was set to the value used by G09b ($Q_H^0 = 5 \times 10^9$ ionising photons

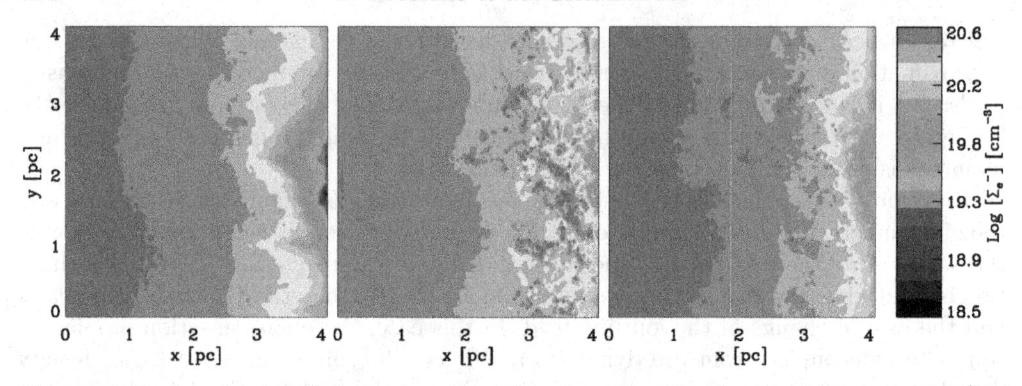

Figure 1. Surface density of electrons projected in the z-direction for the G09b turbulent ISM simulation at t = 0.5 Myr. *Left:* iVine; *Middle:* MOCASSIN H-only; *Right:* MOCASSIN nebular abundances.

per second) and a blackbody spectrum of 40kK was assumed. We run H-only simulations (referred to as "H-only") and simulations with typical HII region abundances (referred to as "Metals"). The elemental abundance are as follows, given as number density with respect to Hydrogen: He/H = 0.1, C/H = 2.2e-4, N/H = 4.0e-5, O/H = 3.3e-4, Ne/H = 5.0e-5, S/H = 9.0e-6.

The resulting MOCASSIN temperature and ionisation structure grids were compared to those obtained by IVINE in order to address the following questions:

(*a*) Are the global ionisation fractions accurate?
(*b*) How accurate is the gas temperature distribution?
(*c*) What is the effect of the diffuse field?
(*d*) How can the algorithm be improved?

3.1. *Global Properties*

Figure 1 shows the surface density of electrons projected in the z-direction for the G09b turbulent ISM simulation at t = 0.5 Myr. The figure shows that no significant differences are noticeable in the integrated ionisation structure, implying that the global ionisation structure is correctly determined by IVINE. This is also confirmed by the comparison of the total ionised mass fractions: at t = 0.5 Myr, iVine obtains a total ionised mass of 13.9%, while MOCASSIN "H-only" and "Metals" obtain 15.6% and 14.0%, respectively. The agreement at other time snapshots is equally good (e.g. at t = 250kyr IVINE obtains 9.1% and MOCASSIN "Metals" 9.5%).

It may at first appear curious that the agreement should be better between IVINE and MOCASSIN "Metals", rather than MOCASSIN "H-only", given that only H-ionisation is considered in IVINE. This is however simply explained by the fact that IVINE adopts a "ionised gas temperature" (T_{hot}) of 10kK, which is close to a *typical* HII region temperature, with *typical* gas abundances. The removal of metals in the "H-only" simulations causes the temperature to rise to values close to 17kK, due to the fact that cooling becomes much less efficient without collisionally excited lines of oxygen, carbon etc. The hotter temperatures in the "H-only" models directly translate to slower recombinations, as the recombination coefficient is proportional to the inverse square root of the temperature. As a result of slower recombination the "H-only" grids have a slightly larger ionisation degree.

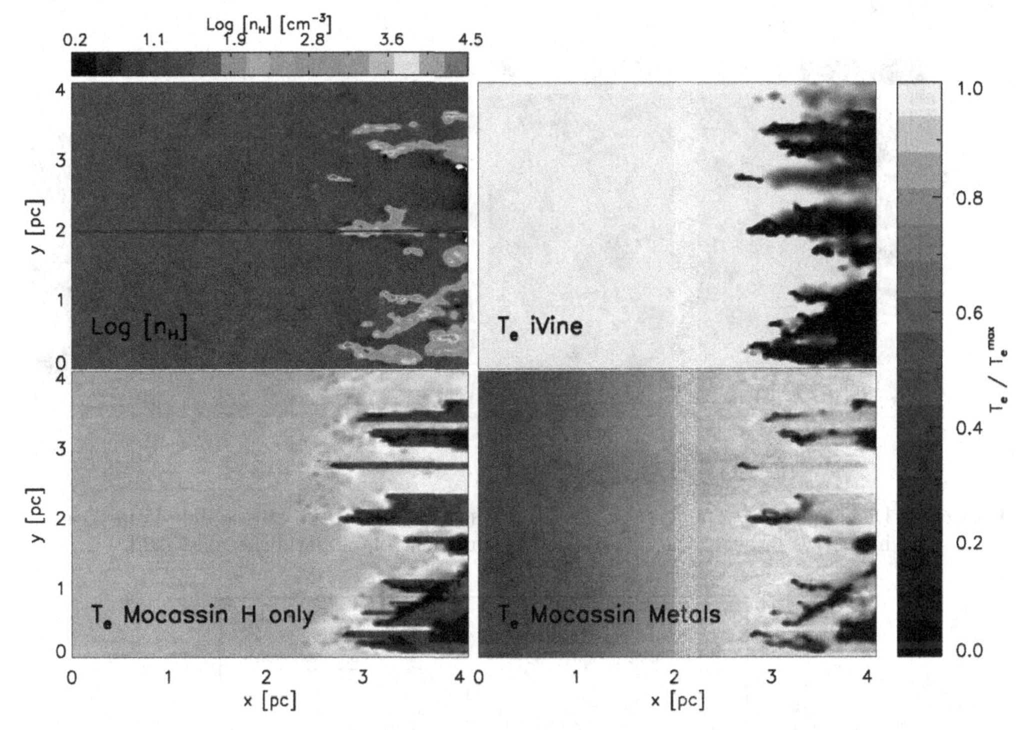

Figure 2. Density and temperature maps for the z = 25 slice of the G09 turbulent ISM simulation at t = 0.5 Myr. *Top left:* Gas density map; *Top right:* electron temperature, T_e as calculated by iVine; *Bottom left:* electron temperature, T_e as calculated by MOCASSIN with H-only; *Bottom right:* electron temperature, T_e as calculated by MOCASSIN with nebular abundances.

3.2. *Ionisation and temperature structure*

Accurate gas temperatures are of prime importance as this is how feedback from ionising radiation impacts on the hydrodynamics of the system. In Figure 2 we compare the electron temperatures T_e calculated by IVINE and MOCASSIN ("H-only" and "Metals") in a z-slice of the t = 0.5 Myr grid. The top-right panel shows the number density [cm^{-3}] map for the selected slice. The large shadow regions behind the high density clumps are immediately evident from both figures. These shadows are largely reduced in the MOCASSIN calculations as a result of diffuse field ionisation. The diffuse field is softer than the stellar field and therefore temperatures in the shadow regions are lower. The higher temperatures in the shadow regions of the MOCASSIN "Metals" model are a consequence of the Helium Lyman radiation and the heavy elements free-bound contribution to the diffuse field. The rise in gas temperature shown in the MOCASSIN results at larger distances from the star is not surprising and a simple consequence of radiation hardening and the recombining of some of the dominant cooling ions.

4. Towards more realistic algorithms

As IVINE solves the transfer along plane parallel rays, it has currently no means of bringing ionisation (and hence heating) to regions that lie behind high density clumps. This creates a large temperature (pressure) gradient between neighbouring direct and diffuse-field dominated regions, which may have important implications for the dynamics,

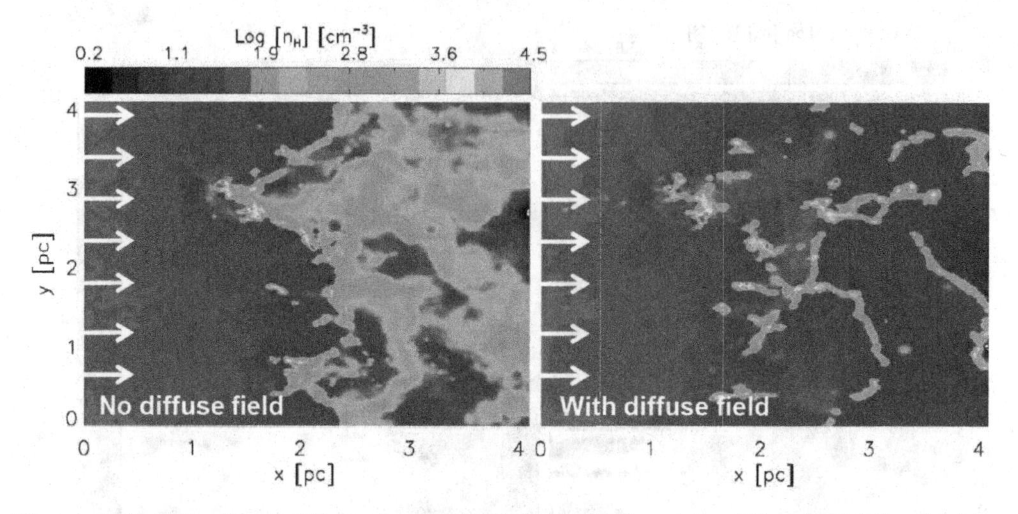

Figure 3. Density slice at 250 kyr for the OTS iVINE (left) and the diffuse field iVine (right). The arrows indicate the direction of the incident plane parallel stellar field.

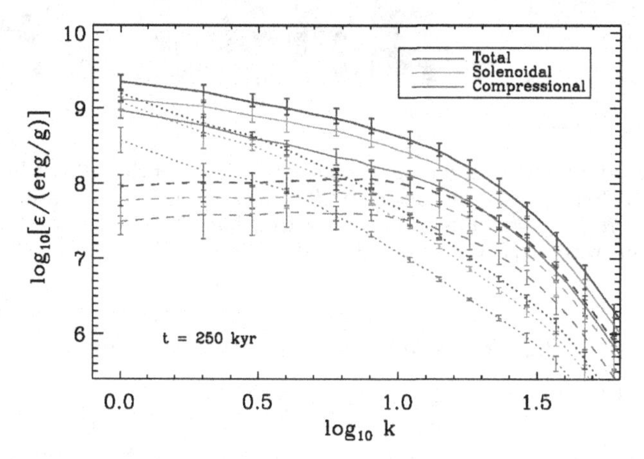

Figure 4. Turbulence spectrum obtained for the standard OTS iVINE (solid lines), the control run with no ionising radiation (dotted line) and the diffuse field iVINE (dashed lines).

particularly with respect to turbulence calculations. The same problem is faced by all codes that employ the OTS approximation and thus ignore the diffuse field contributions.

In order to investigate whether the error introduced by OTS approximation actually bears any consequence on the dynamical evolution of the system and on the turbulence spectrum, we propose here a simple zeroth order strategy to include the effects of the diffuse field in iVINE and which can be readily extended to other codes. It consists of the following steps: (i) identify the diffuse field dominated regions (shadow); (ii) study the realistic temperature distribution in the shadow region using fully frequency resolved three-dimensional photoionisation calculations performed with MOCASSIN and parameterise the gas temperature in the shadow regions as a function of (e.g.) gas density; (iii) implement the temperature parameterisation in iVINE and update the gas temperatures in the shadow regions at every dynamical time step accordingly.

We note that this approach allows for environmental variables, such as the hardness of the stellar field and the metallicity of the gas to be accounted for in the SPH calculation, since their effect on the temperature distribution is folded in the parameterisation obtained with MOCASSIN.

Figure 3 shows a slice of the density structure snapshots at 250k year for a standard IVINE (left panel) compared to a first attempt at a diffuse field implementation in IVINE (right panel). The effects of the diffuse field in this calculation are purposely exaggerated to highlight possible consequences. In this toy calculation much of the lower density gas has already been expelled from the grid at this time by diffuse field ionisation, indicating a clear divergence of the dynamical evolution of the system with and without the OTS approximation. The turbulence spectrum obtained in the two cases are also rather different, as shown in Figure 4, where the specific kinetic energy is plotted as a function of wave number in the case of the control run with no ionisation at all (dotted lines), OTS IVINE (solid lines) and diffuse field IVINE (dashed lines). The suppression of the larger scales is probably due to the fact that much of the gas has been removed from the region, however it also appears that the small scale turbulence is not as efficiently driven when the diffuse field is considered. The latter is due to the fact that the large temperature gradients created by the OTS at the shadow regions are removed when the diffuse field is considered.

We stress that the results presented here are to be considered only a first exploratory step to establish whether diffuse field effects are likely to play a role in the dynamical evolution of a turbulent medium. While the above suggests that this may indeed be the case, it is important to note here that our current crude implementation overestimates the effects of diffuse fields. More detailed comparisons will be presented in a forthcoming article (Ercolano & Gritschneder 2010, in prep)

5. Conclusions

We have presented a review of the current implementations of photoionisation algorithms in star formation hydrodynamical simulation, highlighting some of the most common approximation that are employed in order to simplify the radiative transfer and photoionisation problems.

We discuss the robustness of the temperature fields obtained by such methods in light of recent tests against detailed 3D photoionisation calculations for complex density distributions typical of star forming regions. We conclude that while the global ionised mass fractions obtained by the simplified methods are roughly in agreement, the temperature fields are poorly represented. In particular, the assumption of the OTS approximation may lead to unrealistic shadow regions and extreme temperature gradients that affect the dynamical evolution of the system and its turbulence spectrum.

We propose a simple strategy to provide a more realistic description of the temperature distribution based on parameterisations obtained with a dedicated photoionisation code, MOCASSIN, which includes frequency resolved 3D radiative transfer and all the microphysical process needed for an accurate calculation of the temperature distribution of ionised regions. This computationally inexpensive method allows to include the thermal effects of diffuse field, as well as accounting for environmental variables, such as gas metallicity and stellar spectra hardness.

References

Bertoldi, F. 1989, *ApJ*, 346, 735
Bisbas, T. G., Wünsch, R., Whitworth, A. P., & Hubber, D. A. 2009, *A&A*, 497, 649
Dale, J. E., Bonnell, I. A., Clarke, C. J., & Bate, M. R. 2005, *MNRAS*, 358, 291

Dale, J. E., Ercolano, B., & Clarke, C. J. 2007, *MNRAS*, 382, 1759
Elmegreen, B. G., Kimura, T., & Tosa, M. 1995, *ApJ*, 451, 675
Ercolano, B., Barlow, M. J., Storey, P. J., & Liu, X.-W. 2003, *MNRAS*, 340, 1136
Ercolano, B., Barlow, M. J., & Storey, P. J. 2005, *MNRAS*, 362, 1038
Ercolano, B. & Storey, P. J. 2006, *MNRAS*, 372, 1875
Ercolano, B., Young, P. R., Drake, J. J., & Raymond, J. C. 2008, *ApJS*, 175, 5345, 165
Garcia-Segura, G. & Franco, J. 1996, *ApJ*, 469, 171
Goodwin, S. P. & Bastian, N. 2006, *MNRAS*, 373, 752
Gritschneder, M., Naab, T., Walch, S., Burkert, A., & Heitsch, F. 2009, *ApJL*, 694, L26
Gritschneder, M., Naab, T., Burkert, A., Walch, S., Heitsch, F., & Wetzstein, M. 2009, *MNRAS*, 393, 21
Kessel-Deynet, O. & Burkert, A. 2000, *MNRAS*, 315, 713
Klessen, R. S., Krumholz, M. R., & Heitsch, F. 2009, Advanced Science Letters (ASL), Special Issue on Computational Astrophysics, edited by Lucio Mayer, arXiv:0906.4452
Landi, E., Del Zanna, G., Young, P. R., Dere, K. P., Mason, H. E., & Landini, M. 2006, *ApJS*, 162, 261
Mac Low, M.-M. 2007, proceedings of "Massive Star Formation: Observations Confront Thoery", ASP conference series, eds. H. Beuther *et al.*, arXiv:0711.4047
Mellema, G., Arthur, S. J., Henney, W. J., Iliev, I. T., & Shapiro, P. R. 2006, *ApJ*, 647, 397
Miao, J., White, G. J., Nelson, R., Thompson, M., & Morgan, L. 2006, *MNRAS*, 369, 143
Osterbrock, D. E., & Ferland, G. J. 2006, Astrophysics of gaseous nebulae and active galactic nuclei, 2nd. ed. by D.E. Osterbrock and G.J. Ferland. Sausalito, CA: University Science Books, 2006,
Peters, T., Mac Low, M.-M., Banerjee, R., Klessen, R. S., & Dullemond, C. P. 2010, *ApJ*, 719, 831
Verner, D. A., Ferland, G. J., Korista, K. T., & Yakovlev, D. G. 1996, *ApJ*, 465, 487
Yorke, H. W., Tenorio-Tagle, G., Bodenheimer, P., & Rozyczka, M. 1989, *A&A*, 216, 207

Computational Star Formation
Proceedings IAU Symposium No. 270, 2011
J. Alves, B.G. Elmegreen, J. M. Girart & V. Trimble, eds.

doi:10.1017/S1743921311000561

Supernova Feedback on the Interstellar Medium and Star Formation

Gerhard Hensler

Institute of Astronomy, University of Vienna, Tuerkenschanzstr. 17, 1180 Vienna, Austria
email: gerhard.hensler@univie.ac.at

Abstract. Supernovae are the most energetic stellar events and influence the interstellar medium by their gasdynamics and energetics. By this, both also affect the star formation positively and negatively. In this paper, we review the development of the complexity of investigations aiming at understanding the interchange between supernovae and their released hot gas with the star-forming molecular clouds. Commencing from analytical studies the paper advances to numerical models of supernova feedback from superbubble scales to galaxy structure. We also discuss parametrizations of star-formation and supernova-energy transfer efficiencies. Since evolutionary models from the interstellar medium to galaxies are numerous and apply multiple recipes of these parameters, only a representative selection of studies can be discussed here.

Keywords. stars: formation, ISM: kinematics and dynamics, ISM: bubbles, (ISM:) supernova remnants ISM: structure, galaxies: evolution, galaxies: ISM

---

1. Introduction

Since stars are formed within the coolest molecular material of the interstellar medium (ISM), the star-formation rate (SFR) should be determined simply by the gas reservoir and by the free-fall time τ_{ff} of molecular clouds (Elmegreen 2002). This, however, raises a conflict between the ISM conditions and observed SFRs in the sense that τ_{ff} for a typical molecular cloud density of 100 cm^{-3} amounts to 10^{14} sec, i.e. $3 \cdot 10^6$ yrs. For the total galactic molecular mass of $10^9 - 10^{10}\, M_\odot$ the SFR should then amount to about 100 to 1000 $M_\odot yr^{-1}$, what is by orders of magnitudes larger than observed and the gas reservoir within the Milky Way would have been used up today. This means, that the SF timescale must be stretched with respect to collapse or dynamical timescale by introducing a SF efficiency (SFE) ϵ_{SF} and its definition could read: $\tau_{SF} = \epsilon_{SF}^{-1} \cdot \tau_{ff}$.

Already in 1959 Schmidt argued that the SFR per unit area is related to the gas column density Σ_g by a power law with exponent n. Kennicutt (1998) derived from the Hα luminosity of spiral galaxies $\Sigma_{H\alpha}$ a vertically integrated and azimuthally averaged SFR (in $M_\odot yr^{-1} pc^{-2}$), i.e. of gas disks in rotational equilibrium, and found a correlation with $n = 1.4 \pm 0.15$ holding over more than 4 orders of magnitude in Σ_g with a drop below a density threshold at 10 $M_\odot/pc^2$. While this relation establishes an equilibrium SFR, it is not surprising that the slope varies for dynamically triggered SF, as in starburst and merger galaxies and for high-z galaxies, when the disks form by gas infall, for the latter reaching $n = 1.7 \pm 0.05$ (Bouché *et al.* 2007).

The ordinary Kennicutt-Schmidt (KS) relation can be understood by the simple analytical assumption allowing for the $\tau_{ff} - \tau_{SF}$ relation and for a uniform state of the ISM. Its equilibrium on disk scales requires that heating processes counteract to the natural cooling of plasmas. Besides the possible heating processes from dissipation of dynamical effects, as there are the differential rotation of the disk, gas infall, tidal interactions, shocks, etc., to the feedback by freshly produced stars, not for all of them it is obvious, how effectively they influence the SF by the expected self-regulation or vice versa trigger it.

Unfortunately, the issue of a general KS-law is confused by the similarity of slopes under various stellar feedback strengths (see e.g. sect. 2 in Hensler (2009)). Köppen, Theis, & Hensler (1995) demonstrated already that the SFR achieves a dependence on ρ_g^2, if the stellar heating is compensated by collisional-excited cooling radiation (e.g. Böhringer & Hensler 1989). The coefficient of this relation determines τ_{SF}. Obviously, the SF in galaxies, therefore, depends on both the gas content and the energy budget of the ISM.

Since the most efficient stellar energy power is exerted by supernovae (SNe), and here particularly by the explosions of the shortly living massive stars as type II SNe, their feedback to the ISM is of fundamental relevance for the SF. In this paper, we overview and discuss the regulation of the ISM by the SF feedback thru SNe and more pronounced by their cumulative effect as superbubbles. Although the expression *feedback* of SNe also includes their release of freshly produced elements, here we only focus on the dynamical and energetic issues and refer the reader interested on the chemical evolutionary consequences to Hensler & Recchi (2010).

2. Supernova feedback

2.1. *Energy release and the Interstellar Medium*

As SN explosions were since long known to expel vehemently expanding hot gas, Spitzer (1956) predicted this hot ISM phase to expand from the galactic disk where it cannot be bound or pressure confined to form a hot halo gas. Not before the 70's and with the aid of observations which made formerly unaccessible spectral ranges (from the FIR/submm to X-rays) available, the existence a hot gas component within our Milky Way was manifested and later also perceived in other disk galaxies. Although the ISM in its cool molecular component is conditioned for SF, hot gas regulates its dynamics as driver of shocks and turbulence as well as its energetics by heating thru cooling radiation and heat conduction, by this, exerting negative stellar feedback.

After these facts became internalized, a significant amount of mostly analytical explorations were dedicated to a first understanding of the ISM structure, its temporal behaviour with respect to the hot ISM (comprehensively reviewed by Spitzer (1990)), to volume filling factors and mass fractions of the gas phases. For this purpose, randomly distributed and temporally exploding SNe were considered with cooling and according expansion (Cioffi & Shull 1991) and with transitions to/from the warm/cool gas (McKee & Ostriker 1977) or as non-dynamically interchanging gas phases (see e.g. Habe, Ikeuchi, & Takaka 1981, Ikeuchi & Tomita 1983). In general, the action of SNe on the ISM is multifacetted (Chevalier 1977) and to a significant part affecting the SF. Because SNe type II happen on much shorter timescales than type Ia, all the ISM and SF feedback studies focus mostly on their energy deposit and dynamical issues.

While these latter studies include arbitrary SN rates with a description of the temporal size evolution of individual supernova remnants (SNRs), the SF dependence on the physical state is not yet properly treated. With a toy model consisting of 6 ISM components and at least 10 interchange processes Ikeuchi, Habe, & Tanaka (1984) also included the formation of stars from giant molecular clouds. Those can only form from cool clouds which, on the other hand, are swept-up and condensed in SN shells, so that SF and SN explosions together with different gas phases and interchange processes form a consistent network. As a natural effect of this local consideration the SF can oscillates with the corresponding timescales.

First dynamical approaches to the structure evolution of the ISM and gas disks, aiming at understanding the excitation of turbulence, were performed by Rosen & Bregman (1995). As heating sources they took the energy of massive stellar winds only into account, which is at the lower range of stellar power to the ISM because of their very low energy transfer efficiencies (Hensler 2007). Nevertheless, their models already demonstrate the compression of gas filaments and the expulsion of gas vertically from the disk even with the inclusion of self-gravity. Indeed, the SF recipe is not self-consistent and the vertical boundary conditions seem to affect the vertical gas dynamics artificially.

In more recent numerical investigations of the ISM evolution in the context of SF and SN explosions Slyz *et al.* (2005) performed studies of the influence of SN feedback, self-gravity, and stars as an additional gravitational component on the SFR. The main issues can be summarized as that feedback enhances the ISM porosity, increases the gas velocity dispersion and the contrasts of T and ρ, so that smaller and more pronounced structures form. Since the SFR depends on most of these variables, more importantly, the SN feedback models reach SFRs by a factor of two higher than without feedback.

At the same time, de Avillez & Breitschwerdt (2004) simulated the structure evolution of the solar vicinity in a box of $1 \times 1 \times 10\, kpc^3$ size and identified the Local Bubble and its neighbouring Loop I in their models. In addition, filamentary neutral gas structures, resolved down to 0.625 pc, become visible and the vertical matter cycle. However, because of the lack of self-consistent SF they applied an analytical recipe of random SN events (de Avillez 2000) to their models that obviously underestimates the production of superbubbles from massive OB associations and, moreover, cannot reasonably achieve issues on the feedback of SNe on the SF. The ISM processes also do not include self-regulation effects by radiation or heat conduction, which lead to negative SF feedback.

2.2. *Star-formation triggering*

SN and stellar wind-driven bubbles sweep up surrounding gas, condense it, and are, by this, expected to excite SF in a self-propagating manner (abbreviated SPSF) as a positive feedback. In a cornerstone paper Gerola & Seiden (1978) have demonstrated by a parameter study, how SNR expansion can lead to structure formation in galactic disks by SPSF. Their main aim, however, to reproduce the general spiral arm structure thru SPSF by means of the coherence of different timescales cannot be reached and most galaxies develop patchy structures randomly.

Such SF trigger by the condensation of swept-up gas in SN or more efficiently in superbubbles (see fig. 1) can be explored in detail by the investigation of the fragmentation timescale (see e.g. Ehlerova *et al.* 1997, Fukuda & Hanawa 2000). Ehlerova *et al.* compared a self-similar analytical solution with the results of 3D numerical simulations of superbubble expansions in homogeneous media. The amount of energy supply from the final number of young stars in an OB association, the value of the sound speed, the stratification and density of the ambient medium, the galactic differential rotation, and the vertical gravitational force in the galactic disk, all these influence the fragmentation. The typical superbubble radius at which shells start to fragment decreases from almost 700 pc at an ambient gas density n of 1 cm^{-3} to 200 pc at 10 cm^{-3}. While in thick disks like they exist in dwarf galaxies (DGs) nearly the whole shell fragments, in thin disks it is restricted to the galactic equator only. The SF process itself cannot be resolved in these studies and the assumption follows the line, that unstable fragments may become molecular and trigger the formation of molecular clouds in which new stars are formed. The main conclusion is that in DGs the SF may propagate in all directions possibly turning the system as the whole into a starburst, while in spiral galaxies the SF propagates within a thin strip near the symmetry plane only. Since the applied thin shell

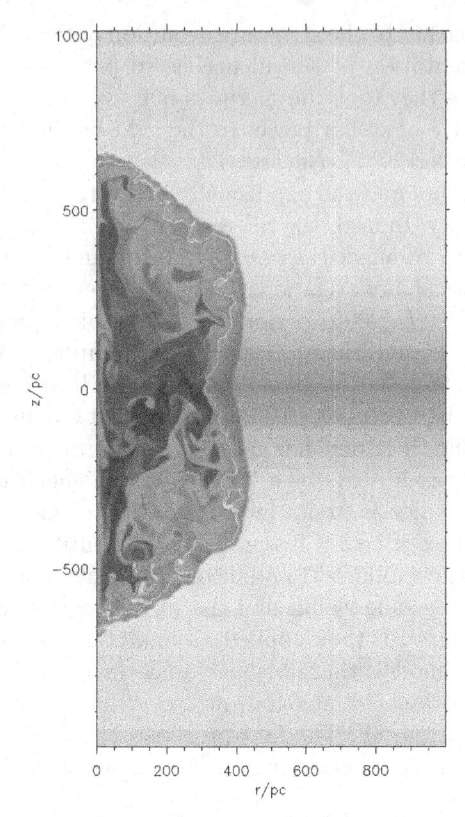

Figure 1. Density distribution of a superbubble after 20 Myrs. The superbubble results from 100 supernova typeII explosions at the origin of the coordinate system. The temporal sequence of explosions happens according to the lifetimes of stars in the mass range between 10 and 100 $M_\odot$ with a Salpeter IMF. The galactic disk is vertically composed of the three-phase interstellar medium, cool, warm and hot phase, respectively, with (central density ρ_0 [*in* $g\,cm^{-3}$]; temperature T [*in K*]; vertical scaleheight H [*in pc*]) of (2×10^{-24}; 150; 100), (5×10^{-25}; 9000; 1000), (1.7×10^{-27}; 2×10^6; 4000). The density varies from almost $10^{-23}\,g\,cm^{-3}$, in the densest part of the shell to $2 \times 10^{-28}\,g\,cm^{-3}$ in the darkest bubble interiors. *(from Gudell 2002)*

approximation is reasonably only a 0th-order approximation, in a recent paper Wünsch *et al.* (2010) (see also this volume) clarify that the shell thickness and the environmental pressure influences the fragments in the sense that their sizes become smaller for higher pressure. Nevertheless, the deviations from the thin-shell approximations are not large. Yet as a drawback the influences of magnetic fields is not taken into account.

The perception of SF trigger in SN or superbubble shells sounds reasonable from the point of view of numerical models because (as shown in fig. 1) sufficient ambient ISM mass is swept up, preferably in the gas disk itself, and is capable to cooling and gravitational instabilities. Shell-like distributions of young stars, are e.g. found in G54.4-0.3, called *sharky* (Junkes, Fürst, & Reich 1992), in the Orion-Monoceros region (Wilson *et al.* 2005), more promising in the Orion-Eridanus shell (Lee & Chen 2009), and in several superbubbles in the Large Magellanic Cloud, as e.g. Henize 206 (Gorjian *et al.* 2004). Also the formation of Gould's Belt stellar associations in the shell of a superbubble is most probable (Moreno, Alfaro, & Franco 1999), however, still debated (Comeron & Torra 1994).

Another possible feedback effect by SN is caused, when the ultra-fast SNR shock overruns a dense interstellar cloud, so that the clouds are quenched and stars are expected to be formed instantaneously. Although such cloud crushing is numerically modelled its effect on the necessary collapse of sub-clumps is not resolved simultaneously (Orlando *et al.* 2005).

Since in several DGs excessively high SFRs are observed (Hunter *et al.* 1998, van Zee, Skillman, & Salzer 1998, Stil & Isreal 2002 and further more), it is a matter of study whether starbursts are the result of a SF self-trigger mechanism or the issue of an external process because the objects are obviously linked to large enveloping gas reservoirs from which gas infall must be invoked (Mühle *et al.* 2005, Hensler *et al.* 2004). How the hot SN gas that transits from superbubbles to galactic winds interacts with infalling cold gas is yet unexplored but an attractive challenge from the present to the early universe (Recchi & Hensler 2007). High-velocity clouds (HVCs) in our Milky Way on their passage through the hot halo gas are expected to be easily disrupted due to Kelvin-Helmholtz (KH) instability. In contrast, model clouds of self-gravitating clouds including the effect of saturated heat conduction survive over almost 100 Myrs because they are mostly stabilized against KH instability (Vieser & Hensler 2007). While their compression in the stratified halo gas should not be able to avoid SF, its absence in HVCs is still a mistery, so that its understanding would provide a further insight into the positive vs. negative feedback effect of SN gas on SF in clouds.

2.3. *Supernova energy impact*

Because of their enormous power in various forms, the majority of galaxy evolution models usually take the energy deposit of SNeII explosions into account as the only heating source for the ISM. Although it is generally agreed that the explosive energy of an individual SN lies around 10^{51} ergs with significant uncertainties (or an intrinsic scatter), however, of one order of magnitude, the energy deposit as turbulent and consequently as thermal energy to the ISM is still more than unclear, but it is one of the most important ingredients for our understanding of galaxy formation (e.g. Efstathiou 2000, Silk 2003).

As a similar study, the energy release of massive stars as radiation-driven and wind-blown HII regions can be considered. Analytical estimates for purely radiative HII regions yielded an energy transfer efficiency ϵ of the order of a few percent (Lasker 1967). Although the additional stellar wind power L_w can be easily evaluated from models and observations, its fraction that is transferred into thermal and turbulent energy is not obvious from first principles. Transfer efficiencies for both radiative and kinetic energies remain much lower from detailed numerical simulations than analytically derived and amount to only a few per mil (Hensler 2007, and references therein). Nevertheless, as valid for HII regions and because massive stars do not disperse from the SF site, SNeII explode within the stellar associations. While massive stars contribute significantly to the ISM structure formation as e.g. by cavities and holes, in the HI gas and chimneys of hot gas, on large scales the energy release by massive stars triggers the matter circulations via galactic outflows from a gaseous disk and galactic winds. By this, also the chemical evolution is affected thru the loss of metal-enriched gas from a galaxy (for observations see e.g. Martin, Kobulnicki & Heckman (2002), for models e.g. Recchi & Hensler (2006b)).

SN explosions as an immediate consequence of SF stir up the ISM by the expansion of hot bubbles, deposit turbulent energy into the ISM, thereby, heat the ISM and regulate the SF again (Hensler & Rieschick 2002). This negative energy feedback is enhanced at low gravitation because the SN energy exceeds easily the galactic binding energy and drives a galactic wind. Although investigations have been performed for the heating (or energy transfer) efficiency ϵ_{SN} of SNe (Thornton *et al.* 1998), superbubbles (e.g. Strickland *et al.* 2004), and starbursts (Melioli & de Gouveia Dal Pino 2004) they are yet too simplistic and mostly spatially poorly resolved to account for quantitative results.

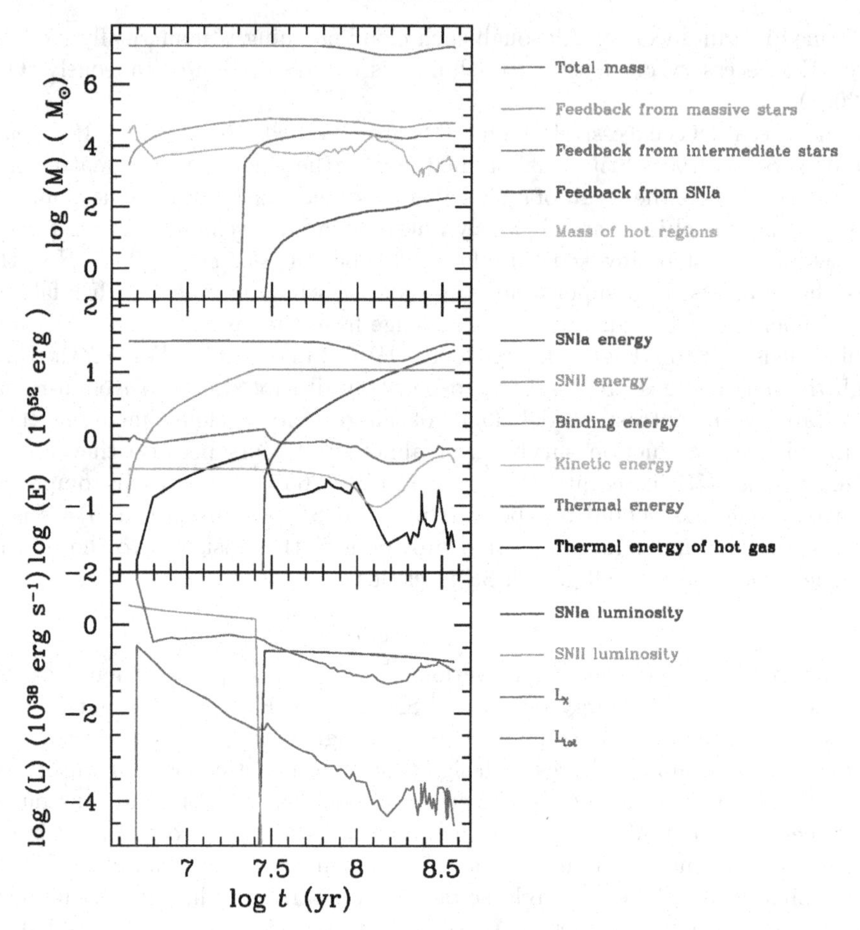

Figure 2. Temporal evolution of masses energies, and luminosities, respectively, of contained and released components for a single star-formation burst in a dwarf galaxy like I Zw 18 (Recchi *et al.* (2002). The green (light) curves refer to the supernovae typeII contribution. Comparison of the thermal energy (red line) and the accumulated supernova typeII energy (middle panel) shows that the thermal energy is almost not enhanced by SNeII.

Thornton *et al.* derived an efficiency ϵ_{SN} of 0.1 from 1D SN simulations as already applied by chemo-dynamical galaxy models (Samland, Hensler & Theis 1997), while unity is also used in some galaxy models (see sect. 3), but seems far too large.

Although numerical experiments of superbubbles and galactic winds are performed, yet they only demonstrate the destructive effect on the surrounding ISM but lack of self-consistency and a complex treatment. Simulations of the chemical evolution of starburst DGs by Recchi *et al.* (2002, 2006), that are dedicated to reproduce the peculiar abundance patterns in these galaxies by different SF episodes, found, that ϵ_{SN} can vary widely. A superbubble expanding from a stellar association embedded in a thick HI disk has, at first, to act against the surrounding medium, by this, is cooling due to its pressure work and radiation, but compresses the swept-up shell material and implies turbulent energy to the ISM. Here the superbubble expansion is efficiently hampered, but depends on the HI disk thickness (Recchi *et al.* 2009) and the energy loss by radiative cooling.

In fig.2 the temporal behaviour of the above-mentioned starburst DG model by Recchi *et al.* (2002) is displayed. Comparison of the various kinds of energy contents reveal an heating efficiency ϵ_{SN} of to about 18%. While the accumulated SNII energy release reaches

10^{53} ergs after 20 Myrs, the thermal energy content starts at 10^{52} drops and increases successively but not above 1.8×10^{52} ergs. For the subsequent SNIa explosions, always single events, the accumulated energy release is clearly discernible while the thermal energy decreases again and varies below 10^{52} ergs so that here $\epsilon_{SN} \to 0$. Moreover, if a closely following SF episode pushes its SNeII into the already existing chimney of a preceding superbubble, the hot gas can easily escape without any hindrance and thus affects the ISM energy budget much less. Recchi & Hensler (2006a) found that depending on the external HI density the chimneys do not close before a few hundred Myrs.

3. Galaxy evolution

Numerical simulations of galaxy evolution are only feasible with intermediate-scale spatial resolution so that SF cannot be resolved and must already be prescribed by reasonable recipes. Numerous papers have implemented SF criteria, such as e.g. threshold density ρ_{SF} with SF if $\rho_g \geqslant \rho_{SF}$, excess mass in a specified volume with respect to the Jeans mass, i.e. $M_g \geqslant M_J$, convergence of gas flows ($div \cdot \underline{\mathbf{v}} < 0$), cooling timescale $t_{cool} \leqslant t_{dyn}$, temperature limits $T_g \leqslant T_{lim}$, Toomre's Q parameter, temperature dependent SFR, but to some extent also already under the assumption of the KS law (Dalla Vecchia & Schaye 2008). This latter, however, is unjustified for non-equilibrium situations, because the SF should converge to this relation due to self-regulation. If at least some of these conditions are fulfilled, as a further step, the gas mass which is converted into stars, i.e. the SFE, has to be set e.g. by fixing an empirical value ϵ_{SF} from observations, i.e. $\Delta m_{SF} = \epsilon_{SF} \cdot \rho_g \cdot \Delta x^3$, where Δx^3 is e.g. the mesh volume in a grid code. If the numerical timestep Δt is smaller than the dynamical timestep τ_{ff}, Δm_{SF} has to be weighted by this time ratio (Tasker & Bryan 2006). Since ϵ_{SF} must inherently depend on the local conditions so that it is high in bursting SF modes, as requested for the Globular Cluster formation, but of percentage level in the self-regulated SF mode, numerical simulations often try to derive the realistic SFE by comparing models of largely different ϵ_{SF} with observations, as e.g. to reproduce gas structures in galaxy disks and galactic winds (e.g. $\epsilon_{SF} = 0.05$ and 0.5 in Tasker & Bryan 2006, 2008). In addition, ϵ_{SN} by them is fixed to 10^{51} erg per 55 $M_\odot$ of formed stars (1.8×10^{49} erg $M_\odot^{-1}$ by Dalla Vecchia & Schaye 2008), so that these results can be treated indicatively but not yet quantitatively, since they also mismatch with the KS relation.

Theoretical studies by Elmegreen & Efremov (1997) achieved a dependence of ϵ_{SF} on the external pressure, while Köppen, Theis, & Hensler (1995) explored a temperature dependence of the SFR. Furthermore, most galaxy evolutionary models at present lack of the appropriate representation of the different ISM phases allowing for their dynamics and their direct interactions by heat conduction, dynamical drag, and dynamical instabilities thru forming interfaces, not to mention resolving the turbulence cascade.

4. Conclusions

The dominating influence of SN explosions and superbubbles on structure, dynamics, and energy budget of the ISM are obvious and agreed. Signs and strengths of these feedback effects are, however, widely uncertain. Whether the feedback is positive (trigger) or negative (suppression) can be understood analyticly from first principles, but because of the non-linearity and the complexity level of the acting plasmaphysical processes clear results cannot be quantified reliably. In addition, the temporal behaviour varies by orders of magnitude because of the changing conditions. In summary, the energy transfer efficiencies of SNe and superbubbles to the ISM are much below unity and depend on the temporal and local conditions, but must not be overestimated. Spatially and temporarily resolved simulations of SN and superbubbles in an extended environment with varying conditions of the ISM are necessary in order to connect large-scale effects on SF clouds with the existing detailed simulations on the star-forming scales.

Acknowledgements

The author is grateful to Simone Recchi for substantial contributions to this topic and providing fig. 2. The attendance of the symposium was funded by the key programme "Computational Astrophysics" of the University of Vienna under project no. FS538001.

References

Böhringer, H. & Hensler, G. 1989, *A&A*, 215, 147
Bouché, N., Cresci, G., Davies, R., *et al.* 2007, *ApJ*, 671, 303
Chevalier, R.A. 1977, *ARA&A*, 15, 175
Cioffi, D.F. & Shull, J.M. 1991, *ApJ*, 367, 96
Comeron, F. & Torra, J. 1994, *A&A*, 281, 35
Dalla Vecchia, C. & Schaye, J. 2008, *MNRAS*, 387, 1431
de Avillez, M.A. 2000, *MNRAS*, 315, 479
de Avillez, M.A. & Breitschwerdt, D. 2004, *A&A*, 425, 899
Efstathiou, G. 2000, *MNRAS*, 317, 697
Ehlerova, S., Palous, J., Theis, C., & Hensler, G. 1997, *A&A*, 328, 111
Elmegreen, B.G. 2002, *ApJ*, 577, 206
Elmegreen, B.G. & Efremov, Y.N. 1997, *ApJ*, 480, 235
Fukuda, N. & Hanawa, T. 2000, *ApJ*, 533, 911
Gerola, H. & Seiden, P.E. 1978, *ApJ*, 223, 129
Gorjian, V., Werner, M.W., Mould, J.R., *et al.* 2004, *ApJS*, 154, 275
Gudell, A. 2002, diploma thesis, University of Kiel
Habe, A., Ikeuchi, S., & Tanaka, Y.D. 1981, *PASJ*, 33, 23
Hensler, G. 2007, in eds. E. Ensellem *et al.*, *Chemodynamics: from first stars to local galaxies* Proc. CRAL-Conference Series I, EAS Publ. Ser. No. 7, p. 113
Hensler, G. 2009, in: J. Andersen, J. Bland-Hawthorn, & B. Nordstroem (eds.), *The Galaxy Disk in Cosmological Context*, Proc. IAU Symposium No. 254, p. 269
Hensler, G. & Recchi, S. 2010, in: K. Cunha, M. Spite & B. Barbuy, (eds.), *Chemical Abundances in the Universe: Connecting First Stars to Planets*, Proc. IAU Symp. No. 265, p. 325
Hensler, G. & Rieschick, A. 2002, in: E. Grebel & W. Brandner (eds.), *Modes of Star Formation and the Origin of Field Populations*, ASP Conf. Ser., 285, 341
Hensler, G., Köppen, J., Pflamm, J., & Rieschick, A. 2004, in: P.-A. Duc, J. Braine, E. Brinks (eds.), *Recycling intergalactic and interstellar matter*, Proc. IAU Symp. No. 217, p. 178
Hunter, D.A., Wilcots, E.M., van Woerden, H., *et al.* 1998, *ApJ*, 495, L47
Kennicutt, R.J. 1998, *ApJ*, 498, 541
Ikeuchi, S. & Tomita, H. 1983, *PASJ*, 35, 56
Ikeuchi, S., Habe, A., & Tanaka, Y.D. 1984, *MNRAS*, 207, 909
Junkes, N., Fürst, E., & Reich, W. 1992, *A&A*, 261, 289
Köppen, J., Theis, C., & Hensler, G. 1995, *A&A*, 296, 99
Lasker, B.M. 1967, *ApJ*, 149, 23
Lee, H.-T. & Chen, W.P. 2009, *ApJ*, 694, 1423
Martin, C.L., Kobulnicki, H.A., & Heckman, T.M. 2002, *ApJ*, 574, 663
McKee, C.F. & Ostriker, J.P. 1977, *ApJ*, 218, 148
Melioli, C. & de Gouveia Dal Pino, E.M. 2004, *A&A*, 424, 817
Moreno, E., Alfaro, E.J., & Franco, J. 1999, *ApJ*, 522, 276
Mühle, S., Klein, U.,Wilcots, E. M., & Hüttermeister, S. 2005, *AJ*, 130, 524
Orlando, S., Peres, G., Reale, F., *et al.* 2005, *A&A*, 444, 505

Recchi, S. & Hensler, G. 2006a, *A&A*, 445, L39
Recchi, S. & Hensler, G. 2006b, *Rev. Mod. Astronomy*, 18, 164
Recchi, S. & Hensler, G. 2007, *A&A*, 476, 841
Recchi, S., Hensler, G., & Anelli, D. 2009, arXiv:0901.1976
Recchi, S., Hensler, G., Angeretti, L., & Matteucci, F. 2006, *A&A*, 445, 875
Recchi, S., Matteucci, F., D'Ercole, A., & Tosi, M. 2002, *A&A*, 384, 799
Rosen, A. & Bregman, J.N. 1995, *ApJ*, 440, 634
Samland, M., Hensler, G., & Theis, C. 1997, *ApJ*, 476, 544
Schmidt, M. 1959, *ApJ*, 129, 243
Slyz, A.D., Devriendt, J.E.G., Bryan, G., & Silk, J. 2005, *MNRAS*, 356, 737
Silk, J. 2003, *MNRAS*, 343, 249
Spitzer, L. 1956, *ApJ*, 124, 20
Spitzer, L. 1990, *ARA&A*, 28, 71
Stil, J. M. & Israel, F. P. 2002, *A&A*, 392, 473
Strickland, D.K., Heckman, T.M., Colbert, E.J.M., *et al.* 2004, *ApJ*, 606, 829
Tasker, E.J. & Bryan, G.L. 2006, *ApJ*, 641, 878
Tasker, E.J. & Bryan, G.L. 2008, *ApJ*, 673, 810
Thornton, K., Gaudlitz, M., Janka, H.-Th., & Steinmetz, M. 1998, *ApJ*, 500, 95
van Zee, L., Skillman, E.D., & Salzer, J.J. 1998, *AJ*, 116, 1186
Vieser, W. & Hensler, G. 2007, *A&A*, 472, 141
Wilson, B.A., Dame, T.M., Masheder, M.R.W., & Thaddeus, P. 2005, *A&A*, 430, 523
Wünsch, R., Dale, J.E., Palous, J., & Whitworth, A.P. 2010, *MNRAS*, in press

Matthias Gritschneder and Johanna Malinen

Computational Star Formation
Proceedings IAU Symposium No. 270, 2011
J. Alves, B.G. Elmegreen, J. M. Girart & V. Trimble, eds.

doi:10.1017/S1743921311000573

Pillars, Jets and Dynamical Features

Matthias Gritschneder[1,2], **Andreas Burkert**[2,3], **Thorsten Naab**[2,4], **Stefanie Walch**[5]

[1]Kavli Institute for Astronomy and Astrophysics, Peking University, 100871 Beijing, China
email: `gritschneder@pku.edu.cn`

[2]Universitäts-Sternwarte München, 81679 München, Germany

[3]Max Planck Institut für Extraterrestrische Physik, 85748 Garching bei München, Germany

[4]Max Planck Institut für Astrophysik, 85740 Garching bei München, Germany

[5]School of Physics & Astronomy, Cardiff University, Cardiff CF24 3AA, United Kingdom

Abstract. We present high resolution simulations on the impact of ionizing radiation on turbulent molecular clouds. The combination of hydrodynamics, gravitational forces and ionization in the tree-SPH code iVINE naturally leads to the formation of elongated filaments and clumps, which are in excellent agreement with the pillars observed around HII regions. Including gravity the formation of a second generation of low-mass stars with surrounding protostellar disks is triggered at the tips of the pillars, as also observed. A parameter study allows us to determine the physical conditions under which irregular structures form and whether they resemble large pillars or a system of small, isolated globules.

1. Introduction

Stars are known to form in turbulent, cold molecular clouds. When massive stars ignite, their UV-radiation ionizes and heats the surrounding gas, leading to an expanding HII bubble. As soon as the HII region breaks through the surface of the molecular cloud a low-density, optically thin hole is formed which reveals the otherwise obscured interior. At the interface between the HII region and the molecular gas peculiar structures, often called pillars, are found. In general, the pillars point like fingers towards the ionizing source (see Figure 1d) and show a common head-to-tail structure. Most of the mass is concentrated in the head which has a bright rim facing the young stars. Thin, elongated pillars connect the head with the main body of the molecular cloud. They have typical widths of 0.1 – 0.7 pc and are 1 – 4 pc long (Gahm *et al.* 2006, Schuller *et al.* 2006). Observations show that the pillars are not smooth, but built of small substructures, filaments and clumps (Pound 1998). Some filaments run diagonal across the pillars, suggesting a complex twist into a helical structure (Carlqvist *et al.* 2003). This is also supported by spectroscopic measurements of the line-of-sight (LOS) gas velocity: the pillars show a bulk motion away from the ionizing stellar sources with a superimposed complex shear flow that could be interpreted as corkscrew rotation (Gahm *et al.* 2006). Occasionally, close to the tip of the head small spherical gas clumps are observed to break off and float into the hot HII region. These so called evaporating gaseous globules (EGGs) have been found with HST e.g. in the Eagle Nebula (McCaughrean & Andersen 2002). If stars with surrounding gas discs happen to form in these clumps they transform into evaporating proto-planetary discs, so called proplyds. Direct signatures of star formation are also found in the head of the pillars, e.g. through jets from obscured proto-stars piercing through the surface into the HII region (e.g. in Eta Carina, Smith *et al.* 2000).

Early models of pillar formation suggested that the pillars form by Rayleigh-Taylor instabilities when the expanding hot, low-density HII region radially accelerates the cold, dense gas (Frieman 1954). This has been ruled out by the observations of the complex flows inside the pillars. Another possibility is the radiation-driven implosion (e.g. Bertoldi 1989, Lefloch & Lazareff 1994, Miao *et al.* 2009) of preexisting dense cores. This scenario results in cometary-shaped structures. However, the overall shape of the pillars in M16 would require several cores with fine tuned positions and relative velocities. Especially the veil between the two smaller pillars in Figure 1d poses a real challenge. A third scenario is the collect and collapse model. Here, the radiation sweeps up a large shell, which then fragments to form stars and pillars (e.g. Elmegreen & Lada 1977, Klein *et al.* 1980). However, the timescales (>5 Myr) and masses (>1000 $M_{\odot}$) involved (Elmegreen *et al.* 1995, Wünsch & Palouš 2001) are much larger than in M16. Therefore, this scenario is more likely i.e. in supernova-driven shells.

We demonstrate that the observed pillars form naturally from the ionization of a turbulent molecular cloud, representing substructure that is excavated by the stellar UV-radiation and then compressed by the high-pressure HII environment. This process has already been studied on large scales (Mellema *et al.* 2006, Dale *et al.* 2007, Mac Low *et al.* 2007). We present, for the first time, simulations with a high enough resolution (100 AU spatial and $2 \times 10^{-4} M_{\odot}$ in mass) to follow the formation of individual pillars.

2. Results

Because of its non-linearity and high complexity, a deeper insight into the evolution of HII regions and their interaction with the surrounding gas can only be gained with numerical simulations. For this purpose we developed a new parallelized 3-dimensional hydrodynamical code to simulate radiative ionization of a turbulent molecular cloud from a nearby star cluster with so far unprecedented resolution (Gritschneder *et al.* 2009a, Gritschneder *et al.* 2009b, Gritschneder *et al.* 2010). Figure 1(a,b,c,e) shows the results of two typical simulations, corresponding to the ionization of a cloud with a temperature of $T = 10K$ and a mean number density of $n = 300\,\mathrm{cm}^{-3}$ at turbulent velocities of Mach 5 and 7, respectively. This is the range of random velocities that are typically measured spectroscopically in molecular clouds. The HII region extends north towards the massive stars that are located outside of the computational domain. Along channels of low gas density, the ionizing radiation penetrates deeply into the molecular cloud, generating channels of hot, pressurized gas. Ionization is however less efficient along line-of sights with a density enhancement near the cloud surface. These regions are the seeds of the pillars. The enhancement is compressed into a clump that becomes the head of the pillar. Heated gas evaporates off the surface and streams into the direction of the ionizing star cluster. Momentum conservation drives a compression wave into the clump, leading to a flattened, optically thick "cap" that casts its shadow radially away from the ionizing source. The gas in the shadow remains cold while the environment is ionized and heated. An isolated, elongated pillar forms that is surrounded by hot and diffuse gas. The structure in the pillar is very clumpy and filamentary with an irregular and twisted velocity field which still traces the previous molecular cloud turbulence.

Figure 1e shows a zoom on the pillars that form in our simulations on the same scale as the HST observation of the pillars of creation (Figure 1d). The size and geometry of the simulations matches very well the observed structures. We also find excellent agreement with respect to kinematics and density substructure. The pillars surface densities in our simulations are $N(H_2) \approx 3 \times 10^{20}\,\mathrm{cm}^{-2}$ on average. In total, the mass contained in Figure 1e is $\approx 45\,M_{\odot}$, the most prominent pillar has a mass of $\approx 15\,M_{\odot}$, similar to

Figure 1. The evolutionary stages of two cloud ionization simulations, $t = 500$ kyr after the ignition of the ionizing source, are compared with the observed pillars of creation in M16 by Hester *et al.* 1996. We show colour coded gas surface density, integrated along the z-axis. The upper left **(b)** and upper right **(c)** panels correspond to clouds with initial turbulent velocities of Mach 7 and Mach 5, respectively. The box size is 4 pc × 4 pc. The lower right panel **(e)** shows a zoom into the upper right picture with a box size of 1 pc × 1 pc. Here and for the inset of the upper left panel **(a)** the colour-coding is shifted by an order of magnitude to higher surface densities. The observation in the lower left panel (d) (Hester *et al.* 1996) has the same physical length scale as the zoomed simulated pillars to the right (e).

the observed $9.2\,M_{\odot}$ in the Dancing Queen trunk in NGC 7822 (Gahm *et al.* 2006). In addition, the simulated line widths with a standard mean deviation of $\approx 0.38\,\mathrm{km\,s^{-1}}$, corresponding to a FWHM $0.94\,\mathrm{km\,s^{-1}}$, are in excellent agreement with the observed FWHM $1.2\,\mathrm{km\,s^{-1}}$. Furthermore, these lines resemble a cork-screw pattern, as also observed (for details see Gritschneder *et al.* 2010).

The pillars heads continuously change due to this internal turbulence seen in the line-widths and their streaming velocity. An example is the small spherical gas globule that is about to break off from the tip of the head in the lower right corner of the snapshot shown in Figure 1e. It results from a small sub-clump that separates from the turbulent head due to its relative velocity, drifts into the hot HII region and forms an isolated EGG. The irregular, elongated vertical gas filament in Figure 1a with a veil of cold gas in its shadow results from the head's tangential motion perpendicular to the impinging radiation. A similar structure is observed in M16 (Figure 1d). At the same time the head is pushed by gas evaporation into the pillar, sweeping up additional cold and clumpy gas. The formation and lifetime of a pillar depends strongly on the local velocity field of the parental molecular cloud, prior to ionization. Parameter studies (Gritschneder *et al.* 2010) show that the turbulent gas velocity of the molecular cloud has to exceed Mach numbers $M \sim 2$ at a density of $n = 300\,\mathrm{cm}^{-3}$ for substructures to be dense and massive enough in order to survive the ionization and cast long shadows. If the turbulent velocity is on the other hand too large ($M \geqslant 10$), the tangential velocities are in general high enough to destroy the coherence of forming pillars quickly, leading to larger numbers of isolated globules and rarely to massive structures. We find that the size of the pillars is typically 2.5% of the wavelength of the dominant mode in the turbulent power-spectrum. Applying these results to the M16 pillars, we conclude that their parental molecular cloud was characterized by irregular motion in the range of $M \approx 5 - 7$, corresponding to velocities of $1.0 - 1.4\,\mathrm{km\,s^{-1}}$ with a turbulent driving scale of order 4 pc.

3. Conclusions

In summary, it is remarkable that the simple model of a turbulent, cold cloud being ionized by a nearby star cluster can reproduce the observed structures around H II regions so well. For the first time the complex morphology and kinematics can be matched simultaneously. We conclude that ionization, momentum exchange between the evaporating hot gas and the cold surrounding as well as shadowing effects dominate the formation and evolution of structures in ionization fronts around star-forming regions. Large, coherent pillars form only under certain conditions that provide interesting information about the initial molecular cloud environment. Low-mass star formation can be triggered in the dense heads of the pillars, facing the young stellar cluster.

Acknowledgements

This research was supported financially by the Cluster of Excellence "Origin and Structure of the Universe" which also partly funded the SGI Altix 3700 Bx2 supercomputer where all simulations were performed.

References

Bertoldi, F. 1989, *ApJ*, 346, 735
Carlqvist, P., Gahm, G. F., & Kristen, H. 2003, *A&A*, 403, 399
Dale, J. E., Clark, P. C., & Bonnell, I. A. 2007, *MNRAS*, 377, 535
Elmegreen, B. G., Kimura, T., & Tosa, M. 1995, *ApJ*, 451, 675+
Elmegreen, B. G. & Lada, C. J. 1977, *ApJ*, 214, 725
Frieman, E. A. 1954, *ApJ*, 120, 18
Gahm, G. F., Carlqvist, P., Johansson, L. E. B., & Nikolić, S. 2006, *A&A*, 454, 201
Gritschneder, M., Naab, T., Burkert, A., Walch, S., Heitsch, F., & Wetzstein, M. 2009a, *MNRAS*, 393, 21
Gritschneder, M., Naab, T., Walch, S., Burkert, A., & Heitsch, F. 2009b, *ApJ Letters*, 694, L26
Gritschneder, M., Burkert, A., Naab, T., & Walch, S. 2010, ArXiv e-prints 1009, arXiv:1009.0011
Hester, J. J. *et al.* 1996, *AJ*, 111, 2349
Klein, R. I., Sandford, II, M. T., & Whitaker, R. W. 1980, Space Science Reviews, 27, 275
Lefloch, B. & Lazareff, B. 1994, *A&A*, 289, 559
Mac Low, M.-M., Toraskar, J., Oishi, J. S., & Abel, T. 2007, *ApJ*, 668, 980
McCaughrean, M. J. & Andersen, M. 2002, *A&A*, 389, 513
Mellema, G., Arthur, S. J., Henney, W. J., Iliev, I. T., & Shapiro, P. R. 2006, *ApJ*, 647, 397
Miao, J., White, G. J., Thompson, M. A., & Nelson, R. P. 2009, *ApJ*, 692, 382
Pound, M. W. 1998, *ApJ Letters*, 493, L113+
Schuller, F., Leurini, S., Hieret, C., Menten, K. M., Philipp, S. D., Güsten, R., Schilke, P., & Nyman, L. 2006, *A&A*, 454, L87
Smith, N., Egan, M. P., Carey, S., Price, S. D., Morse, J. A., & Price, P. A. 2000, *ApJ Letters*, 532, L145
Wünsch, R. & Palouš, J. 2001, *A&A*, 374, 746

Computational Star Formation
Proceedings IAU Symposium No. 270, 2011
J. Alves, B.G. Elmegreen, J. M. Girart & V. Trimble, eds.

doi:10.1017/S1743921311000585

The interaction of an HII region with a fractal molecular cloud

Steffi Walch[1], **Ant Whitworth**[1], **Thomas Bisbas**[1,2], **Richard Wünsch**[1,2] and **David Hubber**[3]

[1]School of Physics & Astronomy, Cardiff University, UK
emails: Stefanie.Walch@astro.cf.ac.uk, Anthony.Whitworth@astro.cf.ac.uk

[2]Czech Academy of Sciences, Prague, Czech Republic
emails: Thomas.Bisbas@astro.cf.ac.uk, richard@wunsch.cz

[3]Department of Physics & Astronomy, Sheffield University, UK
email: D.Hubber@sheffield.ac.uk

Abstract. We describe an algorithm for constructing fractal molecular clouds that obeys prescribed mass and velocity scaling relations.The algorithm involves a random seed, so that many different realisations corresponding to the same fractal dimension and the same scaling relations can be generated. It first generates all the details of the density field, and then position the SPH particles, so that the same simulation can be repeated with different numbers of particles to explore convergence. It can also be used to initialise finite-difference simulations. We then present preliminary numerical simulations of HII regions expanding into such clouds, and explore the resulting patterns of star formation. If the cloud has low fractal dimension, it already contains many small self-gravitating condensations, and the principal mechanism of star formation is radiatively driven implosion. This results in star formation occurring quite early, throughout the cloud. The stars resulting from the collapse and fragmentation of a single condensation are often distributed in a filament pointing radially away from the source of ionising radiation; as the remainder of the condensation is dispersed, these stars tend to get left behind in the HII region. If the cloud has high fractal dimension, the cloud does not initially contain dense condensations, and star formation is therefore delayed until the expanding HII region has swept up a sufficiently massive shell. The shell then becomes gravitationally unstable and breaks up into protostars. In this collect-and-collapse mode, the protostars are distributed in tangential arcs, they tend to be somewhat more massive, and as the expansion of the shell stalls they move ahead of the ionisation front.

Keywords. HII regions, molecular clouds, ionisation fronts, triggered star formation

---

1. Introduction

This project is aimed at understanding the role of feedback from massive stars in star-forming molecular clouds, i.e. how, and under what circumstances, massive stars may trigger, or accelerate, or inhibit, or terminate star formation. In particular, we are concerned here with the effect of an HII region expanding into a molecular cloud, and the relative importance of (i) *radiatively-driven implosion* (i.e. where the action of the ionisation front, and the shock front that precedes it, is to compress pre-existing condensations and trigger their collapse) and (ii) *collect-and-collapse* (i.e. where the HII region sweeps up a dense shell, which eventually becomes sufficiently massive to fragment and collapse). These two mechanisms are expected to deliver rather distinct patterns of star formation, and there may also be differences in the properties of the stars they produce (e.g. mass function, binary statistics, velocity dispersion). We are therefore investigating, by means of SPH simulations, how the expansion of an HII region into a molecular cloud is influenced by the cloud having pre-existing fractal substructure. This is a timely project because, in the aftermath of SPITZER, and with the advent of HERSCHEL, we now have a wealth of observational data on star formation at the boundaries of HII regions, against which to test the credibility of our models (e.g. Churchwell *et al.* 2004; Deharveng *et al.* 2009; Koenig *et al.* 2008; Smith *et al.* 2010).

2. Initial conditions

We model molecular clouds with a fractal structure, i.e. a nested, self-similar hierarchy of clumps within clumps. This hierarchy is characterised by two parameters, D and C. The fractal dimension D determines, for a clump on level ℓ of the hierarchy, what fraction f of its volume is occupied by clumps on the next level $(\ell + 1)$. For an octal fractal structure, i.e. one in which each clump can be divided into a maximum of eight subclumps,

$$f = 2^{3-D} \tag{2.1}$$

Thus, if $D = 3$, $f = 1$ and the density distribution inside the clump is smooth. As D is reduced, f decreases, and hence the fraction of a clump's volume that is occupied by subclumps is reduced. We consider values of D in the interval $2.0 < D < 2.8$.

The density contrast C determines the factor by which clumps on level $\ell+1$ are denser than those on level ℓ. It follows that mass, M, scales with linear size, L, according to

$$M \propto L^{\chi_{ML}}, \qquad \chi_{ML} = 3 - \log_2(C). \tag{2.2}$$

Kaufmann *et al.* (2010) find $\chi_{ML} = 1.3 \pm 0.1$, and hence we put

$$C = 2^{3-\chi_{ML}} = 3.25 \pm 0.25. \tag{2.3}$$

Once the fractal dimension, D, and the density contrast, C, are specified, the mass spectrum of clumps is given by

$$\frac{dN}{dM} \propto M^{-\zeta_{NM}}, \qquad \zeta_{NM} = \frac{D}{\chi_{ML}} \tag{2.4}$$

(cf. Stutzki *et al.* 1998); and the volume-weighted logarithmic density PDF is given by

$$\frac{dV}{d\ell og(\rho)} \propto \rho^{-\zeta_{V\rho}}, \qquad \zeta_{V\rho} = \frac{(3-D)}{\log_2(C)} = \frac{(3-D)}{(3-\chi_{ML})}. \tag{2.5}$$

Note that we can also give the clumps bulk velocities, so that the internal velocity dispersion of a clump on level ℓ derives from the bulk velocities of the smaller clumps it contains. In this case we have to introduce a third parameter characterising the velocity scaling law (Larson 1981). As with the density contrast, C, this parameter is constrained by the observed velocity scaling law. However, in the simulations presented here these velocities are unimportant, and so we neglect them.

We construct fractal clouds by starting with a cubic root cell (the whole computational domain) in which the density is set to a minimum value, B. We then divide the root cell into eight approximately, but not exactly, equal subcells, by splitting the root cell first with a single surface orthogonal to one of the Cartesian axes, then splitting the two resulting parts with two surfaces orthogonal to one of the other Cartesian axes, and finally splitting the four resulting parts with four surfaces orthogonal to the remaining Cartesian axis. Next we pick, using random numbers, a subset of 2^D of these subcells. These are the fertile subcells, and we therefore increase their densities by a factor of C. The remaining $8 - 2^D$ subcells are infertile, and so their densities are unchanged. We then repeat this process of dividing fertile cells into subcells and picking a subset of the subcells to be fertile; once a cell is identified as being infertile, nothing further happens to it. Since 2^D is not in general an integer, we carry forward surplus fertility to the next cell, or the next generation of cells. The process is repeated recursively through a user-specified number of levels (specified with the proviso that the clumps on the lowest level should be resolved by $\gtrsim 50$ SPH particles). In order to get rid of alignments with the Cartesian axes, the subcells within a given cell are rotated about the centre of the parent cell, through three random angles; where necessary particles are wrapped periodically. This algorithm works well for clouds with low fractal dimension, $D \lesssim 2.8$, but breaks down as the fractal dimension approaches 3, because – in this limit – the random rotations create a significant amount of spurious (non-fractal) sub-structure.

The cloud mass and initial radius are fixed at $M_{\rm CLOUD} = 780\,{\rm M}_\odot$ and $R_{\rm CLOUD} = 1\,{\rm pc}$. By fixing the cloud mass and radius, we can easily generate a single-parameter family of clouds by only varying the fractal dimension. However, we should be aware that the density PDF then extends to higher densities in the clouds with lower fractal dimension.

A constant isotropic source of ionising photons, $\dot{\mathcal{N}}_{\rm LyC} = 10^{49}\,{\rm s}^{-1}$, is switched on instantaneously at the centre of the cloud. The temperature of the ionised gas is set to $10^4\,{\rm K}$, and the temperature of the neutral gas is set to $10\,{\rm K}$; the temperature discontinuity across the ionisation front is smoothed over a few smoothing lengths.

3. Numerical method

We use the state-of-the-art SPH code SEREN (Hubber *et al.* 2010) to follow the self-gravitating gas dynamics, with between 10^6 and 10^7 particles. We use the HEALPix-based algorithm of Bisbas *et al.* (2009) to treat the transport of ionising radiation, with up to 9 levels of refinement, and hence a finest resolution of ~ 100 rays per square degree. We neglect the diffuse ionising radiation field, by invoking the On-The-Spot Approximation. Hence, along each ray (represented by the unit vector $\hat{\mathbf{k}}$) the location of the ionisation front ($\mathbf{r}_{\rm IF} = R\hat{\mathbf{k}}$) is given by

$$\int_{r=0}^{r=R} n^2\left(r\hat{\mathbf{k}}\right) r^2\,dr = \frac{\dot{\mathcal{N}}_{\rm LyC}}{4\,\pi\,\alpha_{\rm B}}. \tag{3.1}$$

Here n is the number-density of hydrogen, and $\alpha_{\rm B}$ is the recombination coefficient into excited states only. Sink particles are introduced if the density exceeds $\rho_{\rm SINK} = 10^{-11}\,{\rm g\,cm}^{-3}$ (by which stage a protostellar condensation should be well into its Kelvin-Helmholtz contraction phase), provided various other conditions are met (see Hubber *et al.* 2010). The simulations are extremely well converged, in the sense that almost exactly the same evolution is seen with 10^6 and 10^7 SPH particles. Magnetic effects are not included; nor is radiation pressure.

4. Results and Conclusions

Fractal clouds appear to offer a promising way of modelling observed molecular clouds with a relatively small number of parameters, and hence of initialising numerical simulations of cloud evolution and star formation. As the fractal dimension decreases from $D = 3$, we move from a regime in which the cloud density is initially uniform to one in which the cloud is, from the outset, highly structured, with many dense compact condensations.

For low fractal dimension, star formation occurs mainly by radiatively driven implosion. Because there is no delay whilst a dense massive shell is swept up, star formation

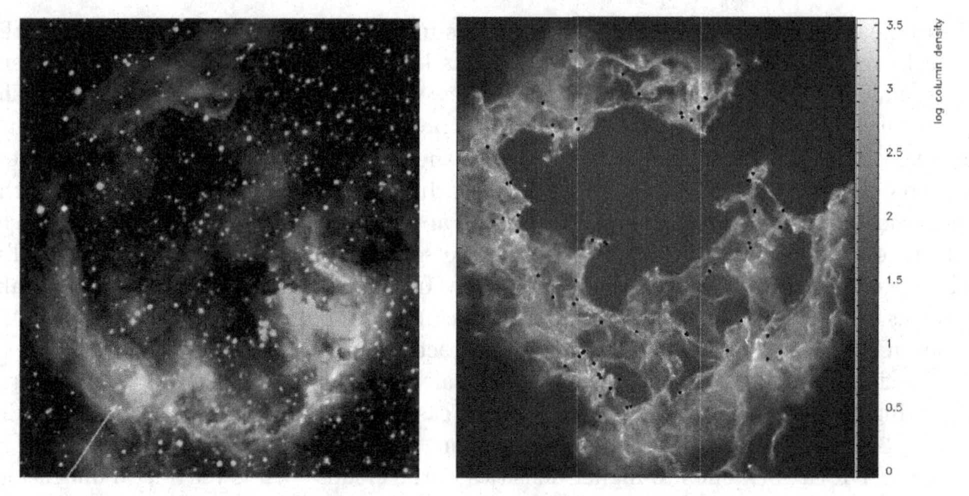

Figure 1. Left: RCW 79 observed with Spitzer IRAC 8 μm image (orange) superimposed on a SuperCOSMOS Hα image (turquoise) (credit: Zavagno *et al.* 2006). Right: Column density plot showing the dense shell structure formed when ionising a cloud with $D = 2.8$ for 0.5 Myr.

occurs early (at $t \gtrsim 0.2$ Myr for the cloud parameters given in section 2), and at many different radii, more or less simultaneously. The resulting protostars tend to have random velocities. As their natal envelope is ablated, they get left behind in the HII region.

For high fractal dimension, star formation occurs mainly by the collect-and-collapse mode, therefore later (at $t \gtrsim 0.4$ Myr for this setup) and at larger radius. It creates stars arranged in tangential arcs, and with systematic outwards velocities, so that they tend to move ahead of the ionisation front. Figure 1 (righthand frame) shows one example of a dense, swept-up shell formed from a cloud with fractal dimension $D = 2.8$.

We have, as yet, only considered a small part of the relevant parameter space. However, the morphological features of the HII regions generated by these simulations bear a striking resemblance to observed HII regions, with an abundance of bright arcs, bright rims and elephants' trunks. A fuller report on these results is in preparation but Figure 1 gives a first impression. The morphology of RCW 79 (Fig.1, righthand frame), a region where the collect and collapse process seems to be responsible for triggering star formation, can be well matched with a model of an HII region expanding into a rather uniform cloud, where bright rims and trunks dominate the final picture.

References

Bisbas, T. G., Wünsch, R., Whitworth, A. P., & Hubber, D. A. 2009, *A&A*, 497, 649
Churchwell, E., Whitney, B. A., & Babler, B. L., *et al.* 2004, *ApJS*, 154, 322
Deharveng, L., Zavagno, A., Schuller, F., Caplan, J., Pomarès, M., & De Breuck, C. 2009, *A&A*, 496, 177
Zavagno, A., Deharveng, L., Comerón, F., Brand, J., Massi, F., Caplan, J., & Russeil, D. 2006, *A&A*, 446, 171
Hubber, D. A., Batty, C. P., McLeod, A., & Whitworth, A. P. 2010, submitted to *MNRAS*
Kauffmann, J., Pillai, T., Shetty, R., Myers, P. C., & Goodman, A. A. 2010, *ApJ*, 716, 433
Koenig, X. P., Allen, L. E., Gutermuth, R. A., Hora, J. L., Brunt, C. M., & Muzerolle, J. 2008, *ApJ*, 688, 1142
Larson, R. B. 1981, *MNRAS*, 194, 809
Stutzki, J., Bensch, F., Heithausen, A., Ossenkopf, V., & Zielinsky, M. 1998, *A&A*, 336, 697
Smith, N., Povich, M. S., Whitney, B. A., Churchwell, E., Babler, B. L., Meade, M. R., Bally, J., Gehrz, R. D., Robitaille, T. P., & Stassun, K. G. 2010, *MNRAS*, in press

Computational Star Formation
Proceedings IAU Symposium No. 270, 2011
J. Alves, B.G. Elmegreen, J. M. Girart & V. Trimble, eds.

doi:10.1017/S1743921311000597

Scaling Relations between Gas and Star Formation in Nearby Galaxies

Frank Bigiel[1], Adam Leroy[2] and Fabian Walter[3]

[1]Department of Astronomy, Radio Astronomy Laboratory, University of California, Berkeley, CA 94720, USA
email: bigiel@astro.berkeley.edu

[2]National Radio Astronomy Observatory, 520 Edgemont Road, Charlottesville, VA 22903, USA

[3]Max-Planck-Institut für Astronomie, Königstuhl 17, 69117 Heidelberg, Germany

Abstract. High resolution, multi-wavelength maps of a sizeable set of nearby galaxies have made it possible to study how the surface densities of H I, H_2 and star formation rate ($\Sigma_{\rm HI}$,$\Sigma_{\rm H2}$,$\Sigma_{\rm SFR}$) relate on scales of a few hundred parsecs. At these scales, individual galaxy disks are comfortably resolved, making it possible to assess gas-SFR relations with respect to environment within galaxies. $\Sigma_{\rm H2}$, traced by CO intensity, shows a strong correlation with $\Sigma_{\rm SFR}$ and the ratio between these two quantities, the molecular gas depletion time, appears to be constant at about 2 Gyr in large spiral galaxies. Within the star-forming disks of galaxies, $\Sigma_{\rm SFR}$ shows almost no correlation with $\Sigma_{\rm HI}$. In the outer parts of galaxies, however, $\Sigma_{\rm SFR}$ does scale with $\Sigma_{\rm HI}$, though with large scatter. Combining data from these different environments yields a distribution with multiple regimes in $\Sigma_{\rm gas} - \Sigma_{\rm SFR}$ space. If the underlying assumptions to convert observables to physical quantities are matched, even combined datasets based on different SFR tracers, methodologies and spatial scales occupy a well define locus in $\Sigma_{\rm gas} - \Sigma_{\rm SFR}$ space.

Keywords. galaxies: evolution, galaxies: ISM, radio lines: ISM, radio lines: galaxies

---

1. Introduction and the Global Star Formation Law

Great progress has been made towards an understanding of star formation (SF) on small scales in the Milky Way, but many open questions remain about its connection to large scale processes: what sets where SF occurs in galaxies and how efficiently gas is converted into stars? How important are global, galaxy-scale environmental parameters as opposed to small-scale properties of the interstellar medium (ISM)? What is the role of feedback in regulating SF? To address such questions, theoretical modeling and simulations need to be constrained by comprehensive observations.

Both observations and theory have focused on the relationship between the star formation rate (SFR) and the gas density, for which a tight power-law relationship was observed in a large number of galaxies by Kennicutt(1998). Such a relationship was first suggested many decades ago by Schmidt(1959), who studied the distributions of atomic gas and stars in the Galaxy. Over the following decades, similar studies targeted individual Local Group galaxies, e.g., M33 (Madore *et al.* (1974), Newton(1980)), the Large Magellanic Cloud (Tosa & Hamajima(1975)), and the Small Magellanic Cloud (Sanduleak(1969)). Kennicutt(1989) carried out the first comprehensive extragalactic study targeting a large sample of nearby galaxies and Kennicutt(1998) followed up this work, focusing on measurements averaged across galaxy disks. In a sample of 97 nearby normal and starburst galaxies, he found a close correlation between the galaxy-average total gas surface density ($\Sigma_{\rm gas} = \Sigma_{\rm HI} + \Sigma_{\rm H2}$) and the galaxy-average SFR surface density ($\Sigma_{\rm SFR}$). Following this work, it has become standard to study the relationship between gas and

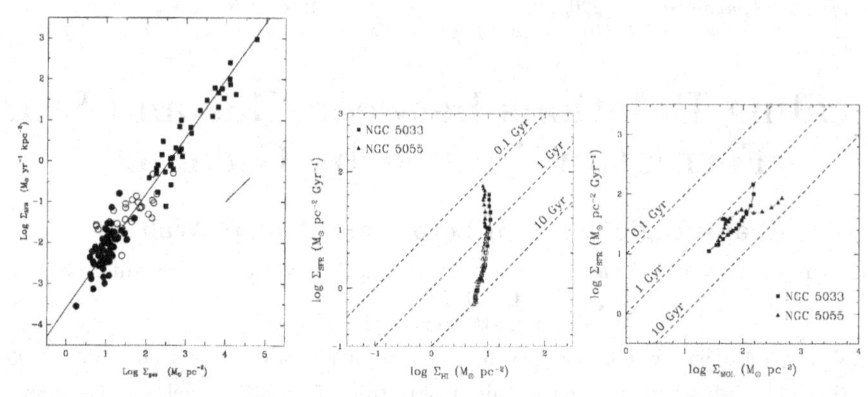

Figure 1. *Left:* Kennicutt(1998) found a strong correlation over many orders-of-magnitude between global averages of $\Sigma_{\rm SFR}$ and $\Sigma_{\rm gas} = \Sigma_{\rm HI} + \Sigma_{\rm H2}$ for a large sample of nearby normal and starburst galaxies. He derived a power law index $N \approx 1.40$, implying more efficient SF for galaxies with higher average gas columns. *Middle and Right:* Surface densities of gas and SFR measured in radial profiles by Wong & Blitz(2002) for two exemplary nearby spirals. The middle panel shows $\Sigma_{\rm SFR}$ versus $\Sigma_{\rm HI}$, the right panel $\Sigma_{\rm SFR}$ versus $\Sigma_{\rm H2}$. Wong & Blitz(2002) found no correlation between atomic gas and SFR, whereas molecular gas and SFR scale with one another.

star formation via surface densities, which are observationally more easily accessible than volume densities.

Kennicutt(1998) found $\Sigma_{\rm SFR} = A \times \Sigma_{\rm gas}^N$, with intercept A and power law index N — a relationship that is variously referred to as the "star formation law," "Schmidt-Kennicutt law," or "Schmidt Law." Kennicutt(1998) derived $N \approx 1.40$. Because the ratio $\Sigma_{\rm SFR}/\Sigma_{\rm gas}$ describes how efficiently gas is converted into stars (and is thus often referred to as the star formation efficiency, SFE), this super-linear power law index implies that systems with higher average gas surface densities more efficiently convert gas into stars (left panel, Figure 1). This measured value is close to $N = 1.5$, which is expected if the free-fall time in a fixed scale height gas disk is the governing timescale for SF on large scales. Other studies working with disk-averaged, global measurements found N to be in the range of $\sim 0.9 - 1.7$ (e.g., Buat *et al.* (1989), Buat(1992), Deharveng *et al.* (1994)).

The availability of high-resolution maps of CO emission, the standard tracer of molecular gas, made it possible to follow up the work of Kennicutt(1998) with studies focusing on azimuthally-averaged radial profiles of gas and SF. Resolving galaxies in this way makes it possible to look at how gas and SF relate within individual galaxy disks, opening up a wide range of environmental factors to explore. Wong & Blitz(2002) used BIMA SONG data (Helfer *et al.* (2003)) to study 6 nearby spirals, Boissier *et al.* (2003) explored a larger sample of nearby spirals, Heyer *et al.* (2004) studied the Local Group galaxy M33, and Schuster *et al.* (2007) explored the gas-SF relation in M51. These studies derived power law indices in the range $N \approx 1 - 3$, leaving it unclear whether a single relation relates gas and SF when galaxy disks are spatially resolved. Further disagreement centered on the relationship of SF to different types of gas — H I, H_2, and total gas. Intuitively, one might expect a stronger correlation between SF and the cold, molecular phase, rather than the atomic phase. Wong & Blitz(2002) indeed found a much stronger correlation of $\Sigma_{\rm SFR}$ with the molecular gas, $\Sigma_{\rm H2}$ (compare Figure 1). However, Kennicutt(1998) and Schuster *et al.* (2007) both found a better correlation of $\Sigma_{\rm SFR}$ with the total gas, $\Sigma_{\rm gas}$, than with $\Sigma_{\rm H2}$.

2. Recent Advances: Gas and Star Formation on sub-kiloparsec Scales

One reason that different studies returned such different results were the wide range of SFR tracers employed in the various analyses. Furthermore, different studies employed widely varying methods to correct the observed UV and Hα intensities for the effects of extinction by dust. Because the correction factor is usually $\gtrsim 2$, the adopted methodology makes a large difference. A large step forward in this field came from the *Spitzer* space telescope. As part of SINGS (*Spitzer* Infrared Nearby Galaxies Survey, Kennicutt *et al.* (2003)), *Spitzer* obtained high-resolution IR maps of a large sample of nearby galaxies. Calzetti *et al.* (2005) and Calzetti *et al.* (2007) demonstrated the utility of these maps to trace recently formed stars obscured on small scales, particularly when used in combination with Hα — a tracer of unobscured star formation.

At the same time the VLA large program THINGS (Walter *et al.* (2008)) obtained the first large set of high resolution, high sensitivity 21-cm line maps for the same sample of galaxies. Following shortly thereafter, the IRAM large program HERACLES mapped CO emission for an overlapping sample of nearby galaxies (first maps in Leroy *et al.* (2009)). The result was, for the first time, a matched set of sensitive, high spatial resolution maps of atomic gas, molecular gas, embedded and unobscured star formation for a large sample of nearby galaxies. The resolution of the maps allowed hundreds of independent measurements per galaxy, leading to significantly improved statistics and the ability to carefully isolate regions with specific physical conditions.

Bigiel *et al.* (2008) combined these data to compare H I, H_2, and SFR across a sample of 7 nearby spiral galaxies. Figure 2 shows the results of this analysis: the left panel shows $\Sigma_{\rm HI}$, the middle panel $\Sigma_{\rm H2}$, and the right panel $\Sigma_{\rm gas} = \Sigma_{\rm HI} + \Sigma_{\rm H2}$ versus $\Sigma_{\rm SFR}$ (derived from a combination of far UV and 24μm emission). Galex far UV emission was chosen to trace the recent, unobscured SF because of the low background in the FUV channel and the large field-of-view of the GALEX satellite. In these plots, H I and H_2 show distinct behaviors: the atomic gas shows no clear correlation with the SFR, whereas the molecular gas exhibits a strong correlation. As a result, the composite total gas-SFR relation is more complex than a single power law. In the combined (total gas) plot in the right panel, one can clearly distinguish where the ISM is H I dominated (low gas columns, steep relation) form where it is H_2 dominated (high gas columns, roughly linear correlation).

If a power law is fit to the molecular gas distribution in the middle panel, one obtains $N \approx 1.0$. This can be restated as a constant ratio $\Sigma_{\rm SFR}/\Sigma_{\rm H2}$, which means that on average each parcel of H_2 forms stars at the same rate. Leroy *et al.* (2008) searched for correlations between $\Sigma_{\rm SFR}/\Sigma_{\rm H2}$ and a number of environmental variables — ISM pressure, dynamical time, galactocentric radius, stellar and gas surface density — and found little or no variation across the disks of 12 nearby spirals. On the other hand many of these same environmental variables do correlate strongly with the H_2-to-H I ratio. The combined conclusion of these two studies was that the average depletion time in the molecular gas (i.e., $\Sigma_{\rm H2}/\Sigma_{\rm SFR}$) of nearby spirals is fairly constant at ~ 2.0 Gyr, while the abundance of molecular gas is a strong function of environment.

Blanc *et al.* (2009) carried out a similar experiment to Bigiel *et al.* (2008). They sampled the inner part of M51 with 170 pc diameter apertures and estimated local SFR surface densities from Hα emission. Their integral field unit observations allowed for accurate estimates of internal extinction and corrections for contamination by the AGN and diffuse ionized gas. They used a Monte-Carlo approach to incorporate upper limits into their power law fit. Their results are in good agreement with Bigiel *et al.* (2008) regarding M51 in particular as well as regarding the general conclusions they reached: a

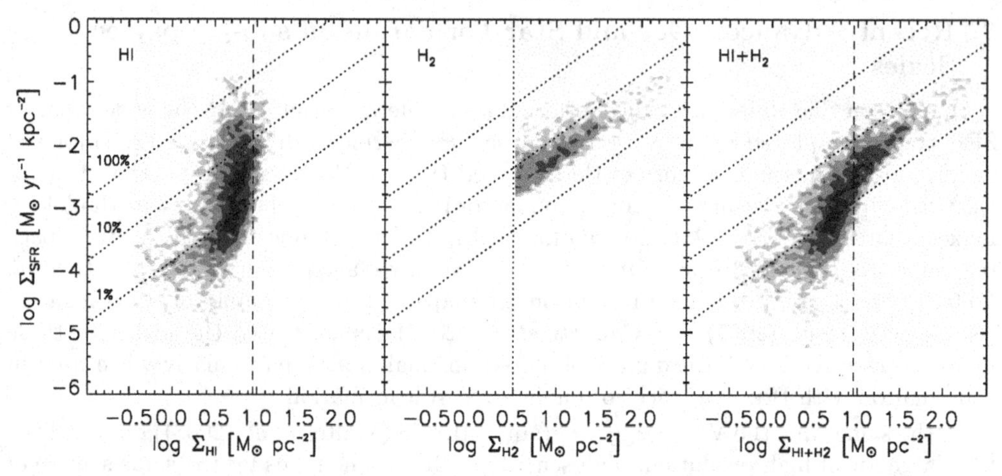

Figure 2. $\Sigma_{\rm SFR}$ versus $\Sigma_{\rm HI}$ (left), $\Sigma_{\rm H2}$ (middle) and $\Sigma_{\rm gas}$ (right panel) for pixel-by-pixel data from 7 nearby spirals at 750 pc resolution. The contours represent the density of sampling points (pixels), where darker colors indicate a higher density. The H I distribution saturates at about 10 $M_\odot$ pc^{-2} and shows no correlation with the SFR (left panel). Gas in excess of this surface density is predominantly molecular. The middle panel illustrates the correlation between H_2 and SFR, which can be described by a power law with slope $N \approx 1.0$. This implies a constant H_2 depletion time of ~ 2 Gyr. The total gas plot (right panel) illustrates the different behavior of H I and H_2 dominated ISM at low and high gas column densities, respectively.

virtual absence of correlation between SFR and H I, a strong correlation between SFR and H_2 and a molecular gas depletion time that shows little variation with molecular gas column.

Recently, Rahman *et al.* (2010) explored the impact of different SFR tracers and the role of possible contributions from diffuse emission and different sampling and fitting strategies on the relationship between $\Sigma_{\rm H2}$ and $\Sigma_{\rm SFR}$. They found that the SFR derived for low surface brightness regions is extremely sensitive to the underlying assumptions, but that at high surface brightness the result of a roughly constant H_2 depletion time is robust.

Even more recently, Schruba *et al.* (in prep.) combined the HERACLES and THINGS data to coherently average CO spectra as a function of radius. With this approach they are able to trace molecular gas out to 1.2 r_{25}, allowing them to study the relation between H_2 and SFR where H I dominates the ISM. They demonstrate that the tight correlation between H_2 and SFR crosses seamlessly into the H I-dominated outer disk (left panel Figure 3).

With so much effort expended measuring SFRs and gas densities over the years, it is interesting to ask whether the literature largely agrees. The right panel in Figure 3 shows a collection of literature measurements along with the data from Bigiel *et al.* (2008). After matching underlying assumptions about how to derive physical quantities from the observables, the literature data populate a well-defined locus in $\Sigma_{\rm SFR}$–$\Sigma_{\rm gas}$ space.

With some consensus emerging on the broad distribution of data in SFR-H_2 space, attention is turning to the origin of the intrinsic scatter in the SFR-H_2 ratio. Schruba *et al.* (2010) looked at this as a function of spatial scale in M33 and showed that scatter in the CO-to-Hα ratio increases dramatically once a resolution element contains only a single star-forming region (i.e., H II region or giant molecular cloud). This occurs at scales of $\sim$ 150 pc in M33 but should be a function of the environment studied. They interpreted

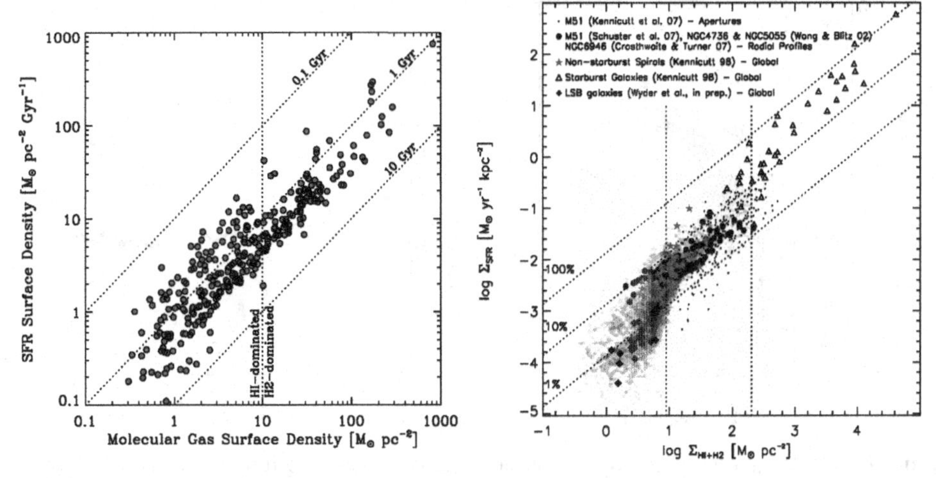

Figure 3. *Left:* $\Sigma_{\rm SFR}$ versus $\Sigma_{\rm H2}$ from Schruba *et al.* (in prep.). They apply a stacking analysis to the HERACLES CO data to probe the H_2-SFR relation far into the regime where $\Sigma_{\rm HI} > \Sigma_{\rm H2}$. The correlation between $\Sigma_{\rm SFR}$ and $\Sigma_{\rm H2}$ extends smoothly into the H I-dominated parts of galaxies out to $1.2\,r_{25}$. *Right:* Comparison between different datasets using different methodologies from Bigiel *et al.* (2008). The contours are identical to the right panel in Figure 2 and the overplotted symbols come from studies using a variety of SFR tracers and methodologies. All datasets have been adjusted to match the same set of assumptions when converting observables to physical quantities. The composite sample occupies a well-defined locus in $\Sigma_{\rm SFR} - \Sigma_{\rm gas}$ space.

this finding as a result of the evolution of these regions, so that information on the life cycle of giant molecular clouds is embedded in the $\Sigma_{\rm SFR}$-$\Sigma_{\rm H2}$ relation, especially at high resolutions.

Another avenue of investigation is the role of the host galaxy in driving the scatter in the SFR-H_2 ratio. Leroy *et al.* (2008) and Bigiel *et al.* (2008) found little environmental dependence of this quantity in large spiral galaxies, but with the completion of the HERACLES survey (47 galaxies spanning from low-mass dwarfs to large spirals) we can now apply a similar analysis to a much wider sample of galaxies. Bigiel *et al.* (in prep.) and Schruba *et al.* (in prep.) explore how scatter in the SFR-to-H_2 ratio breaks into scatter *within* galaxies and scatter *among* galaxies. They find that scatter among galaxies dominates the relation, with a clear trend evident so that less massive, more metal poor, later type galaxies show systematically higher ratios of SFR-to-H_2. Young *et al.* (1996) saw this trend using integrated FIR-to-CO ratios. These new analyses show that once these galaxy-to-galaxy variations are removed, galaxies exhibit a very tight internal H_2-SFR relation.

3. Scaling Relations beyond the Optical Disks

H I maps reveal atomic gas out to many optical radii in spiral galaxies and over the past few years, GALEX UV observations have revealed widespread SF in the outer parts of many galaxies (e.g., Thilker *et al.* (2005), Gil de Paz *et al.* (2007), Bigiel *et al.* (2010b)). These outer disks have fewer heavy elements, less dust, and lower stellar and gas surface densities than the inner parts of spiral galaxies. Contrasting the gas-SFR relationship in outer disks with that in the inner parts of normal galaxies gives a chance to assess the impact of these parameters on SF.

In the left panel of Figure 4 we show the results of a pixel-by-pixel analysis of outer galaxy disks: the open contours show $\Sigma_{\rm SFR}$ versus $\Sigma_{\rm gas}$ for a sample of 17 spiral galaxies

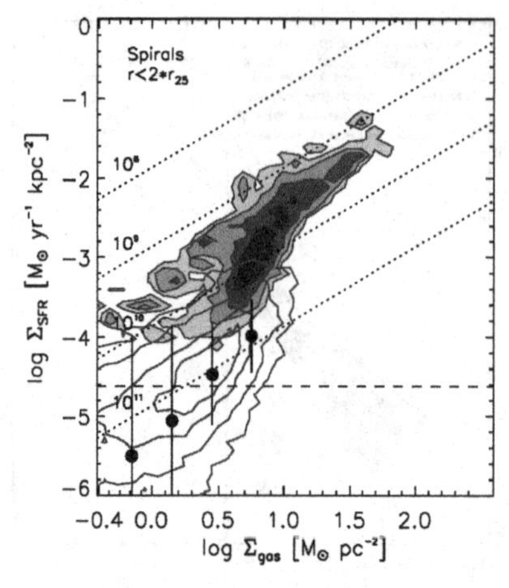

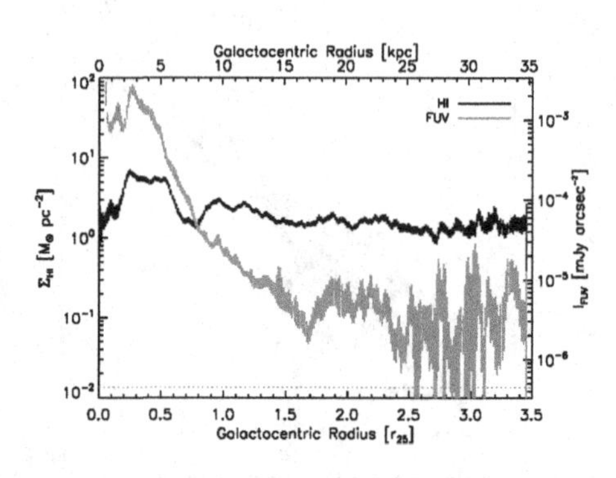

Figure 4. *Left:* $\Sigma_{\rm SFR}$ versus $\Sigma_{\rm gas}$ for the outer (open contours) and inner (filled contours) parts of nearby spiral galaxies (Bigiel *et al.* (2010a)). The combined distribution reveals multiple regimes: at large radii and low surface densities, $\Sigma_{\rm SFR}$ scales with $\Sigma_{\rm gas} \approx \Sigma_{\rm HI}$. Over most of the area in disks, $\Sigma_{\rm SFR}$ is a very steep function of $\Sigma_{\rm gas}$, with the H_2-to-H I ratio being the key determinant of $\Sigma_{\rm SFR}$. At high column densities, the gas is predominantly molecular and correlates well with $\Sigma_{\rm SFR}$. At even higher column densities, a steepening of this relation, meaning and increasing efficiency of SF, may accompany the transition from galaxy disks to starbursts. *Right:* H I and far UV radial profiles for M83 out to almost 4 optical radii r_{25}. The inferred H I depletion time is about a Hubble time at large radii, much longer than the molecular gas depletion times measured in many nearby spirals.

at 15" resolution (corresponding to physical scales between 200 pc and 1 kpc) from Bigiel *et al.* (2010a). SFRs are estimated from GALEX far UV emission and $\Sigma_{\rm gas}$ is estimated from HI emission alone, assuming a negligible contribution from molecular gas on ~kpc scales in outer galaxy disks. For comparison, the filled contours show sampling data from the star forming disks of 7 spiral galaxies from Bigiel *et al.* (2008) (compare the right panel in Figure 2 above). In the outer disks (open contours) $\Sigma_{\rm SFR}$ scales with $\Sigma_{\rm gas}$, i.e., $\Sigma_{\rm HI}$, though the scatter in $\Sigma_{\rm SFR}$ for a given H I column is large. This is an interesting difference compared to the inner parts of spiral galaxies, where the H I showed no clear correlation with $\Sigma_{\rm SFR}$ (compare Section 2).

The H I-FUV relation observed for outer disks suggests two things. First, that at large radii the availability of H I may be a bottleneck for star formation. Even if stars form directly from H_2, molecular clouds must be assembled from H I and this will only be possible in regions with enough H I to assemble these clouds. Second, in the inner parts of galaxies many physical conditions important to the H I-H_2 conversion change while $\Sigma_{\rm HI}$ remains approximately fixed, but in the outer parts $\Sigma_{\rm HI}$ varies while other environmental conditions show comparatively little variation. As a result, $\Sigma_{\rm HI}$ transitions from being a relatively unimportant driver for SF in the inner parts of galaxies to a key quantity at large radii. This is apparent from the right panel of Figure 4, where we use extremely deep FUV data from GALEX and H I data from THINGS to trace SF and H I out to almost four optical radii in the nearby spiral M83 (Bigiel *et al.* (2010b)). The H I depletion time (H I-to-SFR ratio) inferred from the radial profiles in this plot is approximately constant at large radii: it is about a Hubble time, i.e., much longer than the ~ 2 Gyr molecular gas depletion time scale observed in the inner parts of galaxies. This suggests relatively fixed conditions for molecular cloud formation and that assembling H_2, rather than forming stars out of H_2, is the rate-limiting process for SF in outer disks.

4. The Composite Star Formation Law

The combined distribution (inner and outer disks) in the left panel of Figure 4 can be divided into different parts according to gas column density, each part describing the relation between gas and SFR in a particular regime in a typical spiral galaxy disk. For low gas columns (outer disks), the SFR scales with gas column, though with significant scatter. At smaller radii and higher gas columns — corresponding to much of the area inside the star-forming disk — the distribution becomes much steeper. In this regime, knowing $\Sigma_{\rm gas}$ alone is not enough to predict $\Sigma_{\rm SFR}$ with any accuracy. Across this regime the H_2-to-H I ratio is varying steadily as a function of other environmental quantities. At yet smaller radii and higher gas columns, the dominant phase of the ISM transitions from atomic to molecular and a strong correlation emerges between $\Sigma_{\rm gas}$ and $\Sigma_{\rm SFR}$.

There is good observational evidence (e.g., Kennicutt(1998), Gao & Solomon(2004), Greve *et al.* (2005), Bouché *et al.* (2007)) that at higher gas columns, the relation steepens further, so that the SFR-per-H_2 ratio is higher in starburst galaxies than in normal galaxy disks. This may drive the frequent observation of $N \approx 1.5$ in starburst galaxies. However, it must be emphasized that there is currently a lack of data probing normal disk galaxies and starbursts in a self-consistent way, so the details at the high end of this relation remain uncertain.

5. Conclusions

With vast improvements in the data available for nearby galaxies some consensus is beginning to emerge on how different parts of galaxies populate the $\Sigma_{\rm SFR}$-$\Sigma_{\rm gas}$ parameter space. The role of environmental quantities other than gas surface density alone are beginning to become clear and different relations are emerging for different types and parts of galaxies. When viewed in detail the composite relation may not be a simple power law, but it contains key information to constrain theories and to benchmark simulations.

Challenges remain, too. The determination of star formation rates at low surface brightness is still difficult. The use of CO to trace H_2 underpins almost all of this work but the CO-to-H_2 conversion factor remains imprecisely calibrated as a function of environment. Finally, the fundamental units of star-formation, individual molecular clouds, remain largely observationally inaccessible beyond the Local Group — a situation that will not change until ALMA begins its full operations.

Acknowledgements

We thank the SOC and LOC for organizing this stimulating and productive meeting. F.B. gratefully acknowledges financial support from the IAU. We thank Andreas Schruba for providing us with the SFR-H_2 plot before publication.

References

Bigiel, F., Leroy, A., Walter, F., Brinks, E., de Blok, W. J. G., Madore, B., & Thornley, M. D. 2008, *AJ*, 136, 2846

Bigiel, F., Leroy, A., Walter, F., Blitz, L., Brinks, E., de Blok, W. J. G., & Madore, B. 2010a, arXiv:1007.3498

Bigiel, F., Leroy, A., Seibert, M., Walter, F., Blitz, L., Thilker, D., & Madore, B. 2010b, *ApJL*, 720, L31

Blanc, G. A., Heiderman, A., Gebhardt, K., Evans, N. J., & Adams, J. 2009, *ApJ*, 704, 842

Boissier, S., Prantzos, N., Boselli, A. & Gavazzi, G. 2003, *MNRAS*, 346, 1215

Bouché, N., *et al.* 2007, *ApJ*, 671, 303

Buat, V., Deharveng, J. M., & Donas, J. 1989, *A&A*, 223, 42
Buat, V. 1992, *A&A*, 264, 444
Calzetti, D., *et al.* 2005, *ApJ*, 633, 871
Calzetti, D., *et al.* 2007, *ApJ*, 666, 870
Deharveng, J.-M., Sasseen, T. P., Buat, V., Bowyer, S., Lampton, M., & Wu, X. 1994, *A&A*, 289, 715
Gao, Y. & Solomon, P. M. 2004, *ApJ*, 606, 271
Gil de Paz, A., *et al.* 2007, *ApJS*, 173, 185
Greve, T. R., *et al.* 2005, *MNRAS*, 359, 1165
Helfer, T. T., Thornley, M. D., Regan, M. W., Wong, T., Sheth, K., Vogel, S. N., Blitz, L., & Bock, D. C.-J. 2003, *ApJS*, 145, 259
Heyer, M. H., Corbelli, E., Schneider, S. E. & Young, J. S. 2004, *ApJ*, 602, 723
Kennicutt, R. C. 1989, *ApJ*, 344, 685
Kennicutt, R. C. 1998a, *ApJ*, 498, 541
Kennicutt, R. C., Jr., *et al.* 2003, *PASP*, 115, 928
Kennicutt, R. C., Jr., *et al.* 2007, *ApJ*, 671, 333
Leroy, A. K., Walter, F., Brinks, E., Bigiel, F., de Blok, W. J. G., Madore, B., & Thornley, M. D. 2008, *AJ*, 136, 2782
Leroy, A. K., *et al.* 2009, *AJ*, 137, 4670
Madore, B. F., van den Bergh, S., & Rogstad, D. H. 1974, *ApJ*, 191, 317
Newton, K. 1980, *MNRAS*, 190, 689
Onodera, S., *et al.* 2010, arXiv:1006.5764
Rahman, N., *et al.* 2010, arXiv:1009.3272
Sanduleak, N. 1969, *AJ*, 74, 47
Schmidt, M. 1959, *ApJ*, 129, 243
Schruba, A., Leroy, A. K., Walter, F., Sandstrom, K., & Rosolowsky, E. 2010, arXiv:1009.1651
Schuster, K. F., Kramer, C., Hitschfeld, M., Garcia-Burillo, S. & Mookerjea, B. 2007, *A&A*, 461, 143
Thilker, D. A., *et al.* 2005, *ApJL*, 619, L79
Tosa, M. & Hamajima, K. 1975, *PASJ*, 27, 501
Walter, F., Brinks, E., de Blok, W. J. G., Bigiel, F., Kennicutt, R. C., Thornley, M. D., & Leroy, A. 2008, *AJ*, 136, 2563
Wong, T. & Blitz, L. 2002, *ApJ*, 569, 157
Young, J. S., Allen, L., Kenney, J. D. P., Lesser, A., & Rownd, B. 1996, *AJ*, 112, 1903

Computational Star Formation
Proceedings IAU Symposium No. 270, 2011
J. Alves, B.G. Elmegreen, J. M. Girart & V. Trimble, eds.

doi:10.1017/S1743921311000603

Observational Comparison of Star Formation in Different Galaxy Types

Eva K. Grebel[1]

[1]Astronomisches Rechen-Institut, Zentrum für Astronomie der Universität Heidelberg, Mönchhofstr. 12–14, D-69120 Heidelberg, Germany
email: grebel@ari.uni-heidelberg.de

Abstract. Galaxies cover a wide range of masses and star formation histories. In this review, I summarize some of the evolutionary key features of common galaxy types. At the high-mass end, very rapid, efficient early star formation is observed, accompanied by strong enrichment and later quiescence, well-described by downsizing scenarios. In the intermediate-mass regime, early-type galaxies may still show activity in low-mass environments or when being rejuvenated by wet mergers. In late-type galaxies, we find continuous, though variable star formation over a Hubble time. In the dwarf regime, a wide range of properties from bursty activity to quiescence is observed. Generally, stochasticity dominates here, and star formation rates and efficiencies tend to be low. Morphological types and their star formation properties correlate with environment.

Keywords. galaxies: formation, galaxies: evolution, galaxies: elliptical and lenticular, galaxies: spiral, galaxies: dwarf, galaxies: stellar content

---

1. Introduction

Star formation differs widely in different galaxy types (e.g., Kennicutt 1998), ranging from slow, low-efficiency events that may be long-lasting to intense, short-duration starbursts. Along the Hubble sequence, typical global present-day star formation rates range from ~ 0 M$_\odot$ yr^{-1} in giant ellipticals to ~ 20 M$_\odot$ yr^{-1} in gas-rich, late-type spirals. Starburst galaxies show star formation rates of up to $\sim$100 M$_\odot$ yr^{-1}, and ultra-luminous infrared galaxies (ULIRGs) appear to form up to $\sim$1000 M$_\odot$ yr^{-1} in stars. Star formation may be localized or encompass a large fraction of the baryonic gas mass of a galaxy. It may be continuous, declining or increasing in intensity, or episodic. It may be triggered by internal processes within a galaxy or by interactions with other galaxies. Typical sites of present-day star formation in galaxies are located in the extended disks of spirals and irregulars, in the dense gas disks in galaxy centers (circumnuclear star formation), and in regions of compressed gas in starbursts, interacting galaxies, or tidal tails. The star formation histories of galaxies vary with galaxy type, mass, gas content, and environment as well as with time, and are tightly coupled with their chemical evolution.

2. Very early star formation

Giant elliptical galaxies are found to have experienced the bulk of their star formation at early times and to have undergone enrichment very rapidly. A similar evolution appears to have taken place in the bulges of spiral galaxies. But how early is “early”? Age dating resolved old stellar field populations in nearby galaxies remains difficult due to crowding, extinction, the superposition of stellar populations of different ages, and the general difficulty of associating individual red giants with a specific age. These problems are exacerbated in galaxies where only integrated-light properties can be analyzed, which makes it very difficult to date stellar populations older than a few Gyr.

2.1. *Quasars at redshift* $\geqslant 6$

An alternative approach, limited to very luminous and hence presumably very massive, actively star-forming objects, is the analysis of galaxies observed at high redshift. Surprisingly, even the very young quasars discovered at a redshift of $z \sim 6$ (age of the Universe: $\sim$ 900 Myr) reveal metal absorption lines that translate into supersolar metallicities, indicating very rapid, early enrichment. These objects may be the precursors of giant ellipticals (Fan et al. 2001). The masses of the central black holes in $z \sim 6$ QSOs are estimated to range from several 10^8 to several 10^9 $M_\odot$. This suggests formation redshifts of more than 10 for putative 100 $M_\odot$ seed black holes if continuous Eddington accretion is assumed. Other, more rapid black hole formation mechanisms may also play a role (e.g., Fan 2007).

For one of the highest-redshift quasars known, SDSS J114816.16+525150.3 at $z = 6.42$, sub-millimeter and radio observations suggest a molecular gas mass (CO) of $\sim 5 \times 10^{10}$ $M_\odot$ within a radius of 2.5 kpc around the central black hole if the gas is bound. The star formation rate in this object is $\sim$1000 $M_\odot$ yr^{-1}, akin to what has been derived for star-bursting ultraluminous infrared galaxies. Even within 20 kpc, the inferred mass estimate for the QSO's host is still comparatively low, $\sim 10^{11}$ $M_\odot$. A central massive 10^{12} $M_\odot$ bulge has evidently not yet formed (Walter *et al.* 2004; Fan 2007).

In several $z \gtrsim 6$ QSOs dust has been detected. One of the primary sources for dust in the present-day Universe are low- and intermediate-mass asymptotic giant branch stars, but these need at least some 500 Myr to 1 Gyr to begin generating dust in their envelopes. In quasars in the early Universe, other mechanisms are believed to be responsible, including dust production in supernovae of type II (Maiolino *et al.* 2004) and in quasar winds (Elvis *et al.* 2002). Recently two $z \sim 6$ QSOs *without dust* were detected, which indicates that these objects are probably first-generation QSOs forming in an essentially dust-free environment (Jiang *et al.* 2010). This discovery illustrates that even at these very early times of less than 1 Gyr after the Big Bang, galactic environments differ in the onset of massive star formation and in the degree of heavy-element pollution, with some regions already having experienced substantial enrichment. One may speculate that the densest regions are the first ones to start massive star formation, and that the dusty high-redshift QSOs trace these regions.

Magnesium, an α element, is produced in supernovae of type II and hence expected to be generated soon after massive star formation commences. Iron is predominantly produced in supernovae of type Ia, requiring a minimum delay time of $\sim$300 Myr when instantaneous starbursts are considered (e.g., Matteucci & Recchi 2001). Fe is detected in $z \sim 6$ QSOs despite their young age. The Fe II/Mg II ratio (which is essentially a proxy for Fe/α) is found to be comparable to that of lower-redshift QSOs (e.g., Barth *et al.* 2003; Freudling *et al.* 2003; Kurk *et al.* 2007) and to be around solar or super-solar metallicity. The near-constancy of the Fe II/Mg II ratio as a function of redshift implies a lack of chemical evolution in QSOs since $z \sim 6$ and suggests a formation redshift of the SN Ia progenitors of $z \gtrsim 10$. Extremely rapid enrichment on a time scale of just a few hundred Myr must have occurred in these QSOs, much faster than the slow enrichment time scales of spiral galaxies. Nonetheless, also here evolutionary differences are becoming apparent: Some of the $z \sim 6$ QSOs are less evolved, *not* showing strong Fe II emission lines (Iwamuro *et al.* 2004) and hence no significant SN Ia contributions yet.

2.2. *Galaxies at redshift* $\geqslant 7$

Moving to even higher redshifts, the analysis of spectral energy distributions obtained from near-infrared imaging data of galaxies at redshifts of 7 (age of the Universe $\sim$770 Myr) and 8 ($\sim$640 Myr) revealed median ages of $\sim$200 Myr for their stellar populations, but even younger ages are not excluded (Finkelstein *et al.* 2010). The typical stellar masses of the galaxies at $z \sim 7$ are $<10^9$ $M_\odot$; they may be as low as only 10^7 $M_\odot$ at $z \sim 8$. While some galaxies are consistent with having no internal dust extinction, the median value is $A_V \sim 0.3$ mag. Finkelstein *et al.* (2010) estimate that some 10^6 $M_\odot$ of dust may have been produced in these galaxies through massive (12 – 35 $M_\odot$) stars that turned into SNe II, a process expected to take less than 20 Myr (Todini & Ferrara 2001).

Finkelstein *et al.* (2010) derive metallicities of $0.005\,Z_\odot$ for the majority of their galaxies, while some may have somewhat higher values ($0.02\,Z_\odot$). Inferring maximal stellar masses of a few times 10^9 $M_\odot$ for the $z \sim 7$ and 8 galaxies, the authors emphasize that these masses are still considerably lower than suggested for $L_\star$ counterparts at $z < 6$. These distant, luminous galaxies whose colors resemble those of local, metal-poor star-bursting dwarf galaxies trace the earliest times of high-redshift galaxy formation accessible to us with the current instrumentation.

3. Global evolutionary trends

Global evolutionary trends in galaxy evolution over cosmic times (e.g., Madau *et al.* 1996) are best inferred from individual or combined galaxy surveys covering a wide redshift range. Marchesini *et al.* (2009) derive the redshift evolution of the global stellar mass density for galaxies with stellar masses in the range of $10^8 < M_\star/M_\odot < 10^{13}$. Galaxies with lower stellar masses do not appear to contribute significantly to the mass density budget. Marchesini *et al.* find that approximately 45% of the present-day stellar mass was generated from $3 > z > 1$ (within 3.6 Gyr). From $z \sim 1$ to the present, i.e., during the last ~7.5 Gyr, the remaining 50% were produced.

3.1. *Downsizing*

Considering cosmic star formation histories, there is compelling evidence for "downsizing" (Cowie *et al.* 1996): The stars in more massive galaxies usually formed at earlier cosmic epochs and over a shorter time period. For a pedagogical illustration, see Fig. 9 in Thomas *et al.* (2010). High-redshift galaxies tend to have star formation rates higher than those found in the local Universe. Also, these galaxies are typically more massive than low-redshift star-forming galaxies (Seymour *et al.* 2008).

Juneau *et al.* (2005) show that the star formation rate density depends strongly on the stellar mass of galaxies. For massive galaxies with $M_\star > 10^{10.8}$ $M_\odot$ the star formation rate density was about six times higher at $z = 2$ than at the present, remaining approximately constant since $z = 1$. The star formation rate density in their "intermediate-mass" bin, $10^{10.2} \leqslant M_\star/M_\odot \leqslant 10^{10.8}$ $M_\odot$, peaks at $z \sim 1.5$, whereas since $z < 1$ most of the activity was in lower-mass galaxies (Juneau *et al.* 2005).

For quasar host galaxies observed with the Herschel satellite, Serjeant *et al.* (2010) find that high-luminosity quasars have their peak contribution to the star formation density at $z \sim 3$, while the maximum contribution of low-luminosity quasars peaks between $1 < z < 2$. The authors suggest that this indicates a decrease in both the rate of major mergers and in the gas available for star formation and black hole accretion.

3.2. *Mass-metallicity relations*

From an analysis of ~160,000 galaxies in the local Universe observed by the Sloan Digital Sky Survey (SDSS), Gallazzi et al. (2008) conclude that approximately 40% of the total amount of metals contained in stars is located in bulge-dominated galaxies with predominantly old populations. Disk-dominated galaxies contain <25%, while both types of galaxies contribute similarly to the total stellar mass density.

Panter *et al.* (2008) derived the mass-fraction-weighted galaxian metallicity as a function of present-day stellar mass based on the analysis of $> 300,000$ galaxies from the SDSS. Panter *et al.* find a flat relation with little scatter ($\sim$0.15 dex) around $\sim$1.1 $Z_\odot$ for galaxies with masses $\gtrsim 10^{10.5}$ M$_\odot$. Below $\sim 10^{10}$ M$_\odot$ there is a clear trend of decreasing metallicity of about 0.5 dex per dex decline in mass, although the dispersion is relatively large ($\sim$0.5 dex). When considering only (the very few) galaxies in which more than 50% of the light comes from populations younger than 500 Myr, only systems with stellar masses $<10^{10}$ M$_\odot$ contribute significantly – another indication of downsizing.

3.3. *Environmental trends*

Galaxies in high-density regions are generally found to have higher metallicities than those in low-density regions (see, e.g., Panter *et al.* 2008). Sheth *et al.* (2006) find that galaxies with above-average star formation rates and high metallicities at high redshifts are situated mainly in galaxy clusters in the present-day Universe. Interestingly, Poggianti *et al.* (2010) infer that high-redshift clusters were denser environments with respect to both galaxy number and mass than contemporary clusters, which might have fostered the intense star-formation activity and growth in their most massive galaxies. The clustering strength of star-forming galaxies decreases with decreasing redshift (Hartley *et al.* 2010). Galaxies that are passively evolving at the present time (typically galaxies with halo masses $> 10^{13}$ M$_\odot$) are twice as strongly clustered than present-day star-forming galaxies (which are typically at least a factor of 10 less massive).

Sheth *et al.* (2006) point out that at lower redshifts, star formation (either in terms of mass or fraction) is anti-correlated with environment such that dense environments have shown lower star formation rates than low-density regions during the past $\sim$5 Gyr. Interestingly, in the present-day Universe star-forming galaxies appear to evolve fairly independently of their environment, with intrinsic properties playing a determining role (Balogh *et al.* 2004; Poggianti *et al.* 2008). Overall, environment seems to have had little influence on the cosmic star formation history since $z < 1$ (Cooper *et al.* 2008).

The famous morphology-density relation in clusters and groups (Oemler 1974; Dressler 1980; Postman & Geller 1984) describes the increase in the fraction of ellipticals with galaxy density, while the fraction of spirals declines. Similarly, there is a pronounced correlation of morphological types with cluster-centric radius (Whitmore et al. 1993) such that in the innermost regions of a cluster the fraction of ellipticals shows a strong increase, while the fraction of spirals drops sharply. The S0 fraction rises less steeply with decreasing radius and also drops in the innermost regions. These relations are interpreted as suggestive of the growth of ellipticals (and S0s) in high-density regions at the expense of spirals. Goto *et al.* (2003) find three different regimes depending on galaxy density: For densities <1 Mpc^{-2}, the morphology-density and morphology-radius relations become rather weak. For 1 – 6 Mpc^{-2}, the fraction of late-type disks decreases with cluster-centric radius, while early-type spiral and S0 fractions increase. For the highest-density clusters with > 6 Mpc^{-2}, also these intermediate-type fractions decrease with cluster-centric radius, while the early-type fractions increase.

For dwarf galaxies, similar global morphology-density and morphology-radius trends are observed in the sense that gas-deficient early-type dwarfs tend to concentrate within $\sim$300 kpc around massive galaxies in groups or are predominantly found in the inner regions of clusters, while gas-rich late-type dwarfs are found in the outskirts and in the field (e.g., Grebel 2000; Karachentsev *et al.* 2002a, 2003a; Lisker *et al.* 2007). The poorest low-density groups are dominated by late-type galaxies both among their massive and their dwarf members (Karachentsev *et al.* 2003a, 2003b), while in richer, more compact groups the early-type fractions increase also among the dwarfs, as does the morphological segregation (see, e.g., Karachentsev *et al.* 2002a, 2002b). Various physical mechanisms including ram pressure and tidal stripping are discussed to explain apparent evolutionary connections of dwarf galaxies with environment (e.g., van den Bergh 1994; Vollmer *et al.* 2001; Grebel *et al.* 2003; Dong *et al.* 2003; Hensler *et al.* 2004; Kravtsov *et al.* 2004; Mieske *et al.* 2004; Lisker *et al.* 2006; Mayer et al. 2006; D'Onghia *et al.* 2009).

4. Massive early-type galaxies

Analyzing a volume-limited sample of $> 14,000$ early-type galaxies from the SDSS, Clemens *et al.* (2009a) find that their ages, metallicities, and α-element enhancement increase with their mass (using velocity dispersion, σ, as indicator). For galaxies with $\sigma > 180$ km s^{-1}, the mean age decreases with decreasing galactocentric radius, while the metallicity increases. Clemens *et al.* suggest that the massive early-type galaxies were assembled at $z \lesssim 3.5$, merging with low-mass halos that began to form at $z \sim 10$. These subhalos contributed older, metal-poor stars that are still distributed over large radii. Gas-rich mergers, very frequent at early times, contributed fuel for intense star formation in the central regions of the galaxies, while mergers at later times were increasingly gas-poor or dry. Clemens et al. find these radial age and metallicity gradients in early-type galaxies regardless of environment, although massive ellipticals in clusters are on average ~2 Gyr older than those in the field, supporting the trends expected for downsizing.

4.1. *Environment and rejuvenation*

From an analysis of Spitzer Space Telescope (SST) data of 50 early-type galaxies in the Coma cluster, Clemens *et al.* (2009b) find that while the majority is passive, some ~30 % of the galaxies are either younger than 10 Gyr or were rejuvenated in the last few Gyr.

Combining near-UV photometry from the Galaxy Evolution Explorer (GALEX) satellite with SDSS data of a volume-limited sample of 839 luminous early-type galaxies, Schawinski *et al.* (2007) conclude that ~30 % of these objects show evidence of recent (<1 Gyr) star formation (~29 % ellipticals, ~39 % lenticulars). Moreover, they show that that low-density environments contain ~25 % more UV-bright early-type galaxies.

Thomas *et al.* (2010) analyze low-redshift > 3000 early-type galaxies from the SDSS and infer that intermediate-mass and low-mass galaxies show evidence for a secondary peak of more recent star formation around ~2.5 Gyr ago. They find that the fraction of these rejuvenated galaxies becomes larger with decreasing galaxy mass and with decreasing environmental density, reaching up to 45 % at low masses and low densities. Thomas *et al.* conclude that the impact of environment increases with decreasing galaxy mass via mergers and interactions and has done so since $z \sim 0.2$.

4.2. *E+A galaxies*

An interesting class of rejuvenated early-type galaxies are the so-called "E+A" galaxies, ellipticals that show the typical K-star spectra with Mg, Ca, and Fe absorption lines as well as strong Balmer lines akin to A-stars (Dressler & Gunn 1983), indicating that in addition to the usual passive evolution, they experienced star formation within the last Gyr. The absence of [O II] and Hα emission lines shows that there is no ongoing star formation. These post-starburst galaxies are observed both in clusters and in the field.

SDSS studies support suggestions that the E+A phenomenon is created by interactions and/or mergers. About 30 % of the E+A galaxies show disturbed morphologies or tidal tails (Goto 2005). The analysis of 660 E+A galaxies revealed that these objects have a 54 % higher probability of having close companion galaxies than normal galaxies (~8 % vs. ~5 %; Yamauchi *et al.* 2008).

5. Spirals and irregulars

5.1. *Stellar halos of disk galaxies*

Spiral galaxies usually have extended, low-density stellar halos (Zibetti *et al.* 2004) whose density decreases with $\sim r^{-3}$. In the Milky Way the stellar halo consists of old, metal-poor stars and globular clusters on eccentric prograde or retrograde orbits (see Freeman & Bland-Hawthorn 2002 and Helmi 2008 for recent reviews of the Galactic halo). As many of half of the field stars in the halo may have originated in disrupted globular clusters (Martell & Grebel 2010; see Odenkirchen *et al.* 2001a for an example).

ΛCDM simulations suggest that in galaxies with few recent mergers the fraction of halo stars formed *in situ* amounts to 20 % to 50 % (Zolotov *et al.* 2009). Johnston *et al.* (2008) propose that halos dominated by very early accretion show higher [α/Fe] ratios, whereas those that accreted mainly high-luminosity satellites should exhibit higher [Fe/H].

The detection of substructure (e.g., Newberg *et al.* 2002; Yanny *et al.* 2003; Bell et al. 2008) as well as chemical and kinematic signatures (Carollo *et al.* 2007; Geisler *et al.* 2007) support the scenario that part of the Galactic halo was accreted. The individual stellar element abundance patterns suggest that such accretion may have mainly occurred at very early times, since the [α/Fe] ratios of metal-poor halo stars match the ones found in similar stars in the Galactic dwarf spheroidal and irregular companions (e.g., Koch *et al.* 2008a). Prominent morphological evidence of ongoing dwarf galaxy accretion has been found not only in the Milky Way (e.g., Ibata *et al.* 1994), but also around other spiral galaxies (e.g., Ibata *et al.* 2001; Zucker *et al.* 2004; Martínez-Delgado *et al.* 2008).

5.2. *Bulges of disk galaxies and formation scenarios*

Early-type spirals show prominent bulges, which become less pronounced and ultimately vanish in late-type spirals and irregulars. Classical bulges (found in early-type to Sbc spirals) resemble elliptical galaxies in their properties, are dominated by old, mainly metal-rich stars with a large metallicity spread, show hot stellar kinematics, and follow a de Vaucouleurs surface brightness profile just like typical elliptical galaxies. Pseudobulges (in disk galaxies later than Sbc) resemble disk galaxies, have similar exponential profiles, are rotation-dominated, and may contain a nuclear bar, ring, or spiral. They are believed to form from disk material via secular evolution (Kormendy & Kennicutt 2004). A recent analysis combining data from SST and GALEX found that all bulges show some amount of ongoing star formation, regardless of their type (Fisher *et al.* 2009), with small bulges having formed 10 to 30 % of their mass in the past 1 to 2 Gyr (Thomas & Davies 2006). Extracting a sample of > 3000 nearby edge-on disk galaxies from the SDSS, Kautsch *et al.* (2006) showed that approximately 30 % of the edge-one galaxies are bulge-less disks.

Noguchi (1999) suggested that massive clumps forming at early times in galactic disks move towards the galactic center due to dynamical friction, merge, and form the galactic bulge. This scenario leads to the observed trend of increased bulge-to-disk ratios with increased total galactic masses. van den Bergh (2002) noted that while most of the galaxies observed in the Hubble Deep Fields at $z < 1$ have disk-like morphologies, most galaxies at $z > 2$ look clumpy or chaotic. Analyzing such "clump clusters" and "chain galaxies" in the Hubble Ultra Deep Field, Elmegreen *et al.* (2009) find that the masses of the star-forming clumps are of the order to 10^7 to 10^8 $M_{\odot}$. Bournaud *et al.* (2007) argue that clusters of such massive, kpc-sized clumps can form bulges in less than 1 Gyr, while the system as a whole evolves from a violently unstable disk into a regular spiral with an exponential or double exponential disk profile on a similarly rapid time scale. While the coalescence of these clumps resembles a major merger with respect to orbital mixing, the resulting bulge has no specific dark-matter component, which distinguishes it from bulges formed via galaxy mergers (Elmegreen *et al.* 2008).

5.3. *Disks and long-term evolutionary trends for disk galaxies*

Disks are the primary sites of present-day star formation in spiral galaxies, and it seems likely that they have continued to form stars for a Hubble time. Disks show ordered rotation, and their stars move around the galactic center on near-circular orbits. The rotational velocities greatly exceed the velocity dispersion by factors of 20 or more.

Gas-deficient early-type disk galaxies show little activity at the present time, while gas-rich late-type disks experience wide-spread, active star formation. Star formation occurs mainly in the midplane of the thin disks, in particular along spiral arms, where recent events are impressively traced by giant H II regions. Spiral density waves may induce star formation (see Martínez-García *et al.* 2009; and references therein), although it has been suggested that this mechanism may contribute less than 50 % to the overall star formation rate (Elmegreen & Elmegreen 1986).

In the Milky Way, the star formation in the disk was not constant, but shows extended episodes of increased and reduced activity (e.g., Rocha-Pinto *et al.* 2000), a radial metallicity gradient, a G-dwarf problem, and a large metallicity scatter at all ages (Nordström *et al.* 2004). The thin disk is embedded in a lower-density, kinematically hotter stellar population consisting of older, more metal-poor stars – the thick disk (Gilmore & Reid 1983; Bensby *et al.* 2005). The chemical similarity of Galactic bulge and thick disk stars might suggest that the Milky Way does not have a classical bulge (Meléndez *et al.* 2008).

Dalcanton & Bernstein (2002) showed that thick disks are ubiquitous also in bulge-less late-type disk galaxies, which indicates that their formation is a universal property of disk formation independent from the formation of a bulge. A variety of mechanisms for the formation of thick disks has been proposed, including formation from accreted satellites, gas-rich mergers, heating of an early thin disk by mergers, heating via star formation processes, and radial migration (e.g., Wyse *et al.* 2006; Brook *et al.* 2004; Kroupa 2002; Bournaud *et al.* 2009; Roškar *et al.* 2008). Sales *et al.* (2009) suggest that the eccentricity distribution of thick disk stars may permit one to distinguish between these scenarios.

In cosmological simulations disk galaxies may form, for example, via major, wet mergers (see, e.g., Barnes 2002; Governato *et al.* 2009) or without mergers via inside-out and vertical collapse in a growing dark matter halo (Samland & Gerhard 2003).

As noted by van den Bergh (2002) based on an analysis of the Hubble Deep Fields, roughly one third of the objects at $z > 2$ seem to be experiencing mergers. He suggests that from $1 < z < 2$ a transition from merger-dominated to disk-dominated star formation occurred. Moreover, he finds that at $z > 0.5$, there are fewer and fewer barred spirals. While early-type galaxies assume their customary morphologies relatively early on, 46 % of the spirals at $0.6 < z < 0.8$ are still peculiar, and with higher redshift, the spiral arm patterns become increasingly chaotic. Also within the class of spiral galaxies there are trends: Only ∼5 % of the Sa and Sab galaxies are peculiar at $z \sim 0.7$, while almost 70 % of the Sbc and Sc types are still peculiar.

Elmegreen *et al.* (2007) suggest that the formation epoch of clumpy disk galaxies may extend up to $z \sim 5$. The ones experiencing major mergers may form red spheroidals at $2 \lesssim z \lesssim 3$, whereas the others evolve into spirals. Elmegreen *et al.* propose that the the star formation activity in clumpy disks is caused by gravitational collapse of portions of the disk gas without requiring an external trigger.

Regarding environment, Poggianti *et al.* (2009) find that the fraction of ellipticals remains essentially constant below $z = 1$, while the spiral and S0 fractions continue to evolve, showing the most pronounced evolution in low-mass galaxy clusters. They attribute this to secular evolution and to environmental mechanisms that are more effective in low-mass environments. At low redshifts, the declining spiral fraction with density is driven by late-type spirals (Sc and later; Poggianti *et al.* 2008).

5.4. *Irregulars*

Irregular galaxies are gas-rich, low-mass, metal-poor galaxies without spiral density waves, which show recent or ongoing star formation that appears to have extended over a Hubble time (Hunter 1997). Many studies found the H I gas to be considerably more extended than the stellar component in irregulars (e.g., Young & Lo), but more recent, deep optical surveys show that the optical extent of at least some of these galaxies has been underestimated (e.g., Kniazev *et al.* 2009). All nearby irregulars and dwarf irregulars have been found to contain old populations, although their fractions differ (Grebel & Gallagher 2004). The old populations tend to be more extended than the more recent star formation (e.g., Minniti & Zijlstra 1996; Kniazev *et al.* 2009) and show a more regular distribution (e.g., Zaritsky *et al.* 2000; van der Marel 2001).

Irregulars are usually found in the outskirts of groups and clusters or in the field, thus interactions with other galaxies are likely to be rare. Their star formation appears to be largely governed by internal processes and seems to be stochastic. Rather than experiencing brief, intense starbursts, irregulars typically show extended episodes of star formation interrupted by short quiescent periods – so-called gasping star formation (e.g., Cignoni & Tosi 2010). The long-term star formation amplitude variations amount to factors of 2 to 3 (Tosi *et al.* 1991). For a review of irregulars and dwarf irregulars in the Local Group, for which we have the most detailed data to date, see Grebel (2004).

While the more massive irregulars are rotationally supported and show solid-body rotation, low-mass dwarf irregulars are dominated by random motions. Star formation ceases at lower gas density thresholds than in spirals (e.g., Parodi & Binggeli 2003), and the global gas density of the highly porous interstellar medium has been found to lie below the Toomre criterion for star formation (van Zee *et al.* 1997). Turbulence may create local densities exceeding the star formation threshold (e.g., Stanimirovic *et al.* 1999). Low-mass dwarf irregulars without measurable rotation show less centrally concentrated star formation and have lower star formation rates (Roye & Hunter 2000; Parodi & Binggeli 2003). In most irregulars, star formation occurs within the galaxies' Holmberg radius and within three disk scale lengths (Hunter & Elmegreen 2004).

In contrast, in blue compact dwarf (BCD) galaxies the highest star formation rates are found, and star formation occurs mainly in the central regions (Hunter & Elmegreen). (We do not discuss BCDs and other gas-rich dwarfs in more detail here. For an overview of different dwarf types and their properties, see Grebel (2003)).

Once believed to be chemically homogeneous, there is now evidence of metallicity variations at a given age in several irregulars (e.g., Kniazev *et al.* 2005; Glatt *et al.* 2008). This suggests that local processes dominate the enrichment and that mixing is not very efficient. Irregulars follow a fairly well-defined metallicity-luminosity relation, which however is offset from that of early-type dwarfs covering the same luminosity range. Surprisingly, the offset is such that the continuously star-forming irregulars and dwarf irregulars have lower metallicities at a given luminosity than the inactive early-type dwarfs (e.g., Richer *et al.* 1998), a discrepancy that holds even when comparing stellar populations of the same age (Grebel et al. 2003). Taken at face value, this may imply that the enrichment of irregulars was less efficient and slower than that of early-type dwarfs. BCDs and in particular extremely metal-deficient galaxies continue this trend and appear to be too luminous for their present-day, low metallicities even when compared to normal irregulars (Kunth & Östlin 2000; Kniazev *et al.* 2003).

5.5. *Star formation "demographics"*

Lee *et al.* (2007) investigate the star formation "demographics" of star-forming galaxies out to 11 Mpc combining Hα and GALEX UV fluxes. Their sample includes spirals, irregulars, and BCDs. Lee *et al.* identify three different star formation regimes:

1. Galaxies with maximum rotational velocities $V_{max} > 120$ km s^{-1}, total B-band magnitudes of $M_B \lesssim -19$, and stellar masses $\gtrsim 10^{10}$ M$_\odot$ are mainly bulge-dominated galaxies with relatively low specific star formation rates and increased scatter in these rates. Also the mass-metallicity relation changes its slope in this regime (Panter *et al.* 2007), and supernova ejecta can be retained (Dekel & Woo 2003). Bothwell *et al.* (2009) find that the H I content of these massive galaxies decreases faster than their star formation rates, leading to shorter H I consumption time scales and making the lack of gas a plausible reason for the observed quenching of star formation activity.

2. Galaxies with ~120 km s^{-1} $> V_{max} >$ 50 km s^{-1} and $-19 < M_B < -15$ comprise mainly late-type spirals and massive irregulars. Lee at al. (2007) suggest that spiral structure acts as an important regulatory factor for star formation. They find that the galaxies in this intermediate-mass regime exhibit a comparatively tight, constant relation between star formation rate and luminosity (or rotational velocity). The star formation rates show fluctuations of 2 to 3, and the current star formation activity is about half of its average value in the past. Bothwell *et al.* (2009) argue that the galaxies in this regime evolve secularly. They show that the star formation rates decrease with the galaxies' H I mass, and that the H I consumption time scales increase with decreasing luminosity.

3. Below $V_{max} = 50$ km s^{-1} and $M_B > -15$, dwarf galaxies, particularly irregulars, dominate. At these low masses, the star formation rates exhibit much more variability ranging from significantly higher (e.g., in BCDs) to significantly lower (e.g., in so-called transition-type dwarfs with properties in between dwarf irregulars and dwarf spheroidal galaxies, see Grebel *et al.* 2003) star formation activity than in the higher-mass regimes. Overall, there is a general trend towards lower star formation rates. Stochastic intrinsic processes, feedback, and the ability to retain gas play an important role here. Bothwell et al. (2009) find that for many of the galaxies in the low-mass regime the H I consumption time scale exceeds a Hubble time (in good agreement with the results of Hunter 1997).

Bothwell *et al.* show that the H I consumption time scales have a minimum duration of more than 100 Myr. They argue that this minimum duration corresponds to the gas mass divided by the minimum gas assembly time, i.e., the free-fall collapse time.

6. Early-type dwarfs

We end this review by summarizing the star formation properties of early-type dwarfs, most notably of dwarf ellipticals (dEs) and dwarf spheroidals (dSphs; see Grebel 2003; Grebel et al. 2003). Typically located in high-density regions such as the immediate surroundings of massive galaxies or in galaxy clusters, this dense environment may have affected the evolution of these now gas-deficient galaxies (Section 3.3). Structural and kinematic studies suggest that early-type dwarfs are strongly dark-matter dominated (e.g., Odenkirchen *et al.* 2001b; Klessen *et al.* 2003; Wilkinson *et al.* 2004; Koch *et al.* 2007a, 2007b; Walker *et al.* 2007; Gilmore *et al.* 2007; Wolf et al. 2010), and there are even indications of a constant dark-matter halo surface density from spirals to dwarfs (Donato *et al.* 2009). However, it is not yet clear whether the apparently constant total mass regardless of a dwarf's baryonic luminosity is universal (e.g., Adén *et al.* 2009).

All of dEs and dSphs studied in detail so far reveal varying fractions of old populations (Grebel & Gallagher 2004; Da Costa *et al.* 2010) that become dominant at low galactic masses, while intermediate-age populations (> 1 Gyr) are prominent at higher galactic masses. Still, no two dwarfs share the same evolutionary history or detailed abundance properties (Grebel 1997). Population gradients are found in many early-type dwarfs. Where present, younger and/or more metal-rich populations tend to be more centrally concentrated (e.g., Harbeck *et al.* 2001; Lisker *et al.* 2006; Crnojević *et al.* 2010).

The Galactic dSphs reach solar [α/Fe] ratios at much lower [Fe/H] than typical Galactic halo stars, which suggests low star formation rates, the loss of metals and supernova ejecta, and/or a larger contribution from SNe Ia (e.g., Shetrone *et al.* 2001). Abundance spreads of 1 dex in [Fe/H] and more are common. The scatter in α element abundance ratios at a given metallicity underlines the inhomogeneous, localized enrichment in the early-type dwarfs, another characteristic expected of slow, stochastic star formation and low star formation efficiencies (Koch *et al.* 2008a, 2008b; Marcolini *et al.* 2008).

References

Adén, D., *et al.* 2009, *ApJ*, 706, 150
Balogh, M. L., *et al.* 2004, *ApJ*, 615, L101
Barnes, J. E. 2002, *MNRAS*, 333, 481
Barth, A. J., Martini, P., Nelson, C. H., & Ho, L. C. 2003, *ApJ*, 594, L95
Bell, E. F., *et al.* 2008, *ApJ*, 680, 295
Bensby, T., Feltzing, S., Lundström, I., & Ilyin, I. 2005, *A&A*, 433, 185
Bothwell, M. S., Kennicutt, R. C., & Lee, J. C. 2009, *MNRAS*, 400, 154
Bournaud, F., Elmegreen, B. G., & Elmegreen, D. M. 2007, *ApJ*, 670, 237
Bournaud, F., Elmegreen, B. G., & Martig, M. 2009, *ApJ*, 707, L1
Brook, C. B., Kawata, D., Gibson, B. K., & Freeman, K. C. 2004, *ApJ*, 612, 894
Bush, S. J., *et al.* 2010, *ApJ*, 713, 780
Carollo, D., *et al.* 2007, *Nature*, 450, 1020
Cignoni, M. & Tosi, M. 2010, *Adv. Ast.*, Vol. 2010, 1
Clemens, M. S., Bressan, A., Nikolic, B., & Rampazzo, R. 2009a, *MNRAS*, 392, L35
Clemens, M. S., *et al.* 2009b, *MNRAS*, 392, 982
Cooper, M. C., *et al.* 2008, *MNRAS*, 383, 1058
Cowie, L. L., Songaila, A., Hu, E. M., & Cohen, J. G. 1996, *AJ*, 112, 839
Crnojević, D., Grebel, E. K., & Koch, A. 2010, *A&A*, 516, A85
Da Costa, G. S., Rejkuba, M., Jerjen, H., & Grebel, E. K. 2010, *ApJ*, 708, L121
Dalcanton, J. J. & Bernstein, R. A. 2002, *AJ*, 124, 1328
Dekel, A. & Woo, J. 2003, *MNRAS*, 344, 1131
Dressler, A. 1980, *ApJ*, 236, 351
Dressler, A. & Gunn, J. E. 1983, *ApJ*, 270, 7
Donato, F., *et al.* 2009, *MNRAS*, 397, 1169
Dong, S., Lin, D. N. C., & Murray, S. D. 2003, *ApJ*, 596, 930
D'Onghia, E., Besla, G., Cox, T. J., & Hernquist, L. 2009, *Nature*, 460, 605
Elvis, M., Marengo, M., & Karovska, M. 2002, *ApJ*, 567, L107
Elmegreen, B. G. & Elmegreen, D. M. 1986, *ApJ*, 311, 554
Elmegreen, D. M., Elmegreen, B.G., Ravindranath, S., & Coe, D. A. 2007, *ApJ*, 658, 763
Elmegreen, B. G., Bournaud, F., & Elmegreen, D.M. 2008, *ApJ*, 688, 67
Elmegreen, B. G., Elmegreen, D. M., Fernandez, M. X., & Lemonias, J. J. 2009, *ApJ*, 692, 12
Fan, X., *et al.* 2001, *AJ*, 122, 2833
Fan, X. 2006, *New Astron. Revs*, 50, 665
Finkelstein, S. L., *et al.* 2010, *ApJ*, 719, 1250
Fisher, D. B., Drory, N., & Fabricius, M. H. 2009, *ApJ*, 697, 630
Freeman, K. & Bland-Hawthorn, J. 2002, *Ann. Rev. Astron. Astroph.*, 40, 487
Freudling, W., Corbin, M. R., & Korista, K. T. 2003, *ApJ*, 587, L67
Gallazzi, A., Brinchmann, J., Charlot, S., & White, S. D. M. 2008, *MNRAS*, 383, 1439

Geisler, D., Wallerstein, G., Smith, V. V., & Casetti-Dinescu, D. I. 2007, *PASP*, 119, 939
Gilmore, G. & Reid, N. 1983, *MNRAS*, 202, 1025
Gilmore, G., *et al.* 2007, *ApJ*, 663, 948
Glatt, K., *et al.* 2008, *AJ*, 136, 1703
Governato, F., et al. 2009, *MNRAS*, 398, 312
Goto, T., *et al.* 2003, *MNRAS*, 346, 601
Goto, T. 2005, *MNRAS*, 357, 937
Grebel, E. K. 1997, *Rev. Mod. Astron.*, 10, 29
Grebel, E. K. 2000, in Star formation from the small to the large scale, ESLAB Symposium 33, ESA SP 445, eds. F. Favata, A. Kaas, & A. Wilson (ESA: Noordwijk), 87
Grebel, E. K. 2003, *Ap&SSS*, 284, 947
Grebel, E. K. 2004, in Origin and Evolution of the Elements, Carnegie Observatories Astrophysics Series, Vol. 4, eds. A. McWilliam & M. Rauch (Cambridge: CUP), 237
Grebel, E. K., Gallagher, J. S., III, & Harbeck, D. 2003, *AJ*, 125, 1926
Grebel, E. K. & Gallagher, J. S., III 2004, *ApJ*, 610, L89
Harbeck, D., et al. 2001, *AJ*, 122, 3092
Hartley, W. G., *et al.* 2010, *MNRAS*, 1089
Helmi, A. 2008, *A&ARv*, 15, 145
Hensler, G., Theis, C., Gallagher, J. S., III 2004, *A&A*, 426, 25
Hunter, D. A. 1997, *PASP*, 109, 937
Hunter, D. A. & Elmegreen, B. G. 2004, *AJ*, 128, 2170
Ibata, R. A., Gilmore, G., & Irwin, M. J. 1994, *Nature*, 370, 194
Ibata, R., Irwin, M., Lewis, G., Ferguson, A. M. N., & Tanvir, N. 2001, *Nature*, 412, 49
Iwamuro, F., *et al.* 2004, *ApJ*, 614, 69
Jiang, L., *et al.* 2010, *Nature*, 464, 380
Juneau, S., *et al.* 2005, *ApJ*, 619, L135
Johnston, K. V., *et al.* 2008, *ApJ*, 689, 936
Karachentsev, I. D., *et al.* 2002a, *A&A*, 385, 21
Karachentsev, I. D., *et al.* 2002b, *A&A*, 383, 125
Karachentsev, I. D., *et al.* 2003a, *A&A*, 398, 467
Karachentsev, I. D., *et al.* 2003b, *A&A*, 404, 93
Kautsch, S. J., Grebel, E. K., Barazza, F. D., & Gallagher, J. S., III 2006, *A&A*, 445, 765
Kennicutt, R. C., Jr. 1998, *Ann. Rev. Astron. Astroph.*, 36, 189
Klessen, R. S., Grebel, E. K., & Harbeck, D. 2003, *ApJ*, 589, 798
Kniazev, A. Y., *et al.* 2003, *ApJ*, 593, L73
Kniazev, A. Y., *et al.* 2005, *AJ*, 130, 1558
Kniazev, A. Y., *et al.* 2009, *MNRAS*, 400, 2054
Koch, A., *et al.* 2007a, *AJ*, 134, 566
Koch, A., *et al.* 2007b, *ApJ*, 657, 241
Koch, A., *et al.* 2008a, *AJ*, 135, 1580
Koch, A., McWilliam, A., Grebel, E. K., Zucker, D. B., & Belokurov, V. 2008b, *ApJ*, 688, L13
Kravtsov, A. V., Gnedin, O. Y., & Klypin, A. A. 2004, *ApJ*, 609, 482
Kormendy, J. & Kennicutt, R. C., Jr. 2004, *Ann. Rev. Astron. Astroph.*, 42, 603
Kroupa, P. 2002, *MNRAS*, 330, 707
Kunth, D. & Östlin, G. 2000, *A&ARv*, 10, 1
Kurk, J. D., *et al.* 2007, *ApJ*, 669, 32
Lee, J. C., *et al.* 2007, *ApJ*, 671, L113
Lisker, T., Grebel, E. K., & Binggeli, B. 2006a, *AJ*, 132, 497
Lisker, T., Glatt, K., Westera, P., & Grebel, E. K. 2006b, *AJ*, 132, 2432
Lisker, T., Grebel, E. K., Binggeli, B., & Glatt, K. 2007, *ApJ*, 660, 1186
Madau, P., *et al.* 1996, *MNRAS*, 283, 1388
Maiolino, R., *et al.* 2004, *Nature*, 431, 533
Marchesini, D., *et al.* 2009, *ApJ*, 701, 1765
Marcolini, A., D'Ercole, A., Battaglia, G., & Gibson, B.K. 2008, *MNRAS*, 386, 2173

Martell, S. L. & Grebel, E. K. 2010, *A&A*, in press (arXiv:1005.4070)
Martínez-Delgado, D., *et al.* 2008, *ApJ*, 689, 184
Martínez-García, E. E., González-Lópezlira, R. A., & Bruzual, G. 2009, *ApJ*, 694, 512
Matteucci, F., & Recchi, S. 2001, *ApJ*, 558, 351
Mayer, L., Mastropietro, C., Wadsley, J., Stadel, J., & Moore, B. 2006, *MNRAS*, 369, 1021
Meléndez, J., *et al.* 2008, *A&A*, 484, L21
Mieske, S., Hilker, M., & Infante, L. 2004, *A&A*, 418, 445
Minniti, D. & Zijlstra, A. A. 1996, *ApJ*, 467, L13
Newberg, H. J., *et al.* 2002, *ApJ*, 569, 245
Noguchi, M. 1999, *ApJ*, 514, 77
Nordström, B., *et al.* 2004, *A&A*, 418, 989
Odenkirchen, M., et al. 2001a, *ApJ*, 548, L165
Odenkirchen, M., *et al.* 2001, *AJ*, 122, 2538
Oemler, A., Jr. 1974, *ApJ*, 194, 1
Panter, B., Jimenez, R., Heavens, A. F., & Charlot, S. 2008, *MNRAS*, 391, 1117
Parodi, B. R. & Binggeli, B. 2003, *A&A*, 398, 501
Poggianti, B. M., et al. 2008, *ApJ*, 684, 888
Poggianti, B. M., et al. 2009, *ApJ*, 697, L137
Poggianti, B. M., et al. 2010, *MNRAS*, 405, 995
Postman, M. & Geller, M. J. 1984, *ApJ*, 281, 95
Richer, M., McCall, M. L., & Stasinska, G. 1998, *A&A*, 340, 67
Rocha-Pinto, H. J., Scalo, J., Maciel, W. J., & Flynn, C. 2000, *A&A*, 358, 869
Roškar, R., Debattista, V. P., Quinn, T. R., Stinson, G. S., & Wadsley, J. 2008, *ApJ*, 684, L79
Roye, E. W. & Hunter, D. A. 2000, *AJ*, 119, 1145
Sales, L. V., *et al.* 2009, *MNRAS*, 400, L61
Samland, M. & Gerhard, O. E. 2003, *A&A*, 399, 961
Schawinski, K., et al. 2007, *ApJS*, 173, 512
Serjeant, S., *et al.* 2010, *A&A*, 518, L7
Seymour, N., *et al.* 2008, *MNRAS*, 386, 1695
Sheth, R. K., Jimenez, R., Panter, B., & Heavens, A. F. 2006, *ApJ*, 650, L25
Shetrone, M. D., Côté, P., & Sargent, W. L. W. 2001, *ApJ*, 548, 592
Stanimirovic, S., *et al.* 1999, *MNRAS*, 302, 417
Thomas, D. & Davies, R. L. 2006, *MNRAS*, 366, 510
Thomas, D., Maraston, C., Schawinski, K., Sarzi, M., & Silk, J. 2010, *MNRAS*, 404, 1775
Todini, P. & Ferrara, A. 2001, *MNRAS*, 325, 726
Tosi, M., Greggio, L., Marconi, G., & Focardi, P. 1991, *AJ*, 102, 951
Yamauchi, C., Yagi, M., & Goto, T. 2008, *MNRAS*, 390, 383
Yanny, B., *et al.* 2003, *ApJ*, 588, 824
Young, L. M. & Lo, K. Y., 1997, *ApJ*, 490, 710
van den Bergh, S. 1994, *ApJ*, 428, 617
van den Bergh, S. 2002, *PASP*, 114, 797
van der Marel, R. P. 2001, *AJ*, 122, 1827
van Zee, L., Haynes, M. P., Salzer, J. J., & Broeils, A. H. 1997, *AJ*, 113, 1618
Vollmer, B., Cayatte, V., Balkowski, C., & Duschl, W. J. 2001, *ApJ*, 561, 708
Walker, M. G., *et al.* 2007, *ApJ*, 667, L53
Walter, F., *et al.* 2004, *ApJ*, 615, L17
Whitmore, B. C., Gilmore, D. M., & Jones, C. 1993, *ApJ*, 407, 489
Wilkinson, M. I., *et al.* 2004, *ApJ*, 611, L21
Wolf, J., *et al.* 2010, *MNRAS*, 406, 1220
Wyse, R. F. G., *et al.* 2006, *ApJ*, 639, L13
Zaritsky, D., Harris, J., Grebel, E. K., & Thompson, I. B. 2000, *ApJ*, 534, L53
Zibetti, S., White, S. D. M., Brinkmann, J. 2004, *MNRAS*, 347, 556
Zolotov, A., *et al.* 2009, *ApJ*, 702, 1058
Zucker, D. B., *et al.* 2004, *ApJ*, 612, L117

Computational Star Formation
Proceedings IAU Symposium No. 270, 2011
J. Alves, B.G. Elmegreen, J. M. Girart & V. Trimble, eds.

doi:10.1017/S1743921311000615

Theory of the Star Formation Rate

Paolo Padoan[1] **and Åke Nordlund**[2]

[1]ICREA & ICC, University of Barcelona, Marti i Franquès 1, E-08028 Barcelona, Spain
email: ppadoan@icc.ub.edu

[2]Niels Bohr Institute, University of Copenhagen, Juliane Maries Vej 30, DK-2100, Copenhagen, Denmark
email: aake@nbi.dk

Abstract. This work presents a new physical model of the star formation rate (SFR), tested with a large set of numerical simulations of driven, supersonic, self-gravitating, magneto-hydrodynamic (MHD) turbulence, where collapsing cores are captured with accreting sink particles. The model depends on the relative importance of gravitational, turbulent, magnetic, and thermal energies, expressed through the virial parameter, $\alpha_{\rm vir}$, the rms sonic Mach number, $\mathcal{M}_{\rm S,0}$, and the ratio of mean gas pressure to mean magnetic pressure, β_0. The SFR is predicted to decrease with increasing $\alpha_{\rm vir}$ (stronger turbulence relative to gravity), and to depend weakly on $\mathcal{M}_{\rm S,0}$ and β_0, for values typical of star forming regions ($\mathcal{M}_{\rm S,0} \approx$ 4-20 and $\beta_0 \approx$ 1-20). The star-formation simulations used to test the model result in an approximately constant SFR, after an initial transient phase. Both the value of the SFR and its dependence on the virial parameter found in the simulations agree very well with the theoretical predictions.

Keywords. ISM: kinematics and dynamics, MHD, stars: formation, turbulence

1. Introduction

The star-formation process is slow. Only a small fraction of the mass of cold gas is converted into stars in a free-fall time, $\tau_{\rm ff}$, both on Galactic scale (Zuckerman & Palmer 1974; Williams & McKee 1997) and on the scale of individual clouds (Krumholz & Tan 2007, Evans *et al.* 2009). Several authors have proposed that the observed supersonic turbulence may be the reason for the low star formation rate (SFR). The turbulence is responsible for much of the complex and filamentary density structure observed in molecular clouds, and prestellar cores are likely assembled as the densest regions in this turbulent fragmentation process (Padoan *et al.* 2001). But even if supersonic turbulence intermittently creates dense regions that are gravitationally unstable, it does it inefficiently; its net effect on the large scale is to suppress star formation when the total turbulent kinetic energy exceeds the total gravitational energy.

Due to the importance of turbulence, the SFR depends primarily on the ratio of the turbulent kinetic energy, $E_{\rm K}$, and the gravitational energy, $E_{\rm G}$, of a star-forming region. This ratio is expressed by the virial parameter introduced by Bertoldi & McKee (1992),

$$\alpha_{\rm vir} \sim \frac{2E_{\rm K}}{E_{\rm G}} = \frac{5\sigma_{\rm v,1D}^2 R}{GM}, \tag{1.1}$$

where $\sigma_{\rm v,1D}$ is the one-dimensional rms velocity, R and M the cloud radius and mass respectively, and G the gravitational constant, and it has been assumed the cloud is a sphere with uniform density.

Krumholz & McKee (2005) derived a theoretical model where the SFR is primarily controlled by the virial parameter. In this model, it is assumed that the gas mass above some critical density, $\rho_{\rm cr}$, is gravitationally unstable, and the fraction of this unstable mass is computed assuming the gas density obeys a Log-Normal pdf (Nordlund & Padoan 1999). The idea of relying on the density pdf was also exploited in Padoan & Nordlund (2002) and in Padoan & Nordlund (2004) to explain the stellar IMF and the origin of brown dwarfs, and by Padoan (1995) to model the SFR.

The model of Krumholz & McKee (2005) did not include magnetic fields and was calibrated and tested using low-resolution SPH simulations by Vazquez-Semadeni *et al.* (2003). In this work we propose a new model of the SFR that includes magnetic fields and that is tested with an unprecedented set of large numerical simulations of driven, supersonic, self-gravitating, MHD turbulence, where collapsing cores are represented by accreting sink particles. To model the process of star formation we must include gravitational, turbulent, magnetic, and thermal energies. Their relative importance is expressed by the virial parameter, $\alpha_{\rm vir}$, the rms sonic Mach number, $\mathcal{M}_{\rm S,0}$, and the mean gas pressure to mean magnetic pressure, β_0, and we derive a model that depends explicitly on all three non-dimensional parameters. In the non-magnetized limit of $\beta_0 \to \infty$, our definition of the critical density for star formation has the same dependence on $\alpha_{\rm vir}$ as in the model of Krumholz & McKee (2005), but our derivation does not rely on the concepts of local turbulent pressure support and sonic scale.

2. Critical density in MHD Turbulence

Including both thermal and magnetic pressures, the pressure balance condition for magneto-hydrodynamic (MHD) shocks is:

$$\rho_{\rm MHD}(c_{\rm S}^2 + v_{\rm A}^2/2) = \rho_0(v_0/2)^2, \tag{2.1}$$

where $c_{\rm S}$ is the sound speed, ρ_0 and $\rho_{\rm MHD}$ the preshock and postshock gas densities, $v_0/2$ the shock velocity, and $v_{\rm A}$ is the Alfvén velocity in the postshock gas defined by the postshock magnetic field perpendicular to the direction of compression. Because the field is amplified only in the direction perpendicular to the compression, the postshock perpendicular field is comparable to the total postshock field, and we can write, $v_{\rm A} \approx B/(4\pi\rho)^{1/2}$, where B is the postshock magnetic field and ρ the postshock gas density. The characteristic gas density and thickness of postshock layers are thus given by:

$$\rho_{\rm MHD} = \rho_0(\mathcal{M}_{\rm S,0}^2/4)\left(1+\beta^{-1}\right)^{-1}, \tag{2.2}$$

$$\lambda_{\rm MHD} = (\theta\, L_0)(\mathcal{M}_{\rm S,0}^2/4)^{-1}\left(1+\beta^{-1}\right), \tag{2.3}$$

where L_0 is the size (e.g. the diameter for a sphere) of the system, $\theta\, L_0$, with $\theta \leqslant 1$, is the turbulence integral scale, and we have introduced the ratio of gas to magnetic pressure in the postshock gas, $\beta = 2\,c_{\rm S}^2/v_{\rm A}^2$. In the limit of $\beta \to \infty$, these expressions reduce to the corresponding HD ones.

In numerical simulations of supersonic and super-Alfvénic turbulence, it is found that, although $v_{\rm A}$ has a very large scatter for any given density, its mean value is nearly density independent, corresponding to a mean relation approaching $B \propto \rho^{1/2}$ for a very weak mean magnetic field (Padoan & Nordlund 1999). In the specific MHD simulation used in this work, the mean value of $v_{\rm A}$ is almost exactly constant for any density $\rho \gtrsim 2\rho_0$. Zeeman splitting measurements of the magnetic field strength in molecular cloud cores are also consistent with an average value of $v_{\rm A}$ nearly independent of density (Crutcher 1999).

We define the critical density as the density above which a uniform sphere of radius $\lambda_{\rm MHD}/2$ is gravitationally unstable. To account for both thermal and magnetic support, we adopt the approximation of the critical mass for collapse, $M_{\rm cr}$, introduced by McKee

(1989),

$$M_{\rm cr} \approx M_{\rm BE} + M_{\phi}, \tag{2.4}$$

where M_{ϕ} is the magnetic critical mass for a sphere of radius R, mean density equal to the postshock density ρ, and constant mass-to-flux ratio,

$$M_{\phi} = 0.17\pi R^2 B/G^{1/2} = 0.387 v_{\rm A}^3/(G^{3/2}\rho^{1/2}) \tag{2.5}$$

where the numerical coefficient 0.17 is from Tomisaka *et al.* (1988) (see also Nakano & Nakamura (1978) for the case of an infinite sheet, and McKee & Ostriker 2007 for a discussion of ellipsoidal clouds and other geometries), and $M_{\rm BE}$ is the Bonnor-Ebert mass (Bonnor 1956; Ebert 1957) with external density equal to the postshock density ρ,

$$M_{\rm BE} = 1.182\, c_{\rm S}^3\, /(G^{3/2}\rho^{1/2}) \tag{2.6}$$

The critical density is then defined by the condition,

$$M_{\rm MHD}(\rho_{\rm cr,MHD}) = M_{\rm BE}(\rho_{\rm cr,MHD}) + M_{\phi}(\rho_{\rm cr,MHD}), \tag{2.7}$$

where $M_{\rm MHD}(\rho) = (4/3)\pi\lambda_{\rm MHD}^3\,\rho$. Equation (2.7) results in the following expression for the critical density as a function of the three non-dimensional parameters, $\alpha_{\rm vir}$, $\mathcal{M}_{\rm S,0}$, and β:

$$\frac{\rho_{\rm cr,MHD}}{\rho_0} = 0.067\,\theta^{-2}\alpha_{\rm vir}\,\mathcal{M}_{\rm S,0}^2\,\frac{(1+0.925\beta^{-\frac{3}{2}})^{\frac{2}{3}}}{(1+\beta^{-1})^2}, \tag{2.8}$$

Based on numerical simulations, we adopt the value of $\beta = 0.39$ for $\mathcal{M}_{\rm S,0} \approx 10$ and $0.2 \lesssim \beta_0 \lesssim 20$ (see details in Padoan & Nordlund 2010, in preparation). The compilations of OH and CN Zeeman measurements by Troland & Crutcher (2008) and Falgarone *et al.* (2008) give an average value of $\beta = 0.34$ and 0.28 respectively, very close to the value adopted here.

3. Gas Density PDF and Star Formation Rate

We estimate the gas mass fraction that is turned into stars by computing the mass fraction above the critical density, as in Krumholz & McKee (2005). For given values of $\alpha_{\rm vir}$, $\mathcal{M}_{\rm S,0}$, and β (or β_0), the critical density is fixed, and the mass fraction above the critical density is determined by the density pdf. In the HD case, the density pdf is known to be Log-Normal, with a standard deviation depending on the rms Mach number. We assume that the pdf can be approximated by a Log-Normal also in the MHD case,

$$p_{\rm MHD}(x)dx = \frac{x^{-1}}{(2\pi\sigma_{\rm MHD}^2)^{1/2}}\exp\left[-\frac{(\ln x + \sigma_{\rm MHD}^2/2)^2}{2\,\sigma_{\rm MHD}^2}\right]dx \tag{3.1}$$

with

$$\sigma_{x,\rm MHD} \approx (1+\beta^{-1})^{-1/2}\mathcal{M}_{\rm S,0}/2, \tag{3.2}$$

corresponding to

$$\sigma_{\rm MHD}^2 \approx \ln\left[1+\left(\frac{\mathcal{M}_{\rm S,0}}{2}\right)^2(1+\beta^{-1})^{-1}\right]. \tag{3.3}$$

In the non-magnetized case ($\beta = \infty$), these equations reduce to the result of Padoan *et al.* (1997).

Assuming that a fraction ϵ of the mass fraction above the critical density is turned into stars in a free-fall time of the critical density, $\tau_{\rm ff,cr} = (3\pi/(32G\rho_{\rm cr,MHD}))^{1/2}$, the star formation rate per free-fall time (the mass fraction turned into stars in a free-fall time) is given by

$$\mathrm{SFR_{ff}} = \epsilon \, \frac{\tau_{\mathrm{ff},0}}{\tau_{\mathrm{ff,cr}}} \int_{x_{\mathrm{cr}}}^{\infty} x \, p_{\mathrm{MHD}}(x) \, dx = \epsilon \, \frac{x_{\mathrm{cr}}^{1/2}}{2} \left(1 + \mathrm{erf} \left[\frac{\sigma^2 - 2 \ln(x_{\mathrm{cr}})}{2^{3/2} \, \sigma} \right] \right) \tag{3.4}$$

where $\tau_{\mathrm{ff},0} = (3\pi/(32G\rho_0))^{1/2}$ is the free-fall time of the mean density, $x_{\mathrm{cr}} = \rho_{\mathrm{cr,MHD}}/\rho_0$ given by equation (2.8), $\sigma = \sigma_{\mathrm{MHD}}$ given by equation (3.3), and the expression is valid also in the HD limit of $\beta \to \infty$. The value of ϵ is 1 in the HD case and 0.5 in the MHD case (see details in Padoan & Nordlund 2010, in preparation).

4. SFR in Simulations of Driven MHD Turbulence

In order to test the SFR model, we have run a set of simulations of driven supersonic turbulence, on meshes with 500^3–$1,000^3$ computational zones. Using the same methods and setup as in Padoan & Nordlund (2002) and Padoan & Nordlund (2004), we adopt periodic boundary conditions, isothermal equation of state, and random forcing in Fourier space at wavenumbers $1 \leqslant k \leqslant 2$ ($k = 1$ corresponds to the computational box size). The simulations are based on two initial snapshots of fully developed turbulence, one for HD and one for MHD. These snapshots are obtained by running the HD and the MHD simulations from initial states with uniform initial density and magnetic field, and random initial velocity field with power only at wavenumbers $1 \leqslant k \leqslant 2$, for approximately 5 dynamical times, on meshes with $1,000^3$ computational zones, with the driving force keeping the rms sonic Mach number at the approximate value of $\mathcal{M}_{\mathrm{S},0} = \sigma_{\mathrm{v,3D}}/c_{\mathrm{S}} \approx 9$ (except for one HD run with $\mathcal{M}_{\mathrm{S},0} \approx 4.5$). The initial pressure ratio is $\beta_0 = 22.2$ in all MHD runs, and $\beta_0 = \infty$ in the HD runs.

The star formation simulations start when the gravitational force is included. The computational mesh is downsized from $1,000^3$ to 500^3 zones for the 500^3 runs, or kept the same for the $1,000^3$ runs. The driving force is still active during the star-formation phase of the simulations, in order to achieve a stationary value of α_{vir} to correlate with the SFR. The virial parameter varies in the range $\alpha_{\mathrm{vir}} = 0.22$–$2.04$. To define the virial parameter of the simulations, we have chosen to use equation (1.1), with $R = L_0/2$, where L_0 is the box size, and M equal to the total mass in the box, M_0. The virial parameter is then $\alpha_{\mathrm{vir}} = 5\, v_0^2 \, L_0/(6\, G M_0)$, where v_0 is the three-dimensional rms velocity in the box. A collapsing region is captured by the creation of an accreting sink particle if the density exceeds a certain density threshold (8,000 times the mean density in both 500^3 and 1000^3 runs). Further accretion (defined as density exceeding the density threshold) is collected onto the nearest sink particle if the distance is less than four grid zones. Sink particles are never merged. An example of a projected density field from a star formation simulation is shown in Figure 1.

5. Models versus Numerical Results

Figure 2 compares the SFR model with the numerical results. The HD simulations follow almost exactly the theoretical prediction with $\epsilon = 1$, suggesting that all the gas with density above the critical value collapses in a timescale of order $\tau_{\mathrm{ff,cr}}$, as assumed in the model. The dependence of $\mathrm{SFR_{ff}}$ on α_{vir} is too shallow to be consistent with the parametrization in Krumholz & McKee (2005). Only the run with the highest α_{vir} deviates significantly ($\approx$50%) from the theoretical prediction. This is probably only a numerical resolution effect, because the corresponding higher resolution run yields a higher

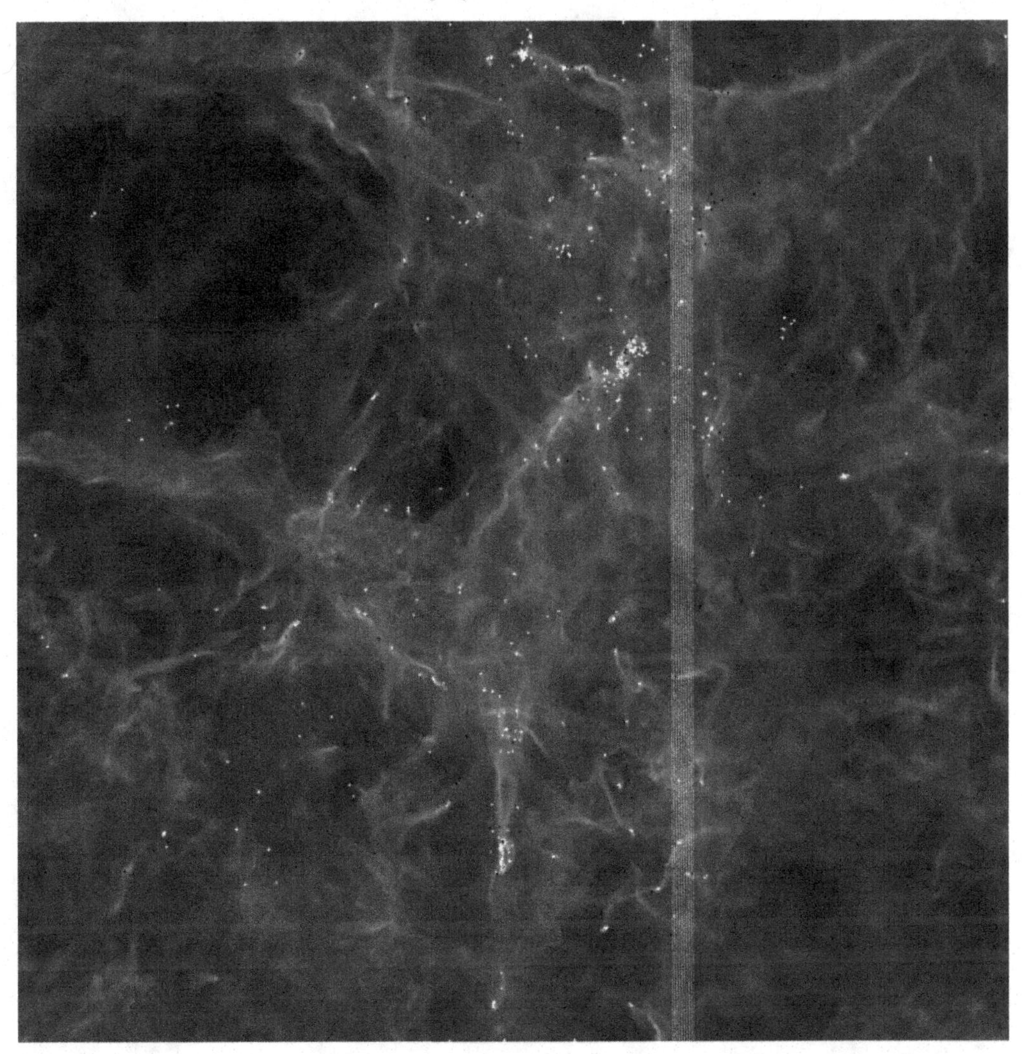

Figure 1. Logarithm of projected density from a snapshot of an exploratory $1,000^3$ run with $\beta_0 = 22.2$, $\mathcal{M}_{S,0} = 18$, and $\alpha_{vir} = 0.9$, at a time when approximately 10% of the mass has been converted into stars. Bright dots show the positions of the stars (sink particles), while black dots are for brown dwarfs (some of which are still accreting and may later grow to stellar masses).

value of $\mathrm{SFR_{ff}}$, nearly identical to the theoretical prediction. At $\alpha_{vir} = 0.95$, instead, the 500^3 run is already converged to the SFR of the corresponding $1,000^3$ run (HD5 and HD8 respectively). The HD runs also confirm the theoretical prediction that $\mathrm{SFR_{ff}}$ should increase with increasing $\mathcal{M}_{S,0}$ (the opposite of the prediction in Krumholz & McKee (2005), as shown by the comparison of the runs with $\mathcal{M}_{S,0} = 4.5$ and 9, respectively. The lower Mach number run fits very well the theoretical prediction, confirming our choice of $\tau_{ff,cr}$ for the timescale of star formation.

There is good agreement between the MHD simulations and the theoretical model with $\epsilon = 0.5$ as well, though the model predicts a significantly higher SFR than the 500^3 simulation with the largest value of α_{vir}. This discrepancy may be entirely due to the insufficient numerical resolution of the simulation, because the $1,000^3$ run with the same virial parameter, $\alpha_{vir} = 2.04$, yields a value of $\mathrm{SFR_{ff}}$ almost identical to the theoretical prediction. Like in the HD simulations, the case with $\alpha_{vir} = 0.95$ seems to be already

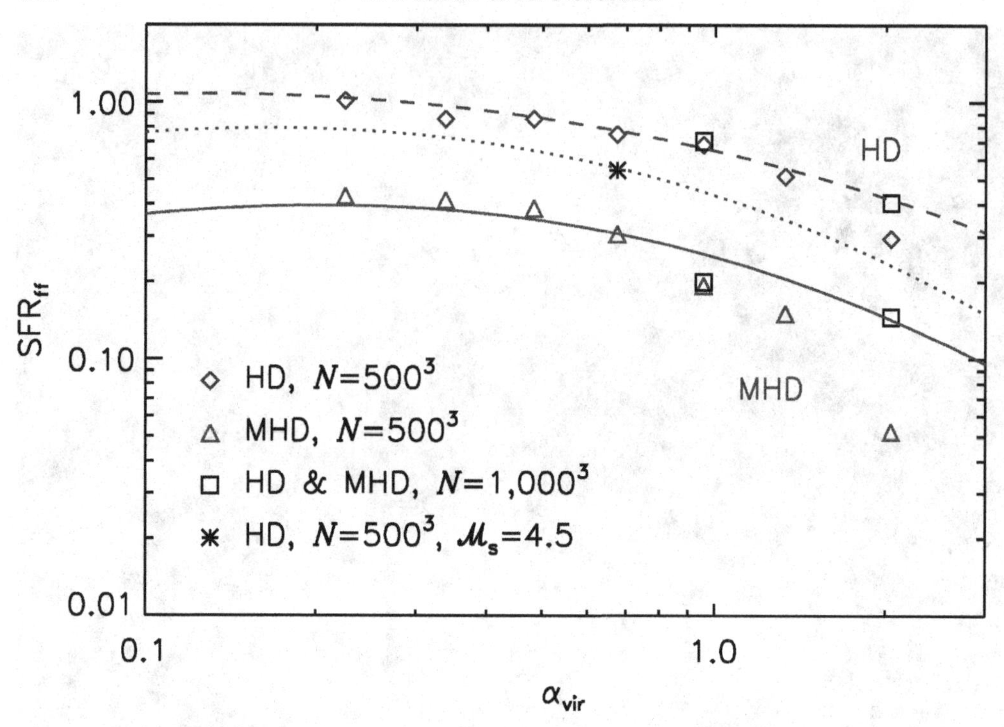

Figure 2. Star formation rate per free-fall time versus virial parameter for the 500^3 MHD simulations (triangles) and for the 500^3 HD simulations (diamonds) with $\mathcal{M}_{S,0} = 9$. The squares are for the $1,000^3$ runs, and the asterisk for the 500^3 HD run with $\mathcal{M}_{S,0} = 4.5$. The MHD model with $\mathcal{M}_{S,0} = 9$, $\beta_0 = 22.2$, and $b = 0.63$ is shown by the solid line. The HD model ($\beta_0 = \infty$) is shown by the dashed line for $\mathcal{M}_{S,0} = 9$, and by the dotted line for $\mathcal{M}_{S,0} = 4.5$.

converged at a resolution of 500^3 computational zones, as its $\mathrm{SFR_{ff}}$ is nearly identical to that of the corresponding $1,000^3$ run (and only approximately 20% below the predicted value).

6. Comparison with SFR in Molecular Clouds

Evans *et al.* (2009) have estimated values of $\mathrm{SFR_{ff}}$ in giant molecular clouds (GMCs) and within some of the dense cloud cores. They find values of $\mathrm{SFR_{ff}} = 0.03$ to 0.06 for GMCs with mean densities distributed around a mean value of $\langle n \rangle = 390$ cm^{-3}, and $\mathrm{SFR_{ff}} = 0.05$ to 0.25 for dense cores with mean densities 50-200 times those of the GMCs. These values are computed by assuming that all the stars detected (by their infrared excess) have been formed in the last 2 Myr. The authors report a best estimate of 2±1 Myr for the lifetime of the Class II phase, so the $\mathrm{SFR_{ff}}$ could be 50% lower, or 100% higher than the values given above. Accounting for this uncertainty, one gets $\mathrm{SFR_{ff}} = 0.02$ to 0.12 for GMCs, and $\mathrm{SFR_{ff}} = 0.03$ to 0.5 for dense cores, suggesting a characteristic value of order 0.1.

For a range of values of $\mathcal{M}_{S,0}$ characteristic of MCs, we predict $\mathrm{SFR_{ff}} \approx 0.12$ to 0.28 at $\alpha_{\mathrm{vir}} = 2$. These values should be reduced by a factor of two or three (Matzner & McKee 2000; André *et al.* 2010), to account for mass loss from stellar outflows and jets, not included in the model and in the simulations. With this reduction, our results are consistent with the relatively high values of $\mathrm{SFR_{ff}}$ found by Evans *et al.* (2009). The definition of the virial ratio for our periodic box (Eq. 1.1) and for observed star forming regions is an important source of uncertainty. Heyer *et al.* (2009) have recently studied a large subset of the GMCs sample of Solomon *et al.* (1987). For each cloud, they compute LTE masses based on the J=1-0 emission lines of ^{13}CO and ^{12}CO. They find masses smaller by a factor of 2 to 5 than the virial masses derived by Solomon *et al.* (1987). Their revised velocity dispersion are also somewhat smaller than in Solomon *et al.* (1987), but their resulting virial parameters are still a factor of approximately 2-3 larger, with a mean value of $\alpha_{\rm vir} = 2.8 \pm 2.4$. If the LTE-derived mass underestimates the real mass by a factor up to two, as argued by the authors, then the values of $\alpha_{\rm vir}$ should be reduced by a factor of two. The mean value is therefore likely to lie in the range $\alpha_{\rm vir} = 1.4$ to 2.8, with a very large scatter. If GMCs have a characteristic value of $\alpha_{\rm vir} \approx 2$, as suggested by this observational sample, the SFR predicted by our model for a reasonable range of values of $\mathcal{M}_{\rm S,0}$, and accounting for a factor of two or three reduction due to mass-loss in outflows and jets, is then consistent with the recent observational estimates by Evans *et al.* (2009).

7. Summary and Concluding Remarks

This work presents a new physical model of the SFR that could be implemented in galaxy formation simulations. The model depends on the relative importance of gravitational, turbulent, magnetic, and thermal energies, expressed through the virial parameter, $\alpha_{\rm vir}$, the rms sonic Mach number, $\mathcal{M}_{\rm S,0}$, and the ratio of the mean gas pressure to mean magnetic pressure, β_0. The value of SFR$_{\rm ff}$ is predicted to decrease with increasing $\alpha_{\rm vir}$, and to increase with increasing $\mathcal{M}_{\rm S,0}$, for values typical of star forming regions ($\mathcal{M}_{\rm S,0} \approx$ 4-20. In the complete absence of a magnetic field, SFR$_{\rm ff}$ increases typically by a factor of three, proving the importance of magnetic fields in star formation, even when they are relatively weak (super-Alfvénic turbulence). The model predictions have been tested with an unprecedented set of large numerical simulations of supersonic MHD turbulence, including the effect of self-gravity, and capturing collapsing cores as accreting sink particles. The SFR in the simulations follow closely the theoretical predictions.

This work illustrates how the turbulence controls the SFR. It does not address how the turbulence is driven to a specific value of $\alpha_{\rm vir}$. Because much of the turbulence driving is likely due to SN explosions, the turbulent kinetic energy and the value of $\alpha_{\rm vir}$ are coupled to the SFR in a feedback loop. The feedback determines the equilibrium level of the SFR (and hence also the equilibrium level of $\alpha_{\rm vir}$) at large scales. If $\alpha_{\rm vir}$ were to decrease (increase) relative to the equilibrium, the SFR would increase (decrease), according to the results of this work, resulting in an increased (decreased) energy injection rate by SN explosions, thus restoring a higher (lower) value of $\alpha_{\rm vir}$. The dependence of the SFR on $\alpha_{\rm vir}$ found in this work suggests that this self-regulation may work quite effectively.

Cosmological simulations of galaxy formation provide the rate of gas cooling and infall, which sets the gas reservoir for the star formation process and thus ultimately controls the SFR. They also include prescriptions for the star formation feedback, known to be essential to recover observed properties of galaxies (Gnedin *et al.* 2009; Gnedin *et al.* 2010). Future galaxy formation simulations should adopt a physical SFR law with an explicit dependence on $\alpha_{\rm vir}$, $\mathcal{M}_{\rm S,0}$, and β as derived in this work, in order to correctly reflect specific conditions of protogalaxies at different redshifts. This requires a treatment of the star formation feedback capable of providing an estimate of $\alpha_{\rm vir}$ on scales of order 10-100 pc, not far from the spatial resolution currently achieved by the largest cosmological simulations of galaxy formation.

Acknowledgements

This research was supported in part by the NSF grant AST-0908740, and a grant from the Danish Natural Science Research Council. We utilized computing resources provided by the San Diego Supercomputer Center, by the NASA High End Computing Program, and by the Danish Center for Scientific Computing.

References

André, P., *et al.* 2010, arXiv:1005.2618
Beetz, C., Schwarz, C., Dreher, J., & Grauer, R. 2008, *Physics Letters A*, 372, 3037
Bertoldi, F. & McKee, C. F. 1992, *ApJ*, 395, 140
Bonnor, W. B. 1956, *MNRAS*, 116, 351
Brunt, C. M., Federrath, C., & Price, D. J. 2010, *MNRAS*, 403, 1507
Crutcher, R. M. 1999, *ApJ*, 520, 706
Ebert, R. 1957, *Zeitschrift fur Astrophysik*, 42, 263
Elmegreen, B. G. 2000, *ApJ*, 530, 277
Elmegreen, B. G. 2007, *ApJ*, 668, 1064
Evans, N. J., *et al.* 2009, *ApJ*, 181, 321
Falgarone, E., Troland, T. H., Crutcher, R. M., & Paubert, G. 2008, *A&A*, 487, 247
Federrath, C., Klessen, R. S., & Schmidt, W. 2008, *ApJL*, 688, L79
Gnedin, N. Y., & Kravtsov, A. V. 2010, *ApJ*, 714, 287
Gnedin, N. Y., Tassis, K., & Kravtsov, A. V. 2009, *ApJ*, 697, 55
Heyer, M., Krawczyk, C., Duval, J., & Jackson, J. M. 2009, *ApJ*, 699, 1092
Kritsuk, A. G., Norman, M. L., Padoan, P., & Wagner, R. 2007, *ApJ*, 665, 416
Kritsuk, A. G., Ustyugov, S. D., Norman, M. L., & Padoan, P. 2009, *Journal of Physics Conference Series*, 180, 012020
Kritsuk, A. G., Ustyugov, S. D., Norman, M. L., & Padoan, P. 2009b, arXiv:0912.0546
Krumholz, M. R. & McKee, C. F. 2005, *ApJ*, 630, 250
Krumholz, M. R. & Tan, J. C. 2007, textitApJ, 654, 304
Li, P. S., Norman, M. L., Mac Low, M.-M., & Heitsch, F. 2004, *ApJ*, 605, 800
Matzner, C. D. & McKee, C. F. 2000, *ApJ*, 545, 364
McKee, C. F. 1989, *ApJ*, 345, 782
McKee, C. F. & Ostriker, E. C. 2007, *ARAA*, 45, 565
Nakano, T. & Nakamura, T. 1978, *PASJ*, 30, 671
Nordlund, Å. & Padoan, P. 1999, in: J. Franco & A. Carramiñana (eds.), *Interstellar Turbulence* (Cambridge University Press), p. 218
Padoan, P. 1995, *MNRAS*, 277, 377
Padoan, P., Juvela, M., Goodman, A. A., & Nordlund, Å. 2001, *ApJ*, 553, 227
Padoan, P. & Nordlund, Å. 1999, *ApJ*, 526, 279
Padoan, P. & Nordlund, Å. 2002, *ApJ*, 576, 870
—. 2004, *ApJ*, 617, 559
Padoan, P., Nordlund, Å., & Jones, B. 1997, *MNRAS*, 288, 145
Solomon, P. M., Rivolo, A. R., Barrett, J., & Yahil, A. 1987, *ApJ*, 319, 730
Tomisaka, K., Ikeuchi, S., & Nakamura, T. 1988, *ApJ*, 335, 239
Troland, T. H. & Crutcher, R. M. 2008, *ApJ*, 680, 457
Vazquez-Semadeni, E. 1994, *ApJ*, 423, 681
Vázquez-Semadeni, E., Ballesteros-Paredes, J., & Klessen, R. S. 2003, *ApJL*, 585, L131
Vázquez-Semadeni, E., Kim, J., & Ballesteros-Paredes, J. 2005, *ApJL*, 630, L49
Williams, J. P. & McKee, C. F. 1997, *ApJ*, 476, 166
Zuckerman, B. & Palmer, P. 1974, *ARAA*, 12, 279

Computational Star Formation
Proceedings IAU Symposium No. 270, 2011
J. Alves, B.G. Elmegreen, J. M. Girart & V. Trimble, eds.

doi:10.1017/S1743921311000627

Collisions of supersonic clouds

Andrew McLeod[1,2], Jan Palouš[1] and Anthony Whitworth[2]

[1] Astronomical Institute, Academy of Sciences of the Czech Republic
[2] Dept. of Physics and Astronomy, Cardiff University,
email: Andrew.McLeod@astro.cf.ac.uk

Abstract. We present simulations of supersonic collisions between molecular clouds of mass $500\,M_\odot$ and radius 2.24 pc. The simulations are performed with the SEREN SPH code. The code treats the energy equation and the associated transport of heating and cooling radiation. The formation of protostars is captured by introducing sink particles. Low velocity collisions form a shock-compressed layer which fragments to form stars. For high-velocity collisions, $v_{rel} \gtrapprox 5\,\mathrm{km\,s^{-1}}$, the non-linear thin shell instability strongly disrupts the shock-compressed layer, and may inhibit the formation of stars.

Keywords. instabilities, shock waves, ISM:clouds, stars: formation, stars: mass function

---

1. Introduction

Cloud–cloud collisions may be important in understanding the star-formation process. The supersonic collision of clouds can compress gas causing it to become gravitationally unstable and collapse into stars. Clouds may also merge or disperse in collisions. Feedback from stellar clusters created by cloud–cloud collisions may cause the recoil of the original clouds, which can then go on to collide with other clouds.

Whitworth *et al.* (1994) analysed the effects of the gravitational instability in a shocked layer. They found that fragments in the layer could be Jeans-unstable and collapse gravitationally, with the fastest modes being those of approximately the Jeans length. They derive the typical fragment separation

$$L_{\mathrm{fragment}} \sim c_s \times t_{\mathrm{fragment}} \sim \mathcal{M}^{1/2} \times \frac{c_s}{\left(G\,\bar{\rho}_{\mathrm{layer}}\right)^{1/2}}, \tag{1.1}$$

where $\mathcal{M}$ is the Mach number of the collision relative to the sound speed in the layer c_s.

Vishniac (1994) describes for the first time the non-linear thin shell instability (NTSI). He studies the growth of the instability in an initially perturbed shock-compressed slab. Ram pressure acts along the collision axis, however thermal pressure is always normal to the slab. In the NTSI, this creates a shear in the outer layers of a perturbed slab, driving material towards the extremities of the perturbation and strengthening it.

Vishniac derives the growth rate of the NTSI as

$$\tau^{-1}(k) \propto k^{3/2} \qquad \text{over unstable wavenumbers} \qquad L^{-1} > k \gg \left(\frac{V_s}{c_s}\right) R^{-1}, \tag{1.2}$$

where L is the thickness of the slab, V_s is the velocity of the shock, c_s is the sound speed in the shock, and R is the propagation distance over which zeroth order shock properties evolve. For a stationary slab, $R = V_e \times t$, the velocity of the external gas multiplied by the time since the slab was formed. The NTSI should be strongest for high Mach number collisions, as these will have a higher density contrast.

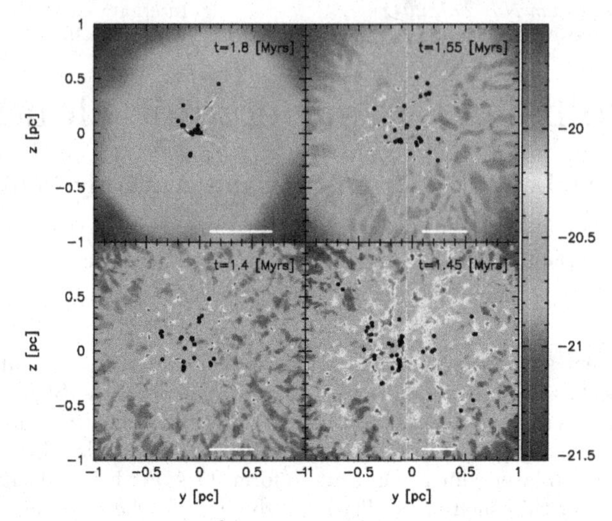

Figure 1. Projected densities ($\mathrm{g\,cm^{-3}}$) in the y–z plane. The solid white line indicates the expected length scale of the gravitational instability. Solid black dots indicate the presence of a sink particle (not to scale). From top left to bottom right: Mach 4, 8, 12 and 16.

2. Method

We set up spheres of radius 2.24 pc and mass 500 $\mathrm{M_\odot}$ containing six million settled SPH particles. The spheres are initially touching. We collide our clouds head-on at Mach 4, 8, 12, 16 and 40, assuming a sound speed of $0.25\,\mathrm{km\,s^{-1}}$. We also superimpose a weak turbulent velocity field of purely solenoidal turbulence with r.m.s. velocity $0.1\,\mathrm{km\,s^{-1}}$. This is too weak to significantly support the cloud but may seed instabilities.

We use an MPI-parallelized version of the SEREN (Hubber *et al.*, 2010) code to simulate cloud-cloud collisions with smoothed particle hydrodynamics (SPH). We include self-gravity. We use sink particles to represent protostellar objects with a sink creation density of $10^{-12}\,\mathrm{g\,cm^{-3}}$. We use the Monaghan (1997) method of artificial viscosity; we use the Morris & Monaghan (1997) switch to reduce viscosity away from shocks. As in Price & Federrath (2010), we use $\beta = 4$ instead of the standard $\beta = 2$ to enhance viscosity in high Mach number shocks.

We use the Stamatellos *et al.* (1997) method to follow the energy equation and capture the effects of heating and cooling radiation. One limitation of this method is that it assumes the gas and dust are well-coupled and so below a density of $\sim 10^5\,\mathrm{cm^{-3}}$ it is likely to overestimate the cooling and underestimate the gas temperature.

3. Results

3.1. *Low velocity collisions (Mach 4, 8, 12 & 16)*

Figure 1 shows column densities along the collision axis. In the Mach 4 and Mach 8 collisions the layer appears to have fragmented into a filamentary structure. For the Mach 12 and Mach 16 collisions the network of linear filaments has been largely replaced by a more 'clumpy' structure. Although this is also organized into a filamentary network, in many cases clumps exist that are completely separate from these networks. We suggest that this is due to the combination of the gravitational instability and the NTSI as the Mach number increases.

The expected length scale of fragmentation is larger than that observed in the filamentary structure by approximately a factor of two. The exception to this is the slowest

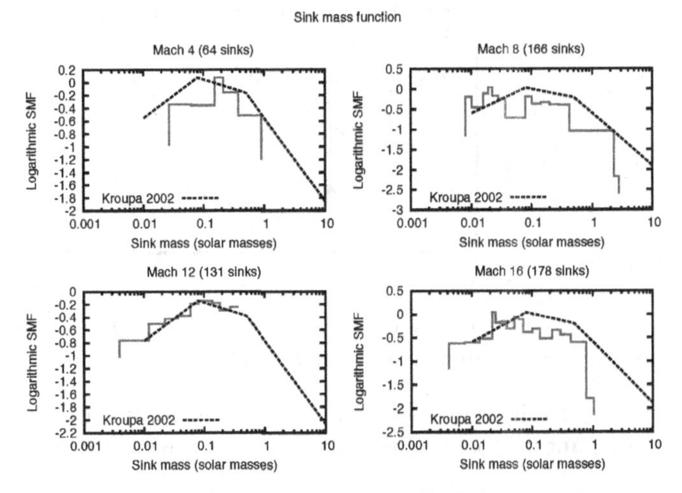

Figure 2. Sink mass function at end of simulation. Each bin contains an equal number of sinks. The Kroupa (2002) IMF is overlaid, fitted to match the highest bin.

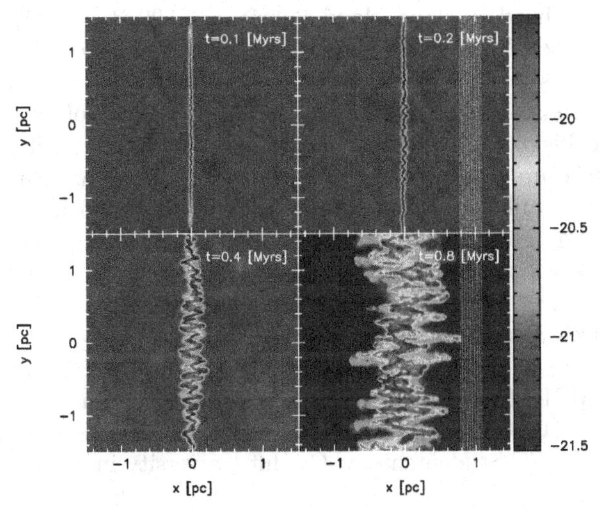

Figure 3. Cross-sections of density ($\mathrm{g\,cm^{-3}}$) in the x–y plane of the Mach 40 cloud–cloud collision.

(Mach 4) collision, but this snapshot is taken at a later time when the pattern of fragments has begun to collapse.

All four low-velocity simulations produce sink particles. Figure 2 shows the sink mass functions; these are noisy but compatible with the Kroupa (2002) IMF.

3.2. *High velocity collision (Mach 40)*

This collision produces no sink particles. Instead, the shocked layer is unstable to the NTSI and eventually breaks up completely. This can be seen in figure 3 which shows cross-sections of density in the x–y plane. The initially weak perturbations grow non-linearly until they saturate and the layer becomes bloated and chaotic. This collision does not form stars.

In figure 4 we show the growth rates of the NTSI as a function of wavenumber $|\boldsymbol{k}|$, for a number of time intervals. At early times the NTSI is not fully developed, but at intermediate times the growth rate obeys the expected relation $\tau^{-1} \propto |\boldsymbol{k}|^{3/2}$ up to the

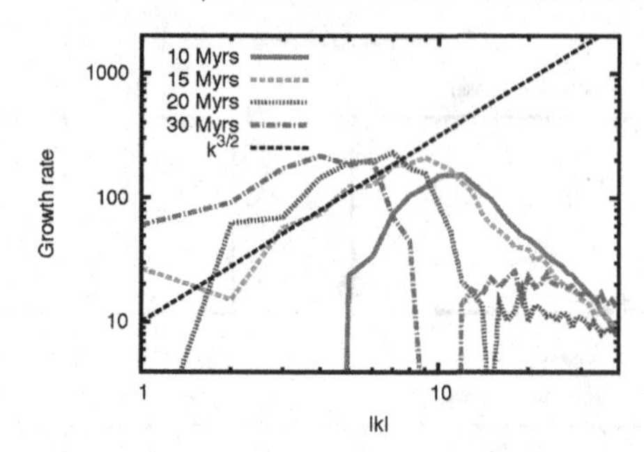

Figure 4. Growth rates of the NTSI as a function of wavenumber $|\boldsymbol{k}|$ at different times. The dashed black line shows the solution of Vishniac, $\tau^{-1} \propto |\boldsymbol{k}|^{3/2}$.

wavenumber equivalent to the thickness of the layer $|\boldsymbol{k}| \sim 10$. At later times than shown the NTSI is saturated and no longer obeys the Vishniac growth rate.

4. Conclusions

We have simulated head-on cloud–cloud collisions of 500 $M_{\odot}$ clouds with uniform density and radius 2.24 pc. We have simulated collision velocities of Mach 4, 8, 12, 16 and 40. All these collisions produce a shock-compressed layer. The lower velocity collisions form a network of filaments. Stars form in the densest parts of these filaments. Intermediate velocity collisions are affected by the non-linear thin shell instability, causing the filamentary structure in the layer to be overlaid with a more clumpy texture. Stars then form out of these filaments and clumps. The highest velocity collision produces a layer which is strongly affected by the NTSI and does not form stars. We reproduce the expected growth rate of the NTSI for its early evolution.

Plots of SPH data in this work were produced with SPLASH, written by Daniel Price (2007). This research has made use of NASA's Astrophysics Data System. Simulations were performed using the Cardiff University ARCCA (Advanced Research Computing CArdiff) cluster (MERLIN). Andrew McLeod is funded as an Early-Stage Researcher by the EC-funded CONSTELLATION Marie Curie training network MRTN-CT-2006-035890 and has an STFC Studentship at Cardiff University in abeyance.

References

Bate, M., Bonnell, I., & Price, N. 1995 *MNRAS*, 277, 362
Barnes, J. & Hut, P. 1986 *Nature*, 324, 446
Hubber, D. A., Batty, C. P., McLeod, A., & Whitworth, A. P. 2010, *A&A*, submitted
Kroupa, P. 2002 *Science*, 295, 82
Monaghan, J. 1997 *JCoPh*, 136, 298
Morris, J. & Monaghan, J. 1997 *JCoPh*, 136, 41
Price, D. 2007, *PASA*, 24, 159
Price, D. & Federrath, C. 2010 *MNRAS*, MNRAS, 406, 1659
Stamatellos, D., Whitworth, A., Bisbas, T., & Goodwin, S. 2007 *MNRAS*, 475, 37
Vishniac, E. T. 1994, *ApJ*, 428, 186
Whitworth, A., Bhattal, A., Chapman, S., Disney, M., & Turner, J. 1994 *A&A*, 290, 421

Computational Star Formation
Proceedings IAU Symposium No. 270, 2011
J. Alves, B.G. Elmegreen, J. M. Girart & V. Trimble, eds.

doi:10.1017/S1743921311000639

Star Formation in the Central Molecular Zone of the Milky Way

Sungsoo S. Kim,[1,2] Takayuki R. Saitoh,[3] Myoungwon Jeon,[1,5] David Merritt,[6] Donald F. Figer,[2] and Keiichi Wada[4]

[1]Dept. of Astronomt & Space Science, Kyung Hee University, Yongin, Kyungki 446-701, Korea

[2]Chester F. Carlson Center for Imaging Science, Rochester Institute of Technology, Rochester, NY 14623, USA

[3]Division of Theoretical Astronomy, National Astronomical Observatory of Japan, Mitaka, Tokyo 181-8588, Japan

[4]Graduate School of Science and Engineering, Kagoshima University, Kagoshima 890-8580, Japan

[5]Dept. of Astronomy, University of Texas, Austin, TX 78712, USA

[6]Centre for Computational Relativity and Gravitation, Rochester Institute of Technology, Rochester, NY 14623, USA

Abstract. Gas materials in the inner Galactic disk continuously migrate toward the Galactic center (GC) due to interactions with the bar potential, magnetic fields, stars, and other gaseous materials. Those in forms of molecules appear to accumulate around 200 pc from the center (the central molecular zone, CMZ) to form stars there and further inside. The bar potential in the GC is thought to be responsible for such accumulation of molecules and subsequent star formation, which is believed to have been continuous throughout the lifetime of the Galaxy. We present 3-D hydrodynamic simulations of the CMZ that consider self-gravity, radiative cooling, and supernova feedback, and discuss the efficiency and role of the star formation in that region. We find that the gas accumulated in the CMZ by a bar potential of the inner bulge effectively turns into stars, supporting the idea that the stellar cusp inside the central 200 pc is a result of the sustained star formation in the CMZ. The obtained star formation rate in the CMZ, 0.03–0.1 $M_\odot$, is consistent with the recent estimate based on the mid-infrared observations by Yusef-Zadeh *et al.* (2009).

1. Introduction

The CO emission survey along the Galactic plane shows that molecular gas is abundant down to a Galactocentric radius R_G of $\sim$3 kpc. Inside this radius, the gas content is mostly in forms of atoms and there are no noticeable star formation activities. However, a significant amount of molecular gas, as well as various evidence of recent and current star formation, appears again inside a projected R_g of $\sim$200 pc. This distribution of molecular gas is called the Central Molecular Zone (CMZ). The observed longitude-velocity diagram of the molecular emission in the CMZ is generally interpreted as a result of torus-like distribution of molecules with an outer radius of $\sim$200 pc. The total gas mass in the CMZ is estimated to be $\sim 5 \times 10^7\ M_\odot$ (Pierce-Price *et al.* 2000).

The gas content in the CMZ is thought to have migrated inward to the current location from the Galactic disk. Serabyn & Morris (1996) enumerate mechanisms for the inward transport of gas: shear viscosity due to the differential rotation of gas disk, compression and shocks associated with the elongated or cusped stable orbits in a non-axisymmetric potential, dynamical friction with the field stars, magnetic field viscosity, and the dilution

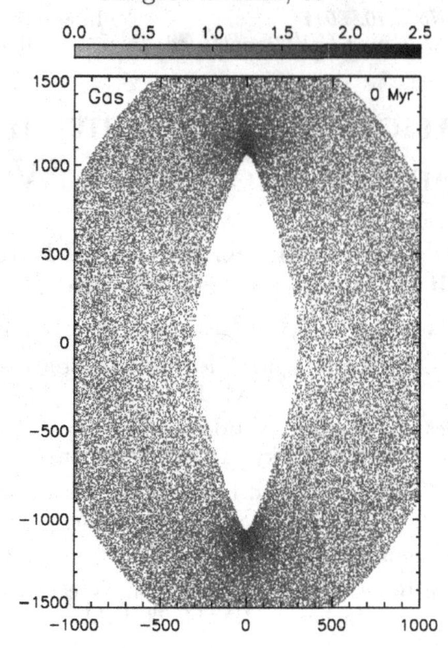

Figure 1. Initial gas particle distribution of our standard run. Length units are in parsecs.

of specific angular momentum by stellar mass loss material from outer bulge (Jenkins & Binney 1994).

The transition from atomic to molecular status around 200 pc is believed to be responsible for the characteristics of the stable orbits in a bar potential (Binney *et al.* 1991). Bar potentials have several distinct families of stable orbits, among which X_1 and X_2 orbit families are related to the discussion here: X_1, the outermost orbit family, is elongated along the bar's major axis, whereas X_2 is elongated along the bar's minor axis deeper inside the potential. Hydrodynamic effects cause gas particles to generally move along stable orbits, but the innermost X_1 orbits are sharply cusped or even self-intersecting at the apocenters. Gas is compressed or even undergoes a shock in these regions, loses its orbital energy, and falls inward to settle onto an X_2 orbit inside. Such compression and subsequent cooling will transform the mostly atomic gas to molecular clouds.

Some of the accumulated molecular gas will keep moving inward and reach the Circum Nuclear Disk (CND) of molecular clouds at R_G of few parsecs and/or eventually sink to the central supermassive black hole, while some will collapse and form stars in the CMZ. If the non-sphericity of the Galactic potential in the inner bulge has been significant enough to give rise to X_1, X_2 orbit families for the lifetime of the Galaxy, a fair amount of stars must have formed in the CMZ so far. Serabyn & Morris (1996) argue that the sustained star formation in the CMZ has resulted in a cusp in the stellar number density profile with R_G of 100–200 pc.

In the present Proceedings paper, we present 3-D hydrodynamic simulations of gas particles in the inner bulge of the Galaxy that consider the effects of star formation, supernova feedback, and realistic gas cooling and heating. We will show that a bar potential that has a X_1–X_2 transition at ~200 pc can indeed compress gas sufficiently enough to form stars and that the obtained star formation rates are consistent with observations.

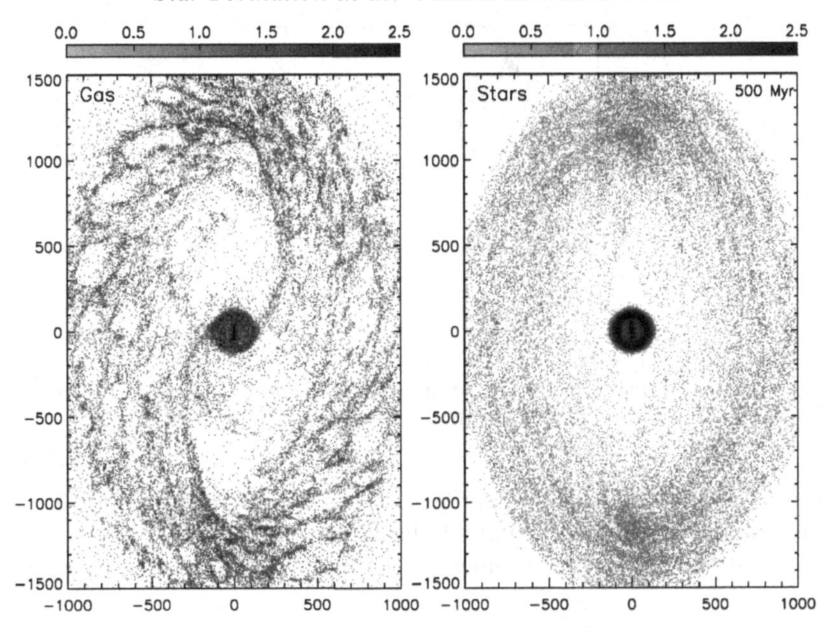

Figure 2. Distribution of gas (*left*) and star (*right*) particles in our standard run at $T = 500$ Myr. Length units are in parsecs. Stars are predominantly formed in the central 200 pc.

2. The Simulations

We use a parallel tree SPH code ASURA (Saitoh, in preparation) that can utilize the special-purpose hardware GRAPE or an optimally-tuned gravity calculation library, Phantom-GRAPE. We use an opening angle of $\theta = 0.5$ for a cell opening criterion, and the kernel size of an SPH particle is determined by imposing the number of neighbors to be 32 ± 2. We use a cooling function by Spaans & Norman (1997) for gas with the solar metallicity for a temperature range of 10–10^8 K. A uniform heating from far-UV radiation is considered with a value observed in the Solar neighborhood (Wolfire *et al.* 1995).

A star particle is spawned when a gas particle satisfies all three following conditions: 1) the hydrogen number density is larger than a threshold value ($n_H > n_{th}$), 2) the temperature is lower than a threshold value ($T < T_{th}$), and 3) the flow is converging ($\nabla \cdot v < 0$). We adopt $n_{th} = 100$ cm^{-3} and $T_{th} = 100$ K. The effect of supernova feedback is implemented in a probabilistic manner. We assume that stars more massive than $8\,M_\odot$ explode as Type II supernovae and that each explosion outputs 10^{51} ergs of thermal energy into the surrounding 32 SPH particles. Detailed discussion on the choice of n_{th} and T_{th}, spawning of star particles, and supernova feedback is given in Saitoh *et al.* (2008).

For the Galactic potential, we adopt a power-law density distribution with a slope of -1.75 and a nonaxisymmetry of $m = 2$ for the bulge, and a Miyamoto-Nagai model for the disk.

Initially, the simulation has gas particles only. Our standard run has the following initial distribution of particles: They are on one of the X_1 orbits whose semi-minor axes range from 300 to 1200 pc. The number of particles on each orbit is proportional to the length of the orbit, and on a given orbit, the particles are spaced over the same time-interval (see Fig. 1). The vertical distribution follows a Gaussian function with a

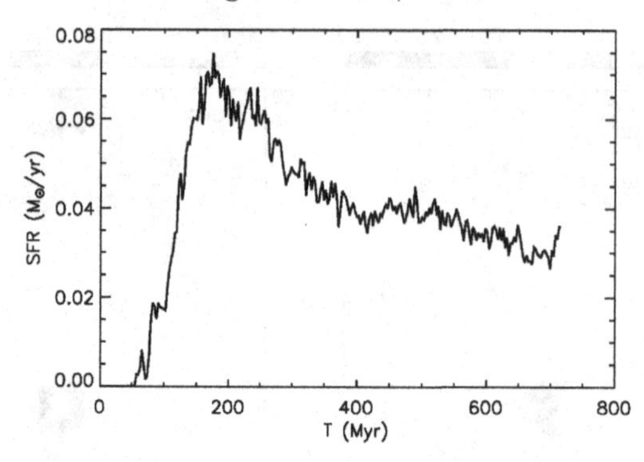

Figure 3. Star formation rate in our standard run. The obtained SFRs are very similar to the values estimated from mid-infrared observations by Yusef-Zadeh *et al.* (2009), 0.04–0.08 $M_\odot$ yr^{-1}.

scale height of 40 pc. The standard run has 10^5 gas particles, and the total gas mass is 5×10^7 $M_\odot$, thus each gas particle initially has a mass of 500 $M_\odot$.

We evolve the system without cooling for the first 50 Myr to have a relaxed particle distribution.

3. Results

Figure 2 shows our standard run at $T = 500$ Myr. About a half of the gas particles that were initially on X_1 orbits have migrated inward to the CMZ, and the majority of newly formed stars remain in the central 200 pc. This implies that if gas has been supplied down to the inner bulge region from the disk throughout the lifetime of the Galaxy, then indeed a significant amount of stars would have been born in the CMZ and a resulting stellar population would form a central cusp that is distinctive from the larger bulge.

Figure 3 shows the evolution of overall star formation rate (SFR) in our standard run. The SFR increases rather steeply during the first 200 Myr because there are not yet many supernova explosions that increase the temperature of nearby gas particles. During the later half of the simulation, most of the star formation takes place in the CMZ, thus the obtained SFR values of 0.03–0.04 $M_\odot$ yr^{-1} can be regarded as the SFR in the CMZ. These values are very close to the recent SFR estimated from mid-infrared observations by Yusef-Zadeh *et al.* (2009), 0.04–0.08 $M_\odot$ yr^{-1}.

References

Binney, J., Gerhard, O. E., Stark, A. A., Bally, J., & Uchida, K. I. 1991, *MNRAS*, 252, 210

Jenkins, A., & Binney, J. 1994, *MNRAS*, 270, 703

Pierce-Price, D., Richer, J. S., Greaves, J. S., Holland, W. S., Jenness, T., Lasenby, A. N., White, G. J., Matthews, H. E., Ward-Thompson, D., Dent, W. R. F., Zylka, R., Mezger, P., Hasegawa, T., Oka, T., Omont, A., & Gilmore, G. 2000, *ApJ*, 545, L121

Saitoh, T. R., Daisaka, H., Kokubo, E., Makino, J., Okamoto, T., Tomisaka, K., Wada, K., & Yoshida, N. 2008, *PASJ*, 60, 667

Serabyn, E. & Morris, M., 1996, *Nature*, 382, 602

Spaans, M. & Norman, C. A. 1997, *ApJ*, 483, 87

Yusef-Zadeh, F., Hewitt, J. W., Arendt, R. G., Whitney, B., Rieke, G., Wardle, M., Hinz, J. L., Stolovy, S., Lang, C. C., Burton, M. G., & Ramirez, S. 2009, *ApJ*, 702, 178

Wolfire, M. G., Hollenbach, D., McKee, C. F., Tielens, A. G. G. M., & Bakes, E. L. O. 1995, *ApJ*, 443, 152

Computational Star Formation
Proceedings IAU Symposium No. 270, 2011
J. Alves, B.G. Elmegreen, J. M. Girart & V. Trimble, eds.

doi:10.1017/S1743921311000640

Galactic scale star formation: Interplay between stellar spirals and the ISM

Keiichi Wada[1], Junichi Baba[2], Michiko Fujii[1] and Takayuki R. Saitoh[2]

[1]Graduate School of Science and Engeering, Kagoshima University, Kagoshima, Japan
email: wada@cfca.jp

[2]National Astronomical Observatory of Japan

Abstract. Spiral structures in the disk galaxies have been extensively studied by many theoretical papers, but conventional steady-state models are not consistent with what we observe in time-dependent, multi-dimensional numerical simulations and also in real galaxies. Here we review recent progress in numerical modeling of stellar and gas spirals in disk galaxies. The spiral arms excited in a stellar disk can last for 10 Gyrs without the ISM, but each spiral arm is short-lived and is recurrently formed. The stellar spirals are not waves propagating with a single pattern speed. The ISM is concentrated in local potential minima, which roughly follow the galactic rotation together with the stellar arms, therefore galactic dust lanes are not the classic 'galactic shocks'.

Keywords. N-body simulation, SPH, hydrodynamics

1. Introduction

What we probably learned from conventional textbooks and papers on galactic spirals would be, for example:

- The 'winding dilemma' is tricky.
- Galactic stellar spiral arms are stationary waves propagating with a single pattern speed, and they reflect at resonances or Q-barrier.
- The ISM (or some kind of dissipation) is necessary to keep the long-lived stellar spirals.
- Gas spirals (or dust lanes) are caused by 'galactic shocks', which are offset against the stellar arms.
- GMCs are formed through gravitational instability in the shocked layers, and thereby star formation is triggered.

The concept of galactic shock was originally proposed by Fujimoto (1968) and extensively studied by many followers, e.g. Roberts (1969) and Shu (1973). Fujimoto found a stationary solution that represents a standing shock in a tightly wrapped spiral potential. When a supersonic flow of the gas passes through a spiral potential which is caused by a stellar density wave, the flow is decelerated, and often makes a shock. The subsonic flow behind the shock is then accelerated again toward the other spiral potential (Fig. 1). Due to the radiative cooling, the ISM is strongly compressed at shocked layer, then molecular clouds are formed through gravitational and thermal instabilities followed by star formation (e.g. Hartmann *et al.* 2001; Vazquez-Semadeni 2006; Heitsch *et al.* 2009). As a result, spiral arms in galaxies are illuminated by massive stars and HII regions. This is a conventional, and widely accepted picture, but it is not clear whether the stationary shocks predicted by theories are consistent with the star formation process, which is a basically unstable phenomena. In fact, recent time-dependent simulations suggest that

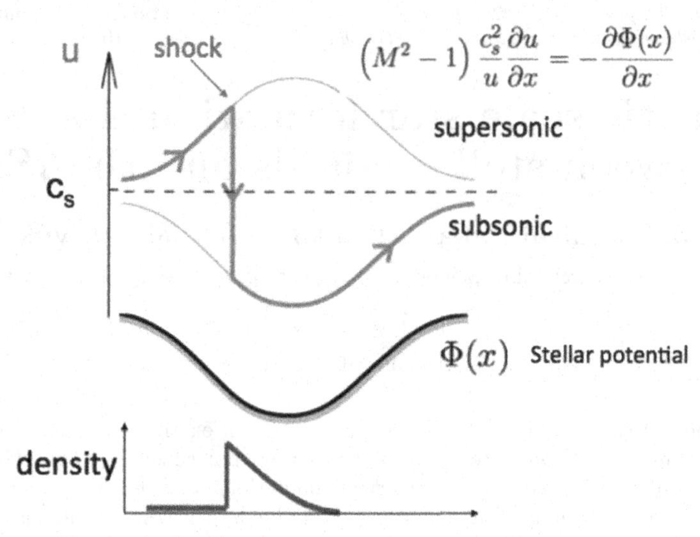

Figure 1. An one-dimensional image of 'galactic shock'.

spiral shocked layers are not always stable (e.g. Shetty & Ostriker 2006; Dobbs *et al.* 2008). Wada & Koda (2004) showed that spiral shocks are unstable ('wiggle instability'), if the Mach number and pitch angles are large enough. The substructures in the inter-arm regions (i.e. 'spurs') are naturally formed as a result of formation of clumps in the layers†. A similar phenomenon can be seen even in 'cloud-fluid' simulations by e.g. Tomisaka (1986). The wiggle instability seen in the hydrodynamic simulations is essentially the Kelvin-Helmholtz instability, but this is not the only process to generate complex structures in the ISM. Various physics, such as magnetic field, self-gravity of the gas, radiative cooling, heating by stars could affect stability, structures and dynamics of the gas in a spiral potential (see e.g. Kim *et al.* 2010 and references therein).

Using three-dimensional hydrodynamic simulations, taking into account the self-gravity of the gas and radiative cooling/heating and supernova feedback, Wada (2008) found that the classic galactic shocks are unstable and transient, and they shift to a globally quasi-steady, inhomogeneous arms due to the nonlinear development of gravitational, thermal, and hydrodynamical instabilities (Fig. 2). As a result, the arms with many GMC-like condensations are formed, but those substructures are not steady. The clumps eventually move into the inter-arm regions, and they are tidally stretched, eventually turn into spurs. In the quasi-steady state, the density enhancement tends to be associated with the spiral potential. This chaotic state is no longer a 'shock', but more resembles the spiral arms in real galaxies, in a sense that high density regions like GMCs coexist in the quasi-steady arms.

However, Wada (2008) as well as most other previous hydrodynamic and magneto- hydrodynamic simulations in spiral potentials still did not deal with an important aspect in spiral galaxies. The spiral potentials in these previous simulations are time-independent, therefore gravitational interaction between stars and gas is not properly considered. This is an essential point on dynamics of spirals, especially if spiral potentials are not steady, which is actually the case as will be discussed in the next section.

† Note that spurs are not 'waves'. Clumps formed by any kind of instabilities near the potential trough, such as gravitational instability, can be sheared off, and as a result elongated structures in the inter-arm regions are formed.

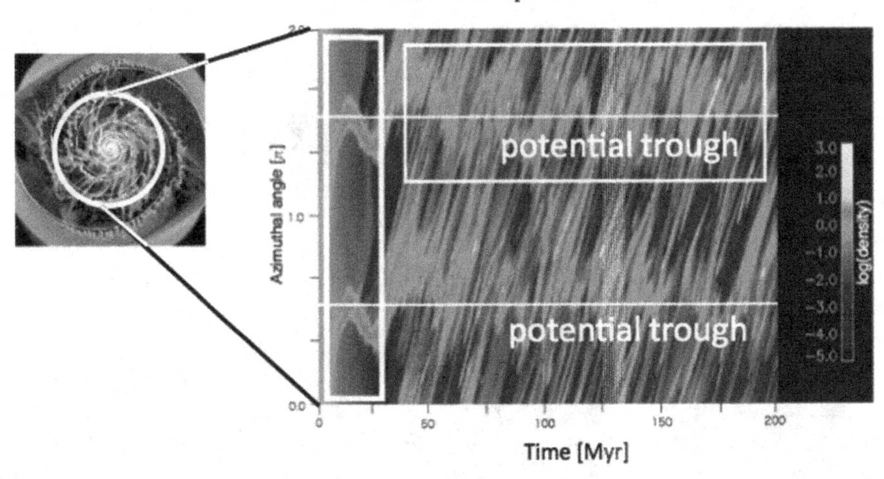

Figure 2. Time evolution of the azimuthal density profile. The horizontal lines show the positions of the minimum of the spiral potential. The initially appeared shocks turn to be stochastic arms consisting of many substructures after $t \sim 50$ Myrs (Wada 2008).

2. Evolution of Pure Stellar Disks

Using 3-D N-body (tree-GRAPE) simulations, Fujii *et al.* (2010) showed that stellar disks can maintain spiral features for several tens of rotations without the help of cooling mechanisms, such as dissipational effects of the ISM. In the simulations, multi-arm spirals developed in an isolated disk can survive for more than 10 Gyrs, if the number of particles is sufficiently large, e.g., $N > 3 \times 10^6$ (Fig. 3). They claimed that there is a self-regulating mechanism that maintains the amplitudes of the spiral arms. Interestingly, they found that spiral arms developed in the disk are not steady, but rather recurrently formed. Locally developed arms often merge into other high density regions, and form grand-design spiral arms. However, the grand-design arms are short-lived structures, and break into local arms. The process is repeated, and as a result the dominant Fourier mode is also time-dependent, and it radially changes (see the bottom panels of Fig. 3).

In fact, the non-steady spiral is a common feature seen in previous N-body simulations. In other words, it is hard to produce 'steady' spirals that can be understood by Lin and Shu's density wave theory, in time-dependent simulations of collisionless disks. For example, Sellwood & Carlberg (1984), using 2D, particle-mesh simulations, pointed out that spirals become faint in about ten rotations ($\sim$ a few Gyrs) †. In order to keep spirals for many rotations, they claimed that some mechanisms to reduce the velocity dispersion of stars are necessary. They added 'dynamically cold' stars, that is, stars with small velocity dispersion, and showed that spirals can last for more rotational periods. Since dynamically cold stars could be formed from the cold ISM, one might think that the ISM is essential to keep spirals long-lived. However, as mentioned above, the recent results by Fujii *et al.* (2010) suggest that this is not actually true.

3. Interplay between gas and stellar spirals

Even if the N-body models suggest that spiral arms can survive without help of the ISM, the ISM is actually present in galactic disks, and it is still not clear how the ISM reacts with the time-dependent, transient stellar spirals. Stellar spirals could be also affected by the ISM. In order to see the interplay between stellar spirals and the ISM,

† See also the recent review by Sellwood (2010) on the life time of spirals.

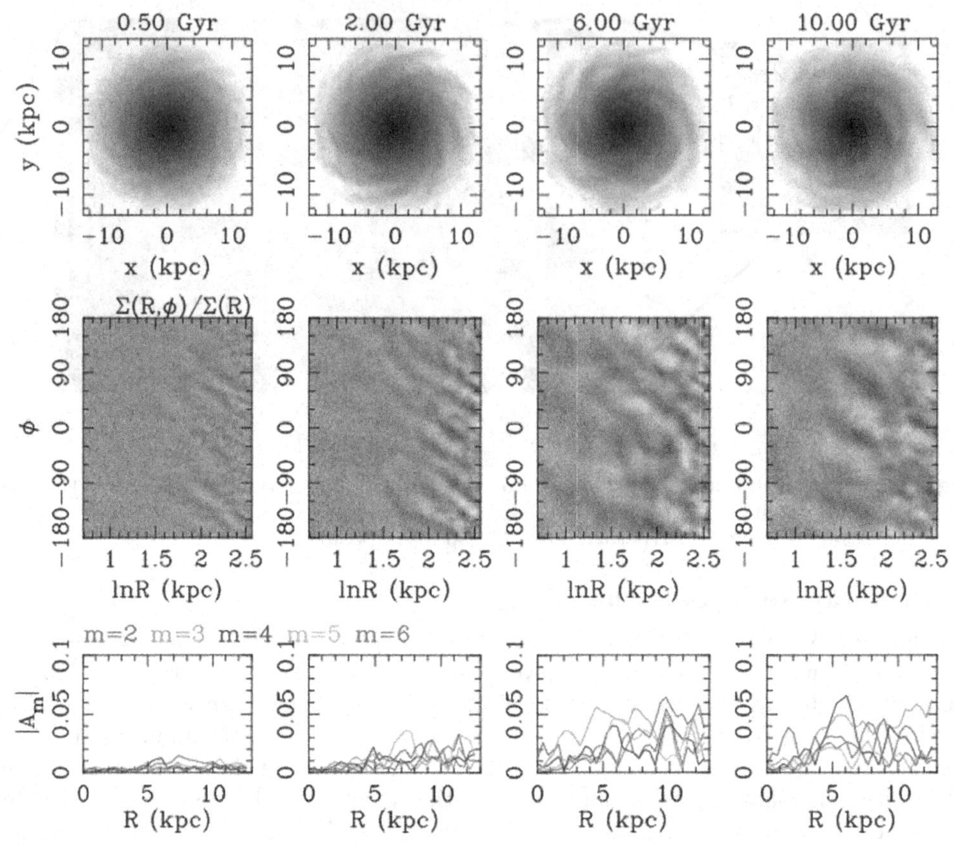

Figure 3. Evolution of spiral arms in a pure N-body disk with $N = 9 \times 10^6$. Top panels show the surface density, middle panels shows the surface density normalized at each radius, and bottom panels show the Fourier amplitudes. (Fujii *et al.* 2010).

and to see how spiral structures and star forming regions are formed, we performed N-body/SPH simulations of a disk in a fixed spherical potential. The initial profile of the disk is exponential, and 1×10^6 SPH particles and 3×10^6 N-body particles are used. The numerical code used here is ASURA (Saitoh *et al.* 2008, 2009; see also Baba *et al.* 2009). We adopted a fixed, spherical dark matter (DM) halo as a host of a stellar disk. The density profile of the DM halo follows the Navarro-Frenk-White (NFW) profile. The treatment of star formation and the heating due to the SN feedback are the same as those in Saitoh *et al.* (2008).

Figure 4 shows a snapshot of stars and gas in a typical model. Depending on the initial mass profile, either mullti-arm spirals with large pitch angle as seen in Fig. 4 or a central bar associated with spiral arms as shown in Baba *et al.* (2009) is formed. Cold gas forms filamentary and clumpy structures, in contrast to the relatively smooth distribution of stellar spirals. Young stars formed from gas in a cold, dense phase are roughly distributed along stellar spirals.

One should note that these stellar and gas spiral arms do not propagate in the stellar disk with a single pattern speed. In other words, they do NOT show a rigid rotation. Figure 5 shows motions of spiral arms at three different radii. It is clear that the stellar arms roughly follow the radial change of the galactic rotation, represented by the solid lines, at each radius. This means that spiral arms are in fact 'wound up', therefore each spiral arm should not be long-lived. This kinematical feature seems to be consistent with

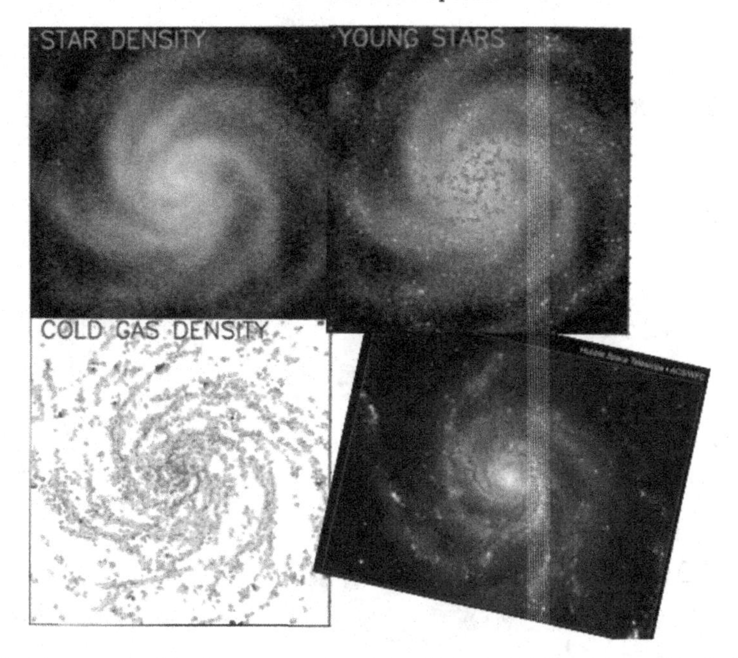

Figure 4. Multi-arm spiral 'galaxy' reproduced by N-body/SPH code: ASURA (Saitoh *et al.* 2008). Stars formed form cold and dense gas are represented by dots (upper right panel). Cold gas roughly follow the stellar spirals, but it has complicated structures. The morphology at this snapshot incidentally looks similar to M101 (bottom right: HST image. ©NASA/ESA).

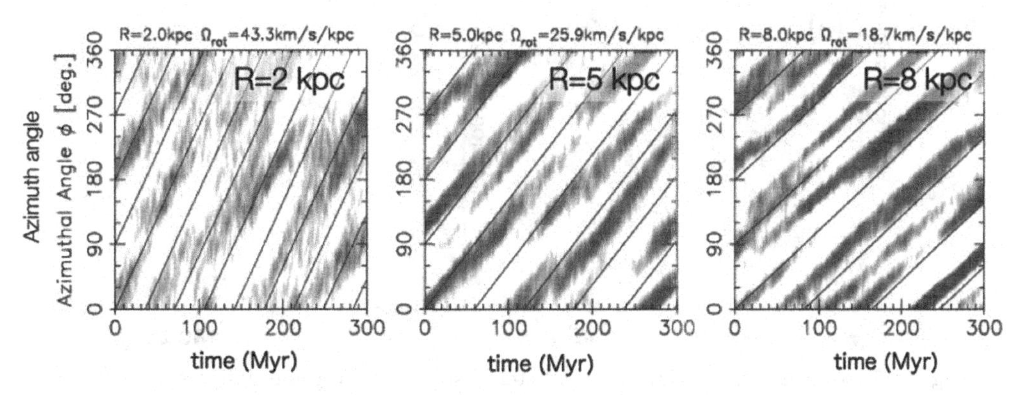

Figure 5. Kinematics of stellar arms at three radii. Stripes represent how the position of high density regions of stars move. Solid orthogonal lines represent galactic rotation at the radius.

some observations in M51 and NGC 1068, in which a radially declining 'pattern' speed is reported using the Tremmaine-Weinberg method (Merrifield *et al.* 2006; Meidt *et al.* 2008).

Figure 6 is cross-sections at five different galactic radii, showing amplitudes of stellar spirals and gas density. Relatively smooth curves represent stellar density; for example there are four arms at $R = 5$ kpc . Gas density, on the other hand, is more spiky, and there are many peaks. However, high density regions on average are roughly associated with stellar arms. Moreover, the complicated structure of the gas is not steady at all, and cannot be simply explained by the standing galactic shock solution. It is also clear that there is no apparent offset between stellar arms and high density regions of the gas.

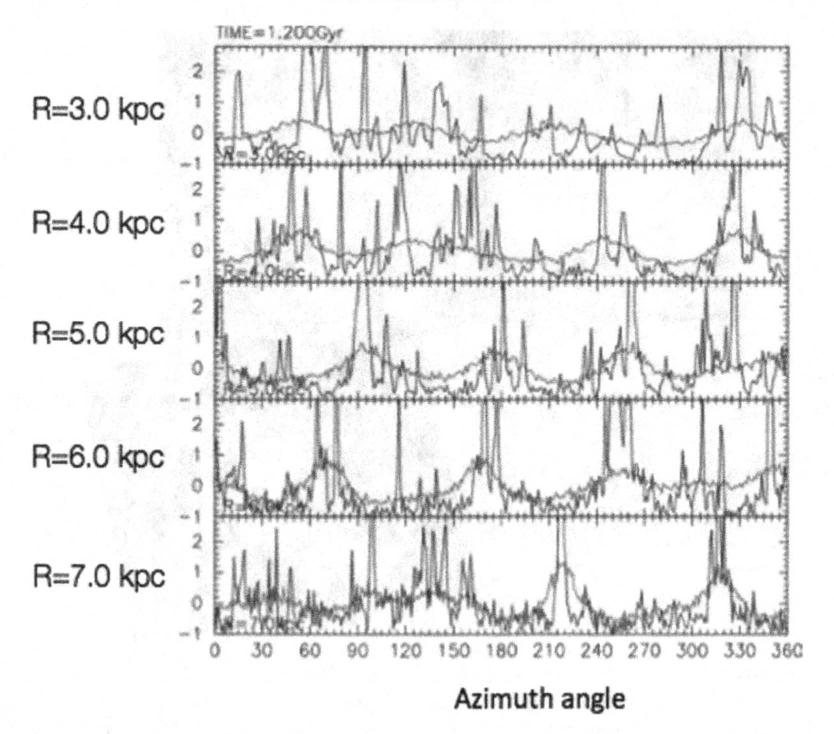

Figure 6. Gas and stellar density at different radii.

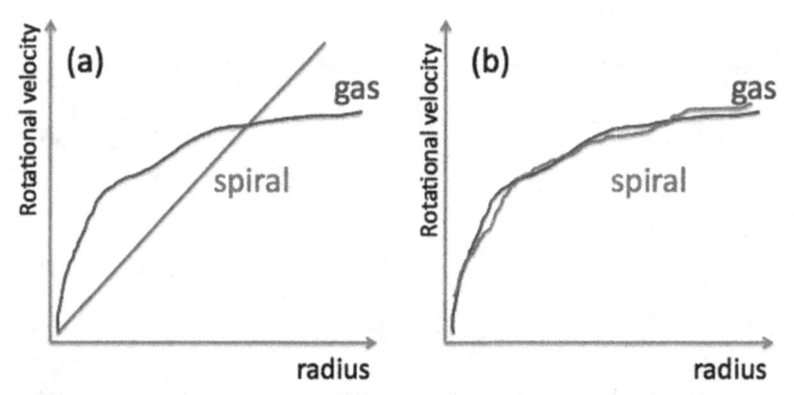

Figure 7. Rotation curve of the gas and stellar spiral. (a) A grand-design spiral with a constant pattern speed. (b) More realistic case.

A similar conclusion was also proposed by Dobbs & Bonnell (2008). They take a time-dependent gravitational potential from N-body simulations of Sellwood & Carlberg (1984), and run 3-D, two-phase (i.e. cold and warm) SPH simulations. They pointed out that "*c gas generally falls into a developing potential minimum and is released only when the local minimum dissolves. In this case, the densest gas is coincident with the spiral potential, rather than offset as in the grand-design spirals*". This is indeed what we see in more self-consistent (i.e. in terms of interaction between stars and ISM, thermal properties of the ISM, and star formation) N-body/SPH simulations.

Figure 7 shows schematically the difference between the rotation curves of spiral and gas (a) in the classical picture and (b) in the N-body/SPH simulations. If the multi-arm spiral structures are naturally developed in the galactic disk, the rotational velocity of the

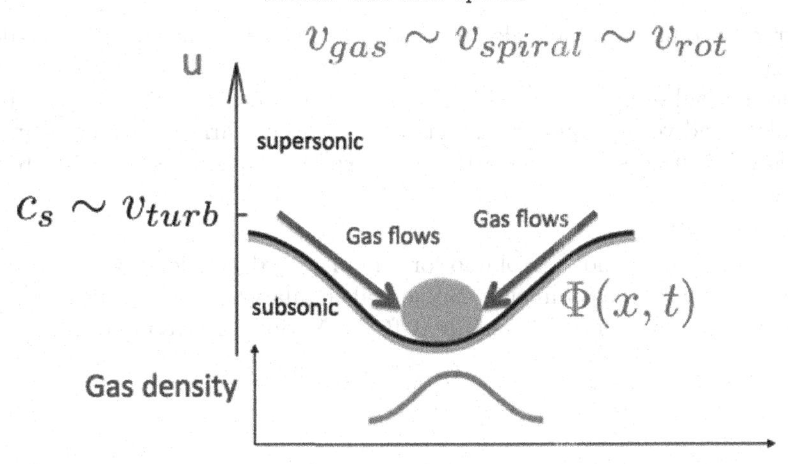

Figure 8. A modern picture of galactic spirals.

gas and stars should not be very different as shown in Fig. 7(b). Therefore, in contrast to the classical 'galactic shock' (Fig. 1), the ISM is always 'subsonic'† relative to the spiral potential, since the ISM basically moves together with the stellar spirals at any radii. In this case, as schematically shown in Fig. 8, the ISM falls into the spiral potential from both sides, and forms condensations at the bottom of the potential. Depending on the gas density, the condensation fragments and forms sub-structures. As a result, it is natural that there is no clear offset between stellar arms and gas arms, as seen in Fig. 6.

We also found that the ISM is not dynamically essential to keep stellar spirals, but it temporarily enhances the amplitude of local stellar arms. This is simply because the ISM tends to concentrate near the bottom of the potential as mentioned above. One should note that the local stellar potential is not steady, but time-dependent. Each stellar arm becomes faint or merges into other arms on a time scale of the galactic rotation, or even shorter. This local change of the potential can make the gas concentrations, like Giant Molecular Associations (GMA), gravitationally unbound. Therefore the time-evolution of the perturbation in the stellar potential should be considered when we discuss life-time of GMAs. The interaction between non-steady spirals and the ISM also causes large non-circular motions of the ISM. This is in fact observed in our Galaxy as the peculiar motions of the star forming regions (Baba *et al.* 2009).

4. Summary

The 'standard' picture of galactic spirals should now be replaced with a modern picture, which has been revealed by hydrodynamic and N-body simulations. Although some of them were already recognized by theorists since 1960s, recent progress in numerical techniques more clearly showed the dynamical nature of the galactic spirals in a more self-consistent manner. Some important features in the modern picture can be summarized as follows:

- There is NO winding dilemma!
- Spiral dust-lanes are not shocks.
- Galactic stellar spiral arms follow the galactic rotation . They are wound up locally,

† Here we assume the effective sound velocity of the ISM is comparable to its random motion, i.e. $\sim$ 10 km s^{-1}.

and recurrently connect to form global spirals, therefore they are highly time-dependent phenomena.

- The ISM is 'subsonic' to the stellar spiral potential, and no global shocks are formed†.
- Gas arms and young stars are associated with stellar arms, without clear offset.
- Gas is NOT necessary to keep the long-lived stellar spirals. It could temporarily enhance them.

We thank J. Makino and E. Kokubo for stimulating discussions. We are also grateful to the referee for valuable comments. Numerical simulations were performed by facilities at Center for Computational Astronomy (CfCA), National Astronomical Observatory of Japan.

References

Baba, J., Asaki, Y., Makino, J., Miyoshi, M., Saitoh, T. R., & Wada, K. 2010, *ApJ*, 706, 471
Dobbs, C. L. 2008, *391*, 844
Dobbs, C. L. & Bonnell, I. A. 2008, *MNRAS*, 385, 1893
Fujii, M. S., Baba, J., Saitoh, T.R., Makino, J., Kokubo, E., Wada, K., 2010, arXiv:1006.1228
Fujimoto, M. 1968, Non-stable Phenomena in Galaxies, proceedings of IAU Symposium No. 29, Byurakan, May 4-12, 1966. Published by "The Pubblishing House of the Academy of Sciences of Armenian SSR Yerevan, 1968 (translated from Russian)., p.453
Hartmann, L., Ballesteros-Paredes, J., & Bergin, E. A. 2001, *ApJ*, 562, 852
Heitsch, F. & Hartmann, L. 2008, *ApJ*, 689, 290
Kim, C.-G., Kim, W.-T., & Ostriker, E. C. 2010, arXiv:1006.4691
Meidt, S. E., Rand, R. J., Merrifield, M. R., Shetty, R., & Vogel, S. N. 2008, *ApJ*, 688, 224
Merrifield, M. R., Rand, R. J., & Meidt, S. E. 2006, *MNRAS*, 366, L17
Roberts, W. W. 1969, *ApJ*, 158, 123
Saitoh, T. R., Daisaka, H., Kokubo, E., Makino, J., Okamoto, T., Tomisaka, K., Wada, K., & Yoshida, N. 2008, *PASJ*, 60, 667
Saitoh, T. R., Daisaka, H., Kokubo, E., Makino, J., Okamoto, T., Tomisaka, K., Wada, K., & Yoshida, N. 2009, *PASJ*, 61, 481
Sellwood, J. A. & Carlberg, R. G. 1984, *ApJ*, 282, 61
Sellwood, J. A. 2010, arXiv:1008.2737
Shetty, R. & Ostriker, E. C. 2006, *ApJ*, 647, 997
Shu, F. H., Milione, V., & Roberts, W. W., Jr. 1973, *ApJ*, 183, 819
Tomisaka, K. 1986, *PASJ*, 38, 95
Vázquez-Semadeni, E., Ryu, D., Passot, T., González, R. F., & Gazol, A. 2006, *ApJ*, 643, 245
Wada, K. 2008, *ApJ*, 675, 188
Wada, K. & Koda, J. 2004, *MNRAS*, 349, 270

† Even in this picture, 'local' shocks associated with collision between clouds, where thermal temperature is smaller than 100 K, can be formed.

Computational Star Formation
Proceedings IAU Symposium No. 270, 2011
J. Alves, B.G. Elmegreen, J. M. Girart & V. Trimble, eds.

doi:10.1017/S1743921311000652

Control of galactic scale star formation by gravitational instability or midplane pressure?

Mordecai-Mark Mac Low[1]

[1]Department of Astrophysics, American Museum of Natural History, Central Park West at 79th Street, New York, NY, 10024-5192, USA. email: mordecai@amnh.org

Abstract.
Star formation in galaxies has been suggested to depend on large-scale gravitational instability or on the pressure required to form molecular hydrogen. I present numerical models and analysis of observations in support of the gravitational instability hypothesis. I also consider whether the correlation between the surface densities of molecular hydrogen and star formation implies causation, and if so in which direction.

Keywords. stars: formation, galaxies: evolution, ISM: molecules, ISM: clouds, ISM: structure

---

1. Context

The importance of understanding galactic-scale star formation has steadily increased as it has become a major stumbling block to confirming the fundamental picture of galaxy evolution in a universe dominated by dark matter and dark energy. The observational study of the question has progressed from the measurements integrated over galaxies of Kennicutt (1989, 1998) to the radial profiles of Martin & Kennicutt (2001) and the sub-kiloparsec scale pixels of Bigiel *et al.* (2008). These works have found strong correlations between the surface density of either molecular gas Σ_{H_2} or total gas Σ_{gas} and that of star formation Σ_{SFR}. The limits of these averaged approaches have been found, however, with the work by Evans (see this volume) showing that individual star forming molecular clouds have star formation rates well above the values derived from integrating over larger scales.

Attempts to predict the star formation rate can broadly be described as taking either global or local approaches to the question, depending on what they propose to be the rate-limiting step in the star formation process. For example, a major class of local models relies on the formation and destruction of molecular clouds as the balance between H_2 formation on grains and H_2 dissociation by far UV shifts (McKee 1989; Krumholz & McKee 2005; Shu *et al.* 2007; Krumholz *et al.* 2009). Two global models that I will focus on here are global gravitational instability (Martin & Kennicutt 2001; Rafikov 2001; Kravtsov 2003; Li *et al.* 2005, 2006; Tasker & Bryan 2006), and molecular cloud formation determined by the midplane pressure of galactic disks (Elmegreen 1989; Wong & Blitz 2002; Blitz & Rosolowsky 2006)

Although proposals for global and local models often present them as opposed to each other, careful examination of the models at each scale reveals that they generally rely on parameters predicted by their counterparts at the other scale, usually through the assumption of empirically determined parameters. For example, global models assume the observed velocity dispersion of the gas and the local star formation efficiency of molecular clouds, while local models assume the distribution of cloud properties such as masses, sizes, and locations.

2. Global Models

2.1. *Midplane Pressure*

After evaluation of a large set of coordinated observations using The H I Nearby Galaxies Survey (Walter *et al.* 2008), the Spitzer Infrared Nearby Galaxies Survey (Kennicutt *et al.* 2003) and the HERACLES CO survey (Leroy *et al.* 2009), Leroy *et al.* (2008) concluded that the best predictor of star formation was disk midplane pressure. They relied on two observations. The first, by Wong & Blitz (2002); Blitz & Rosolowsky (2004) and Blitz & Rosolowsky (2006) is that the molecular fraction $R_{\rm mol} = \Sigma_{\rm H_2}/\Sigma_{\rm H\,I}$ correlates near linearly with the midplane pressure (Elmegreen 1989; Elmegreen & Parravano 1994)

$$P_h \simeq \frac{\pi}{2} G \Sigma_{\rm tot} \left(\Sigma_{\rm tot} + \frac{\sigma_{\rm gas}}{\sigma_{*,z}} \Sigma_* \right). \quad (2.1)$$

where σ_*, z is the stellar velocity dispersion in the vertical direction, and Σ_* is the surface density of the stellar disk. The second is that $\Sigma_{\rm SFR} \propto \Sigma_{\rm H_2}$ (Rownd & Young 1999; Wong & Blitz 2002; Gao & Solomon 2004). The star formation rate can thus be re-expressed as a function of total gas as

$$\Sigma_{\rm SFR}(\Sigma_{\rm tot}) = \Sigma_{\rm SFR}(\Sigma_{\rm H_2}) \frac{R_{\rm mol}}{R_{\rm mol}+1}. \quad (2.2)$$

The molecular fraction was evaluated by Elmegreen (1993) as being dependent on P_h and the FUV radiation field j as $R_{\rm mol} \propto P_h^{2.2} j^{-1}$. If $j \propto \Sigma_{\rm SFR} \propto \Sigma_{\rm H_2}$, then $R_{\rm mol} \propto P_h^{1.2}$, reproducing the observed correlation.

This raises the question though, of why molecular hydrogen formation should be so important to star formation. The observations of the correlation appear to be on solid ground, at least for galaxies within an order of magnitude or so of the Milky Way in mass. However, the question can be raised of whether correlation implies causation? If so, in which direction does the causation run? Rather than H_2 formation being the gate to star formation, perhaps large scale gravitational collapse leads to both H_2 formation, and, not much later, to star formation.

Indeed, simulations of H_2 formation in a self-gravitating, magnetized, periodic, turbulent box show that (Glover & Mac Low 2007, 2010) the density enhancements produced by supersonic turbulence can lead to substantial molecular fractions within a few million years. The fraction of molecular hydrogen in these simulations shows a clear dependence on the density of the gas in the box. At a roughly constant temperature (such as the 60K characteristic of the cold neutral medium), this also would yield a dependence on pressure.

The evaluation of the star formation law by Leroy *et al.* (2008) relied on the assumption that the stellar radial velocity dispersion $\sigma_{*,r}$ declined exponentially with radius, based on the apparent constant scale height of stellar disks, and thus exponential decline of $\sigma_{*,z}$. If a constant value of $\sigma_{*,r}$ is instead assumed, however, they find that the next model that I will discuss, gravitational instability, is equally predictive.

2.2. *Gravitational Instability*

Global disk instability models postulate that star formation happens wherever gravitational instability (Gammie 1992; Rafikov 2001) of the combined collisionless stars

(Toomre 1964) and collisional gas (Goldreich & Lynden-Bell 1965) in the disk sets in. The criteria for linear instability leading to radial collapse of axisymmetric rings for gas and stellar disks are

$$Q_g \equiv \frac{\kappa c_g}{\pi G \Sigma_{\rm tot}} < 1, \quad Q_* \equiv \frac{\kappa \sigma_{*,r}}{\pi G \Sigma_*} < 1.07, \tag{2.3}$$

where κ is the epicyclic frequency, c_g the speed of sound, σ_* the radial stellar velocity dispersion, and Σ_* is the stellar surface density. (Note that following Rafikov (2001) we use a factor of π in the definition of Q_* rather than 3.36, shifting the instability criterion to slightly higher value.) Define the dimensionless quantities $q = k\sigma_*/\kappa$ and $s = c_g/\sigma_*$. Then the instability criterion for the combined disk of gas and stars is given by

$$\frac{2}{Q_*}\frac{1}{q}\left[1 - e^{-q^2} I_0(q^2)\right] + \frac{2}{Q_g} q \frac{s}{1+q^2 s^2} > \frac{1}{Q_{sg}}, \tag{2.4}$$

where I_0 is the Bessel function of order 0.

3. Simulations

Numerical experiments on the behavior of the gravitational instability in disks have been done by Li *et al.* (2005), using isothermal gas, collisionless stars, and live dark matter halos computed with GADGET (Springel *et al.* 2001). They controlled the initial gravitational instability of the disk, and then computed its subsequent behavior. Using sink particles, they measured the amount of gas that collapsed as a function of time, and related it to the minimum radial value of the initial instability $Q_{sg,\rm min}$. Care was taken to resolve the local Jeans length at all densities below the sink particle threshold to avoid artificial fragmentation (Truelove *et al.* 1997; Bate & Burkert 1997), requiring as many as five million particles in their largest simulations.

Kravtsov (2003) and Li *et al.* (2006) demonstrate that the Schmidt law (Kennicutt 1998) is a natural consequence of a gravitationally unstable galactic disk. Kravtsov (2003) computed a cosmological volume and followed the star formation in individual disks, using a star formation law $\dot{M}_* \propto \rho_g$ deliberately chosen to not automatically reproduce the Schmidt law, as compared to the frequently chosen $\dot{M}_* \propto \rho_g^{1.5}$. The sink particles used by Li *et al.* (2006) effectively give a similar star formation law, as they measure collapsed gas above a fixed threshold.

The threshold for sink particle formation corresponds to a pressure $P/k > 10^7$ cm^{-3}, corresponding to densities high enough for molecule formation to proceed in under a megayear (Glover & Mac Low 2007). If the sink particles are interpreted as consisting of mostly molecular gas (the fraction chosen was 70%), while the SPH particles are treated as atomic gas, the characteristic radial profiles of molecular and atomic gas fond by Wong & Blitz (2002) are recovered, as is the relation $\Sigma_{\rm SFR} \propto \Sigma_{\rm H_2}$.

Both Kravtsov (2003) and Li *et al.* (2006) demonstrated the dropoff from the Kennicutt-Schmidt Law found by Bigiel *et al.* (2008) at $\Sigma_{\rm tot} < 10$ M$_\odot$ pc^{-2}. This reproduction of a major observational result must be counted as a success for the gravitational instability model.

4. Observational Comparisons

The first local stability analysis of an external galaxy with resolution below 50 pc was performed by Yang *et al.* (2007). To measure $\Sigma_{\rm H\,I}$ and $\Sigma_{\rm H_2}$ they used a combined data set of the Australia Telescope Compact Array and the Parkes multibeam receiver (Kim *et al.* 2003) for the H I, and the NANTEN CO survey of the LMC done by Fukui *et al.* (2001). The rotation curve used to derive κ is a fit to the H I and the carbon star (Kunkel *et al.* 1997) measurements inside and outside of about 3.2 kpc, respectively (Kim *et al.* 1998). To estimate the stellar surface density Σ_s, they used the number density of red giant branch (RGB) and asymptotic giant branch (AGB) stars, follow a procedure similar to that outlined by van der Marel (2001) but use only the Two Micron All Sky Survey Point Source Catalogue (Skrutskie *et al.* 2006) and different color criteria. They assumed constant values of both σ_g and $\sigma_{*,r}$, the latter after considering the velocity dispersions of carbon stars, red supergiants, and young globular clusters.

Considering the gas alone, Yang *et al.* (2007) found that 38% of the YSOs identified by Gruendl & Chu (2009) from the Spitzer LMC survey lay in Toomre stable regions. However, when the stellar contribution was included, only 15% of the YSOs were in stable regions, showing that the stability analysis using the combination of gas and stars provides a reliable way of predicting star formation.

The apparent stability of the M33 disk was raised by Padoan in his talk at this conference as a counterargument to the theory of gravitational instability, with a citation to Corbelli (2003). Examination of that paper shows that the star forming disk of M33 is indeed stable if only the gas is considered. However, Corbelli (2003) also performed a simplified analysis including the stellar surface density, and concluded that the entire star-forming disk is indeed Toomre unstable if the stellar contribution is accounted for (see her Fig. 8).

5. Conclusions

Why does midplane pressure appear to work so well at predicting the star formation efficiency of galaxies? The key driver appears to be the linear correlation between Σ_{H_2} and Σ_{SFR}. But why should H_2 exert such a strong influence on star formation? Although H_2 is a coolant, interstellar gas can already cool down to below 60 K with atomic fine structure lines. The remaining cooling is probably not a hugely limiting factor in star formation.

However, gravitational instability produces dense gas that quickly forms H_2 (Glover & Mac Low 2007). Thus, H_2, and other molecules such as CO that form with it, may primarily just act to trace dense gas that is already gravitationally unstable and collapsing. Ostriker (this volume) offers an analytic model for how the combination of dynamical and thermodynamical equilibrium may lead to this situation.

Galaxies clearly form stars at widely varying efficiency. Including that in cosmological models seems to be one way forward to understanding the evolution of galaxies over cosmic time (Kravtsov 2010). Varying global efficiency of star formation is a natural consequence of gravitational instability in galaxies with varying properties. The galaxies modeled by Li *et al.* (2006) show orders of magnitude variation in global efficiency, even with the assumption of constant local efficiency of star formation in any individual collapsing region (an assumption that must ultimately be demonstrated with something looking a lot like a local model, to be sure.)

Acknowledgements

I thank the organizers for their invitation to speak and partial support of my attendance, and Paolo Padoan for provoking me to check the stability of M33. This work was partly funded by NASA/SAO grant TM0-11008X, by NASA/STScI grant HST-AR-11780.02-A, and by NSF grant AST08-06558.

References

Bate, M. R. & Burkert, A. 1997, *MNRAS*, 288, 1060
Bigiel, F., Leroy, A., Walter, F., Brinks, E., de Blok, W. J. G., Madore, B., & Thornley, M. D. 2008, *AJ*, 136, 2846
Blitz, L. & Rosolowsky, E. 2004, *ApJ (Letters)*, 612, L29
Blitz, L. & Rosolowsky, E. 2006, *ApJ*, 650, 933
Corbelli, E. 2003, *MNRAS*, 342, 199
Elmegreen, B. G. 1989, *ApJ*, 338, 178
Elmegreen, B. G. 1993, *ApJ*, 411, 170
Elmegreen, B. G. & Parravano, A. 1994, *ApJ (Letters)*, 435, L121
Fukui, Y., Mizuno, N., Yamaguchi, R., Mizuno, A., & Onishi, T. 2001, *PASJ*, 53, L41
Gammie, C. F. 1992, Ph.D. Thesis, Princeton University, Princeton, NJ, USA
Gao, Y. & Solomon, P. M. 2004, *ApJ*, 606, 271
Glover, S. C. O. & Mac Low, M.-M. 2007, *ApJ*, 659, 1317
Glover, S. C. O. & Mac Low, M.-M. 2010, *MNRAS*, submitted (arXiv:1003.1340)
Goldreich, P. & Lynden-Bell, D. 1965, *MNRAS*, 130, 97
Gruendl, R. A. & Chu, Y.-H. 2009, *ApJ Suppl.*, 184, 172
Kennicutt, R. C. 1989, *ApJ*, 344, 685
Kennicutt, R. C. 1998, *ARAA*, 36, 189
Kennicutt, R. C., Jr. *et al.* 2003, *PASP*, 115, 928
Kim, S., Staveley-Smith, L., Dopita, M. A., Freeman, K. C., Sault, R. J., Kesteven, M. J., & McConnell, D. 1998, *ApJ*, 503, 674
Kim, S., Staveley-Smith, L., Dopita, M. A., Sault, R. J., Freeman, K. C., Lee, Y., & Chu, Y.-H. 2003, *ApJ Suppl.*, 148, 473
Kravtsov, A. V. 2003, *ApJ (Letters)*, 590, L1
Kravtsov, A. 2010, *Adv. Astron.*, 2010, 1
Krumholz, M. R. & McKee, C. F. 2005, *ApJ*, 630, 250
Krumholz, M. R., McKee, C. F., & Tumlinson, J. 2009, *ApJ*, 699, 850
Kunkel, W. E., Demers, S., Irwin, M. J., & Albert, L. 1997, *ApJ (Letters)*, 488, L129
Leroy, A. K., Walter, F., Brinks, E., Bigiel, F., de Blok, W. J. G., Madore, B., & Thornley, M. D. 2008, *AJ*, 136, 2782
Leroy, A. K. *et al.* 2009, *AJ*, 137, 4670
Li, Y., Mac Low, M.-M., & Klessen, R. S. 2005, *ApJ (Letters)*, 620, L19
Li, Y., Mac Low, M.-M., & Klessen, R. S. 2006, *ApJ*, 639, 879
Martin, C. L. & Kennicutt, R. C. 2001, *ApJ*, 555, 301
McKee, C. F. 1989, *ApJ*, 345, 782
Rafikov, R. R. 2001, *MNRAS*, 323, 445
Rownd, B. K. & Young, J. S. 1999, *AJ*, 118, 670
Shu, F. H., Allen, R. J., Lizano, S., & Galli, D. 2007, *ApJ (Letters)*, 662, L75
Skrutskie, M. F. *et al.* 2006, *AJ*, 131, 1163
Tasker, E. J. & Bryan, G. L. 2006, *ApJ*, 641, 878
Truelove, J. K., Klein, R. I., McKee, C. F., Holliman, J. H., Howell, L. H., & Greenough, J. A. 1997, *ApJ (Letters)*, 489, L179
Toomre, A. 1964, *ApJ*, 139, 1217
Walter, F., Brinks, E., de Blok, W. J. G., Bigiel, F., Kennicutt, R. C., Thornley, M. D., & Leroy, A. 2008, *AJ*, 136, 2563
Wong, T. & Blitz, L. 2002, *ApJ*, 569, 157
van der Marel, R. P. 2001, *AJ*, 122, 1827
Yang, C.-C., Gruendl, R. A., Chu, Y.-H., Mac Low, M.-M., & Fukui, Y. 2007, *ApJ*, 671, 374

Mordecai Mac Low

Computational Star Formation
Proceedings IAU Symposium No. 270, 2011
J. Alves, B.G. Elmegreen, J. M. Girart & V. Trimble, eds.

doi:10.1017/S1743921311000664

Star Formation and the Properties of Giant Molecular Clouds in Global Simulations

Elizabeth J. Tasker[1] & Jonathan C. Tan[2]

[1]Department of Physics & Astronomy, McMaster University,
Hamilton, ON, Canada
email: taskere@mcmaster.ca
[2]Department of Astronomy, University of Florida,
Gainesville, FL, USA

Abstract. We simulated an isolated quiescent Milky Way-type galaxy with a maximum effective resolution of 7.8 pc. Clouds formed in the interstellar medium through gravitational fragmentation and became the sites for star formation. We tracked the evolution of the clouds through 300 Myr in the presence of star formation, photoelectric heating and feedback from Type II supernovae. The cloud mass distribution agreed well with observational results. Feedback suppressed star formation but did not destroy the surrounding cloud. Collisions between clouds were found to be sufficiently frequent to be a significant factor in determining the star formation rate.

Keywords. hydrodynamics, stars: formation, ISM: clouds, Galaxy: disk, ISM: structure.

1. Introduction

Almost all star formation in galaxies occurs in cold, dense clumps within giant molecular clouds (GMCs), whose properties thus set the environment for the star formation process. Inevitably, GMC evolution must be a determining factor for galactic scale star formation rates.

GMCs in the Milky Way are observed to have masses up to 6×10^6 $M_\odot$, with an equally massive atomic envelope (Williams & McKee, 1997, Blitz *et al.*, 1990). The mass profile is well fitted by a power-law distribution:

$$\frac{dN}{dM} = M^{-\alpha} \tag{1.1}$$

where the exponent α is observed to take a value in the range $1.5 - 2.5$ for GMCs in the Milky Way and M33 (Rosolowsky *et al.*, 2006).

How connected these objects are to their galactic environment is a subject of debate. It is possible that GMCs are isolated entities whose properties are independent of the galaxy's potential and each other. Alternatively, they could be coupled, leading to a link between galaxy-scale evolution on 10s kpcs and that of star formation on the sub-parsec scale.

Evidence that such a coupling should exist can be found in the Schmidt-Kennicutt relation (Kennicutt, 1998).This empirical law links the surface density of star formation rate to the surface density of gas:

$$\sum_{\rm sfr} \propto \sum_{\rm gas}^{1.4\pm0.15} \tag{1.2}$$

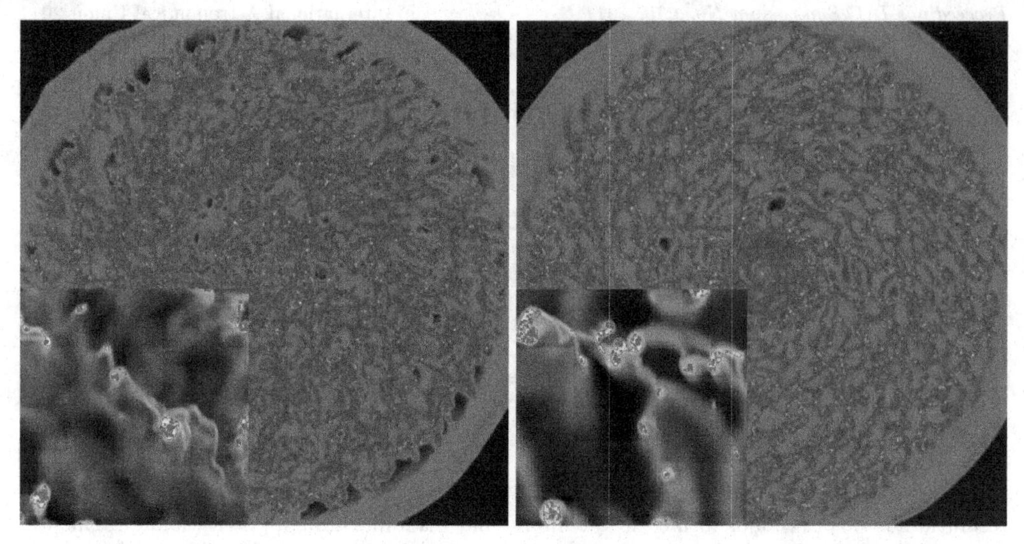

Figure 1. Surface density of the simulation with star formation (left) and the simulation with star formation and photoelectric heating. The whole disc image is 20 kpc across, while the small cut-outs in the bottom left show a 2 kpc slice of the density. Blue contours mark our cloud threshold density of $n_{H,c} = 100\,\mathrm{cm}^{-3}$ and red points show the star particles that are less than 1 Myr old.

or alternatively, also to the orbital angular frequency at the outer radius, Ω_{out}: $\Sigma_{\mathrm{sfr}} \propto \Sigma_{\mathrm{gas}}\Omega_{\mathrm{out}}$.

That the global orbital angular frequency should be related to the star formation rate suggests that star formation is strongly influenced by its large scale environment.

One method for this link is via collisions of GMCs. In his paper, Tan (2000) performed analytical calculations to show that cloud collisions could trigger star formation and produce the Schmidt-Kennicutt relation if they occurred at a frequency of a fixed fraction of their orbital period.

2. Simulation details

We used the adaptive mesh refinement code *Enzo* (Bryan & Norman, 1997) to model the isolated galaxy disc. Details of the disc initial conditions for simulations without star formation are described in Tasker & Tan (2009). Our highest resolution simulation has a smallest cell size of 7.8 pc and star particles were formed with typical masses of $10^3\,\mathrm{M}_\odot$ from gas with $n_{H,c} > 100\,\mathrm{cm}^{-3}$ (i.e. our definition of GMCs) with an efficiency of 2% per local free-fall time (Krumholz & Tan, 2007). Radiative cooling is allowed down to 300 K and the gas is purely atomic, with the dense cloud gas assumed to have a high molecular component. Simulations with star formation and FUV feedback are presented here and described in more detail by Tasker & Tan (in prep).

Clouds were defined as continuous structures with densities greater than $100\,\mathrm{cm}^{-3}$, a typical mean density of observed GMCs. They were tracked through the simulation by comparing outputs at 1 Myr intervals. Details of the cloud identification and tracking scheme are also described in Tasker & Tan (2009).

The surface density of the galaxy disc at 200 Myr is shown in Figure 1. The left-hand image shows the face-on disc for a simulation that includes star formation but no form of feedback. The panel in the bottom left of the image is a 2 kpc slice through the disc mid-plane. The blue contours mark out the $100\,\mathrm{cm}^{-3}$ cloud threshold and the red particles mark the stars born in the last 1 Myr. As expected, the stars are born in the identified

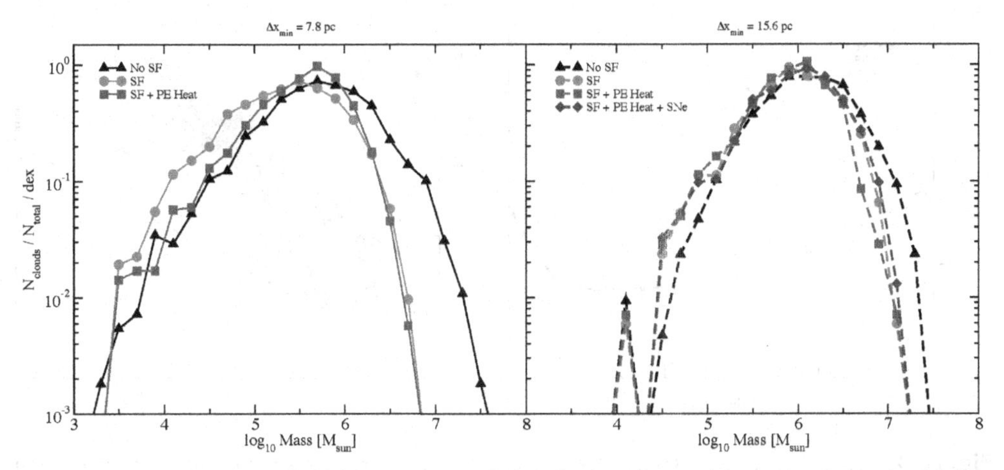

Figure 2. Distributions of GMC masses. Left-hand plot shows the results from the highest resolution (7.8 pc) simulations with and without star formation and with the inclusion of photoelectric heating. The right plot shows the same profile for a resolution of 15.6 pc and also includes results from the run with supernovae feedback.

GMCs. The right-hand side of Figure 1 shows the same images for the simulation with feedback from photoelectric heating. Smaller scale structures appear to be suppressed by the presence of FUV heating.

3. Cloud properties

Figure 2 shows the distributions of GMC masses. The left-hand plot shows clouds in the simulations with 7.8 pc resolution. Without star formation, there is no destructive mechanism and the clouds can reach artificially large masses. With star formation, this mass reaches a maximum of $M \sim 10^{6.7}$ $M_\odot$. This value is unaffected by the presence of photoelectric heating. Observed GMCs in the Milky Way have masses up to 6×10^6 $M_\odot$ (Williams & McKee, 1997) which, allowing for an atomic envelope of roughly equal size, is in good agreement with what we find.

The right-hand plot shows the mass profile at a lower resolution of 15.6 pc and includes the simulation with supernovae feedback. We can see that the shape of the mass profile is relatively independent of the physics included. Supernovae feedback acts to greatly suppress star formation, but the relatively constant profile shape suggests that supernova feedback has only a modest influence on GMC properties.

4. Star formation

As mentioned in § 1, cloud collision could potentially play a significant role in driving star formation. The left-hand plot in Figure 3 shows the average collision time of clouds as a fraction of their orbital time plotted against their radial position in the disc. Regardless of the physics included, the cloud collision rate is small and approximately constant fraction of the orbital time, with clouds typically experiencing a collision every quarter of an orbit. This may be sufficient for such interactions to be important for global star formation rates, if GMCs have lifetimes of this order (Tan, 2000).

The right-hand image of Figure 3 shows the Schmidt-Kennicutt relation. Squares show the gas surface density versus surface density of star formation rate averaged over disc annuli. Open circles show this relation for the individual GMCs. Over-plotted are the results from the THINGS survey (Bigiel *et al.*, 2008) for local galaxies. Diagonal lines mark lines mark constant star formation efficiency, denoting the rate of surface star

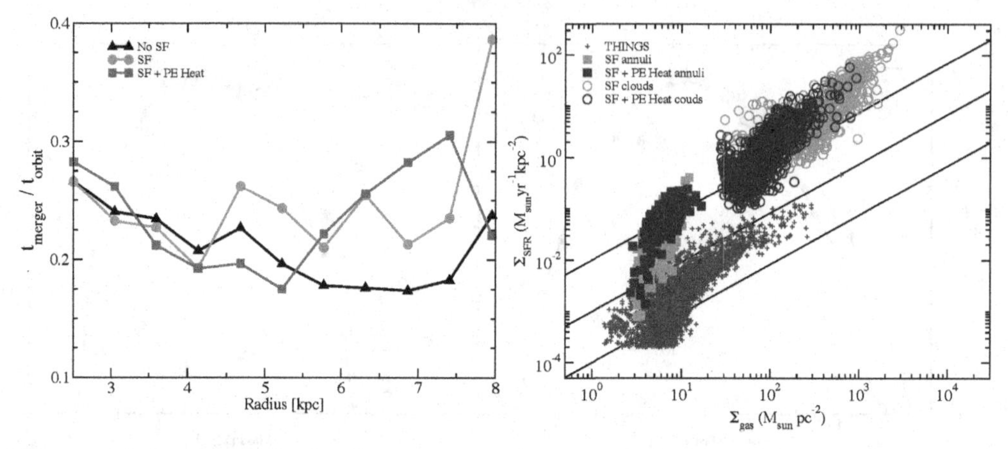

Figure 3. Left: average cloud collision rate as a function of radius for runs at 7.8 pc resolution. Clouds from simulation times 200 ± 25 Myr were included. Right: the local Schmidt-Kennicutt relation for annuli in the disc and individual GMCs at 250 Myr. Solid lines mark constant star formation efficiencies and observational data from the THINGS survey is shown.

formation needed to consume 1%, 10% and 100% of the gas in 100 Myr. Averaging on scales of the disk annuli, the simulations form stars at a rate about a factor of 10 higher than the observed values for a given Σ_{gas}. This is likely due to other feedback processes, such as ionization, and magnetic fields which are not yet included in our simulations.

Averaging the Schmidt-Kennicutt relation on the scale of GMCs, we find an approximate power law behavior, similar to 1.2 but with a higher normalisation. There is some evidence that on GMC scales the star formation law does have such an elevated normalization (Heiderman, Evans & Huard, in prep.). This could be accounted for both by the physics we are not including and by the observations being averaged over a larger surface area than the individual clouds in our simulation.

5. Summary

We modeled isolated galaxy discs as gas orbiting in fixed axisymmetric potentials, investigating GMC formation and evolution and its dependence on star formation and feedback from diffuse background photoelectric heating and localized supernovae. Models including star formation produce GMC mass functions similar to those observed in the Milky Way, including a truncation at masses $\sim 10^7$ $M_\odot$. We find cloud collisions occur approximately every 1/4 of an orbital period, indicating that this process could be a link between the global and local scales of star formation.

References

Bigiel, F., Leroy, A., Walter, F., Brinks, E., de Blok, W. J. G., Madore, B., & Thornley, M. D. 2008, *AJ*, 136, 2846

Blitz, L., Bazell, D., & Desert, F. X. 1990, *ApJL*, 352, L13

Bryan, G. L. & Norman, M. L. 1997, *ASP Conf. Ser.* 123: Computational Astrophysics; 12th Kingston Meeting on Theoretical Astrophysics, 363

Kennicutt, R. C., Jr. 1998, *ApJ*, 498, 541

Krumholz, M. R. & Tan, J. C. 2007, *ApJ*, 654, 304

Rosolowsky, E., Engargiola, G., Plambeck, R., & Blitz, L. 2003, *ApJ*, 599, 258

Tan, J. C. 2000, *ApJ*, 536, 173

Tasker, E. J. & Tan, J. C. 2009, *ApJ*, 700, 358

Williams, J. P. & McKee, C. F. 1997, *ApJ*, 476, 166

Computational Star Formation
Proceedings IAU Symposium No. 270, 2011
J. Alves, B.G. Elmegreen, J. M. Girart & V. Trimble, eds.

doi:10.1017/S1743921311000676

Modeling Formation of Globular Clusters: Beacons of Galactic Star Formation

Oleg Y. Gnedin

University of Michigan, Department of Astronomy, Ann Arbor, MI 48109, USA
email: ognedin@umich.edu

Abstract. Modern hydrodynamic simulations of galaxy formation are able to predict accurately the rates and locations of the assembly of giant molecular clouds in early galaxies. These clouds could host star clusters with the masses and sizes of real globular clusters. I describe current state-of-the-art simulations aimed at understanding the origin of the cluster mass function and metallicity distribution. Metallicity bimodality of globular cluster systems appears to be a natural outcome of hierarchical formation and gradually declining fraction of cold gas in galaxies. Globular cluster formation was most prominent at redshifts $z > 3$, when massive star clusters may have contributed as much as 20% of all galactic star formation.

Keywords. galaxies: formation — galaxies: star clusters — globular clusters: general

---

1. Clues from Old and Young Star Clusters

A self-consistent description of the formation of globular clusters remains a challenge to theorists. Most of the progress is driven by observational discoveries. The Hubble Space Telescope observations have convincingly demonstrated one of the likely routes for the formation of massive star clusters today – in the mergers of gas-rich galaxies. These observations have also shown the differences between the mass function of young clusters (power-law $dN/dM \propto M^{-2}$) and old clusters (log-normal or broken power-law).

Surveys of the globular cluster systems of galaxies in the Virgo and Fornax galaxy clusters have solidified the evidence for bimodal, and even multimodal, color distribution in galaxies ranging from dwarf disks to giant ellipticals (Peng *et al.* 2008). This color bimodality likely translates into a bimodal distribution of the abundances of heavy elements such as iron. We know this to be the case in the Galaxy as well as in M31, where accurate spectral measurements exist for a large fraction of the clusters. The two most frequently encountered modes are commonly called *blue* (metal-poor) and *red* (metal-rich).

Detailed spectroscopy reveals a significant spread of ages of the red clusters in the Galaxy, up to 6 Gyr (Dotter *et al.* 2010). The spread increases with metallicity and distance from the center. The age spread of the blue clusters is smaller, in the range 1-2 Gyr, and is consistent with the measurement errors.

2. Modeling the Formation of Globular Clusters is Hard

The first attempt to model the formation of globular clusters within the framework of hierarchical galaxy formation was by Beasley *et al.* (2002). Their semi-analytical model could reproduce the metallicity bimodality only by assuming two separate prescriptions for the blue and red clusters: blue clusters formed in quiescent disks with an efficiency of 0.002 relative to field stars, whereas red clusters formed in gas-rich mergers with a higher efficiency of 0.007. The formation of blue clusters also had to be artificially halted after $z = 5$, so as not to dilute the bimodality.

Moore *et al.* (2006) considered an idealized scenario for the formation of blue globular clusters at high redshift, inside dark matter halos that would eventually merge into the Galaxy, one cluster per halo. They used the observed spatial distribution of the Galactic clusters to constrain the formation epoch and found that the clusters would need to form by $z \sim 12$, in relatively small halos. Such an early formation is inconsistent with the simultaneous requirements of high mass and density for the parent molecular clouds to produce such dense ($\rho_* > 10^4$ M$_\odot$ pc^{-3}) and massive ($M > 10^5$ M$_\odot$) clusters as observed. This scenario also places stringent constraints on the age spread of blue clusters to be less than 0.5 Gyr, which may already be inconsistent with the available age measurements. The tension with observations of this scenario, and of its several variants in the literature, probably indicates that globular clusters cannot be simply associated with early dark matter halos and, instead, must be studied as an integral part of galactic star formation.

3. Globular Clusters Could Form in Protogalactic Disks

Kravtsov & Gnedin (2005) used a hydrodynamic simulation of a Galactic environment at redshifts $z > 3$ and found dense, massive gas clouds within the protogalactic clumps. These clouds assemble within the self-gravitating disk of progenitor galaxies after gas-rich mergers. The disk develops strong spiral arms, which further fragment into separate molecular clouds located along the arms as beads on a string. A working assumption, that the central high-density region of these clouds formed a star cluster, results in the distributions of cluster mass, size, and metallicity that are consistent with those of the Galactic metal-poor clusters. The high stellar density of Galactic clusters restricts their parent clouds to be in relatively massive progenitors, with the total mass $M_h > 10^9$ M$_\odot$. The mass of the molecular clouds increases with cosmic time, but the rate of mergers declines steadily. Therefore, the cluster formation efficiency peaks during an extended epoch, $5 < z < 3$, when the Universe is still less than 2 Gyr old. The parent molecular clouds are massive enough to be self-shielded from UV radiation, so that globular cluster formation should be unaffected by the reionization of cosmic hydrogen at $z > 6$. The mass function of model clusters is consistent with a power law $dN/dM \propto M^{-2}$, similar to the local young star clusters. The total mass of clusters formed in each progenitor is roughly proportional to the available gas supply and the total mass, $M_{GC} \sim 10^{-4}\ M_h$.

Prieto & Gnedin (2008) showed that subsequent mergers of the progenitor galaxies would ensure the present distribution of the globular cluster system is spheroidal, as observed, even though initially all clusters form on nearly circular orbits. Depending on the subsequent trajectories of their host galaxies, clusters form three main subsystems at present time. *Disk clusters* formed in the most massive progenitor that eventually hosts the present Galactic disk. These clusters are scattered into eccentric orbits by perturbations from accreted galactic satellites. *Inner halo clusters* came from the now-disrupted satellite galaxies. Their orbits are inclined with respect to the Galactic disk and are fairly isotropic. *Outer halo clusters* are either still associated with the surviving satellite galaxies, or were scattered away from their hosts during close encounters with other satellites and consequently appear isolated.

Following the scenario outlined above, Muratov & Gnedin (2010) developed a semi-analytical model that aims to reproduce statistically the metallicity distribution of the Galactic globular clusters. The formation of clusters is triggered during a merger of gas-rich protogalaxies with the mass ratio 1:5 or higher, and during very early mergers with any mass ratio when the cold gas fraction in the progenitors is close to 100%. Model clusters are assigned the mean metallicity of their host galaxies, which is calculated using the observed galaxy stellar mass-metallicity relation.

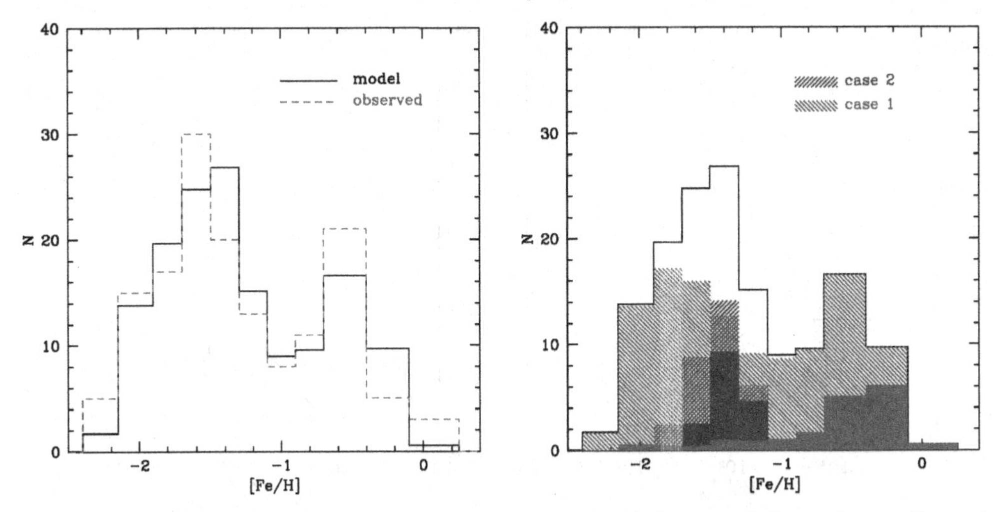

Figure 1. *Left:* Metallicities of model clusters that survived dynamical disruption until $z = 0$, compared to the observed distribution of Galactic globular clusters. *Right:* Model metallicity distribution split by the formation criterion: major mergers (`case-1`) and early mergers (`case-2`). Filled histograms show clusters formed in the main Galactic disk. From Muratov & Gnedin (2010).

4. Metallicity Bimodality

Figure 1 shows the metallicity bimodality in the model of Muratov & Gnedin (2010). Note that the model imposes the same formation criteria for all clusters, without explicitly differentiating between the two modes. The only variables are the gradually changing amount of cold gas, the growth of protogalactic disks, and the rate of merging. Yet, the model produces two peaks of the metallicity distribution, centered at [Fe/H] ≈ -1.6 and [Fe/H] ≈ -0.6, matching the Galactic globular clusters.

The red peak is not as pronounced as in the observations but is still significant. Early mergers of low-mass progenitors contribute only blue clusters. Interestingly, later major mergers contribute both to the red and blue modes, in about equal proportions. They are expected to produce a higher fraction of red clusters in galaxies with more active merger history, such as in massive ellipticals.

In this scenario, bimodality results from the history of galaxy assembly (rate of mergers) and the amount of cold gas in protogalactic disks. Early mergers are frequent but involve relatively low-mass protogalaxies, which produce preferentially blue clusters. Late mergers are infrequent but typically involve more massive galaxies. As the number of clusters formed in each merger increases with the progenitor mass, just a few late supermassive mergers can produce a significant number of red clusters. The concurrent growth of the average metallicity of galaxies between the late mergers leads to an apparent "gap" between the red and blue clusters.

Our prescription links cluster metallicity to the average galaxy metallicity in a one-to-one relation, albeit with random scatter. Since the average galaxy metallicity grows monotonically with time, the cluster metallicity also grows with time. The model thus encodes an age-metallicity relation, in the sense that metal-rich clusters are younger than their metal-poor counterparts by several Gyr. However, clusters of the same age may differ in metallicity by as much as a factor of 10, as they formed in the progenitors of different mass.

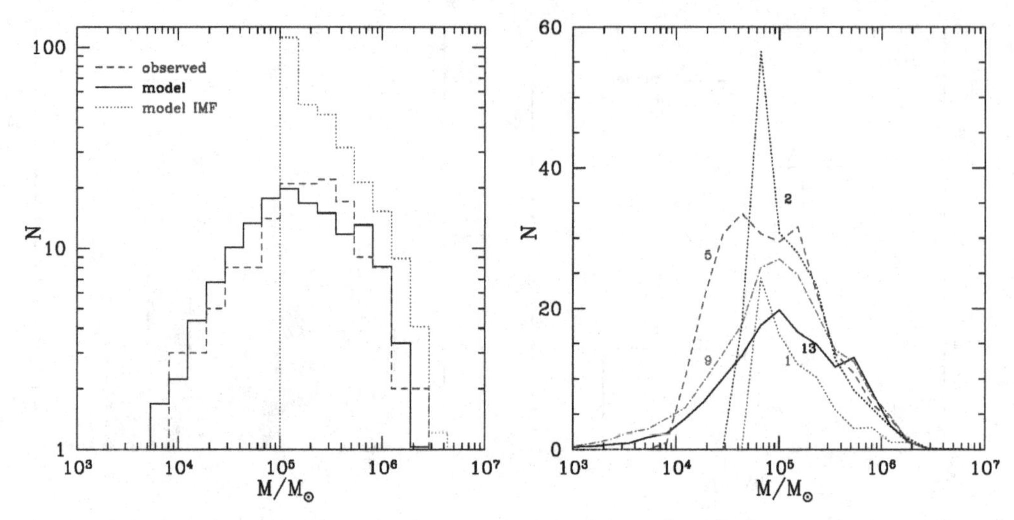

Figure 2. *Left:* Dynamically evolved model clusters at $z = 0$ (*solid*), compared to the Galactic globular clusters (*dashed*). Dotted histogram shows the combined initial masses of model clusters with $M > 10^5$ M$_\odot$ formed at all epochs, including those that did not survive until the present. *Right:* Evolution of the mass function at cosmic times of 1 Gyr ($z \approx 5.7$, *dotted*), 2 Gyr ($z \approx 3.2$, *dotted*), 5 Gyr ($z \approx 1.3$, *dashed*), 9 Gyr ($z \approx 0.5$, *dot-dashed*), and 13.5 Gyr ($z = 0$, *solid*).

5. Evolution of the Mass Function

Some of the old and low-mass clusters will be disrupted by the gradual escape of stars and will not appear in the observed sample. Figure 2 shows the dynamical evolution of the cluster mass function, as a result of stellar mass loss, tidal truncation, and two-body evaporation. Even though the model parameters were tuned to reproduce the metallicity, not the mass distribution, the mass function at $z = 0$ is consistent with the observed in the Galaxy. Majority of the disrupted clusters were blue clusters that formed in early low-mass progenitors.

Right panel of Figure 2 illustrates the interplay between the continuous buildup of massive clusters ($M > 10^5$ M$_\odot$) and the dynamical erosion of the low-mass clusters ($M < 10^5$ M$_\odot$). Expecting that most clusters below 10^5 M$_\odot$ would eventually be disrupted, we did not track their formation in the model. Instead, the low end of the mass function is built by the gradual evaporation of more massive clusters. Note that most of the clusters were not formed until the universe was 2 Gyr old, corresponding to $z \approx 3$. The fraction of clusters formed before $z \approx 6$, when cosmic hydrogen was reionized, is small.

An exciting prediction of the model is a high fraction of galaxy stellar mass locked in star clusters at $z > 3$: $M_{GC}/M_* \approx 10 - 20\%$. This fraction declines steadily with time and reaches 0.1% at the present epoch.

References

Beasley, M. A., Baugh, C. M., Forbes, D. A. *et al.* 2002, *MNRAS*, 333, 383

Dotter, A., Sarajedini, A., Anderson, J. *et al.* 2010, *ApJ*, 708, 698

Kravtsov, A. V. & Gnedin, O. Y. 2005, *ApJ*, 623, 650

Moore, B., Diemand, J., Madau, P., Zemp, M., & Stadel, J. 2006, *MNRAS*, 368, 563

Muratov, A. L. & Gnedin, O. Y. 2010, *ApJ*, 718, 1266

Peng, E. W., Jordán, A., Côté, P. *et al.* 2008, *ApJ*, 681, 197

Prieto, J. L. & Gnedin, O. Y. 2008, *ApJ*, 689, 919

Computational Star Formation
Proceedings IAU Symposium No. 270, 2011
J. Alves, B.G. Elmegreen, J. M. Girart & V. Trimble, eds.

doi:10.1017/S1743921311000688

The formation of super-star clusters in disk and dwarf galaxies

Carsten Weidner[1], Ian A. Bonnell[1] and Hans Zinnecker[2,3]

[1] SUPA, School of Physics & Astronomy, University of St Andrews, North Haugh, St Andrews, Fife KY16 9SS, UK
email: cw60@st-andrews.ac.uk, iab1@st-andrews.ac.uk

[2] Astrophysikalisches Institut Potsdam, An der Sternwarte 16, D-14482 Potsdam, Germany
[3] SOFIA Science Center, University of Stuttgart, Institut für Raumfahrtsyteme, Pfaffenwaldring 31, D-70569 Stuttgart, Germany
email: zinnecker@dsi.uni-stuttgart.de

Abstract. Super-star clusters are probably the largest star-forming entities in our local Universe, containing hundreds of thousands to millions of young stars usually within less than a few parsecs. While no such systems are known in the Milky Way (MW), they are found especially in pairs of interacting galaxies but also in some dwarf galaxies like R 136 in the Large Magelanic Cloud (LMC). With the use of SPH calculations we show that a natural explanation for this phenomenon is the presence of shear in normal spiral galaxies which facilitates the formation of low-density loose OB associations from giant molecular clouds (GMC) instead of dense super-star clusters. In contrast, in interacting galaxies and in dwarf galaxies, regions can collapse without having a large-scale sense of rotation. This lack of rotational support allows the giant molecular clouds to concentrate into a single, dense and gravitationally bound system.

Keywords. ISM: clouds – open clusters and associations: general – galaxies: star clusters: general – galaxies: star formation

---

1. Introduction

The formation of stars does not take place in isolation but in loose groups or embedded clusters in Giant Molecular Clouds, each cluster containing a dozen to many million of stars (Lada & Lada 2003). Such embedded clusters need not be bound structures and actually ~90% of them will dissolve within about 10 Myr (Lada & Lada 2003) due to gas expulsion and internal N-body effects.

Interestingly, many so-called super-star clusters ($M_{\rm cluster} \geqslant 10^5\ M_{\odot}$) are not found in star-forming disk galaxies but in relatively small dwarf galaxies. Especially, when excluding merger and interacting galaxies, there seems to be a trend with galaxy type and mass of the most-massive young (< 50 Myr) star cluster (Fig. 1).

Besides the total mass, one main difference between dwarf galaxies, disk galaxies and interaction regions of galaxies is the amount of shear acting on GMCs in them. Disk galaxies are rotationally supported systems, resulting in relatively large shear forces on GMCs. Dwarf galaxies show far less amounts of shear. Interacting galaxies with tidally driven structure also have low shear in the tidal arms and the interaction region.

The presence of retrograde rotating GMCs in spiral galaxies is sometimes seen as evidence that the formation and evolution of GMCs is not or only weakly connected to the shear/rotation of spiral arms. But Dobbs (2008) has shown that spiral shock models of GMC formation produce both prograde and retrograde rotating GMCs due to the random nature of the coagulation process.

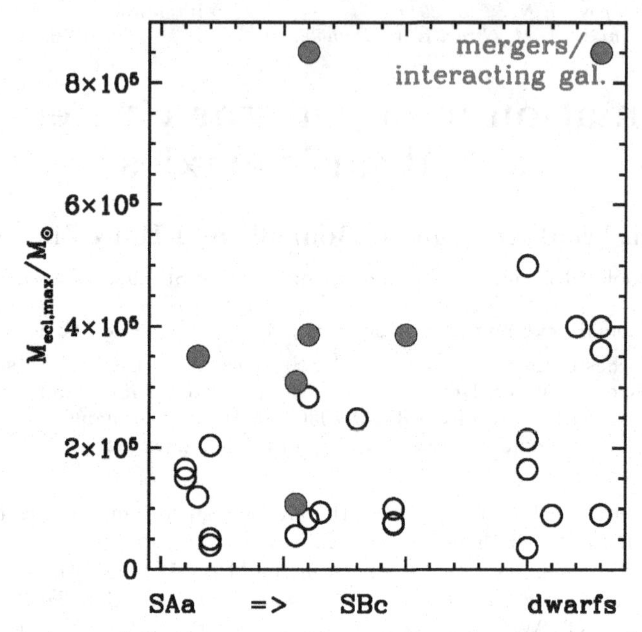

Figure 1. Dependence of the most-massive young cluster in galaxies on the galaxy type.

2. The Model

The simulations of the gravitational collapse of GMCs were carried out using a 3D-SPH code. The thermal and turbulent energies are initially significantly subvirial and therefore the clouds collapse rapidly due to gravitation. The initial temperature of the gas is about 100 K and the gas is allowed to cool down to $\approx$ 10 K due to a Larson equation-of-state with $\gamma = 0.75$ (Larson 2005). 10^6 SPH-particles are used to model the 10^6 $M_\odot$ gas within 50 pc cloud radius. Star formation is modelled by sink particles, which can grow through accretion of infalling gas (SPH particles) and interact gravitationally with the rest of the simulation.

The calculations are evolved for about one free-fall time of the GMC. With a radius of $R_{\rm GMC} = 50$ pc and a mass $M_{\rm GMC} = 10^6$ $M_\odot$ the free-fall time is $t_{\rm ff} = 5.9$ Myr in all cases considered here.

Four different models are calculated with identical initial conditions but four different levels of shear, corresponding to solid body rotation with angular velocities of $\Omega = 0$, $\Omega = 2 \times 10^{-15}, 5 \times 10^{-15}$ and 10^{-14} rad s^{-1}.

For MW GMCs rotations rates of $\Omega_{\rm MW} \sim 3 - 6 \times 10^{-15}$ rad s^{-1} are observed (Bissantz *et al.* 2003) while for the LMC the rate is significantly lower with $\Omega_{\rm LMC} \sim 6 \times 10^{-16}$ rad s^{-1} (Alves & Nelson 2000).

Models 2-4 represent typical conditions expected in spiral galaxies.

The simulations presented here do not include any feedback from supernovae, radiation or stellar winds. Though, the effects of ionizing radiation (Dale *et al.* 2005) and stellar winds (Dale & Bonnell 2008) have been studied before. We note that the inclusion of these sources of feedback in the above models did not have a significant effect on the star formation rate or efficiency.

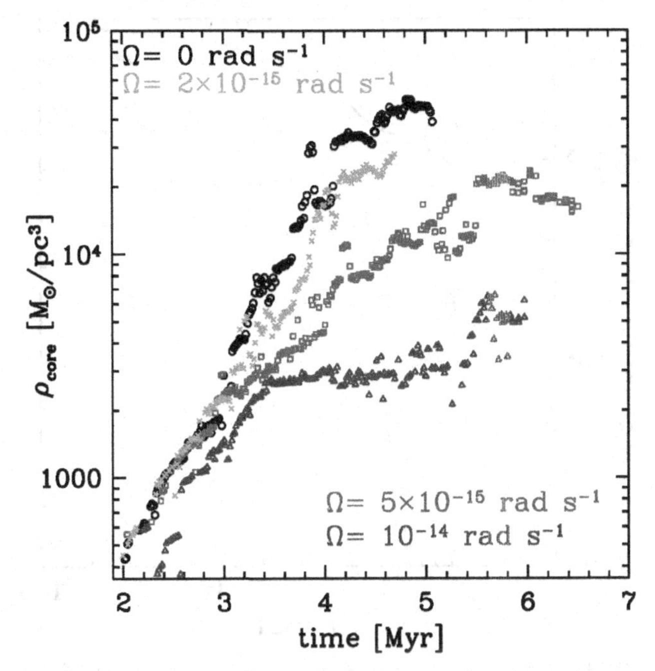

Figure 2. Temporal evolution of mass density in a core with $r_{\rm core}$ = 1.0 pc. The black dots mark model 1, the turquoise crosses model 2, the red boxes model 3 and the blue triangles model 4. The final densities are about 3 to 15 times larger for the no shear run than those in the runs with shear.

3. Results

In the models with low or no shear, more sinks are formed and they are much more highly concentrated towards the center of the cloud. In Figure 2 is shown how the density of mass in sinks evolves with time. The mass density is calculated in a sphere of $r_{\rm core}$ = 1.0 pc around the center of mass of all sinks. After about 3 Myr the models start to diverge as rotational support in the shear models halts the central collapse while in the absence of shear the formation of a dense core occurs. The no-shear model reaches average central densities in excess of $6 \times 10^4 M_\odot$ pc^{-3}, up to 15 times larger than the runs with shear.

An other indicator for the difference between models with and without shear can be seen in Figure 3. Here the radial distribution of the mass in sinks is shown at a time of 4.3 Myr. In the model without shear, a cluster forms with nearly $10^5 M_\odot$ inside 0.1 pc whereas the models with shear have a much more distributed population and much less significant central condensations. Though the total stellar masses formed in all four cases does not differ drastically between the models. Therefore the star-formation efficiencies (SFE = $M_{\rm sinks}/M_{\rm GMC}$) are relatively similar, with values between 20 and 40%. It is simply the distribution of the resultant stellar populations which are very different.

4. Conclusions

The models show that the formation of super-star clusters depends crucially on the shear content in the pre-collapse giant molecular cloud. Massive clouds in slowly and non rotating environments, as to be expected in dwarf galaxies like the LMC and in interaction regions of colliding galaxies, tend to form a massive, centrally condensed cluster, in agreement with the results by Escala & Larson (2008). GMC's in disk galaxies

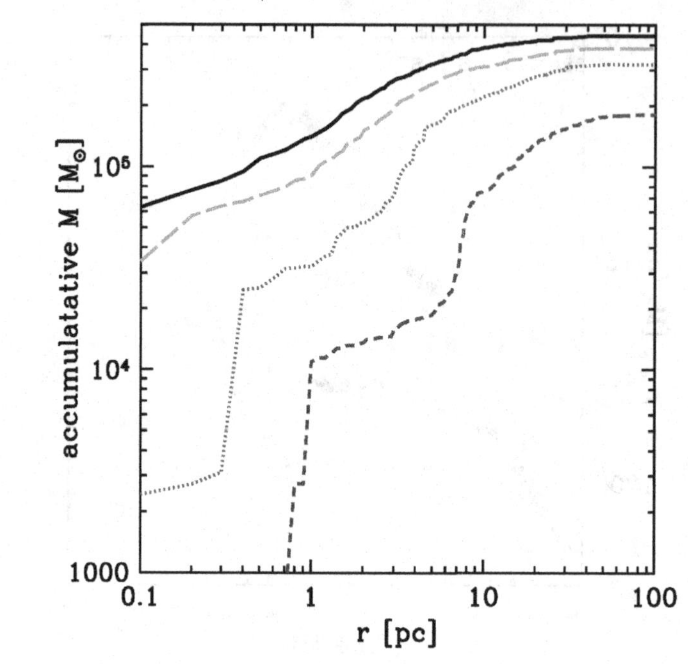

Figure 3. The radial dependence of the accumulated mass for all four simulations after 4.3 Myr for the no shear model (solid black line), the light shear model (long-dashed turquoise line), the intermediate shear one (dotted red line) and the high shear model (dashed blue line).

are more likely to fragment and form a system of smaller clusters or structures more like OB associations. Globular clusters are therefore unlikely to have formed in the disk of galaxies. They may originate from an initial monolithic collapse which formed the bulge of a galaxy, a major merging event, or they have formed in dwarf galaxies which have then been accreted later on (Zinnecker *et al.* 1988).

References

Alves, D. R. & Nelson, C. A. 2000, *ApJ*, 542, 789

Bissantz, N., Englmaier, P., & Gerhard, O. 2003, *MNRAS*, 340, 949

Dale, J. E. & Bonnell, I. A. 2008, *MNRAS*, 391, 2

Dale, J. E., Bonnell, I. A., Clarke, C. J., & Bate, M. R. 2005, *MNRAS*, 358, 291

Dobbs, C. L. 2008, *MNRAS*, 391, 844

Escala, A. & Larson, R. B. 2008, *ApJ*, 685, L31

Lada, C. J. & Lada, E. A. 2003, *ARA&A*, 41, 57

Larson, R. B. 2005, *MNRAS*, 359, 211

Zinnecker, H., Keable, C. J., Dunlop, J. S., Cannon, R. D., & Griffiths, W. K. 1988, in IAU Symposium, Vol. 126, The Harlow-Shapley Symposium on Globular Cluster Systems in Galaxies, ed. J. E. Grindlay & A. G. D. Philip, 603–+

Computational Star Formation
Proceedings IAU Symposium No. 270, 2011
J. Alves, B.G. Elmegreen, J. M. Girart & V. Trimble, eds.

doi:10.1017/S174392131100069X

GRAPE Accelerators

Junichiro Makino[1,2,3]

[1]Division of Theoretical Astronomy, National Astronomical Observatory of Japan, 2–21–1 Osawa, Mitaka-shi, Tokyo 181–8588.
[2]Center for Computational Astrophysics, National Astronomical Observatory of Japan, 2–21–1 Osawa, Mitaka-shi, Tokyo 181–8588
[3]School of Physical Sciences, Graduate University of Advanced Study (SOKENDAI), 2–21–1 Osawa, Mitaka-shi, Tokyo 181–8588
email: `makino@cfca.jp`

Abstract. I'll overview the past, present, and future of the GRAPE project, which started as the effort to design and develop specialized hardware for gravitational N-body problem. The current hardware, GRAPE-DR, has an architecture quite different from previous GRAPEs, in the sense that it is a collection of small, but programmable processors, while previous GRAPEs had hardwired pipelines. I'll discuss pros and cons of these two approaches, comparisons with other accelerators and future directions.

Keywords. methods: n-body simulations — methods: numerical

1. Introduction

In many simulations in astrophysics, it is necessary to solve gravitational N-body problems. In some cases, such as the study of formation of galaxies or stars, it is important to treat non-gravitational effects such as the hydrodynamical interaction, radiation, and magnetic fields, but in these simulations calculation of gravity is usually the most time-consuming part.

To solve the gravitational N-body problem, one needs to calculate the gravitational force on each body (particle) in the system from all other particles in the system. There are many ways to do so, and if relatively low accuracy is sufficient, one can use the Barnes-Hut tree algorithm (Barnes & Hut 1986) or FMM(Greengard and Rokhlin 1987). Even with these schemes, the calculation of the gravitational interaction between particles (or particles and multipole expansions of groups of particles) is the most time-consuming part of the calculation. Thus, one can greatly improve the speed of the entire simulation, just by accelerating the speed of the calculation of particle-particle interaction. This is the basic idea behind GRAPE computers.

The basic idea is shown in figure 1. The system consists of a host computer and special-purpose hardware, and the special-purpose hardware handles the calculation of gravitational interaction between particles. The host computer performs other calculations such as the time integration of particles, I/O, and diagnostics.

2. History

GRAPE Project was started in 1988. The first machine completed, the GRAPE-1 (Ito *et al.* 1990), was a single-board unit on which around 100 IC and LSI chips were mounted and wire-wrapped. The pipeline processor of GRAPE-1 was implemented using commercially available IC and LSI chips This choice was a natural consequence of the fact

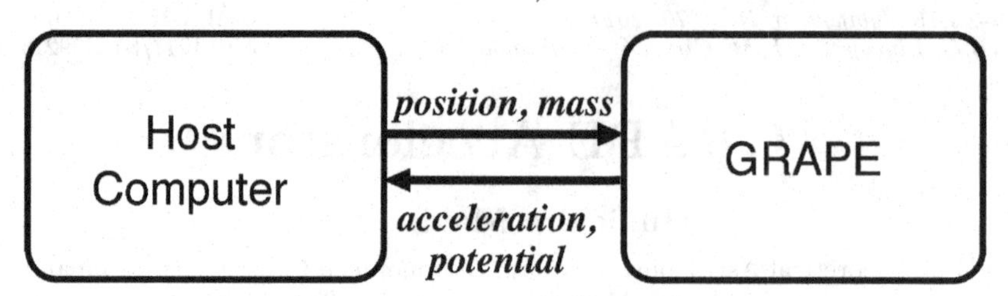

Figure 1. Basic structure of a GRAPE system.

that project members lacked both money and experience to design custom LSI chips. In fact, none of the original design and development team of GRAPE-1 had the knowledge of electronic circuits more than what was learned in basic undergraduate courses for physics students.

For GRAPE-1, an unusually short word format was used, to make the hardware as simple as possible. Except for the first subtraction of the position vectors (16-bit fixed point) and final accumulation of the force (48-bit fixed point), all operations are done in 8-bit logarithmic format, in which 3 bits are used for the "fractional" part. This choice simplified the hardware significantly. The use of extremely short word format in GRAPE-1 was based on the detailed theoretical analysis of error propagation and numerical experiment (Makino *et al.* 1990).

GRAPE-2 was similar to GRAPE-1A, but with much higher numerical accuracy. In order to achieve higher accuracy, commercial LSI chips for floating-point arithmetic operations such as TI SN74ACT8847 and Analog Devices ADSP3201/3202 were used. The pipeline of GRAPE-2 processes the three components of the interaction sequentially. So it accumulates one interaction in every three clock cycles. This approach was adopted to reduce the circuit size. Its speed was around 40 Mflops, but it is still much faster than workstations or minicomputers at that time.

GRAPE-3 was the first GRAPE computer with custom LSI chip. The number format was the combination of the fixed point and logarithmic format similar to what were used in GRAPE-1. The chip was fabricated using 1μm design rule by National Semiconductor. The number of transistors on a chip was 110K. The chip operated at 20MHz clock speed, offering a speed of about 0.8 Gflops. Printed-circuit board with 8 chips were mass-produced, for a speed of 6.4 Gflops per board. Thus, GRAPE-3 was also the first GRAPE computer to integrate multiple pipelines into a system. Also, GRAPE-3 was the first GRAPE computer to be manufactured and sold by a commercial company. Nearly 100 copies of GRAPE-3 have been sold to more than 30 institutes (more than 20 outside Japan).

With GRAPE-4, a high-accuracy pipeline was integrated into one chip. This chip calculates the first time derivative of the force, so that a fourth-order Hermite scheme (Makino & Aarseth 1992) can be used. Here, again, a serialized pipeline similar to that of GRAPE-2 was used. The chip was fabricated using 1μm design rule by LSI Logic. Total transistor count was about 400K.

The completed GRAPE-4 system consisted of 1728 pipeline chips (36 PCB boards each with 48 pipeline chips). It operated on 32 MHz clock, delivering the speed of 1.1 Tflops. Technical details of machines from GRAPE-1 through GRAPE-4 can be found in our book (Makino & Taiji 1998) and reference therein.

GRAPE-5 (Kawai *et al.* 2000) was an improvement over GRAPE-3. It integrated two full pipelines which operate on 80 MHz clock. Thus, a single GRAPE-5 chip offered a speed 8 times more than that of the GRAPE-3 chip, or the same speed as that of an

Table 1. History of GRAPE project

GRAPE-1	(89/4 — 89/10)	310 Mflops, low accuracy
GRAPE-2	(89/8 — 90/5)	50 Mflops, high accuracy(32bit/64bit)
GRAPE-1A	(90/4 — 90/10)	310 Mflops, low accuracy
GRAPE-3	(90/9 — 91/9)	18 Gflops, high accuracy
GRAPE-2A	(91/7 — 92/5)	230 Mflops, high accuracy
HARP-1	(92/7 — 93/3)	180 Mflops, high accuracy Hermite scheme
GRAPE-3A	(92/1 — 93/7)	8 Gflops/board some 80 copies are used all over the world
GRAPE-4	(92/7 — 95/7)	1 Tflops, high accuracy Some 10 copies of small machines
MD-GRAPE	(94/7 — 95/4)	1Gflops/chip, high accuracy programmable interaction
GRAPE-5	(96/4 — 99/8)	5Gflops/chip, low accuracy
GRAPE-6	(97/8 — 02/3)	64 Tflops, high accuracy

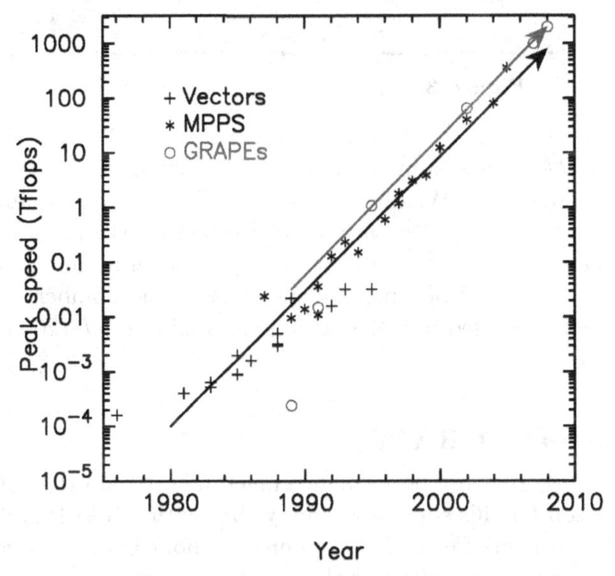

Figure 2. The evolution of GRAPE and general-purpose parallel computers. The peak speed is plotted against the year of delivery. Open circles, crosses and stars denote GRAPEs, vector processors, and parallel processors, respectively.

8-chip GRAPE-3 board. GRAPE-5 was awarded the 1999 Gordon Bell Prize for price-performance. The GRAPE-5 chip was fabricated with 0.35μm design rule by NEC.

Table 1 summarizes the history of GRAPE project. Figure 2 shows the evolution of GRAPE systems and general-purpose parallel computers. One can see that evolution of GRAPE is faster than that of general-purpose computers.

The GRAPE-6 was essentially a scaled-up version of GRAPE-4(Makino *et al.* 1997), with the peak speed of around 64 Tflops. The peak speed of a single pipeline chip was 31 Gflops. In comparison, GRAPE-4 consists of 1728 pipeline chips, each with 600 Mflops. The increase of a factor of 50 in speed was achieved by integrating six pipelines into one chip (GRAPE-4 chip has one pipeline which needs three cycles to calculate the force from one particle) and using 3 times higher clock frequency. The advance of the device technology (from 1μm to 0.25μm) made these improvements possible. Figure 3 shows the processor chip delivered in early 1999. The six pipeline units are visible.

Figure 3. The GRAPE-6 processor chip.

The completed GRAPE-6 system consisted of 64 processor boards, grouped into 4 clusters with 16 boards each. Within a cluster, 16 boards are organized in a 4 by 4 matrix, with 4 host computers. They are organized so that the effective communication speed is proportional to the number of host computers. In a simple configuration, the effective communication speed becomes independent of the number of host computers. The details of the network used in GRAPE-6 is in Makino *et al.* (2003).

3. LSI economics and GRAPE

GRAPE has achieved the cost performance much better than that of general-purpose computers. One reason for this success is simply that with GRAPE architecture one can use practically all transistors for arithmetic units, without being limited by the memory wall problem. Another reason is the fact that arithmetic units can be optimized to their specific uses in the pipeline. For example, in the case of GRAPE-6, the subtraction of two positions is performed in 64-bit fixed point format, not in floating-point format. Final accumulation is also done in fixed point. In addition, most of the arithmetic operations to calculate the pairwise interactions are done in single precision. These optimizations made it possible to pack more than 300 arithmetic units into a single chip with less than 10M transistors. The first microprocessor with fully-pipelined double-precision floating-point unit, Intel 80860, required 1.2M transistors for two (actually one and half) operations. Thus, the number of transistors per arithmetic unit of GRAPE is smaller by more than a factor of 10. When compared with more recent processors, the difference becomes even larger. The Fermi processor from NVIDIA integrates 512 arithmetic units (adder and multiplier) with 3G transistors. Thus, it is five times less efficient than Intel 80860, and nearly 100 times less efficient than GRAPE-6.

However, there is another economical factor. As the silicon semiconductor technology advances, the initial cost for the design and fabrication of custom chips increases. In 1990, the initial cost for a custom chip was around 100K USD. By the end of the 1990s, it has become higher than 1M USD. By 2010, the initial cost of a custom chip is around

10M USD. Thus, it has become difficult to get a budget large enough to make a custom chip, which has rather limited range of applications.

There are several possible solutions. One is to reduce the initial cost by using FPGA (Field-Programmable Gate Array) chips. An FPGA chip consists of a number of "programmable" logic blocks (LBs) and also "programmable" interconnections. A LB is essentially a small lookup table with multiple inputs, augmented with one flip-flop and sometimes full-adder or more additional circuits. The lookup table can express any combinatorial logic for input data, and with flip-flop it can be part of a sequential logic. Interconnection network is used to make larger and more complex logic, by connecting LBs. The design of recent FPGA chips has become much more complex, with large functional units like memory blocks and multiplier (typically 18×18 bits) blocks.

Unfortunately, because of the need for the programmability, the size of the circuit that can fit into an FPGA chip is much smaller than that for a custom LSI, and the speed of the circuit is also slower. In order to be competitive, it is necessary to use much shorter word length. GRAPE architecture with reduced accuracy is thus an ideal target for FPGA-based approach. Several successful approaches have been reported (Hamada *et al.* 1999, Kawai & Fukushige 2006).

4. GRAPE-DR

Another solution for the problem of the high initial cost is to widen the application range by some way to justify the high cost. With GRAPE-DR project Makino *et al.*(2007), we followed this approach.

With GRAPE-DR, the hardwired pipeline processor of previous GRAPE systems were replaced by a collection of simple SIMD programmable processors. The internal network and external memory interface were designed so that it could emulate GRAPE processors efficiently and could be used for several other important applications, including the multiplication of dense matrices.

GRAPE-DR is an acronym of "Greatly Reduced Array of Processor Elements with Data Reduction". The last part, "Data Reduction", means that it has an on-chip tree network which can do various reduction operations such as summation, max/min and logical and/or.

The GRAPE-DR project was started in FY 2004, and finished in FY 2008. The GRAPE-DR processor chip consists of 512 simple processors, which can operate at the clock cycle of 500MHz, for 512 Gflops of single precision peak performance (256 Gflops double precision). It was fabricated with TSMC 90nm process and the size is around 300mm^2. The peak power consumption is around 60W. The GRAPE-DR processor board (figure 4) houses 4 GRAPE-DR chips, each with its own local DRAM chips. It communicates with the host computer through Gen1 16-lane PCI-Express interface.

This card gives the theoretical peak performance of 819 Gflops (in double precision) at the clock speed of 400 MHz. The actual performance numbers are 640 Gflops for matrix-matrix multiplication, 430 Gflops for LU-decomposition, and 500 Gflops for direct N-body simulation with individual timesteps (figure 5). These numbers are typically a factor of two or more better than the best performance number so far reported with GPGPUs.

In the case of parallel LU decomposition, the measured performance was 24 Tflops on 64-board, 64-node system. The average power consumption of this system during the calculation was 29KW, and thus performance per Watt is 815 Mflops/W. This number is listed as No. 1 in the Little Green 500 list of June 2010. Thus, from a technical point of

Figure 4. The GRAPE-DR processor board.

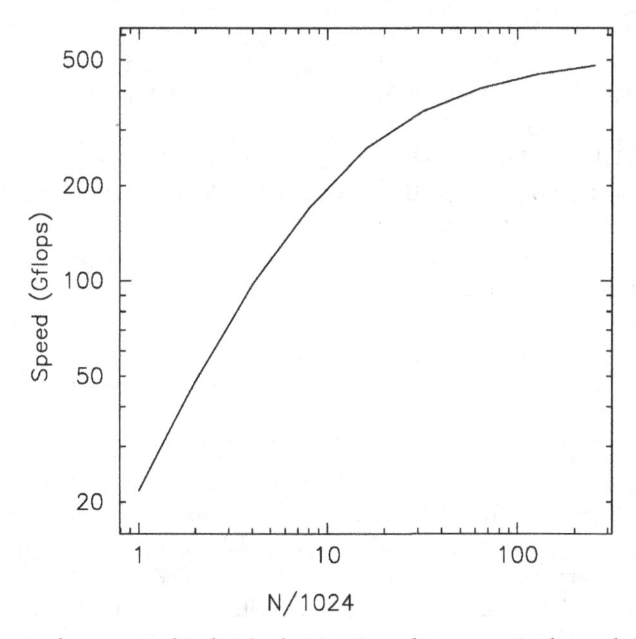

Figure 5. The performance of individual-timestep scheme on single-card GRAPE-DR in Gflops, plotted as a function of the number of particles.

view, we believe the GRAPE-DR project is highly successful, in making multi-purpose computers with highest single-card performance and highest performance-per-watt.

Whether or not the approach like GRAPE-DR will be competitive with other approaches, in particular GPGPUs, is at the time of writing rather unclear. The reason is simply that the advantage over GPGPUs is not quite enough, primarily because of the low production cost of GPGPUs. On the other hand, the transistor efficiency of general-purpose computers, and that of GPUs, have been decreasing for the last 20 years and probably will continue to do so for the next 10 years or so. GRAPE-DR can retain its

Figure 6. The GRAPE-DR cluster.

efficiency when it is implemented with more advanced semiconductor technology, since, as in the case of GRAPE, one can use the increased number of transistors to increase the number of processor elements. Thus, it might remain competitive.

5. Future directions

In hindsight, 1990s was a very good period for the development of special-purpose architecture such as GRAPE, because of two reasons. First, the semiconductor technology reached to the point where many floating-point arithmetic units can be integrated into a chip. Second, the initial design cost of a chip was still within the reach of fairly small research projects in basic science.

By now, semiconductor technology reached to the point that one could integrate thousands of arithmetic units into a chip. On the other hand, the initial design cost of a chip has become too high.

The use of FPGAs and the GRAPE-DR approach are two examples of the way to tackle the problem of increasing initial cost. However, unless one can keep increasing the budget, GRAPE-DR approach is not viable, simply because it still means exponential increase in the initial, and therefore total, cost of the project.

On the other hand, such increase in the budget might not be impossible, since the field of computational science as a whole is becoming more and more important. Even though a supercomputer is expensive, it is still much less expensive compared to, for example, particle accelerators or space telescopes. Of course, computer simulation cannot replace the real experiments of observations, but computer simulations have become essential in many fields science and technology.

In addition, there are several technologies available in between FPGAs and custom chips. One is what is called "structured ASIC". It requires customization of typically just one metal layer, resulting in large reduction in the initial cost. The number of gates one can fit into the given silicon area falls between those of FPGAs and custom chips. We are currently working on a new fully-pipelined system, based on this structured ASIC.

The price of the chip is not very low, but in the current plan it gives extremely good performance for very low energy consumption.

References

Barnes, J. & Hut, P. 1986, *Nature*, 324, 446

Greengard, L. & Rokhlin, V. 1987, Journal of Computational Physics, 73, 325

Hamada, T., Fukushige, T., Kawai, A., & Makino, J. 1999, PROGRAPE-1: A Programmable, Multi-Purpose Computer for Many-Body Simulations, submitted to *PASJ*

Ito, T., Makino, J., Ebisuzaki, T., & Sugimoto, D. 1990, Computer Physics Communications, 60, 187

Kawai, A. & Fukushige, T. 2006, $158/GFLOP Astrophysical N-Body Simulation with a Reconfigurable Add-in Card and a Hierarchical Tree Algorithm

Kawai, A., Fukushige, T., Makino, J., & Taiji, M. 2000, *PASJ*, 52, 659

Makino, J. & Aarseth, S. J. 1992, *PASJ*, 44, 141

Makino, J., Hiraki, K., & Inaba, M. 2007, in Proceedings of SC07, ACM, (Online)

Makino, J., Ito, T., & Ebisuzaki, T. 1990, *PASJ*, 42, 717

Makino, J., Fukushige, T,. Koga, M., & Namura, K., T. 2003, *PASJ*, 55, 1163

Makino, J. & Taiji, M. 1998, Scientific Simulations with Special-Purpose Computers — The GRAPE Systems (Chichester: John Wiley and Sons)

Makino, J., Taiji, M., Ebisuzaki, T., & Sugimoto, D. 1997, *ApJ*, 480, 432

Makino, J., Fukushige, T,. Koga, M., & Namura, K., T. 2003, *PASJ*, 55, 1163

Computational Star Formation
Proceedings IAU Symposium No. 270, 2011
J. Alves, B.G. Elmegreen, J. M. Girart & V. Trimble, eds.

doi:10.1017/S1743921311000706

Numerical Cosmology powered by GPUs

Dominique Aubert[1]

[1]Observatoire Astronomique, Universite de Strasbourg, CNRS,UMR 7550
11 rue de l'Universite, 67000 Strasbourg France
email: `dominique.aubert@astro.unistra.fr`

Abstract. Graphics Processing Units (GPUs) offer a new way to accelerate numerical calculations by means of on-board massive parallelisation. We discuss two examples of GPU implementation relevant for cosmological simulations, an N-Body Particle-mesh solver and a radiative transfer code. The latter has also been ported on multi-GPU clusters. The range of acceleration (x30-x80) achieved here offer bright perspective for large scale simulations driven by GPUs.

Keywords. methods: n-body simulations, numerical, cosmology

---

1. GPUs for scientific applications

The last few years have seen the rise of a new technique for parallel calculations which relies on graphics processing units (GPUs hereafter). This type of hardware was originally designed and optimized for applications related to graphics display such as 3D rendering. Compared to the classical CPUs, the GPU architecture favors 'calculation units' against cache and control flow units. It originates from the fact that graphical calculations are numerically intensive but the same set of operations is identically applied to a large number of data (like e.g. polygons rotations). Ideally, the lack of cache is compensated by the intensity of calculations and control flow is unnecessary since the same operations are applied to all data. Furthermore, graphics boards typically contain a few hundreds of multicore units, implying that a large number of clone calculations can be dealt in parallel. If a scientific application can fit in this model, one can expect a significant acceleration.

Due to their original field of applications, GPUs were at first limited to single-float calculations but the latest generation boards are also able to perform double-precision calculations. Single boards can achieve a peak performance of 1.5 Tflop/s in single precision and 500 Gflop/s in double precision. Theoretical bandwidth can be as high as 180 GB/s. Applications which have been ported on GPUs experience typically a x10 to x100 acceleration compared to a single CPU-core calculation. The final improvement rate is dependent on several factors: the suitability of a given application to the GPU programming model, the amount of communication between the host PC and the board and of course the level of optimisation such as memory access patterns.

Several options exists to interact with this hardware for scientific calculations. Originally, the scientists literally wrote their applications in order to mimic graphics calculations, with e.g. shaders. Hopefully, more user friendly solutions exists today. Currently, the most popular one is CUDA, written by the Nvidia company exclusively for its devices. CUDA is a toolkit which contains a compiler plus several libraries and development tools. In practice, one writes a regular C or Fortran code coupled to a set of libraries which provide primitives to interact with the devices, mostly for communications, launching calculations and organizing the parallel task among the data. More recently, the OpenCL standard has been defined by the Khronos group: this standard allows to program on any multicore architecture in an unified way and is not limited to the devices designed by a specific company.

2. Particle-Mesh N-Body integrator

The particle-mesh (PM) technique relies on a grid-based description of the gravitational field created by a distribution of massive particles. This method is fast but suffers from a low spatial resolution and is usually coupled to a more accurate technique such as direct n-body (PPM) or tree-based calculations (TPM). Once the initial particle distribution is known, the density is calculated on a fixed grid. The Poisson equation is solved usually by means of Fourier-space technique or relaxation and the resulting potential provides the force field at the grid nodes. The force is interpolated at particle positions and the latter can be moved and prepared for the next timestep.

In Aubert, Amini & David 2009, we described a GPU implementation of this technique using CUDA and further technical details can be found in this publication. In our implementation, the Poisson equation can be solved either by FFT (using the cuFFT library) or by our own implementation of the multigrid relaxation technique. Interestingly most of the steps of a PM calculation consist in identical operations performed independently on a large set of data. For instance, updating the velocity of particles can easily be done in parallel. Also the restriction or prolongation operations in the multigrid solver can be performed cell-by-cell without any communication. Even the relaxation can be performed on the cells of the grid completely in parallel. Overall, any operation restricted to the particle-based description or the grid-based description can be fully executed by independent and parallel threads. As a consequence large accelerations (x50 -x100) can be achieved on these operations on GPUs. This is illustrated by the Figure 1.

However, the PM technique implies that several switches between the particle based and the grid based description should be performed. An example is the histogramming step, where particles are projected on a grid using e.g. a CIC interpolating scheme. A first issue is related to the fact that particles can hit the same cell, implying that the number of hits in a given cell cannot be updated in parallel. Moreover particles are distributed 'randomly' in the computational volume leading to non ordered memory accesses in order to update the grid, killing the GPU performances. The same issues are encountered when interpolating the force computed on a grid back on particles. Interestingly, these points were already tackled when vector-based architectures were the norm and a way to circumvent these problems are described e.g. in Ferrel & Bertschinger (1994). Using brute force with atomic operations which serialize

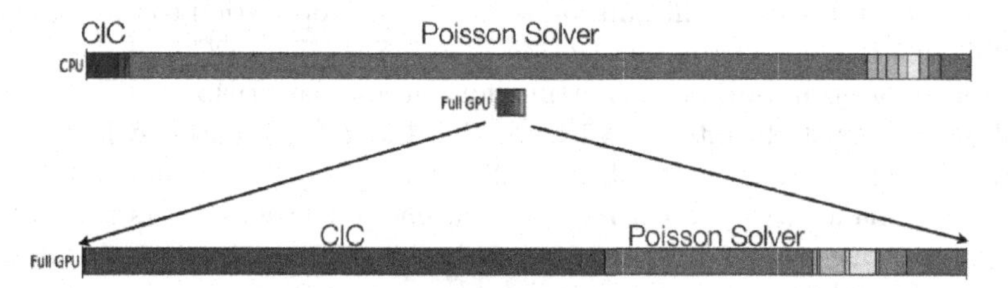

Figure 1. A typical PM timestep on CPU and GPU. Dark blue stand for the time spent in the histogramming step while light blue stands for the resolution of the Poisson equation. The other colors stand for force calculation and position update and are negligible. The bars are on scale and the GPU is globally 30 times faster than CPU, but the relative importance of CIC and Poisson resolution is different.

accesses when required is also an option. In any case the operation remains quite GPU-unfriendly.

As a consequence, the relative importance of the tasks involved in PM calculations differs from classic CPU calculations. Figure 1 compare the two cases: overall a x30 acceleration is achieved on GPU but the importance of the Poisson resolution is reduced compared to the CPU implementation, whereas the CIC-histogramming step becomes dominant. It is due to the fact that the Poisson resolution experiences a better benefit from the GPU implementation than the CIC step. The Poisson resolution fits into the GPU programming model while the histogramming is intrinsically more difficult to parallelise.

3. Cosmological Radiative transfer on multi-GPUs

The second application is described in full details in Aubert & Teyssier (2008) and Aubert & Teyssier (2010). It consists in a cosmological radiative transfer code, called ATON. Given a gas density distribution and a collection of sources, this code solves the radiative transfer using the moment-based M1 technique (Levermore 1984) and computes the chemistry and heating created by the ionizing radiation. Radiation is described as a fluid on a fixed grid and its transport is computed using Godunov-like techniques in an *explicit fashion*. Thanks to the latter choice, the radiative state of cell can be updated in parallel and using the same set of operations, fitting perfectly in the GPU programming model. Chemistry and photo-heating process are purely local and can also be computed in a fully parallel fashion. More generally all hyperbolic solvers are perfectly adequate for GPU-based calculations. For instance, using CUDA, ATON is 80x faster on a GeForce 8800 GTX than on a Opteron 2.7 GHz. It should be noted that such an acceleration is mandatory to use ATON: because of the explicit resolution, the Courant condition on the speed of light forces the calculation to operate on a large number of timesteps. Without GPUs, such a technique would be extremely limited in practice.

Another feature of ATON is that it runs on more than one GPU using an additional MPI layer. Practically, the GPUs are seen only by a single host but the host communicate through regular MPI communications. Each GPU is assigned a sub-volume of the global computational volume and the exchanges between subvolumes are implemented by exchanging ghost layers between the GPUs: the graphics boards send the data on their hosts, the hosts communicates and once ready they send the exchanged data back on their GPUs.

As an application we performed a study of the cosmological reionisation in a set 1024^3 cosmological simulations (see Aubert & Teyssier 2010). The number of required timesteps to evolve the system from $z \sim 15$ to $z = 5.5$ lie between 50 000 and 150 000 depending on the physical size of the simulated volume. These calculations were performed on 128 Tesla GPUs hosted by the Titane Supercomputer of the CEA. Thanks to the acceleration provided by GPUs, such calculations are completed in 3-10 hours. It allowed us to assess the issue of communication costs: since all the exchanged data must go through the PCI-bus, a legitimate question would be to understand how much is lost in communications. Globally, it should be noted that the speedup remains reasonably linear up to 128 devices. Furthermore, our measurements for different parallel configurations show that 10 - 15% of a calculation is spent in MPI+PCI communications. While significant, it remains reasonable and demonstrates the usability of GPUs for large scale multi-device calculations.

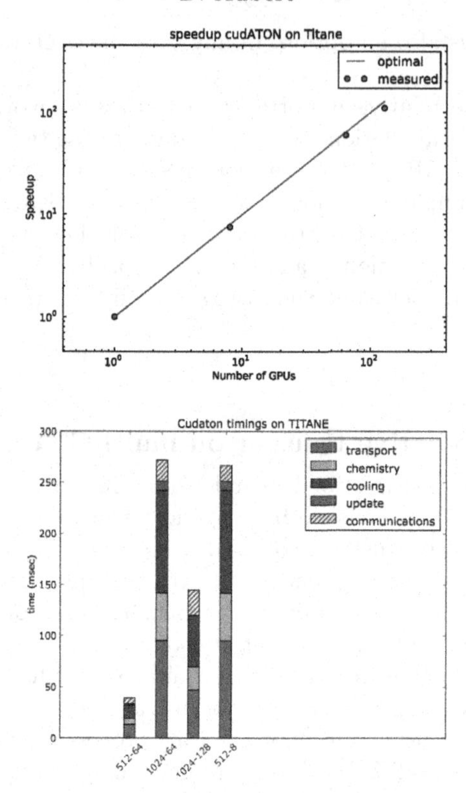

Figure 2. Top: speedup curve achieved by the radiative transfer code ATON as a function of the number of GPUs. Bottom: time spent in the different stages of radiative timestep of aton. The communications are limited to 15% fraction.

4. Conclusion

In addition to these two applications (PM and radiative transfer) we also implemented on GPUs a standard hydrodynamical solver on fixed grid, achieving an acceleration close to x100. It should be noted that such accelerations are by no means exceptional for GPU codes and are obtained in a large range of applications from graphics rendering, to medical applications or genomics. It should also be emphasized that GPU-programming is fairly easy to learn, however the difficulty lies in fitting existing applications into this specific programming model, which can be fairly different than the usual CPU programming, less flexible and closer to the hardware. These difficulties set aside, significant accelerations can be quickly obtained for a moderate developing cost.

References

Aubert, D., Amini, M., & David, R. 2009, *LNCS*,5544,874
Ferrel, E. & Bertschinger, E. 1994, *Int. J. Mod. Phys.*, 933
Aubert, D. & Teyssier, R. 2008, *MNRAS*, 387, 295
Aubert, D. & Teyssier, R. 2010, submitted to *ApJ*, arxiv:1004.2503
Levermore, C. 1984, *Journal of Quantitative Spectroscopy and Radiative Transfer*, 31, 149

Computational Star Formation
Proceedings IAU Symposium No. 270, 2011
J. Alves, B.G. Elmegreen, J. M. Girart & V. Trimble, eds.

doi:10.1017/S1743921311000718

GAMER with out-of-core computation

Hsi-Yu Schive[1,2,3], Yu-Chih Tsai[1], & Tzihong Chiueh[1,2,3]

[1]Department of Physics, National Taiwan University, 106, Taipei, Taiwan, R.O.C.
email: b88202011@ntu.edu.tw (Hsi-Yu Schive)

[2]Center for Theoretical Sciences, National Taiwan University, 106, Taipei, Taiwan, R.O.C.
[3]Leung Center for Cosmology and Particle Astrophysics, National Taiwan University, 106, Taipei, Taiwan, R.O.C.

Abstract. *GAMER* is a GPU-accelerated Adaptive-MEsh-Refinement code for astrophysical simulations. In this work, two further extensions of the code are reported. First, we have implemented the MUSCL-Hancock method with the Roe's Riemann solver for the hydrodynamic evolution, by which the accuracy, overall performance and the GPU versus CPU speed-up factor are improved. Second, we have implemented the out-of-core computation, which utilizes the large storage space of multiple hard disks as the additional run-time virtual memory and permits an extremely large problem to be solved in a relatively small-size GPU cluster. The communication overhead associated with the data transfer between the parallel hard disks and the main memory is carefully reduced by overlapping it with the CPU/GPU computations.

Keywords. gravitation, hydrodynamics, methods: numerical

---

1. Introduction

Novel use of the graphic processing unit (GPU) has becoming a promising technique in the computational astrophysics. The applications that have been reported include purely hydrodynamics, magnetohydrodynamics, gravitational lensing, radiation transfer, direct N-body calculation, particle-mesh method, hierarchical tree algorithm, and etc. Typically, one to two order of magnitudes performance improvements were reported (Schive *et al.* 2008, for example).

Schive *et al.* (2010) presented the first multi-GPU-accelerated Adaptive-MEsh-Refinement (AMR) code, named *GAMER*, which is dedicated for high-resolution astrophysical simulations. A GPU hydrodynamic solver and a GPU Poisson solver have been implemented in the code, while the AMR data structure is still manipulated by CPUs. An overall performance speed-up up to 12x was reported. In this work, two further extensions are implemented into *GAMER*, namely, the MUSCL-Hancock method (Toro 2009) with the Roe's Riemann solver (Roe 1981) for the hydrodynamic evolution, and the out-of-core computation. The latter uses parallel hard disks to increase the total amount of available virtual memory. By integrating the high computation performance of GPUs and the out-of-core technique, it provides an extremely efficient solution to increase both the simulation problem size and performance of the AMR simulations.

2. Extension I: hydrodynamic solver

In the previous work (Schive *et al.* 2010), the second-order relaxing total variation diminishing (TVD) method (Trac & Pen 2003) has been adopted, in which the three-dimensional evolution is achieved by the dimensional splitting method. Figure 1 shows the performance speed-up versus the number of GPUs, in which we compare the performance using the same number of GPUs (Tesla T10 GPU) and CPU cores (Xeon E5520).

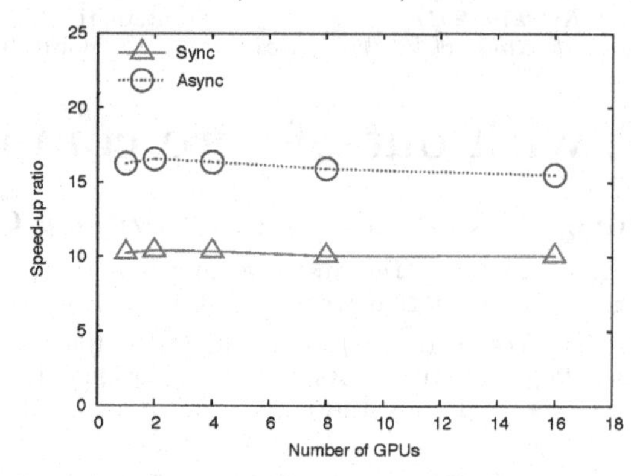

Figure 1. Overall performance speed-up versus the number of GPUs using the relaxing TVD scheme. The circles and triangles show the results with and without the concurrent execution between CPU and GPU, respectively.

The simulations are conducted in the GPU cluster installed in the National Astronomical Observatories, Chinese Academy of Sciences. We also compare the results with and without the concurrent execution between CPU and GPU, and a maximum speed-up up to 16.5x is demonstrated when the concurrency is enabled. We also notice that the speed-up factor only decreases slightly to 15.5x in the 16 GPUs/CPUs test, indicating that the network time is nearly negligible. Timing measurements show that the MPI data transfer takes less than 2% of the total simulation time.

To further enhance the capability of the code, in this work we have implemented a new GPU hydrodynamic solver based on the MUSCL-Hancock method. This approach contains four steps, namely, the spatial data reconstruction, the half-step prediction, the Riemann solver, and the full-step update. We apply the unsplit finite volume method for the three-dimensional evolution, and the Roe's solver is adopted for the Riemann problem. Comparing with the second-order relaxing TVD scheme, the MUSCL-Hancock method has two main advantages for the AMR+GPU implementation. First, it requires only a five-point stencil in each spatial direction, while the relaxing TVD scheme requires seven points. Since the operation of preparing the ghost-zone data is conducted by CPU, which has been shown to be more time expensive than the GPU hydrodynamic solver (Schive *et al.* 2010), reducing the size of stencil can directly lead to significant improvement of the overall performance. Second, the MUSCL-Hancock method has higher arithmetic intensity, and hence is more GPU-friendly. Factor of 55x performance speed-up is measured by comparing the GPU and CPU versions of this method.

Figure 2 compares the overall performance speed-up in purely baryonic cosmological simulations using the two different hydrodynamic schemes. The same gravity solver is adopted in the two cases. The performance is measured by using one NVIDIA GeForce 8800 GTX GPU and one AMD Athlon 64 X2 3800 CPU core. Clearly, the MUSCL-Hancock scheme achieves a superior performance improvement, in which a speed-up of 19.2x is demonstrated. Moreover, although in the CPU-only runs the MUSCL-Hancock method is more time-consuming, it is not the case when the GPU-acceleration is activated. Timing experiments show that the total execution time is actually reduced when we replace the GPU relaxing-TVD solver by the GPU MUSCL-Hancock solver, which results from the less computing time required for the ghost-zone preparation in CPU and the more efficient GPU kernel.

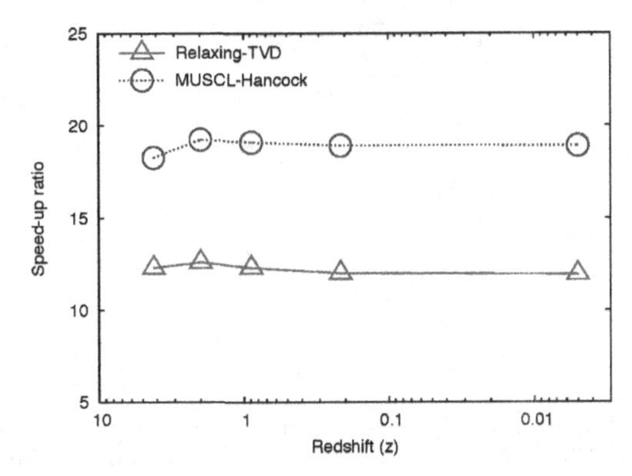

Figure 2. Overall performance speed-up in purely baryonic cosmological simulations using the MUSCL-Hancock scheme (circles) and the relaxing-TVD scheme (triangles), respectively.

3. Extension II: out-of-core computation

In *GAMER*, we have demonstrated that the performance of the AMR simulations can be highly improved by using GPUs. However, the simulation size is still limited by the total amount of main memory. To alleviate this limitation, we further implement the out-of-core technique, by which only a small portion of the simulation data needs to be loaded into the main memory while the rest of data remain stored in the hard disks. To increase the total I/O bandwidth in a single node, we evenly distribute the data in eight hard disks and perform the data transfer between the main memory and the eight disks concurrently. By doing so, a maximum bandwidth of 750 MB/s is achieved.

The parallelization in *GAMER* is based on the rectangular domain decomposition. To perform the out-of-core computation in a multi-node system, we let each computing node to work on a group of nearby sub-domains, and each of which will be assigned a different out-of-core rank that is similar to the concept of the MPI rank. Figure 3 shows a two-dimensional example of the domain decomposition. Different sub-domains within the same node are always evaluated sequentially, while sub-domains in different nodes can be evaluated in parallel. In each node, only the data of the sub-domain being advanced are loaded from the hard disks to the main memory. After the targeted sub-domain is advanced by one time-step, the updated data will be stored back to the hard disks, and the data of the next targeted sub-domain will be loaded into the main memory. Furthermore, to improve the efficiency of the out-of-core computation, the hard disk I/O time for one out-of-core rank is arranged to be overlapped with the computation for a different out-of-core rank in the same node.

Updating the buffer data of each sub-domain requires transferring data in between adjacent sub-domains. However, since different out-of-core ranks within the same computing node are calculated sequentially, we need a data transfer mechanism different from the MPI implementation. To this end, we have implemented two functions named *OOC_Send* and *OOC_Recv*, which are similar to the MPI functions *MPI_Send* and *MPI_Reve*, but use the hard disks as the data exchange buffer. For example, to send data from the out-of-core rank A to rank B, the former first invokes the function OOC_Send to store the transferring data in the hard disks. Afterward, rank B can invoke the function OOC_Recv with the correct data tag to load the transferring data from the hard disks, thus completing a single data transferring operation. On the other hand, the data transfer between different computing nodes is still accomplished by using the MPI functions.

To test the performance, we conduct single-node simulations with the 512^3 root level and five refinement levels, giving $16,384^3$ effective resolution. The total memory

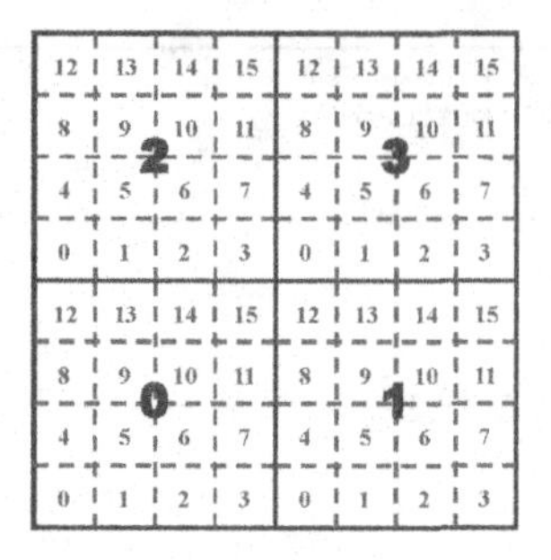

Figure 3. Domain decomposition of the parallelized out-of-core computation. The solid blue lines represent the sub-domain boundaries between different MPI ranks in different computing nodes, and the dashed red lines represent the sub-domain boundaries between different out-of-core ranks in the same computing node. The blue and red numbers stand for the MPI ranks and the out-of-core ranks, respectively.

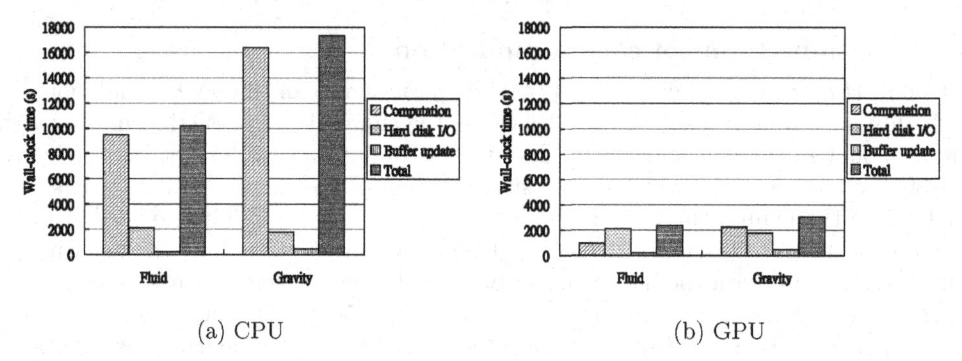

Figure 4. Performance of the fluid and gravity solvers in the out-of-core AMR simulations. The right and left panels show the results with and without the GPU-acceleration, respectively.

requirement is about 100 GB. By dividing the simulation domain into 64 sub-domains, the amount of memory actually allocated is only 3 GB. The performance is measured by using one NVIDIA GeForce 8800 GTX GPU and one Intel i7-920 CPU core. Figure 4 shows the timing measurements of the hydrodynamic and gravity solvers with and without the GPU acceleration. In the CPU-only case, the data I/O time is always much shorter than the computation time. In the case with GPU acceleration, the performance of the hydrodynamic solver is dominated by the data I/O, while that of the gravity solver is still dominated by the CPU/GPU computation. Also note that in each case, the total elapsed time is significantly shorter than the sum of the computation time and the I/O time, indicative of efficient overlap between computation and data I/O. We conclude that the out-of-core computation is reviving and can potentially be a powerful vehicle to deliver the optimal performance of a GPU cluster.

References

Roe, P. L. 1981, *J. Comput. Phys.*, 43, 357

Schive, H., Chien, C., Wong, S., Tsai, Y., & Chiueh, T. 2008, *New Astron.*, 13, 418

Schive, H., Tsai, Y., & Chiueh, T. 2010, *ApJS*, 186, 457

Toro E. F. 2009, *Riemann Solvers and Numerical Methods for Fluid Dynamics. A Practical Introduction.* (3rd ed; Heidelberg: Springer)

Trac, H. & Pen, U. 2003, *PASP*, 115, 303

João Alves and Bruce Elmegreen

Computational Star Formation
Proceedings IAU Symposium No. 270, 2011
J. Alves, B.G. Elmegreen, J. M. Girart & V. Trimble, eds.

doi:10.1017/S174392131100072X

Future Trends in Computing

Bruce G. Elmegreen[1]

[1]IBM Research Division, T.J. Watson Research Center,
1101 Kitchawan Road, Yorktown Hts., NY 10598 USA
email: `bge@us.ibm.com`

Abstract. According to a Top500.org compilation, large computer systems have been doubling in sustained speed every 1.14 years for the last 17 years. If this rapid growth continues, we will have computers by 2020 that can execute an Exaflop (10^{18}) per second. Storage is also improving in cost and density at an exponential rate. Several innovations that will accompany this growth are reviewed here, including shrinkage of basic circuit components on Silicon, three-dimensional integration, and Phase Change Memory. Further growth will require new technologies, most notably those surrounding the basic building block of computers, the Field Effect Transistor. Implications of these changes for the types of problems that can be solved are briefly discussed.

Keywords. methods: n-body simulations, methods: numerical

1. Introduction

We can make progress in numerical star formation only as fast as we make progress in its three main components: our understanding of the important physical processes, our ability to program these processes for accurate simulations in a computer, and the capabilities of the computers that are used. This talk describes the expected progress of computer capabilities in the coming decade.

Our first consideration should be the scale of computations that are being done today. The recent SPH simulation of star formation in a cluster by Bate (2009) took $\sim 10^{17}$ floating point instructions and resulted in 2 TeraBytes of movie-format output. The AMR simulations by Kritsuk *et al.* (2007) and Norman *et al.* (2009) used $\sim 10^{17}$ flops and generated ~20 TBytes of data. On the fastest machine in the world, which can run a well-tuned problem at ~1 Petaflop (10^{12} floating point operations per second), a simulation with 10^{17} flops would take 28 hours. In fact these simulations were run on smaller machines for longer times. Generally we choose to run problems that finish in a reasonable time, such as a day or a week or a month, depending on how important and unique the problem is. Computer centers rarely give their full machine power to a single user. If these two things remain true, the time we are willing to wait for results and our willingness to share, then the scale of our computation will progress with the overall power of our computers.

Figure 1 shows the sustained speed of the fastest computers in the world over the last 17 years (middle squares), and the sustained speed of the 500th-fastest computers (bottom squares) (reference: Top500.org). The sums of all computers in the Top500.org lists are the top squares. All three speeds increase exponentially with time, and they follow each other well. It takes about 7 years for a system speed that is number one in the world to be bypassed by increasingly faster new computers until it is number 500 in the world. The whole field of computation scales approximately in proportion to the fastest computers.

Scaling for an individual user is determined also by the availability of funds. If a person accustomed to 1% of the time on a public system were suddenly given 100% of the time

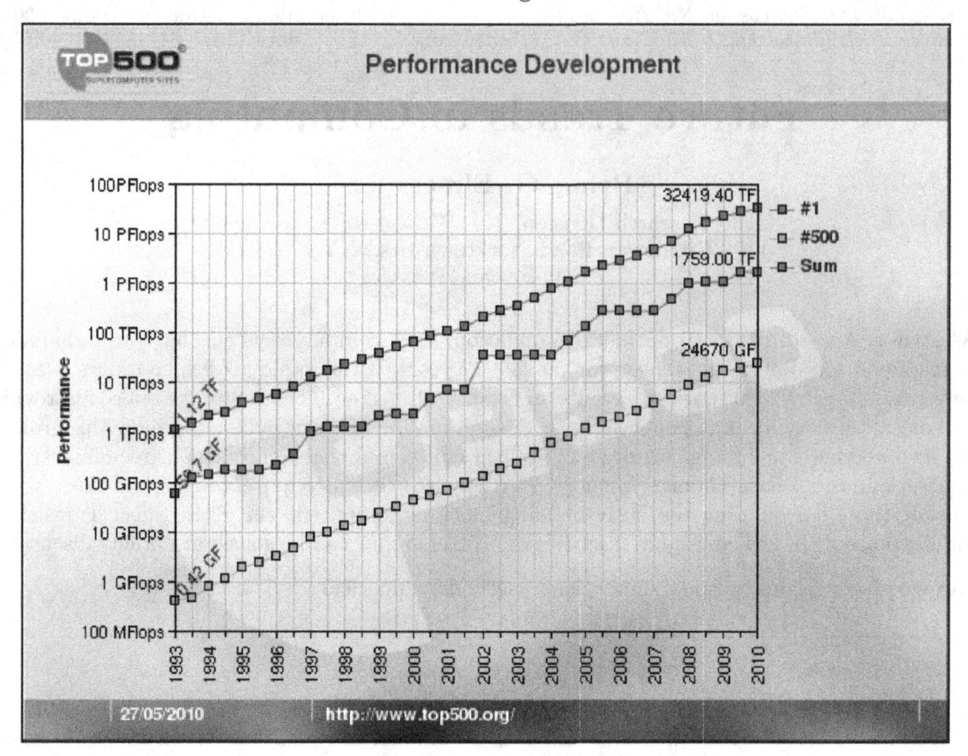

Figure 1. The sustained speed in floating point operations per second on the software package LINPACK of the Top 500 supercomputers in the world as a function of the year. The fastest computer is indicated by the middle curve and the sum of all is the top curve. From the URL http://www.top500.org/lists/2010/06/performance_development. This trend suggests an Exaflop computer (10^{18} flops per second) might be achieved by the year 2020.

on a dedicated machine of the same total size, then the jump in productivity would be a factor of 100 – much greater than the scaling factor of the technology. This boost from machine ownership is one of the strong points of dedicated hardware like GRAPE or GPU clusters. Generally the user base for these systems is much smaller than for all-purpose hardware on a campus or government network, so in addition to the speed factor from the pipelining or multicore technologies, respectively, there is also a speed factor from dedicated use. Even specialized hardware should scale with the technology, however, just like the number 1 computer.

This concept of scaling assumes that the user can program his or her algorithm to a reasonably high efficiency on any of these machines. That may not be true for all algorithms and systems. Since 2004, increasing peak speed has been the result of increasing parallelism with an approximately constant clock speed. The clock speed has been constant to keep the power consumption about constant, given that the voltage of Field Effect Transistors (FETs), has reached its minimum useful value. Power consumption scales with the square of the voltage multiplied by the clock frequency.

Increasing parallelism means that there are more separate arithmetic units, or cores, in a processor chip, and also more processor chips in a system. For example, the three fastest computers in the world, all Peta-flop scale, contain about 10^5 cores each. To make use of all this capacity, a problem has to be very finely divided into many separate computations, one set for each core. Thus scaling to larger-size problems, more grid points, for example, or more particles, is relatively easy as computers grow with increasing parallelism, but scaling to shorter run times for the same size simulation is more difficult. Adding more physics to the same-size problem and expecting the same run time in a larger system is also difficult. If a simulation requires a very long relative run time today, such as thousands of crossing times, and is impractical to run because of that, then it may be impractical to run for a long time in the future too, if technology scaling is only by increased parallelism. This is a very different situation than what we had 20 years ago, when clock speed was increasing at an exponential rate and all problems were run on single cores. The current era of increasing parallelism, rather than increasing clock speed, is a major challenge for algorithmic development and programmers.

Along with increased parallelism has come a decrease in the memory associated with each processor. Memory includes cache for ready use by the core, and more distant on-chip memory for use several tens of clock cycles away. There is also memory on a typical processor board that is separate from the processor, and several hundred clock cycles away. Clock cycle distance means how long it takes, in step-by-step instruction cycles, to fetch a number from memory and deliver it to the core floating point register where it can be ready for multiplication or addition. If the numbers used in a sequence of calculations are random or dependent on the results of the calculation, i.e., unpredictable for future clock cycles, then the processors will spend much of their time waiting for data. Good programming means high predictability for the numbers fetched from memory. Vectors are highly predictable, for example, because the calculations can be done in the order of the vector index. In a well written program, compilers that convert user language into machine language can hide almost all of the memory latency (clock cycles to memory) inside other useful work that takes place at the same time. Generally a programmer has to iterate with the compiler, trying little tricks like unwrapping do-loops and ordering numbers in memory, in order to coax even the best compilers to compile efficient code.

The trade off between memory space and processing space on a chip depends on the use of that chip. Graphics Processing Units (GPUs), for example, were designed for very rapid graphics processing, which involves algorithms that stream low-bit data from calculations or video into numbers used by a display screen. GPUs need a lot of computation speed but not much memory per processor. If an algorithm matches the balance between the data input and output rate, the memory access rate, and the computation rate in a chip or system, then the calculation can be efficient for that hardware. Otherwise there will be a bottleneck somewhere. Most of the processors and systems that have grown rapidly in the last few years by extreme parallelism have migrated toward low ratios of memory to computation per core, often limiting their use to problems that can be very finely divided.

The trend toward greater parallelism implies that coding for algorithms has to change continuously. Professional programmers and astronomers who develop new algorithms will become an increasingly important part of team research efforts. This suggests a change toward increased specialization within astronomy subfields and increased division between programmers and users. Ten to 20 years ago, a programmer could write a code and have it scale for many years simply by increasing clock speeds, but this era is gone for a while. It may come back before the end of the decade with the advent of storage-class memory, as discussed below, but for a while, programs will have to be rewritten or significantly retuned for each new generation of hardware.

2. Future Speeds and Storage

Figure 1 suggests that with further improvements in system architecture and basic technologies, by 2020, the fastest computer in the world will be ~1000 times faster than the fastest today. It would run well-tuned jobs at an Exaflop per second. The trend in Figure 1 for the number 1 system is: speed = Petaflop$\times 2^{(Year-2009)/1.14}$. Thus the doubling time is 1.14 year.

The same improvements can be expected for smaller systems too. By 2020, something close to one million dollars (in today's currency) should buy a Petaflop computer with a general purpose design. This would be the scale of a university data center or a small government lab. A lap-top computer for ~$1000 might run at a Teraflop. There should also be more specialized computers too, like GPUs and GRAPEs, which even today can be purchased with Teraflop speeds for ~$1000. Special hardware like this might run with Petaflop speeds by 2020 and be available for private use ($1000 price).

Magnetic disk storage is increasing in capacity at an exponential rate too. For a given cost, storage capacity has increased by a factor of ~2 every year for the last 30 years. This is faster than the increase in computation rate per dollar, which corresponds to a doubling every 1.5 years or so. The rapid growth of storage capabilities is the primary reason for the current "information age," where enormous quantities of data are available to us on the internet. By 2020, a single magnetic disk could hold ~100 TeraBytes of data (Walter 2005) and a PetaByte could cost only $200 (Komorowski 2009), with the current trends. Exponential growth of storage means that we never have to erase anything (although we should for sanity reasons). With exponential growth, the sum of everything ever stored in a unit of a certain cost equals what can be stored in a fraction of the next generation system for the same cost.

Ray Kurzweil (2001) has considered many of the implications of increasing computer speeds if the trends continue for the next several decades. He thinks we will achieve the Human Brain capacity, which is 10^{16} computations per second, for around $1,000 by the year 2023, and for one cent by 2037. We will also achieve the Human Race capacity (10^{26} cps) for $1,000 around the year 2049, and for one cent around the year 2059. We cannot imagine how the world will differ in this era, but fortunately most graduate students today will still be around to see it.

3. Technology Improvements

Scaling for the next 10 years is not as difficult to imagine as scaling for the next 40 years because the next decade is part of the planned technology road map for many computer and chip companies. Basically, it will result from continued shrinkage of each component size, allowing more and more to fit on a single chip (increased parallelism and increased functionality), combined with a few important one-shot improvements that provide a new boost every few years. Currently, the best available technology has a smallest design scale of 45 nanometers, which means that 200 Billion design structures can be put on a 2-centimeter-square chip. For example, the most complex chips today contain around 2 Billion transistors, each of which has considerable internal structure that has to be designed on the surface of a Silicon crystal. In a few years, the target size for a minimum scale will be 32 nm, in around 5 years it will be 22 nm, and by 2017 or so, it could be 15 nm. Chip complexity and component count in two dimensions increases as the inverse square of this minimum size.

3.1. *Packaging*

Further gains are being made with new technologies in packaging. Currently, with conventional packaging, individual Silicon chips are attached to printed circuit cards with a density for communication between the chip and the card of several thousand input/output terminals per cm^2. Multiple chips can be packaged into a single module with

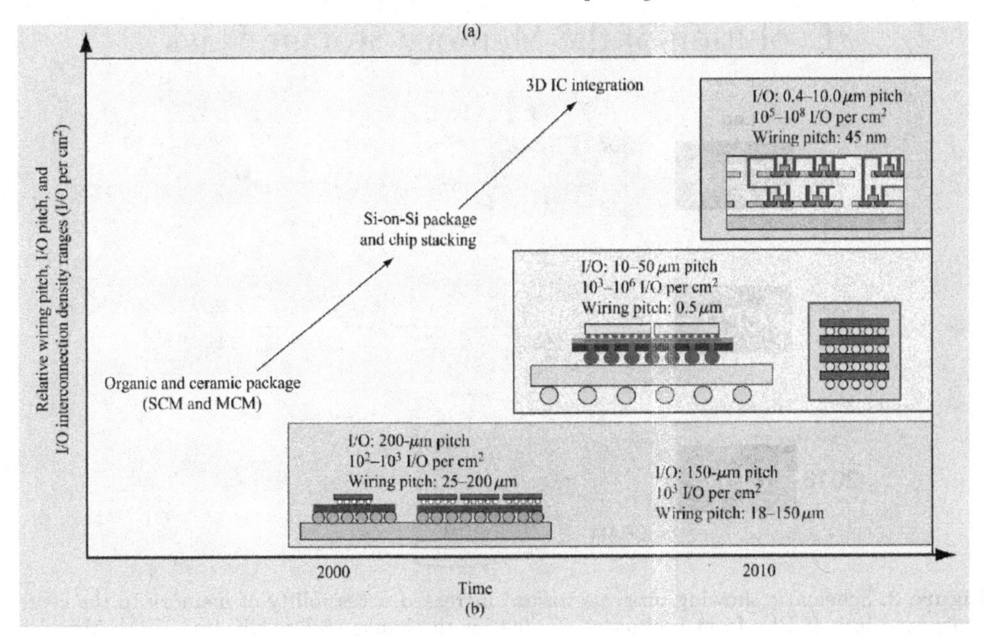

Figure 2. Roadmap of packaging showing progress toward more 3D integration and higher IO density (Knickerbocker *et al.* 2008). The abbreviations are: SCM, single-chip module; MCM, multichip module.

communication bandwidth like this from each chip to an underlying Silicon carrier layer, and slightly lower bandwidth between the Silicon carrier and a substrate. Newer technology can stack Silicon layers vertically using "through Silicon vias," which are thin metal rods that carry currents vertically from one layer to another. Vertical stacking is important because all of the functional parts of Silicon are close to the surface, built up in layers with masking, using vapor and other deposition techniques, and etching or other selective removal techniques. Stacking of two layers doubles the density of useful components in a device. A timeline for packaging improvements is shown in Figure 2.

Input/output densities are currently increasing to several 10's of thousands per cm^2. This can be done, for example, with a 2D array of microsolder bumps 25 μm in size and 50μm apart. In the next 5 years, three-dimensional Silicon integration and packaging should become even more advanced, with IO densities per chip in the millions per cm^2. Problems with cooling the middle components, alignment, and assembly will have to be overcome. At the end of this decade, 3D integration should be able to combine processors, memory, accelerators, and dense, probably optical, IO, into complex structures, enabling greater functionality, increased proximity of memory to processors, and greater IO per compute cycle.

Packaging of boards into racks and racks into systems is improving too. The current family of IBM BlueGene computers, which ranked number 1 in the world from 2004 to 2007 in terms of speed on the benchmark software LINPACK, has a 3D torus design with copper cables for all communications, board-to-board and rack-to-rack. The number 1 ranked computer in 2008 and 2009, the IBM Roadrunner at Los Alamos, has copper cables between the boards inside a rack and optical fiber connections between the racks.

3.2. *New Memory Technology*

One of the oldest bottlenecks in a computer is the memory bandwidth, which is the rate at which the processor can put and retrieve data to the memory. Current chips

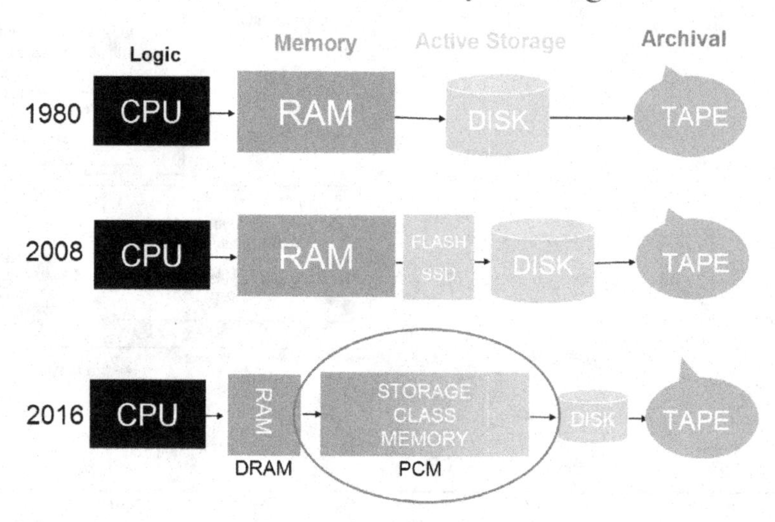

Figure 3. Schematic showing progress toward increased accessibility of memory to the central processing unit (CPU; from Freitas *et al.* 2009, with thanks to Dr. KK Rao, IBM Almaden). Random access memory (RAM) is the first step. Recently, Storage Class Memory has been inserted between the RAM and the Disk to increase the speed of high volume data access. This intermediate step should grow over time until it contains much of the memory currently in RAM and Disk.

have memory inside them ("Embedded DRAM"), which helps by decreasing the distance the current pulses have to travel, but most of the memory is still on the board that contains the chip. Standard memory is DRAM, which stands for Dynamic Random Access Memory. DRAM is very simple, consisting of a Field Effect Transistor (FET) and a capacitor. The capacitor can be a trench in the Silicon that holds electrons on the inner surface, or it can be a stack of plates above the FET gate. The value of a bit (0 or 1) depends on the level of charge on the capacitor. Charging the capacitor is done by passing a current through the FET, which acts as a switch (http://en.wikipedia.org/wiki/Dynamic_random_access_memory). This simple design was invented by Robert Dennard at IBM in 1968 and has been at the core of computer memory since 1970 when Intel released the 1103 chip (http://inventors.about.com/od/rstartinventions/a/Ram.htm). The advantage of DRAM is its simplicity, high density, and scalability with new generations of technology. The disadvantage of DRAM is that the electrons leak out of the capacitor and so all the memory units have to be recharged according to their state every 64 millisecond or so. This takes power – far too much power for continued scaling into the coming decade. DRAM also loses its state when the computer is turned off.

There are several options for "non-volatile" memory, which is the designation for memory that retains its state without power, although it still takes power for changing states. Flash is non-volatile memory used in cameras, cell-phones and in some large-scale computer memories, but Flash is slow and the number of writes before degradation is small, $\sim 10^5$ (compared to DRAM, which is $\sim 10^{15}$).

A promising new type of memory that recently came to market is Phase-Change Memory (PCM), which uses the crystalline structure of Germanium-Antimony-Tellurium compounds, or other chalcogenide glasses, to store information. In these compounds, the crystalline phase passes electricity and light, while the amorphous phase has high electrical resistance and is opaque. It has been used in re-writable Compact Disks and DVDs since the mid-1990's. The original patent for these materials was granted to Stanford R.Ovshinsky in 1961 (http://en.wikipedia.org/wiki/Stanford_R._Ovshinsky). In September 2009, Samsung announced a 512 Mbit memory chip based on PCM and in April 2010, Numonyx BV announced a 128 Mbit memory chip.

Compared to Silicon DRAM, PCM has about the same read/write time, bandwidth, and power, but 100 times the density and is non-volatile. Compared to Flash memory, PCM has the same power, but PCM is 1000 times faster and PCM can write more before degradation (10^5 for Flash, $10^8 - 10^{12}$ times for PCRAM, 10^{12} for disks, and 10^{15} for DRAM). Compared to Disks, PCM is 10^5 times faster, 1% of the power, and has about the same number of writes before degradation. With this potential, PCM could replace disks as a storage medium, greatly increasing the speed of data storage and greatly decreasing the power consumption. With greater speed to stored data, somewhat approaching the speed to DRAM, it might be possible to return to systems with enormous memory capacity, even virtual memory or shared memory among processors. Figure 3 shows the possible evolution of logic, memory, active storage, and archival storage over the next few years. Storage class memory, based on PCM or other non-volatile types, would allow less DRAM and less disk storage, and act as an intermediate step replacing some of the function of both.

3.3. *New Transistor Technology*

The metal - oxide - semiconductor field-effect transistor (MOSFET) is the basis for much of computer technology including logic and memory. It was proposed in 1925 by Julius Edgar Lilienfeld. Today, the metal contact at the gate has been replaced by polycrystalline silicon, but newer, smaller technologies are sometimes returning to metal.

The FET has source and drain electrodes at two ends of a doped semiconductor layer like Silicon, and a gate electrode in the middle. A voltage between the gate and the substrate beneath the Silicon layer induces a field in the Silicon that opens a channel for current to flow from the source to the drain. Unlike bipolar transistors that pass a current through the gate, the gate of an FET acts like a capacitor and changes only the internal field structure. This saves power.

A problem with FETs at very small scales is that the ultrathin (~2 nm) insulator between the gate electrode and the doped Silicon passes a small current by quantum mechanical tunneling. There is also a small leakage current between the source and the drain. These currents are a net energy loss and a source of chip heating. The insulator has been improved considerably over the last few years from the most common Silicon Dioxide to new materials with high dielectric constants. A high dielectric constant allows the gate to maintain a high capacitance as the area gets smaller, without requiring the capacitor to get any thinner, thereby avoiding serious tunneling losses. Since 2007, an important material for FETs with high dielectric constant contains Hafnium.

Higher performance FETs could come from Silicon nanowires, which are fairly easy to grow and have properties that can be used in transistors (Schmidt *et al.* 2006; Yoon *et al.* 2006), and Carbon nanotubes, which are also commonly available and can make transistors (Martel *et al.* 2001; Chen *et al.* 2008). Assembly of these nanoscale objects into useful circuitry is a challenge. An interesting new technique is to use lithography and etching to pattern artificial DNA nanostructures on a Silicon surface, which can then bind Carbon nanotubes and Silicon nanowires into useful shapes (Kershner *et al.* 2009). Commercial products with these and other novel technologies are not likely to be seen for at least 10 years.

A bigger problem for FETs is the inability to reduce the supply voltage, V. Recall that FET power depends on $V^2 f$ for clock frequency f, and that the power dissipation limit has already been reached for today's V and f; higher f requires lower V. Theis (2010) reviews two emerging concepts for FET-like switches that might operate at lower voltage. One is the Tunnel FET (Banerjee *et al.* 1987; Appenzeller *et al.* 2004), in which the source-to-drain current depends sensitively on tunneling through a barrier that is regulated by small changes in gate voltage. So far this has too small an on-state current for use in common devices. Another uses a layer of ferroelectric material in the gate dielectric that can switch polarization abruptly with small changes in the gate voltage (Salahuddin & Datta 2008). The energy barrier for the source-to-drain current depends on the polarization of this ferroelectric, so the current is controlled by the gate again. Unfortunately, ferroelectric switching is too slow for useful devices at the present time. Nevertheless, these and other emerging concepts suggest that today's limits on clock frequency may be temporary.

4. Conclusions

Exponential growth of computer capabilities should continue into the near future, driven by the needs of scientists, financiers, industries, and the general public, but the rate of growth will depend on the state of the world economy and the ability of research, development, and manufacturing labs to overcome technology challenges. Exponential growth implies that all that computers have ever done throughout history can be repeated in the next e-folding time. The same would be true for astronomical observations, considering that detectors are following the same technology curve. One wonders whether our ability to understand the results of our computations and observations can keep up with such a rapid pace in technology development.

Acknowledgements: Helpful comments and information were provided by Drs. Thomas Theis and KK Rao, of IBM Research in Yorktown, NY, and Almaden, CA, respectively.

References

Appenzeller, J., Lin, Y.-M., Knoch, J., & Avouris, Ph., 2004, *Phys. Rev. Lett.*, 93, 19

Banerjee, S., Richardson, W., Coleman, J., & Chatterjee, A. 1987, *IEEE Electron Device Lett.*, 8, 347

Bate, M.R. 2009, *MNRAS*, 392, 590

Chen, Z. *et al.*, 2008, *IEEE EDL*, 29, 183

Freitas, R., Wilcke, W., Kurdi, B., & Burr, G. 2009, FAST 2009 Tutorial T3, http://www.usenix.org/events/fast09/tutorials/T3.pdf

Kershner, R.J., *et al.* 2009, *Nature Nanotechnology*, 4, 557

Knickerbocker, J.U. *et al.* 2008, *IBM J. Res.& Dev.*, 52, no. 6, p 553

Komorowski, M. 2009, http://www.mkomo.com/cost-per-gigabyte

Kritsuk, A. G. Norman, M. L., Padoan, P., & Wagner, R. 2007, *ApJ*, 665, 416

Kurzweil, R. 2001, Lifeboat Foundation Special Report, Law of Accelerating Returns, http://lifeboat.com/ex/law.of.accelerating.returns

Martel, R., Wong, H.-S.P., Chan, K., & Avouris, Ph. 2001, *IEDM Tech. Dig.*, 159

Norman, M. L., Paschos, P. & Harkness, R. 2009, *J. Phys., Conf. Ser.*, 180 012021

Salahuddin, S. & Datta, S. 2008, *Nano Lett.*, 8, 405

Schmidt, V., Riel, H., Senz, S., Karg, S., Riess, W., & Gösele, U. 2006, *Small*, 2, 85

Theis, T. N. 2010, *Science*, 327, 1600

Walter, C. 2005, *Scientific American*, August Issue

Yoon, C., *et al.* 2006, *Nanotech. Materials Devices Conference*, IEEE, 1, 424

Computational Star Formation
Proceedings IAU Symposium No. 270, 2011
J. Alves, B.G. Elmegreen, J. M. Girart & V. Trimble, eds.

doi:10.1017/S1743921311000731

A Fast Explicit Scheme for Solving MHD Equations with Ambipolar Diffusion

Jongsoo Kim

Korea Astronomy and Space Science Institute, Daejeon 305-348, Republic of Korea
email: jskim@kasi.re.kr

Abstract. We developed a fast numerical scheme for solving ambipolar diffusion MHD equations with the strong coupling approximation, which can be written as the ideal MHD equations with an additional ambipolar diffusion term in the induction equation. The mass, momentum, magnetic fluxes due to the ideal MHD equations can be easily calculated by any Godunov-type schemes. Additional magnetic fluxes due to the ambipolar diffusion term are added in the magnetic fluxes, because of two same spatial gradients operated on the advection fluxes and the ambipolar diffusion term. In this way, we easily kept divergence-free magnetic fields using the constraint transport scheme. In order to overcome a small time step imposed by ambipolar diffusion, we used the super time stepping method. The resultant scheme is fast and robust enough to do the long term evolution of star formation simulations. We also proposed that the decay of alfv́en by ambipolar diffusion be a good test problem for our codes.

Keywords. methods:numerical, MHD, stars: formation

---

1. Introduction

Ambipolar diffusion (AD), which arises in partially ionized plasmas, causes the relative drift of ions coupled to magnetic fields with respect to neutrals. AD enables molecular cloud cores to collapse gravitationally, so is one of important processes for star formation (e.g., Mestel & Spitzer 1956; Mouschovias 1987; Shu *et al.* 1987).

Several numerical methods have been proposed in the studies of the dynamics of partially ionized plasmas within the frame of single or two fluid formulations. An incomplete list of them is Tóth (1994), Mac Low *et al.* (1995), Mac Low & Smith (1997), Stone (1997), Li *et al.* (2006), and Tilley & Balsara (2008). Implicit schemes for the multifluid treatment of the Hall term and ambipolar diffusion have also been suggested by Falle (2003) and O'Sullivan & Downes (2006, 2007). In this work, we describe a fully explicit method for incorporating ambipolar diffusion with the strong coupling approximation into a multidimensional MHD code based on the total variation diminishing scheme. The divergence-free condition of magnetic fields is ensured by a flux-interpolated constrained transport scheme, and a super time stepping method is used in order to considerably accelerate the otherwise painfully short diffusion-driven time steps. More detailed information on this work can be found in Choi, Kim, & Wita (2009).

2. AD MHD Equations and Numerical Methods

The isothermal MHD equations including ambipolar diffusion with the strong coupling approximation can be written as

$$\frac{\partial \rho}{\partial t} + \nabla \cdot (\rho \boldsymbol{v}) = 0, \tag{2.1}$$

$$\frac{\partial \boldsymbol{v}}{\partial t} + \boldsymbol{v} \cdot \nabla \boldsymbol{v} + \frac{a^2}{\rho} \nabla \rho - \frac{1}{\rho} (\nabla \times \boldsymbol{B}) \times \boldsymbol{B} = 0, \tag{2.2}$$

$$\frac{\partial \boldsymbol{B}}{\partial t} - \nabla \times (\boldsymbol{v} \times \boldsymbol{B}) = \nabla \times \left\{ \left[\frac{1}{\gamma \rho_i \rho} (\nabla \times \boldsymbol{B}) \times \boldsymbol{B} \right] \times \boldsymbol{B} \right\}, \tag{2.3}$$

$$\nabla \cdot \boldsymbol{B} = 0, \tag{2.4}$$

where a is an isothermal sound speed, γ is the collisional coupling constant between ions and neutrals, and ρ_i is the ion density. The other variables ρ, $\boldsymbol{v}$, and $\boldsymbol{B}$ denote neutral density, neutral velocity, and magnetic field, respectively. We further assume that the ion density is constant in this work.

The above equations except the term in the right hand side of equation (2.3) are same as the ideal MHD equations. So our strategy of solving the above AD MHD equations is first to update the fluxes of ρ, $\boldsymbol{v}$, and $\boldsymbol{B}$ using any Godunov type schemes for solving the isothermal MHDs, then to make a correction to the fluxes of $\boldsymbol{B}$ due to AD with the divergence free condition. We take a total variation diminishing scheme in Kim *et al.* (1999) for the flux calculations due to the idea MHD.

Let's now look at the induction equation in details. The B_x component, for example, of the equation can be written as

$$\frac{\partial B_x}{\partial t} + \frac{\partial}{\partial y} (B_x v_y - B_y v_x) - \frac{\partial}{\partial z} (B_z v_x - B_x v_z) = \frac{\partial S_z}{\partial y} - \frac{\partial S_y}{\partial z}, \tag{2.5}$$

where S_y and S_z are defined by

$$\begin{aligned} S_y &= \frac{1}{\gamma \rho_i \rho} \left[\left(\frac{\partial B_z}{\partial y} - \frac{\partial B_y}{\partial z} \right) B_y B_x + \left(\frac{\partial B_z}{\partial x} - \frac{\partial B_x}{\partial z} \right) (B_x^2 + B_z^2) \right. \\ &\quad \left. + \left(\frac{\partial B_y}{\partial x} - \frac{\partial B_x}{\partial y} \right) B_y B_z \right], \end{aligned} \tag{2.6}$$

$$\begin{aligned} S_z &= \frac{1}{\gamma \rho_i \rho} \left[\left(\frac{\partial B_x}{\partial z} - \frac{\partial B_z}{\partial x} \right) B_z B_y + \left(\frac{\partial B_x}{\partial y} - \frac{\partial B_y}{\partial x} \right) (B_y^2 + B_x^2) \right. \\ &\quad \left. + \left(\frac{\partial B_z}{\partial y} - \frac{\partial B_y}{\partial z} \right) B_z B_x \right]. \end{aligned} \tag{2.7}$$

Note that the first and second terms on the right-hand side of equation (2.5) have the same gradients as the second and third terms do on the left-hand side, respectively. Applying the second-order finite difference operators to equations (2.6) and (2.7), we calculate their values at every grid center, $S_{y,i,j,k}$, and $S_{z,i,j,k}$. And their values at every face center can be simply calculated as the half of the sum of two nearby grid centered values. Then we can define a new flux for B_x along y- and z-directions at face centers as

$$f^{(5)}_{y,i,j+1/2,k} = \bar{f}^{(5)}_{y,i,j+1/2,k} - \frac{1}{2} (S_{z,i,j,k} + S_{z,i,j+1,k}), \tag{2.8}$$

$$f^{(5)}_{z,i,j,k+1/2} = \bar{f}^{(5)}_{z,i,j,k+1/2} + \frac{1}{2} (S_{y,i,j,k} + S_{y,i,j,k+1}), \tag{2.9}$$

respectively, where the first term in right hand side is the conventional flux due to the advection and the second term is due to ambipolar diffusion. Similarly, we can define $f^{(6)}_{x,i+1/2,j,k}$, $f^{(6)}_{z,i,j,k+1/2}$ $f^{(7)}_{x,i+1/2,j,k}$, and $f^{(7)}_{y,i,j+1/2,k}$, where the first two are fluxes for B_y

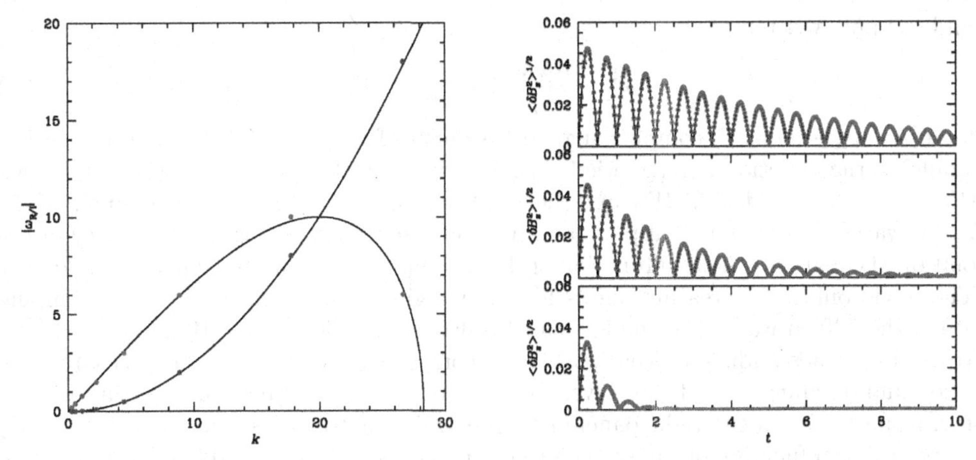

Figure 1. Left panel: A dispersion relation of the Alfv́en waves in a partially ionized medium. The solid lines represent the real and imaginary (monotonically increasing one) parts of the complex angular frequency. Circles are measurements from numerical experiments. Right panel: Time evolution of the root-mean-square of the B_z component for the cases of $\gamma\rho_i = 100$ (top), 50 (middle), 10 (bottom). Solid lines are from theoretical predictions and open circles are from numerical simulations.

along x- and z-directions, and the latter two are fluxes for B_z along x- and y-directions. These combined fluxes at face centers, $f^{(n)}_{x,i+1/2,j,k}$, $f^{(n)}_{y,i,j+1/2,k}$, and $f^{(n)}_{z,i,j,k+1/2}$, are used to enforce $\nabla \cdot \boldsymbol{B} = 0$ as well as to update the magnetic field components to the next time step, as the advection only fluxes for the ideal MHD case do. In fact, we used the flux-interpolated CT (Constraint Transport) scheme developed in Balsara & Spicer(1999).

The time step for the ambipolar diffusion term is proportional to the square of the grid size, so the explicit treatment of ambipolar diffusion terms leads to very small time steps (Mac Low *et al.* 1995). In this work we adopt the "super time stepping" approach (Alexiades *et al.* 1996) to increase the effective time interval and allow much faster computations for ambipolar diffusion. O'Sullivan & Downes (2006,2007) also used this strategy in their multifluid MHD models. The super time stepping technique considerably accelerates the explicit schemes for parabolic problems (Alexiades *et al.* 1996). The key advantage of this approach is that it demands stability over large compound time steps, rather than over each of the constituent substeps. In addition to allowing larger effective time steps, the super time stepping approach offers relatively simple implementation. Readers who are interested in the technical details may look up Alexiades *et al.* (1996) and Choi *et al.*(2009).

3. A New Test Problem

Probably, the most popular test problem for AD or two fluid codes is oblique C (continuous) shocks, whose steady state solutions can be easily obtained (Mac Low *et al.* 1995). In fact, we also tested our code with this problem and presented its results in Figure 1 in Choi *et al.* (2009). One drawback of this test problem is the steady-state nature, which doesn't enable us to check any states in between from an initial state to the final steady state.

We proposed a new test problem that is standing Alfv́en waves in a weakly ionized plasma. In the strong coupling approximation, a dispersion relation for the Alfv́en waves can be simply written as

$$\omega^2 - i\frac{c_A^2 k^2}{\gamma\rho_i}\omega - c_A^2 k^2 = 0, \tag{3.1}$$

where $\omega = \omega_R + i\omega_I$ is the complex angular frequency of a wave and k is a real wavenumber parallel to the direction of magnetic fields. The first and third terms give the well-known Alfv́en waves of the ideal MHDs, and the additional second term gives the damping of the Alfv́en waves by AD. The real part and imaginary parts of the angular frequency of the solution of equation (3.1) as a function of the wavenumber are plotted with two solid lines (The monotonically increasing one is for the imaginary part, which gives the damping rate of the Alfv́en waves.) in the left panel of Figure 1, where $\gamma\rho_i = 10$ and $c_A = 1/\sqrt{2}$. We setup up a standing Alfv́en wave in a computational domain and measured the period and damping rate of the wave, which correspond to the real and imaginary parts of ω, respectively. In the right panel of Figure 1, the root-mean-square values of the B_z component as a function of time are plotted for the cases of $\gamma\rho_i = 100$ (top), 50 (middle), 10 (bottom). As the couple of neutrals and ions becomes weaken, the damping of the Alfv́en waves becomes stronger. We did several experiments with different wavenumbers for the case of $\gamma\rho_i = 10$, measured the periods and decay rates, and put circles in the left panel of Figure 1. It shows good agreements between the numerical measurements and the theoretical prediction.

We also did decay and forced turbulence simulations with AD. The successful simulation results show the flexibility of our method as well as its ability to follow complex MHD flows in the presence of ambipolar diffusion. Readers who are interested in these results may look up Choi *et al.* (2009).

4. Conclusion

We described a method for incorporating ambipolar diffusion in the strong coupling approximation into a multidimensional magnetohydrodynamics code based on the total variation diminishing scheme. Contributions from ambipolar diffusion terms are included by explicit finite difference operators in a fully unsplit way, maintaining second order accuracy. The divergence-free condition of magnetic fields is exactly ensured at all times by a flux-interpolated constrained transport scheme. The super time stepping method is used to accelerate the timestep in high resolution calculations and/or in strong ambipolar diffusion. The test results of the decay of Alfvén waves in this paper and the steady-state oblique C-type shocks in Choi *et al.* (2009) showed the accuracy and robustness of our numerical approach.

References

Alexiades, V., Amiez, G., & Gremaud, P.-A. 1996, *Comm. Num. Meth. Eng.*, 12, 31
Balsara, D. S. & Spicer, D. S. 1999, *J. Comput. Phys.*, 149, 270
Choi, E., Kim, J., & Wiita, P. J. 2009, *ApJS*, 181, 413
Falle, S. A. E. G. 2003, *MNRAS*, 344, 1210
Kim, J., Ryu, D., Jones, T. W., & Hong, S. S. 1999, *ApJ*, 514, 506
Li, P. S., McKee, C. F., & Klein, R. I. 2006, *ApJ*, 653, 1280
Mac Low, M.-M., Norman, M. L., Königl, A., & Wardle, M. 1995, *ApJ*, 442, 726
Mac Low, M.-M. & Smith, M. D. 1997, *ApJ*, 491, 596
Mestel, L. & Spitzer, L., Jr. 1956, *MNRAS*, 116, 503
Mouschovias, T. Ch. 1987, in: G. E. Morfill & M. Scholer (eds.), *The Origin of Stars and Planetary Systems*, (Dordrecht: Reidel), p. 453

O'Sullivan, S. & Downes, T. P. 2006, *MNRAS*, 366, 1329
O'Sullivan, S. & Downes, T. P. 2007, *MNRAS*, 376, 1648
Shu, F. H., Adams, F. C., & Lizano, S. 1987, *ARAA*, 25, 23
Smith, M. D. & Mac Low, M.-M. 1997, *A&A*, 326, 801
Stone, J. M. 1997, *ApJ*, 487, 271
Tilley, D. A. & Balsara, D. S. 2008, *MNRAS*, 389, 1058
Tóth, G. 2000, *J. Comput. Phys.*, 161, 605

Maite Beltrán and Josep Miquel Girart at the banquet

Computational Star Formation
Proceedings IAU Symposium No. 270, 2011
J. Alves, B.G. Elmegreen, J. M. Girart & V. Trimble, eds.

doi:10.1017/S1743921311000743

Ambipolar Diffusion Effects on Weakly Ionized Turbulence Molecular Clouds

Pak Shing Li[1], Christopher F. McKee[2], and Richard I. Klein[3]

[1]Astronomy Department, University of California, Berkeley, CA 94720, USA
email: psli@astro.berkeley.edu

[2]Physics Department and Astronomy Department, University of California, Berkeley, CA 94720, USA
email: cmckee@astro.berkeley.edu

[3]Astronomy Department, University of California, Berkeley, CA 94720, USA; and Lawrence Livermore National Laboratory, P.O.Box 808, L-23, Livermore, CA 94550, USA
email: klein@astron.berkeley.edu

Abstract. Ambipolar diffusion (AD) is a key process in molecular clouds (MCs). Non-ideal MHD turbulence simulations are technically very challenging because of the large Alfvén speed of ions in weakly ionized clouds. Using the Heavy-Ion Approximation method (Li, McKee & Klein 2006), we have carried out two-fluid simulations of AD in isothermal, turbulent boxes at a resolution of 512^3, to investigate the effect of AD on the weakly ionized turbulence in MCs. Our simulation results show that the neutral gas component of the two-fluid system gradually transforms from an ideal MHD turbulence system to near a pure hydrodynamic turbulence system within the standard AD regime, in which the neutrals and ions are coupled over a flow time. The change of the turbulent state has a profound effect on the weakly ionized MCs.

Keywords. turbulence, ISM: magnetic fields, ISM: clouds, methods: numerical.

1. Introduction

For an ideal MHD fluid, the magnetic field is assumed to be frozen into the ionized gas. From observations, the ionization fraction of dense molecular clouds (MCs) is very low, on the order of 10^{-7} or less. Here we treat the MC with a highly simplified 2-fluid model composed of only neutral and ionized molecules. The gyro-frequency of an ion is $\omega_{\rm ci} = eB/m_i c = 9.58 \times 10^{-9} ZBm_p/m_i$ rad s^{-1}, which is still very large compared to the ion-neutral collision frequency, $t_{\rm in}^{-1} = \gamma_{\rm AD}\rho_n \sim 10^{-6}$ s inside typical MCs. At the MC densities, the ions are tied to the magnetic field. The neutral gas feels the presence of the magnetic field indirectly through collisions with the ions. With such a low ionization fraction, treating the weakly ionized MCs using ideal MHD is inappropriate because the coupling between ions and neutrals is weak. The result is that the majority of the MC gas will slowly drift relative to the magnetic field. This slow ambipolar diffusion (AD) process is important because it allows gravitational collapse even in the cloud that is initially magnetically sub-critical.

However, many observations show that MCs are in a supersonically turbulent state. If the AD time scale is much longer than the characteristic time scale of the turbulence, AD is unimportant. The observed line-width size relation implies that the dynamical time in a dense clump with size $\lesssim 1$ pc is comparable to the AD time scale. Can we ignore AD and use ideal MHD to treat these small regions in MCs? We use non-ideal MHD numerical simulations with AD to answer this question.

2. Numerical Method and Simulation Model Parameters

We have implemented a two-fluid semi-implicit method into ZEUS-MP code using Mac Low & Smith(1997) approach. We assume that the MCs are isothermal. As a result, the MHD equations for the two fluids, ions and neutrals, with AD are:

$$\frac{\partial \rho_n}{\partial t} = -\nabla \cdot (\rho_n \mathbf{v}_n), \tag{2.1}$$

$$\frac{\partial \rho_i}{\partial t} = -\nabla \cdot (\rho_i \mathbf{v}_i), \tag{2.2}$$

$$\rho_n \frac{\partial \mathbf{v}_n}{\partial t} = -\rho_n(\mathbf{v}_n \cdot \nabla)\mathbf{v}_n - \nabla P_n - \gamma_{\rm AD}\rho_i\rho_n(\mathbf{v}_n - \mathbf{v}_i), \tag{2.3}$$

$$\rho_i \frac{\partial \mathbf{v}_i}{\partial t} = -\rho_i(\mathbf{v}_i \cdot \nabla)\mathbf{v}_i - \nabla P_i - \gamma_{\rm AD}\rho_i\rho_n(\mathbf{v}_i - \mathbf{v}_n) + \frac{1}{4\pi}(\nabla\times\mathbf{B})\times\mathbf{B}, \tag{2.4}$$

$$\frac{\partial \mathbf{B}}{\partial t} = \nabla\times(\mathbf{v}_i\times\mathbf{B}), \tag{2.5}$$

$$\nabla \cdot \mathbf{B} = 0, \tag{2.6}$$

where ρ is the density, $\mathbf{v}$ is the velocity, $\mathbf{B}$ is the magnetic field strength, and $\gamma_{\rm AD}$ is the ion-neutral collisional coupling constant. The subscripts i and n denote ions and neutrals, respectively. Note that our simulations do not include gravity because we want to understand the complex interaction between AD and turbulence first. The two momentum equations with a stiff drag term are solved implicitly. Using this approach, the time step size is proportional to the grid size Δx (Mac Low & Smith 1997).

We use the Heavy-Ion Approximation (Li, McKee & Klein 2006) to speed up the computation by a substantial factor. The principle of the Heavy-Ion Approximation (HIA) is to increase the mass of the ions by a large factor in order to reduce the frequency of ion Alfven waves so that the numerical time step size can be significantly increased. At the same time, $\gamma_{\rm AD}$ is reduced by the same amount to keep the drag force term the same. In Li, McKee & Klein (2006), we show that there is a numerical criterion when using HIA: the inertial force term for the ions must be much smaller than the Lorentz force or the drag force term in the ion momentum equation. That leads to the condition for the accuracy of the approximation:

$$R_{\rm AD}(\ell_{v_i}) \gg \mathcal{M}_{{\rm A}i}{}^2, \tag{2.7}$$

where ℓ_{v_i} is the velocity length scale, $\mathcal{M}_{{\rm A}i}$ is the ion Alfvén Mach number, and $R_{\rm AD}$ is the AD Reynolds number,

$$R_{\rm AD}(\ell) \equiv \frac{\ell v}{t_{\rm ni} v_A^2} = \frac{4\pi\gamma_{\rm AD}\rho_i\rho_n\ell v}{\langle B^2\rangle} = \frac{\ell}{\ell_{\rm AD}}; \tag{2.8}$$

where $\ell_{\rm AD}$ is the AD length scale. $R_{\rm AD}$ is found to be a good parameter to measure the coupling strength between the ions and the neutral gas (Li, McKee & Klein 2006, Li *et al.* 2008). When $R_{\rm AD} \gg 1$, coupling is strong, and when $R_{\rm AD} \lesssim 1$, coupling is weak.

We have five 512^3 AD models driven at Mach 3 at the largest scale, $k = \ell_0/\lambda = 1 \sim 2$, where ℓ_0 is the box length. If the line-width size relation is satisfied, the size of the box is ~ 0.41 pc. The simulation box is periodic and the temperature is at 10K. The plasma $\beta = 0.1$ so that the magnetic field is strong. All simulations ran for 3 crossing time. $R_{\rm AD}(\ell_0)$ spans 4 orders of magnitude in these models, from 0.12 to 1200, enough to cover the range from weak to strong AD. We also have an ideal MHD and a pure hydrodynamic run to compare with the strong and weak AD coupling cases.

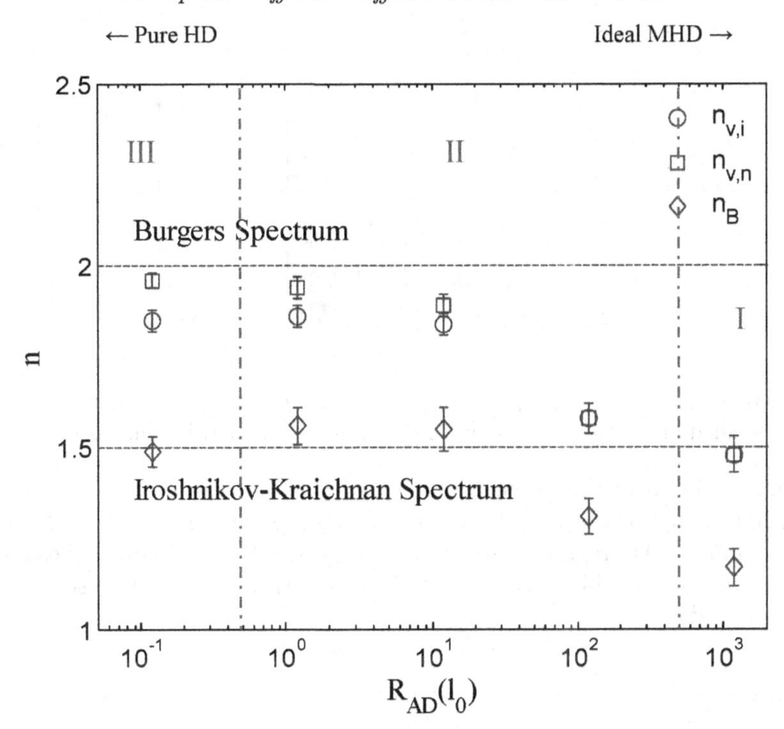

Figure 1. Change of the velocity and magnetic field power spectral indexes of ions and neutrals (n_{vi}, n_{vn}, and n_B) as functions of $R_{\rm AD}$. The transition from the ideal MHD regime (I) to the pure hydro regime begins in the standard AD regime (II). The strong AD regime (III) is also shown. The ion velocity spectral index diverges from the neutral velocity spectral index when AD becomes strong.

3. Simulation Results: Velocity Power Spectrum

We present only the AD effect on the velocity power spectrum of a weakly ionized MC in this short report. For the low Alfvén Mach number in our simulations, the 1D velocity power spectrum of ideal MHD turbulence will be an Iroshnikov-Kraichnan spectrum, with an index of 3/2. For a pure hydrodynamic system, the velocity spectral index is 2, a shock dominated Burgers spectrum. In Figure 1, we plot the ion and neutral velocity spectral indexes and the magnetic field spectral indexes versus $R_{\rm AD}$ for the five turbulent AD models. The neutral velocity spectral indexes of our five AD models trace a transition between the ideal MHD and the pure hydrodynamic regimes. We can categorize five different regimes for AD (McKee, Li, & Klein 2010) by comparing $R_{\rm AD}$ to the ion and neutral Alfvén Mach numbers. Figure 1 shows only the ideal MHD (regime I: $R_{\rm AD} \to \infty$), standard AD (regime II: $R_{\rm AD} > \mathcal{M}_{\rm A}$), and strong AD (regime III: $\mathcal{M}_{\rm A} > R_{\rm AD} > \mathcal{M}_{{\rm A}i}$) regimes. The change in the spectral index occurs mainly inside the standard AD regime, where the neutrals and ions are coupled together over a dynamic time scale and the damping of Alfvén waves of wavelength comparable to the size of the region is weak. Note that the ion power spectral index diverges from the neutral index when the coupling becomes weak. The magnetic field spectral indexes show a similar behavior.

Future observations could verify the change of the velocity power spectral index predicted by our simulations. Crutcher (1999) reports 27 molecular clumps with sensitive magnetic field measurements. Using the cloud clump diameter as the length scale and other measured parameters, we computed $R_{\rm AD}$ for these 27 clouds as shown in Figure 2 (unshaded histogram), including the 12 clouds with only upper limits on the magnetic

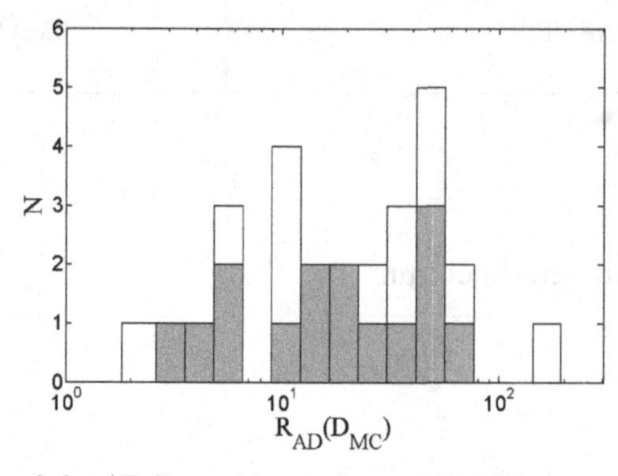

Figure 2. Histograms of the AD Reynolds numbers of 27 MC clumps observed by Crutcher (1999). The shaded histogram shows only the 15 clumps with detected magnetic field strength.

field strength. The logarithmic mean of $R_{\rm AD}$ of the 15 clouds with detected magnetic fields (shaded histogram) is 17 (McKee, Li, & Klein 2010). All these clouds have $R_{\rm AD}$ inside the standard AD regime. Once the velocity spectral indexes of these clouds are determined, we shall be able to determine if the spectral indexes of these clouds change with $R_{\rm AD}$ as predicted.

4. Conclusion

We report our non-ideal MHD turbulence simulations of the effects of AD on weakly ionized MCs. We found that the neutral velocity spectral index in a strong magnetic field environment changes from the Iroshnikov-Kraichnan spectrum expected for a strong B-field system to a Burgers spectrum when AD is strong, as the result of the decoupling of neutrals from ions. Other effects of AD on weakly ionized MCs, such as the clump mass function, the mass-flux ratio, and the line-width size relation are reported in Li *et al.* (2008), McKee, Li & Klein (2010), Li, McKee & Klein (2010). Our simulations show that AD is important in changing the turbulent state of weakly ionized MCs.

Acknowledgements

This research has been supported by NASA under an ATFP grant NNX09AK31G, the NSF under grant AST-0908553 (CFM and RIK), and the high performance computing resources from the NCSA through the grant TG-MCA00N020. CFM also acknowledges the support of the Groupement d'Intérêt Scientifique (GIS) "Physique des deux infinis (P2I)." RIK received support for this work provided by the US Department of Energy at Lawrence Livermore National Laboratory under contract DE-AC52-07NA 27344.

References

Crutcher, R. M. 1999, *ApJ*, 520, 706
Li, P. S., McKee, C. F., & Klein, R. I. 2006, *ApJ*, 653, 1280
Li, P. S., McKee, C. F., Klein, R. I., & Fisher, R. T. 2008, *ApJ*, 684, 380
Li, P. S., McKee, C. F., & Klein, R. I. 2010, in preparation
McKee, C. F., Li, P. S., & Klein, R. I. 2010, *ApJ*, accepted, arXiv:1007.2032
Mac Low, M.-M. & Smith, M. D. 1997, *ApJ*, 491, 596

Computational Star Formation
Proceedings IAU Symposium No. 270, 2011
J. Alves, B.G. Elmegreen, J. M. Girart & V. Trimble, eds.

doi:10.1017/S1743921311000755

Implementing and comparing sink particles in AMR and SPH

Christoph Federrath[1], Robi Banerjee[1], Daniel Seifried[1], Paul C. Clark[1], and Ralf S. Klessen[1]

[1]Zentrum für Astronomie der Universität Heidelberg,
Institut für Theoretische Astrophysik, Albert-Ueberle-Str. 2, 69120 Heidelberg, Germany
email: chfeder@ita.uni-heidelberg.de

Abstract. We implemented sink particles in the Adaptive Mesh Refinement (AMR) code FLASH to model the gravitational collapse and accretion in turbulent molecular clouds and cores. Sink particles are frequently used to measure properties of star formation in numerical simulations, such as the star formation rate and efficiency, and the mass distribution of stars. We show that only using a density threshold for sink particle creation is insufficient in case of supersonic flows, because the density can exceed the threshold in strong shocks that do not necessarily lead to local collapse. Additional physical collapse indicators have to be considered. We apply our AMR sink particle module to the formation of a star cluster, and compare it to a Smoothed Particle Hydrodynamics (SPH) code with sink particles. Our comparison shows encouraging agreement of gas and sink particle properties between the AMR and SPH code.

Keywords. accretion, accretion disks, hydrodynamics, ISM: kinematics and dynamics, ISM: jets and outflows, methods: numerical, shock waves, stars: formation, turbulence

---

1. Introduction

Stars form in turbulent, magnetized molecular clouds by local gravitational collapse of dense gas cores (Mac Low & Klessen 2004). Modeling this process in computer simulations is extremely difficult. It is necessary to follow the freefall collapse of each individual star, while keeping track of the global evolution of the entire cloud at the same time. The fundamental numerical difficulty is that the freefall timescale decreases with increasing gas density: $t_{\rm ff} = \sqrt{3\pi/32G\rho}$. Modeling each individual collapse and following the large-scale evolution of the cloud over several global freefall times in a single magneto-hydrodynamical calculation is beyond the capabilities of modern numerical schemes and supercomputers. Thus, if one wants to model the evolution of such a cloud, the individual runaway collapse must be cut-off in a controlled way and replaced by a subgrid model.

There are two such subgrid models: heating the gas or using sink particles. In the first approach the gas is heated up above a given density threshold to prevent artificial fragmentation beyond the resolution limit for collapse (Truelove *et al.* 1997). There are two problems with the heating approach. First, the Courant timestep decreases, because the sound speed increases, and second, the gas equation of state is changed above the density threshold. For molecular clouds, heating of the gas may only occur for densities $\rho \gtrsim 10^{-14}\,\mathrm{g\,cm^{-3}}$ in the optically thick regime (Larson 1969; Penston 1969). However, gas can become denser than the threshold value in shocks that do not necessarily lead to the formation of a gravitationally bound structure. Thus, shocked gas not going into freefall collapse will be heated up artificially at least in the case where the threshold density for heating is in the isothermal regime.

The alternative subgrid model is to use Lagrangian sink particles, a method invented by Bate, Bonnell, & Price (1995) for Smoothed Particle Hydrodynamics (SPH), and

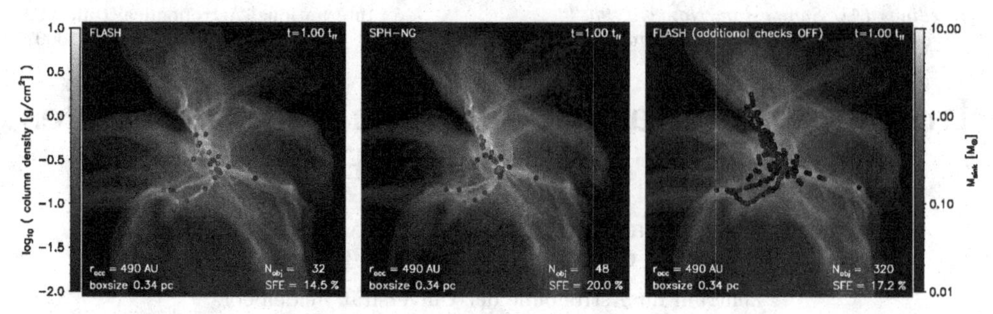

Figure 1. The formation of a stellar cluster from a $100\,M_\odot$ turbulent cloud after one global freefall time. *Left:* FLASH (default settings), *middle:* SPH-NG, and *right:* FLASH (first four checks switched off). See Federrath *et al.* (2010) for further details.

first adopted for Eulerian, Adaptive Mesh Refinement (AMR) by Krumholz, McKee, & Klein (2004). If the gas reaches a given density, a Lagrangian, accreting sink particle is introduced. However, sink particles are supposed to represent bound objects, and thus, a density threshold for their creation is insufficient. Compression in shocks can temporarily create densities larger than the threshold without triggering gravitational collapse. Previous grid-based implementations of sink particles are mostly based on a density threshold criterion. Here, we present an implementation of sink particles for the Eulerian, AMR code FLASH (Fryxell *et al.* 2000) that uses a series of checks, such that only bound and collapsing gas is turned into sink particles. We show that the star formation efficiency and the number of fragments is overestimated, if additional, physical checks in addition to the density threshold are ignored in the isothermal regime.

The main results presented here were previously published in Federrath *et al.* (2010). However, we discuss two new tests for linear and angular momentum conservation in §3. In §2 we present the physical checks necessary to avoid spurious sink particle creation and present the main results of the AMR–SPH comparison of sink particles.

2. Sink particle creation checks and AMR–SPH comparison

We refer the reader to Federrath *et al.* (2010, §2.2) for a detailed discussion and for the implementation of the sink particle creation checks. In summary, for successful sink creation the gas exceeding the density threshold must also

(*a*) be converging (along each cardinal direction individually),
(*b*) have a central gravitational potential minimum,
(*c*) be Jeans-unstable (including magnetic pressure),
(*d*) be bound (including magnetic energy),
(*e*) be on the highest level of the AMR (i.e., Jeans length resolved),
(*f*) not be within accretion radius of existing sinks (then accretion checks apply).

During testing it turned out that the first two checks are particularly important to avoid spurious sink particle creation, however, the relative importance of each of the individual checks should be investigated in more detail (Wadsley *et al.*, in prep.).

Fig. 1 shows the comparison of FLASH (all checks on; left panel) against FLASH (first four checks switched off; right panel), clearly demonstrating the importance of the physical sink creation checks in the isothermal regime. The middle panel of Fig. 1 shows the SPH-NG run, exhibiting some differences to the FLASH run, which can be attributed to the slightly faster collapse of the cloud core in the SPH run. This is most likely a consequence of the faster decay of the supporting initial turbulence due to the slightly higher viscosity in SPH (Price & Federrath 2010). After correcting for this and comparing at times when 26% of the gas has been accreted onto sinks, the FLASH

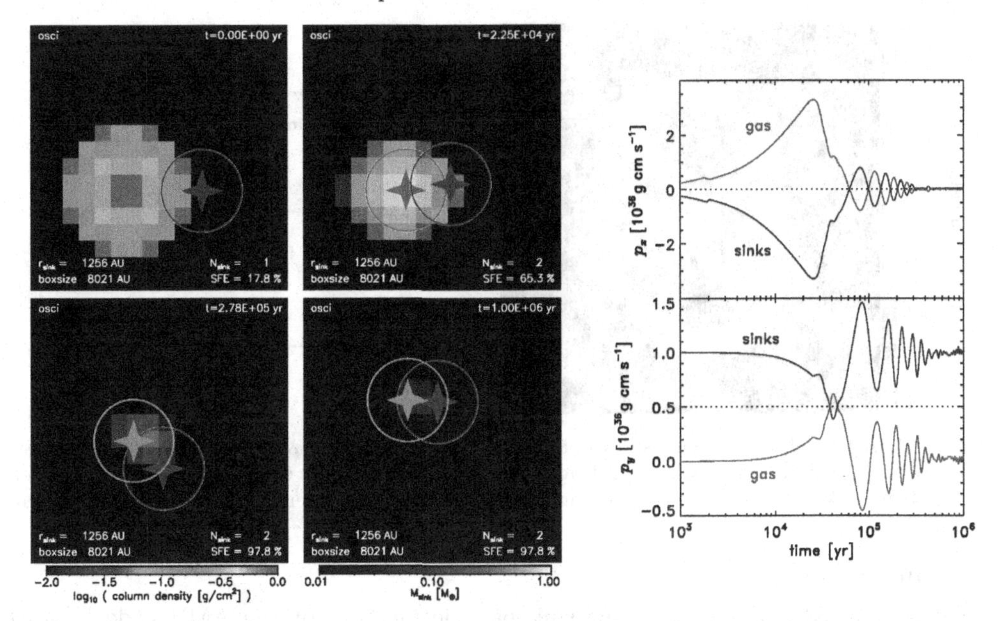

Figure 2. *Left panels:* time evolution of the column density in the momentum conservation test. *Right panels:* time evolution of the momenta of sink particles and gas in x- (*top*) and y-direction (*bottom*). For momentum conservation to hold, the momenta of sinks and gas must be symmetric about the dotted line, which indicates half the initial total momentum.

and SPH-NG runs are in very good agreement with 49 and 50 sink particles formed, respectively, and having similar mass distributions (see §4 in Federrath *et al.* 2010).

3. Momentum and angular momentum conservation test

In Federrath *et al.* (2010) we performed a suite of test simulations for the new sink particle module in FLASH, including circular and highly eccentric orbits, the collapse of a Bonnor-Ebert sphere and a singular isothermal sphere, and a rotating cloud core fragmentation test. Here we add a momentum conservation test shown in Fig. 2, where we initialized a $0.46\,\mathrm{M}_\odot$ core (with very low resolution for testing purposes) at rest and a $0.1\,\mathrm{M}_\odot$ sink particle with an initial momentum of $p_y = 10^{36}\,\mathrm{g\,cm\,s^{-1}}$ in y-direction. Both gas core and sink particle are offset from the center. Fig. 2 (left panels) show column density snapshots of the time evolution of the system. The initial gas core collapses and a second sink particle forms and accretes almost all the initial gas mass, while the initial sink particle only has time to accrete about 2% of the gas. The two sink particles are then followed for 20 orbits up to $t_{\mathrm{end}} = 10^6\,\mathrm{yr}$. The right panels of Fig. 2 show the time evolution of the momentum. Momentum is conserved to within 3% over the whole duration even in this very low resolution run, shown by the symmetry of sink and gas momenta. Kinks in the momentum evolution indicate strong accretion events.

For testing angular momentum conservation, we use a high-resolution model of the Boss & Bodenheimer (1979) test. Figure 3 (left) shows the column density of the disk (face on in the x-y-plane) for the last time frame computed, where 12 fragments have formed in the disk, which turned into a ring-like structure at late times. The right panel shows the time evolution of the total angular momentum, L_z. Angular momentum is conserved to within 0.5% (see Commerçon *et al.* 2008, for a comparison of angular momentum conservation in SPH and AMR).

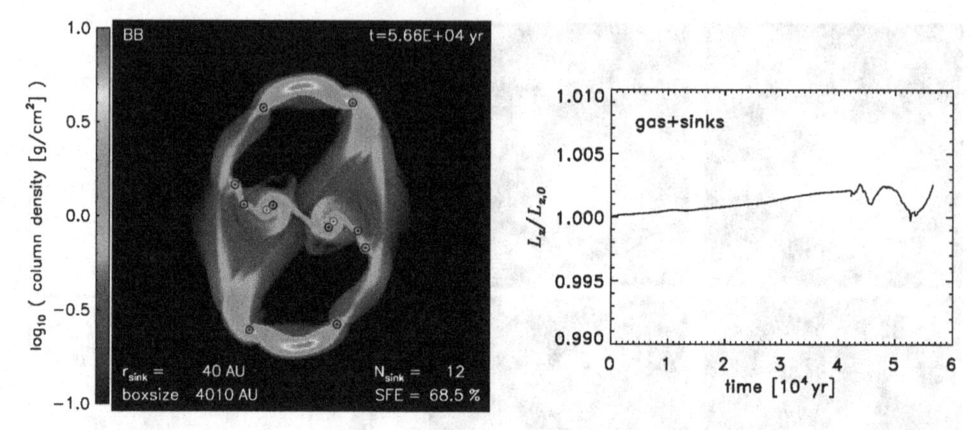

Figure 3. *Left:* Column density snapshot of a variant of the Boss & Bodenheimer (1979) test, followed to very late times when 12 fragments have formed and 68.5% of the gas has been accreted by sinks. *Right:* Shows that angular momentum is conserved to within 0.5% for all times.

4. Conclusions

We introduced and tested a new sink particle method for the AMR code FLASH in Federrath *et al.* (2010). In addition to the tests shown there we presented a linear and angular momentum conservation test in §3. A comparison of gas and sink particle properties showed encouraging agreement with the SPH-NG code (Bate *et al.* 1995). More recently, another SPH code, GASOLINE (Wadsley *et al.* 2004), also showed very good agreement, if the sink particle creation checks outlined in §2 are used (Wadsley *et al.*, in prep.).

Some open issues remain concerning the modeling of magnetic fields in combination with sink particles in SPH (Price & Bate 2008; Price 2010). The problem is that SPH particles are accreted inside the sink particle radius and are thus lost as resolution elements for the magnetic field. The advantage of using a grid-based implementation of sink particles is that the stencil for the magnetic field remains when gas is accreted. The geometry of the magnetic field thus remains intact. We also performed simulations of the collapse of magnetized, rotating cloud cores, which self-consistently produce bipolar outflows (Duffin *et al.*; Seifried *et al.*, in prep.), showing that our sink particle approach works in combination with magnetic fields.

References

Bate, M. R., Bonnell, I. A., & Price, N. M. 1995, *MNRAS*, 277, 362
Boss, A. P. & Bodenheimer, P. 1979, *ApJ*, 234, 289
Commerçon, B., Hennebelle, P., Audit, E., Chabrier, G., & Teyssier, R. 2008, *A&A*, 482, 371
Federrath, C., Banerjee, R., Clark, P. C., & Klessen, R. S. 2010, *ApJ*, 713, 269
Fryxell *et al.* 2000, *ApJS*, 131, 273
Krumholz, M. R., McKee, C. F., & Klein, R. I. 2004, *ApJ*, 611, 399
Larson, R. B. 1969, *MNRAS*, 145, 271
Mac Low, M.-M. & Klessen, R. S. 2004, *Reviews of Modern Physics*, 76, 125
Penston, M. V. 1969, *MNRAS*, 144, 425
Price, D. J. 2010, *MNRAS*, 401, 1475
Price, D. J. & Bate, M. R. 2008, *MNRAS*, 385, 1820
Price, D. J. & Federrath, C. 2010, *MNRAS*, 406, 1659
Truelove *et al.* 1997, *ApJ*, 489, L179
Wadsley, J. W., Stadel, J., & Quinn, T. 2004, *New Astron.*, 9, 137

Computational Star Formation
Proceedings IAU Symposium No. 270, 2011
J. Alves, B.G. Elmegreen, J. M. Girart & V. Trimble, eds.

doi:10.1017/S1743921311000767

Convergence of SPH and AMR simulations

David A. Hubber[1,2], Sam A. E. G. Falle[3] and Simon P. Goodwin[1]

[1]Department of Physics and Astronomy, University of Sheffield,
Hicks Building, Hounsfield Road, Sheffield, S3 7RH, UK

[2]School of Physics and Astronomy, University of Leeds, Leeds, LS2 9JT, UK

[3]Department of Applied Mathematics, University of Leeds, Leeds, LS2 9JT, UK

Abstract. We present the first results of a large suite of convergence tests between Adaptive Mesh Refinement (AMR) Finite Difference Hydrodynamics and Smoothed Particle Hydrodynamics (SPH) simulations of the non-linear thin shell instability and the Kelvin-Helmholtz instability. We find that the two methods converge in the limit of high resolution and accuracy. AMR and SPH simulations of the non-linear thin shell instability converge with each other with standard algorithms and parameters. The Kelvin-Helmholtz instability in SPH requires both an artificial conductivity term and a kernel with larger compact support and more neighbours (e.g. the quintic kernel) in order converge with AMR. For purely hydrodynamical problems, SPH simulations take an order of magnitude longer than the grid code when converged.

Keywords. hydrodynamics, methods: numerical

1. Introduction

Most simulations of self-gravitating hydrodynamics in astrophysics are performed using either Adaptive Mesh Refinement (AMR) Finite Difference hydrodynamics or Smoothed Particle Hydrodynamics (SPH; Lucy 1997; Gingold & Monaghan 1977). Both methods attempt to solve the fluid equations using very different paradigms, each with its own advantages and disadvantages (e.g. grid vs. particles, or Eulerian vs. Lagrangian). Ultimately they should both give the same result in the limit of high resolution and accuracy. Only in the limit of low resolution and accuracy should their limitations become important enough to warrant special attention.

One particular problem with SPH that has been highlighted in recent years is its inability to correctly resolve mixing with particular attention given to the Kelvin-Helmholtz instability (Agertz *et al.* 2007). As well as this problem, there are other examples where SPH and AMR may disagree. We have begun a widespread suite of tests using both SPH and AMR in order to determine i) if SPH and AMR give the same results at high resolution and accuracy; ii) at what resolution(s) SPH and AMR begin to diverge and iii) determine how SPH and AMR behave when under-resolved. We present here the first results of simulations of the non-linear thin-shell instability and the Kelvin-Helmholtz instability using the SPH code SEREN (Hubber *et al.* 2010; in press) SPH code and the AMR-MHD grid code MG (van Loo, Falle & Hartquist 2006).

2. Non-linear thin-shell instability

The non-linear thin-shell instability (hereafter NTSI; Vishniac 1994) occurs when two colliding streams of gas form a shock along a non-planar boundary. If the interface between the two flows is some form of sinusoidal boundary, then this shape can effectively 'funnel' material towards the extrema of the sinusoid. This leads to the growth of density enhancements as more material flows into the shock. We model the NTSI with two

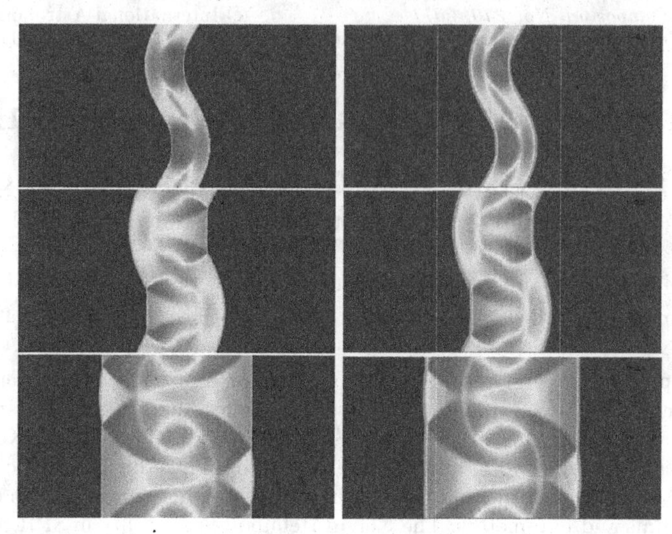

Figure 1. The development of the non-linear thin-shell instability at times $t = 0.1$ (top rown), $t = 0.2$ (middle row) and $t = 0.4$ (bottom row) modelled with MG (left-hand column) and SEREN (right-hand column) in the region $-1 < x < 1$, $0 < y < 1$. Dark colours indicate the pre-shock density and lighter shades the higher-density shocked regions.

uniform density gas flows with the same initial density ($\rho = 1$), constant pressure $P = 1$ and ratio of specific heats $\gamma = 5/3$. The initial velocity profile is

$$v_x(x, y) = \begin{cases} +2 & x < A \sin(k\,y) \\ -2 & x > A \sin(k\,y) \end{cases} \tag{2.1}$$

where $A = 0.1$, $k = 2\,\pi/\lambda$, $\lambda = 1$ and $v_y = 0$ everywhere initially. The computational domain extends between the limits $-3 < x < 3$ and $0 < y < 1$ with open boundaries in the x-dimension and periodic boundaries in the y-dimension. We follow the evolution of the instability until $t = 0.4$ using MG and SEREN. Both codes use standard algorithms and parameters for this test. The resolution of the grid simulation is 1200×200 and the number of SPH particles is 240,000, initially placed on a uniform lattice.

In Figure 1, the AMR results (left-hand column) and the SPH results (right-hand column) are shown at times $t = 0.1$, $t = 0.2$ and $t = 0.4$. This is beyond the growth of the initial density enhancements and towards the generation of more complicated density structures. It can be seen in Figure 1 at $t = 0.1$ that the growth of the initial density perturbation still closely follows the initial sinusoidal boundary. At later times, the flow pattern becomes more complex with a more elongated density structure forming. The AMR and SPH results are nearly identical with only small noticable deviations, possibly due to smoothing in SPH. We note that the SPH simulations are performed with the standard M4 kernel (Monaghan & Lattanzio 1985) and without any additional algorithms such as artificial conductivity (e.g. Price 2008).

3. Kelvin-Helmholtz instability

The Kelvin-Helmholtz instability (hereafter KHI) is a classical hydrodynamical instability generated at the boundary between two shearing fluids which leads to vorticity and mixing between the two fluids. Analysis of the linear growth is given in many classical textbooks and papers (e.g. Chandrasekhar 1961; Junk *et al.* 2010). We model a 2 : 1 density contrast following Agertz *et al.* 2007 and other authors. The two fluids are separated

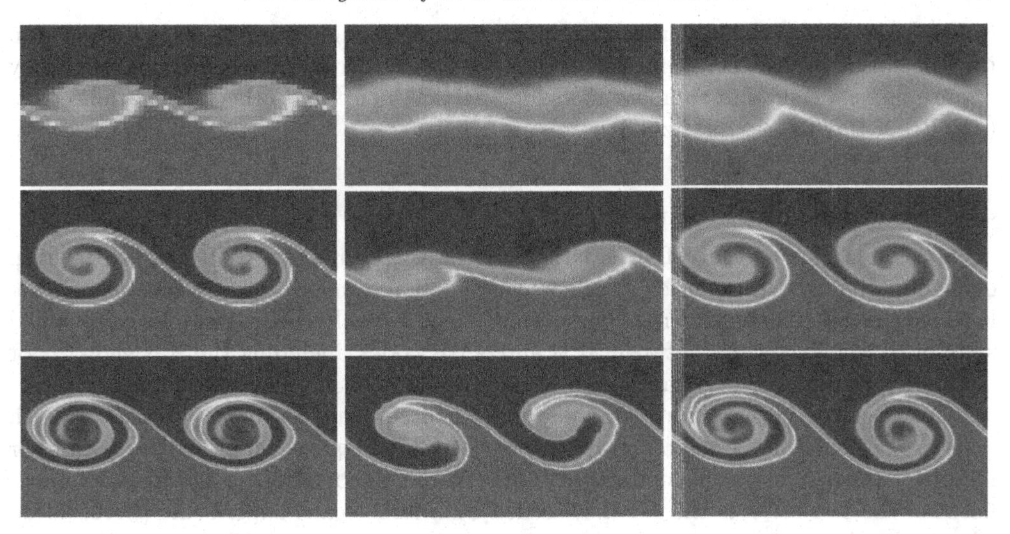

Figure 2. The Kelvin-Helmholtz instability for a 2:1 density contrast at a time $t = 2$ modelled with MG (left-hand column), SEREN with the M4 kernel (middle column) and SEREN with the quintic kernel (right-hand column). The resolution of the grid simulation is 64×32 (top row), 128×64 (middle row) and 256×128 (bottom row). The no. of particles in the SPH simulations are $3,274$ (top row), $24,484$ (middle row) and $97,936$ (bottom row).

along the x-axis and have a x-velocity shear, but are in pressure balance with $P = 2.5$. The ratio of specific heats is $\gamma = 7/5$. Fluid 1 ($y > 0$) has a density $\rho_1 = 1$ and x-velocity $v_x = 0.5$. Fluid 2 ($y < 0$) has a density $\rho_2 = 2$ and x-velocity $v_x = -0.5$. The velocity perturbation in the y-direction with wavelngth $\lambda = 0.5$ is given by

$$v_y(x, y) = \begin{cases} A \sin(k\,x) \exp(-k\,y) & y > 0 \\ A \sin(k\,x) \exp(k\,y) & y < 0 \end{cases} \tag{3.1}$$

where $k = 2\,\pi/\lambda$. The computational domain is $-0.5 < x < 0.5$ and $-0.25 < y < 0.25$ with periodic and mirror boundaries in the x- and y-dimensions respectively. We follow the evolution of the KHI until a time of $t = 2$ using both MG and SEREN, beyond the linear growth of the instability and into the non-linear regime where vorticity develops.

Figure 2 shows the KHI for AMR (column 1), SPH with the M4 kernel (column 2; Monaghan & Lattanzio 1985) and SPH with the quintic kernel (column 3; Morris 1996). For the AMR simulations, the resolution of each row from top to bottom is 64×32, 128×64 and 256×128. For the SPH simulations, the number of particles, from top to bottom, and are $3,274$, $24,484$ and $97,936$. Figure 2 shows that these choices of resolution give very similar results between the AMR simulations and SPH simulations with the quintic kernel, For the lowest resolution (Figure 2, row 1), the two layers only just start to mix together, but there is insufficient resolution for a true vortex to be modelled. As we increase the resolution in both cases (Figure 2, rows 2 and 3), we can see the two methods develop in a similar way, with more vorticity in the spiral patterns being realised.

The SPH simulations with the M4 kernel (middle row) do not agree very well with the AMR simulations in terms of the development of the vorticity. As we increase the resolution, the results with the M4 kernel improve somewhat, but even for very high resolution, they do not agree very well with the grid code at lower resolution. It should be noted that the simulations with the M4 and quintic kernels contain exactly the same numbers of particles. However, since the quintic kernel has a larger compact support ($3\,h$ instead of $2\,h$), its formal resolution is lower. However, it has higher accuracy than the M4 kernel since the larger average number of SPH neighbours means a reduction in particle noise in the SPH summation approximation. Other authors have noted that they require larger neighbour numbers to reduce particle noise and other summation errors (e.g. Read, Hayfield & Agertz 2010).

4. Discussion and Future work

Our simulations of the NTSI and the KHI have shown that AMR and SPH, given enough resolution and accuracy, can show very good agreement in hydrodynamical problems with complex flow patterns. These simulations are only the first in a larger planned suite of convergence tests between SPH and AMR codes, including the Rayleigh-Taylor instability, radiative shocks, advection of polytropes, orbiting binary polytropes.

It is noticeable in our simulations that the NTSI simulations converge very well with standard options and parameters, whereas the SPH KHI required additional algorithms or modifications (e.g. artificial conductivity) plus more neighbours. One principle difference between the two cases is the NTSI is seeded by a large scale, super-sonic perturbation where particle noise and errors are not important, whereas the KHI is the growth of a seeded, low-amplitude velocity perturbation where noise and errors can corrupt the instability early on before it can grow. The accuracy of the SPH method (controlled somewhat by the number of neighbours) required to converge on the same results as the AMR code is therefore dependent somewhat on the problem studied. This is clear from the KHI convergence tests where the M4 kernel does not appear to converge no matter how high the resolution. An important consequence of this is SPH convergence studies should consider varying both the total particle number and neighbour number.

One important practical implication of these convergence tests is that where convergence is achieved, the run times of the SPH simulations are more than an order of magnitude longer than the AMR simulations. This suggests that SPH is not the optimal method for pure-hydrodynamical simulations. The only principal advantage that SPH maintains over grid codes in such scenarios is its Lagrangian nature, which may be important when large advection velocities are present (e.g. Springel 2010). However, preliminary simulations of tests involving self-gravity (e.g. polytrope tests) suggest that the disparity in run-times of astrophysics simulations will not be so great (see also, Federrath *et al.* 2010).

References

Agertz, O., Moore, B., Stadel, J., *et al.*, 2007, *MNRAS*, 380, 963
Chandrasekhar, S., 1961, 'Hydrodynamic and Hydromagnetic Stability', *Oxford University Press*
Federrath, C., Banerjee, R., Clark, P. C., Klessen, R. S., 2010, *AJ*, 713, 269
Gingold, R. A., & Monaghan, J. J., 1977, *MNRAS*, 181, 375
Hubber, D. A., Batty, C. P., McLeod, A., Whitworth, A. P., 2010, *A&A*, submitted
Lucy, L., 1977, *AJ*, 82, 1013
Junk, V., Walch, S., Heitsch, F., *et al.*, 2010, *MNRAS*, submitted
Monaghan, J. J., Lattanzio, J. C., 1985, *A&A*, 149, 135
Morris, J. P., 1996, PhD Thesis - 'Analysis of Smoothed Particle Hydrodynamics with Applications', Monash University
Price, D. J., 2008, *JCoPh*, 227, 10040
Read, J. I., Hayfield, T., Agertz, O., 2010, *MNRAS*, 405, 1513
Springel, V., 2010, *MNRAS*, 401, 791
van Loo, S., Falle, S. A. E. G., Hartquist, T. W., 2006, *MNRAS*, 370, 975
Vishniac, E. T., 1994, *ApJ*, 428, 186

Computational Star Formation
Proceedings IAU Symposium No. 270, 2011
J. Alves, B.G. Elmegreen, J. M. Girart & V. Trimble, eds.
doi:10.1017/S1743921311000779

Radiative Transfer Modeling of Simulation and Observational Data

Jürgen Steinacker[1], Thomas Henning[1] & Aurore Bacmann[2]

[1]Max-Planck-Institut für Astronomie, Königsstuhl 17, 69117 Heidelberg, Germany
email: stein@mpia.de, henning@mpia.de

[2]Université Joseph Fourier - Grenoble 1/CNRS, Laboratoire d'Astrophysique de Grenoble (LAOG) UMR 5571, BP 53, 38041 Grenoble Cedex 09, France
email: aurore.bacmann@obs.ujf-grenoble.fr

Abstract. Radiative Transfer (RT) is considered to be one of the four Grand Challenges in Computational Astrophysics aside of Astrophysical Fluid Dynamics, N-Body Problems in Astrophysics, and Relativistic Astrophysics. The high dimensionality (7D instead of 4D for MHD) and the underlying integro-differential transport equation have forced coders to implement approximative RT methods in order to fit spectra and images or to treat RT in their HD and MHD codes.

The central role of RT in star formation (SF) is based on several facts: a) The dense dusty gas in SF regions alters the radiation substantially making SF one of the most complex applications of RT. b) Radiation transports energy within the object and is therefore an essential part of any dynamical SF model. c) RT calculations tell us which of the processes/structures are visible at what wavelength by which telescope/instrument. Hence, RT is the central tool to analyze simulation results or to explore the scientific capabilities of planned instruments. d) With inverse RT, we can obtain the 1D-3D density and temperature structure from observations, completely decoupled from any (M)HD modeling (and the approximations made within).

In this review, we summarize the main difficulties and the currently used computational techniques to calculate the RT in SF regions. Recent applications of 3D continuum RT in molecular clouds and disks around young massive stars are discussed to illustrate the capabilities and limits of current RT modeling.

Keywords. accretion, accretion disks, radiative transfer, scattering, methods: data analysis, methods: numerical, stars: formation, ISM: clouds, ISM: dust, extinction, ISM: globules, infrared: ISM

1. Introduction

Since scientist have started to tackle the problem of star formation with the aid of computational facilities, the two main questions are: a) Which are the essential physical processes we have to model? and b) which processes can we afford to model given the current computer power? As illustrated in the historical part of these proceedings, the goal of the computational models run on older computers was to understand the combined action of the three basic processes gravitation, gas kinematics, and magnetic fields (see, e.g., the review by Norman). With the use of adaptive meshes and smooth particle codes it became possible to handle the six orders-of-magnitude variation in density as documented in the method section. More recently, codes have been able to study special aspects of the three basic processes, namely jets as a combined action of gas kinematics and magnetic fields, winds, or turbulence within the potentially magnetized gas (see star formation feedback session). Meanwhile, it is also considered to include basic chemical networks in order to determine the ionization degree of the gas for magnetic field coupling purposes, or to compute the abundances of molecules with observable line transitions. While chemical network solvers are no computational problem nowadays, their consideration slows down the codes depending on the sophistication of the network.

In this perspective, the inclusion of radiative transfer effects appears to be a further refinement after the major ingredients to form a star have been dealt with. This is by far not the case, and this misleading concept has been discussed controversially during the conference.

Radiation is a basic ingredient of the star formation process. It is due to its high dimensionality that it was neglected in early calculations or crudely approximated by some heating/cooling functions. With the currently available computer power, the star formation community now is starting to return to more realistic descriptions of how the radiation is transported within the gas, as discussed in this book in the radiative dynamics and feedback sessions. For example, the question where and when gas is starting to collapse to form protostars strongly dependents on a correct treatment of the cooling. In the dense regions of cloud cores, the cooling is dominated by the dust, and it takes complex radiative transfer calculations to determine the heating and cooling in each cell of the computational grid.

There are two main obstacles when performing RT calculations. The intensity of the 3D radiation field depends on the wavelength, three spatial, two directional coordinates, and time. To get an impression of the required resolution, one can assume a decent resolution of 100 points in each variable (or the equivalent number of spatial grid cells when using adaptive grids), In this case, the solution vector has 10^{14} entries. Beside this enormous requirement for the internal memory size of the used computer, the RT equation is an integro-differential equation including a scattering integral, making it difficult to apply common solvers. And 100 grid points might not be enough depending on the problem, e.g., to resolve the strongly peaked UV radiation of the stars in the direction space. Steinacker *et al.* (1996) calculated optimized equally-spaced direction grid points on the unit sphere, but still 100×100 directional grid points correspond to a mean resolution of about 2.7° only.

Aside of its role to transport energy within the object, the transport of the radiation from the enshrouded object to the observer is one of the most fundamental processes in astrophysics, and crucial for our interpretation whenever radiation is altered on its way from the source to the telescope. Within computational star formation, it remains the greatest challenge to link the observational data to the simulation findings. The alteration of the radiation can range from small perturbations like the reddening of stellar light due to interstellar extinction or the change of polarization due to the Faraday effect, to almost complete shielding at short wavelengths and thermal re-emission at infrared and sub-millimeter wavelengths in the case of deeply-embedded star-forming regions.

2. Radiative transfer: solvers and improvements

Comparing continuum and line radiative transfer. For the physical interpretation of observed images, often both sets of line and continuum data have to be considered simultaneously. Here, line transfer is numerically more complex as well as in terms of modeling.

Line RT generally has to be performed for atomic or molecular effects of photon absorption, scattering, or emission. As the atoms and molecules are moving and the lines appear in a narrow wavelengths range, the emission is strongly influenced by the velocity field within the astrophysical structure to investigate. Furthermore, the abundance of the atom or molecule within the gas has to be known or to be determined often depending on a complex network of chemical reactions. This has several effects on calculating and modeling with line RT. First, parameter numbers of the order of several hundreds are needed to describe the model of the gas in its morphology, chemical composition, and kinematics. As the images and SEDs contain line-of-sight-integrated information only, modeling line data means to deal with possible model ambiguities. The model parameters are also strongly coupled so that it is hard to disentangle them just from the line data. On the other side, there is hope the richness of the data will enable us to improve our understanding of the astrophysical processes substantially once 3D line transfer codes are able to determine all parameters. Second, to know the emission of a cell of gas, the level population of the molecule needs to be determined. Given the often complex level structure of the important molecules, this will slow down the line RT calculation (e.g. compared to continuum codes by a factor 10 to 100). Thirdly, the Doppler shift term appears in the equation system which is not present in the continuum case.

Continuum RT is considered, e.g., for dust particles as small solid bodies which absorb, scatter, and re-emit radiation. The micron-sized particles which commonly dominate the RT are well-mixed in the gas, and follow a size and shape distribution (Min *et al.* 2003) as well as a distribution of chemical compositions. This again introduces a strong ambiguity in the modeling process of projected images and SEDs. The time-dependent re-emission of smaller particles down to a size of a few atomic layers again requires more computational effort to calculate the local mean dust temperature (Guhathakurta & Draine(1989) or Siebenmorgen *et al.* (1992)).

In the following, we will concentrate on the simple case of continuum RT of dust particles which have the size of typical interstellar dust particles around the tenth of a micron.

Solution methods. Each of the solution algorithms used sofar has its advantages and drawbacks.

In *Monte-Carlo methods*, a photon is propagated through the calculation domain and its scattering, absorption, and re-emission are tracked in detail (Wolf 2003, Wood *et al.* 2004, Ercolano *et al.* 2005, Jonsson 2006, Pinte *et al.* 2006). This allows to treat very complicated spatial distributions (Juvela & Padoan(2003)), arbitrary scattering functions (Mattila(1970), Witt & Stephens(1974)), and polarization (Whitney *et al.* 2003, Bethell *et al.* 2007). Monte-Carlo methods encountered difficulties when covering re-emission in all directions over many events, and for very small or very high optical depths (Juvela 2005). Meanwhile, several algorithms have been proposed to deal with high optical depths. The major drawback, however, is that there is no global error control when using Monte-Carlo schemes.

Ray-tracing solvers can treat arbitrary density configuration with full error control. General purpose solvers for ordinary differential equations can be used to overcome the problem of strongly varying optical depth when using the ray-tracing solution method. These schemes are available in high-order accuracy and with adaptive step size control (e.g. advanced 5th-order Runge Kutta solvers). But they require the implicit recalculation of the step size and become very time-consuming when the optical depth is high, so that more sophisticated solvers are required.

Grid-based solvers, in combination either with finite differencing or short characteristics, have the advantage of error control on the grid. The drawback is the stiff grid so that the resolution of complex 3D structures needs an appropriate grid generation algorithm (Steinacker *et al.* 2003). Adaptive grids became standard in lower-dimensional problems like hydrodynamical calculations - for RT, the refinement criterion is less clear. An interpolation between the grids for the temperature iteration is numerically costly though and gives rise to interpolation errors in the obtained solution. In most papers presenting grid-based 3D RT codes, numerical diffusion has not been considered and taken into account (Steinacker *et al.* 2002a). The effect is well-known in the course of discretizating hyperbolic equations.

Moment methods are well-posed to treat the optically thick regime and are usually applied to radiation fields with a moderately varying direction dependency. They encounter problems when describing a strongly peaked radiation field arising in the parts of the computational domain where the optical depths is of the order of unity.

Further methods are now in use based on triangulation, or hybrid approaches combining methods (a collection of approaches can be found under www.mpia.de/RT08/).

Radiative transfer on adaptively refined grids. With a finite differencing scheme, the ray equation can be discretized. But the linear grids will not work to resolve the large density gradients. We have proposed to calculate an adaptively refined grid for each wavelength separately as the optical depths strongly varies with wavelength (Steinacker *et al.* 2002b). We have generated grids which are minimizing the 1st-order discretization error in the scattered radiation intensity and provided global error control for solutions of RT problems on this grid. In order to reduce the grid point number in regions where the optical depth becomes large we have proposed to use the concept of penetration depth. The proposed grid generation algorithm is easy to implement, allows pre-calculation of the grids and storage in integer arrays, making a fast solution of the 3D RT equation possible. The drawback of this method is the cost of interpolation calculations and the introduced interpolation errors when it comes to calculating the temperatures using the different grids. *Ray-tracing through regions of very high optical depth.* Especially in massive star formation, the optical depth can vary 6 orders of magnitude or more (Beuther & Steinacker 2007). Commonly, most solvers, be it a Monte-Carlo approach or ray-tracing, are forced to perform small steps slowing down the code. A new ray-tracing technique has been proposed by Steinacker *et al.* (2006) to use the optically thick approximation in order to speed up the calculations. Assuming radial powerlaw dependencies for the density and temperature distribution, we have calculated the absolute solution errors, the numerical effort, and the stepsize variation for a given accuracy. We have shown that advanced ordinary partial equation solvers like 5th-order Runge Kutta with adaptive stepsize control are too expensive to be applied to the inverse 3D RT modeling problem. Instead we suggested a 2nd-order ray-tracing scheme controlling the relative change of the intensity and making use of the diffusion approximation in the regions of high optical depth. The method is designed to cross optical thick regions quickly, to resolve the important regions with an optical depth around unity, and to have a moderate computational expense mandatory for inverse 3D RT modeling. We apply the method to calculate a far-infrared image of a dense molecular cloud core which represents the initial configuration of a star formation process with speed-ups of the order of several hundreds.

3. Applications

We illustrate the requirements and capabilities of RT using a few recent applications. *Dust Absorption and Emission: An evolving molecular cloud core seen in space.* The results of cloud core collapse simulations have been analyzed with RT in Steinacker *et al.* (2004). In the simulations, it was assumed that initially low-mass condensations pass through a stage of turbulence-dominated condensation where they accumulate mass and merge together to form extended prestellar core-like objects. The typical density structures in the cores were non-spherical throughout their evolution. Using a 3D continuum RT program, images at 7 μm, 15 μm, 175 μm, and 1.3 mm were generated for different evolutionary times and viewing angles. As an example, Fig. 1 shows images at these wavelengths for an evolution time of 56 000 years after the onset of the gravitation. It

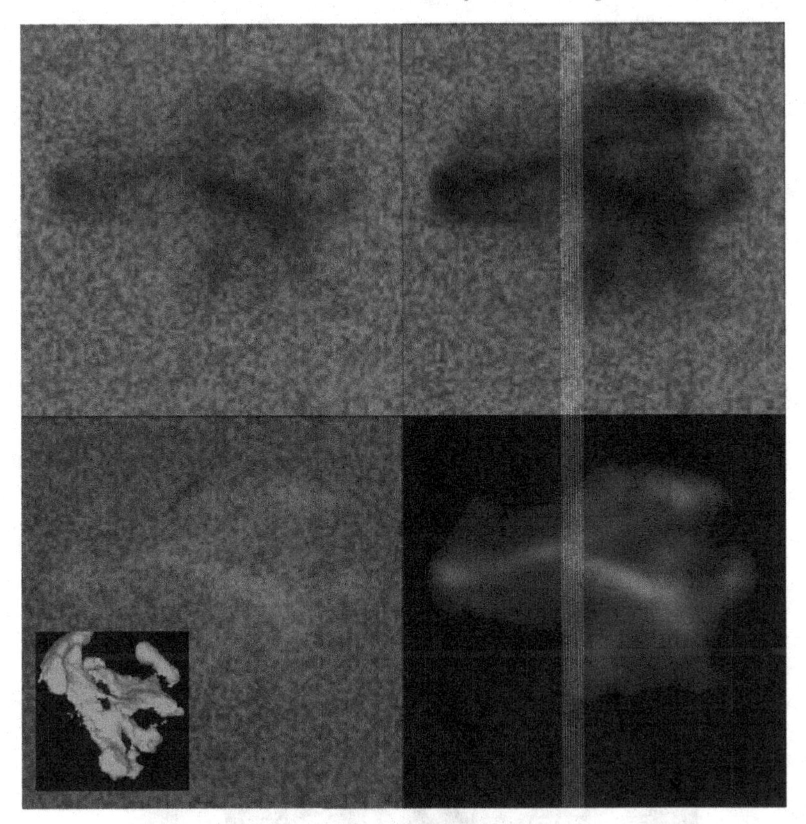

Figure 1. Images of the cloud core fragment at the wavelengths 7 (top left), 15, 175, and 1300 μm (bottom right), at a simulation time of 56 000 years after the start of the start of the gravitational influence. The inlet shows the iso-density surfaces of the used averaged SPH density distributions of the cloud core fragment corresponding to a density of 4×10^{-17} kg m^{-3}. (from Steinacker *et al.* (2004))

was shown that projection effects can lead to a severe misinterpretation of images: A 1D analysis of the vicinity of the density maxima would suggest density profiles in agreement with 1D-core collapse models. The underlying density structure, however, is intrinsically 3D and deviates strongly from the obtained 1D model distribution.

The massive disk candidate SO1 in M17. While observations show that young low- and intermediate-mass stars are surrounded by a circumstellar disk which are often actively accreting matter from it, only a few candidates for such a circumstellar disk around a massive star are known (see Chini *et al.* 2004 for a list of candidates).

Steinacker *et al.* (2006) have analyzed the prominent silhouette structure in M 17 showing a symmetric large-scale pattern in absorption against a bright background, with a central emission region, an hourglass-shaped reflection nebula perpendicular to the extinction bar, a complex outflow over and below the dark extinction lane, and signatures for accretion of matter. Due to the large scale and the strong symmetry of the structure, as well as its presence within a massive star formation region, it attracts special attention as a candidate for a massive disk around a star that might be massive or has the potential to reach such a mass. While the estimate of the disk mass for most massive disk candidates comes from low resolution FIR/mm measurements, they used the advantage that due to the background illumination, the column density can be determined at $\lambda = 2.2$ μm from a high-resolution NAOS/CONICA image.

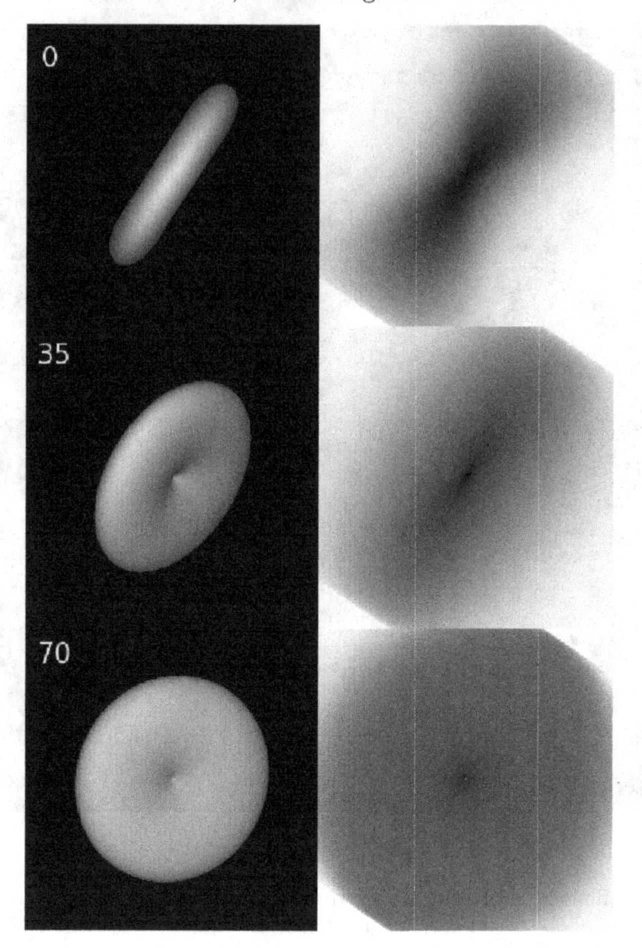

Figure 2. Theoretical absorption image of the circumstellar disk using the optimized fit parameters and varying the inclination angle against the edge-on case. The left panel shows an iso-density surface of the disk distribution (the inclination angle against the edge-on case is given in degrees). The right panel shows the corresponding 2.2 μm absorption image . (from Steinacker *et al.* (2006))

It was investigated whether the observed extinction structure is consistent with a model of a circumstellar rotationally-symmetric flattened density distribution. Applying a commonly used analytical disk model with a powerlaw in radius and a vertical Gaussian distribution, they have fitted a 7 parameter model to the about 8000 pixels of the image. The PSF of a point source was covered by about 3.3 pixels, corresponding to about 2400 independent data points. It was found that the derived optical depth is consistent with a rotationally-symmetric distribution of gas and dust around the central emission peak. The extent of the axisymmetric disk part is about 3000 AU, with a warped point-symmetrical extension beyond that radius, and therefore larger than any circumstellar disk detected sofar. The resulting theoretical absorption image of the disk is shown in Fig. 2 for different inclination angles, along with the corresponding 2.2μm-image.

The mass of the entire disk estimated from the column density was discussed depending on the assumed distance and the dust model and ranges between 0.02 and 5 $M_\odot$. The derived disk mass range is smaller than the mass range of 110 to 330 $M_\odot$ derived from ^{1}3CO data (Chini *et al.* 2004).

Figure 3. Two layers of constant density (solid and semi-transparent) illustrating of the full 3D configuration of the molecular cloud core Rho Oph D. The density was determined by fitting a continuum map at 1.3 mm and two continuum maps at 7 and 15 micron. (from Steinacker *et al.* (2005))

The next step: Inverse 3D RT modeling of the molecular cloud core Rho Oph D. Calculating images from a given density and temperature distribution using a RT code can quickly reach the capabilities of current computers. The finally desired approach would be, however, to determine the density and temperature structure in 3D from the images. This so-called *inverse* RT problem features several difficulties, most prominently the loss of information due to the line-of-sight integration performed during the observations. It was shown in Steinacker *et al.* (2005) that inverse RT transfer is possible nevertheless already for the analysis of molecular cloud cores. A new method was proposed to model the 3D dust density and temperature structure of such a cores. It is based on the fits of multi-wavelengths continuum images in which the core is seen in absorption and emission and requires only a few computations with a 3D CRT code. The large parameter space of a 3D structure is covered using simulated annealing as optimization algorithm.

The method was applied to model the dense molecular cloud core ρ Oph D. In the MIR, the core is seen in absorption against a bright background from the photo-dissociation region of the nearby B2V star, illuminating the cloud from behind. Two ISOCAM images at 7 and 15 μm have been fitted simultaneously by representing the dust distribution in the core with a series of 3D Gaussian density profiles. The background emission behind the core was interpolated from nearby regions of low extinction. Using simulated annealing, a 2D column density map of the core was obtained. The column density of the core has a complex elongated pattern with two peaks, with the southern peak being more

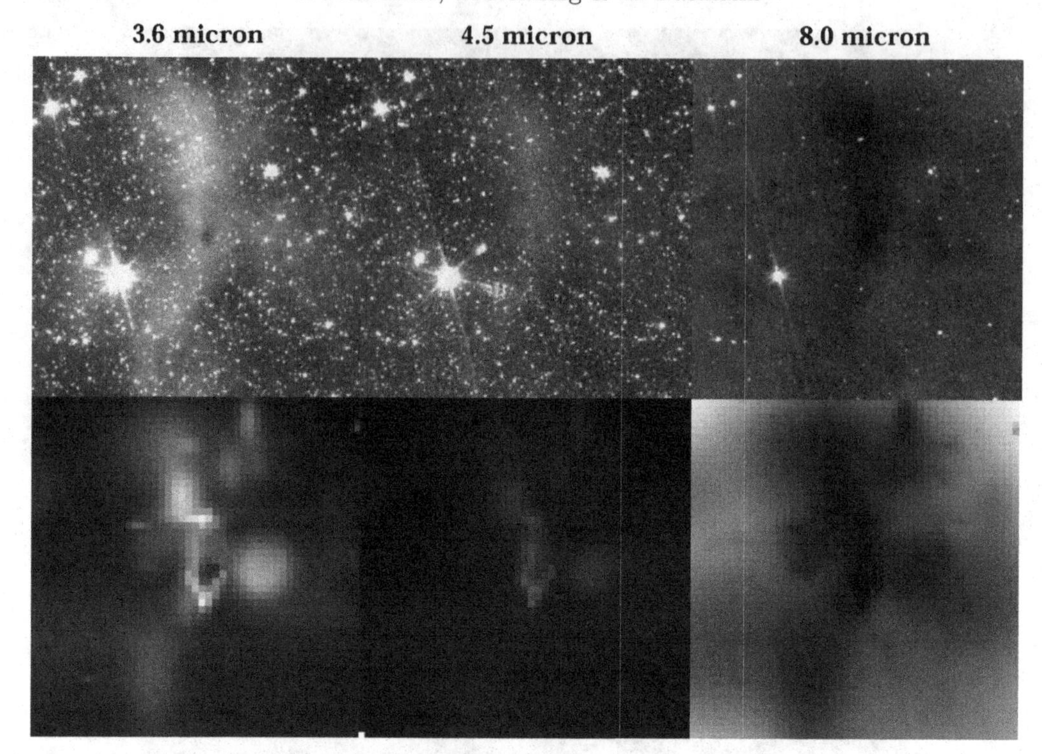

Figure 4. Comparison of the three Spitzer images at 3.6, 4.5, and 8 m of the inner 66000 AU of L183 (top) with scattered light models based on grains growing as a function of density (bottom). The underlying 3D structure of the model images is consistent with the measured Av map. The general pattern of the modeled diffuse emission is similar while, the flux is about a factor of 2 lower in the model. (from Steinacker *et al.* (2010))

compact. Fig. 3 illustrates the resulting 3D density data cube by showing two iso-density layers (semi-transparent and solid).

Coreshine: the new window to explore dust growth in dense molecular cloud cores. Theoretical arguments suggest that dust grains should grow in the dense cold parts of molecular clouds. Evidence of larger grains has so far been gathered in near/mid infrared extinction and millimeter observations. Interpreting the data is, however, aggravated by the complex interplay of density and dust properties (as well as temperature for thermal emission). In Steinacker *et al.* (2010), new Spitzer data of L183 have been presented in bands that contain substantial and marginal PAH emission. A visual extinction map of L183 was fitted by a series of 3D Gaussian distributions. For different dust models, the scattered MIR radiation images of structures have been calculated that agree with the A_V map and they were compared to the Spitzer data. The Spitzer data of L183 show emission in the 3.6 and 4.5 μm bands, while the 5.8 μm band shows slight absorption. This emission was interpreted to be MIR scattered light from grains located further inside the core, and was called "coreshine". Models with grains growing with density yield images with a flux and pattern comparable to the Spitzer images in the bands 3.6, 4.5, and 8.0 μm.

References

Bethell, T. J., Chepurnov, A., Lazarian, A., & Kim, J. 2007, *ApJ*, 663, 1055
Beuther, H. & Steinacker, J. 2007, *ApJL*, 656, L85

Bianchi, S. 2007, *A&A*, 471, 765
Chini, R., Hoffmeister, V., Kimeswenger, S., Nielbock, M., Nürnberger, D., Schmidtobreick, L., & Sterzik, M. 2004, *Nature*, 429, 155
Ercolano, B., Barlow, M. J., & Storey, P. J. 2005, *MNRAS*, 362, 1038
Guhathakurta, P. & Draine, B. T. 1989, *ApJ*, 345, 230
Jonsson, P. 2006, *MNRAS*, 372, 2
Juvela, M. & Padoan, P. 2003, *A&A*, 397, 201
Juvela, M. 2005, *A&A*, 440, 531
Mattila, K. 1970, *A&A*, 9, 53
Min, M., Hovenier, J. W., & de Koter, A. 2003, *A&A*, 404, 35
Pinte, C., Ménard, F., Duchêne, G. & Bastien, P. 2006, *A&A*, 459, 797
Siebenmorgen, R., Kruegel, E., & Mathis, J. S. 1992, *A&A*, 266, 501
Steinacker, J., Thamm, E. & Maier, U. 1996, *JQSRT*, 97, 56
Steinacker, J., Hackert, R., Steinacker, A., & Bacmann, A. 2002, *JQSRT*, 73, 557 (a)
Steinacker, J., Bacmann, A., & Henning, T. 2002, *JQSRT*, 75, 765 (b)
Steinacker, J., Henning, T., Bacmann, A., & Semenov, D. 2003, *A&A*, 401, 405
Steinacker, J., Lang, B., Burkert, A., Bacmann, A., & Henning, T. 2004, *ApJL*, 615, L157
Steinacker, J., Bacmann, A., Henning, T., Klessen, R., & Stickel, M. 2005, *A&A*, 434, 167
Steinacker, J., Bacmann, A., & Henning, T. 2006, *ApJ*, 645, 920
Steinacker, J., Chini, R., Nielbock, M., Nürnberger, D., Hoffmeister, V., Huré, J.-M., & Semenov, D. 2006, *A&A*, 456, 1013
Steinacker, J., Pagani, L., Bacmann, A., & Guieu, S. 2010, *A&A*, 511, A9
Whitney, B. A., Wood, K., Bjorkman, J. E., & Wolff, M. J. 2003, *ApJ*, 591, 1049
Witt, A. N. & Stephens, T. C. 1974, *AJ*, 79, 948
Wolf, S. 2003, *Computer Physics Communications*, 150, 99
Wood, K., Mathis, J. S., & Ercolano, B. 2004, *MNRAS*, 348, 1337

Eve Ostriker, Åke Nordlund, and Paolo Padoan

Computational Star Formation
Proceedings IAU Symposium No. 270, 2011
J. Alves, B.G. Elmegreen, J. M. Girart & V. Trimble, eds.

doi:10.1017/S1743921311000780

Radiative Transfer in Molecular Clouds

M. Juvela
Department of Physics, FI-00014 University of Helsinki, Finland

Abstract. Information of astronomical objects is obtained mainly through their radiation. Thus, the radiative transfer problem has a central role in all astrophysical research. Basic radiative transfer analysis or more complex modeling is needed both to interpret observations and to make predictions on the basis of numerical models. In this paper I will discuss radiative transfer in the context of interstellar molecular clouds where the main scientific questions involve the structure and evolution of the clouds and the star formation process. The studies rely on the analysis of spectral line and dust continuum observations. After a discussion of the corresponding radiative transfer methods, I will examine some of the current challenges in the field. Finally, I will present three studies where radiative transfer modeling pays a central role: the polarized dust emission, the Zeeman effect in emission lines, and the continuum emission from dense cloud cores.

Keywords. radiative transfer, ISM: clouds, ISM: lines and bands, ISM: molecules, dust, infrared: ISM, radio lines: ISM

---

1. Introduction

The radiative transfer problem includes both the formation and transport of radiation. Because the radiation field affects the way the medium emits, one must know the radiation field to calculate the emission and vice versa. All parts of an interstellar cloud can interact with each other through radiation and, when optical depths are high, a self-consistent solution may require a lengthy iterative process. This makes the radiative transfer problem hard and computationally demanding.

Interstellar dust interacts with radiation through absorption, emission, and scattering. Monte Carlo (MC) still remains the most common tool for these calculations. After the simulation of the radiation field, the dust temperatures are solved for and, if necessary, the process is repeated until convergence is reached. When small grains are included the calculation of their stochastic heating becomes the dominant cost that limits the size of the models that can be handled. With radiative transfer modeling one can obtain a possible although usually not a unique solution for the cloud structure, the dust properties, and the strength of radiation sources. It is the only way to reliably study more complex phenomena like the spatial variations of dust grain properties.

Line transfer is different in two respects. Firstly, the calculations need to resolve the line profiles and consider the effect of doppler shifts. Secondly, the scattering (induced emission excluded) can usually be ignored so that it is enough to solve the radiative transfer equation along a fixed set of lines. Monte Carlo is often used although non-random ray-tracing methods should be more efficient. Compared to the continuum case, the need for fine discretization of the line profiles tends to make the calculations more time consuming. A large number of energy levels also means large memory requirements that can limit the spatial resolution of the models. The fractional abundances of the studied species are a crucial and often a poorly known parameter. Conversely, radiative transfer modeling of observations is the best way to estimate the abundances and, at the same time, to draw conclusions on the structure and kinematics of the clouds.

2. Radiative transfer methods

I will present a brief review of radiative transfer methods. This is in part biased towards MC methods and the reader is advised to look up the references for a broader view.

2.1. *Methods for continuum radiative transfer*

The Monte Carlo (MC) method simulates actual processes in a cloud: Photons are emitted at random locations (but weighted according to the real emission) and interact with the medium at random positions (but in accordance with the optical depths). This makes MC intuitive and easy to implement, even in the case of arbitrary scattering functions. MC was used already in the 1970's to study the scattering in interstellar clouds (Mattila 1970, Witt 1977) and later also to calculate dust emission (e.g. Bernard *et al.* 1992). Although spherically symmetric models are still used, 2D and 3D models have become common during the past decade (e.g., Wood *et al.* 1998, Dullemond & Turolla 2000, Juvela & Padoan 2003, Pascucci *et al.* 2004). When the radiation field is sampled with random numbers, the noise decreases slowly, $\sim 1/\sqrt{N}$, as the number of photon packages increases. The efficiency can be improved in many ways. When calculating images of scattered light, the method of forced first scattering guarantees that every photon package scatters at least once (Mattila 1970) and the peel-off scheme (Yusef-Zadeh *et al.* 1984) reduces the noise further by registering a contribution from every scattering. In dust temperature calculations, the noise can be decreased by calculating the absorptions explicitly along the photon path (Lucy 1999). Weighting can be applied to the spatial, angular, and spectral distribution of the packages (Watson & Henney 2001, Juvela 2005, Jonsson 2006). For example, in a spherically symmetric model the innermost shells are rarely hit by any of photon packages unless weighted sampling is used. If the same number of packages representing background photons is sent towards each annulus defined by the spatial discretization, the reduced randomness may even lead to a convergence of $\sim 1/N$. Similar improvement can in some cases be obtained with the use of quasi random numbers. Spatial weighting can also be a solution to sampling problems caused by high optical depths. By creating photon packages preferentially close to cell boundaries, one can guarantee that a fixed fraction of the packages crosses the cell boundary and radiative couplings can be determined for any optical depth (Juvela 2005).

When dust emission has a significant contribution to the radiation field, iterations are needed between the radiative transfer and the re-evaluation of dust temperatures. In optically thin clouds this is not necessary but optically thick dust shells around (proto)stars may require hundreds of iterations. The convergence can be improved by the use of the accelerated MC (AMC) methods (Juvela 2005), sometimes by orders of magnitude. During the iterations only the emission from the medium needs repeated simulation. All the other components of the radiation field can be included as a fixed reference field. When also the dust emission is included in the reference field, simulations are used only to make small corrections to the previously estimated field. As a result, the sampling noise is smaller and is further reduced as the iterations progress (Juvela 2005).

A completely new Monte Carlo scheme was presented by Bjorkman & Wood (2001). In this 'immediate re-emission' method a photon package is initiated in a radiation source. When the package is absorbed, the local temperature is updated and a package is re-emitted from the same location. The package is generated according to a probability distribution that reflects the change in the local SED. The generation of packages follows the natural flow of energy and the method provides automatic weighted sampling. In basic MC, a large number of packages is simulated before the temperatures of all cells are updated at the same time. The immediate re-emission mechanism requires one temperature update after each interaction but, in the case of grains at an equilibrium temperature, the method is quite competitive with the traditional MC even when the latter is enhanced with weighted sampling, AMC, and the use of a reference field (see Juvela 2005). The situation can change if the cost of temperature updates is increased, for example, by the inclusion of several grain populations, grain size distributions, or stochastically heated grains.

In MC the scattering function is taken into account in a statistical way by drawing scattering angles from the appropriate probability distribution. If the angle is discretized, non-stochastic methods can be used with the expense of somewhat larger memory requirements. Examples of such codes can be found in Steinacker *et al.* (2003), Semionov & Vansevičius (2005), and Ritzerveld & Icke (2006) (see also Pascucci *et al.* 2004 and Steinacker in this volume).

2.2. *Line transfer methods*

In line transfer it is usually enough to solve absorption, emission, and stimulated emission along straight rays, ignoring general scattering processes. However, the optical depths are often high so that the solution needs iterations between the estimation of the radiation field and updates of the level populations.

In Monte Carlo the locations and directions of the rays are selected randomly. The first implementations were made in the 1960s but the history of many of the current codes can be traced back to the paper of Bernes (1979). Bernes employed a fixed reference field to reduce the Monte Carlo noise. Choi *et al.* (1995) introduced an adaptive version where, as iterations progress, the reference field approaches the true field. Other improvements included the handling of overlapping lines (Gonzáles-Alfonso & Cenicharo 1993) and the explicit handling of emission and absorption along the rays (Juvela 1997). The first 3D implementations were also made already in 1990's (Park & Hong 1995, Juvela 1997).

Although the $1/\sqrt{N}$ convergence of Monte Carlo could be improved by using quasi random numbers, it is equally easy to implement a regular grid of rays. In such 'ray-tracing codes' one usually follows intensity instead of photon numbers. In the short characteristic method, the intensity is propagated one layer of cells at a time and interpolation is used to derive the values at exact grid positions (e.g., Kunasz & Auer 1988, Auer & Paletou 1994). Long characteristics are continuous rays from the cloud boundary to a cell. They avoid diffusing the radiation field but cause some duplication of calculations. The ray tracing codes estimate the intensity at grid positions while the Monte Carlo versions estimate an average quantity for a cell volume. By adopting the latter approach, the grid of rays becomes independent of the spatial discretization and the local intensity can be estimated with any rays that cross a cell. The result is essentially a long characteristic method without the need for redundant work (Juvela & Padoan 2005).

The similarity of Monte Carlo and ray tracing codes is evident also by looking at the methods used to accelerate the convergence. If emitted photons are absorbed locally, the flow of energy between cells becomes small. This is reflected in the iterations as slow convergence. The core saturation method improves the situation by a specific handling of the optically thick part of the line. The method that was originally developed by Rybicki 1972 was first implemented in the Monte Carlo context by Hartstein & Liseau (1998). In accelerated Λ-iteration (ALI) one explicitly takes into account the emission-absorption cycle within each cell (diagonal Λ operator) or even the radiative couplings between neighboring cells (Scharmer 1981, Olson *et al.* 1986). On the Monte Carlo side the term Accelerated Monte Carlo (AMC) is used for essentially identical methods (Juvela & Padoan 1999; Hogerheijde & van der Tak 2000).

3. Current challenges

Better integration with models of other physical processes and the quest for higher spatial resolution through the use of adaptive grids are some of the current challenges in the radiative transfer modeling.

3.1. *Models of chemistry and interstellar dust*

A complete model of an interstellar cloud should include self-consistent calculations of cloud dynamics, thermal equilibrium, and chemistry. Molecular clouds are cooled through atomic and molecular emission lines and dust continuum emission and, with the exception of the densest cores, heated mainly through the photoelectric effect. Therefore, radiative transfer plays a central core in the thermal balance. Similarly, the chemical photo-reaction rates can be estimated only if the radiation field is known.

The chemical abundances are a major source of uncertainty for a quantitative analysis of line observations. The abundances can vary by orders of magnitude and depend on many parameters like the density, the temperature, and the local radiation field. The large spread in the time constants of the chemical reactions means that a chemical equilibrium may be reached only after several million years. In a dynamic environment the equilibrium is never attained and the prediction of line emission would require the tracing of the chemical history of each gas element. In the setting of turbulent clouds, the dissipation of turbulent energy may produce very localized abundance variations for species like HCO+ (Falgarone *et al.* 2006) while even for the most important observational tracer molecule, CO, the time dependence leads to a large scatter in the abundances found for given density and temperature (Glover *et al.* 2010). These could affect our view of the large scale statistics of the density and velocity fields and need to be studied further, also with high resolution simulations.

In dense regions of the clouds, depletion poses additional problems. For CO, the complementarity of the gas phase phase and solid state abundances has been demonstrated (e.g. Whittet *et al.* 2010) but the general modeling of the depletion process still contains many uncertainties. This is particularly relevant for pre-stellar cores where the depletion is eventually almost complete for most of the commonly observed species and where the depletion time scales are not very different from those of the dynamic core evolution. Although some emission lines can be observed towards even the coldest of cores (e.g., H_2D+, N_2H+, and ammonia) one cannot be quite sure how they probe the different regions of the cores (see, e.g., Walmsley *et al.* 2004; Sipilä *et al.* 2010). This makes it difficult to draw definite conclusions even on the basic core kinematics.

In the case of dust, dense cores are again associated with the strongest effects with an almost complete disappearance of small grains, the accumulation of ice mantles, and the possible formation of fluffy, very large grains (e.g., Ossenkopf & Henning 1994, Stepnik *et al.* 2003). These are used to explain the variations in the ratio of mid-infrared and far-infrared emission and the sub-millimeter emissivity that, at least in some dense clouds, appear to have increased several fold (Bernard *et al.* 1999, Lehtinen *et al.* 2007). The temperature dependence of the spectral index (e.g., Dupac *et al.* 2003) has also been seen in laboratory measurements (Boudet *et al.* 2005) and can be explained by properties of the grain material itself (Meny *et al.* 2007). When these effects are included in dust models (e.g., DUSTEM, Compiègne *et al.* 2010) their effect on the interpretation of far-infrared and sub-millimetre observations must be studied with radiative transfer models. From a technical point of view, the temperature dependence of the dust opacities is a fundamental change that requires modifications in many algorithms and may prove impractical in connection of some previously used methods.

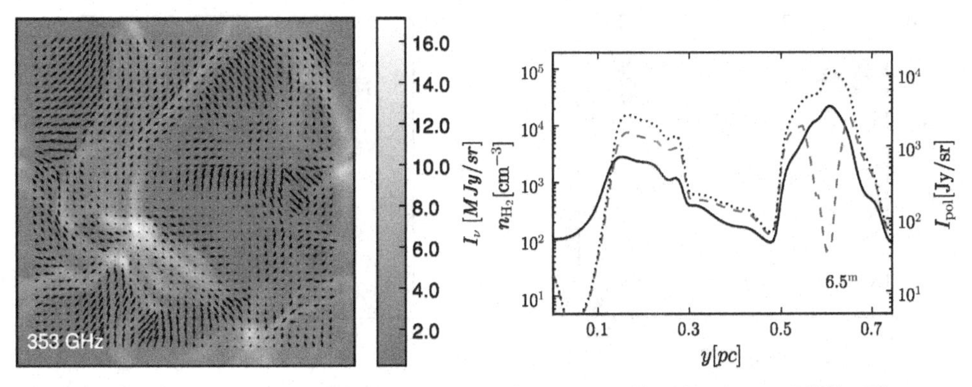

Figure 1. *Left:* Simulated dust polarization map of an interstellar cloud at 353 GHz. The vectors show the polarization degree on a map of total intensity (Pelkonen *et al.* 2007). *Right:* The gas density (solid line) and the polarized intensity at 353 GHz for normal (dashed line) and doubled dust grain size (dotted line) along a line of sight. Without significant grain size increase, the densest core does not contribute significantly to the polarized intensity (see Pelkonen *et al.* 2009).

3.2. *High spatial resolution*

The star formation process involves size scales spanning more than seven orders of magnitude. It is possible to study separately phenomena at large scales (clouds) and at small scales (cores or protostars) but there is clear need for methods with adaptive spatial resolution. This possibility already exists in many codes (e.g., Steinacker *et al.* 2003, Juvela & Padoan 2005, Ritzerveld & Icke 2006, Niccolini & Alcolea 2006).

The implementation of variable resolution using a hierarchy of grids provides advantages beyond the improved effective resolution. Each grid can be processed relatively independently providing a basis of parallel implementations and sequential programs with very low memory usage. However, more interesting is the way the grid structure can help the organization of the calculations. Most effort is usually spent on optically thick regions that fill a small fraction of the full model volume. Optical depth causes problems for the sampling of the radiation field (especially for MC) and for the convergence. When the spatial resolution only changes between well defined grids, simple ray splitting/joining at the grid boundaries ensures that the sampling of the radiation field always corresponds to the spatial discretization. In MC weighted sampling of the emission from the medium can be implemented with any discretization. However, for external radiation this is easier if the incoming energy is first stored on the grid boundary. The effect of the external field on the grid or a subtree in a grid hierarchy can then be estimated by sampling this information with any number of photon packages.

If the information of the external radiation is stored (explicitly on the grid boundary or, in MC, in the form of a reference field) it is trivial to carry out sub-iterations on arbitrary parts of the grid hierarchy. This can results in very large savings in the modeling of inhomogeneous clouds. The overhead of basic ALI/AMC is small enough so that these methods can be used throughout the models. However, the radiative couplings between neighboring cells already require a significant amount of storage. Because higher order ALI/AMC methods are not helpful in optically thin medium, their use can be safely limited to the grids with the highest optical depths. In dust continuum calculations, if a subgrid contains only a few thousand cells, it becomes feasible to store the couplings between most or even all the cells within the grid. So far as the external field remains unchanged, the temperatures can then be solved for that grid almost without any iterations. Only if the dust opacities are temperature dependent, iterations cannot be avoided.

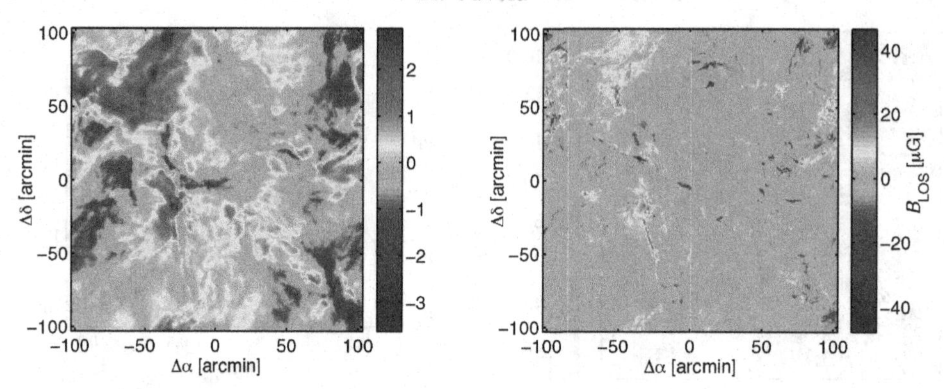

Figure 2. *Left:* Mean line-of-sight magnetic field in the MHD model used for simulations of the Zeeman splitting of OH lines. *Right:* Line-of-sight magnetic field strength determined from the simulated Zeeman observations. The derived magnetic field estimates are strongly weighted towards the densest regions (Lunttila et al. 2009).

4. Examples of radiative transfer studies

Below are given three examples of studies of interstellar clouds where radiative transfer modeling plays a major role.

Polarized dust emission. If dust grains are aligned with the magnetic field in interstellar clouds, they will polarize background starlight (excess intensity along the magnetic field), and will themselves emit polarized thermal radiation (excess intensity perpendicular to the magnetic field). Different grain alignment mechanisms have been debated since the 1950's (see Lazarian 2007). The current favorite is alignment by radiative torques (RATs) where a grain that reacts differently to left- and right-handed circularly polarized light is spun up even when subjected to unpolarized but anisotropic light. Draine & Weingartner (1997) demonstrated with numerical simulations that radiative torques are able to spin up and, importantly, also align the dust grains with respect to the magnetic field. When polarized dust emission of interstellar clouds is modelled, radiative transfer calculations are needed not only to determine the total intensity of the dust emission but also to estimate the anisotropic illumination of the grains and, thus, the degree of grain alignment. Using spherically symmetric model clouds, Cho and Lazarian (2005) showed that RATs lead to the observed drop in the polarization degree as a function of intensity. The conclusion has been confirmed by the subsequent studies with more realistic cloud models (Pelkonen *et al.* 2007) and, in addition, with more realistic grain alignment modeling (Bethell *et al.* 2007, Pelkonen *et al.* 2009) that employs radiative transfer calculations to derive the radiation field inside clumpy clouds.

The grain growth may be of critical importance for dense cores to contribute to polarized emission (see Fig.1). Pelkonen *et al.* (2009) added two caveats. Firstly, grain growth is a slow process and thus not all dense cores might be observable in polarization. Secondly, previous studies assumed that the magnetic field direction and the anisotropic radiation field direction were identical. Hoang & Lazarian (2009) showed that when the angle between these two directions increases, the efficiency of the grain alignment by RATs weakens. When this effect is included the ability of polarization to trace the magnetic field is limited to much lower A_V (see Pelkonen *et al.* 2009).

Zeeman effect in emission lines. Measurements of the Zeeman effect provide information on the absolute value of the line-of-sight component of the magnetic field. This is crucial input for theories of cloud dynamics and prestellar core formation but the complex structure of molecular clouds makes the interpretation of observations difficult. Lunttila

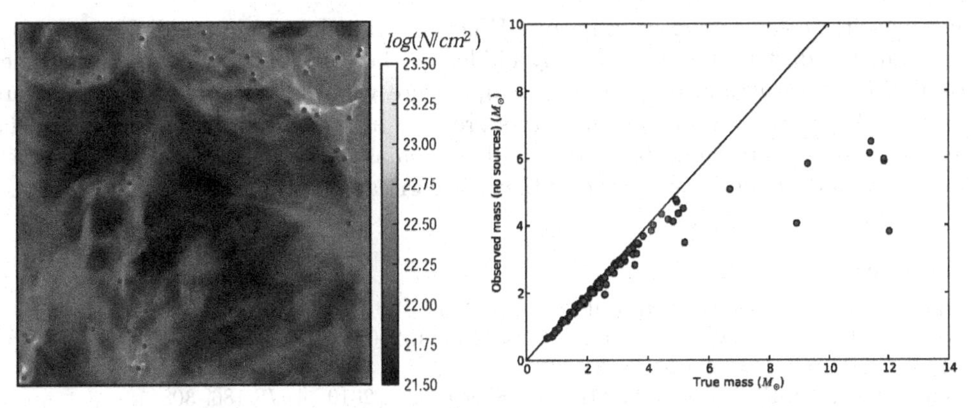

Figure 3. *Left:* Column density in a MHD run that is used to study the accuracy of the mass estimates of cloud cores. The mean extinction through the cloud is $A_V = 13^m$. The locations of the self-gravitating cores are marked with blue circles. *Right:* Comparison of the observed and real masses within a fixed radius of 0.05 pc around the core positions. The masses are derived from observations of dust emission at 250 and 500 μm. The cores contain no internal heating sources. The mass estimates are found to be biased only for cores that are more compact than stable Bonnor-Ebert spheres (Malinen *et al.* 2010).

et al. (2009) used radiative transfer calculations to simulate Zeeman splitting observations of a MHD model cloud. By closely following the procedures used by observers, some key parameters of the models were compared with observations. The cores were selected with the clumpfind algorithm from simulated maps of OH line intensity. The Zeeman splitting of OH 1665 and 1667 MHz lines was calculated by integrating the coupled radiative transfer equations along the line of sight. The simulated observations were done using a beam size and noise level similar to recent OH Zeeman surveys (Troland & Crutcher 2008). The OH Zeeman splitting observations were found to be sensitive to the magnetic field strength in dense cores, where the magnetic field is strong (see Fig. 2). The observations significantly overestimate the mean magnetic field, a conclusion that might be further strengthened if variations in the fractional abundance of OH were considered. Despite the low mean magnetic field of that particular model, $< B >= 0.34\ \mu$G, the properties of the simulated cores were found to agree with observations.

Dust emission from pre-stellar cores. Star formation takes place in the dense cores of molecular clouds. The mass spectrum and internal structure of the cores can be determined with far-infrared and sub-millimetre observations. The accuracy of such determinations can be examined with radiative transfer modeling. Malinen *et al.* (2010) are studying these issues using adaptive mesh refinement (AMR) MHD simulations (Collins et al. 2010) with effective resolutions up to 4096^3. The radiative transfer calculations are conducted on the same AMR grids (for the methods, see Lunttila *et al.* 2010). For the type of observations made by the Herschel satellite, the extracted core masses are usually quite accurate and the possible errors result mainly from the uncertainty in the dust properties, its mass absorption coefficient and spectral index. This is true for stable, Bonnor-Ebert type cores. However, in the 3D simulations some of the gravitationally bound cores are more opaque and have densities that significantly exceed the densities found in stable cores. This will lead to strong temperature gradients in the cores. The observed color temperatures may overestimate the average dust temperature and, consequently, lead to underestimation of the core masses (see Fig. 3). If cloud collapse can proceed without any internal heating, the observed masses will eventually be underestimated by one order of magnitude. It still remains open if such extremely dense and at the same time cold cores exist in nature. When a core is heated internally by a formed protostar, the dust that was previously hidden in the cold central core again becomes visible. In that situation, despite large temperature variations, the mass estimates are not strongly biased. Although some errors are seen in the case of individual cores, the core mass spectra are found to be quite robust against such observational effects, even in the case of dust property variations that are correlated with density.

References

Bernard, J. P., Abergel, A., Ristorcelli, I., *et al.* 1999, *A&A* 347, 640

Bethell, T. J., Chepurnov, A., Lazarian, A., & Kim, J. 2007, *ApJ* 663, 1055

Cho, J. & Lazarian, A. 2005, *ApJ* 631, 361

Collins, D. C., Xu, H., Norman, M., Hui, L., Shengtai, L. 2010, *ApJS* 186, 308

Compiègne, M., Verstraete, L., Jones, A. *et al.* 2010, *A&A, submitted*

Draine, B. T. & Weingartner, J. 1997, *ApJ* 480, 633

Dullemond, C. P., Turolla, R. 2000, *A&A* 360, 1187

Falgarone, E., Pineau Des Forêts, G., Hily-Blant, P., Schilke, P. 2006, *A&A* 452, 511

Glover, S. C. O., Federrath, C., Mac Low, M.-M., Klessen, R. S. 2010, *MNRAS* 404, 2

Hoang T. & Lazarian, A. 2009, *ApJ* 697, 1316

Hogerheijde, M. R. & van der Tak, F. F. S. 2000, *A&A*, 362, 697

Jonsson, P. 2006, *MNRAS* 372, 2

Juvela, M. 1997, *A&A* 322, 943

Juvela, M. 2005, *A&A* 440, 531

Juvela, M. & Padoan, P. 2003, *A&A* 397, 201

Juvela, M. & Padoan, P. 2005, *ApJ* 618, 744

Juvela, M. & Padoan, P. 1999, in *Science with the Atacama Large Millimeter Array, ASP Conference Series, Vol. A. Wootten, ed.*, 28

Lehtinen, K., Juvela, M., Mattila, K., Lemke, D., & Russeil, D. 2007, *A&A* 466, 969

Lucy, L. B. 1999, *A&A*, 344, 282

Lunttila T., Padoan P., Juvela M., & Nordlund Å 2009, *ApJ* 702, L37

Lunttila, T. & Juvela, M. 2010, *in preparation*

Malinen, J., Juvela, M., Collins, D., Lunttila, T., & Padoan, P. 2010, *submitted*, arXiv:1009.4580

Meny, C., Gromov, V., Boudet, N., *et al.* 2007, *A&A* 468, 171

Niccolini, G. & Alcolea, J., 2006 *A&A* 456, 1

Olson, G. L., Auer, L. H., & Buchler, J. R. 1986, *J. Quant. Spectrosc. Radiat. Transfer* 35, 431

Ossenkopf, V. & Henning, T. 1994, *A&A* 291, 943

Pascucci, I., Wolf, S., Steinacker, J. *et al.* 2004, *A&A* 417, 793

Pelkonen, V.-M., Juvela, M., & Padoan, P. 2009, *A&A* 502, 833

Ritzerveld, J. & Icke, V. 2006, *Phys.Rev.* E74, 026704

Scharmer, G. B. 1981, *ApJ* 249, 720

Semionov, D. & Vansevičius V. 2005, *Balt. A* 14, 543

Steinacker, J., Henning, Th., Bacmann, A., & Semenov, D. 2003, *A&A* 401, 405

Troland, T. H. & Crutcher, R.M. 2008, *ApJ* 680, 457

Walmsley, C. M., Flower, D. R., & Pineau des Forêts, G. 2004, *A&A* 418, 1035

Watson, A. M. & Henney, W. J. 2001, *Rev. Mex. Astron. Astrofis.* 37, 221

Whittet, D., Goldsmith, P., & Pineda, J. 2010, *ApJ, submitted,*

Wood, K., Kenyon, S. J., Whitney, B., & Turnbull, M. 1998, *ApJ*, 497, 404

Yusef-Zadeh, F., Morris, M., & White, R. L. 1984, *ApJ* 278, 186

Computational Star Formation
Proceedings IAU Symposium No. 270, 2011
J. Alves, B.G. Elmegreen, J. M. Girart & V. Trimble, eds.

doi:10.1017/S1743921311000792

Adaptable Radiative Transfer Innovations for Submillimeter Telescopes (ARTIST)

Marco Padovani[1] and Jes K. Jørgensen[2]
on behalf of the ARTIST team:
Frank Bertoldi[3], Christian Brinch[4], Pau Frau[1], Josep Miquel Girart[1], Michiel Hogerheijde[4], Attila Juhasz[4], Rolf Kuiper[3], Reinhold Schaaf[3], Wouter H. T. Vlemmings[3]

[1]Institut de Ciències de l'Espai (CSIC-IEEC)
Universitat Autònoma de Barcelona, Spain
email: [padovani,girart,frau] @ice.cat

[2]Centre for Star and Planet Formation
University of Copenhagen, Denmark
email: jes@snm.ku.dk

[3]Argelander Insitute for Astronomy
University of Bonn, Germany
email: [wouter,bertoldi,rschaaf,kuiper] @astro.uni-bonn.de

[4]Leiden Observatory
University of Leiden, the Netherlands
email: [michiel,brinch,juhasz] @strw.leidenuniv.nl

Abstract. Submillimeter observations are a key for answering many of the big questions in modern-day astrophysics, such as how stars and planets form, how galaxies evolve, and how material cycles through stars and the interstellar medium. With the upcoming large submillimeter facilities ALMA and Herschel a new window will open to study these questions. ARTIST is a project funded in context of the European ASTRONET program with the aim of developing a next generation model suite for comprehensive multi-dimensional radiative transfer calculations of the dust and line emission, as well as their polarization, to help interpret observations with these groundbreaking facilities.

Keywords. radiative transfer, methods: numerical, stars: formation, submillimeter, polarization

---

1. Introduction

The Atacama Large Millimeter Array is the largest ground based project in astronomy. It will be the world's most powerful instrument for millimeter and submillimeter astronomy, providing enormous improvements in sensitivity, resolution and imaging fidelity in these wavelength bands. The main focus of the early use of ALMA, compared to present-day facilities, will be its high angular resolution and sensitivity. For studies of star formation, for example, ALMA will zoom in to AU scales in circumstellar disks in nearby star forming regions and thereby address some of the key questions in disk formation. It will also open up new possibilities in the study of planet formation, resolving the effects of planets on the disks around young stars and thus providing direct observational constraints on planet formation models. At the same time, the Herschel Space Observatory is providing high spatial and spectral resolution observations at wavelengths unobservable from the ground, in particular, of lines of H_2O and high excitation transitions of molecules tracing the dense warm gas, e.g., in the innermost regions of protostars or in shocks.

The simultaneous observation of dust and of a multitude of spectral lines opens new horizons for the physical and chemical analysis of the objects, e.g., to determine the excitation conditions (density, temperature, radiation), the chemical abundances and chemical network, and through polarization measurements, the magnetic field. It will be necessary to take all these observational constraints into account for a realistic quantitative description. ALMA will offer a new chance to study magnetic fields: due to receiver constraints, full polarization calibration and imaging will be the norm rather than the exception, making it essential that a self-consistent polarization modeling tool is available to the ALMA users.

With the novelty of these observations it will be critical to have an efficient, flexible and state-of-the-art modeling package that can provide a direct link between the theoretical predictions and the quantitative constraints from the submillimeter observations. The new observational opportunities require a new generation of modeling tools that can model the full multi-dimensional structure of, e.g., a low-mass protostar, including its envelope, disk, outflow and magnetic field, and their time evolution. Current tools are inadequate for the modeling of such complex structures because of their speed and inaccessibility, while tools to model polarization are completely lacking. Both ALMA and Herschel will provide us with large samples of sources, observed homogeneously as part of large key and legacy programs. This makes it prudent to have easily accessible and efficient tools, which with high convergence speed incorporate all observational constraints, for large source samples, and in a systematic fashion. It is the aim of this program to provide such an innovative suite of model tools and test it with existing submillimeter data and with new data from ALMA and Herschel.

2. Objectives

The goal of this project is to deliver a next generation radiative transfer modeling package that provides a self-consistent model for the emission of a multi-dimensional source observed at submillimeter wavelengths. For this project we are specifically motivated, without any loss of more general applicability, by low-mass star formation. Our modeling package shall be able to provide a self-consistent modeling tool for the line and continuum as well as polarization emission from, e.g., a young stellar object, incorporating an infalling large-scale envelope, rotationally decoupled protoplanetary disk, outflow cavity, and magnetic field - with no restrictions of the intrinsic geometry. With such an innovative model tool it will be possible to provide quantitative constraints on the relation between the large-scale angular momentum in the core and the disk evolution, on the direct impact of the outflows and their launching close to the disk surface, and on the importance of the magnetic field. Two important issues need to be addressed in this particular example: (*i*) young stellar objects are characterized by structure on a wide range of spatial scales and with complex geometries, but current radiative transfer tools are locked to fixed linear or logarithmic grids and can therefore model multi-dimensional source structures only with great computational time expense; (*ii*) while most current observations indicate that magnetic fields play an important role in various stages of the star formation process, their relation to the physical source structure is yet poorly constrained. ALMA observations will study the magnetic fields through resolved line and continuum polarization observations that can not be properly analyzed with current models. The modeling tool will also be useful for tackling scientific questions relating to ALMA observations of, e.g., evolved stars, planetary nebulae or extragalactic starbursts. Providing radiative transfer tools that take complex source structures into account will also be of great help to interpret Herschel observations, e.g., to understand the origin of

H_2O and high excitation CO lines in the interface between protostellar cores and outflows or jets. There is a high degree of coupling between different modeling aspects: e.g., to understand the polarization of a given molecule's emission it is necessary to understand the physical conditions and chemical network that leads to the molecule's formation and excitation. For complex source structures it is necessary to develop an approach to time-dependent multi-dimensional modeling, e.g., through libraries of theoretical model prescriptions that can readily be incorporated and expanded for comparison to the data.

3. Tools

The planned model suite will have the following three components: (*a*) an innovative radiative transfer code using adaptive gridding that allows simulations of sources with arbitrary multi-dimensional and time-dependent structures ensuring a rapid convergence and thus allowing an exploration of parameters; (*b*) unique tools for modeling the polarization of the line and dust emission, information that will come with standard ALMA observations; (*c*) a comprehensive Python-based interface connecting these packages, thus with direct link to, e.g., ALMA data reduction software (CASA). A schematic overview of the program is shown in Fig. 1.

(a) – 3D Line Radiative Transfer: LIME. ALMA's high resolution data will produce the need to model phenomena with non-symmetric structures, such as spiral-waves, proto-planet resonances in evolving circumstellar disks, close protostellar binaries etc. Conventional radiative transfer tools use simple linear or logarithmic spatial grids. To model complex source structures in higher dimensions thus requires increasingly finer grid scaling, which becomes very difficult to handle computationally. As an alternative, the SimpleX algorithm (Ritzerveld & Icke 2006) uses a Poisson method to define a grid based on the density distribution. The cornerstone of ARTIST is the 3D line radiative transfer code, LIME (Brinch & Hogerheijde 2010), which utilizes the SimpleX gridding algorithm in a 3D extension to the RATRAN radiative transfer code (Hogerheijde & van der Tak 2000). LIME is currently being applied to for example modeling new H_2O observations of protostars and disks from the Herschel Space Observatory. For these models the adaptive gridding method ensures rapid convergence, for lines from molecules such as H_2O that are far from LTE excitation.

(b) – Dust and Line Polarization. Various theoretical studies predict that magnetic fields, turbulence or/and magneto-hydrodynamic waves may be the main agents controlling both the evolution of molecular clouds and the star-formation process (e.g., Bertoldi & McKee 1992; Mac Low & Klessen 2004; Mouschovias *et al.* 2006; van Loo *et al.* 2007). Unfortunately, the magnetic field is the least-known observable in star formation, due to the inherent difficulty to measure it with present telescopes (mainly through polarimetric observations of dust and molecular emission; e.g., Girart, Rao & Marrone 2006). This situation will change dramatically with ALMA, which will provide such an improvement in sensitivity that polarization observations of dust and molecular lines can be done in many sources at a very good angular resolution. However, the interpretation of the polarized data is difficult and appropriate tools are needed to scientifically harvest the wealth of polarization data expected from ALMA. Note that the ALMA band-7 (275 - 373 GHz) receivers will require full polarization calibration and imaging even for projects with no primary interest in polarization. The ARTIST package consists of a set of modeling tools for polarization and magnetic fields in different molecular and dust environments (e.g., low- and high-mass protostars, envelopes around evolved stars).

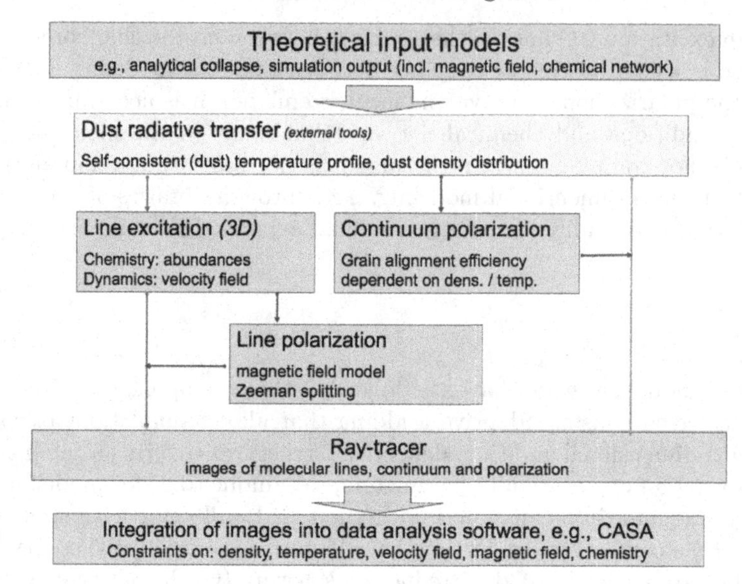

Figure 1. Schematic overview of the components in the ARTIST program.

(c) – Model interface and library. An important component of ARTIST is a common interface for the codes used for radiative transfer modeling of typical data produced by submillimeter telescopes such as ALMA and Herschel. The model interface will include: (*i*) a library of standard input models, e.g., for collapsing protostars, circumstellar disks and evolved stars; (*ii*) a wrapper package that links the existing dust and line radiative transfer codes; (*iii*) tools to analyze model output, e.g., to extract information about molecular excitation and its deviation from LTE, or optical depth surfaces, and to import this in existing visualization packages; (*iv*) a ray tracing backend that readily provides data cubes that can be used in data reduction packages.

4. Current Status

The aim of the ARTIST project is to supply the community with the described tools in one coherent modeling package. The tools will be made publicly available as they are finished: we expect LIME to be released medio-2010 with the remaining components of the package to follow. The tools will be distributed and supported through the ALMA regional center nodes in Bonn and Leiden as well as the Danish initiative for Far-infrared and Submillimeter Astronomy (DFSA; Copenhagen, Denmark). For more information see `http://www.astro.uni-bonn.de/ARC/artist`.

References

Bertoldi, F. & McKee, C. F. 1992, *ApJ*, 395, 140
Brinch, C. & Hogerheijde, M. 2010, *A&A*, in press (arXiv:1008.1492)
Girart, J. M., Rao, R. & Marrone, D. P. 2006 *Science*, 313, 812
Hogerheijde, M. M. & van der Tak, F. F. S. 2000, *A&A*, 362, 697
MacLow, M.-M. & Klessen, R. S. 2004, *RvMP*, 76, 125
Mouschovias, T. Ch., Tassis, K. & Kunz, M. W. 2006, *ApJ*, 646, 1043
Ritzerveld, J. & Icke, V. 2006, *PhRvE*, 74, 26704
van Loo, S., Falle, S. A. E. G., Hartquist, T. W. & Moore, T. J. T. 2007 *A&A*, 471, 213

Computational Star Formation
Proceedings IAU Symposium No. 270, 2011
J. Alves, B.G. Elmegreen, J. M. Girart & V. Trimble, eds.

doi:10.1017/S1743921311000809

Radiative Transfer Simulations of Infrared Dark Clouds

Yaroslav Pavlyuchenkov[1], **Dmitry Wiebe**[1], **Anna Fateeva**[1], **& Tatiana Vasyunina**[2]

[1]Institute of Astronomy, Russian Academy of Sciences,
Pyatniskaya str. 48, 119017 Moscow, Russia
email: pavyar@inasan.ru

[2]Max Planck Institute for Astronomy,
Koenigstuhl 17, D-69117 Heidelberg, Germany
email: vasyunina@mpia.de

Abstract. The determination of prestellar core structure is often based on observations of (sub)millimeter dust continuum. However, recently the Spitzer Space Telescope provided us with IR images of many objects not only in emission but also in absorption. We developed a technique to reconstruct the density and temperature distributions of protostellar objects based on radiation transfer (RT) simulations both in mm and IR wavelengths. Best-fit model parameters are obtained with the genetic algorithm. We apply the method to two cores of Infrared Dark Clouds and show that their observations are better reproduced by a model with an embedded heating source despite the lack of 70 μm emission in one of these cores. Thus, the starless nature of massive cores can only be established with the careful case-by-case RT modeling.

Keywords. radiative transfer; ISM: clouds, dust, extinction; methods: data analysis, numerical;

1. Objectives

Parameters of prestellar and protostellar molecular cores are usually derived from millimeter and sub-millimeter observations of their thermal dust emission. However, the interpretation of these observations is ambiguous as the same spectrum can be produced by different density and temperature distributions. In some cases, millimeter continuum data are supplemented by molecular line observations to derive the temperature. But the interpretation of molecular line observations is also not trivial because of the non-LTE conditions of line formation and complex chemical structure of the cores. The Spitzer Space Telescope provided us with infrared images of many dense cores, seen not only in emission but also in absorption. These data can be used to derive the core density and temperature distributions by fitting absorption and emission simultaneously.

We present a framework to reconstruct the density and temperature distributions of protostellar objects based on detailed RT simulations and on the comparison of model intensity distributions with observations both in emission and absorption at mm and IR wavelengths. We apply the method to two Infrared Dark Cloud cores IRDC-320.27+029 (subcore P2) and IRDC-321.73+005 (subcore P2) which are believed to represent the initial stage for the massive star formation (Fig. 1). For brevity, they are further referred to as IRDC 320 and IRDC 321.

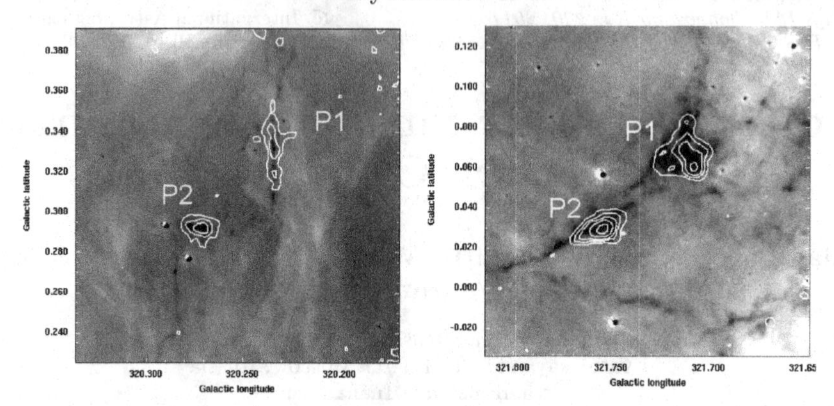

Figure 1. Composite images of dark cores IRDC-320.27+029 and IRDC-321.73+005. The grey map is for 8 μm emission provided by the Spitzer Space Telescope while contours show 1.2 mm emission from the SEST Telescope.

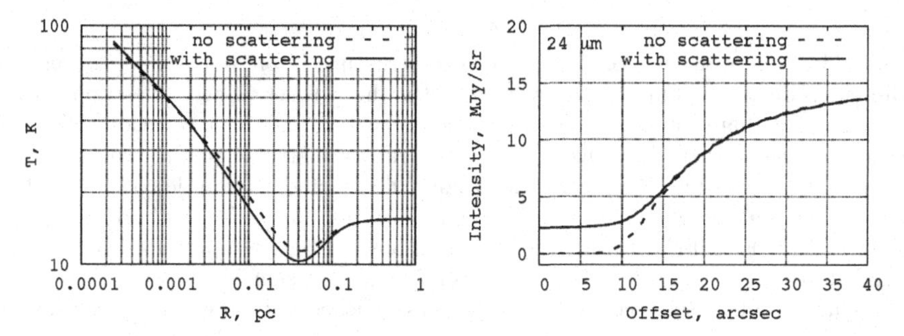

Figure 2. Temperature and intensity distributions for a representative model of a protostellar core. The central heating source is responsible for the inner temperature gradient while the external radiation field heats the envelope. The solid and dashed lines correspond to the models with and without scattering, correspondingly.

2. Model of a protostellar core

We assume that a core is spherically symmetric and its density distribution can be represented by following law, see Tafalla *et al.* (2002):

$$n(\mathrm{H}_2) = \frac{n_0}{1 + (r/r_0)^\beta}, \tag{2.1}$$

where n_0 is the central density, r_0 is the radius of inner plateau, β describes the density decrease in the envelope. We also set the inner core radius of 50 AU and the outer radius of 1 pc. To take into account the possible spread in evolutionary stages, e.g. the presence of a (proto)star, we assume that there is a blackbody source in the core center with the temperature of T_* and radius of 5 $R_\odot$. The core is illuminated by the diffuse interstellar field with the color temperature 10000 K and dilution $D_{\rm bg}$. This field controls the temperature in the envelope. We also postulate the isotropic infrared background emission for those wavelengths where the comparison with observations has to be made. The intensity of infrared background field is equal to the observed intensity at the edge of the core. Note that it is much stronger than the diluted interstellar field.

Temperature distribution and emergent intensity distribution are found as a result of RT simulation with the NATA(LY) code by Pavlyuchenkov *et al.* (2010) for the given set of parameters. Dust emission, absorption and isotropic scattering are taken into account. The accelerated Λ-iteration method is used where the mean intensity is calculated using

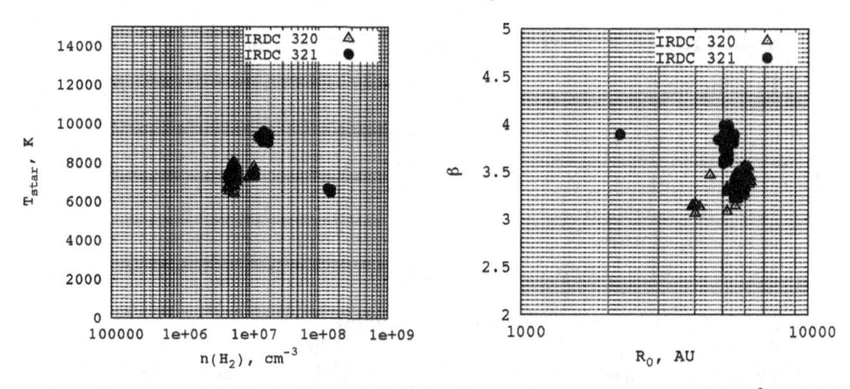

Figure 3. Localization of IRDC 320 and IRDC 321 model parameters with $\chi^2 < 11$ found by PIKAIA code. In the case of IRDC 320 the parameters are well localized while for IRDC 321 there are two groups of parameters with nearly the same χ^2.

the set of selected rays. The frequency dependent dust opacities are calculated from the Mie theory. Temperature and intensity distributions for a representative core model are shown in Fig. 2. Note that scattering does not significantly affect the thermal structure but can strongly affect the emergent intensity distribution.

3. Search for best-fit parameters

For each set of free parameters we perform RT simulations and calculate the model intensity distributions for 1.2 mm, 70 μm, 24 μm, and 8 μm. Then we calculate the χ^2-criterion which characterizes the agreement between the model and observed intensity distributions. Millimeter data are taken from Vasyunina *et al.* (2009). Infrared data are downloaded from NASA/IPAC Infrared Science Archive (http://irsa.ipac.caltech.edu/).

The search for best-fit parameters i.e. the minimization of χ^2 is performed using the genetic code PIKAIA, see Charbonneau (1995). This algorithm allows not only to identify the global minimum of χ^2 in the multi-dimensional parameter space but also to study the degeneracy of the solution. The localization of the optimal parameters (with $\chi^2 < 11$) for IRDC 320 and IRDC 321 in two parameter subspaces is shown in Fig. 3. In Table 1 the best-fit, derived and fixed parameters for the studied cores are shown.

4. Main Results

In general, we achieved a good agreement between synthetic and observed intensity distributions both for IRDC 320 and IRDC 321 (Fig. 4). In the case of IRDC 320 we are able to fit 1.2 mm emission profile and to reproduce the flat intensity distribution at 70 μm and absorption profiles at 24 μm and 8 μm. In the model of IRDC 321 the emission appears not only at 1.2 mm but also at 70 μm in agreement with observations. That is a result of higher central density and higher temperature of the inner heating source.

In both cases our results are consistent with the presence of internal heating sources (proto-stars) in cores' interiors. While for IRDC 321 this assertion seems to be robust as it is seen in emission at 70 μm, for the particular case of IRDC 320 it should only be considered as tentative. In particular, we did not try to distinguish between the foreground and background radiation. Second, available molecular data do not show any typical hot core tracers in this object. So, this case deserves further study, in particular, with chemical models. The important conclusion of this study is that 70 μm band seems to be most promising to discriminate between starless and protostellar massive cores.

Table 1. Best-fit, derived and fixed parameters for template cores

Parameter	Name	IRDC 320	IRDC 321
Varied parameters			
H_2 central density, cm^{-3}	n_0	1.1×10^7	1.8×10^7
Radius of plateau, AU	r_0	4×10^3	5×10^3
Index for envelope	β	3.1	3.8
Star temperature, K	T_*	7300	9300
Dilution of background field	D_{bg}	1.3×10^{-13}	6.7×10^{-14}
Derived parameters			
Cloud mass, $M_\odot$	M	170	230
H_2 surface density, cm^{-2}	N	8.1×10^{23}	1.6×10^{24}
Star luminosity, $L_\odot$	L	60	160
Fixed parameters			
Radius of inner hole	R_{in}	50 AU	
Cloud radius	R_{out}	1 pc	
Star radius	R_*	5 $R_\odot$	
Temperature of background field	T_{bg}	10^4 K	

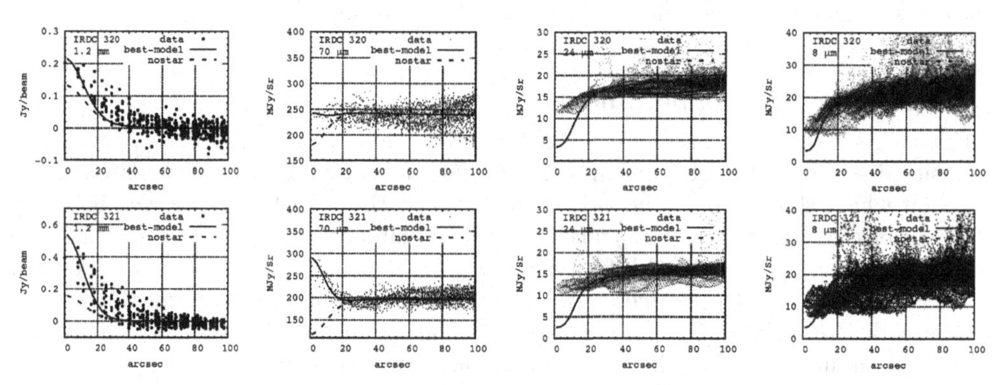

Figure 4. Observed (dots) and modeled (lines) intensity distributions at selected wavelengths for IRDC 320 (top row) and IRDC 321 (bottom row). The model distributions correspond to the model with lowest χ^2. With dashed lines, we show results for the same core model, but without the central heating source. It is noteworthy that not only 70 μm but also millimeter emission is sensitive to the presence of the heating source.

However, observations in this band should not be interpreted without careful modeling to avoid simplified conclusions. In particular, the mere absence of 70 μm emission is not a solid evidence that a core is starless.

This study has been supported by RF President Grant MK-4713.2009.2 and by RFBR grant 10-02-00612.

References

Charbonneau, P.; 1995, Astroph. J. Suppl., 101, 309
Pavlyuchenkov, Ya.; Wiebe, D.; Fateeva, A.; Vasyunina, T.; 2010, Astron. Reports, in press
Tafalla, M., Myers, P. C., Caselli, P., Walmsley, C. M., & Comito, C.; 2002, ApJ, 569, 815
Vasyunina, T.; Linz, H.; Henning, T. *et al.*; 2009, A&A, 499, 149

Computational Star Formation
Proceedings IAU Symposium No. 270, 2011
J. Alves, B.G. Elmegreen, J. M. Girart & V. Trimble, eds.

doi:10.1017/S1743921311000810

Gas dynamics in whole galaxies: SPH

Clare Dobbs[1,2]

[1]Max-Planck-Institut für extraterrestrische Physik,
Giessenbachstraße, D-85748 Garching, Germany
email: cdobbs@mpe.mpg.de

[2]Universitäts-Sternwarte München, Scheinerstraße 1, D-81679 München, Germany

Abstract. I review the progress of SPH calculations for modelling galaxies, and resolving gas dynamics on GMC scales. SPH calculations first investigated the response of isothermal gas to a spiral potential, in the absence of self gravity and magnetic fields. Surprisingly though, even these simple calculations displayed substructure along the spiral arms. Numerical tests indicate that this substructure is still present at high resolution (100 million particles, $\sim$ 10 pc), and is independent of the initial particle distribution. One interpretation of the formation of substructure is that smaller clouds can agglomerate into more massive GMCs via dissipative collisions. More recent calculations have investigated how other processes, such as the thermodynamics of the ISM, and self gravity affect this simple picture. Further research has focused on developing models with a more realistic spiral structure, either by including stars, or incorporating a tidal interaction.

Keywords. ISM: clouds, galaxies: ISM, galaxies: spiral, stars: formation, hydrodynamics

---

1. Introduction

The last decade has seen massive progress in the use of numerical calculations to model the ISM, both on galactic and subgalactic scales. Early particle simulations (Levinson & Roberts 1981; Kwan & Valdes 1983, 1987; Hausman & Roberts 1984; Tomisaka 1984, 1985; Roberts & Stewart 1987) were largely limited to modelling individual clouds as ballistic particles. Now, hydrodynamic simulations of galaxies are able to resolve GMCs. Furthermore they can capture the dynamical evolution of molecular (and atomic) clouds, as they form, interact and disperse.

To maximise resolution, hydrodynamic simulations frequently apply an external potential, which includes the dark matter halo and stellar disc. Numerous calculations have also included a spiral component to the stellar potential in order to simulate a grand design galaxy. Surprisingly, calculations which model the gas response to a spiral potential show the presence of substructure in the gas, even without magnetic fields or self gravity (Wada & Koda 2004, Dobbs & Bonnell 2006). This substructure becomes more clearly visible as the spiral arm clumps are sheared out into spurs, or feathers, in the interarm regions. Such structure is found to occur in both grid (Wada & Koda 2004, Kim & Ostriker 2006) and SPH (Wada & Koda 2004, Dobbs & Bonnell 2006) codes.

Wada & Koda 2004 interpreted the formation of substructure along the spiral arms in terms of Kelvin Helmholtz instabilities. However Dobbs *et al.* 2006 provided a different explanation for the formation of spiral arm clumps. They likened the passage of gas through the spiral shock to a queue of traffic, where gas particles (or clumps) become bunched together. This occurs because the gas particles undergo dissipative collisions in the spiral arms. This process resembles prior models of cloud coalescence, but the accretion and dispersal of gas into and from clumps is more continuous. Also, there is no need for the gas to be molecular, although for some environments e.g. M51 it may well be. The main requirement (Dobbs & Bonnell 2006) is that the gas is cold, so that the spiral shock is strong, and the gas pressure (which would cause clouds to diffuse) is minimal. The spacing of the clumps is $\sim \sigma_v T$ where σ_v is the turbulent velocity dispersion of the gas, and T the time spent in the spiral arms (Dobbs, Bonnell & Pringle 2006, Dobbs 2008). In simulations with higher spiral potentials (and consequently higher arm to interarm contrasts), T increases and the spacing between the clumps increases (Dobbs 2008).

2. Resolution studies

A potential issue with the results of these calculations is that the structure may be resolution dependent, or even disappear at high resolution. In Fig. 1, we show this is not the case. Similar structure is apparent with 1, 20 and 100 million particles, the latter corresponding to a mass resolution of 10 $M_\odot$ and a spatial resolution of < 1 pc. It is difficult to quantify the spacing between the features along the spiral arms - partly because there is no exact periodicity. Nevertheless in Fig. 2, we show the auto-correlation function, taken from the distribution of particles in $log(r)$ versus θ space. The position of the first peak indicates the spacing which provides the best correlation. We see that the peak is in a similar location for each resolution, and also that the typical spacing between subsequent peaks is similar. Unfortunately the spacing of the clumps has not

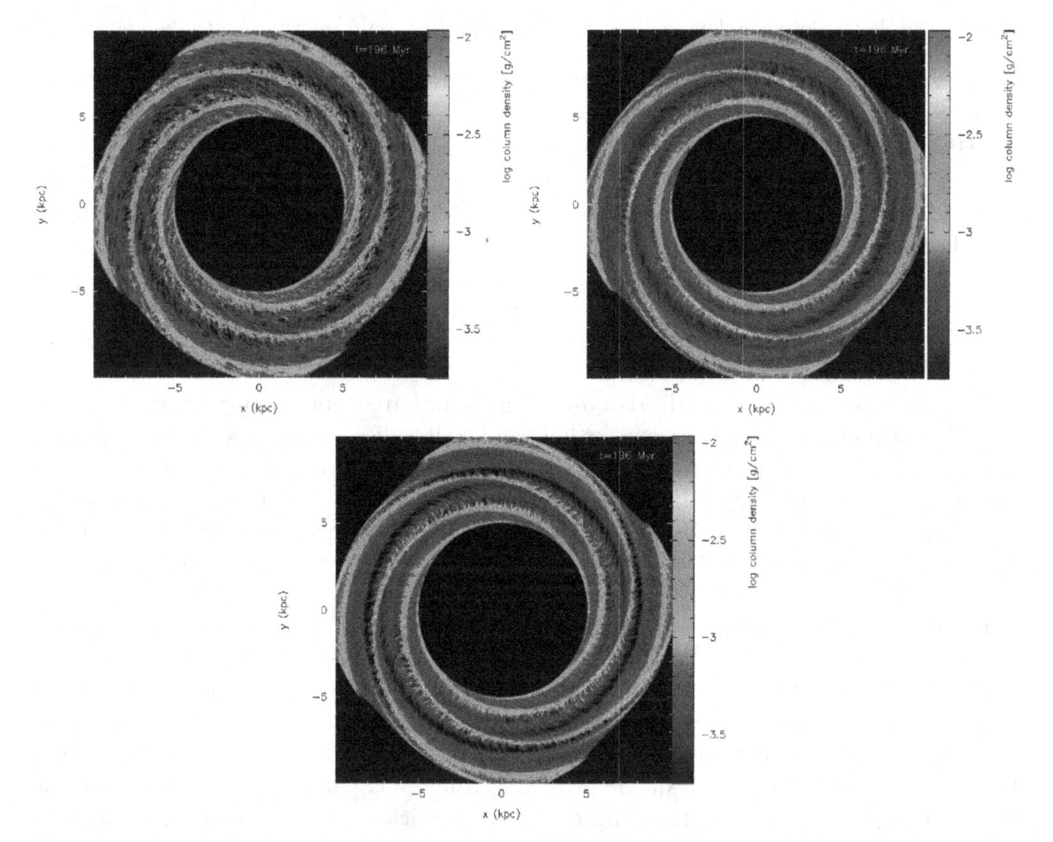

Figure 1. Resolution test for hydrodynamic calculations with a spiral potential alone. The number of particles in each panel is 1 million (top left), 20 million (top right) and 100 million (lower).

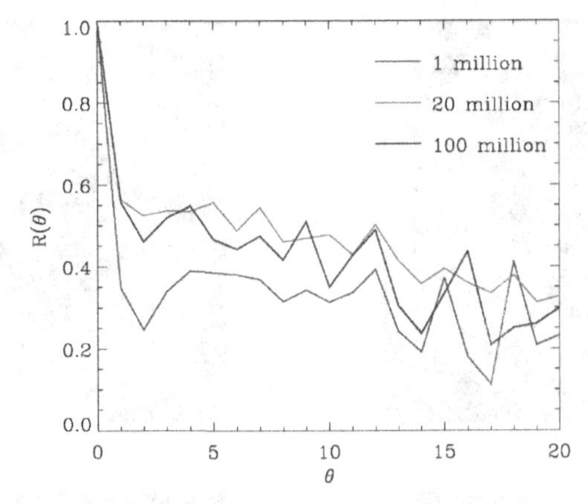

Figure 2. The auto-correlation function (R) is plotted for the distribution of particles along a spiral arm. θ is the angle, or distance along the spiral arm. The correlation function is the sum of the distribution of particles, multiplied by the distribution shifted by θ. Thus the first peak represents the best-fit separation of the spurs. The similarity in location of the first peaks, and spacing of subsequent peaks implies that the spacing is not resolution dependent. The spacing is around 200 pc, but has not yet reached a maximum (which is closer to 700 pc (Dobbs, Bonnell & Pringle 2006)).

reached a maximum at this time (200 Myr) as it was too time consuming to run the 100 million particle simulation further.

In Fig. 3, we show further tests, varying the viscosity and the initial particle distribution. We use 1 million particles in each calculation. The interarm spurs occur in all the cases, and in particular using an initially uniform particle distribution (rather than a random distribution, which is used for all the other calculations) makes negligible difference to the results.

3. Thermodynamics and molecular hydrogen formation

Early calculations adopted an isothermal medium (Slyz *et al.* 2003, Wada & Koda 2004, Dobbs & Bonnell 2006), but more recent calculations have included a much more sophisticated treatment of the ISM (Dobbs *et al.* 2008, Peluppessy & Papadopoulos 2009). In Dobbs *et al.* 2008, we used the thermodynamics and chemistry of Glover & MacLow 2007. The results of these calculations indicate that GMC formation occurs in the same way as isothermal simulations of cold gas, by the coalescence of clouds in the spiral shock, as most of the gas (70%) tends to be cold.

By including H_2 formation, we can start to make estimates on cloud formation timescales, and cloud lifetimes. In Dobbs *et al.* 2008, we found H_2 forms on timescales of Myrs, whilst once formed, H_2 lasts 10's of Myrs. However these calculations do not include any feedback processes, only H_2 photodissociation. Lifetimes based around the cloud dynamics, rather than H_2 fractions, are difficult to determine, since the clouds evolve dynamically on relatively short timescales.

The inclusion of H2, and CO also allows the opportunity of comparison with observations. Pelupessy & Papadopoulos 2009 (Fig. 5) calculate synthetic UV and CO maps of their galaxy models. In Douglas *et al.* 2010, we present synthetic HI maps, which show abundant HI self absorbtion features (also Acreman *et al.*, these proceedings).

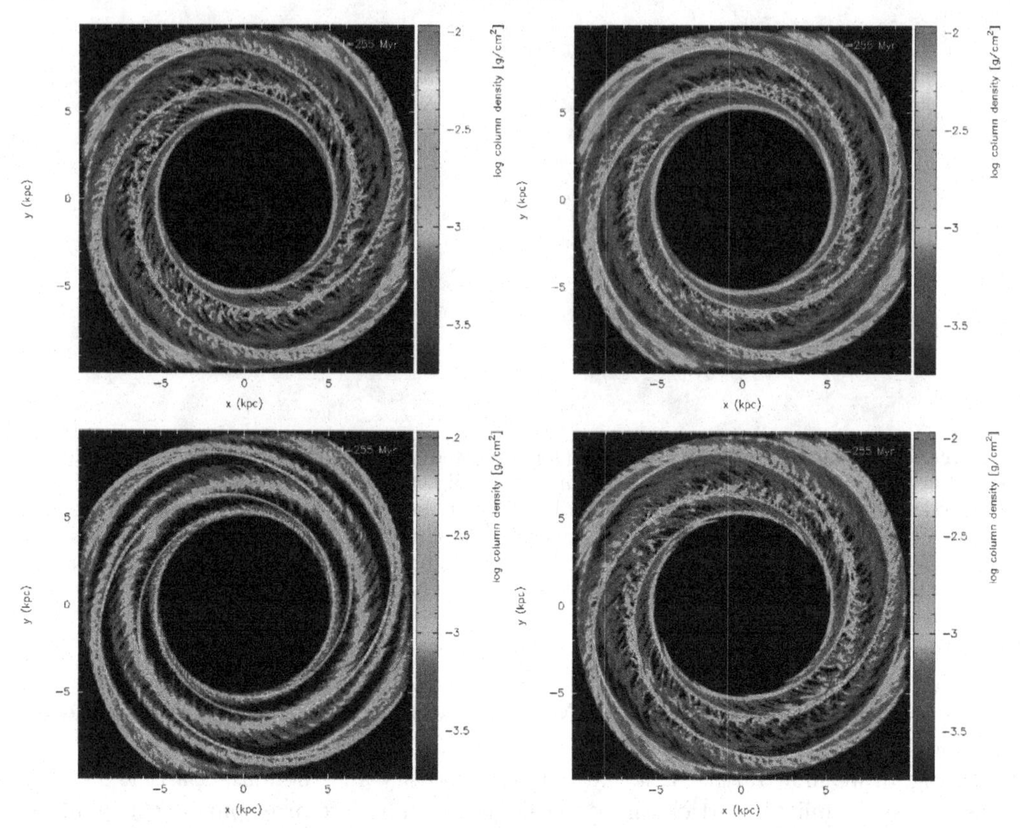

Figure 3. Column density images from SPH simulations of isothermal gas subject to a spiral potential. Three plots show different artificial viscosity settings: the standard case, $\alpha = 1$, $\beta = 2$ (top left); $\alpha = 2$, $\beta = 4$ (top right); Balsara switch with $\alpha = 1$, $\beta = 2$ (lower left). Finally the lower right panel shows the column density when the particles are initially set up uniformly, on concentric circles. The difference for the Balsara switch is probably because the shock is not captured so well.

4. Magnetic fields

Magnetic fields have recently been implemented in simulations of galaxies (Dobbs & Price 2008, Dolag & Stasyszyn 2008, Kotarba *et al.* 2009) using the Euler potential method (Price, these proceedings). Although the current Euler potential implementation is known to have limitations, in particular amplification of the field due to winding is not captured, this method does allow a first order study of the effects of magnetic fields.

In Dobbs & Price 2008 we find that the main effect of the magnetic fields is to contribute pressure to the ISM. The result is similar to increasing the thermal pressure, as the spiral arms and spurs are less dense and more diffuse. However spurs are still visible for plasma $\beta > 0.1$. Kotarba *et al.* 2009 study the amplification of magnetic fields in SPH simulations with stars and gas, where the spiral arms arise consistently. They compare calculations with the Euler method, and direct calculation of B, finding that even though the amplification is likely underestimated with the Euler method, the large value of *div B* when using the direct calculation of B render this method much less reliable.

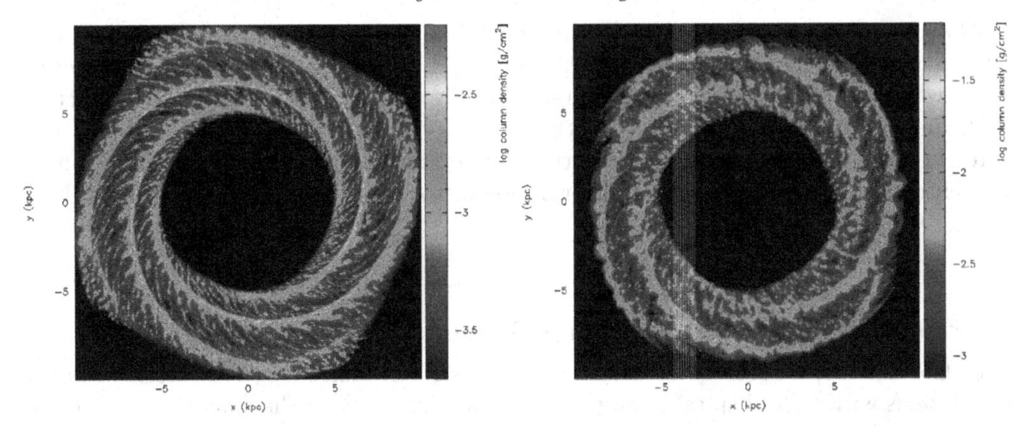

Figure 4. Column density from simulations with surface densities of 4 $M_{\odot}pc^{-2}$ (left) and 20 $M_{\odot}pc^{-2}$ (right). The simulations include self gravity and magnetic fields, adopting an isothermal two-phase medium of cold and warm gas. Self gravity has little effect on the global structure in the low surface density case, but aids the formation of more massive complexes in the higher density case.

5. Self gravity

In Dobbs 2008, we perform simulations with self gravity and magnetic fields. Self gravity enhances the formation of molecular clouds, by increasing the frequency and success (i.e. the likelihood that clouds merge rather than get disrupted) of collisions. Self gravity also leads to gravitational instabilities in the gas. The net result is that clouds reach higher densities compared to the ambient medium, and are easier to distinguish along the spiral arms. The difference in structure becomes more noticeable with increasing surface densities - the gas distribution when $\Sigma = 4$ Mpc^{-2} is similar whether or not self gravity is included, but self gravity has a much stronger impact on the global structure when $\Sigma = 20$ Mpc^{-2} (Fig. 4). The surface density in the solar neighbourhood lies in between these regimes (Wolfire *et al.* 2003).

6. Feedback

So far the effect of feedback on molecular cloud formation, and the evolution of spurs has not been studied for grand design galaxies with SPH. This has been studied with grid codes, e.g. Wada 2008 find that supernovae have a secondary effect, and spurs are still evident.

7. SPH simulations without a spiral potential

Several authors have also modelled flocculent galaxies, where the structure is solely due to gravitational and thermal instabilities, and is presented elsewhere in these proceedings (Wada).

Pelupessy & Papadopoulos (2006,2009) also model flocculent spirals (Fig. 5), and show that they are able to reproduce characteristics of the ISM such as PDFs and the pressure H_2 relation (Blitz & Rosolowsky 2004) reasonably well. They also use the amount of H_2 formation to estimate the star formation efficiency, which is then used in their models of feedback. One of their main results is that for gas rich, and metal poor galaxies, they find strong deviations from the star formation rates in their models and those predicted from the Schimdt-Kennicutt relation (Pelupessy & Papadopoulos 2010).

Alternatively a grand design spiral pattern can be obtained more realistically by modelling an interaction with another galaxy (see Struck, these proceedings). In Dobbs *et al.* 2010, we modelled the grand design spiral M51 (Fig. 6), and showed that the interaction with the companion can reproduce both the observed spiral pattern, and detailed features in the gas. We also found the pattern highly time dependent. Other studies have concentrated on reproducing the star formation histories of interacting galaxies, e.g. Karl *et al.* 2010.

8. Properties of molecular clouds formed in the simulations

In Dobbs 2008, we started to determine properties of molecular clouds formed in the simulations with a fixed spiral potential. The mass functions of the clouds have an index of 1.7-2.0, in line with observations. The cloud angular momenta also appear in good agreement with clouds in both the Milky Way (Phillips 1999) and M33 (Rosolowsky *et al.* 2003). Clumps regularly undergo collisions or interactions with other clumps, which can lead to clumps rotating in a retrograde direction, i.e. in the opposite sense to galactic rotation. Properties of molecular clouds have also been deteremined in calculations which used the AMR code ENZO (Tasker & Tan 2009), and show remarkable similarity with SPH calculations, although there are a number of differences between the models (e.g. presence of magnetic field, spiral potential).

Alves (these proceedings) derive a remarkably tight mass radius relation for molecular clouds. This very tight correlation is not apparent in the results of Heyer *et al.* 2009,

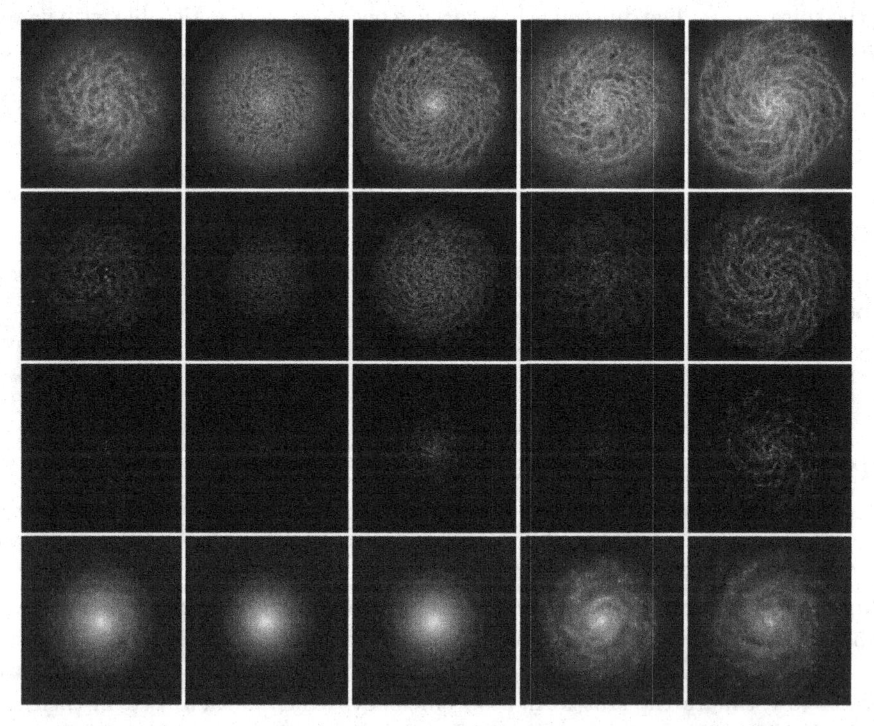

Figure 5. Column density images from Pelupessy & Papadopoulos 2009, for simulations of flocculent spirals, which include the gas and stars. The images show HI (top), H_2 (second line), CO (third line) and UBV (lowest line) from models with varying amounts of gas and metallicites.

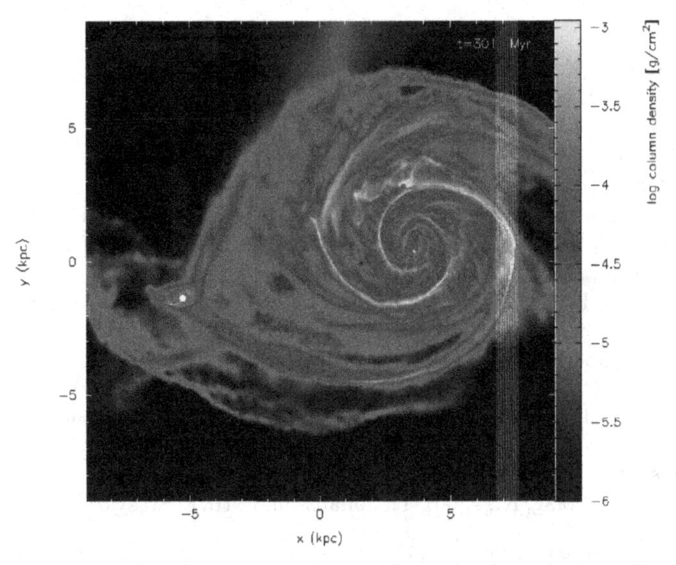

Figure 6. This figure is taken from a calculation designed to reproduce the spiral structure of M51 (Dobbs *et al.* 2010). The companion galaxy, NGC 5193, is modelled as a point mass (seen as a white dot in the figure), and the two galaxies are given inital velocities and positions derived using N-body calculations (Theis & Spinneker 2003).

who re-examined the properties of clouds observed by Solomon *et al.* 1987. Numerical simulations may help explain the discrepancies in these results, and why the clouds found by Heyer *et al.* 2009 tend to be slightly unbound.

9. Conclusion

SPH calculations allow the ISM and molecular clouds to be studied on galactic scales. Unlike previous particle methods, these calculations can model individual clouds, and thus capture changes in the size, shape, and constituent gas of the clouds. Hydrodynamic simulations show that when a global grand design spiral pattern is present, GMCs form by the agglomeration of gas clouds in the spiral shock. This process requires the gas is clumpy (or cold), but appears independent of resolution down to at least 10 $M_\odot$ per particle. It is also independent of the treatment of artificial viscosity, and the initial particle distribution. Magnetic fields do not strongly affect this picture, and the detailed thermodynamics do not make a significant difference, provided there is cold gas. Self gravity can alter the large scale distribution, both enhancing cloud agglomeration and gravitational instabilities, the importance of self gravity increasing with surface density.

10. Acknowledgments

I would like to thank Ian Bonnell and Jim Pringle for their help throughout this research. I also thank Daniel Price for providing his SPH code 'phantom' which was used to perform the resolution test in Section 2. The calculations reported here were performed using the University of Exeter's SGI Altix ICE 8200 supercomputer, and the HLRB-II: SGI Altix 4700 supercomputer at the Leibniz supercomputer centre, Garching. Images included in this review were produced using SPLASH (Price 2007), a visualisation tool for SPH that is publicly available at http://www.astro.ex.ac.uk/people/dprice/splash.

References

Blitz, L. & Rosolowsky, E. 2004, *ApJL*, 612, 29
Dobbs, C. L. & Bonnell, I. A. 2006, *MNRAS*, 367, 873
Dobbs, C. L., Bonnell, I. A., & Pringle, J. E. 2006, *MNRAS*, 371, 1663
Dobbs, C. L. 2008, *MNRAS*, 391, 844
Dobbs, C. L. & Price, D. J. 2008, *MNRAS*, 383, 497
Dobbs, C. L., Glover, S. G., Clark, P. C., & Klessen, R. S. 2008, *MNRAS*, 389, 1097
Dobbs, C. L., Theis, C., Pringle, J. E., & Bate, M. R. 2010, *MNRAS*, 403, 625
Dolag, K. & Stasyszyn, F. 2009, *MNRAS*, 398, 1678
Douglas, K., Acreman, D., Dobbs, C., & Brunt, C. 2010, *MNRAS*, tmp, 896
Glover, S. G. & Mac Low, M. M. 2007, *ApJ*, 169, 239
Hausman, M. A. & Roberts, W. W. Jr 1984, *ApJ*, 282, 106
Heyer, M., Krawczyk, C., Duval, J., & Jackson, J. M. 2009, *ApJ*, 699, 1092
Karl, S. J., Naab, T., Johansson, P. H., Kotarba, H., Boily, C. M., Renaud, F., & Theis, C. 2010, *ApJ*, 715, 88
Kim, W.-T. & Ostriker, E. C. 2006, *ApJ*, 646, 213
Kotarba, H., Lesch, H., Dolag, K., Naab, T., Johansson, P. H., & Stasyszyn, F. A. 1983, *MNRAS*, 397, 733
Kwan, J. & Valdes, F. 1983, *ApJ*, 271, 604
Kwan, J. & Valdes, F. 1987, *ApJ*, 315, 92
Levinson, F. H. & Roberts, W. W. Jr 1981, *ApJ*, 245, 465
Papadopoulos, P. P. & Pelupessy, F. I.. 2010, *ApJ*, 717, 1037
Pelupessy, F. I., Papadopoulos, P. P., & van der Werf, P. 2006, *ApJ*, 645, 1024
Pelupessy, F. I. & Papadopoulos, P. P. 2009, *ApJ*, 707, 954
Phillips, J. P. 1999, *A&AS*, 134, 241
Price, D. J. 2007, *PASA*, 24, 159
Roberts, W. W. Jr & Stewart, G. R. 1987, *ApJ*, 282, 106
Rosolowsky, E., Engargiola, G., Plambeck, R., & Blitz, L. 2003, *ApJ*, 599, 258
Slyz, A., Kranz, T., & Rix, H.-W. 2003, *MNRAS*, 346, 1162
Solomon, P. M., Rivolo, A. R., Barrett, J., & Yahil, A. 1987, *ApJ*, 319, 730
Tasker, E. J. & Tan, J. C. 2009, *ApJ*, 700, 358
Theis, C. & Spinneker, C. 2003, *PASJ*, 284, 495
Tomisaka, K. 1984, *PASJ*, 36, 457
Tomisaka, K. 1986, *PASJ*, 38, 95
Wada, K. & Koda, J. 2004, *PASJ*, 38, 95
Wada, K. 2008, *ApJ*, 675, 188
Wolfire, M. G., McKee, C. F., Hollenbach, D., & Tielens, A. G. G. M. 2003, *ApJ*, 587, 278

Computational Star Formation
Proceedings IAU Symposium No. 270, 2011
J. Alves, B.G. Elmegreen, J. M. Girart & V. Trimble, eds.

doi:10.1017/S1743921311000822

Star Formation and Gas Dynamics in Galactic Disks: Physical Processes and Numerical Models

Eve C. Ostriker

Department of Astronomy, University of Maryland, College Park, MD 20742;
email: ostriker@astro.umd.edu

Abstract. Star formation depends on the available gaseous "fuel" as well as galactic environment, with higher specific star formation rates where gas is predominantly molecular and where stellar (and dark matter) densities are higher. The partition of gas into different thermal components must itself depend on the star formation rate, since a steady state distribution requires a balance between heating (largely from stellar UV for the atomic component) and cooling. In this presentation, I discuss a simple thermal and dynamical equilibrium model for the star formation rate in disk galaxies, where the basic inputs are the total surface density of gas and the volume density of stars and dark matter, averaged over $\sim$ kpc scales. Galactic environment is important because the vertical gravity of the stars and dark matter compress gas toward the midplane, helping to establish the pressure, and hence the cooling rate. In equilibrium, the star formation rate must evolve until the gas heating rate is high enough to balance this cooling rate and maintain the pressure imposed by the local gravitational field. In addition to discussing the formulation of this equilibrium model, I review the current status of numerical simulations of multiphase disks, focusing on measurements of quantities that characterize the mean properties of the diffuse ISM. Based on simulations, turbulence levels in the diffuse ISM appear relatively insensitive to local disk conditions and energetic driving rates, consistent with observations. It remains to be determined, both from observations and simulations, how mass exchange processes control the ratio of cold-to-warm gas in the atomic ISM.

---

1. Introduction

Disk galaxies are gas-rich systems, with a multi-phase, highly structured interstellar medium (ISM). Within the ISM, star formation takes place in giant molecular clouds (GMCs), sometimes concentrated in spiral arms. The rate and character of star formation are influenced by physical processes from sub-pc to multi-kpc scales (McKee & Ostriker 2007). In spite of the complexity of the ISM and star formation at small scales, there are nevertheless clear correlations between the large-scale rate at which stars are born, and the properties of the ISM and (intra-)galactic environment on large ($\sim$ kpc) scales.

As discussed by Frank Bigiel at this meeting (see also Bigiel *et al.* 2008, and references therein), in regions of galaxies where the gaseous surface density $\Sigma \lesssim 100$ M$_\odot$ pc^{-2}, the star formation rate closely follows the surface density of molecular gas. This can be understood in terms of the gas having an essentially constant star formation timescale, $t_{\rm SF} \sim 2 \times 10^9$ yr, within molecular gas (which is observed to be in organized in gravitationally bound clouds with properties that are similar in different galaxies). As a consequence, $\Sigma_{\rm SFR} \propto \Sigma$ in regions where the molecular gas dominates the atomic gas. For regions where atomic gas dominates (primarily in the outer parts of galaxies), $\Sigma_{\rm SFR}$ instead varies as a steeper power of Σ. In addition to this superlinear behavior, there is considerable scatter in the relation between $\Sigma_{\rm SFR}$ vs. Σ at low surface density, suggesting that one or more other parameters, in addition to Σ, controls the star formation rate.

Indeed, recent examination of the correlation of $\Sigma_{\rm SFR}$ with "non-interstellar" galactic environmental properties has revealed interesting dependences, indicating that in the outer parts of galaxies, both the specific star formation rate and the ratio of molecular-to-atomic gas increase roughly linearly with the *stellar* surface density Σ_s (Leroy *et al.* 2008). Previously, Blitz & Rosolowsky (2006) found an approximately linear increase of the molecular content with the estimated dynamic pressure of the ISM, and this is evident in the sample analyzed by Leroy *et al.* (2008) as well. The physical reason for the relationship between molecular content (and star formation) and pressure has not, however, been clear from these empirical studies.

Observations of star formation pose a number of challenges: Why is there an increase in the slope of $\Sigma_{\rm SFR} \propto \Sigma^{1+p}$ in going from molecular- to atomic-dominated regions? What is the physical reason for the empirical relation between ISM pressure and star formation; more generally, how do galactic parameters such as Σ_s, the velocity dispersions of stars and of gas, and spiral structure affect $\Sigma_{\rm SFR}$? Is it possible to explain the observed behavior of $\Sigma_{\rm SFR}$ using simplified theoretical models, and what is required in numerical simulations in order to reproduce observed star formation relationships? Recent theoretical work has taken on these challenges with increasing success; a key to these advances has been a more sophisticated treatment of both the ISM and the galactic environment. For example, Koyama & Ostriker (2009a) found, using numerical simulations of the ISM and a cooling function allowing multiple phases, star formation rates and proportions between self-gravitating and diffuse gas similar to the observations of Blitz & Rosolowsky (2006) and Leroy *et al.* (2008) provided that turbulent driving is included; for non-turbulent models, the proportion of self-gravitating gas was found to be much too high.

2. A thermal/dynamical equilibrium model for $\Sigma_{\rm SFR}$

Motivated by recent observations as well as simulations and earlier theory, Ostriker *et al.* (2010) (hereafter OML) have developed a simple model for star formation regulation in multiphase, turbulent ISM disks. In essence, the OML model combines three basic principles: (1) the diffuse (atomic) component of the ISM is in approximate thermal equilibrium, with a density (and pressure) proportional to the heating rate; (2) the diffuse component of the ISM is in approximate dynamical equilibrium, with the pressure at any height above the galactic midplane given by the weight of the overlying gas; (3) UV from young stars provides most of the heating for the atomic component of the ISM, with star formation taking place only within the gravitationally-bound component of the ISM. These principles have been individually established and extensively studied (over several decades) in the astrophysical literature. Field *et al.* (1969) combined (1) and (2) to conclude that the diffuse atomic gas in the local Milky Way must consist of a two-phase cloud/intercloud medium. In this and subsequent treatments of thermal and dynamical equilibrium, the heating rate has generally been treated as an independent (empirical) parameter. But, by including (3) together with (1) and (2), OML obtained a closed system representing a local patch of a disk galaxy. For this closed system, the partition of gas into phases and the star formation rate are obtained self-consistently.

In the OML model, the (simplified) ISM is treated as having two components, one consisting of diffuse gas (including both high-density cold atomic cloudlets and a low-density warm atomic intercloud medium), and the other consisting of gravitationally-bound clouds (GBCs). Although hot gas is also present in the ISM, it is a tiny fraction of the mass, and fills $\lesssim 20\%$ of the volume (Heiles 2001) (OML describe how to correct for this effect). For galaxies with normal metallicity, the GBCs would represent giant molecular clouds, including their atomic shielding layers. Averaged over $\sim$ kpc scales (which may contain many or few individual GBCs), the total surface density of the GBC component is $\Sigma_{\rm GBC}$, and the total surface density of the diffuse component is $\Sigma_{\rm diff}$.

The diffuse component is assumed to be in vertical dynamical equilibrium (as has been verified by numerical simulations; e.g. Piontek & Ostriker 2007; Koyama & Ostriker 2009b), with the vertical gravity (from the diffuse gas, the GBC component, the stellar disk, and the dark matter halo) balanced by the difference between midplane and external values of thermal pressure $P_{\rm th}$, turbulent pressure ρv_z^2, and magnetic stresses $(8\pi)^{-1}(B^2 - 2B_z^2)$. Because cooling times are short compared to other timescales, the diffuse gas is assumed to be in thermal equilibrium, with the additional provision that both warm and cold phases are present. This allows a range of pressures between $P_{\rm min,cold}$ and $P_{\rm max,warm}$; for the model of OML, it is assumed that the pressure is equal to the geometric mean of these limits, $P_{\rm two-phase} \equiv (P_{\rm min,cold} P_{\rm max,warm})^{1/2}$. For atomic gas, heating is generally dominated by the UV and cooling by collisionally-excited lines (Wolfire *et al.* 2003), which yields $P_{\rm two-phase} \propto J_{\rm UV}$. (Note that other heating – e.g. cosmic rays and shocks – can be more important for very dense, shielded cores and very hot gas, respectively.) Finally, the OML model assumes that the rate of star formation is proportional to the total surface density $\Sigma_{\rm GBC}$ of gas in the GBC component, $\Sigma_{\rm SFR} = \Sigma_{\rm GBC}/t_{\rm SF} = (\Sigma - \Sigma_{\rm diff})/t_{\rm SF}$.

Vertical dynamical equilibrium within the diffuse layer is expressed as $P_{\rm tot} \equiv \alpha P_{\rm th} = \Sigma_{\rm diff}\langle g_z\rangle/2$, where the mean vertical gravity is

$$\langle g_z\rangle \approx \pi G(\Sigma_{\rm diff} + 2\Sigma_{\rm GBC}) + 2(2G\rho_{\rm sd})^{1/2}\sigma_z; \tag{2.1}$$

$\rho_{\rm sd}$ is the midplane density of stars plus dark matter, σ_z is the total vertical velocity dispersion of the diffuse gas, and the total pressure is larger than the thermal pressure by a factor α (see below). The GBC component contributes more strongly (per unit mass) to the gravity because its scale height is smaller than that of the diffuse gas.

If $n^2\Lambda(T)$ is the cooling rate per unit volume and $n\Gamma$ is the heating rate per unit volume, then the two-phase pressure is given by

$$\begin{aligned} \frac{P_{\rm two-phase}}{k} &\equiv (n_{\rm min,cold}T_{\rm min,cold}n_{\rm max,warm}T_{\rm max,warm})^{1/2} \\ &= \Gamma\frac{(T_{\rm min,cold}T_{\rm max,warm})^{1/2}}{[\Lambda(T_{\rm min,cold})\Lambda(T_{\rm max,warm})]^{1/2}}, \end{aligned} \tag{2.2}$$

where we have used the equilibrium condition $\Gamma = n\Lambda$ for both phases. Cooling of the cold atomic medium is dominated by metals (in particular, C II) so that $\Lambda \propto Z_{\rm gas}$, while heating is dominated by the photoelectric effect with $\Gamma \propto Z_{\rm dust}J_{\rm UV}$; since $T_{\rm min,cold}$ and $T_{\rm max,warm}$ are relatively independent of the heating rate (Wolfire *et al.* 1995), this yields $P_{\rm two-phase} \propto J_{\rm UV}$ if $Z_{\rm dust}/Z_{\rm gas} = const$. The terms $Z_{\rm gas}$ and $Z_{\rm dust}$ represent the ratios of metals and dust to hydrogen, respectively. The mean UV intensity is affected by radiative transfer through the diffuse gas, but for modest optical depth in the diffuse gas the relation $J_{\rm UV} \propto \Sigma_{\rm SFR}$ is expected to hold. In addition, a larger fraction of the UV escapes from GBCs if Z_d is very sub-Solar, which increases $J_{\rm UV}$ for a given $\Sigma_{\rm SFR}$ (this effect is quite uncertain, but might increase $J_{\rm UV}$ by a factor ~ 2). In thermal equilibrium with $P_{\rm th} \sim P_{\rm two-phase}$, the midplane pressure is therefore expected to vary roughly as $P_{\rm th} \propto \Sigma_{\rm SFR}$, with a somewhat larger coefficient for very low-metallicity regions.

Combining the thermal equilibrium relation $P_{\rm th} = P_{\rm th,0}\Sigma_{\rm SFR}/\Sigma_{\rm SFR,0}$ (normalized using the Solar neighborhood thermal pressure $P_{\rm th,0}$ and star formation rate $\Sigma_{\rm SFR,0}$) with the dynamical equilibrium relation $P_{\rm th} = \Sigma_{\rm diff}\langle g_z\rangle/(2\alpha)$ and the star formation relation

$\Sigma_{\rm SFR} = \Sigma_{\rm GBC}/t_{\rm SF}$, we obtain

$$\frac{\Sigma_{\rm GBC}}{\Sigma_{\rm diff}} = \frac{\langle g_z \rangle}{g_*} \propto \pi G(\Sigma_{\rm diff} + 2\Sigma_{\rm GBC}) + 2(2G\rho_{\rm sd})^{1/2}\sigma_z. \tag{2.3}$$

Here, $g_* = 2\alpha P_{\rm th,0}/(\Sigma_{\rm SFR,0}t_{\rm SF})$; for fiducial parameters, this acceleration is $g_* \sim$ pc Myr^{-2}.

It is interesting to compare outer and inner disks. In outer disks (similar to the Solar neighborhood and beyond, in galaxies like the Milky-Way), diffuse gas dominates the total so that $\Sigma_{\rm GBC} \ll \Sigma_{\rm diff} \approx \Sigma$; in addition, the term depending on $\rho_{\rm sd}$ dominates the gravity g_z. In this regime, the relation $\Sigma_{\rm SFR} \propto \Sigma_{\rm GBC} \propto \Sigma\sqrt{\rho_{\rm sd}}$ is therefore expected to hold. Physically, this regime may be thought of as the result of star formation increasing until the heating it provides is sufficient to balance cooling at the (dynamically-imposed) midplane pressure. If there is too little gas in the GBC component, the star formation rate would be extremely low, and the UV field would be very weak. A very low heating rate could not maintain a warm medium at the pressure imposed by the local gravitational field, so that (a portion of the) warm gas would condense out and become cold clouds. These cold clouds would collect to create more GBCs, which would then initiate star formation, raising the local UV radiation field until heating balances cooling. Given the low gravity and pressure of outer disks, cooling rates are moderate, and relatively low levels of star formation are needed to produce enough UV that heating balances cooling.

For outer disks where the stars and dark matter dominate gravity, the vertical oscillation time is $t_{\rm osc} = \pi^{1/2}/(G\rho_{\rm sd})^{1/2}$; a dense cloud settles to the midplane in $\sim t_{\rm osc}/4$. In this regime, the conversion time from gas to stars, $t_{\rm con} \equiv \Sigma/\Sigma_{\rm SFR}$, is given by

$$t_{\rm con} = t_{\rm osc}\frac{\sigma_z P_{\rm th,0}}{(2\pi)^{1/2}\langle v_{\rm th}^2\rangle \Sigma_{\rm SFR,0}}, \tag{2.4}$$

where $\langle v_{\rm th}^2\rangle \equiv \tilde{f}_w c_w^2 \approx c_w^2 M_{\rm diff,warm}/M_{\rm diff,total}$ is the mean thermal dispersion in the diffuse medium (here $c_w \sim 8$ km s^{-1} is the thermal speed in the warm ISM). Using $P_{\rm th,0} \sim \langle v_{\rm th}^2\rangle P_{\rm gas,0}/\sigma_z^2$ and defining a star formation energy conversion efficiency $\varepsilon_{\rm rad} \equiv 4\pi J_{\rm rad,0}/(c^2\Sigma_{\rm SFR,0})$ for $P_{\rm rad,0} = 4\pi J_{\rm rad,0}/(3c)$,

$$t_{\rm con} = t_{\rm osc}\frac{c}{3(2\pi)^{1/2}\sigma_z}\frac{P_{\rm gas,0}}{P_{\rm rad,0}}\varepsilon_{\rm rad}. \tag{2.5}$$

That is, the gas conversion time (or depletion time) is set by the time for gas to settle to the midplane, scaled by factors for the ratio of gas-to-radiation pressure in the Solar neighborhood, the mass-to-energy conversion efficiency, and c/σ_z.

In inner disks, unlike outer disks, we have $\Sigma_{\rm diff} \ll \Sigma_{\rm GBC} \approx \Sigma$, so that $\Sigma_{\rm SFR} \propto \Sigma$. In inner disks, it is straightforward to show that there is an upper limit on the diffuse gas surface density $\Sigma_{\rm diff}$. Physically, the reason for this limit is that the diffuse-gas cooling rate per particle increases with higher density and pressure in the inner parts of disks at least as $n\Lambda \propto \Sigma_{\rm diff}\Sigma_{\rm GBC}$ (since $n\Lambda \propto \Sigma_{\rm diff}/H \propto \Sigma_{\rm diff}g_z/\sigma_z^2 \propto \Sigma_{\rm diff}\Sigma_{\rm GBC}[1 + g_{\rm sd}/g_{\rm GBC}]/\sigma_z^2$), whereas the heating rate per particle varies as $\Gamma \propto \Sigma_{\rm SFR} \propto \Sigma_{\rm GBC}$. Thus, cooling will exceed heating (causing mass to drop out of the diffuse component) unless $\Sigma_{\rm diff}$ is sufficiently low. Enhanced cooling and mass "dropout" is likely responsible at least in part for the "saturation" of HI surface densities at $\lesssim 10$ M$_\odot$ pc^{-2} that has been observed in the inner parts of galaxies (Bigiel *et al.* 2008).

Based on the relations described above, the star formation law is expected to steepen from $\Sigma_{\rm SFR} \propto \Sigma$ in inner disks to $\Sigma_{\rm SFR} \propto \Sigma\sqrt{\rho_{\rm sd}}$ in outer disks. A reduction of the specific star formation rate $\Sigma_{\rm SFR}/\Sigma$ is indeed observed in galaxies starting at $\Sigma \lesssim 10$ M$_\odot$ pc^{-2} (Bigiel *et al.* 2008; Leroy *et al.* 2008). For some galaxies, a further power-law relation

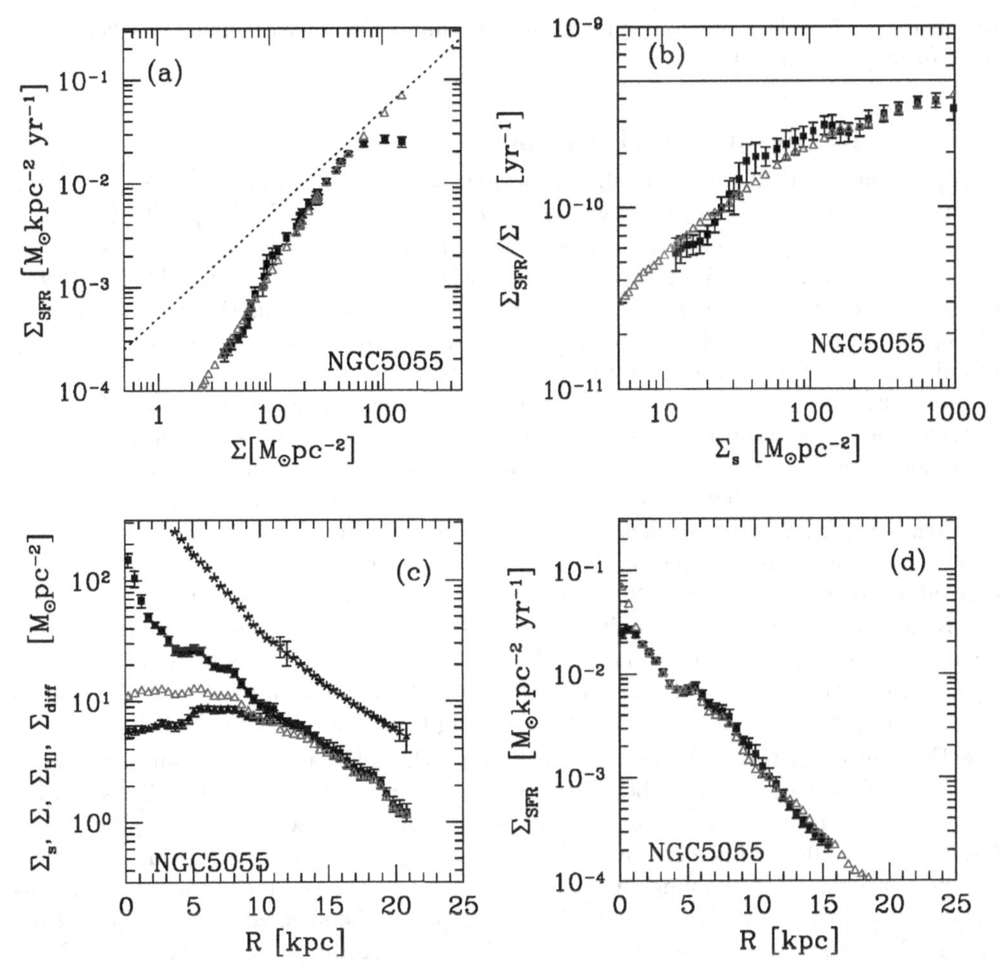

Figure 1. Comparison between annular averages of the data (*squares*) for NGC 5055 (Leroy *et al.* 2008), and the thermal/dynamical equilibrium model (*triangles*) developed in OML. Both the star formation rates as a function of radius in the galaxy (panel d), and star formation rates as a function of gas and stellar density (panels a and b) agree well with the model predictions.

$\rho_{sd} \propto \Sigma^{2p}$ may hold such that $\Sigma_{SFR} \propto \Sigma^{1+p}$ in outer disks, but this need not be the case in general – that is, integrated "Schmidt"-type relations need not hold.

In OML, the full solution for Σ_{SFR} is obtained as a function of Σ, ρ_{sd}, and the parameters $\alpha \equiv P_{tot}/P_{th}$ and $\tilde{f}_w \equiv \langle v_{th}^2 \rangle / c_w^2$, under the assumptions of thermal and dynamical equilibrium described above. It is also shown that the theoretical solution for Σ_{SFR} agrees well overall with a sample of disk galaxies analyzed in Leroy *et al.* (2008), with especially close correspondence for the large flocculent galaxies NGC 7331 and NGC 5055. Figure 1 shows an example of the comparison between the model and data, for NGC 5055.

Given the promising comparisons between the analytic theory and observations, it will be quite interesting to develop numerical simulations that fully test the assumptions and results of the thermal/dynamical equilibrium model. Encouragingly, the poster presented by C.-G. Kim at this meeting shows that initial numerical tests support the assumptions of thermal and dynamical equilibrium adopted in the analysis of OML. As discussed above, the OML theory contains parameters that must be set from either observations or detailed simulations. In the remainder of this contribution, we review what is known in this regard based on previous numerical work, and what measurements will be needed from future modeling efforts.

3. Numerical evaluation of parameters

From equations (2.1) and (2.3), the star formation rate in outer-disk regions is expected to vary as $\Sigma_{\rm SFR} \propto \Sigma\sqrt{2G\rho_{\rm sd}}\sigma_z/\alpha$, where $\alpha \equiv \sigma_z^2/v_{\rm th}^2$ and $\sigma_z^2 = v_{\rm th}^2 + v_{\rm turb}^2 + (1/2)\Delta(v_A^2 - 2v_{A,z}^2)$, with $v_{\rm th}^2$, $v_{\rm turb}^2$, and v_A^2 the (mass-weighted) mean thermal, turbulent, and Alfvén speeds in the diffuse gas (we now omit angle brackets denoting averaging). The coefficient σ_z/α can also be written as $v_{\rm th}^2/\sigma_z = c_w^2\tilde{f}_w/\sigma_z$. Thus, the star formation rate is expected to depend on the total velocity dispersion σ_z (or the ratio σ_z/c_w, where c_w is fixed by atomic physics), and on the fraction of diffuse gas in the warm phase $\approx \tilde{f}_w$

The ratios σ_z/c_w and $\tilde{f}_w \approx M_{\rm diff,warm}/M_{\rm diff,total}$ depend on the details of gas dynamics in the diffuse ISM. Important effects include warm and cold phase exchange via thermal instability; turbulence (with the associated shock heating and adiabatic temperature changes, as well as turbulent mixing); conversion of diffuse gas to GBCs via midplane settling, self-gravity, and turbulence-induced cloudlet collisions; return of GBC gas to the diffuse phase by photodissociation and by "mechanical" destruction processes (including expanding HII regions, winds, SNe, and radiation pressure). Turbulence in the diffuse gas can be driven by stellar energetic inputs as well as spiral shocks, the magnetorotational instability, large-scale gravitational instabilities in the disk, and cosmic infall.

Numerical studies to understand the various effects involved are very much a work in progress, but some consensus is already beginning to emerge on a number of points:

- For a medium with a bistable cooling curve, the midplane thermal pressure tends to evolve, by exchange of mass between cold and warm components of the diffuse phase, such that the mean value is comparable to, or slightly below, $P_{\rm two-phase}$ (Piontek & Ostriker 2005, 2007). Since out-of equilibrium effects depend on the heating time from shocks compared to the cooling time, the mean value of the thermal pressure, as well as the breadth of the pressure distribution, must in general be affected by the scale and the amplitude of turbulence (see Gazol *et al.* 2005, 2009; Audit & Hennebelle 2005, 2010; Hennebelle & Audit 2007; Joung & Mac Low 2006; Joung *et al.* 2009). Realistic numerical evaluations of the mean thermal pressure (for a given radiative heating rate) therefore will require numerical simulations in which the vertical box size is comparable to the true scale height of the diffuse ISM, and in which the turbulent amplitude is $\sim 5-10$ km s^{-1}.
- Magnetic fields in differentially-rotating multiphase disks are amplified by the magnetorotational instability until the magnetic pressure becomes comparable to the thermal gas pressure, with $B_z^2 \ll B^2$ (Piontek & Ostriker 2005, 2007; Wang & Abel 2009). Supernova-driven turbulence also contributes to amplifying the magnetic field (de Avillez & Breitschwerdt 2005; Mac Low *et al.* 2005).
- The energy input from supernovae yield ISM velocity dispersions $\sim 5 - 10$ km s^{-1} for models with a wide range of supernova driving rates and disk properties (e.g. Kim 2004; de Avillez & Breitschwerdt 2005; Dib *et al.* 2006; Shetty & Ostriker 2008; Agertz *et al.* 2009; Joung *et al.* 2009). These values are comparable to those observed in the HI gas. Simulations have also shown that the turbulent amplitudes decrease at smaller scales and for higher densities. With this range of turbulent velocity dispersions, the turbulent pressure in simulations of the diffuse ISM is comparable to the thermal pressure.
- The interaction between self-gravity and rotational shear also drives turbulence at significant levels ($\gtrsim 10$ km s^{-1}) in galactic disks (Kim & Ostriker 2001, 2007; Wada *et al.* 2002; Shetty & Ostriker 2008; Tasker & Tan 2009; Agertz *et al.* 2009; Aumer *et al.* 2010; Bournaud *et al.* 2010). However, the turbulent power is much larger at the large

($\sim$ kpc) scales that dominate the swing amplifier than at scales below the disk scale height, and in-plane velocities (which do not contribute to vertical support of the disk) are much larger than vertical velocities. Thus, turbulence driven by instabilities on large scales is likely of limited importance in regulating the effective midplane pressure (for a given local gas surface density Σ), and hence the star formation rate. (Gravitational instabilities would, however, enhance Σ and thus $\Sigma_{\rm SFR}$ locally.) Flapping associated with non-steady spiral shocks also drives turbulence in the diffuse ISM (Kim *et al.* 2006; Kim *et al.* 2010), but again, vertical motions are small compared to horizontal motions

Although numerical results have shown that the total turbulent velocity dispersion σ_z is relatively insensitive to the disk properties and the supernova driving rate (consistent with observations), it is much less certain how the warm fraction, or $v_{\rm th}^2 = \tilde{f}_w c_w^2 \approx c_w^2 M_{\rm diff,warm}/M_{\rm diff,total}$, depends on disk conditions and/or the star formation rate. Assessing this dependence, including a full exploration of parameter space, is an important task for future numerical studies. The fraction of diffuse atomic gas in different phases is not well known empirically, either, although observations of C II with *Herschel* potentially afford a means to separate cold and warm components of the atomic medium (which both contribute to 21 cm emission).

Finally, it remains important to understand more fully how spiral structure develops, and in particular, whether it is possible to characterize in a simple way the fraction of gas in a given annulus that is found in "arm" vs. "interarm" conditions, and what the compression factor is for the gas surface density. Numerical simulations have begun to marry spiral structure with an increasingly realistic treatment of the ISM (including multiple phases, turbulence, and magnetic fields); much more, however, remains to be done on this front. It also remains to be determined how well models like that of OML apply locally for galaxies with strong spiral structure. More generally, it is important to assess which equilibria (thermal, dynamical, star formation) still apply locally even in galaxies with large-scale transient structure in the ISM (due to spiral arms, tidal interactions, mergers, cosmic inflows, etc.).

4. Conclusion

Gas is the raw material for star formation, but the detailed state of the ISM, which depends in turn on the internal galactic environment, determines the rate at which this material is processed to create new stars. Recent observations have begun to explore the correlation between gas content and star formation at increasingly high spatial resolution, revealing changes in star formation "laws" between inner and outer disks; other environmental dependences of star formation have also been explored, including intriguing correlations between molecular and stellar content of galactic disks.

Although the simplest recipes for star formation (such as a rate that depends inversely on the free-fall time at the mean ISM density) have difficulty matching the data, models that account for feedback and the multiphase character of the ISM are more successful. In particular, recent work suggests that the empirical correlation between molecular content and estimated midplane pressure can be understood as reflecting a state of simultaneous thermal and dynamical equilibrium in the diffuse ISM. The thermal/dynamical equilibrium model of OML develops the idea that UV from OB stars provides a feedback loop that regulates the star formation rate: the proportions of diffuse and self-gravitating gas adjust themselves so that the heating rate (proportional to the mass of self-gravitating gas) matches the cooling rate (proportional to the mass of diffuse gas and to the vertical gravitational field). The model formulated in OML is promising in terms of explaining star-forming behavior in observed systems. With numerical simulations, it will be possible to appraise – and potentially revise – the simplifying assumptions and parameterizations adopted by this equilibrium model. Time-dependent simulations will also lead to a much clearer understanding of how GBCs form and disperse, and how their properties and the formation/destruction timescales relate to galactic environment. This will aid in defining limits for applying equilibrium relations, while also pointing the way towards non-equilibrium theories of star formation.

Acknowledgements: This work was supported by grant AST-0908185 from the National Science Foundation, and by a fellowship from the John Simon Guggenheim Foundation. The author thanks the referee for a helpful report.

References

Agertz, O., Lake, G., Teyssier, R., Moore, B., Mayer, L., & Romeo, A. B. 2009, *MNRAS*, 392, 294
Audit, E. & Hennebelle, P. 2005, *A&A*, 433, 1
—. 2010, *A&A*, 511, A76+
Aumer, M., Burkert, A., Johansson, P. H., & Genzel, R. 2010, *ApJ*, 719, 1230
Bournaud, F., Elmegreen, B. G., Teyssier, R., Block, D. L., & Puerari, I. 2010, arXiv:1007.2566
Bigiel, F., Leroy, A., Walter, F., Brinks, E., de Blok, W. J. G., Madore, B., & Thornley, M. D. 2008, *AJ*, 136, 2846
Blitz, L. & Rosolowsky, E. 2006, *ApJ*, 650, 933
de Avillez, M. A. & Breitschwerdt, D. 2005, *A&A*, 436, 585
Dib, S., Bell, E., & Burkert, A. 2006, *ApJ*, 638, 797
Field, G. B., Goldsmith, D. W., & Habing, H. J. 1969, *ApJL*, 155, L149
Gazol, A., Luis, L., & Kim, J. 2009, *ApJ*, 693, 656
Gazol, A., Vázquez-Semadeni, E., & Kim, J. 2005, *ApJ*, 630, 911
Heiles, C. 2001, Tetons 4: Galactic Structure, Stars and the Interstellar Medium, 231, 294
Hennebelle, P. & Audit, E. 2007, *A&A*, 465, 431
Joung, M. K. R. & Mac Low, M. 2006, *ApJ*, 653, 1266
Joung, M. R., Mac Low, M., & Bryan, G. L. 2009, *ApJ*, 704, 137
Kim, C.-G., Kim, W.-T., & Ostriker, E. C. 2006, *ApJL*, 649, L13
Kim, C., Kim, W., & Ostriker, E. C. 2010, ArXiv e-prints
Kim, J. 2004, Journal of Korean Astronomical Society, 37, 237
Kim, W. & Ostriker, E. C. 2007, *ApJ*, 660, 1232
Kim, W.-T. & Ostriker, E. C. 2001, *ApJ*, 559, 70
Koyama, H. & Ostriker, E. C. 2009a, *ApJ*, 693, 1316
—. 2009b, *ApJ*, 693, 1346
Leroy, A. K., Walter, F., Brinks, E., Bigiel, F., de Blok, W. J. G., Madore, B., & Thornley, M. D. 2008, *AJ*, 136, 2782
Mac Low, M., Balsara, D. S., Kim, J., & de Avillez, M. A. 2005, *ApJ*, 626, 864
McKee, C. F., & Ostriker, E. C. 2007, *ARAA*, 45, 565
Ostriker, E. C., McKee, C. F., & Leroy, A. K. 2010, *ApJ*, 721, 975 (OML)
Piontek, R. A. & Ostriker, E. C. 2005, *ApJ*, 629, 849
—. 2007, *ApJ*, 663, 183
Shetty, R. & Ostriker, E. C. 2008, *ApJ*, 684, 978
Tasker, E. J. & Tan, J. C. 2009, *ApJ*, 700, 358
Wada, K., Meurer, G., & Norman, C. A. 2002, *ApJ*, 577, 197
Wang, P. & Abel, T. 2009, *ApJ*, 696, 96
Wolfire, M. G., Hollenbach, D., McKee, C. F., Tielens, A. G. G. M., & Bakes, E. L. O. 1995, *ApJ*, 443, 152
Wolfire, M. G., McKee, C. F., Hollenbach, D., & Tielens, A. G. G. M. 2003, *ApJ*, 587, 278

Computational Star Formation
Proceedings IAU Symposium No. 270, 2011
J. Alves, B.G. Elmegreen, J. M. Girart & V. Trimble, eds.

doi:10.1017/S1743921311000834

Star Formation in Interacting Galaxies

Curtis Struck
Dept. of Physics and Astronomy, Iowa State University
Ames, IA 50011 USA
email: curt@iastate.edu

Abstract. This brief review emphasizes the wide range of environments where interaction induced star formation occurs. In these environments we can study the numerous elaborations of a few basic physical processes, including: gravitational instability, accretion and large-scale shocks.

Keywords. galaxies: interactions, galaxies: evolution, stars: formation

1. Introduction and Background

This has become a huge topic in the last few decades, with many specific areas of active research. Induced star formation in major mergers produces the most luminous galaxies in the universe, and its cumulative contribution to the net star formation (SF) of typical galaxies is substantial. The amount and type of star formation induced in collisions and mergers depends on environment and cosmological time, with important corollary phenomena like downsizing and the peak in the cosmological SF rate at redshifts of 2-3. The disequilibrium of disks, and the production of tidal structures like bridges, tails and shells, provide unique environments for star formation, e.g, in tidal dwarf galaxies, and proto-globular clusters. (The former are distinguished from the latter by greater masses, e.g., $\geqslant 10^7 M_\odot$.) Interactions also compress and funnel gas into galaxy cores driving a range of activities therein.

This large range of topics cannot be covered in this brief review. Dynamical forces acting on galaxy scales (typically 3-30 kpc) are primarily responsible for the induced star formation, so I will focus the discussion on these scales. The associated phenomena on the smaller scales of molecular clouds, and the larger scales of cosmological structure formation are considered in other reviews at this meeting. Similarly, I will generally not discuss computational methods, but concentrate on what has been learned from the interplay between observations, models and analytic theory. In some systems with unique SF structures these comparisons can strongly constrain model treatments. The topics that I do want briefly visit include: some special star-forming environments created in collisions between two comparable galaxies, which can be considered playgrounds for the study of induced star formation (ISF), explorations of dynamically complex environments in groups, attempts to automate parts of the dynamical modeling, and the state of analytic theory.

Before probing these specific topics, I would like to review a little background on the big questions in this field. First is the question of how much of the net cosmic star formation can be blamed on interactions and mergers? The last decade has seen a great deal of progress towards answering this from both deep observational surveys, and cosmological structure buildup simulations. Examples of the former include the work of Jogee, *et al.* (2009), who examined data from GEMS, Combo-17 and Spitzer deep fields, with sensitivity out to $z \simeq 0.8$. Their overall conclusion was that mergers major and minor can account for $\leqslant 30\%$ of the star formation in the cosmic time range of 3-7 Gyr ago. As a second example, Shi, *et al.* (2009) in a study of LIRGS in the GOODS survey conclude that the fraction of mergers among these strongly star-forming objects is about 50% up to a redshift of $z \simeq 1.2$. Studies of both the cosmic SFR and the merger rate predict increases out to redshifts of 2-3, so we can expect the interaction induced star formation is even higher at those redshifts. There is little question that interaction/merger ISF is a very important contributor. However, it is also clear that we have some ways to go in order to have accurate, quantitative estimates of the amount in different cosmological epochs.

Other big questions include where, when and how the interaction-induced SF occurs? A significant part of these questions was answered a couple of decades ago with IRAS observations. That data led to the discovery of ULIRGS, to the recognition that most SF occurs in the final phases of wet merging, and that it is buried deep within the nucleus of the remnant. This was summarized in the conference proceedings of Sulentic, Keel, & Telesco (1990), and later explained in part by numerical simulations (e.g., Barnes & Hernquist 1996, Mihos & Hernquist 1996). The high redshift version of that story is being elaborated steadily at the present time.

What about ISF before the merger? A number of surveys have examined this question in the last decade, usually comparing extensive samples of paired versus unpaired galaxies (see the reviews of Struck 2006, and in Smith, *et al.* 2010a), or samples of obviously interacting galaxies to comparable isolated objects. Weaknesses of the former technique include the fact a fraction of the paired galaxies may not be interacting (even with very similar recession velocities), and minor companions of comparison objects may not always be visible. Both effects are probably small in recent studies. In the case of strongly interacting samples, the difficulties are often associated with obtaining data on a large sample of the rather rare objects. Yet studies with the two techniques seem to be coming to similar conclusions. The result is that SF is enhanced in pre-merger interactions, but only by a factor of a few or less, and this enhancement is often dominated by nuclear starbursts in one of the pair, at a time near closest approach. However, there are a number of less typical cases worthy of study.

2. Star-forming Environments in Galaxy Collisions

Among the specific reasons for considering these examples is firstly, to understand differences and similarities between star-forming regions in the Local Group and more extreme environments. Secondly, given that we still must use phenomenological prescriptions for SF in models on these and larger scales, it is worth briefly considering how some these apply over the range of observed conditions. Let's begin with some gaseous shocks with an extent and intensity definitely not seen in our galaxy.

I will pass by one of the most famous, the "Overlap Region" in the Antennae system, if only because of the angle at which we view it is not ideal, making observational interpretations difficult. Several posters were presented at this meeting on Antennae models by Chapon, Karl, Kruijssen, and Renaud (see summaries in this volume). Each of these has a different and interesting angle on the system.

Consider instead the shocks in the ocular ring in the disk of IC 2163, which both observations and models (Struck 2005) suggest has been orbiting around its companion, NGC 2207 for some time. Most of the emission in various bands comes from strong SF in the ocular wave, rather than directly from the large-scale shocks. However, the sharp dust lanes in HST observations, and strong dust emission in the Spitzer Space Telescope observations, are indicative of such shocks. Because of its proximity this galaxy is probably the best system for detailed study of such waves. It apparently has another very large shock to offer. An ongoing study of a unique band of radio continuum emission at the interface between the two disks suggests it may be the result of gentle scraping, as the disks almost or just barely touch (see Fig. 1 Kaufman, *et al.* 2010,). In contrast to the Antennae overlap, this soft encounter evidently induces virtually no SF.

One of the most beautiful shocks in an interacting system is found in Arp 118, investigated by Appleton, *et al.* (2003), see the right panel in Fig. 1. This shock is revealed in optical imagery as a long, multi-kiloparsec dust lane. By a fluke of nature we have amazingly direct evidence that this is a shock. Absorption against a background radio jet allows us to see the 185 km/s shock velocity jump in HI absorption. There is little star formation over most of the length of the shock, but Appleton *et al.* presented a plausible case that the shock compression does induce star formation, with a time delay of about 25 Myr, and which is visible at a considerable distance downstream. The shock itself is a type that is commonly seen to develop at the leading edge of a tidal arm in models (e.g., in the semi-analytical models of Gerber & Lamb 1994). In other systems, star formation does occur in such arms, especially at latter stages when the material is flung out to greater distances, and arm material travels with the feature rather than moving through it.

An extreme example of strong, large-scale shocks in interacting systems is found in the so-called Taffy system (UGC 12914/15, see Gao, Zhu & Seaquist 2003), and its several known sister systems. Here a nearly direct collision between galaxy disks leaves gas and magnetic fields strewn between the two. In the Taffy there is little SF in the debris field, despite the fact that much of the gas there must have been strongly shocked. There is SF in and near the disks, which may include infalling gas.

The assumption that SF is directly induced by shocks has been used as the basis of phenomenological SF laws in simulations. J. Barnes has recently advocated this procedure (see e.g., Chien & Barnes 2009), and has provided numerical examples to show that such a prescription does a better job than gas density-dependent forms in matching the SF

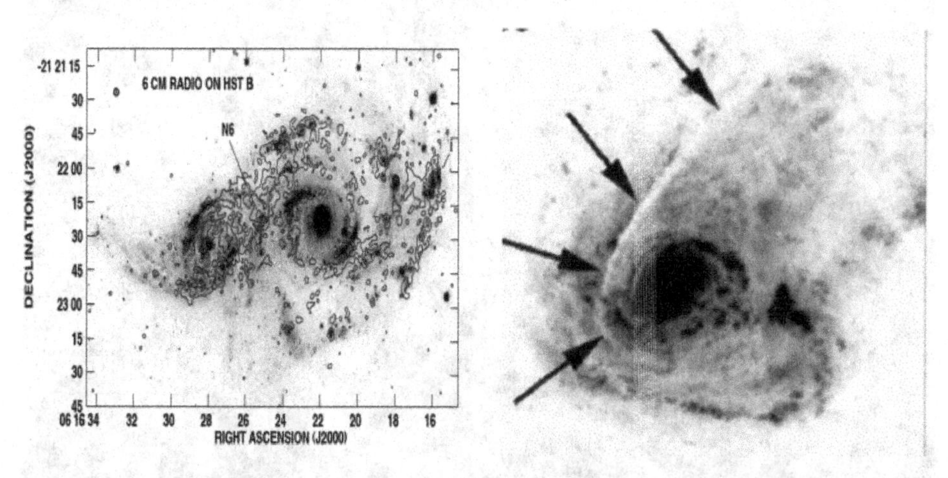

Figure 1. Examples of large-scale shocks with little apparent SF. The NGC 2207/IC 2163 system is shown on the left (from Kaufman, *et al.* 2010). An optical image (HST B-band) is shown in grayscale, and radio continuum emission is shown as contours. Note the continuum emission between the galaxies. An optical (HST) image of Arp 118 is shown on the right (from Appleton, *et al.* 2003). The large scale shock is highlighted by a long thin dust lane. The induced star formation lies off the shock. Note: according to the NASA Extragalactic Database, the scale of the system is 560 pc/arcsec.

in a couple of well-known merger remnants. In light of the above examples I would caution that there may be subtleties in other cases. Some of these examples suggest that there may a threshold compression before shocks can effectively trigger SF, and the Arp 118 example suggests that the time delays may play a role in some cases. Recent high resolution numerical studies shed further light on these questions (see Kim, *et al.* 2009, Saitoh, *et al.* 2009, Teyssier, *et al.* 2010), and some of these works were updated at this meeting.

Another set of processes for triggering SF are gravitational or magneto-gravitional instabilities (see e.g., Elmegreen 1989, Efremov 2010). A characteristic and oft-cited symptom of this disease is regularly spaced clumps of young stars, or molecular clouds, or HI superclouds the so-called beads-on-a-string. The beads are relatively rare in isolated galaxies, but interactions can trigger them on large scales. It is my impression that they are not quite so rare in colliding galaxies. The Spitzer/IRAC images of the NGC 2207/IC 2163 system show them in a number of arm sections. HST reveals optical subclumps in some of them (Elmegreen, *et al.* 2006 and references therein).

They are apparently quite common in M51-type flyby systems, in the spirals or extended tails. This includes M51 itself, but other morphologically similar examples are: Arp 72, Arp 82, and Arp 86. The beaded structure in bridges and tails is often more prominent and extensive in GALEX UV observations (see Fig. 2 from Smith, *et al.* 2010b).

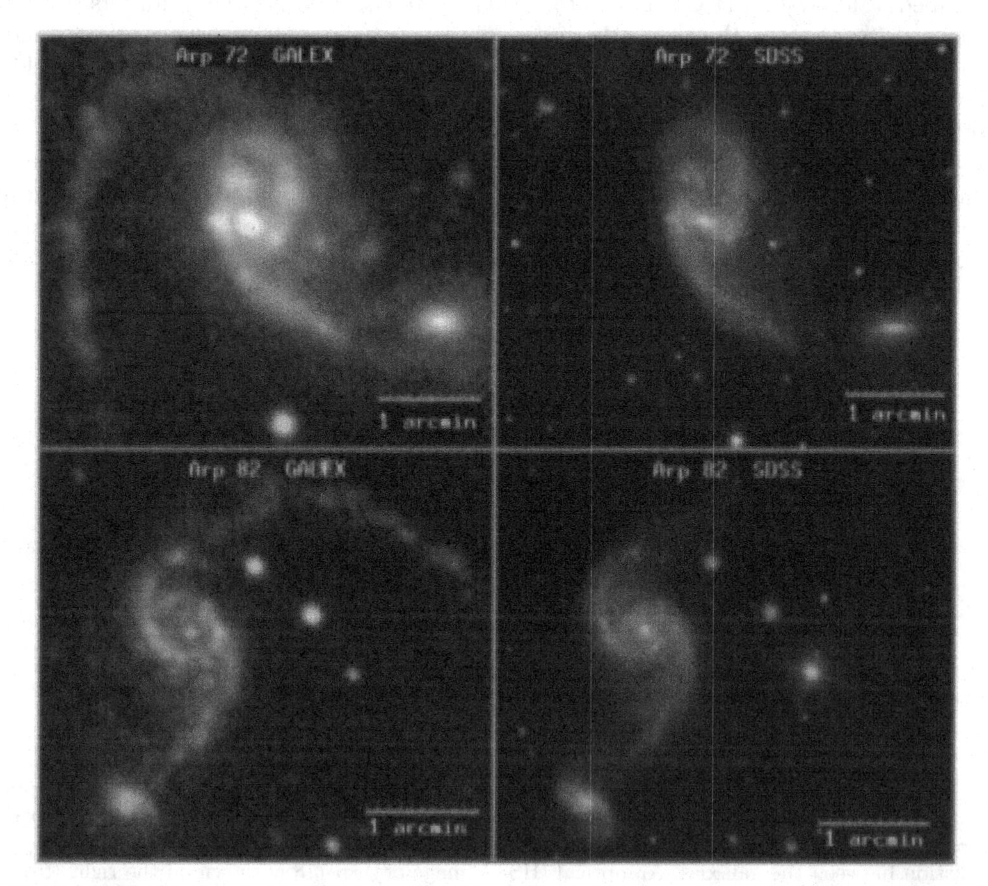

Figure 2. Sample beads-on-a-string in the tidal tails of Arp 72 and Arp 82 (from Smith, *et al.* 2010b). The beads are much more prominent in the GALEX UV images on the left, than in the SDSS images on the right.

Some colliding ring galaxies, like the Cartwheel are also beaded. What these types have in common are strong waves, or as in the case of material tails, material that remains in proximity long enough to pull together under self-gravity.

The beads often have substructure. Thus, they do not result from an instability with a single dominant wavelength. The concepts of "competitive accretion" within the waves, or a tidal wave version of "collect and collapse" may operate on a timescale set by each unique environment. Numerical models now have the resolution to test such notions.

A third process, besides shock compression and gravitational instability, which likely plays an important role in interaction induced star formation is accretion or pile-up. There are many kinds of accretion on galaxy scales. Cosmological accretion has excited a great deal of interest lately as a competitor to ISF in mergers, but is beyond the scope of this review. Gas concentration in wet mergers can also be viewed as accretion, but the topic has been frequently reviewed before. I will survey some examples in ongoing interactions.

The most well known example of recent years is the Duc-Bournaud theory of tidal dwarf galaxy (TDG) formation (see Duc, Bournaud, & Masset 2004, Bournaud, & Duc 2006). Their models demonstrated that TDGs form in a pile-up at the end of tidal tail, where material is unable to push any farther out of the joint dark halo of the merging galaxies. Its hard to get much material out that far, so only dwarfs are made. However, as shown by the models it can accumulate for $\simeq 10^8$ yr., pull together gravitationally, and form stars. According to Lisenfeld, *et al.* (2009) the observed correlation between molecular gas density and SFR obeys the Schmidt Law in these objects. But only a handful of objects to date have observations sufficient to test this.

A similar pileup occurs when bridge material swings around a companion as in Arp 285 (see Fig. 3 and Smith, *et al.* 2008). Like the beads phenomenon, SF occurs in a line of clumps in this region, but simulations suggest that this is random, not systematic.

An equally dramatic situation is the battle of tidal-stretching versus gravitational accumulation, which can occur in the centers of bridges. This is illustrated in the Arp 305 system, where a TDG seems to be forming in the bridge center (see 4 and Hancock, *et al.* 2009). Similarly, there are the objects my collaborators have named Hinge Clumps (see Hancock, *et al.* 2009, and Smith 2010b). These are luminous star forming regions found in disks at the base of tidal tails. Examples are shown in Fig. 5.

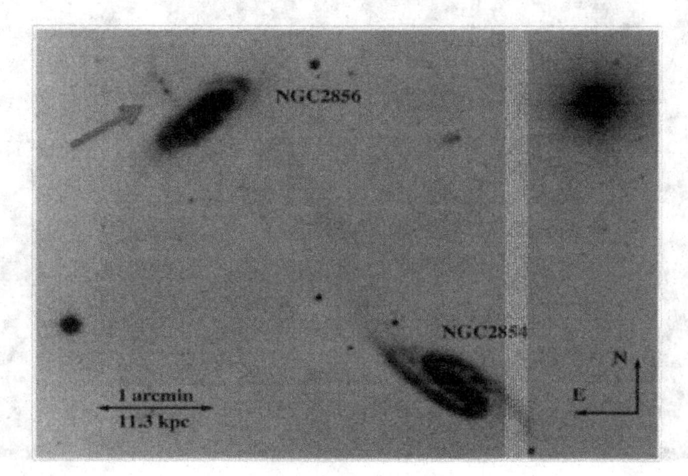

Figure 3. Optical image of the NGC 2854/2856 system (from Smith, *et al.* 2008). The northern galaxy has a rather faint "beads-on-a-string" plume, which models indicate may be a pileup zone of material transferred from the southern galaxy.

In sum, ISF in galaxy collisions comes in a much wider range of variations on themes of compression than in isolated disks. The dominant subthemes are the same: shocks, gravitational collapse/fragmentation and accretion. Another thing that is the same in all the various environments is that star formation occurs in clusters, and often in clusters of clusters. There seems to be wide consensus that the cluster birth function is universal across interacting and isolated galaxies. Currently, there is some debate about cluster disruption processes and timescales. Observations of interacting galaxies have provided input into this debate, but attention has been strongly focused on a few nearby systems (especially M51 and the Antennae), which do not sample the full range of star-forming environmental conditions in interactions. Studies of other nearby systems can provide more input. For example, Peterson, *et al.* (2009) find that some very young groups of star clusters in the Arp 284 system are embedded in a halo of intermediate age stars. This can tell us about the disruption SF/disruption history of such regions.

3. Concluding Miscellany and Prospects

In order to begin to understand the mechanisms of induced SF in the many environments of interacting galaxies, dynamical models of prototypical systems, or if the data are of high enough quality, models of individual systems are needed. In the latter case, such models have traditionally been handcrafted, i.e., fit by trial and error. With aid of some general rules of thumb the art has become quite well developed. Even so, the future promises broader developments.

In the last decade, several groups have attempted automatic modeling in the sense that the simulation code is able to explore models with different collision parameters, seek

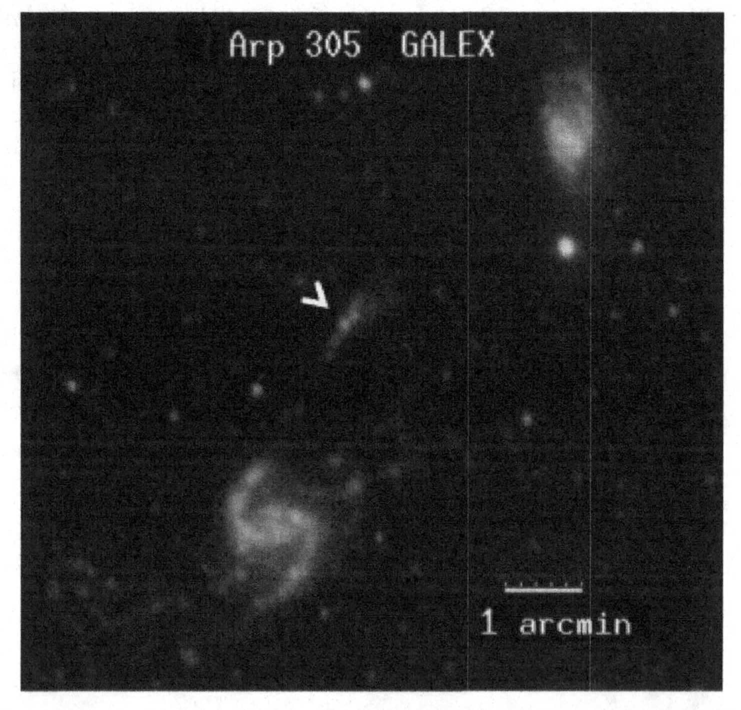

Figure 4. A candidate TDG forming in the bridge between the two galaxies of Arp 305, as revealed in GALEX observations (from Hancock, *et al.* 2009). In published observations HI observations gas is seen to connect the two galaxies, confirming the existence of the bridge.

best fit models, and then use these as the basis for another round of modeling and fitting (see Smith, *et al.* 2010c and references therein, also Gomez, *et al.* 2004). Most of these works use genetic algorithms. Fitting detailed structures of ISF has not been attempted, only morphological and velocity structures. The difficulty of this procedure grows rapidly with the number of the observational constraints and their resolution. Interestingly, the goodness-of-fit measure does not increase smoothly, but rather in jumps. At present, alternatives include "expert-learning" systems like the Barnes/Hibbard Identikit (Barnes & Hibbard 2009) or Chilingarian *et al.*'s GalMer database (Chilingarian, *et al.* 2010).

I also believe that analytic work could be further developed in this field, at least to aid in understanding complex numerical simulations. For example, analytic descriptions of waves have proven feasible and useful in studies of colliding ring galaxies. The study of caustic waveforms can applied more generally (see e.g., Struck-Marcell 1990, Gerber & Lamb 1994). Simple, restricted 3-body modeling is very useful for impulsive encounters, and is nearly semi-analytic in such cases.

Models for multiple interactions in specific groups have not been attempted often. The complexity is great and the collision parameter space is dauntingly large. A significant fraction of cosmic star formation occurs in such environments, and more so at high redshift. This provides motivations for studying local examples in detail. Cosmological structure simulations provide much information about SF in multiple mergers, but not all that we would like. Are there general rules-of-thumb governing the dynamics of such systems?

J.-S. Hwang and I (in collaboration with F. Renaud and P. N. Appleton) are computing models of the evolution of Stephan's Quintet involving multiple collisions to better understand these processes. One simplification: we find that most of the major features can be modeled as the result of a series of 4 close 2-body encounters. A complication: initial conditions of successive 2-body encounters are not in any kind of steady state! These conclusions are probably general.

With work proceeding on so many fronts, and the computational tools to deal with increased complexity, there is good reason to believe that our understanding of interaction induced SF will advance considerably in the next decade. Prominent interacting systems

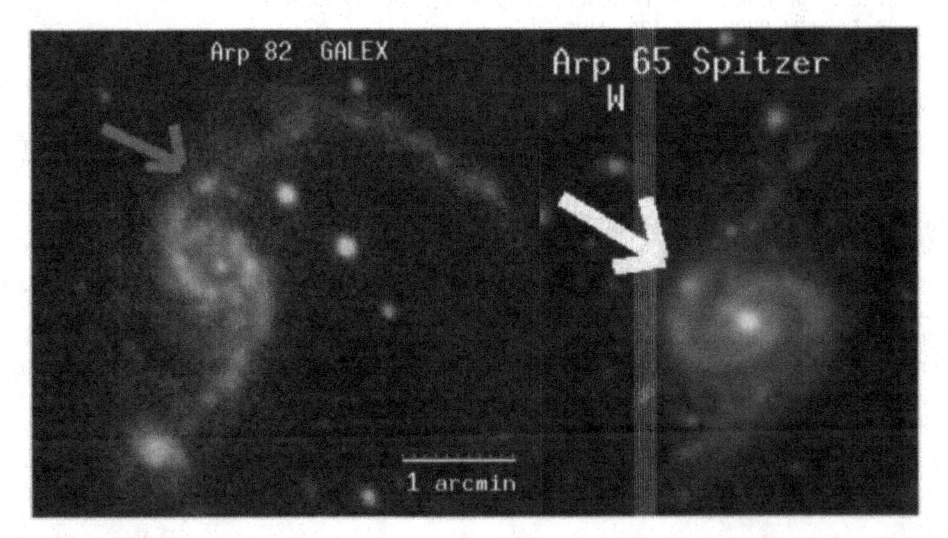

Figure 5. Two examples of "hinge clumps" in Arp 65 and Arp 82 (from Smith 2010b). Hinge clumps are prominent knots of star formation at the base of a tidal tail, but within the galaxy disk.

are not located nearby, so their study is at a more exploratory stage than Galactic or Local Group studies described at this conference. However, the foundation is being laid for more exciting work in the future.

References

Appleton, P. N., Charmandaris, V., Gao, Y., Jarrett, T., & Bransford, M. A. 2003, *ApJ*, 586, 112

Barnes, J. E. & Hernquist, L. 1996, *ApJ*, 471, 115

Barnes, J. E. & Hibbard, J. E. 2009, *AJ*, 137, 3071

Bournaud, F. & Duc, P.-A. 2006, *A&A*, 456, 481

Chien, L.-H. & Barnes, J. E. 2010, *MNRAS*, 407, 43

Chilingarian, I., Di Matteo, P., Combes, F., Melchoir, A.-L., & Semelin, B. 2010, *A&A*, 518, 61

Duc, P.-A., Bournaud, F., & Masset, F. 2004, *A&A*, 427, 803

Efremov, Y. N. 2010, *MNRAS*, 405, 1531

Elmegreen, B. G. 1989, *ApJ* (Letters), 342, L67

Elmegreen, D. M., Elmegreen, B. G., Kaufman, M., Sheth, K., Struck, C., Thomasson, M., & Brinks, E. 2006 *ApJ* 642, 158

Gao, Y., Zhu, M. & Seaquist, E. R. 2003, *AJ*, 126, 2171

Gerber, R. A. & Lamb, S. A. 1994, *ApJ*, 431, 604

Gomez, J. C., Athanassoula, L., Fuentes, O., & Bosma, A. 2004, in F. Ochsenbein, M. G. Allen, & D. Egret (eds.), *Astronomical Data Analysis Software and Systems (ADASS) XIII: A.S.P. Conf. Series, 314* (San Francisco: ASP), p. 629

Hancock, M., Smith, B. J., Struck, C., Giroux, M. L., & Hurlock, S. 2009, *AJ*, 137, 4643

Jogee, S., *et al.* 2009, *ApJ*, 697, 1971

Kaufman, M., Grupe, D., Elmegreen, D. M., Elmegreen, B. G., Struck, C., & Brinks, E. 2010, *AJ*, in preparation

Kim, J., Wise, J. H., & Abel, T. 2009, *ApJ*, 694, L123

Lisenfeld, U., Bournaud, F., Brinks, E., & Duc, P.-A. 2009, *astro-ph*, 0903-0999

Mihos, J. C. & Hernquist, L. 1996, *ApJ*, 464, 641

Peterson, B. W., Struck, C., Smith, B. J., & Hancock, M. 2009, *MNRAS*, 400, 1208

Saitoh, T. R., Daisaka, H., Kokubo, E., Makino, J., Okamoto, T., Tomisaka, K., Wada, K., & Yoshida, N. 2009, *PASJ*, 61, 481

Shi, Y., Rieke, G., Lotz, J., & Perez-Gonzalez, P. G. 2009, *ApJ*, 697, 1764

Smith, B. J., *et al.* 2008, *AJ*, 135, 2406

Smith, B. J., Bastian, N., Higdon, S. J. U., & Higdon, J. L. (eds.) 2010a, *Galaxy Wars: Stellar Populations and Star Formation in Interacting Galaxies: A.S.P. Conf. Series, 423* (San Francisco: ASP)

Smith, B. J., *et al.* 2010c, in Smith, B. J., Bastian, N., Higdon, S. J. U., & Higdon, J. L. (eds.), *Galaxy Wars: Stellar Populations and Star Formation in Interacting Galaxies: A.S.P. Conf. Series, 423* (San Francisco: ASP), p. 227

Smith, B. J., Giroux, M. L., Struck, C., Hancock, M., & Hurlock, S. 2010b, *AJ*, 139, 1212

Struck, C. 2005, *MNRAS*, 364, 69

Struck, C. 2006, in: J. W. Mason (ed.), *Astrophysics Update 2* (Heidelberg: Springer Praxis), p. 115

Struck-Marcell, C. 1990, *AJ*, 99, 71

Sulentic, J. W., Keel, W. C., & Telesco, C. M. (eds.) 1990, *Paired and Interacting Galaxies: I.A.U. Colloq. 124* (Washington, D. C.: NASA)

Teyssier, R., Chapon, D., & Bournaud, F. 2010, *ApJ*, 720, L149

Computational Star Formation
Proceedings IAU Symposium No. 270, 2011
J. Alves, B.G. Elmegreen, J. M. Girart & V. Trimble, eds.

doi:10.1017/S1743921311000846

Shock-induced star cluster formation in colliding galaxies

Takayuki R. Saitoh[1], Hiroshi Daisaka[2], Eiichiro Kokubo[1,3,4], Junichiro Makino[1,3,4], Takashi Okamoto[5], Kohji Tomisaka[1,3,4], Keiichi Wada[3,6], and Naoki Yoshida[7]

[1]Division of Theoretical Astronomy, National Astronomical Observatory of Japan, 2–21–1 Osawa, Mitaka-shi, Tokyo 181–8588
[2]Graduate School of Commerce and Management, Hitotsubashi University, Naka 2–1 Kunitachi-shi, Tokyo 186–8601
[3]Center for Computational Astrophysics, National Astronomical Observatory of Japan, 2–21–1 Osawa, Mitaka-shi, Tokyo 181–8588
[4]School of Physical Sciences, Graduate University of Advanced Study (SOKENDAI), 2–21–1 Osawa, Mitaka-shi, Tokyo 181–8588
[5]Center for Computational Sciences, University of Tsukuba 1–1–1,Tennodai, Tsukuba, Ibaraki 305–8577, Japan
[6]Graduate School of Science and Engineering, Kagoshima University, 1–21–30 Korimoto, Kagoshima, Kagoshima 890–8580
[7]Institute for the Physics and Mathematics of the Universe, University of Tokyo, 5–1–5 Kashiwanoha, Kashiwa, Chiba 277–8568, Japan
email: saitoh.takayuki@nao.ac.jp

Abstract. We studied the formation process of star clusters using high-resolution N-body/smoothed particle hydrodynamics simulations of colliding galaxies. The total number of particles is 1.2×10^8 for our high resolution run. The gravitational softening is 5 pc and we allow gas to cool down to ~ 10 K. During the first encounter of the collision, a giant filament consists of cold and dense gas found between the progenitors by shock compression. A vigorous starburst took place in the filament, resulting in the formation of star clusters. The mass of these star clusters ranges from 10^{5-8} $M_{\odot}$. These star clusters formed hierarchically: at first small star clusters formed, and then they merged via gravity, resulting in larger star clusters.

Keywords. galaxies: star clusters — galaxies: starburst — galaxies: evolution — ISM: evolution — methods: numerical

1. Introduction

Merging galaxies contain many young star clusters (e.g., Whitmore 2003). These star clusters could potentially evolve into the present day metal-rich globular clusters and so they are widely accepted to be a good candidate for globular cluster progenitors. Galaxy-galaxy merger is therefore considered as one of the formation channels of globular clusters (Schweizer 1987).

There have been a number of numerical studies of merging galaxies. There have been, however, only a few numerical studies of star cluster formation in merging galaxies even though it has been shown that resolving a cloudy/multiphase interstellar medium (ISM) and/or clustered star formation can have important consequences for the formation history of early-type galaxies (e.g. Bois *et al.* 2010). Some of the existing studies adopted sub-grid models of star cluster formation (e.g., Bekki & Couch 2001; Li *et al.* 2004), while more recently their formation has been captured directly (e.g. using the sticky particle method in Baurnaud *et al.* 2008). Here we report the result of merger simulations that

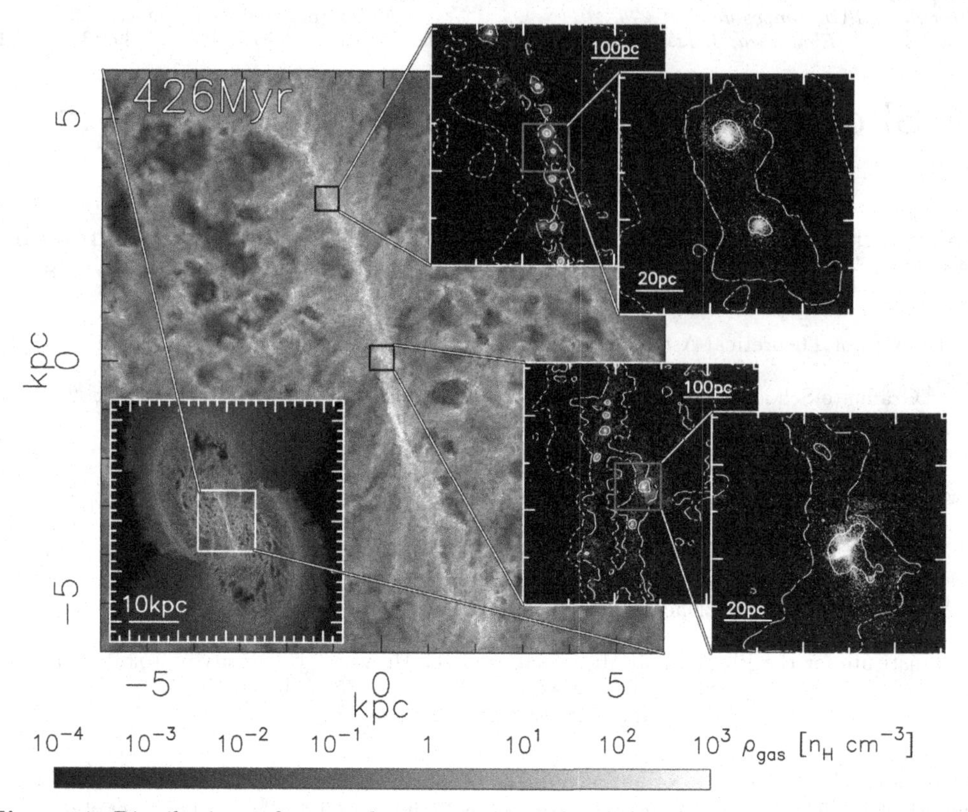

Figure 1. Distributions of gas and stars in the early phase of the star cluster formation (t = 426 Myr). Main panel shows gas density map of the mid-plane in 12 kpc × 12 kpc regions, while the bottom-left inset displays four times larger scale than that of the main panel. Closeup views of star-cluster forming regions show surface stellar density maps with contours of surface gas density in 500 pc × 500 pc and 100 pc × 100 pc.

capture the multiphase nature of the ISM and include realistic models of star formation and feedback.

2. Method

We prepared two identical progenitor galaxies and then let them merge from a parabolic and coplanar configuration. The mass of components in one progenitor galaxy is 10^{11} $M_\odot$ for the dark matter halo, 4.7×10^9 $M_\odot$ for the stellar disk and 1.8×10^9 $M_\odot$ for the gas disk. The galaxies are modeled using both N-body and smoothed particle hydrodynamics (SPH) particles. We employed 1.2×10^8 particles for the two progenitor galaxies for the finest runs where the corresponding mass of each particle is 1.9×10^3 $M_\odot$ for both N-body and SPH particles.

Numerical simulations were performed by our N-body/SPH code, `ASURA`. We adopted the FAST scheme (Saitoh & Makino 2010) that accelerates the time integration of a self-gravitating fluid by asynchronously integrating gravity and hydrodynamical interactions. In addition, we used a time-step limiter for the time-integration of SPH particles with individual time-steps which enforces the differences of time steps in neighboring SPH particles to be small enough so that the SPH particles can correctly evolve under strong shocks (Saitoh & Makino 2009).

The high mass resolution allows us to adopt the ISM model with the wide temperature range down to 10 K and realistic conditions of star formation ($\rho > 100$ cm^{-3}, $T < 100$ K

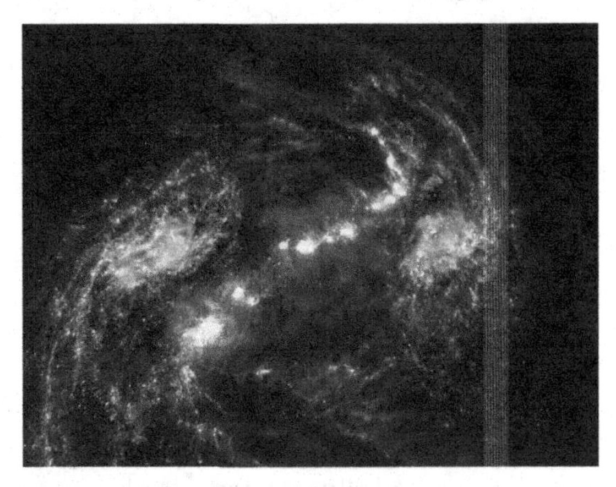

Figure 2. Distributions of star clusters formed during the first encounter at $t \sim 460$ Myr. Blown and blue means cold and hot gas phases, whereas white points are young stars. The visualization was done by Takaaki Takeda.

and $\nabla \cdot v < 0$, where ρ, T and v indicate density, temperature and velocity, respectively). The feedback from type II supernovae (SNe) was also taken into account. These models are the same as those in Saitoh *et al.* (2008; 2009).

3. Results

At the first encounter (after ~ 420 Myr from the beginning of the simulations), strong shocks took place. These shocks induced the formation of a large filament of cold and dense gas between the two galaxies. This filament quickly cooled and fragmented to a number of small, high-density clumps, and a large-scale starburst took place in the filament (see Saitoh *et al.* 2009). This starburst did not take place in previous studies in which gas cooling was inhibited below 10^4 K. This starburst continued for several 10 Myr and was then quenched by the SN feedback. A number of star clusters formed during this starburst.

Figure 1 shows the snapshots of the early phase of the star cluster formation where we can see that a number of small star clusters formed along the gas filament. The late phase of the star cluster formation ($t \geqslant 440$ Myr) was mainly driven by mergers of star clusters without gas, since gas in the filament was blown out by SNe. Hence star clusters grew hierarchically through gravitational mergers of smaller star clusters. There is no star cluster with the mass $\geqslant 10^7$ $M_\odot$ which forms from an instantaneous collapse of a single large cloud. At the final phase (after 40 Myr from the start of the starburst), we found several tens of star clusters between two progenitor galaxies (see figure 2). This is because they were formed only from the gas in the shock-induced filament.

Figure 3 shows cumulative mass functions of star cluster systems. The mass of star clusters ranges from 10^{5-8} $M_\odot$. The mass function for masses larger than 10^7 $M_\odot$ is almost independent of mass and spatial resolutions. The slope of this high mass region is close to a power law function with an index of -2, which is consistent with observations (e.g., Whitmore *et al.* 1999). Since the formation of star clusters is driven by mergers, the shape of the mass function becomes a power law reflecting the scale free nature of gravity (Elmegreen 2009).

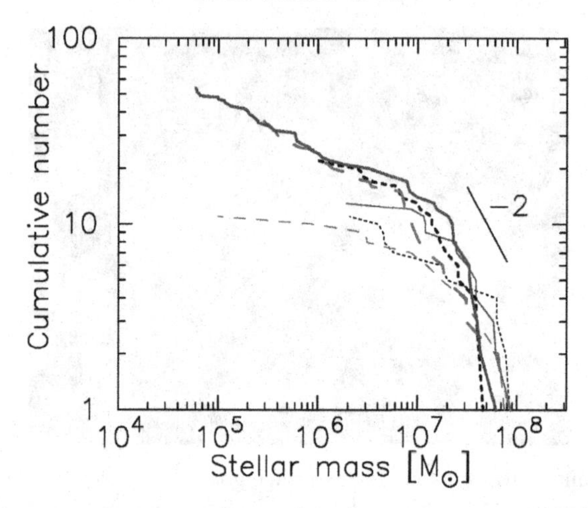

Figure 3. Cumulative mass functions of star cluster systems at $t = 460$ Myr. Thick and thin curves indicate cumulative mass functions of star clusters in runs with $\epsilon = 5$ pc and 20 pc, respectively. Solid, dashed, dotted curves represent high, middle, low mass resolution runs, respectively.

4. Summary

We performed high resolution simulations of merging galaxies that capture the multiphase ISM and the realistic modelings of star formation and of SN feedback. In the simulations, we found that a number of star clusters formed during the starburst at the first encounter. These star clusters grew via hierarchical mergers.

Acknowledgements

Numerical simulations were carried out on Cray XT4 at CfCA of NAOJ. TRS is financially supported by a Research Fellowship from the JSPS for Young Scientists.

References

Bekki, K. & Couch, W. J. 2001, *ApJL*, 557, L19

Bois, M., *et al.* 2010, *MNRAS*, 406, 2405

Bournaud, F., Duc, P.-A., & Emsellem, E. 2008, *MNRAS*, 389, L8

Elmegreen, B. G. 2009, in Globular Clusters, Guide to Galaxies, ed. T. Richtler, *et al.*, ESO: Springer, p. 87

Li, Y., Mac Low, M.-M., & Klessen, R. S. 2004, *ApJL*, 614, L29

Saitoh, T. R., Daisaka, H., Kokubo, E., Makino, J., Okamoto, T., Tomisaka, K., Wada, K., & Yoshida, N. 2008, *PASJ*, 60, 667

Saitoh, T. R., Daisaka, H., Kokubo, E., Makino, J., Okamoto, T., Tomisaka, K., Wada, K., & Yoshida, N. 2009, *PASJ*, 61, 481

Saitoh, T. R. & Makino, J. 2009, *ApJ* (Letters), 697, L99

Saitoh, T. R. & Makino, J. 2010, *PASJ*, 62, 301

Schweizer, F. 1987, in Nearly Normal Galaxies. From the Planck Time to the Present, ed. S. M. Faber, 18–25

Whitmore, B. C. 2003, in A Decade of Hubble Space Telescope Science, ed. M. Livio, K. Noll, & M. Stiavelli, 153–178

Whitmore, B. C., Zhang, Q., Leitherer, C., Fall, S. M., Schweizer, F., & Miller, B. W. 1999, *AJ*, 118, 1551

Computational star formation
Proceedings IAU Symposium No. 270, 2011
J. Alves, B.G. Elmegreen, J. M. Girart & V. Trimble, eds.

doi:10.1017/S1743921311000858

Galactic star formation in parsec-scale resolution simulations

Leila C. Powell[1], Frederic Bournaud[1], Damien Chapon[1], Julien Devriendt[2], Adrianne Slyz[2] and Romain Teyssier[1,3]

[1]SAp, CEA-Saclay, F-91191 Gif-sur-Yvette Cedex, France
email: leila.powell@cea.fr
[2]Oxford Astrophysics, Denys Wilkinson Building, Keble Road, Oxford, OX1 3RH, UK
[3]Institute of Theoretical Physics, University of Zurich Winterhurerstrasse 190, CH-8057 Zurich, Switzerland

Abstract. The interstellar medium (ISM) in galaxies is multiphase and cloudy, with stars forming in the very dense, cold gas found in Giant Molecular Clouds (GMCs). Simulating the evolution of an entire galaxy, however, is a computational problem which covers many orders of magnitude, so many simulations cannot reach densities high enough or temperatures low enough to resolve this multiphase nature. Therefore, the formation of GMCs is not captured and the resulting gas distribution is smooth, contrary to observations. We investigate how star formation (SF) proceeds in simulated galaxies when we obtain parsec-scale resolution and more successfully capture the multiphase ISM. Both major mergers and the accretion of cold gas via filaments are dominant contributors to a galaxy's total stellar budget and we examine SF at high resolution in both of these contexts.

Keywords. methods: numerical, galaxies: evolution, galaxies: interaction, stars: formation

---

By performing adaptive mesh refinement (AMR) simulations in which we resolve the multiphase ISM and the formation of GMCs, we investigate the nature of galactic SF in two important contexts: major mergers (Section 1) and cold accretion via filaments at high-redshift (Section 2).

1. Star formation modes during major mergers

To date, we have performed 8 simulations of 1:1 galaxy mergers ($M_{\rm gal} \approx 8 \times 10^{10} {\rm M}_{\odot}$) with a maximum 5pc spatial resolution and 1.6×10^4 $\rm M_{\odot}$ effective gas mass resolution using the AMR code RAMSES (Teyssier 2002), providing us with a representative sample that contains a variety of orbital parameters. To achieve our high resolution, we use a 'pseudo-cooling' equation of state which dictates the temperature of gas based on its density (see Teyssier *et al.* 2010, for details). This approach has been shown to produce an ISM density power spectrum that agrees well with observations (Bournaud, 2010 in prep.). We choose a moderately high density threshold ($\sim 10^3$ atoms cm^{-3}) and a very low efficiency ($\sim 10^{-4}$) for the SF prescription, and employ the Sedov blastwave model of supernova (SN) feedback (Dubois & Teyssier 2008). This choice of parameters gives a star formation rate (SFR) of $\sim 1{\rm M}_{\odot}{\rm yr}^{-1}$ in the galaxies when they are simulated in isolation, which is reasonable for a $z = 0$ disc galaxy. Work is currently underway to test the sensitivity of the SFR to changes in these parameters.

1.1. *Extended star formation*

In the classical picture, merger-induced starbursts result when tidal torques drive large amounts of gas into the centre of the merging structure, where it is compressed and forms

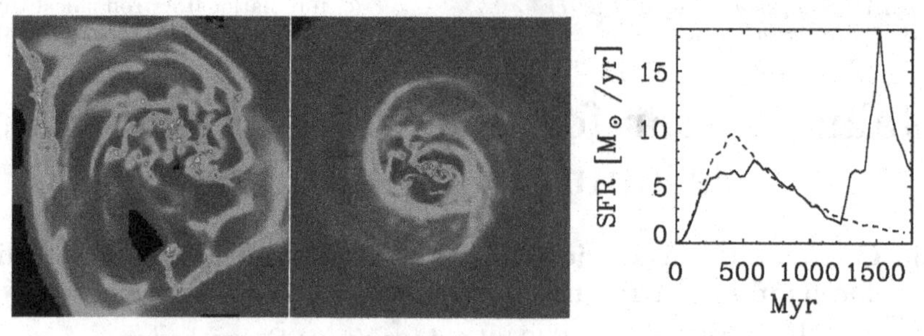

Figure 1. Left and Middle: Gas density images of the merging galaxies, one face-on, one edge-on (left) and of one of the same galaxies when evolved in isolation (middle) at t=1.5 Gyr. Red contours indicate the regions with densities above the SF threshold. Size of image is $\approx$ 13kpc. **Right:** SFR versus time for both of the merging galaxies (solid line) and the sum of the SFRs in the same two galaxies when evolved in isolation (dashed line).

Figure 2. Gas density images from two different mergers. **Left:** Formation of 'beads on a string' of length $\sim$ 100 kpc. **Right:** Two tidal dwarf galaxies (one with $M \sim 10^9 M_\odot$) formed after the first pericenter passage.

stars rapidly (Struck, this meeting). We report the occurrence of another mode of merger-induced SF: clumpy, extended SF. In Fig. 1 we clearly demonstrate this alternative mode by comparing the gas density map of the merging galaxies (left) with that of one of the same galaxies when evolved in isolation (middle) at the same time instant ($t = 1.5$ Gyr). Red contours indicate the high density, star-forming clumps and these are considerably more numerous and significantly less centrally concentrated in the merging galaxies (where the clumps are spread over a region of $\approx$ 7kpc in diameter) than in the isolated galaxy (clumps spread over $\approx$ 3kpc). This is particularly clear when comparing the merging galaxy seen face-on in the left-hand panel of Fig. 1 with the isolated galaxy (middle panel). By reference to the time evolution of the SFR (Fig. 1, right), we can see that $t = 1.5$ Gyr occurs just before the maximum SFR, indicating that this extended, clumpy SF is responsible for the starburst which has increased the SFR in the merger (solid line) by a factor of $\approx$ 10 relative to the SFR in the same two galaxies evolved in isolation (dashed line) i.e. the starburst occurs *before* the final coalescence of the two galaxy centres.

This type of merger-induced star cluster formation over an extended region is also observed in the Antennae galaxies. Teyssier *et al.* (2010) have demonstrated that this is only reproduced in a simulation of the Antennae with RAMSES if a resolution of $\sim$ 10pc is achieved; they find that the star formation is no longer clustered if the gas is thermally supported (as occurs at lower resolutions), highlighting the strong dependence of the mode of simulated SF on resolution.

In Fig. 2 we show some examples of the other varied star-forming features found serendipitously in our merger sample.

2. Star formation during cold-mode accretion and the onset of a galactic wind

The Nut Simulation † is an ultra-high resolution cosmological simulation of a Milky Way like galaxy forming at the intersection of 3 filaments in a ΛCDM cosmology (Slyz *et al.* 2009; Devriendt *et al.* 2010). We perform the simulation with RAMSES (Teyssier 2002) including cooling, a UV background (Haardt & Madau 1996), SF and SN feedback with metal enrichment (Dubois & Teyssier 2008) achieving a maximum physical spatial resolution of $\approx$ 0.5pc in the densest regions at all times. A thin, rapidly rotating disc forms, which is continuously fed by cold gas that streams along the filaments at ~Mach 5. The disc is gravitationally unstable and fragments into star-forming clumps (see Devriendt *et al.* (2010, in prep.) for details of the disc and Slyz *et al.* (2010, in prep.) for a study of the clumpy star formation). SF in the galaxy, its satellites and haloes embedded in the filaments results in many SN explosions, which are individually resolved. The numerous bubbles overlap, creating an extended cavity through which the hot gas escapes the galactic potential in the form of a wind (see Slyz *et al.* 2009; Devriendt *et al.* 2010, for figures illustrating this and additional technical details). By $z = 9$, when the dark matter halo mass has reached $\sim 10^9 \mathrm{M}_\odot$, the wind extends to $\approx 6r_{\mathrm{vir}}$, filling the virial sphere with hot, diffuse gas not unlike the shock-heated gaseous halo found in galaxies above the mass threshold for a stable virial shock (e.g. Birnboim & Dekel 2003). To date, the simulation has reached $z \approx 7.5$ and it is still running.

Here we present results from a study of the interaction between the cold filaments and the hot, far-reaching galactic wind and examine the impact this competition has on the SFR (Powell *et al.*, 2010 in prep.).

2.1. *Inflows and outflows: measuring accretion and winds*

To assess the relative importance of the filaments and the galactic wind we measure the mass inflow and outflow rates within the virial radius of the galaxy. We divide the gas into various phases with the use of temperature and density thresholds, similar to the approach in other simulation studies of hot and cold mode accretion (e.g. Agertz *et al.* 2009). Note, however, that the thresholds are resolution-dependent so ours differ from those in the previous lower resolution studies. The categories we define are: hot diffuse ($T > 2 \times 10^5$K, $n < 10$ atoms cm^{-3}), warm diffuse ($2 \times 10^4\mathrm{K} < T < 2 \times 10^5$K, $n < 10$ atoms cm^{-3}), cold diffuse ($T < 2 \times 10^4$K, $n < 0.1$ atoms cm^{-3}), filaments ($T < 2 \times 10^4$K, $0.1 \leqslant n \leqslant 10$ atoms cm^{-3}) and 'clumpy' ($n > 10$ atoms cm^{-3}). We verify that these categories accurately separate the different gas components using 3D visualisation software.

In Fig. 3 we compare the inflow in a control run with no SN feedback (left) with that in the Nut simulation (middle) and examine the outflow in the Nut simulation (right). We highlight several key results: 1) the inflow of cold gas in both runs (first two panels), is dominated at all radii by the contribution from the filaments (blue dotted line), which supply material right down to the disc, 2) the total inflow rate (first two panels, black solid line) is almost identical in the simulations with SN feedback (middle, the Nut simulation) and without SN feedback (left, control run), revealing that the galactic wind has had a negligible effect on the accretion rate, 3) the only significant mass outflow occurs in the Nut simulation (right) yet this is only around 10 per cent of the mass inflow rate and 4) there is a non-negligible cold diffuse component (blue dashed line) in this outflow, which is gas that was previously in between the filaments that has been entrained by the wind.

† Nut is the Egyptian goddess of the night sky

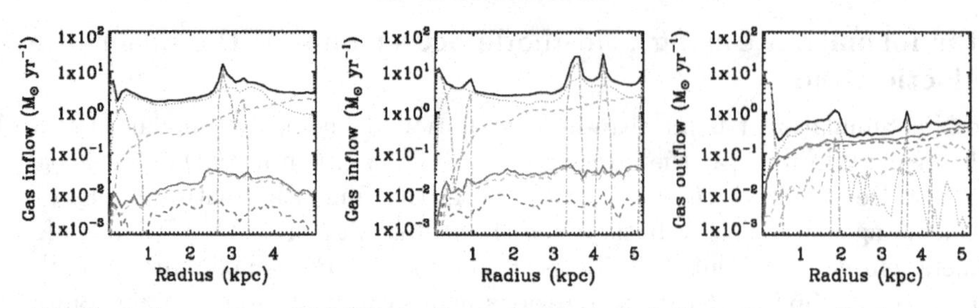

Figure 3. Gas inflow averaged in 100 pc physical spherical shells out to $r_{\rm vir}$ in the control run without supernovae feedback (left) and the Nut simulation (middle) at $z = 9$. Gas outflow in the Nut simulation (right). Flow rate of all gas (solid black line), subdivided into dense (blue dot-dash line), filamentary (blue dotted line), cold diffuse (blue dashed line), hot diffuse (red dashed line), warm diffuse (orange dashed line) and hot+warm diffuse (red solid line).

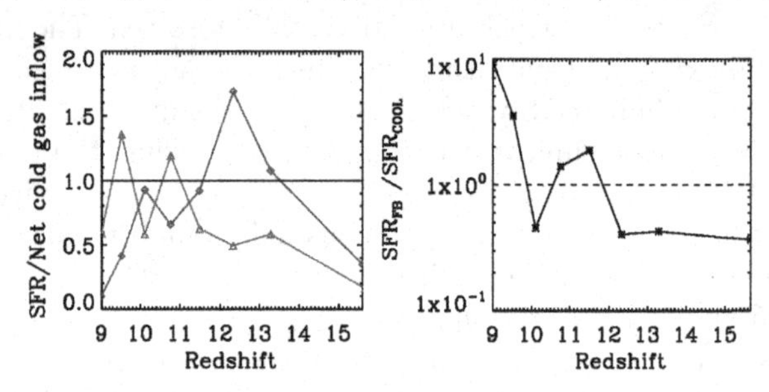

Figure 4. SFR of main galaxy averaged over 10Myr **Left:** SFR divided by cold gas inflow rate for control run (blue) and Nut simulation (red). **Right:** SFR in Nut simulation divided by SFR in control run. Horizontal lines indicate when ratios = 1.

2.2. *What happens to the star formation rate?*

We have established that the high-velocity, far-reaching galactic wind has been unable to impact the supply of cold gas, the 'fuel' for SF, because it is delivered to the galaxy in supersonic, highly collimated, dense filaments. The other major influence on the SFR is the efficiency with which this fuel is converted into stars.

The left-hand panel of Fig. 4 shows the ratio of the SFR over the net cold gas inflow rate for the control run (blue) and the Nut simulation (red). For both runs the ratio oscillates around 1 (the horizontal line), indicating that the conversion of the filament-supplied cold gas to stars is very efficient. The mean efficiency in the central star-forming region is ~ 0.1 in both cases; an order of magnitude higher than the 1 per cent efficiency used in RAMSES to dictate the SFR on a cell by cell basis. The right-hand panel of Fig. 4 shows the ratio of the SFR in the Nut Simulation to that in the control run (the actual SFR is a few $M_{\odot}yr^{-1}$). Approaching $z = 9$, the SFR in the Nut simulation (*with* SN feedback) exceeds that in the control run (*without* SN feedback). We attribute this positive feedback effect to the metal enrichment in the Nut simulation (absent in the control run), which facilitates cooling of the gas, boosting SF.

References

Agertz O., Teyssier R., Moore B., 2009, *MNRAS*, 397, L64
Birnboim Y. & Dekel A., 2003, *MNRAS*, 345, 349
Devriendt J., Slyz A., Powell L., Pichon C., Teyssier R., 2010, *IAU Symposium*, 262, 248
Dubois Y. & Teyssier R., 2008, *A&A*, 477, 79
Haardt F. & Madau P., 1996, *ApJ*, 461, 20
Slyz A., Devriendt J., Powell L., 2009, *ASP conference series*, Astronum-2009
Teyssier R., 2002, *A&A*, 385, 337
Teyssier R., Chapon D., Bournaud F., 2010, arXiv:1006.4757

Computational Star Formation
Proceedings IAU Symposium No. 270, 2011
J. Alves, B.G. Elmegreen, J. M. Girart & V. Trimble, eds.

doi:10.1017/S174392131100086X

Galaxy formation hydrodynamics: From cosmic flows to star-forming clouds

Frédéric Bournaud[1]

[1]CEA, IRFU, SAp. CEA-Saclay, F-91191 Gif-sur-Yvette, France
email: frederic.bournaud@cea.fr

Abstract. Major progress has been made over the last few years in understanding hydrodynamical processes on cosmological scales, in particular how galaxies get their baryons. There is increasing recognition that a large part of the baryons accrete smoothly onto galaxies, and that internal evolution processes play a major role in shaping galaxies – mergers are not necessarily the dominant process. However, predictions from the various assembly mechanisms are still in large disagreement with the observed properties of galaxies in the nearby Universe. Small-scale processes have a major impact on the global evolution of galaxies over a Hubble time and the usual sub-grid models account for them in a far too uncertain way. Understanding when, where and at which rate galaxies formed their stars becomes crucial to understand the formation of galaxy populations. I discuss recent improvements and current limitations in "resolved" modeling of star formation, aiming at explicitly capturing star-forming instabilities, in cosmological and galaxy-sized simulations. Such models need to develop three-dimensional turbulence in the ISM, which requires parsec-scale resolution at redshift zero.

Keywords. galaxies: formation, galaxies: ISM, galaxies : structure, galaxies : high-redshift

---

1. Cosmological hydrodynamics: the growth of galaxies

Dark matter-only simulations have long been the main tool of numerical cosmologists. They have drown a picture of galaxy formation in which mass assembly is mostly driven by the gradual merging of small dark matter halos into larger ones: the galaxies, i.e. the baryons, are a relatively passive component that follows this *hierarchical* growth of dark halos. Cosmological simulations can now directly model the evolution of the baryons, not only in the so-called "zoom" simulations towards a single chosen galaxy, but also in large-scale simulations covering hundreds of Mpc, such as the *Marenostrum* simulation by Teyssier and collaborators. Such simulations have revealed a richer picture of galaxy assembly: not purely hierarchical, and not driven only by the dark matter structures. Baryons do not all lie within galaxies and halos but also form long filaments along the cosmic web, which gather a large fraction of the gas even at moderate redshift (z~1-2).

These reservoirs continuously supply gas into galaxies. For small galaxies in low-mass halos, infalling gas is accelerated but the flow remains subsonic in the hot gas that fills the halo (this hot halo gas, coming for instance from stellar winds, is at $T{\sim}10^{6-7}$K with $c_s \simeq 100-300$ km/s). The infalling gas thus remains relatively cold (10^{4-5}K) and directly fuels the central galactic disk. Around massive galaxies, infalling gas reaches supersonic velocities and is shock-heated to the viral temperature (10^{6-7}K) by this *virial shock*, which typically stabilizes near the virial radius of the dark halo. These cold accretion vs. hot accretion modes have been proposed by Birnboim & Dekel (2003) and detailed in the simulations of (Kereš *et al.* 2005): the cold mode supplies blue star-forming galaxies, while the hot mode could result in red and dead objects.

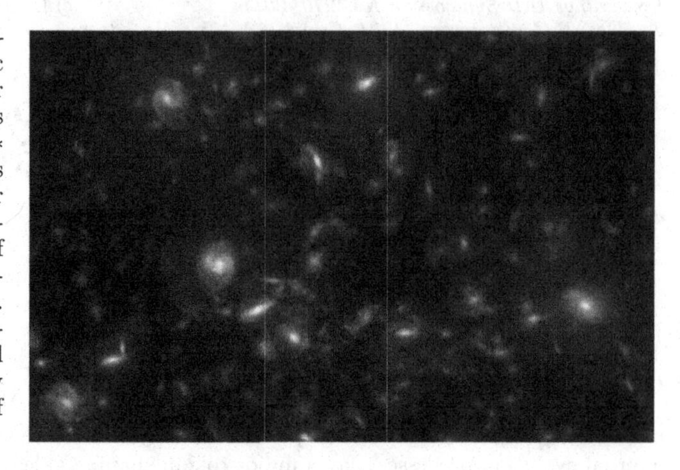

Figure 1. *Marenostrum galaxies* in a true color synthetic image (courtesy from Teyssier & Pichon). The main galaxies in this image are around $L*$ and all have bulge/disk ratios around unity. There is a clear lack of bulgeless or disk-dominated galaxies, but also a lack of real ellipticals even in Marenostrum groups and proto-clusters. Sub-grid recipes that would reduce the bulge fraction in all galaxies are not a satisfactory answer to the whole diversity of galaxy populations.

Gas infall is not limited to pure cold and hot modes. *Cold flows in hot halos* could be a more general mode. Simulations by Kravtsov *et al.* presented in Dekel & Birnboim (2006) suggest that cold gas streams at high redshift are dense enough to penetrate the hot gas halo. They directly reach the central galaxy in the form of cold unshocked gas flows collimated in the hot halo. The ubiquity of this mode for massive galaxies was further pointed out in the Marenostrum simulation (Ocvirk *et al.* 2008). The number density of such stream-fed galaxies in the simulation matches the observed comoving density of star-forming galaxies at $z \sim 2$ (Dekel *et al.* 2009). Cold streams in hot halos could be an important growth mode even for relatively low-mass galaxies, as shown by the *Nut* simulation (see Powell *et al.*, this meeting). This supports an emerging picture in which Milky Way-like galaxies get most of their mass from cold accretion. Major mergers are relatively unfrequent (Genel *et al.* 2008). Even small incoming halos (i.e., minor mergers) account only a small fraction of the total infalling mass, so a substantial part of the infall should be quite smooth (Genel *et al.* 2010). The mass brought in as baryonic clumps in dark halos (i.e. major and minor mergers) probably represents, on average, one third of the baryonic supply onto $\sim L*$ galaxies (Brooks *et al.* 2009; Kereš *et al.* 2009).

The dominance of mergers or smooth accretion should depend on mass. The galaxy mass function roughly as steep as the dark halo mass function at low masses, and becomes much shallower at high masses (Hopkins *et al.* 2010). In the low-mass regime, this implies that galaxy mergers are generally minor ones (having two merging galaxies with similar masses requires two haloes of similar masses, which is unlikely), so cold accretion could dominate. But for very massive galaxies above $L*$, the galaxy masses grow more slowly than the halo masses (because of some debated "quenching" mechanism). When two big halos of different masses merge together, their galaxies have more similar masses, and galaxy mergers thus tend to be relatively major ones: giant elliptical galaxies likely mostly assembled through violent mergers.

The formation of dwarf galaxies with stellar masses around 10^9 $M_\odot$, largely driven by cold gas accretion and feedback processes, can be relatively well understood (Governato *et al.* 2010, and Brook this meeting). Giant elliptical galaxies are also relatively well understood as resulting from numerous early mergers and several possible quenching mechanisms (Naab *et al.* 2007, Johansson *et al.* 2009, Martig *et al.* 2009). Things are much more complicated for galaxies in the $L*$ regime, where the combination of cold accretion, mergers, feedback processes and internal evolution is richer. This regime encompasses a large diversity of galaxy types, from bulgeless rotating disks to slowly-rotating ellipticals. While environmental effects play an important role, they cannot explain the bulk diversity

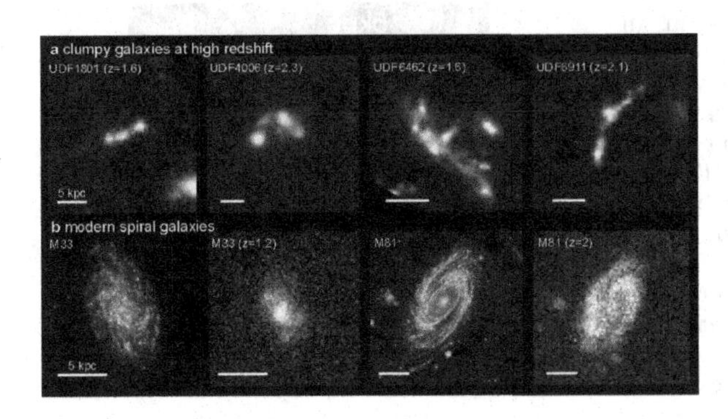

Figure 2. *Star-forming galaxies at $z{\sim}2$ are dominated by kpc-sized clumps of 10^{7-9} $M_\odot$. This is not an artifact from bandshifting (optical imaging traces the UV emission at $z = 2$) or resolution. Modern disk galaxies in the UV have only much smaller star forming regions. They do not show highly contrasted giant clumps but exponential disks and spiral arms instead, even when artificially redshifted (courtesy from Debra Elmegreen).*

of galaxies: there are field ellipticals, and galaxies with very low bulge fractions in groups as well (such as M33 in the Local group). Explaining the diversity of $L*$ galaxies remains a major challenge, the toughest part likely being to explain the survival of bulgeless, pure disk galaxies. For instance, all Marenostrum galaxies at $\sim L*$ are bulgy galaxies, say, S0 or Sa (Fig. 1). It is known that all simulations of this type lack bulgeless galaxies, but it is interesting to note that they lack elliptical galaxies at the same time.

2. Milky Way progenitors in the early Universe

Since the formation of today's $\sim$ $L*$ galaxies cannot be directly understood from cosmological simulations, let us examine the properties of their progenitors at redshift $z \sim 2$, when they contained only half of their present mass, focusing on normally star-forming galaxies rather than strong starbursts: Optical surveys can decently resolve the morphology of $L*$ galaxies at z 2. An extensive study was performed by Elmegreen, Elmegreen and collaborators (2004–2009) in the Hubble Ultra Deep Field (UDF). Most star-forming galaxies around $L*$ at $z > 1$ are very clumpy, with giant clumps that do not resemble the star-forming clouds or GMCs of nearby spiral galaxies (Fig. 2). These high-z galaxies are most often "clumpy disks" rather than multiple mergers. Giant clump masses can reach $\sim 10^{8-9}$ $M_\odot$.Spectroscopic surveys study the resolved kinematics of ionized gas (e.g. Förster Schreiber *et al.* 2009; Epinat *et al.* 2009). They have confirmed that the majority of normally star-forming objects in the young Universe are large rotating disks, with giant Hα clumps, and suggest strong ISM turbulence with 1D dispersions of a few tens of km/s.Molecular gas observations of normally star-forming galaxies at $z > 1$ (Daddi *et al.* 2010; Tacconi *et al.* 2010) have revealed high gas fractions, typically $\sim$50% of the baryons at $z = 2$, with most of the gas most in dense molecular form. Relatively low excitations and Milky Way-like conversion factors are supported by dynamical arguments (Daddi *et al.* 2010) and observations of CO SEDs (Dannerbauer *et al.* 2009).

Gas-rich, turbulent, clumpy disks thus appear to represent the bulk of normally star-forming galaxies at $z > 1$. The giant clumps likely result from gravitational (Jeans) instability in gas-rich disks with a Toomre parameter Q below or close to unity, owing to low spheroid fractions, and strong turbulence maintains a high Jeans mass (typical clump mass) (Bournaud & Elmegreen 2009).

3. Internal evolution in primordial high-redshift galaxies

Models of internal dynamics of clumpy disk galaxies were studied in Bournaud *et al.* (2007a), among others. These clumps are massive enough to undergo tidal torquing and friction and migrate inwards in a few disk rotation periods. They can form a bulge with

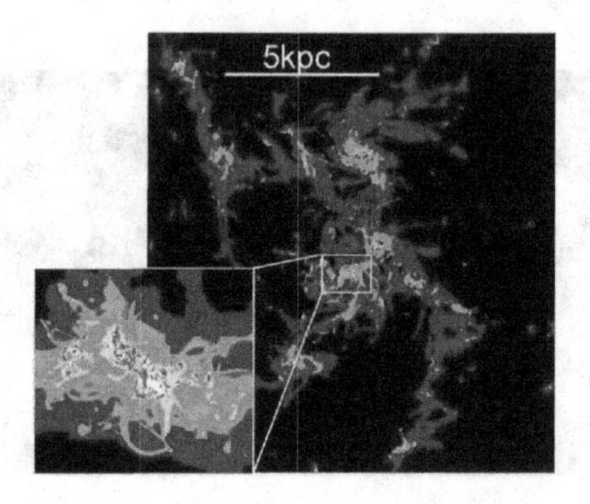

Figure 3. Gas surface density in an AMR model of a gas-rich high-redshift disk with 2 pc resolution, showing a distribution dominated by a few "giant clumps". A zoom on a giant clump shows sub-clumps and shocks. This giant complex is overall supported by fragmentation and internal turbulence, and some internal rotation. This prevents the whole complex from reaching very high average gas densities and high star formation efficiencies, which could leave it long-lived against stellar feedback.

properties typical for classical bulges (at least in massive galaxies). The clumps loose a substantial fraction of their stars via tidal effects, and maybe some gas through feedback processes. The released material can form an exponential disk. The pre-existing stars can be stirred in a very thick stellar disk, which can be hot enough to remain thick down to z=0 even if mass infall continues to fuel the thin disk (Bournaud *et al.* 2009).

Rapid internal evolution with bulge formation by giant clump instabilities in high-redshift disks has now been reproduced with various codes and in cosmological simulations (Agertz *et al.* 2009a; Ceverino *et al.* 2010). A major unknown is the survival of the giant clumps to star formation feedback, in particular radiative pressure from young stars (Murray *et al.* 2010). Krumholz & Dekel (2010) have shown that giant clumps will survive if they do not collapse into small superdense objects that would form stars too rapidly and undergo too strong feedback. This seems to be the case in observations but measurements at the scale of clumps remain uncertain. This is also the case in AMR disk simulations (Fig. 3) where the giant star-forming complexes form many sub-clumps and remain supported by internal turbulence and rotation instead of collapsing. (Burkert *et al.* 2009; Elmegreen & Burkert 2010) further suggest that the strong turbulence in high redshift galaxies is mostly driven by the gravitational instabilities, not by feedback processes. (Burkert 2009) has shown simulations where the main source of turbulence was a strong disruptive feedback, but the result was not to disrupt the giant clumps after they formed, but to completely inhibit their formation, which is not consistent with observations. There are probably observed signatures of feedback, but they may leave the largest and densest clumps relatively unaffected (Lehnert *et al.* 2009).

4. Galaxy mergers and the disk survival issue

The survival of disk galaxies remains a major issue: even if a large part of the mass comes from smooth flows, galaxies should undergo a significant number of mergers. The survival of massive disk galaxies with low bulge fractions remains misunderstood. Even if giant clumps and subsequent bulge formation at high redshift did not affect galaxies much, merger-induced bulge formation alone seems inconsistent with the observed fractions of bulges (Weinzirl *et al.* 2009).

There have been claims that disks can survive or re-form in high-redshift mergers, because these mergers involve very high gas fractions (e.g., Robertson *et al.* 2006). Some results were even called "spiral" galaxies (Springel & Hernquist 2005) because a large rotating gas disk was found, but more than 50% of the baryons were actually in a stellar bulge or spheroid after the merger: this is not typical for a "spiral" galaxy. A more general issue is that mergers have never been studied in realistic high-redshift conditions, because the employed techniques have always modeled a warm, stable and smooth gas.

The real ISM, especially in high-redshift galaxies, is highly unstable, and supported by supersonic turbulence in cold gas. Real mergers at z=2 with high gas fractions may thus be fundamentally different and much more dissipative than existing models. Disk survival is uneasy with a turbulent, heterogeneous ISM (Bournaud *et al.* 2010b).

5. ISM physics in galaxy formation

Star-forming regions are not just substructures that emit most of the visible light. They have major consequences on disks and bulges at high redshift, and a significant impact in mergers, too (Bois *et al.* 2010; Teyssier *et al.* 2010). The crucial role of the heterogeneous, cloudy and turbulent ISM and clustered star formation in global galaxy properties has also been pointed out in recent cosmological simulations. In low-mass dwarf galaxies, this can preserve bulgeless galaxies (Brook, this meeting). More massive galaxies in the $L*$ regime still have too massive bulges and spheroids in cosmological models, but recent work has shown that the physics of ISM turbulence, clustered star formation, and stellar evolution may strongly reduce the predicted bulge fractions (e.g., Semelin & Combes 2005; Piontek & Steinmetz 2009; Martig & Bournaud 2010). Nevertheless, as pointed out by Agertz *et al.* (2010), too many free parameters in sub-grid models prevent robust predictions, and explicitly resolving the main star forming regions is needed.

6. Resolved star formation in cosmological and galaxy-sized models

Galaxy formation models need to directly resolve the largest scales of star formation: a few giant clumps in high-redshift galaxies, and a few tens of GMCs that drive the bulk of the star formation in the Milky Way. In theory, the formation of GMCs is probably initiated by gravitational instabilities in turbulent gas (Elmegreen 2002; Krumholz & McKee 2005), which should be relatively simple to model. However, "resolving" star formation is not achieved so easily in galaxy formation models.

In cosmological simulations, none of the recent and supposedly high-resolution models resolves the main star formation regions at z = 0. Models by Agertz *et al.* or Governato *et al.* show an heterogeneous ISM, which mainly results from supernovae feedback creating empty bubbles and higher-density regions. But high densities of thousands of atoms/cm^3 are not reached and cloud-forming instabilities are absent. Things are a bit easier at redshift $z > 1$ with giant clumps of star formation. They have been directly captured in cosmological simulations (Agertz *et al.* 2009a; Ceverino *et al.* 2010). They have not been "resolved", though: these giant clumps survive feedback because they keep a relatively low density and hence a relatively low star formation efficiency. Maybe this is how a real clump would be, but the resolution limit may induce a severe bias. The smallest cell size in Ceverino *et al.* is 70 pc, and the thermal Jeans length has to be kept larger than a few cells, the total (thermal+turbulent) Jeans length being necessarily even larger, say, around 500 pc. A star-forming clump could not become much more compact, extremely dense, very efficiently star-forming and rapidly disrupted by feedback. It cannot fragment into sub-clumps and become supported by internal turbulence, but remains largely "supported by the resolution limit", instead. Only isolated disk simulations directly resolving the internal fragmentation of such giant clumps is directly resolved and their internal turbulent support, so that their internal density distribution and resulting efficiency in a given star formation scheme can be directly resolved (see Fig. 3).

Resolving star-forming clouds at redshift zero is harder, even in idealized isolated galaxy simulations with box sizes of a few tens of kpc. Recent works have studied the formation of GMCs via cooling and gravitational instabilities in a galactic disk (e.g.

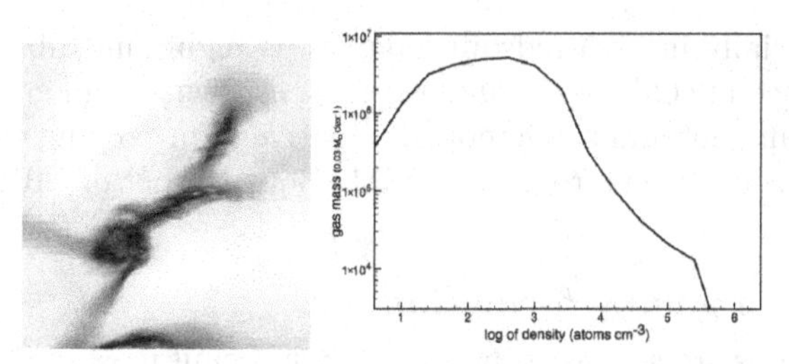

Figure 4. Zoom on a GMC in a simulation of a whole galaxy including gas, stars and dark matter from Bournaud *et al.* (2010a). The left panel shows the gas density over a portion of 300×300 pc in the face-on disk. The whole disk has a log-normal density PDF, the right panel shows the PDF for the single GMC shown to the left: it has a log-normal shape with a significant high-density tail. The high-density tail is where star formation takes place.

Li *et al.* 2005; Agertz *et al.* 2009b; Tasker & Tan 2009). They have highlighted the role of gravity and turbulence in driving GMC formation. However, these simulation do not "resolve" the GMC properties: for instance the typical radius of a GMC in the Tasker & Tan model is around 20 pc, for an AMR cell size of 7 pc. The outer parts of these GMCs are supported by rotation around the central core with a mini-spiral shape (unlike real GMCs) and the internal regions are supported by the resolution limit (i.e., by the thermal pressure floor that maintains the thermal Jeans length larger than a few cells). While the mass of these GMCs could be realistic, their properties such as the gas density distribution or local free-fall time cannot be trusted. Then, assuming a given star formation recipe on small scales (Schmidt law or any other), the resulting star formation activity in each cloud cannot be robustly predicted.

Nevertheless all these simulations have nicely shown that GMC formation is probably mostly driven by ISM turbulence and gravity. We recently went one step further and demonstrated the properties of gas clouds can be directly resolved in simulations of whole galaxies, including their size, density distribution, and internal turbulence (Bournaud *et al.* 2010a and Fig. 4). The average Jeans scale length sets the disk scale height. Starting from this most unstable scale, a 3D turbulence cascade takes place. The turbulence cascade forms denser gas regions, some of which become gravitationally unstable and collapse into dense clouds, with local densities reaching 10^{5-6} cm^{-3}. This occurs mainly in regions already compactified by stellar arms or bars, which traces the coupled instability of gas and stars in a bi-component disk (see also Yang *et al.* 2007). Owing to a very high spatial and mass resolution (0.8 pc and 5×10^3 M$_\odot$), the model shows GMCs that stopped collapsing well before the resolution limit, and inside which dense sub-clouds, bubbles, and numerous structures with chaotic motions are directly resolved. A single GMC is shown on Fig. 4: its PDF is log-normal with a substantial excess of high-density gas, consistently with observations of nearby GMCs (Lombardi *et al.* 2010). Star formation in the model takes place above 5000 cm^{-3}, which is about the high-density tail of the GMC PDF. When collapsing, this GMC evacuated most of its angular momentum by fragmenting and expelling some substructures, and at the observed instant is dominated by internal turbulent motions.

The process is almost entirely driven by turbulence and gravity, but feedback is essential to maintain a steady state distribution. Without feedback, all gas mass will eventually end-up at the smallest scales of the turbulence cascade, in superdense bullets at the resolution limit. Feedback regulates the process by disrupting dense structures on the smallest scales and refueling a turbulent steady state through larger-scale instabilities. This does not seem strongly affected by the hypothesis done on feedback, and adding HII region and radiative pressure feedback to supernovae feedback models could somewhat decrease the lifetime of gas clouds, but would not change the main properties of the ISM density distribution that are primarily controlled by the gravity-driven turbulence cascade. Although the small-scale substructures inside GMCs are probably still far from realistic (magnetic effects should eventually become important), it seems that the main GMC properties can be correctly modeled in simulations of whole galaxies or cosmological simulations, provided that:

– The spatial resolution reaches about 1 pc: the gas disk scale height must be accurately resolved to capture a 3D turbulence cascade in the ISM, because there is no such cascade in 2D. A few resolution elements per scale height are not enough. One must resolve several Jeans length per disk scale height, so that instabilities can arise over-densities above and below the mid-plane and drive a vertical component of turbulent motions. As the thermal Jeans length itself is described by at least a few resolution elements to prevent artificial fragmentation, and the total, thermal+turbulent Jeans length is substantially larger – at least 10-20 resolution elements for supersonic turbulence. This implies that the gas disk scale height has to be resolved with, say, 100-200 resolution elements: the resolution for gravity *and* hydrodynamics for a typical $z = 0$ galaxy needs to be around 1 pc.

– Star formation and energy feedback of some sort is included, so as to maintain a steady turbulent state where clouds do not all collapse down to the resolution limit.

– Long-range gravitational forcings from old stellar populations are included (spiral density waves, bars, etc). Large-scale structures do participate to the turbulence cascade down to small scales, and the contribution of all stars to the disk stability in a combined $Q_{gas+stars}$ parameter is about as important as that from the gas itself.

These requirements could in theory be achieved with any code, but in practice are probably easier to reach with oct-tree AMR codes such as RAMSES (Teyssier 2002): grid refinements over complex systems such as GMCs in a spiral disk are optimized compared to patch-based and multi-grid techniques.

7. Conclusions

While the assembly of baryons into galaxies seems relatively well understood on large scales, the properties of galaxies in the simulations are still far from matching those of observed galaxy populations. There is increasing evidence that small-scale processes such as ISM turbulence, fragmentation, clustered star formation, are crucial to understand galaxy formation, in particular in the primordial high-redshift phases. They can have major effects on the global properties of each galaxy, and even on the global cosmological evolution of baryons. For instance, tentative explanations of the *downsizing* in the λ-CDM paradigm rely on AGN feedback (Scannapieco *et al.* 2005; Di Matteo *et al.* 2008). However, AGN fueling involves on a number processes inside the central kpc of galaxies that are completely unresolved in cosmological models (nuclear bars, resonances, nuclear starbursts, mass loss from clusters, etc: see Combes 2001).

Recent results and objective criteria to capture and correctly resolve the main star-forming regions of galaxies have been presented. They are unfortunately not reached in present-day cosmological simulations, especially at redshift zero, but recent improvements have been made in accounting for the multiphase ISM in galaxy formation models.

References

Agertz, O., *et al.* 2009a, *MNRAS*, 392, 294
Agertz, O., Teyssier, R., & Moore, B. 2009, *MNRAS*, 397, L64
Agertz, O., Teyssier, R., & Moore, B. 2010, *MNRAS* submitted, arXiv:1004.0005
Birnboim, Y. & Dekel, A. 2003, *MNRAS*, 345, 349

Bois, M., *et al.* 2010, *MNRAS* in press, arXiv:1004.4003
Bournaud, F., Elmegreen, B. G., & Elmegreen, D. M. 2007, *ApJ*, 670, 237
Bournaud, F. & Elmegreen, B. G. 2009, *ApJL*, 694, L158
Bournaud, F., Elmegreen, B. G., & Martig, M. 2009, *ApJL*, 707, L1
Bournaud, F., Elmegreen, B. G., Teyssier, R., Block, D. L., Puerari, I. 2010, *MNRAS* in press (arXiv:1007.2566).
Bournaud, F., *et al.* 2010b, submitted to *ApJL*, arXiv:1006.4782
Brooks, A. M., Governato, F., Quinn, T., Brook, C. B., & Wadsley, J. 2009, *ApJ*, 694, 396
Burkert, A. 2009, *IAU Symposium*, 254, 437
Burkert, A., *et al.* 2009, arXiv:0907.4777
Ceverino, D., Dekel, A., & Bournaud, F. 2010, *MNRAS*, 440
Combes, F. 2001, *Advanced Lectures on the Starburst-AGN*, 223. arXiv:astro-ph/0010570
Conselice, C .J., Bershady, M. A., Dickinson, M., & Papovich, C. 2003, *AJ*, 126, 1183
Daddi, E., *et al.* 2010, *ApJ*, 713, 686
Dannerbauer, H., *et al.* 2009, *ApJL*, 698, L178
Dekel, A. & Birnboim, Y. 2006, *MNRAS*, 368, 2
Dekel, A., *et al.* 2009, *Nature*, 457, 451
Di Matteo, T., Colberg, J., Springel, V., Hernquist, L., & Sijacki, D. 2008, *ApJ*, 676, 33
Elmegreen, B. G. 2002, *ApJ*, 577, 206
Elmegreen, D. M., Elmegreen, B. G., & Hirst, A. C. 2004, *ApJL*, 604, L21
Elmegreen, D. M., Elmegreen, B. G., Ravindranath, S., & Coe, D. A. 2007, *ApJ*, 658, 763
Elmegreen, B. G., Elmegreen, D. M., Fernandez, M. X., & Lemonias, J. J. 2009, *ApJ*, 692, 12
Elmegreen, B. G. & Burkert, A. 2010, *ApJ*, 712, 294
Epinat, B., *et al.* 2009, *A&A*, 504, 789
Förster Schreiber, N. M., *et al.* 2009, *ApJ*, 706, 1364
Genel, S., *et al.* 2008, *ApJ*, 688, 789
Genel, S., Bouché, N., Naab, T., Sternberg, A., & Genzel, R. 2010, arXiv:1005.4058
Governato, F., *et al.* 2010, *Nature*, 463, 203
Hopkins, P. F., *et al.*, 2010, *ApJ*, 715, 202
Johansson, P. H., Naab, T., & Ostriker, J. P. 2009, *ApJL*, 697, L38
Kereš, D., Katz, N., Weinberg, D. H., & Davé, R. 2005, *MNRAS*, 363, 2
Kereš, D., Katz, N., Fardal, M., Davé, R., & Weinberg, D. H. 2009, *MNRAS*, 395, 160
Krumholz, M. R. & Dekel, A. 2010, *MNRAS*, 635
Krumholz, M. R. & McKee, C. F. 2005, *ApJ*, 630, 250
Lehnert, M. D., *et al.* 2009, *ApJ*, 699, 1660
Li, Y., Mac Low, M.-M., & Klessen, R. S. 2005, *ApJL*, 620, L19
Lombardi, M., Lada, C. J., & Alves, J. 2010, *A&A*, 512, A67
Martig, M., Bournaud, F., Teyssier, R., & Dekel, A. 2009, *ApJ*, 707, 250
Martig, M. & Bournaud, F. 2010, *ApJL*, 714, L275
Murray, N., Quataert, E., & Thompson, T. A. 2010, *ApJ*, 709, 191
Naab, T., Johansson, P. H., Ostriker, J. P., & Efstathiou, G. 2007, *ApJ*, 658, 710
Ocvirk, P., Pichon, C., & Teyssier, R. 2008, *MNRAS*, 390, 1326
Piontek, F. & Steinmetz, M. 2009, *MNRAS* submitted, arXiv:0909.4167
Scannapieco, E., Silk, J., & Bouwens, R. 2005, *ApJL*, 635, L13
Semelin, B. & Combes, F. 2005, *A&A*, 441, 55
Springel, V. & Hernquist, L. 2005, *ApJL*, 622, L9
Steidel, C. C. *et al.* 2010, *ApJ*, 717, 289
Tacconi, L. J., *et al.* 2010, *Nature*, 463, 781
Tasker, E. J. & Tan, J. C. 2009, *ApJ*, 700, 358
Teyssier, R. 2002, *A&A*, 385, 337
Teyssier, R., Chapon, D., & Bournaud, F. 2010, *ApJL*, arXiv:1006.4757
Robertson, B., *et al.* 2006, *ApJ*, 645, 986
Weinzirl, T., Jogee, S., Khochfar, S., Burkert, A., & Kormendy, J. 2009, *ApJ*, 696, 411
Yang, C.-C., Gruendl, R. A., Chu, Y.-H., Mac Low, M.-M., & Fukui, Y. 2007, *ApJ*, 671, 374

Computational Star Formation
Proceedings IAU Symposium No. 270, 2011
J. Alves, B.G. Elmegreen, J. M. Girart & V. Trimble, eds.

doi:10.1017/S1743921311000871

Fragmentation in turbulent primordial gas

S. C. O. Glover[1], P. C. Clark[1], R. S. Klessen[1] & V. Bromm[2]

[1]Institut für Theoretische Astrophysik, Zentrum für Astronomie der Universität Heidelberg, Albert-Ueberle-Str. 2, 69120 Heidelberg, Germany

[2]Department of Astronomy, University of Texas, Austin, TX 78712

Abstract. We report results from numerical simulations of star formation in the early universe that focus on the role of subsonic turbulence, and investigate whether it can induce fragmentation of the gas. We find that dense primordial gas is highly susceptible to fragmentation, even for rms turbulent velocity dispersions as low as 20% of the initial sound speed. The resulting fragments cover over two orders of magnitude in mass, ranging from $\sim 0.1\ M_\odot$ to $\sim 40\ M_\odot$. However, our results suggest that the details of the fragmentation depend on the local properties of the turbulent velocity field and hence we expect considerable variations in the resulting stellar mass spectrum in different halos.

Keywords. stars: formation – galaxies: formation – cosmology: theory

1. Introduction

Over the course of the last decade, work by a number of groups has lead to the development of a widely accepted picture for the formation of the first stars, the so-called Population III (or Pop. III) stars. In this picture, the very first stars (often termed Population III.1) form within small dark matter halos that have total masses $M \sim 10^6\ M_\odot$, virial temperatures of around a thousand K, and that are assembled at redshifts $z \sim 25$–30 or above (Bromm & Larson 2004; Glover 2005; Bromm *et al.* 2009). Gas falling into one of these small dark matter halos is shock-heated to a temperature close to the virial temperature of the halo, and thereafter cools via H_2 rotational and vibrational line emission. A cold, dense core forms at the center of the dark matter halo, with a characteristic density $n_{\rm ch} \sim 10^4\ {\rm cm}^{-3}$ and temperature $T_{\rm ch} \sim 200$ K; these values are determined by the microphysics of the H_2 molecule, and represent the point at which H_2 cooling becomes inefficient, and the effective equation of state of the gas becomes stiffer than isothermal. Once the mass of this cold dense clump exceeds the critical Bonnor-Ebert mass for this temperature and density, which is approximately $1000\ M_\odot$, it decouples from the larger scale flow and undergoes runaway gravitational collapse. Until recently, numerical simulations that have followed this collapse up to very high densities (e.g. Abel, Bryan, & Norman 2002; Bromm & Loeb 2004) have found no indication of further fragmentation, and so it has typically been assumed that this dense core forms only a single star. In this picture, the final mass of this star will be set only by the efficiency with which it can accrete gas from its natal core, and there are good reasons to believe that for protostellar masses less than ~ 20–$30\ M_\odot$, this efficiency will be close to 100% (McKee & Tan 2008). It is therefore difficult to avoid the conclusion that the final mass of the Pop. III star will be very large.

Over the past few years, however, a new generation of numerical simulations have begun to suggest that the true picture of Pop. III star formation may be rather more complicated than the simple scenario outlined above. Work by a number of groups (Clark, Glover & Klessen 2008; Machida 2008; Machida *et al.* 2009; Turk, Abel & O'Shea 2009; Stacy, Greif & Bromm 2010) has found evidence for fragmentation in high density primordial gas, with predictions of the likelihood that fragmentation occurs ranging from roughly 20% (Turk, Abel & O'Shea 2009) to essentially 100% (Clark, Glover & Klessen 2008; Stacy, Greif & Bromm 2010). If fragmentation, and the formation of binary or multiple systems, is indeed a common outcome of Population III star formation, then this has profound implications for the final masses of the Pop. III stars, their production of ionizing photons and metals, the rate of high redshift gamma ray bursts, and many other issues.

It is therefore important to better understand the physical basis of fragmentation in these systems. Although numerical simulations starting from cosmological initial conditions and following collapse all the way to protostellar densities will give us the most accurate picture of how the first stars formed, these simulations are computationally costly (making it difficult to explore a wide parameter space), and are also difficult to interpret, since many physical effects combine to yield the observed behaviour. Simulations that start from simpler initial conditions and that focus on exploring the importance of a single free parameter therefore play an important role, which is complementary to that of fully cosmological simulations. A good example is the recent work by Machida and collaborators (Machida 2008; Machida *et al.* 2009) which examined the influence of the initial rotational energy of the gas, and found that zero metallicity clouds with sufficient initial rotational energy could fragment into tight binaries. In this contribution, we briefly describe the results of a similar study that we have recently performed that looked at the effects of the initial turbulent energy of the gas (Clark *et al.* 2010).

2. Simulations

2.1. *Numerical method*

We modeled the evolution of the primordial gas using a modified version of the Gadget 2 smoothed particle hydrodynamics code (Springel 2005). Our modifications include the addition of a sink particle implementation (Bate, Bonnell & Price 1995; Jappsen *et al.* 2005), allowing us to follow the evolution of the gas beyond the point at which the first protostar forms, and a detailed treatment of primordial cooling and chemistry (Glover & Abel 2008; Clark *et al.* 2010). Full details can be found in Clark *et al.* (2010).

2.2. *Initial conditions*

We took initial configurations that were unstable Bonnor-Ebert spheres, into which we injected a subsonic turbulent velocity field. In our study, we considered two different sets of initial conditions: one corresponding to the Pop. III.1 scenario described above, and another corresponding to the Pop. III.2 scenario, in which efficient HD cooling leads to a lower $T_{\rm ch}$ and higher $n_{\rm ch}$ for the primordial prestellar core. However, in this contribution we will discuss only the results of the Pop. III.1 simulations. In these, we set the initial gas temperature to be 300 K, and took an initial central number density of 10^5 cm^{-3}. The mass of the Bonnor-Ebert sphere was taken to be $1000\,{\rm M}_{\odot}$, corresponding to roughly three Jeans masses. This mass was represented by 2 million SPH particles, giving us a particle mass of 5×10^{-4} M$_{\odot}$, and a mass resolution of 5×10^{-2} M$_{\odot}$. Within the BE spheres, we imposed a turbulent velocity field that had a power spectrum $P(k) \propto k^{-4}$. The three-dimensional rms velocity in the turbulent field, $\Delta v_{\rm turb}$, was scaled to some fraction of the initial sound speed $c_{\rm s}$. For the simulations presented here, we used four different rms velocities: 0.1, 0.2, 0.4 and 0.8 $c_{\rm s}$. Sink particles were created once the number density of the gas reached 10^{13} cm^{-3}, using the standard Bate, Bonnell & Price (1995) algorithm, with a sink accretion radius of 20 AU.

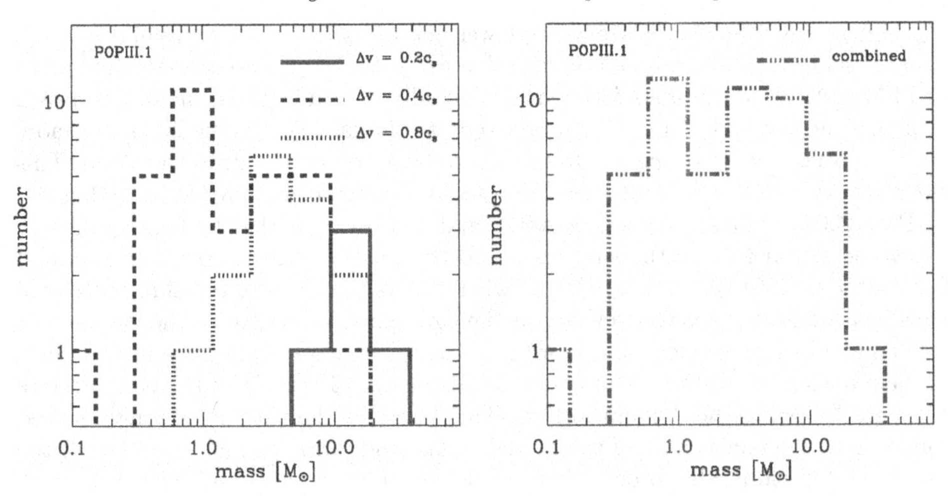

Figure 1. The left panel shows the mass functions from the simulations in which fragmentation occurs. In all cases the mass function is plotted at the point at which the total mass of gas converted to sink particles is $100 M_\odot$. Note that as accretion is ongoing, and the system is still young ($t \sim 1000$ yr), these will often not be the final masses of the sinks. The mass functions in the individual simulations differ substantially, although the combined mass function, shown in the right panel, exhibits a broad and flat distribution between masses of 0.4 and 20 $M_\odot$.

3. Key results

We find that the clouds with $\Delta v_{\rm turb} \geqslant 0.2 c_s$ all fragment into small clusters of sink particles: the runs with $\Delta v_{\rm turb} = 0.2 c_s$, $0.4 c_s$ and $0.8 c_s$ have formed 5, 31, and 15 sink particles, respectively, at the point at which 10% of the cloud mass has been accreted. This clearly demonstrates that turbulent subsonic motions are able to promote fragmentation in primordial gas clouds. Only in the case with $\Delta v_{\rm turb} = 0.1 c_s$ does the cloud form just a single sink particle. In this case, an extended disk builds up around the star, but otherwise the gas contains little structure.

Interestingly, there is no clear trend linking the number of fragments that form and the initial turbulent energy: the 0.4 c_s run fragments more than the 0.8 c_s run, despite containing only one quarter of the initial turbulent energy. The effects at play here are somewhat complex. First, the turbulent velocity field contained in the collapsing region will differ with each value of $\Delta v_{\rm turb}/c_s$, since the different strengths of flow will push the gas around to different degrees. In addition, the nature of the turbulence that survives in the collapsing core will also affect the fragmentation. The cloud with $\Delta v_{\rm turb} = 0.4 c_s$ – the most successful in terms of fragmentation – forms a large disk-like structure, which appears to aid the fragmentation of the infalling envelope. Much of the cloud's ability to fragment therefore seems to depend on the amount of angular momentum that becomes locked up in the collapsing region, and hence on the detailed properties of the turbulent velocity field. Further, the extra delay in the collapse caused by the increased turbulent support also gives the cloud more time to wash out anisotropies. The ability of the cloud to fragment is a competition between these conflicting processes. For the randomly generated velocity field that we used in this study, the ability to fragment is better for $\Delta v_{\rm turb}/c_s = 0.4$, than $\Delta v_{\rm turb}/c_s = 0.8$, but we note that this may not always be the case.

The mass functions of the sink particles from the simulations that undergo fragmentation are shown in Fig. 1. For clarity, we have omitted the single 100 $M_\odot$ sink particle that forms in the 0.1 c_s cloud. For the 0.2 c_s and 0.8 c_s clouds, we see that the sink masses cluster around some central value – roughly 12 $M_\odot$ and 4 $M_\odot$ respectively – while the 0.4 c_s

cloud has a mass function that is skewed to lower masses, with a peak at around 1 $M_\odot$, a sharp fall-off below this, and a broad distribution towards higher masses, extending up to around 13 $M_\odot$. As all of the sinks form with essentially the same initial mass, the spread of sink masses at the end of the simulation reflects the fact that some sinks are more effective than others at accreting gas from the available reservoir within the cloud. This behaviour is very familiar from studies of present-day star formation (see e.g. Bonnell, Vine & Bate 2004; Schmeja & Klessen 2004), and is a result of the the high sensitivity of the mass accretion rate to the sink mass and the relative velocity between sink and gas (Bonnell *et al.* 2001a): $\dot{m}_* \propto \rho m_*^2/v_{\rm rel}^3$, where m_* is the mass of the sink particle, ρ is the gas density, and $v_{\rm rel}$ is the velocity of the sink particle relative to the gas. As the number of sinks increases, more and more get 'kicked' by close dynamical interactions, forcing them onto more distant orbits, and increasing $v_{\rm rel}$. These sinks therefore accrete very little gas following the dynamical kick. The few sinks that are not strongly kicked therefore accrete the lion's share of the available gas, and form the high-mass tail of the resulting mass function. This process is termed 'competitive accretion', (Bonnell *et al.* 2001a) and normally leads to the type of distribution of masses seen in our $0.4c_{\rm s}$ run.

Our simulations therefore demonstrate that if significant levels of subsonic turbulence are present in the primordial gas found within the first dark matter halos, then Pop. III stars forming in these halos are more likely to be born in small stellar groups than as single stars. Moreover, if Pop. III stars do indeed form in this way, then we would expect the resulting initial mass function to be broad, thanks to the influence of competitive accretion within the star cluster. Tests of these expectations will require large, high-resolution studies of the formation of the first dark matter minihalos that are capable of resolving the turbulent flows in the gas at the moment when the baryons become self-gravitating, and that use sink particles (or some comparable technique) to allow the evolution of the self-gravitating system to be followed over multiple dynamical times. Such simulations would help determine which, if any, of the initial conditions presented in our study are actually realized in nature.

References

Abel, T., Bryan, G. L., & Norman, M. L. 2002, *Science*, 295, 93
Bate, M. R., Bonnell, I. A., & Price, N. M. 1995, *MNRAS*, 277, 362
Bonnell, I. A., Bate, M. R., Clarke, C. J., & Pringle, J. E. 2001a, *MNRAS*, 323, 785
Bonnell, I. A., Vine, S. G., & Bate, M. R. 2004, *MNRAS*, 349, 735
Bromm, V. & Larson, R. B. 2004, *ARA&A*, 42, 79
Bromm, V. & Loeb, A. 2004, *New Astron.* 9, 353
Bromm, V., Yoshida, N., Hernquist, L. & McKee, C. F. 2009, *Nature*, 459, 49
Clark, P. C., Glover, S. C. O., & Klessen, R. S. 2008, *ApJ*, 672, 757
Clark, P. C., Glover, S. C. O., Klessen, R. S., & Bromm, V. 2010, *ApJ*, submitted; arXiv:1006.1508
Glover, S. 2005, *Space Sci. Rev.*, 117, 445
Glover, S. C. O. & Abel, T. 2008, *MNRAS*, 388, 1627
Jappsen, A.-K., Klessen, R. S., Larson, R. B., Li, Y., & Mac Low, M.-M. 2005, *A&A*, 435, 611
Machida, M. N. 2008, *ApJ*, 682, L1
Machida, M. N., Omukai, K., Matsumoto, T., & Inutsuka, S.-I. 2009, *MNRAS*, 399, 1255
McKee, C. F. & Tan, J. C. 2008, *ApJ*, 681, 771
Schmeja, S. & Klessen, 2004, *A&A*, 419, 405
Springel, V. 2005, *MNRAS*, 364, 1105
Stacy, A., Greif, T. H., & Bromm, V. 2010, *MNRAS*, 403, 45
Turk, M. J., Abel, T., & O'Shea, B. W. 2009, *Science*, 325, 601

Computational Star Formation
Proceedings IAU Symposium No. 270, 2011
J. Alves, B.G. Elmegreen, J. M. Girart & V. Trimble, eds.

doi:10.1017/S1743921311000883

Cosmological simulations of low-mass galaxies: some potential issues

Pedro Colín[1], Vladimir Avila-Reese[2], and Octavio Valenzuela[2]

[1]Centro de Radioastronomía y Astrofísica, Universidad Nacional Autónoma de México, A.P. 72-3 (Xangari), Morelia, Michoacán 58089, México
[2]Instituto de Astronomía, Universidad Nacional Autónoma de México, A.P. 70-264, 04510, México, D.F., México

Abstract. Cosmological Adaptive Mesh Refinement simulations are used to study the specific star formation rate (sSFR=SSF/M_s) history and the stellar mass fraction, $f_s = M_s/M_T$, of small galaxies, total masses M_T between few $\times 10^{10}$ $M_\odot$ to few $\times 10^{11}$ $M_\odot$. Our results are compared with recent observational inferences that show the so-called "downsizing in sSFR" phenomenon: the less massive the galaxy, the higher on average is its sSFR, a trend seen at least since $z \sim 1$. The simulations are not able to reproduce this phenomenon, in particular the high inferred values of sSFR, as well as the low values of f_s constrained from observations. The effects of resolution and sub-grid physics on the SFR and f_s of galaxies are discussed.

1. Introduction

Numerical N-body + hydrodynamical cosmological simulations have maturated enough as to predict some of the main properties of real galaxies; nevertheless, issues related to resolution and the so-called sub-grid physics are not settled yet. Therefore, inevitably the results of the simulations used here depends on the resolution and on the schemes and parameters used to model the physics of the unresolved scales. The current theoretical background of galaxy formation and evolution is the successful ΛCDM hierarchical clustering scenario. In this work, we simulate 'realistic' sub-L_* galaxies, hereafter low-mass galaxies, by using the hydrodynamics + N-body Adaptive Refinement Tree code (ART; Kravtsov *et al.* 1997; Kravtsov et al 2003) in the context of the ΛCDM scenario (see also Colín *et al.* 2010).

In the recent literature, an unexpected observational trend for low-mass galaxies (stellar masses $M_s < 3\ 10^{10}$ $M_\odot$) has been established: the lower M_s, the higher is on average the specific star formation rate (sSFR=SFR/M_s). This trend, called "downsizing in sSFR", is seen at least out to $z \sim 1$ and it implies a delay in the M_s assembling process. This delay should be larger for lower mass galaxies. The sSFR downsizing is difficult to be reproduced by ΛCDM-based disk galaxy evolutionary models, which predict low values of sSFR for small galaxies at late epochs and, instead of downsizing, present a weak upsizing trend in sSFR (Firmani *et al.* 2010 and references therein). These models include SN-driven outflows, which are important for low-mass systems. In this paper we explore whether our simulated galaxies face such a difficulty.

2. Numerical Setup

Among the physical processes included in the hydrodynamics ART code are: metal-dependent gas cooling, SF, thermal stellar feedback, self-consistent advection of metals, UV heating background source, etc. The cooling of gas takes into account atomic and

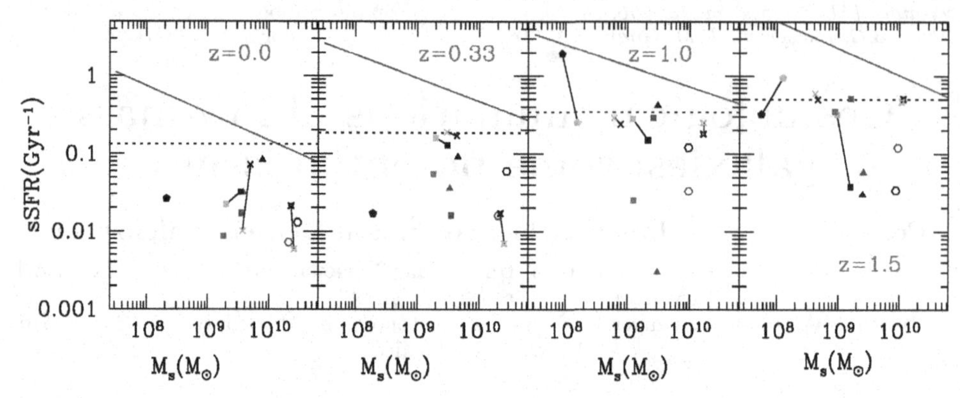

Figure 1. Specific star formation rate (sSFR) as a function of M_s, at four epochs, $z = 0$, 0.33, 1.0, and, 1.5. Different simulated galaxies are denoted by different symbols while galaxies simulated with different resolution are connected by a line (color cyan represents the highest resolution). The symbols of galaxies simulated with different values of the parameters f or Δt (or resolution) have different color. The blue line at $z = 0$ are data from Salim *et al.* (2007) (see their equation 12). The blue lines at other redshifts are the median values for different redshift bins taken from Santini *et al.* (2009): $0.3 < z < 0.6$, $0.6 < z < 1.0$, and $1.0 < z < 1.5$. Dashed lines show the values at each z of $1/[t_H(z)\text{-}1\text{Gyr}](1 - R)$, where R is the gas recycling factor, and it corresponds to galaxies with constant SFR.

molecular cooling. Star formation (SF) takes place in the coldest and densest collapsed regions, defined by $T < T_{\rm SF}$ and $\rho_g > \rho_{\rm SF}$, where T and ρ_g are the temperature and density of the gas, respectively, and $\rho_{\rm SF}$ is a density threshold. A stellar particle of mass $m_* = f m_{gas}$ is placed in a grid cell, of mass m_{gas}, where these conditions are simultaneously satisfied. The parameter $f < 1$ is a measure of the local efficiency with which gas is transformed into stars. Although $n_{\rm SF}$, the hydrogen number density corresponding to $\rho_{\rm SF}$, is in principle a free parameter of the simulations, one can make an educated guess as to which values should be considered most realistic by noting that SF occurs almost exclusively in giant molecular clouds. Thus, it is natural to require that the density in a star-forming cell should be at least such that the cell's column density equals the threshold value for molecular Hydrogen and CO formation, $N \gtrsim 10^{21}\text{cm}^{-2}$ (Franco & Cox 1986; van Dishoeck & Black 1988). For a cell of 100–300 pc on a side, this translates into a $n_{\rm SF}$ = of a few cm^{-3}. Here we set $n_{\rm SF} = 6\ \text{cm}^{-3}$ in all simulations (see Colín *et al.* 2010 for details).

In Table 1, we show with capital letters the simulated galaxies. Model galaxies with the same letter means same galaxy but simulated with different resolution or different values of the parameters f or Δt (see Table 1 for their definitions); for example, the difference between C1 or C2 is resolution, C2 is twice better resolved†. In second column, we show the total mass of the galaxy inside the virial radius $M_T = M_{DM} + M_{bar}$, where DM stands for dark matter. The mass of the DM particle in the highest resolution zone is given in column fifth.

3. Results

Figure 1 shows the sSFR as a function of M_s for all our simulations at different epochs. The SFR is the ratio between the stellar mass, produced during a certain time bin dt,

† Although the number of DM particles inside $R_{\rm vir}$ increases by about a factor of eight in these more resolved galaxies, the number of cells increases by much less because the refinement criteria (DM or gas mass in a cell) was fixed by the corresponding lower resolution simulations.

Table 1. Parameter Simulations

Galaxy	M_T ($M_\odot$)	f[1]	Cooling[2] (10^6 yr)	m_p ($M_\odot$)	l^3_{max} (pc)	Galaxy	M_T ($M_\odot$)	f[1]	Cooling[2] (10^6 yr)	m_p ($M_\odot$)	l^3_{max} (pc)
A1	2.80×10^{10}	0.50	20.0	6.6×10^5	218	C2	6.56×10^{10}	0.50	20.0	8.2×10^4	109
A2	2.48×10^{10}	0.50	20.0	8.2×10^4	109	D1	9.52×10^{10}	off	40.0	7.5×10^5	218
B1	5.94×10^{10}	0.50	40.0	7.5×10^5	218	D2	8.37×10^{10}	0.66	40.0	7.5×10^5	218
B2	5.31×10^{10}	0.50	40.0	9.4×10^4	109	E1	3.04×10^{11}	0.50	40.0	7.5×10^5	218
B3	5.76×10^{10}	0.20	40.0	9.4×10^4	109	E2	3.05×10^{11}	0.50	40.0	9.4×10^4	109
B4	5.11×10^{10}	0.50	10.0	9.4×10^4	109	F1	4.04×10^{11}	off	40.0	7.5×10^5	218
C1	7.79×10^{10}	0.50	20.0	6.6×10^5	218	F2	3.58×10^{11}	0.66	40.0	7.5×10^5	218

Notes:
[1]The parameter f measures the fraction of gas mass, in a star forming cell, that is transformed into stars.
[2]The interval of time Δt during which cooling is turned off after a SF event in order feedback may work.
[3]The cells in the finest grid measures l_{max} on a side.

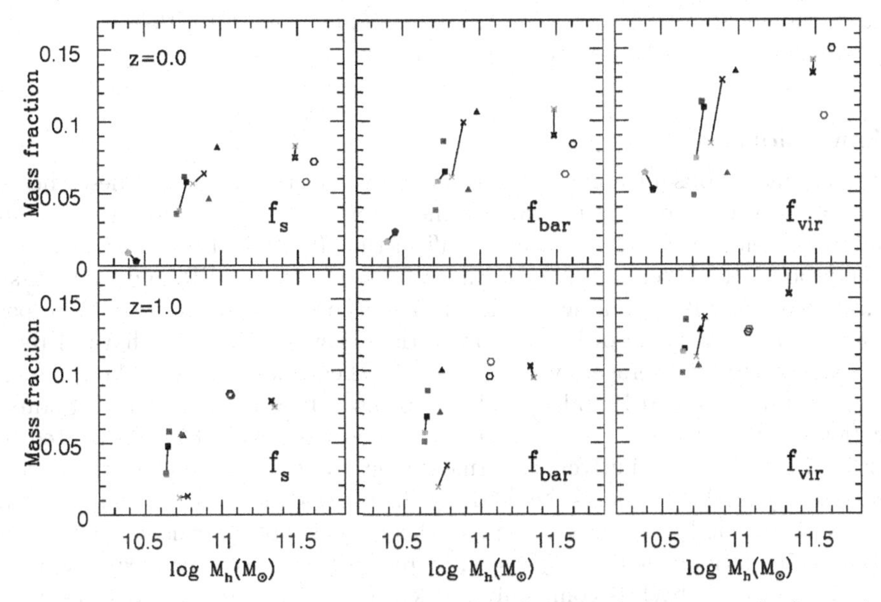

Figure 2. Mass fractions as a function of the mass of the halo, dark matter plus baryonic mass inside the virial radius, at two epochs, $z = 0$ and 1, for the different modeled galaxies, shown in Table 1. Galaxies are denoted by the same symbol and color code as in Figure 1. A trend can be clearly appreciated, the amount of baryons inside the galaxy, $5R_e$ or $R_{\rm vir}$, decreases as the mass of the galaxy gets smaller.

taken here equal to 100 Myr, around z, and dt. The blue line at $z = 0$ is a linear fit to star-forming galaxies (AGN excluded) from the SDSS (Salim *et al.* 2007); the scatter around this fit is large, $1\sigma \approx 0.5$ dex. The blue lines in the other panels are the median values of observational inferences in Santini *et al.* (2009); from left to right, $0.3 < z < 0.6$, $0.6 < z < 1.0$, and $1.0 < z < 1.5$. These inferences are roughly representative of many more observational studies appeared in the last years (see more references therein and in Firmani *et al.* 2010). As in the observational samples, our galaxies have on average a sSFR that increases towards higher redshifts.

The same galaxy simulated with different (reasonable) sub-grid parameters may have different values of sSFR at a given z. We notice also that the SFR has an intrinsic fluctuating behavior (see Colín *et al.* 2010). The increase in resolution affects the sSFR, in most of the cases lowering even more the values of sSFR. Despite all these differences, our results show for all cases that *the lower the mass, the smaller the sSFRs of the simulated galaxies with respect to observations out to $z \sim 1.5$.*

Figure 2 shows the different mass fractions, stellar, $f_s = M_s/M_T$, baryon, $f_{bar=}(M_s + M_g)/M_T$, and, virial, $f_{vir} = M_{bar}/M_T$, computed for the different simulated galaxies, shown in Table 1. Symbol and color code are as in Figure 1. The stellar fraction is computed inside five effective radii, R_e, defined as the half-mass stellar radius of the galaxy. The baryonic mass in f_{bar} considers stars and gas at $T < 10^4$ K inside $5R_e$; f_{vir} uses all baryonic mass within $R_{\rm vir}$. In all three cases, M_T is the total galaxy mass (see Table 1). At the two z's a clear trend is seen: the amount of baryons decreases as M_T decreases. This is clearly a result of the stellar thermal feedback: all of our low-mass simulated galaxies developed in greater or lesser extent gas blow-away episodes, with the strength of this galactic wind being a very sensitive function of potential well of the galaxy. This *trend* is in agreement with several direct and indirect observational estimates, though the *values* for f_s or f_{bar} (e.g. , Guo *et al.* 2010; Behroozi, Conroy & Wechsler 2010), are at least a factor three lower than our results.

4. Conclusions

Our ART simulations are able to produce roughly realistic sub-L_* galaxies. The results depend on the values of the sub-grid parameters and resolution. However, for all the possibilities considered here, our results confirm that (i) ΛCDM-based galaxies are not able to reproduce the observable phenomenon of downsizing in sSFR, and (ii) their stellar mass fractions, in spite of the feedback-driven outflows, are too high with respect to observational inferences. According to observations, low-mass galaxies have SFRs much higher than their past average, even at $z \sim 1$; dashed lines in Fig. 1 show the case if one assumes a constant SFR: galaxies above (below) these lines are forming stars at a higher (lower) rate than the past average. If the observational inferences related to the downsizing in sSFR are definitively confirmed, then in order to reproduce them, galaxies formed in low-mass ΛCDM halos should delay their active SF phase towards later epoch. This delay should be larger for lower mass galaxies. It is not clear how to attain this in models and simulations; new astrophysical ingredients are probably necessary. For high masses, the AGN feedback is commonly invoked to explain mass downsizing. For low-mass galaxies this mechanism does not apply because observations show that the AGN phenomenon is not common in these galaxies.

References

Colín, P., Avila-Reese, V., Vázquez-Semadeni, E., Valenzuela, O., & Ceverino, D. 2010, *ApJ*, 713, 535
Behroozi, P. S., Conroy, C., & Wechsler, R. H. 2010, *ApJ*, 717, 379
Firmani, C., Avila-Reese, V., & Rodríguez-Puebla, A. 2010, *MNRAS*, 404, 1100
Franco, J., & Cox, D. P. 1986, *PASP*, 98, 1076
Guo, Q, White, S., Li, Ch., & Boylan-Kolchin, M. 2010, *MNRAS*, 404, 1100
Kravtsov, A. V., Klypin, A. A., & Khokhlov, A.M., 1997, *ApJS*, 111, 73
Kravtsov, A. V. 2003, *ApJ* (Letters), 590, 1
Salim, S. *et al.* 2007, *ApJS*, 173, 267
Santini P. *et al.*, 2009, *A&A*, 504, 751
van Dishoeck, E. F., & Black, J. H. 1988, *ApJ*, 334, 771

Computational Star Formation
Proceedings IAU Symposium No. 270, 2011
J. Alves, B.G. Elmegreen, J. M. Girart & V. Trimble, eds.

doi:10.1017/S1743921311000895

The impact of metallicity and X-rays on star formation

Marco Spaans[1], Aycin Aykutalp[1], Seyit Hocuk[1]

[1]Kapteyn Astronomical Institute, University of Groningen,
P.O. Box 800, 9700 AV Groningen, the Netherlands
email: **spaans@astro.rug.nl**

Abstract. Star formation is regulated through a variety of feedback processes. In this study, we treat feedback by metal injection and a UV background as well as by X-ray irradiation. Our aim is to investigate whether star formation is significantly affected when the ISM of a proto-galaxxy enjoys different metallicities and when a star forming cloud resides in the vicinity of a strong X-ray source. We perform cosmological Enzo simulations with a detailed treatment of non-zero metallicity chemistry and thermal balance. We also perform FLASH simulations with embedded Lagrangian sink particles of a collapsing molecular cloud near a massive, 10^7 $M_\odot$, black hole that produces X-ray radiation.

We find that a multi-phase ISM forms for metallicites as small as 10^{-4} Solar at $z = 6$, with higher ($10^{-2} Z_\odot$) metallicities supporting a cold (< 100 K) and dense (> 10^3 cm^{-3}) phase at higher ($z = 20$) redshift. A star formation recipe based on the presence of a cold dense phase leads to a self-regulating mode in the presence of supernova and radiation feedback. We also find that when there is strong X-ray feedback a collapsing cloud fragments into larger clumps whereby fewer but more massive protostellar cores are formed. This is a consequence of the higher Jeans mass in the warm (50 K, due to ionization heating) molecular gas. Accretion processes dominate the mass function and a near-flat, non-Salpeter IMF results.

Keywords. stars: formation — hydrodynamics — chemistry — IMF

---

1. Brief Context

The presence of metals and X-rays in star forming regions affect processes like gas heating, ionization degree and cooling time, and do so up to column densities of 10^{24} cm^{-2} that are typical of dense cores (e.g., Hocuk & Spaans 2010; Meijerink & Spaans 2005). At the same time, star formation in the early universe and inside active galactic nuclei may depend strongly on the ambient metallicity and radiation background as well as on mechanical feedback from supernovae (e.g., Greif *et al.* 2010).

2. Low-Metallicity Halos at High Redshift

In the Enzo simulations we use 128^3 grid cells on the top grid with three nested subgrids, each refining by a factor of two. The box size of the simulations is 1-8 Mpc/h and a concordance cosmology is adopted. We run simulations from redshift 99 to 5 for a pre-enriched ISM, using different cooling prescriptions for metallicities of 10^{-4}, 10^{-3}, 10^{-2} and 10^{-1} Solar. In these simulations, we have used a star formation recipe which derives from the metallicity dependent ISM phase structure and we include UV radiation and supernova feedback from PopIII stars. The metallicity dependent cooling is computed with the Meijerink & Spaans (2005) UV chemistry code and includes fine-structure emissions from carbon and oxygen as well as molecular lines from species like

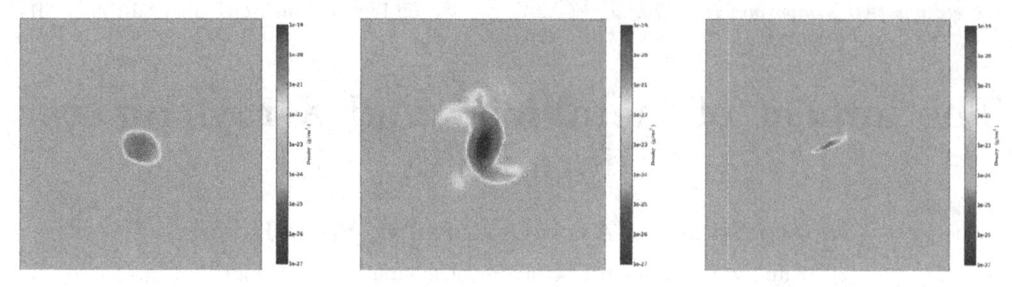

Figure 1. Density profiles of a minihalo at $z \approx 5.4$ for metallicities of $Z/Z_{\odot} = 10^{-4}$ (left), $Z/Z_{\odot} = 10^{-2}$ (middle), and $Z/Z_{\odot} = 10^{-2}$ with a $J_b = 100 J_{-21}$ radiation field (right).

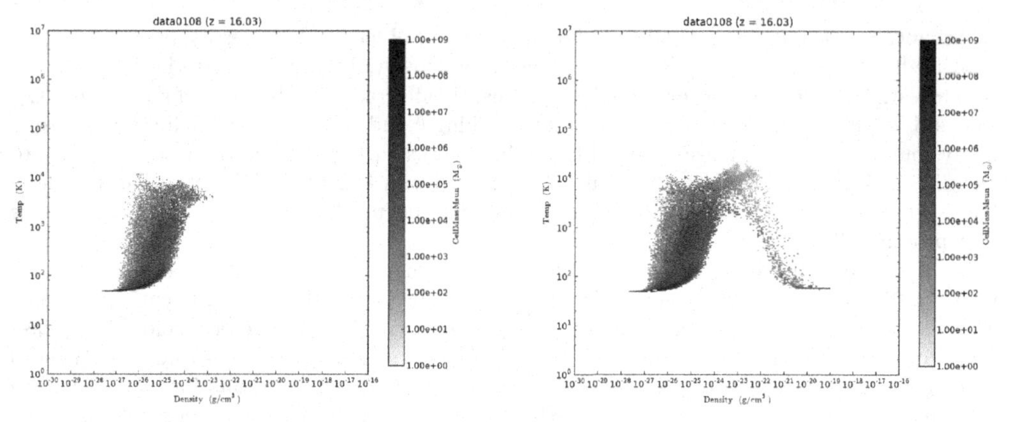

Figure 2. Density-temperature phase diagrams at $z \approx 16$ for $Z/Z_{\odot} = 10^{-4}$ (left) and $Z/Z_{\odot} = 10^{-2}$ (right).

H_2, CO and water. Heating is provided by adiabatic compression, dust grain photoelectric emission and supernovae. All level populations are computed under statistical equilibrium. The chemistry includes gas phase and grain surface formation of H_2 and HD (Cazaux & Spaans 2004, 2009). CPU demands prevent global radiation transfer to be performed. Instead, we work with a precomputed grid of cooling rates that augment the zero metallicity rates already present in Enzo.

A multi-phase ISM develops, at various times, once the metallicity exceeds a value of about 10^{-4} Solar, consistent with Jappsen *et al.* (2009). The key feature in this is the presence of a cold (< 100 K) and dense ($> 10^3$ cm^{-3}) phase. The redshift at which a multi-phase ISM is established depends on the metallicity, see figure 2 for $z \approx 16$. Also, the $Z/Z_{\odot} = 10^{-2}$ case evolves faster dynamically compared to the $Z/Z_{\odot} = 10^{-4}$ case due to the shorter cooling time. The higher metallicity halo is more compact and although mergers are more violent it recovers faster than the $Z/Z_{\odot} = 10^{-4}$ case (see figure 1, left two panels). In a run where we switch to $Z/Z_{\odot} = 10^{-1}$ at redshift 12 we first form more stars due to a stronger presence of a cold dense phase. Consequently, the minihalo experiences severe feedback effects. Due to the efficient cooling ability of the metal-rich gas, the halo maintains its density structure but the cold dense phase is largely destroyed (see figure 3).

As a first attempt to include the effects of UV radiation feedback, we have performed runs with a constant background radiation field of magnitude equal to the mean Galactic value (about $J_b = 100 J_{-21}$). This background dissociates molecules like CO completely and leaves little H_2 despite the self-shielding ability of the latter. It should be emphasized that J_b is merely chosen and that the transfer of radiation is strictly local. Still, these first

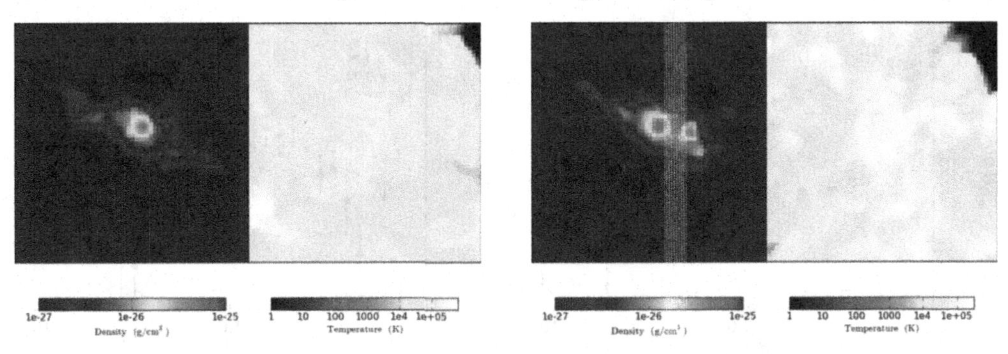

Figure 3. Density-temperature slices at $z \approx 6$, for star formation and feedback on runs. Shown are a constant $Z/Z_{\odot} = 10^{-2}$ run (left) and one for which $Z/Z_{\odot}$ jumps from 10^{-2} to 10^{-1} at $z = 12$ (right).

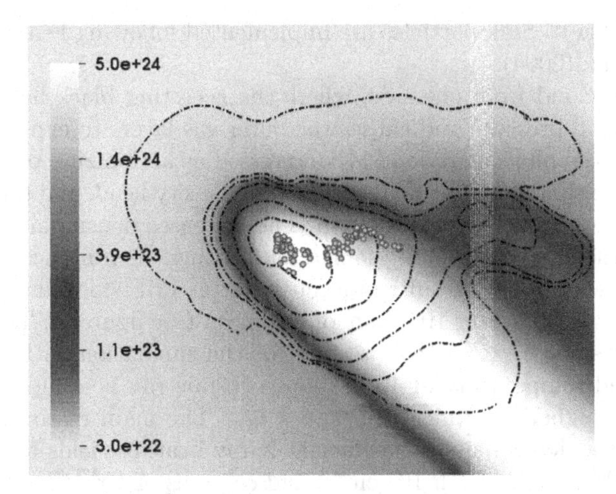

Figure 4. Column density (cm^{-2}) slice of a molecular cloud under the impact of X-rays (at $t = t_{ff}$). The colors represent the column density along the line of sight to the black hole, which is located at the upper left side. The density is shown in black contours. Contour levels are 1, 4, 16, 64, 256, 1024 $\times 10^4$ cm^{-3}. The white spheres indicate the locations of the protostars.

calculations do indicate that the cold dense phase is fragile for metallicities of $Z/Z_{\odot} \leqslant 10^{-2}$ (figure 1, right panel). In all, supernova and radiation feedback act to diminish the presence of a cold dense phase (at a given metallicity) and thus allow self-regulation for a star formation recipe that is based on a cold and dense phase.

3. X-ray Irradiation of a Cloud near a Black Hole

In order to follow the effects caused by X-rays on the temperature of the gas, the X-ray chemistry code of Meijerink & Spaans (2005) is ported into FLASH. This code again includes all relevant atomic and molecular cooling lines and carefully treats secondary ionizations by fast electrons, Coulomb heating and ion-molecule/neutral-neutral reactions. Given an X-ray flux, gas density, and global column density along the line of sight to the source, the code calculates the temperature and chemical abundances self-consistently. This output is fed into the simulation at every iteration yielding a complete X-ray chemistry extension of FLASH. The cloud is located at 10 pc from the black hole. A k^{-4} turbulent spectrum is imposed as well as Keplerian shear. The initial cloud mass is 800 $M_{\odot}$. The X-ray flux is 160 erg s^{-1} cm^{-2} with an $E^{-0.9}$ power-law distribution

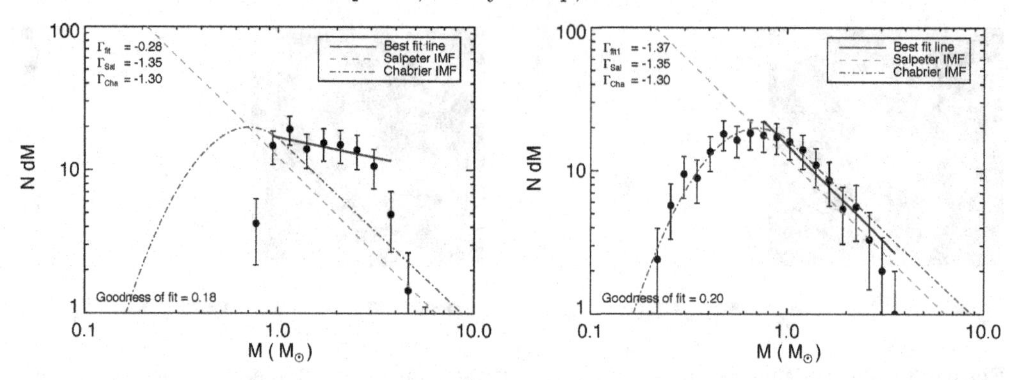

Figure 5. The IMF of two simulations at $t = t_{\rm ff}$. Left panel shows the IMF for the simulation with X-rays, and the right panel shows the IMF for the simulation without X-rays.

between 1 and 100 keV. Sink particles are implemented following Federrath *et al.* (2010) and Krumholz *et al.* (2004).

X-rays heat the cloud from one side, where the accreting black hole is located. This increases the thermal pressure and causes the outer gas layers to expand and evaporate. At the same time, the molecular cloud is compressed as an ionizing pressure flow travels inwards. Finger-like shapes are formed, with a high density head, and the gas that is lying in their shadows is well shielded, see figure 4. The increased density and external thermal pressure, caused the efficient ($\sim 30\%$) ionization heating, then induces star formation.

The simulation with X-rays results in a much flatter IMF slope than the one without radiation, $\Gamma_{\rm X-ray} = -0.28$ after 10^5 years of evolution (see figure 5). This is much flatter than the Salpeter slope of -1.35 that we find for the simulation without X-rays, which has a typical cloud temperature of 10 K due to cosmic ray heating only. These trends are maintained at higher dynamical times, $t = 2t_{\rm ff}$. The main reason for the change in IMF comes from the Jeans mass. The efficient X-ray heating yields temperatures in the cloud of 50 K at densities as high 10^6 cm^{-3} and columns of 10^{23-24} cm^{-2}. We find that the sink particles still grow in mass after $2t_{\rm ff}$ and experience competitive accretion.

References

Cazaux, S. & Spaans, M. 2004, *ApJ*, 611, 40
Cazaux, S. & Spaans, M. 2009, *A&A*, 496, 365
Federrath., C., Banerjee, R., Clark, P. C., & Klessen, R. S. 2010, *ApJ*, 713, 269
Greif, T. M., Glover, S. C. O., Bromm, V., & Klessen, R. S. 2010, *ApJ*, 716, 510
Hocuk, S. & Spaans, M. 2010, *A&A*, 510, A110
Jappsen, A.-K., Klessen, R. S., Glover, S. C. O., & Mac Low, M.-M. 2009, *ApJ*, 696, 1065
Krumholz, M. R., McKee, C. F., & Klein, R. I. 2004, *ApJ*, 611, 399
Meijerink, R. & Spaans, M. 2005, *A&A*, 436, 397

Computational Star Formation
Proceedings IAU Symposium No. 270, 2011
J. Alves, B.G. Elmegreen, J. M. Girart & V. Trimble, eds.

doi:10.1017/S1743921311000901

A Guide to Comparisons of Star Formation Simulations with Observations

Alyssa A. Goodman[1]

[1]Harvard-Smithsonian Center for Astrophysics
60 Garden Street, MS 42
Cambridge, MA 02318, USA
email: agoodman@cfa.harvard.edu

Abstract. We review an approach to observation-theory comparisons we call "Taste-Testing." In this approach, synthetic observations are made of numerical simulations, and then both real and synthetic observations are "tasted" (compared) using a variety of statistical tests. We first lay out arguments for bringing theory to observational space rather than observations to theory space. Next, we explain that generating synthetic observations is only a step along the way to the quantitative, statistical, taste tests that offer the most insight. We offer a set of examples focused on polarimetry, scattering and emission by dust, and spectral-line mapping in star-forming regions. We conclude with a discussion of the connection between statistical tests used to date and the physics we seek to understand. In particular, we suggest that the "lognormal" nature of molecular clouds can be created by the interaction of many random processes, as can the lognormal nature of the IMF, so that the fact that both the "Clump Mass Function" (CMF) and IMF appear lognormal does not necessarily imply a direct relationship between them.

Keywords. star formation, simulations, statistical comparisons.

1. Introduction

Saying that one is working on "comparing observations with simulations" sounds straightforward. Alas, though, comparing astronomical experimental data ("observations") about star formation with relevant theory is much more difficult than [experiment]:[theory] comparisons in almost any other area of physics. The reasons are twofold. First, the theories involved are not typically "clean" analytic ones, but instead require simulation to make predictions. And second, the measurements observers offer involve complicated-to-interpret "fluxes" from various physical processes rather than more direct measures of physical parameters.

Let's say a theory makes a prediction about density. In a laboratory, one fills a container of known interior volume and weight with the substance under study using some kind of simple mechanical device (e.g. a spoon), puts the container on a well-calibrated scale, measures a weight, subtracts the weight of the container, divides by the well-known gravitational constant and the known volume, and *voilà*, density. In a star-forming region...not so simple. One points a telescope of (sometimes poorly) known beam response at some patch of 2D sky and collects all the photons that come to a detector of (sometimes poorly known or variable) efficiency via the telescope during a well-measured period of time. Such an observation gives the "flux" (or lack thereof) of some quantity (e.g. thermal emission from dust, emission from a spectral line in gas, dust extinction). Next, that flux needs to be converted either to a 2D "column density" on the sky, or a 3D volume density (only possible using chemical excitation models for spectral lines). In the case of a column density, the "column" needs to be assigned a length to extract a volume density.

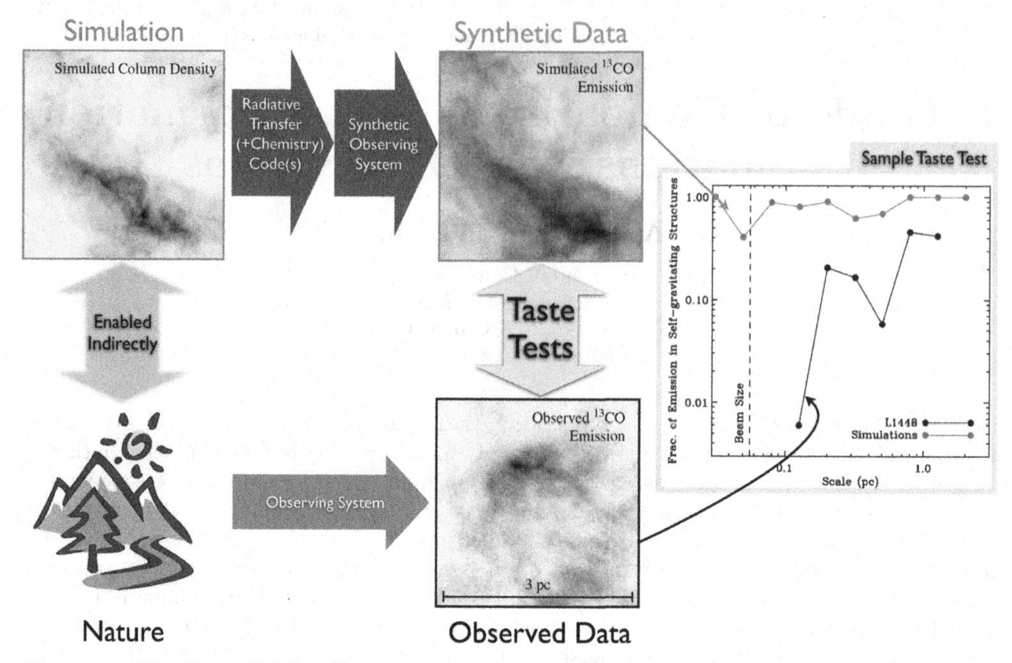

Figure 1. The Taste-Testing Process. The processes leading to measurements are represented by the green arrow for "real" data and by the two purple arrows for synthetic data. The box at right shows a comparison of the fraction of self-gravitating (according to a virial analysis) material as a function of scale in the simulations of Padoan *et al.* (2006a) and in the L1448 region of Perseus (see Rosolowsky *et al.* (2008); Goodman *et al.* (2009b) and §3, below.)

And, in either case, it is not clear what mixture of actual densities conspire along a line of sight to give the measured "representative" volume density.

So, should we give up on observational measurements of physical parameters in star formation research? Of course not. But, we should consider that perhaps comparisons of observations and theory (primarily simulations) are best carried out in an "observational space," where synthetic observations of theoretical output are made in order to enable more direct statistical comparison with "real" (observational) data. This comparison in observational space has been referred to in the past by some (including me) as "Taste-Testing." If one's goal is to reproduce the recipe for a great soup eaten in a restaurant, then a good course of action is to guess the ingredients based on dining experience (observation) and the processes based on cooking experience (physics) create the soup, take a look at it, and then if it seems to look and smell right, then finally "taste" it. Centrifuging the soup–as a theorist might try to do with observations in "theorists'" space–to discern its ingredients might give you a basic chemical breakdown, but it would not tell you how to make it again. Thus, we define a legitimate [observation]:[theory] taste test as one where statistical (taste) tests are applied in observer's (edible) space. An overview of the process is shown in Fig. 1.

2. Synthetic Observations

In creating synthetic observations of star-forming regions, there are two conceptual steps (as shown by the purple arrows in Fig. 1). First, the physical processes producing the spectral-spatial-temporal combination of photons to be observed must be modeled. And, second, the response of the observing system to that flux of photons must be

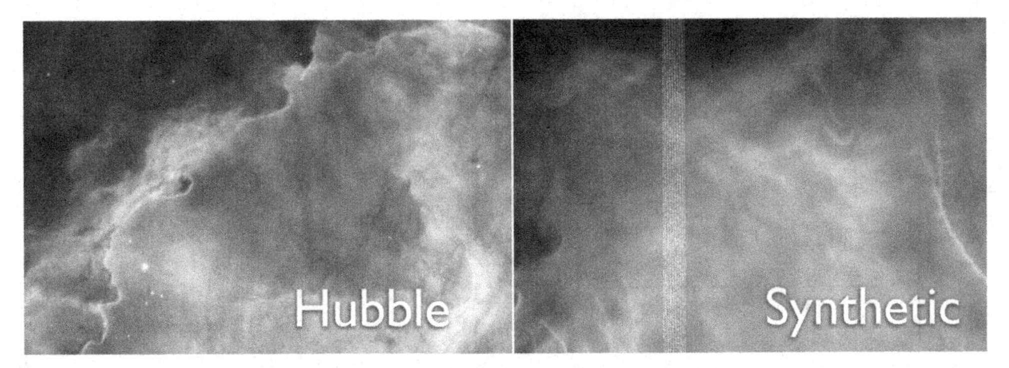

Figure 2. Will the real M17 please stand up?

calculated. In reality, each of these steps can be very difficult, and often one step is taken without the other. For example, it is exceedingly difficult to model the molecular line emission from star-forming gas fully, as chemical, radiative, shock, magnetic, and thermal physics can all come into play. So, researchers like Padoan *et al.* (1999), Ayliffe *et al.* (2007), Offner *et al.* (2008), Rundle *et al.* (2010) are to be forgiven when they create somewhat idealized "synthetic" spectral line maps, not including the exact response of a particular telescope. Other researchers all but skip the first step and carry out only the second, by running only highly idealized analytic models (rather than more complicated simulations) though telescope "simulators," like the ALMA simulator (`http://www.cv.nrao.edu/naasc/alma_simulations.shtml`). Some researchers, for example Krumholz *et al.* (2007), have already gone so far as to "observe" numerical simulations (e.g. of massive disks) with particular (synthetic) telescopes (e.g. ALMA, EVLA). In principle, the ARTIST efforts (`http://www.astro.uni-bonn.de/ARC/artist/`) will be able to automate the process of creating synthetic observations for any "input" theoretical model, once inputting complex numerical models becomes operationally straightforward.

Fig. 2 shows what is certainly one of the most beautiful examples to date of a synthetic observation. The left panel shows the [OIII, blue], H-α [green] and [SII, red] emission-line image (Hester, STScI-PRC03-13) of part of the M17 H II region from the Hubble Space Telescope, and the right panel shows a synthetic view through the same filters of a 512^3 numerical simulation of Mellema *et al.* (2006).

While perhaps less spectacularly beautiful, other synthetic products, such as spectral-line maps, polarization maps, dust emission, scattering, and extinction maps have been generated recently by many groups (see examples in Table 1), and more and more are being created as computational power increases. All of these synthetic data products, created by what some groups call "forward modeling," offer the opportunity for "Taste Testing." It is critical to keep in mind that legitimate comparison requires the taste-testing step: the creation of synthetic observations is a key step toward comparison, but not a comparison itself. Apple juice and iced tea can look very much like each other, but they do *not* taste the same. The images in Fig. 2 may *look* alike to our eyes, but are we being fooled by the matching color scheme and roughly correct morphology? Or, is the similarity really great?

3. Statistical Comparisons (Taste-Tests)

In order to quantify just *how similar* real and simulated observations are, we need powerful statistical measures, preferably of properties that are intuitively and or physically meaningful.

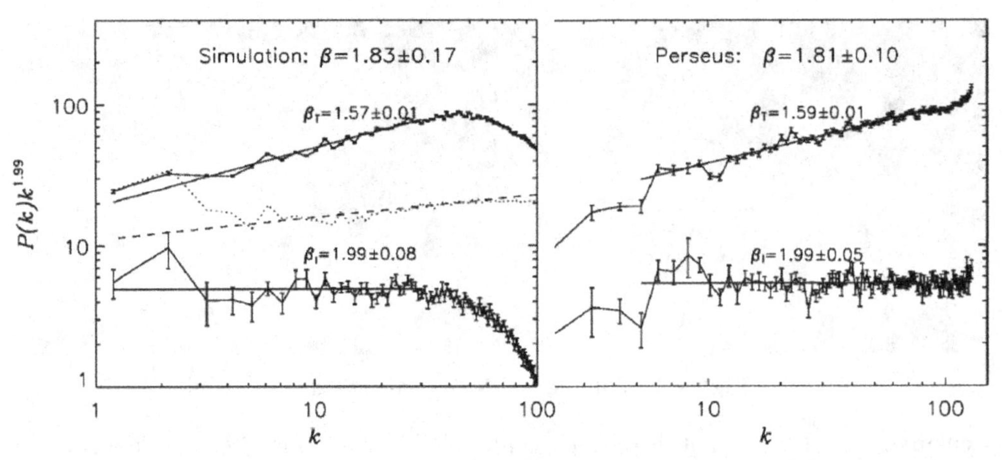

Figure 3. Velocity Power Spectra of Synthetic (left) and Real (right) maps, based on Figures in Padoan *et al.* (2006a)

Some comparative statistical tests are harder to pass than others, and the trick is to find a test that is "hard enough" so that it cannot be fooled, while not being so hard as to require an (unrealistic) level of agreement. For a trivial example of a test "too easy" to pass, think about the intensity distribution (a histogram) of all pixels in a map. Intensity histograms might match for two maps, but so would a histogram for a totally unphysical map created by randomizing the position of every intensity pixel in either map. Too easy.

Many researchers make use of various forms of correlation functions, which take account of the distribution of some measured property at various real or transformed scales, as taste tests (see, for examples, Lazarian (2009); Padoan *et al.* (2003); Brunt *et al.* (2009), and references therein). In our experience, passing any such test is often necessary, but not sufficient, evidence of plausible agreement. By way of example, consider Fig. 3, which shows velocity power spectra for synthetic and real ^{13}CO maps of a star-forming region, from the work of Padoan *et al.* (2006a) The agreement between the slopes of the power spectra shown is excellent, but further work, using harder-to-pass dendrogram-based tests, on the same data shows that the two data cubes used in the comparison are in fact quite different (Rosolowsky *et al.* 2008; Goodman *et al.* 2009b).

The right-most panel of Fig. 1 shows a comparison of the L1448 sub-region of Perseus with a relevant sub-region of the same simulation used by Padoan *et al.* (2006a) to create the left panel of Fig. 3. The hierarchy of both clouds is measured by constructing tree diagrams (dendrograms) that quantify the hierarchy of emission within the position-position-velocity space created by spectral line mapping. A virial parameter can be calculated for every point in the tree, using measured mass, size, and velocity dispersion. The "Sample Taste Test" panel of Fig. 1 shows that the fraction of material with virial parameter below a "self-gravitating" threshold does *not* depend on scale in the same way in both real and synthetic data cubes. (And, the disagreement is particularly off-putting because the majority of structures within the Padoan *et al.* simulation are apparently self-gravitating, even though self-gravity was not included as an input to the simulation.) So, alas, while the synthetic observations of the simulations shown in Padoan *et al.* (2006a) *pass* a velocity power spectrum comparison test, they *fail* the dendrogram-derived virial parameter test. Note that this and any similar failure to pass a taste test can be due to any step along the way through the top half of Fig. 1: the discrepancy can stem from the simulation itself and/or from the radiative transfer and/or chemistry calculations and/or from assumptions about the telescope response.

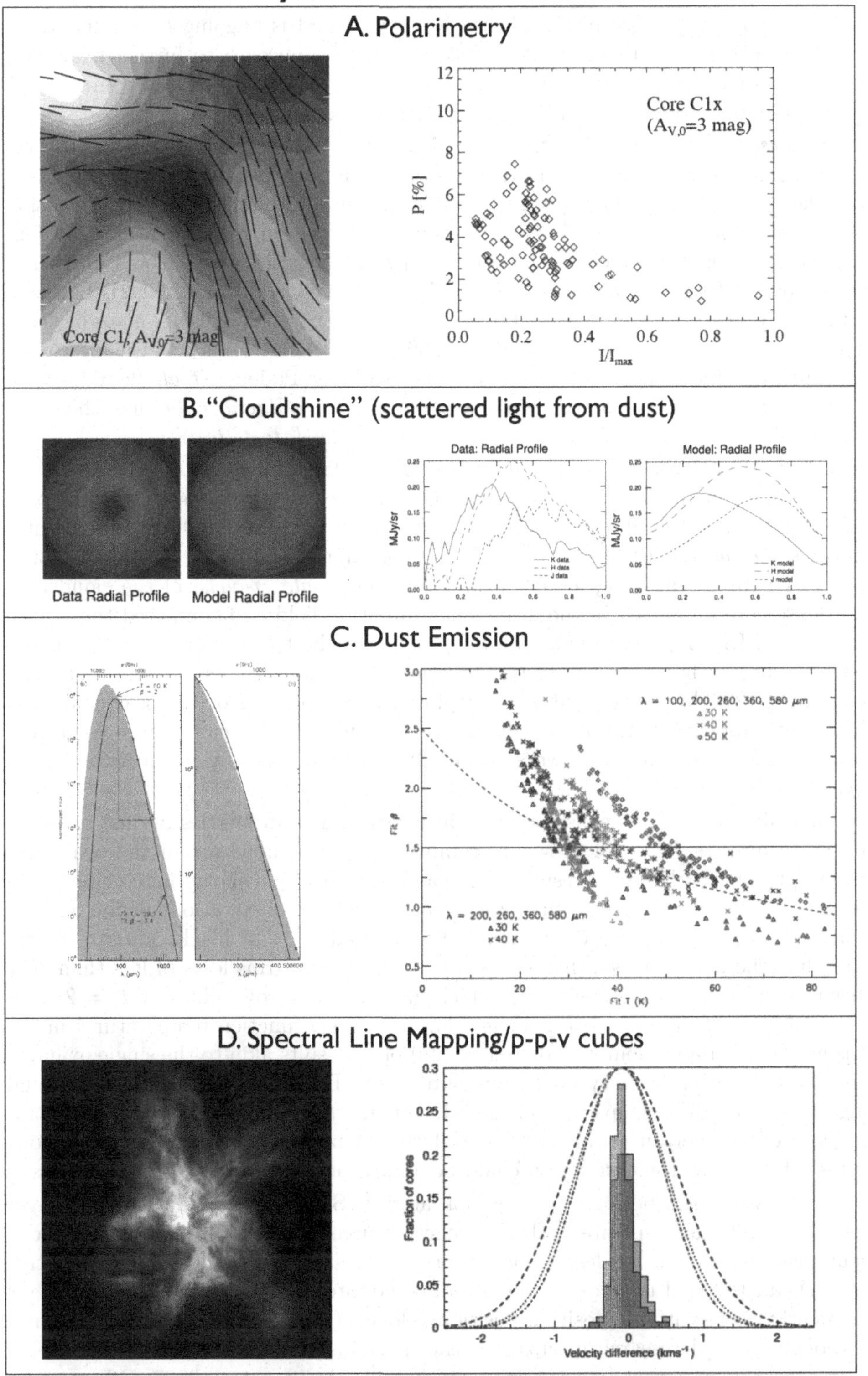

Table 1. Tasting Menu: Please refer to the text on the facing page for references and details.

4. Tasting Menu

Since so much more "Tasting" work has been done and is ongoing than what there is room to mention here, I shall take a "Tasting Menu" approach to further examples, much as a fine restaurant would. Table 1, as a Tasting Menu, offers four courses. The text in the remainder of this section offers descriptions of each course.

For an *hors d'oeuvre*, consider **"A. Polarimetry".** Synthetic polarization observations, giving insight into magnetic field structure, have been created by several authors (cf. Ostriker *et al.* (2001)). The specific example shown in Table 1 shows the re-creation of a so-called "polarization hole" in a sub-mm dust emission map. In nearly all density regimes, polarization efficiency falls as the highest densities probed by a particular tracer are reached (see Goodman *et al.* (1995); Matthews & Wilson (2002); Lazarian *et al.* (1997); Whittet *et al.* (2008) and references therein). When SCUBA maps of dense cores were first made e.g. Matthews *et al.* (2001), the "holes" this inefficiency causes in the center of maps seemed mysterious. The work by Padoan *et al.* (2001) shown here demonstrates that a polarization map (*left panel, synthetic data*) has the same polarization-intensity behavior as the real data (*right panel, taste-test*).

For an appetizer, let's look at **"B. Cloudshine".** Dark clouds are not "dark" when imaged deeply enough. The effect known as "Cloudshine" was first observed in the NIR by Lehtinen & Mattila (1996) then again by Foster & Goodman (2006), and recently by Steinacker *et al.* (2010) in the MIR (a.k.a. "coreshine"). Foster & Goodman (2006) created very simple scattering models (*"model radial profile" panels*) of the cloudshine from a dense core immersed in the interstellar radiation field and compared those with a dense core in L1451 (*"data radial profile" panels*). In the figure shown in Table 1, the taste-test is simply the agreement of the different colors' emission as a function of scale. Padoan *et al.* (2006b) carried out more sophisticated modeling of the scattered light produced within a turbulent box and showed that column density can be recovered from NIR cloudshine at accuracies no worse than 40% off (and typically much better) from reality.

For our main course, let's consider **"C. Emission"**. The properties of dust emission from star-forming regions will grow in prominence as a comparator in the near-term future in light of new Herschel results. The particular example shown here comes from the work of Shetty *et al.* (2009). In the *left panel* synthetic observations of the SED at Herschel-like wavelengths of a 60 K emissivity-modified ($\beta = 2$) blackbody are shown: the solid line shows a fit to synthetic data where noise at an rms level of less than 10% has been included at each wavelength, which gives quite far-off values of $T = 29.3$ K and $\beta = 3.4$. The "taste-test" comparison (showing β as a function temperature) in the *right panel* shows of an ensemble (colored points) of fit results akin to the single example shown at left, for slightly-noisy far-IR and sub-mm SEDs, along with the observational findings (solid curve) of Dupac *et al.* (2003). Dupac *et al.* claim a physical origin for their measured"temperature-emissivity" correlation, while the Shetty *et al.* results show, using taste-tests, that the correlation could be an artifact of SED fitting degeneracies.

And, for dessert (the best part), let's look at **"D. Spectral Line Mapping/*p-p-v* cubes"**. Spectral-line maps are both the richest sources of detailed information about molecular clouds, and the hardest to model correctly. As explained above, since radiative transfer, chemistry, and telescope response all come into play, it is exceedingly difficult to confidently offer synthetic position-position-velocity (*p-p-v*) cubes for testing. Thankfully, though, several research groups have been willing to tread in (or at least near) these treacherous waters of late (Padoan *et al.* 1999; Ayliffe *et al.* 2007; Offner *et al.* 2008; Kirk *et al.* 2009; Tilley & Balsara 2008; Rundle *et al.* 2010). In the example shown here, the *left panel* shows a map of HCO^+ (1-0) emission color-coded by velocity created by Rundle *et al.* (2010). The *right panel* shows a corresponding prediction of how narrow the distribution of core-to-core velocities (filled histograms) is with respect to the widths of various spectral lines in their environs (curves). In particular, these results (cf. Ayliffe *et al.* (2007)) show that even a simulation with significant amounts of competitive accretion can produce the kinds of small velocity offsets amongst features and tracers seen, for example, by Walsh *et al.* (2004) and by Kirk *et al.* (2010).

5. The Future: "Tests of the Tests"

A couple of years ago at a meeting in honor of Steve and Karen Strom, Frank Shu listened to my presentation of an early version of the L1448 results shown in Fig. 1. He astutely commented: "that's nice, but *how about a Test of the Test*"? What Frank meant by this question is critical for the future of observation-simulation comparisons, so allow me to explain. Sure, one can calculate a "virial parameter," and use it as a statistical measure of two *p-p-v* cubes' similarity, but, is that virial parameter really a measure of the same physics (e.g. boundedness) that it would be in real "*p-p-p v-v-v*" space? Similarly, one can use CLUMPFIND to de-compose clouds statistically, but does it measure real structures within cloud, whose physical properties it is worth comparing? Does finding a lognormal column density distribution really uniquely imply anything about turbulence? Let us consider these last three questions in turn.

5.1. *Virial Parameters and p-p-v Space*

Using the virial theorem–even in full, real, 3D (*p-p-p v-v-v*) space–is dangerous, in that key terms are often left out. In recent years, virial analysis has been applied rather cavalierly in *p-p-v* space as well, thanks in part to Bertoldi and McKee (1992), who laid out the ground work for a "virial parameter" that was to be a proxy for the significance of gravity as compared with turbulent pressure. In recent work, Dib *et al.* (2007) have shown that ignoring the subtleties of the virial theorem (e.g. surface energy) leads to incorrect conclusions about the boundedness of particular objects. And, in work directly aimed at understanding the translation from features in *p-p-v* space to *p-p-p v-v-v* space (c.f. Ballesteros-Paredes & Low 2002; Smith *et al.* 2008), Shetty *et al.* have recently shown that the standard virial parameter (e.g. Bertoldi & Mckee 1992) is really only "safe" for very simple, isolated, objects, and that the more crowded a region becomes, the worse an approximation the virial parameter is. Thus, new work seeking perhaps more robust measures–in *p-p-v* space of the role of gravity and fundamental forces/processes in molecular clouds–is sorely needed.

5.2. *Structures, Real or Statistical?*

We heard much at the meeting about "clump mass spectra" and about the "filamentariness" of molecular clouds. However, unless one allows for the possibility of "clumps" *within* "filaments," these measures are topologically inconsistent. Thus, measures of cloud structure that do *not* allow for hierarchy are likely to produce unphysical structures. For example, when CLUMPFIND (Williams *et al.* 1994) is applied in very crowded regions, where hierarchy and overlap are important, not only will spurious, unphysical, structures be cataloged, the exact structures found will depend sensitively on CLUMPFIND's input parameters (Pineda *et al.* 2009). So, while CLUMPFIND can be validly used as a purely statistical comparative test, the naïve *interpretation* of its output as giving a mass spectrum of structures that can directly form individual or small groups of stars, is not physically sensible. (Note that in cases where extended structure is removed before

CLUMPFIND is applied (e.g. Alves *et al.* 2006), the resulting Clump Mass Function (CMF) is then more valid, but it is still reliant on the exact prescription for removing "extended" structure.)

What is needed instead of CLUMPFIND? More unbiased algorithms like the dendrogram (tree) approach blatantly advertised above are a good first step. But, even these more sophisticated segmentations then need measures of the connections between physically-relevant parameters and purely statistical ones. For example, in *p-p-v* space, can we find use new segmentation algorithms like dendrograms to help find a relationship between how "filamentary" a region is, and how "bound" those filaments, or perhaps the cores within them, are?

5.3. *Lognormals are Not Enough*

Many authors have recently shown that when a "large enough" volume of a molecular cloud is sampled, a lognormal column density distribution will be found (Goodman *et al.* (2009a); Wong *et al.* (2009); Froebrich & Rowles (2010); Kainulainen *et al.* (2009); Lombardi *et al.* (2010)). The existence of such lognormals is certainly *consistent* with the predictions of many turbulence simulations (e.g. Vazquez-Semadeni 1994; Ostriker *et al.* 2001; Wada & Norman 2001). However, a Bonnor-Ebert sphere sampled appropriately will also give a lognormal density distribution (Tassis *et al.* 2010)! And, in fact, the central-limit theorem shows that lognormal behavior will result from a confluence of nearly any set of interacting random processes. Thus, measuring lognormal behavior is again "necessary but not sufficient" on its own as a descriptive statistic. It may be, though, that the particular properties of a lognormal (e.g. its width, its mean, and/or over what range it holds) may be more discriminating.

Just because the IMF and CMF are both consistent with lognormals does *not* mean one comes definitively directly from the other. As Swift & Williams (2008) show, assuming very different core-to-star fragmentation/multiplicity schemes can, at the level of detail we can measure them today, produce the "same" IMF from the "same" CMF. It is indeed possible that the CMF could easily be generated by the random (even turbulent!) processes within molecular clouds on large scales, while the IMF is *simultaneously* generated by the randomness (Adams & Fatuzzo 1996) that takes place on smaller scales within clusters and fragmenting cores.

5.4. *Summary*

It is clear that some "taste" tests are harder to fool than others, so clearly, we will seek and prefer the tough ones as discriminators. Our community's next step will be to begin to seriously explore which tests are best for discerning which physical characteristics, as this is not yet very clear. We should think of this process as designing artificial "instruments" capable of measuring the interstellar equivalents of culinary measures like acidity, texture, and salinity: only, our instruments will measure properties like ionization, magnetic beta, vorticity, etc. As the computational power available for simulation and the size of observational databases increase, we will all need to learn how to become tough but fair food critics, and good chefs.

References

Adams, F. & Fatuzzo, M. 1996, The Astrophysical Journal, 464, 256
Alves, J., Lombardi, M., & Lada, C. 2006, Astronomy and Astrophysics, 462, L17
Ayliffe, B. A., Langdon, J. C., Cohl, H. S., & Bate, M. R. 2007, MNRAS, 374, 1198
Ballesteros-Paredes, J. & Low, M.-M. M. 2002, The Astrophysical Journal, 570, 734
Bertoldi, F. & Mckee, C. 1992, The Astrophysical Journal, 395, 140

Brunt, C. M., Heyer, M. H., & Low, M.-M. M. 2009, Astronomy and Astrophysics, 504, 883
Dib, S., Kim, J., Vázquez-Semadeni, E., Burkert, A., & Shadmehri, M. 2007, The Astrophysical Journal, 661, 262
Dupac, X., *et al.* 2003, Astronomy and Astrophysics, 404, L11
Foster, J. & Goodman, A. 2006, The Astrophysical Journal, 636, L105
Froebrich, D. & Rowles, J. 2010, Monthly Notices of the Royal Astronomical Society, 406, 1350
Goodman, A., Jones, T., Lada, E., & Myers, P. 1995, The Astrophysical Journal, 448, 748
Goodman, A. A., Pineda, J. E., & Schnee, S. L. 2009a, The Astrophysical Journal, 692, 91
Goodman, A. A., Rosolowsky, E. W., Borkin, M. A., Foster, J. B., Halle, M., Kauffmann, J., & Pineda, J. E. 2009b, Nature, 457, 63, dOI: 10.1038/nature07609
Kainulainen, J., Beuther, H., Henning, T., & Plume, R. 2009, Astronomy and Astrophysics, 508, L35
Kirk, H., Johnstone, D., & Basu, S. 2009, The Astrophysical Journal, 699, 1433
Kirk, H., Pineda, J., Johnstone, D., & Goodman, A. 2010, eprint arXiv:1008.4527
Krumholz, M., Klein, R., & Mckee, C. 2007, The Astrophysical Journal, 665, 478
Lazarian, A. 2009, Space Science Reviews, 143, 357
Lazarian, A., Goodman, A., & Myers, P. 1997, The Astrophysical Journal, 490, 273
Lehtinen, K. & Mattila, K. 1996, Astronomy and Astrophysics, 309, 570
Lombardi, M., Lada, C., & Alves, J. 2010, Astronomy and Astrophysics, 512, 67
Matthews, B. & Wilson, C. 2002, The Astrophysical Journal, 574, 822
Matthews, B., Wilson, C., & Fiege, J. 2001, The Astrophysical Journal, 562, 400
Mellema, G., Arthur, S., Henney, W., Iliev, I., & Shapiro, P. 2006, The Astrophysical Journal, 647, 397
Offner, S. S. R., Krumholz, M. R., Klein, R. I., & McKee, C. F. 2008, The Astronomical Journal, 136, 404
Ostriker, E. C., Stone, J. M., & Gammie, C. F. 2001, The Astrophysical Journal, 546, 980
Padoan, P., Bally, J., Billawala, Y., Juvela, M., & Nordlund, Å. 1999, The Astrophysical Journal, 525, 318
Padoan, P., Goodman, A., Draine, B. T., Juvela, M., Nordlund, Å., & Rögnvaldsson, Ö. E. 2001, The Astrophysical Journal, 559, 1005
Padoan, P., Goodman, A., & Juvela, M. 2003, The Astrophysical Journal, 588, 881
Padoan, P., Juvela, M., Kritsuk, A., & Norman, M. L. 2006a, The Astrophysical Journal, 653, L125
Padoan, P., Juvela, M., & Pelkonen, V.-M. 2006b, The Astrophysical Journal, 636, L101
Pineda, J. E., Rosolowsky, E. W., & Goodman, A. A. 2009, The Astrophysical Journal Letters, 699, L134
Rosolowsky, E. W., Pineda, J. E., Kauffmann, J., & Goodman, A. A. 2008, The Astrophysical Journal, 679, 1338
Rundle, D., Harries, T., Acreman, D., & Bate, M. 2010, Monthly Notices of the Royal Astronomical Society, -1, 1090
Shetty, R., Kauffmann, J., Schnee, S., & Goodman, A. A. 2009, The Astrophysical Journal, 696, 676
Smith, R., Clark, P., & Bonnell, I. 2008, Monthly Notices of the Royal Astronomical Society, 391, 1091
Steinacker, J., Pagani, L., Bacmann, A., & Guieu, S. 2010, Astronomy and Astrophysics, 511, 9
Swift, J. J. & Williams, J. P. 2008, The Astrophysical Journal, 679, 552
Tassis, K., Christie, D. A., Urban, A., Pineda, J. L., Mouschovias, T. C., Yorke, H. W., & Martel, H. 2010, MNRAS, 1217
Tilley, D. & Balsara, D. 2008, Monthly Notices of the Royal Astronomical Society, 406, 1201
Vazquez-Semadeni, E. 1994, The Astrophysical Journal, 423, 681
Wada, K. & Norman, C. 2001, The Astrophysical Journal, 547, 172
Walsh, A., Myers, P., & Burton, M. 2004, The Astrophysical Journal, 614, 194
Whittet, D., Hough, J., Lazarian, A., & Hoang, T. 2008, The Astrophysical Journal, 674, 304
Williams, J., Geus, E. D., & Blitz, L. 1994, The Astrophysical Journal, 428, 693
Wong, T., *et al.* 2009, The Astrophysical Journal, 696, 370

Computational Star Formation
Proceedings IAU Symposium No. 270, 2011
J. Alves, B.G. Elmegreen, J. M. Girart & V. Trimble, eds.

doi:10.1017/S1743921311000913

Comparison between Simulations and Theory: where we stand

Joe J. Monaghan[1]

[1]School of Mathematical Sciences
Vic 3800, Monash University, Australia
email: joe.monaghan@monash.edu

Abstract. In this paper I relate progress in the simulation of star formation to a remarkable early (1962) paper by Fowler and Hoyle who analyzed star formation in terms of the relevant energies starting from the formation of a large (100 pc) cloud and ending with the final mass distribution of the stars. The way in which modern simulations have clarified and corrected the Fowler-Hoyle picture, and areas where we could improve our simulations, are discussed.

Keywords. Star formation

---

1. Introduction

The aim of a theory of star formation is to predict the mass, internal rotation, composition, and magnetic fields of the observed stars. I take the essentials of a theory to be the equations that describe the system and their subsequent solution. There is general agreement that the relevant equations are those of gas dynamics with gravity and magnetic fields (at the very least in the MHD approximation), the chemistry of transformations between atoms and molecules, radiative transfer and the effects of grain formation and destruction. Most of the gas in the galaxy is turbulent and we could reasonably expect that ultimately we will need a turbulence theory to provide sub-grid effects.

We can gauge the progress made towards achieving our aims by comparing it with the remarkable paper of Willy Fowler and Fred Hoyle (1962). In that paper Fowler and Hoyle recognized the need for a theory to show:

• That a large (100 pc) cloud of gas in our galaxy could collapse despite the shear, magnetic fields, cosmic rays , and turbulence.

• That such a cloud would continue to collapse, and the Jeans mass continue to decrease, allowing fragmentation.

• That the fragmentation would cease when the opacity was sufficiently large for the compressional heat to be retained and, at that stage, the fragments would have masses comparable to the observed masses.

Because the computers available to Fowler and Hoyle were incapable of simulating the gas dynamics relevant to star formation, Fowler and Hoyle used approximate energy arguments to estimate the importance of the various physical effects. Furthermore, they were not aware, or were unable, to include the effects of accretion, and they were not able to predict the formation of binary systems. Nevertheless, much of what they deduced is relevant to the aims of present conference.

The initial step of the Fowler-Hoyle theory was the formation of clouds from the galactic material. They estimated the density required to overcome the rotation of the galaxy, and with this density, and estimates of the magnetic fields and turbulence, they estimated the energy contributions from gravitational, magnetic fields, turbulence and cosmic rays. This enabled them to assert that clouds could collapse and that once they did, further fragmentation was inevitable, though the explanation of the way this occurred was hazy. Contributions to this conference have described simulations of large cloud formation in a galaxy in the absence of magnetic fields along the lines envisaged by Fowler and Hoyle. These simulations have the potential to provide realistic initial conditions for the subsequent fragmentation of the clouds.

In the last several years extensive simulations, some of which have been described at this conference, have followed the evolution of large clouds that fragment to produce stellar systems. The most recent of these, which use both SPH and AMR methods, also include radiative transfer and magnetic fields. This work has clarified the conjectures of Fowler and Hoyle concerning the fragmentation of clouds forming in the galaxy.

In order to follow the dynamics of a fragmenting cloud for as long as possible sink particles are used to replace very high condensations of gas. As a consequence, the full range of phenomena from cloud to stars, have not yet been simulated though it has been possible to show that mass functions similar to those observed are predicted from the simulations. A question that was not taken up by Fowler and Hoyle is the effect of feed-back from newly formed stars sufficiently massive to become supernova. They inject energy and generate turbulence, and they change the metal composition which in turn affects radiative transfer. Studies of this feed-back are in their infancy.

In summary the current simulations allow us first to follow the early stages of clouds forming in the galaxy, second to use a variety of initial conditions to simulate fragmentation, and third to predict the mass function of the stars that are produced by the fragmentation. They do not allow us to start with a galaxy and follow the evolution to stars, and they do include the detailed effects of supernova and other possible feed back mechanisms. There are however, a number of questions that we might ask concerning the accuracy and relevance of simulations, and I will discuss some of these below.

2. Questions concerning the accuracy and relevance of simulations

The following are questions we need to answer in order to establish that the results we get from the simulations of star formation can be believed.

- Do we have the right equations?
- Are we solving them correctly?
- Do we have the right initial conditions?
- Do we have the right boundary conditions?
- Is feed-back from rapidly evolving stars included correctly?
- Are we treating diffusion, mixing and turbulence correctly?
- Is dust being included correctly?

Some of these questions are easy to answer, and they vary in importance. We believe we know the right equations to solve though this is done with varying degrees of approximation. The algorithms currently used fall into two large classes. The first being finite difference methods, conveniently but inaccurately labelled as AMR, the second are particle methods collectively described as SPH. In practice AMR is a handy name to tell us that the mesh of a finite difference scheme is subdivided to achieve better resolution, but the actual finite difference scheme can vary. Some of them are very good, and others rather poor. In this conference Krumholz, Tyssier, Norman and others show impressive results obtained with AMR methods. SPH is a general particle method and the actual algorithm depends on the kernels used, the rules for the spatial gradients, and the time stepping algorithm. Some are better than others. The results of Bate, Price, Bonnell, Springel, Whitworth, Klessen and others at this conference have shown a range of impressive results from carefully tested SPH algorithms.

As an example of the different treatments of the SPH equations we can consider the SPH pressure force term which can be written in the form (Monaghan(2005))

$$\frac{\nabla P}{\rho} \to \sum_b m_b \left(\frac{P_a}{\Omega_a \rho_a^2} + \frac{P_b}{\Omega \rho_b^2} + \Pi_{ab} \right) \nabla_a W_{ab}, \tag{2.1}$$

where m_a, P_a, and ρ_a, are the mass, pressure, density of particle a, Π_{ab} provides viscosity, and W_{ab} is the SPH kernel. The function Ω takes account of the variation of the resolution with density. If this is applied to an ideal gas at constant pressure in a periodic box, and the SPH particles are placed on a grid of squares, it will be found that the particles will move around their equilibrium positions with an energy that is about 0.01 the thermal energy. This occurs because the SPH algorithm has some of the attributes of molecular dynamics and the particles seek out the lowest energy configuration. In two dimensions this low energy state is close to a grid of squares but the preferred positions approximate a grid of hexagons. In practice this initial motion is entirely negligible, and for a star formation problem it is only 0.01 of the thermal energy. However, some authors find it tempting to write the SPH equations so that a constant pressure configuration will remain at rest. One way to do this is to write the previous equation as

$$\frac{\nabla P}{\rho} \to \sum_b m_b \left(\frac{(P_b - P_a)}{\rho_b^2} + \Pi_{ab} \right) \nabla_a W_{ab}, \tag{2.2}$$

where we have dropped the Ω since they do not affect our argument. There are at least two problems with this equation. First it means that gas at constant pressure in a tube with the ends open, and connected to a vacuum, will not move even though it should. If, on the other hand, (2.1) is used the SPH particles near the vacuum will feel a pressure force from their neighbours in the interior of the tube and they will move out of the tube. The same thing happens with a finite difference calculation. The second problem is that the pressure force in (2.2) doesn't conserve momentum. As a consequence two particles initially at rest, but with different pressures will be accelerated in the same direction with increasing speed. The final speed is only limited by collisions or the transfer of heat between the particles which cause the pressures to become equal whereupon the rapidly moving particles will slow down and stop. If (2.1) is used the two SPH particles will move apart with equal and opposite velocities reproducing, though with very poor resolution, the expansion of a parcel of gas in a vacuum.

Because some researchers believe SPH cannot simulate problems where two fluids with very different densities are in contact it is worth noting that SPH can simulate Rayleigh Taylor instabilities and large buoyant bubbles in fluids where the density ratios are 1000:1, Grenier *et al.* (2009) using a momentum conserving algorithm where the pressure term is

$$\frac{\nabla P}{\rho} \to \sum_b m_b \left(\frac{P_a + P_b}{\rho_a \rho_b} + R_{ab} + \Pi_{ab} \right) \nabla_a W_{ab}, \tag{2.3}$$

where R_{ab} is a term which increases the pressure between the particles of different fluids by a fraction 0.08. The algorithm due to Grenier *et al.* (2009)) is rather complicated and simpler algorithms are now available.

2.1. *Initial conditions*

Let us assume then that we can solve the set of equations needed for star formation. The next question is whether or not we have the right initial conditions. One view is that the final results are independent of the initial conditions. The other is that the

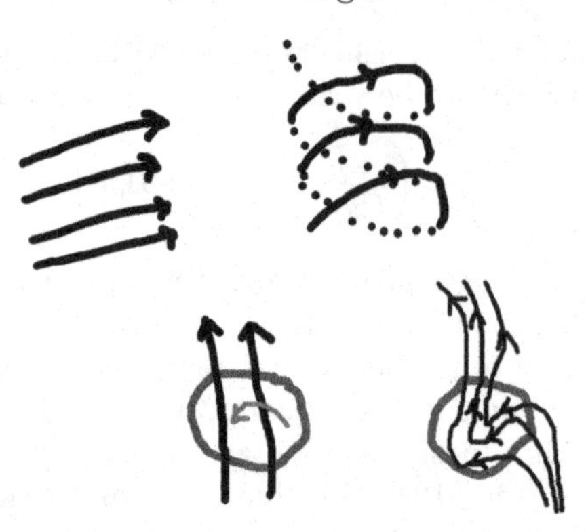

Figure 1. On the upper left is shown a sketch of a typical estimate of the magnetic field from observations, but an alternative magnetic field is that shown on the right. The fields may be twisted by differential rotation as shown in the lower sketches and such a field may be more relevant for the initial conditions for star formation.

results depend significantly on these conditions. Both views have representatives at this conference. My view is that at least some aspects of the initial conditions are important and our knowledge of galaxies should enable us to pin them down. For example, it is clear that molecular clouds forming in the galaxy will have significant rotation, which may or may not be controlled by magnetic fields. Suppose the rotation dominates the local magnetic field. The magnetic field will then be rapidly twisted by the shear (see the lower part of figure 1). At present this combination of rotation and magnetic field is disregarded as an initial condition. Instead the initial conditions are usually chosen to be a turbulent velocity field (see below) with an initial magnetic field which consists of straight field lines.

The observer's also tend to interpret fields in terms of a simple topology as shown in figure 1, though that topology could just as easily be interpreted in terms of a helical field with only one side of the helix observed (as shown in the upper right hand side of figure 1. Where we can observe magnetic fields in turbulent regions such as sunspots we find they are complicated tangles of fields even though their pressure far exceeds the thermal pressure of the gas. The final resolution will come from simulations of the formation of giant molecular clouds along the lines of the work of Claire Dobbs and Ian Bonnell (this conference) but with magnetic fields.

2.2. *Boundary conditions*

Star forming regions such as Orion appear to be independent of neighbouring star forming regions. It is therefore appropriate to consider them as isolated clouds with boundary conditions that are nearly stress free. There is a dangerous tendency however, to replace the universe by periodic boxes without testing how big the box should be before the simulation is affected by the periodicity. Detailed calculations of turbulence in nearly incompressible fluids for example, show that only a small fraction, typically~ 0.01 of the volume sustains isotropic turbulence. Correlation functions are also compromised by the fact that for a periodic box the corners are perfectly correlated. The conclusion is that the periodic box has to be very much larger than the scale of the processes of interest if it is to mimic an infinite fluid with satisfactory accuracy.

It is known from turbulence calculations that the turbulence is strongly affected by whether or not the boundaries are periodic, no slip, or stress free. The left frame of figure 2 shows the vorticity in a two dimensional turbulent liquid in a periodic box as calculated by Clercx *et al.* (2001). The vorticity has one sign in the upper left corner and the lower

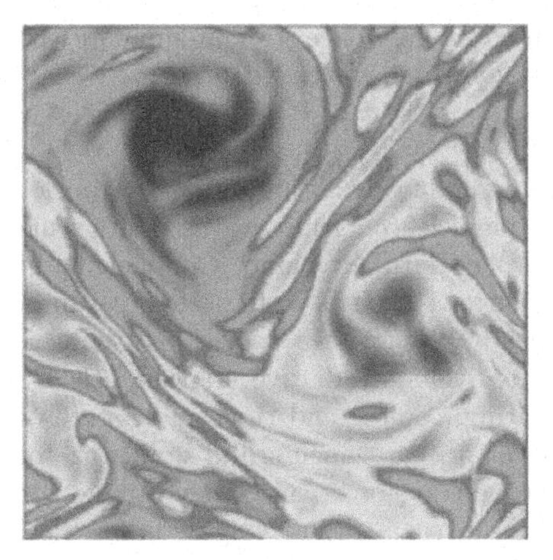

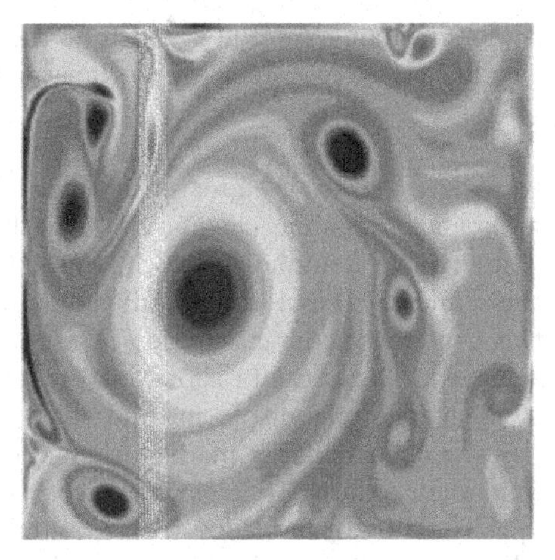

Figure 2. Decaying turbulence in a square box. The left frame shows the vorticity when the boundaries are periodic. There is vorticity of one sign in the upper left hand corner while vorticity of opposite sign occupies the lower right corner. The right frame is for the same initial conditions but with no-slip boundaries. Vorticity of one sign forms a large island surrounded by small islands of opposite vorticity. Note that in black and white the vorticity in the central part of the large circular vortex appears to be the same as that of the small islands, but it is, in fact, the peak value of the vorticity in the large island.

right corner has the opposite sign. In the right frame of figure 2 the simulation uses the same initial conditions but no-slip boundaries. The distribution of vorticity is now completely different. It consists of a large island with vorticity ringed by vorticity of the opposite sign. Similar differences show up with stress-free boundaries and boundaries of different shape e.g. rectangles.

Returning to the initial conditions we can ask what turbulence do we assign to the fluid. The answer will depend on the density distribution and boundary conditions. If they are stress free and the density decreases strongly near the boundary the turbulence will be different from that with periodic boundaries and uniform initial density. In my view periodic boundaries should not be used, or if they are used, they should be tested by comparing results for a range of box sizes.

Another aspect of turbulence which is useful to consider is that while the simulators take the view that the turbulence is largely due to shocks, the situation is much more complicated in reality. The generation of turbulence can come from Kelvin-Helmholtz instabilities, or from explosive outbursts from supernovae or indeed from any unstable fluid process. Furthermore, if there is turbulence at all scales, the shocks themselves will be affected because of turbulent mixing of momentum and thermal energy. I expect to see more work on how turbulence affects the simulation of shocks. One obvious aspect is that the shock profiles will be smoothed and the shock dissipation altered.

2.3. *Mixing*

The extent of mixing becomes important when we need to know the distribution of heavy elements or dust in order to calculate radiative transfer correctly. The extent of mixing depends on the velocity distribution and the scale over which we wish to specify the mixing. The smaller the scale the less likely it is to be well mixed. Furthermore it is known that there can be coherent structures called lagrangian coherent structures within which there is very little mixing. As a consequence rules of thumb like ' the mixing will be nearly complete after two turnover times ', are false. No one at this conference has explored mixing carefully, and it would be useful to do this once we can simulate the ejection of heavy elements into the star forming region from supernova.

2.4. *Radiative transfer*

It was good to see the great strides that have been taken in calculating radiative transfer within gas dynamic calculations whether or not particles or grids are used. In some cases (Bate, Krumholz, Walch, Susa, Offner, Harper-Clark, Ercolano, Commercon in this conference) flux limited diffusion approximations have been used. In others (Norlund this conference) direct solution of radiative transfer by ray tracing algorithms has been achieved. Their results show how important it is to include radiative transfer because it affects the pressure terms and the extent of heating, cooling and the atomic state of the gas. The remarks made earlier about mixing of heavy elements and grains are of fundamental importance for the accuracy of radiative transfer because they change the opacity. If I can have one wish granted it would be to see a ray tracing code for SPH which uses the particles directly without recourse to a grid.

3. A set of final questions

To end here are a set of questions which might be asked if we disappeared and returned in 10 years. Everyone will have their own set of questions.

- Have we got reliable accurate codes that simulate gas magnetic fields, dust, ions, electrons, radiative transfer and chemistry?
- Do they predict almost everything the observers see?
- Do we have agreement on the initial and boundary conditions?
- Do we have a simple, phenomenological rule for star formation in cosmological simulations?
- If Bruce Elmegreen is right about hardware can we simulate seamlessly from GMC to stars?
- Have the observer's stopped observing because there is nothing new to see?

Given the progress shown by the work described at this conference I expect the answers to be at least a qualified yes to each question except, perhaps, the last.

References

Clercx, H. H., Nielson, A. H., Torres, D. J., & Coutsias, E. A. 2001, *Euro. J. Mech-B/Fluids*,20, 557

Fowler, W. A. & Hoyle, F. 1962, *Roy. Obs. Bulletins. Her Majesty's Stationary Office.*, Number 67

Grenier, N., Antuono, M., Colagrossi, A., LeTouzé, , D., & Alessandrini, B. 2009, *J. Computat. Phys*, 228, 8380

Monaghan, J. J. 2005, *Rep. Prog. Phys.*, 68, 1703

Computational Star Formation
Proceedings IAU Symposium No. 270, 2011
J. Alves, B.G. Elmegreen, J. M. Girart & V. Trimble, eds.

doi:10.1017/S1743921311000925

Concluding Remarks: Playing by the Numbers

Virginia Trimble

Department of Physics and Astronomy, University of California Irvine,
Irvine, CA 92697-4575, USA
And
Las Cumbres Observatory
Goleta, CA 93117, USA
vtrimble@uci.edu

Abstract. Advances in available computing facilities, judiciously employed by our colleagues, have undoubtedly enhanced our understanding of the processes by which stars (and planets) form, from very diffuse gas to something you could almost live around. Nevertheless we remain very far from being able to describe (let alone explain) what is going on in many cases. And this is likely to remain true even as Moore's Law growth begins to hit its head against energy and power considerations. A large fraction of long-standing questions appear to have honest answers of the general form, "yes and no", "all of the above", or "some of them are and some of them aren't". This includes many of my favorites, like triggering, formation of binary populations, and the role of magnetic fields. Rather few questions have actually been retired from the universe of discourse in recent years.

1. Introduction

I arrived at the meeting with two lists of questions: a fairly long one of items I had been wondering about before looking at the abstracts, and a considerably longer one (generally addressing details) derived from the titles and abstracts, especially of the posters. Most of these questions have been around for a decade or more. Michael Norman suggested that progress can be gauged by the number and importance of the questions that can be retired from discussion, in the sense, for instance, that we no longer ask (with Simon Newcomb a century ago) what is the primary source of stellar energy. Richard Larson took us briefly back to the era, preceding and surrounding World War II, when it was by no means certain that ongoing star formation was either real or important. This, he pointed out, has definitely been settled, in favor of the process and its early supporters, including Lyman Spitzer.

A good many other classic questions can be retired, but only in the sense that the answer is, for instance, that there is triggering by more than one mechanism, that massive stars grow in several different ways, and that both turbulence and magnetic fields are important, often at the same time. The questions then return in the form of which happens when and where, and why, invariably demanding a good deal more computing power to try to follow everything at once.

The following sections include a few historical remarks, my lists of questions in, at best, semi-ordered form, followed by summaries of a few specific subfields, and a conclusion of a sort.

2. A Bit of History

Early computational tools included the pencil and the abacus (though the ones shown at the symposium were not used either by Jeans - it was an American pencil - or by Hayashi in establishing his tracks - it was a Chinese abacus, and in any case he had access to electronic computing facilities, including those at Goddard Space Flight Center. (Sugimoto 2010). Computers were, of course, once people, generally women with some college education who processed data for the male observers. The best known at Harvard included Henrietta Swan Leavitt (Cepheids), William Fleming, Antonia Maury, and Annie J. Cannon, whose spectral types were, respectively, too simple, too complex (including a criterion that separated out giants to which Pickering did not subscribe), and just right. Cecilia Payne Gaposchkin was their contemporary, but a modern, full-fledged researched astronomer, who asked difficult questions and managed to answer some of them.

Sir James Jeans was a pioneer of the pencil and paper school, who had a mass and a length, and served as president of the Royal Astronomical Society for two year ending in 1927 (keep that date!). He only rather gradually accepted the existence of other galaxies and the "short" time scale (10^{10} yr) for the universe, as against his own $10^{12} - 10^{13}$ yr required for galaxies, star clusters, and binary systems to achieve their current appearance using processes he had thought of. Thus for much of his career, he opined that (a) gaseous spirals became star clusters, (b) rotating gaseous spheres became Maclaurin spheroids, then Jacobi ellipsoids (with the RV Tauri stars as possible examples)and then close binaries, while (c) encounters were needed to account for wide binaries and planetary systems, since the "nebular" hypothesis (to which we all now subscribe) would leave 90% of the angular momentum in the sun. He wrote, first, (Jeans 1933) that by the time "the whole of the gas is condensed into detached globules, calculation shews that each globule would have about the same mass as an actual star, and, second, that if you start with the entire universe as a chaotic mass of gas, finally the whole mass would condense or break up, into detached masses of denser gas (for which) calculations shew that these would be on something like the scale of the actual nebulae, and would form at about the average distance apart of the observed nebulae, all of about the same size, weight, and intrinsic brightness. This is found very approximately to be the case." The observations he had in mind must have been overwhelmingly dominated by selection effects!

Later computational tools included punch cards for both Fortran commands and data, output on oversized, green-lined paper, and the Christy code, mentioned by Richard Larson, which I applied to look for Cepheid-type pulsations of R CorBor stars (yes they do) in 1970-71. Near the beginning of this stage Hayashi (1961) showed that stars condensing out of diffuse gas would be completely convective and so follow tracks in the HR diagram sharply upward and then down onto the main sequence (Hayashi tracks!) rather than sneaking in from the lower right hand corner under radiative equilibrium as had been previously thought.

Abandonment of the encounter hypotheses for planetary system formation and establishment of the Hayashi tracks count as progress in star formation (in Norman's sense) over, roughly, the last 70 years. Table 1 describes progress over about the last 40 years, in the sense that some things just barely calculable then have become routine now, while the topic of galaxy formation has advanced from incomprehensible to, at least, respectable. That item violates my general principle of mentioning very few speakers and almost no poster presenters by name, because those names, in order in their sessions, so conveniently give us the A, B, C of the subject. TLA, in case you weren't sure, is a Three Letter Acronym. In addition, formation of galaxies and stars (at least the first generation)

Table 1. Progress in Computability of Formation and Evolution of Stars and Galaxies.

Topic	New Code, 1968	Now
Galaxy Evolution	Tinsley (1968), top down	Bottom up
Star Evolution	Paczyński (1970)	On a disk near you
Star Formation	Larson (1969)	AMR, SPH, TLA, etc.
Galaxy Formation	——	Abel, Bournaud, Colin, and all

are no longer so separable as we once thought them. The topic of galaxy evolution has gone through a paradigm shift from "top down" to "bottom up." Tinsley's (1968) initial conditions had a gas cloud with mass equal to the eventual galaxy, which was turned into stars in, perhaps, 10 time steps, with fixed IMF, and the nuclear products then dumped back into the remaining gas. It had a "G dwarf problem" but otherwise yielded quite a decent fit to the range of galaxy luminosities, colors, compositions, and residual gas fractions that we see. Galaxy evolution in 2010 proceeds largely through star formation in dwarf entities intermingled with mergers and captures, which tend to trigger more star formation, with the results being required to look not so very different from Tinsley's.

Also more or less historical, I suppose, is a decoding of the hieroglyphic writing of the name of the NUT project, which combines more aspects of Egyptian orthography than one usually sees in one place. The three water jugs and the sky at the beginning associate her with rain. The three jagged lines are N's, giving the plural N-W (which might alternatively have been written with N plus the "weak bird" glyph, whose sound is W). The loaf of bread is a phonetic T, so that we now have her name "spelled out". Then comes the distorted oval shaped womb, indicating her association with childbirth, and finally the figure of a goddess, so that the final ordering is determinatives-alphabetic writing-more determinatives.

3. My Initial Question

Some of these can indeed be phrased as questions (Is star formation triggered? for instance). But most seem to look more natural as parts of a list of things we would like to understand, and are shown that way here.

Physics to be included and initial conditions: Gravity is the sine qua non of structure formation, though my initial list left it out (by accident), and at least one speaker claimed (deliberately) that it wasn't needed. Other items are distributions of temperature and density, ionization (due primarily to cosmic rays inside molecular clouds), the strength and structure of magnetic fields and of turbulence, and the amount and composition of dust.

Molecular clouds, sometimes giant ones (GMCs) are the locus of star formation, at least here and now. Thus "why do stars form?" begins with what assembles GMCs; what causes them to start collapsing or fragmenting; how efficient is the process, and what does this efficiency depend on; and how do stars form in environments very different from a temperature around 10K and a density of 10^{3-5} Hcm^{-3}.

Triggering is generally thought to be part of why collapse and fragmentation begin, but if it happens only some of the time, what difference does it make (to efficiency, the IMF, or whatever). On global scales, mergers, infalling dwarfs, and other encounters are possible triggers. More locally, we might have shock waves at spiral arms (hardly mentioned at the meeting), cloud-cloud collisions, expanding HII regions and supernova remnants, UV radiation, and other shock sources.

Clusters of stars are the natural outcome of fragmenting GMCs, but we are not quite certain that all stars form in clusters, even very short-lived ones. One would like to know the range of final cluster masses (in stars) vs. the initial gas cloud mass, and also, within any one cluster, how co-eval are the stars, is there an ordering of formation by star mass or propagation of star formation in space, and is mass segregation primordial or dynamical, this last a case that demands the answer, both please!

The equivalent of dynamic range remains a problem. A decade ago, it was possible for a theorist to start with a cloud of $50M_\odot$ and follow it fairly honestly down to the scale of binaries and protoplanetary disks (Alves & McCaughrean 2001). Naturally very few $100M_\odot$ stars resulted. I arrived wondering how much the treatable mass had increased. Two different presentations said 500 and $1000M_\odot$ (just right for the number of Moore's law doubling times, though not yet quite an OB association).

Stellar populations formed under very different conditions might well be expected to differ from the disk of the Milky Way. One would particularly like to know just what Population III was like and where their remnants might be lurking (IMBHs in the cores of globular clusters??). Other somewhat mysterious contexts include youngish stars in mostly old galaxies (formed in situ or brought in by mergers?) and the condition under which globular clusters form, both past and present. They aren't making them any more, at least in our galaxy.

The initial mass function looks more similar from place to place than one might have guessed. One would like observations of it in regions of different metallicity, cloud turbulence, magnetic fields, and redshift, and calculations that can match the full range found. A topic of some dispute in recent years has been the extent to which the IMF is already "coded in" at the stage of starless cores in molecular clouds.

Stars of more than about $10M_\odot$ would seem to have difficulty in assembling by accretion (because their expected luminosities rise above the Eddington limit). Thus one asks whether they can fool Mother Nature, whether some other process is required (mergers of less massive cores, presumably), or both. Similar questions arise for the formation of brown dwarfs and orphan planets (not because of Eddington limit, but because there doesn't seem to be anything obvious to stop them accreting). Again, several processes have been suggested, and if they all contribute, one is surprised at the relative uniformity of the IMF from place to place.

Binaries are important (at least to me, since their statistics formed part of "my" territory for a number of years). The problems start at the observational level, where theorists habitually use very incomplete studies of small samples of the incidence of duplicity and of the distributions of primary mass, mass ratio, period (or separation), and eccentricity. One cannot really extract the IMF from observations or define stellar populations by color or integrated line strength without allowing for them. On the theoretical side, one would like to know how they form, their relationship to triples and higher-order multiplicity (hierarchies vs. Trapezia and their stability), implications for numbers of planets, and their role in nucleosynthesis and other "advanced problems".

Disks and jets undoubtedly contribute (positively or negatively!) to planet formation, binary formation and other interactions, the disposal of spare magnetic flux and angular momentum, and a range of observable phenomena. My impression is that theorists have to work harder to make jets than disks.

Global star formation rates are another area where the problems begin with getting hold of the data to be matched. One wants to know the SFR as a function of z, Z, and z, that is time (or redshift), metallicity, and location in galaxies or protogalaxies. General agreement seems to be limited to the rather qualitative - more at $z = 1 - 3$ than now, but less farther back in time. I suspect Redman's theorem would then apply, that any competent theorist could explain the data using any theory. Large redshift is particularly important for reionizing the intergalactic medium so that we can see what is going on. Not that nature had any desire to help us, but there are at least a few clear sight lines back to $z = 7 - 8$.

General vs. special purpose hardware and software. The hardware case was discussed at the symposium. Most star formation codes have been developed specifically for star formation, but I wonder whether there might be material we could steal from the (perhaps better funded) territories of cosmology/galaxy formation/VLSS and high energy astrophysics (which also involve turbulence, magnetic fields, and, of course, gravity!).

4. Questions and Issues Arising from the Abstracts

A large fraction of the titles and abstracts of both talks and posters gave the impression (which usually turned out to be correct) that the presenters intended to address one or more of the issues mentioned in Sect. 3. This is a good thing, and if it had not turned out to be the case, the strong implication would be that I was the wrong person to be given this task! But there were also a good many other items, sometimes more detailed, of which this is an attempted list, meaning that if you look hard in the proceedings you can probably find out something about them.

I have ordered these in the same pattern as the questions of Sect. 3, but as "NASA bullets" since many are fairly short. Interesting items mentioned in the original abstracts that did not find their way into the posters and talks as presented have been omitted.

Initial Conditions

- Measured magnetic fields are clearly important, but if determined from continuum polarization can be badly wrong if dust grains have been oriented by dynamical torques. An anonymous conference participant, acting as a referee, indicates that several theorists have concluded that at least the larger grains will be well aligned with the field in any case. Zeeman broadening of OH lines is an alternative field indicator. This is perhaps the place to mention, after the fact, that I had the great pleasure of being the (non-anonymous) referee of Prof. Larson's contribution, for which my only suggestion (quickly adopted) was full citations of the references.
- One also needs initial conditions for calculations of chemistry (next subtopic).
- Fluctuations in temperature and density of gas and dust can be described by thermodynamic considerations.

Molecular Clouds

- Accurate chemical evolution of the molecular gas is essential if observers' clouds and clumps (traced with CO, NH_3, CH_3OH, etc.) are to be compared with theorists' clouds and clumps (generally identified as density enhancements).
- The gas can be bistable; HI is correlated with total gas and star formation rates, though typically not as well as H_2.
- Formation of GMCs from more extended HI is at least partly a fragmentation process in a multiphase ISM, whose dynamical evolution continues between the initial cloud and final proto-star cluster phases.

• Filamentary clouds may yield a different IMF from spherical ones, and some real clouds are at least non-spherical in a way that may be related to triggering

• Collisions of superbubbles from previous star formation episodes can form molecular cloud complexes between them.

Triggering

• Run-away OB stars are another possibility, as is UV irradiation, from them or other sources (perhaps even AGNs?). Photoevaporative triggering can reveal itself in bright-rimmed molecular clouds.

• Compressive tides could either trigger star formation or shelter gas from star formation.

Star Clusters

• Newly formed clusters are still embedded in dust and gas, and it is possible we may never see the real, unevolved IMF for O stars

• R Associations (with no OB stars but the later types obviously young) exist.

• Clusters can form around an initial very massive star (with triggering in a sort of ring).

• Dynamical segregation occurs gradually on the same time scale as the higher masses die.

• Galaxy mergers (etc.) can both trigger formation of star clusters and destroy existing ones.

• Comparisons of statistics of clusters and molecular clouds is one way of assessing the efficiency of star formation.

Dynamic Range

• Some indication of the difficulty of the problem is given by the extreme messiness of real protostellar envelopes (just think of the Vela and Rosette Nebulae).

Star Populations

• The acronym IRDC takes a moment to translate into InfraRed Dark Clouds. The real curiosity is that some of them actually have protostellar cores in them, and so are not really "starless cores".

The Initial Mass Functions

• No one really claimed that these are all the same, everywhere, though there is some disagreement about whether particular regions with young stars are "top heavy". These include the nucleus of our own galaxy (where one does not see the X-rays expected if there are lots of young, low-mass stars to go with the massive ones in the Arches etc.), 30 Doradus (where a normal IMF would probably tie up more than the total stellar mass of the LMC), and young massive clusters (which will end up as globular clusters in 10 Gyr or so only if they have an appropriate complement of low masses).

• Evolution of cores to stars in a cluster environment may be different enough to affect the IMF. A central black hole (for instance for star formation in disks around AGNs) is likely to inhibit formation of small stars, while having lots of dust should favor extra cooling and a smaller Jeans Mass for the turnover of the IMF.

The Largest and Smallest Masses

• Star growth by collisions in accreting cluster cores is expected.

• Accurate computations of the spectra expected for brown dwarfs as a function of mass and age are essential for recognizing them when they are not in binaries.

Binary Stars

• Accretion can continue from circumbinary disks.

• Binary populations among very low mass stars and brown dwarfs are different from the A-K stars, though portions of this will reflect observational selection effects and dynamical evolution.

Disks and Jets

• Jets must somehow be launched, and get loaded up with molecular gas.

• Disks must disappear on the time scales set by observations (and models exist for making this happen either faster or slower than seen).

• There will be a dust sublimation front where we find dust-free disks.

Global Star Formation Rates

• High Velocity Clouds probably bring in fresh gas to enable star formation to continue. They themselves seem to contain no stars.

• Downsizing - the shift of the bulk of star formation to smaller galaxies as we approach the present - is observationally real and not readily reproduced in bottom-up models of galaxy formation and evolution.

• Something must terminate the formation of stars in early, massive galaxies (arguably the ionization of the gas so that the Jeans mass becomes some enormous number)

• In the case of population III, line pressure radiation on incoming gas (nearly pure H and He) is not sufficient to prevent accretion, which, however, should stop with the UV breakout, ionization of the nearby gas, and radiation pressure from continuum photons.

Hard- and Soft-ware

• A number of presentations featured specific codes and machines on which they run, but my initial thought that we might be able to "borrow" things from the cosmology/galaxy formation community was stretched to include the high energy astrophysics community by a specific poster abstract, which mentioned disks, jets, magnetic fields and all, and was not spotted by the SOC as pertaining to neutron stars rather than YSOs until the program had been largely assembled.

Things I Hadn't Thought of (but Should Have)

• The most important of these is undoubtedly feedback - the effects of protostellar and stellar jets and all the rest on the surrounding gas, where they will certainly be a source of turbulence and, perhaps, of magnetic field amplification.

• Specific YSOs, star formation regions, star clusters, and merging galaxies mentioned in the presentations include the Perseus Arm (a reminder that spiral shock waves can be another trigger for GMC assemblage and collapse), Herbig-Haro objects, FUOr's (and EXor's, which by NOT having EX Ori as their prototype are just there to make things more difficult), the Antennae galaxies, NGC 3603, the Aquila Rift complex, Eta Cha, the Pipe Nebulae, AFGL 5142, Orion KL, NGC 1333, V4046 Sgr, Herbig Ae stars, Sh2-104, IRAS17149-3916, and V 582 Aur. This is not a complete list, which is one of the reasons the editors intend to provide a "object index" to this volume.

5. Some (Partial) Answers from the Symposium

This is a subset. The 43 hours of talks (plus 10 I spent reading posters) would take an equal time to retell, just as you can really simulate a 10^{80} particle universe only with an $N = 10^{80}$ calculation. The ones here are a personal, prejudiced choice.

5.1. *Binary Stars*

These are important for nucleosynthesis and for the use of Type Ia supernovae as distance indicators in cosmology. Whether Population III stars had no binaries or practically nothing but binaries has been determined - theoretically - both ways.

The desired data are numbers as a function of primary mass, mass ratio (my own little territory), period (or separation) and eccentricity, as a function of metallicity and formation epoch. The data change: population II when I was your age had "no" binaries. Globular clusters now have a large excess of X-ray binaries among other things. $N(e)$ is a convolution of formation and dynamical evolution, revealed by distributions of eccentricity vs. orbit periods in clusters of different ages.

$N(M_1)$ is intermediate for OB stars, high (more than 50%) for A-K stars, and low again for M, L, and T types, reflecting formation, survival, and observational selection. One speaker suggested that nearly all stars form as triples, generally ejecting one member to leave binaries and singles about equally numerous. I think I last heard this at a 1979 meeting (Harrington 1982). $N(M_1, M_2/M_1, a)$ is clearly not a separable function. In fact, I suspect that all the distributions are actually sums of about 3 separate curves, arising from 3 different formation mechanisms, and simply blended in most data.

As for $N(Z, z)$, "there are very little data, and they are contradictory, so I will skip this point", quoting Lisa Deharveng on a different topic.

Common proper motion pairs can either be barely bound within clusters, or just drift off together.

One speaker remarked upon the difficulty of calculating the incidence of binaries in dwarf spheroidals (important because they must be allowed for in turning an observed velocity dispersion into a Virial mass to get M/L for these nearly-invisible objects). Luckily observations come to our aid. Minor (2010, a thesis defended the week before the meeting) has found, with about a year of radial velocity data for several Milky Way companions, that there are binaries, some of which indeed have orbital velocities that can disturb σ_v, but that the effect on mass determination is generally modest, not exceeding 15% (but, of course, systematic, not random). A major factor is that large amplitudes come from two stars of nearly equal mass, which show themselves as double lines in the spectrum most of the time.

5.2. *Efficiency*

One speaker described the efficiency of star formation as the elephant in the conference room, because many calculations find a few percent to perhaps 15 %, and yet clusters remain bound when the residual gas leaves. That the average efficiency must be small follows because the Galaxy is about one hundred rotation periods old, so that a given gas atom has had about 100 arm-crossing chances to find its way into a star, and yet gas remains.

Many calculations yield 1-3 %, while cluster survival requires 15-20 %. One poster pointed out that it helps if most of the stars form in the cloud core. Data on numbers of clusters vs. age are not as helpful in defining survival time as you might expect.

Encounter shocks at $z = 8 - 9$ apparently yield quite a high ratio of mass going into stars vs. cold gas coming in.

5.3. *Brown Dwarfs*

These exist (20 years ago they did not, observationally), but are past the peak of the IMF in all contexts so far examined, and so are not a major dark matter component. Binary brown dwarfs also exist, though they are relatively rare and mostly have small separations (surely some combination of formation physics, dynamical evolution, and selection effects). They might form as binaries in accretion disks, be left behind when gas leaves a star formation region, or be gently ejected before they have had time to sweep up their fair share of gas.

One would like to know whether they form a continuous population with the orphan planets or exoplanets in general.

5.4. *Items New to Me*

The concept of sinks is apparently quite new, so I feel no guilt. The same thing happened when I was summarizing a conference on large scale structure soon after the concept of σ_8 had been introduced [it is the rms density fluctuation on a scale of $8Mpc$ for $H = 100km/sec/Mpc$]. Its numerical value is about one.

Other items like this include star formation on filaments (like VLSSS in the cosmological context), halo globular clusters being correlated with their central black hole masses, which in turn are correlated with the mass in bulge stars. If this means all this stuff has come from captures, I am confused.

Some more new stuff - that jets wiggle both observationally and theoretically; that low mass pre-main-sequence stars are a good deal fainter than expected (implying slow or episodic accretion), and that there is the risk of a magnetic braking catastrophe, because some angular momentum must be left to form planets!

5.5. *The Core Mass Function vs. the Stellar Initial Mass Function*

It was suggested that, in different contexts (and so conceivably all true), there is no close relationship (because of ongoing accretion, outflow, etc.), that they differ by a factor three, that they have the same slope when magnetic fields are included (but are otherwise different), that they arrive from the sum of several processes, that the observed IMF is universal, but the CMF is not; or conversely.

5.6. *Extrema*

The prize for the shortest talk went to Andy McLeod. The most popular encounter was NGC 4038/39 and the most popular cluster NGC 3603 (notice that virtually all well-known star clusters and galaxies have names beginning with N or M).

The first star formed at $z = 42$ in 2004 (this is funny if you read certain kinds of low-brow literature), which has perhaps grown to 70 this year.

The identity, mass, and explanation for the most massive stars remain uncertain. Qualitatively, the luminosities of main sequence stars scale as about M^3, so that you hit the Eddington limit at about $100M_\odot$ (which must surely depend on both Z and z). Calculations achieved either 110 or 140 $M_\odot$ in the star from a $480M_\odot$ core, both calculations coming from the same group, and they explained that the simulation is ongoing. Observationally, we certainly don't see many stars in excess of those numbers (when angular resolution is adequate), but it is not certain whether the lack is real or merely having

$N(\geqslant M) = N_0 M^{-2.3}$ go to zero. Formation of these very massive stars may well involve all of accretion, Rayleigh-Taylor instabilities, and mergers.

5.7. *Miscellany*

We are once again reduced to NASA bullets to collect many short (often verbless) items.

H_2 could be just a tracer (vs. the cause) of star formation, in the sense that cold, dense HI would do just as well. Larson admitted off-line that he had once turned off all molecule formation in his 1968 code, and it made very little difference. But of course we need molecules to observe, and when they begin to condense out on ice, the densest cores become very difficult to observe, yielding systematic errors in mass vs. time and temperature.

Bottom up star formation appeared in one presentation, as was popular back in about 1970. It will yield a power law (like the energies of cosmic rays) from a sort of self-organized criticality.

The metallicity to change Pop III to Pop II was 10^{-5} in one poster, but 10^{-2} in another. I would go with the smaller number.

Prototypes are rarely typical, as in the case of Algols, RR Lyrae stars, and Cepheids, so perhaps also for the Pipe Nebula (justifying investigation of other regions).

Gravity can be left out, said one presenter. Not from my universe, despite potentially favorable effects on the masses of astronomers.

Magnetic fields in star formation regions make things smoother, yielding fewer and more massive stars and later star formation, while with turbulence you get finer structure, earlier stars, more stars, and more low mass stars, but measurements of magnetic fields by Chandrasekhar-Fermi method can yield the wrong answer (with the fields coming out smaller than the real values, if there is lots of small-scale structure).

Red and dead? Yes, E and SO galaxies have gas; it is just too hot to make stars.

Do tidal dwarfs get dark matter? They had better if they are to be like the dwarfs in the Local Group, while tidal globular clusters had better not. One generally does not have masses for these young, very bright entities and so cannot tell what they will like in 10 Gyr.

Every trigger or driver of turbulence you have ever heard of is surely important somewhere - mergers, SN(R) shocks, HII expansion, outflow, runaway OB stars, Herbig Haro objects, shocks at arms, photoevaporation, UV radiation, external photoionization, and surely some we have not yet thought of.

6. Looking Ahead

If the past is Sir James Jeans stepping down as president of the Royal Astronomical Society in 1927, and the present is this 2010 conference, then the future means 2093 (which will be my 150th birthday). By then, studies of star formation will probably have bumped against some fairly fundamental limits.

On the observational side, CCDs already have something like 90% quantum efficiency, so "mehr licht" has to mean bigger mirrors. Angular resolution can improve as $\theta = \lambda/D$, but for some phenomena a sort of Olbers' Paradox will set in, with the sky completely covered by the things you want to study (for instance HI fluctuations or even dwarf galaxies at large redshift). Timing is limited by photon bunching on the nano-second scale (but much sooner by atmospheric scintillation, unless you take your 600-meter mirror into space). Spectral resolution is limited by the uncertainty principle and the length of time you are willing to count wave crests! Admittedly, this is a wavelength resolution somewhere around 10^{19} for optical astronomers.

On the computational side, Bruce Elmegreen looked at realistic futures. But, if you wish to look at unrealistic ones you could start with a thermodynamic limit that erasing one Gigabyte adds $S = 8.2 \times 10^{-14} J/K$ of entropy to the universe, and so requires dumping heat of $2.5 \times 10^{-11} J$ at room temperature. Make up your own rules for how many Gbyte per particle your N-body simulation will need, how many time steps are required for your version of the Courant condition, and how long you are willing to let your simulation run. I ended up with 4×10^{26} particles per GWatt, while a mere star has 10^{57} particles. This is apparently not quite the way to go.

The computer of the future will presumably be some sort of quantum picodot, but it will still be I/O limited, as were the very early computers of 1954, when my father worked for NCR, then trying to rival IBM. (It lost.) My theoretical astrophysicist of 2093 is under copyright and cannot be shown here, but he is green and wears glasses over quite large eyes.

7. Last Words

Some of the discussion following papers made me feel that the phrase "star formation community" might be an oxymoron. This is not actually bad - it means that people in the field are sure enough of its importance and permanence to be able to afford to risk not always showing a united front.

"There are", said Anne B. Underhill, "more models that aren't stars than there are stars that aren't models". This still seems to be true, though she said it in 1967 at the Prague IAU (which was attended by at least one person at S270). Miriam Lichtheim was of the same generation and equally tart, but she taught me the basics of reading and writing Middle Egyptian at UCLA in 1962-63.

It is the traditional prerogative of the last speaker to express collective thanks to the local organizers (Josep Miguel Girart and his whole team) and the scientific program organizers (João Alves, Bruce Elmegreen and their team), and this I do both individually and collectively, most heartily.

My last overhead had the words for "farewell" or "until next time" in 20 languages, and it was a special pleasure to be able to add Korean, Turkish, Persian, Finnish, and three other I don't actually recognize to this list during the meeting.

References

Alves, J. F & McCaughrean, J. J. Eds. 2001, *The Origins of Stars and Planets*, Springer ESO Astrophysics Symposia

Harrington, J. P. 1982, in M. H. Hart & B. Zuckerman, Eds., *Where Are They?*, Pergamon Press p. 152 (from a 1979 meeting)

Hayashi, C. 1961, Publ. Astron. Soc. Japan, 13, 450

Jeans, J. 1935, *Through Space and Time:* The Christmas, 1933, Royal Institution Lectures, Macmillan

Larson, R. B. 1969, Mon. Not. Royal Astron. Soc. 145, 271

Minor, Q. 2010, Ph.D. Dissertation, U. California, Irvine, defended May 2010

Paczyński, B. 1970, Acta Astron. 20, 47

Sugimoto, D. 2010, Astron. & Geophys. 51, 3.36, and personal communication

Tinsley, B. M. 1968, Astrophys. J. 151, 547

Historic Building's hall of the University of Barcelona

Sunday's reception at the historic building of the University of Barcelona

Pictures of Sunday's Reception

Coffee Break

The young LOC: Álvaro Sánchez-Monge, Josep M. Masqué, Aina Palau, Pau Frau and Felipe O. Alves

Poster session

Lunch at Mare Nostrum: Shantanu Basu, Patrick Hennebelle, Philippe André, Masahiro Machida and Kengo Tomida

Lunch at Mare Nostrum shared with a cat

Visit to Mare Nostrum

Visit to Caves Codorniu

Visit to Caves Freixenet

“Piscolabis” (snack) at Masia Torreblanca

Banquet

McKee's speech at the Banquet

Human tower by the Colla castellera Xicots de Vilafranca and some meeting's participants volunteers
All pictures are courtesy of Pau Frau, Alyssa Goodman and Bruce Elmegreen

Author Index

Object Index

Subject Index

Printed in the United States
by Baker & Taylor Publisher Services

Printed in the United States
by Baker & Taylor Publisher Services